FUNDAMENTALS
OF CHEMISTRY

FUNDAMENTALS OF CHEMISTRY

THIRD EDITION

FRANK BRESCIA
JOHN ARENTS
HERBERT MEISLICH
AMOS TURK

*The City College of the
City University of New York*

ACADEMIC PRESS, INC.

New York   San Francisco   London

A Subsidiary of
Harcourt Brace Jovanovich, Publishers

Cover illustration: Aspirin under a polarized light microscope. Photograph courtesy of Sidney Braginsky. Reproduced from the cover of American Laboratory, July 1973. © Copyright 1973 by International Scientific Communications, Inc.

ACADEMIC PRESS, INC.
111 Fifth Avenue, New York, New York 10003

United Kingdom Edition published by
ACADEMIC PRESS, INC. (LONDON) LTD.
24/28 Oval Road, London NW1

Library of Congress Cataloging in Publication Data

Brescia, Frank, Date
    Fundamentals of chemistry.

      1.   Chemistry.   I.   Title.
QD31.2.B74 1975   540   74-10192
ISBN 0-12-132372-2

Printed in the United States of America

CONTENTS

PREFACE

The wide acceptance of the earlier editions over a period of eleven years has encouraged us to continue our efforts to improve the effectiveness of this text as a teaching tool. At the same time we have made every effort to preserve the carefully developed expositions for which the previous editions were so favorably cited.

Textbooks are composed of chapters, and chapters are further subdivided into sections. While such categories are pedagogically useful, it is important also to emphasize the progressive development of topics and their interrelationships so that the student sees the subject as a cohesive body of theory and experiment. We have tried to strengthen this aspect in this edition. Some specialized topics have been omitted; others have been condensed and presented in homework problems so that they are available if the instructor wants to assign them. As a result of these efforts, this edition has fewer instructional byways and is somewhat shorter than the previous ones. We hope that this "smoothing out" effort will be pedagogically rewarding.

SPECIFIC CHANGES

The material on separation and purification has been modified and included as part of the first chapter. The chapters on atomic and molecular weights, on stoichiometry, and on the first law have been moved forward. Together with Chapter 2 on gases, this group constitutes an early introduction to the essential arithmetic of chemistry and, additionally, provides the needed background for most of the early laboratory exercises. The structure of atoms and types of chemical bonds are discussed next, but the molecular orbital (MO) theory of bonding is delayed to permit early introduction of topics associated with laboratory work in qualitative and quantitative analysis. The long chapter on galvanic cells and chemical thermodynamics has been split into two chapters, with additional discussion of free energy and entropy.

The various discussions of chemical bonding have been rearranged to create a cohesive unit introduced by general treatment of MO theory, followed by applications to the chemistry of representative elements, transition elements, and organic compounds. The material on the chemistry of the covalent bond has been absorbed in the discussions of Lewis acid–base theory and in the organic chemistry chapter. Separate chapters are devoted to MO theory of covalent and metallic bonding and to orbital hybridization, including molecular geometry. The discussion of intermolecular forces has been shortened and included in the chapter on types of chemical bonds.

HOMEWORK PROBLEMS

We have provided a copious supply that obviates the need to supplement the text with problem books. Each chapter has an initial set of problems which, if assigned *in toto*, will give the student the amount of practice deemed necessary for an understanding of the essential ideas of the chapter. Additional problems are also provided, some of which are suitable for routine drill, some are expected to provide additional challenge to the students, and others serve to extend the content of the chapter. Examples of the latter group are the van der Waals equation, the Schrödinger equation, and the Cannizzaro method for atomic weights. Answers to the numerical problems appear at the end of the text.

ANCILLARY TEXTS THAT ACCOMPANY THE THIRD EDITION

Fundamentals of Chemistry: Laboratory Studies, Third Edition
Student Guide to Fundamentals of Chemistry, Third Edition

ACKNOWLEDGMENT

We wish to express our gratitude to our wives for their patience, to our students who used and made helpful comments on the earlier editions, and to our colleagues at City College and elsewhere for their suggestions. In addition, we wish to thank the many people at Academic Press who helped bring this text to fruition.

Frank Brescia
John Arents
Herbert Meislich
Amos Turk

TO THE STUDENT

Because it is unrealistic that equal attention and effort be given to every paragraph in a textbook, we offer the following suggestions to help you use proper judgment regarding study emphasis.

First, scan the chapter quickly in order to see what topics it covers. (Give it at least a modest level of attention before you attend the lecture covering these topics.) Next, look at the Problems. The first set (before the Additional Problems) attempts to cover at an acceptable level what we think you must understand for a reasonable grasp of the material covered in the chapter, especially topics referred to in subsequent chapters. Now begin to study the chapter, bearing in mind the queries posed in the Problems section. When you come to an illustrative problem, don't look at the solution immediately, but try to solve the problem yourself first. Don't struggle too long with it; even a brief attempt will help you by fixing the question more firmly in your mind. Then go over the answer carefully, making sure you understand all the steps.

Now let us say you come to a section you think requires less emphasis. How should you deal with it? One possibility would be to read it rapidly. You may find that the section gives you new insight into some other topic and thus merits further study. You may even choose to skip it entirely, or perhaps return to it later. Of course your instructor will have preferences for sections to be emphasized, deemphasized, or omitted.

Finally, work out your selected or assigned set of homework problems, referring to the chapter when necessary. You should also try to answer some of the questions in the "Additional Problems" section—they will expand and strengthen your grasp of the subject. Above all, do not fall behind; the study of chemistry builds upon itself, and ensuing chapters require an understanding of prior chapters.

Note that in the text we use SMALL CAPITALS for new terms and *italics* for emphasis. If you think that some additional markings will help you, use them.

Few students make sufficient use of the index. Use it to locate and refer to other topics that will help you better understand the chapter you are studying.

There are appendixes at the end of the book. Consider Appendix I as part of the introduction, and read it with Chapter 1. Also, refer to the mathematical review (Appendixes I.15 and I.16) to refresh your computational techniques.

Finally, the subject is chemistry, but the language is English. Don't guess at the meaning of a word that may be obscure to you—use the dictionary. For example, when we say "discrete" we don't mean "discreet." When we say "resolution" of lines in a spectrum, we're not talking about New Year's Day; the appropriate meaning is "the effect of an instrument in making the separate parts of an image distinguishable." To understand better what you read makes it more enjoyable. And when you enjoy it, you understand it better.

1

INTRODUCTION

Chemistry deals with the properties and transformations of materials. Materials are samples of matter; they include a rock, the secretion from a pituitary gland, the varnish on a table, the odorant from a flower, cobra venom, a helium atom, and a proton. Chemists observe how materials change; they coordinate their observations into useful concepts, and they predict conditions under which specific changes will occur—often to produce materials never previously observed. Chemistry's origins are ancient. Metallurgy, leather tanning, fermentation, and the manufacture of soap, glass, and pigments were all devised before man acquired the concepts that now underlie his understanding of chemical change. Advances in chemical theory and technology during the past two centuries, however, have been spectacular compared with all earlier progress. The most valuable group of ideas held by chemists has proved to be the atomic and molecular hypotheses.

Modern chemistry began with investigations of gases under pressure and vacuum (a discipline then called "pneumatics") in the seventeenth century by Evangelista Torricelli, Blaise Pascal, Otto von Guericke, and Robert Boyle. These studies led to improvements in laboratory techniques that helped another generation of scientists (notably Karl Wilhelm Scheele, Henry Cavendish, Joseph Priestley, and Antoine Lavoisier), about a century later, to formulate a quantitative basis for chemical changes, especially combustion and other reactions involving oxygen. These advances, in turn, set the stage for the chemical pioneers of the nineteenth century (John Dalton, Amedeo Avogadro, Jöns Jakob Berzelius, and Stanislao Cannizzaro) to interpret chemical changes in terms of atoms and molecules, and to devise rational systems of atomic and molecular weights. The latter half of that century witnessed a very fruitful growth of systematizing concepts—the periodic table, the structural theory of organic chemistry, and stereochemistry (the geometry of molecules). In 1896 Henri Becquerel discovered radioactivity, thus initiating a new chain of discoveries that led to a great refinement of our ideas about the atom, and to new understanding of chemical reactions.

The chemist today pursues an ever more intimate understanding of molecular structure and chemical reactions. He wants to know what factors influence chemical reactions—why one molecule behaves differently from another under the same conditions. To answer these questions, he must do more than identify the sequences of atomic linkages in molecules. He needs to know the shapes of molecules in terms of bond distances and angles, the strengths of the various bonds, their vibration frequencies, and the molecular distributions of electrical charge. The chemist also wants to understand how chemical reactions occur. To do this he must elucidate all the intermediate conditions of the molecules undergoing alteration. He will not necessarily demand exact answers from his experiments and theories, however, because he knows that even gross approximations have served chemistry and chemical technology well in the past.

Interactions between observations and ideas are the essence of science. However, if we try to study *all* of the conceptual and historical antecedents of, say, the atomic theory, we find that one year and one book are hardly sufficient. We must, therefore, be selective and, in our selectivity, somewhat arbitrary. This book begins with the consideration of the properties of matter in the gaseous state—the simplest state, in some ways—and then goes on to the study of quantitative chemistry and the concepts of atoms and molecules. This sequence is an approximation of the historical one. There is no rigid order of "facts first, then theory," however, only a continuous interaction between the two. As in the days of Boyle, Lavoisier, and Cannizzaro, as well as in our own time, the facts serve as raw material for the construction of theories, and in turn, theoretical concepts guide us in suggesting what experiments to do next, and thus lead us to new information.

DEFINITIONS OF SOME CHEMICAL TERMS / 1.2

We shall define and discuss briefly some terms which are part of the language of chemists.

The properties of a material may be divided into two general categories: ACCIDENTAL or EXTENSIVE PROPERTIES, on the one hand, depend on the amount of matter present in a sample; SPECIFIC or INTENSIVE PROPERTIES, on the other hand, do not. Thus, the white color of a piece of chalk is a specific property; its length is an accidental one.

A SUBSTANCE is any variety of matter of recognizably definite composition and specific properties. The term is used in distinction to BODY, or OBJECT, which refers to a particular item of matter. Thus, a chair (object) is made of wood (substance). The COMPOSITION of a substance is its makeup of constituent substances, usually expressed in terms of percent by mass.

Some substances have fixed compositions that are associated with their properties; they are said to be PURE SUBSTANCES. For example, red iron rust can be obtained as a pure substance comprising 69.94% iron and 30.06% oxygen. Coal, on the other hand, is not a pure substance; its carbon content ranges approximately from 35 to 85%. Of course, a pure substance may be contaminated by foreign matter. The important point, however, is that the pure substance, when it is recovered from such a MIXTURE, retains its definite composition and specific properties.

It is believed that the constant compositions associated with pure substances are maintained by linkages among elementary units of matter; such linkages are called CHEMICAL BONDS. A transformation accompanied by the making or breaking of chemical bonds is called a CHEMICAL CHANGE or CHEMICAL REACTION. Examples are combustion, corrosion, photosynthesis, and digestion.

A PHYSICAL CHANGE of a substance does not involve a change of definite composition or specific properties. Alterations in the dimensions of objects, or in their physical states (for example, gaseous, liquid, or solid), are considered to be physical changes. Examples are fracture, deformation, pulverizing, drawing (as of a metal wire), thermal expansion or contraction, melting, boiling, and freezing.

The types of behavior that a substance exhibits in chemical reactions are called its CHEMICAL PROPERTIES; other characteristics of a substance are called its PHYSICAL PROPERTIES.

DECOMPOSITION is a chemical reaction in which a substance breaks down into simpler forms. Most of the many substances known to man can undergo decompositions that involve energy changes of moderate magnitude,* and yield two or more decomposition products. A relatively few substances (somewhat over one hundred) do not decompose at all within these ranges of energy change or, if they do, give only one ultimate product (for example, ozone → oxygen). Such substances are considered to be the stuff of which all other substances are made, and are called ELEMENTS. The fundamental unit of the element is the ATOM.

A pure substance that is not an element is called a COMPOUND.

Two or more atoms may be linked together by chemical bonds. When such a collection of bonded atoms is electrically neutral, it is called a MOLECULE. Individual unbonded atoms are also considered to be molecules; for example, the molecules of helium, a typical "monatomic" gas, are individual atoms. There is no absolute upper limit of size for molecules. Viruses (see Fig. 26.6, page 562), which reach dimen-

---

* Up to about 2×10^3 cal/g (released) or 3×10^4 cal/g (absorbed). Processes involving higher energy changes are no longer chemical, but nuclear.

sions of hundreds of angstrom units,* are sometimes called "giant molecules." Molecules of gaseous substances usually have dimensions less than 10 Å; small molecules like those of water and hydrogen chloride are around 2 to 4 Å.

Electrically charged atoms or groups of atoms are called IONS. The smallest ions are individual charged atoms (for example, sodium ion, Na^+ or fluoride ion, F^-). Ions may also be groups of relatively few atoms, such as sulfate ion, SO_4^{2-}.

CLASSIFICATION OF THE STATES OF MATTER / 1.3

If all portions of a body have the same specific properties, the body is said to be HOMOGENEOUS; if not, it is HETEROGENEOUS. In a sample of matter, any portion that is homogeneous and separated from other parts of the sample by a definite surface or boundary is called a PHASE. Thus, ice and water are a two-phase system. A layer of oil floating on water is also a two-phase system (oil and water). If the mixture is shaken until the oil is dispersed as droplets in the water, an EMULSION forms, which still consists of two phases (oil and water).

Molecules may have a greater or lesser tendency to stick together, or cohere. Generally, the stronger the cohesive forces of a substance, the greater are the density, hardness, strength, and ability to resist the effects of heating. The weaker the cohesive forces, the greater is the tendency of a substance to spread out in space.

The arrangement of its molecular particles in space is another important determinant of the properties of a substance. In fact, different materials are known to occur in conditions ranging from complete disorder of their particles to highly ordered states in which only very few particles are out of place. Figure 1.1 shows crystalline shapes, suggesting orderly arrangements among the particles making up the substances.

The degrees of cohesiveness and of order serve as the bases for the classification of phases of matter according to the way the molecules are "put together" or aggregated. For this purpose, matter may be said (a) to be arranged either randomly or in an orderly manner; and (b) to cohere so well that it maintains a rigid shape (like an iron bar), or so poorly that its shape can easily be distorted (like honey), or practically not at all, so that it disperses easily (like air). This classification yields six combinations, four of which are familiar to most of us. These are listed and named in Table 1.1. None of the distinctions given in Table 1.1 is absolute, but some are sharper than others. The distinction between liquid and glass, for example, depends mainly on the degree of rigidity, and is often quite poorly defined. The randomness of molecular arrangements in liquids is far less than it is in gases.

Some substances, especially those composed of large molecules, such

Fig. 1.1 *Quartz crystals. (Courtesy Ward's Natural Science Establishment, Inc., Rochester, N.Y.)*

* One angstrom unit (Å) = 10^{-8} cm = 0.1 nanometer (nm); see Appendix I.4.

Table 1.1
States of Aggregation of Matter

Predominant arrangement of particles	Degree of cohesiveness	Common name of state
Random	Low (no well-defined boundary)	Gas
Random	Intermediate (well-defined boundary, but no rigidity)	Liquid
Random	High (rigid)	Glass (also called non-crystalline solid)
Orderly	Low	Does not ordinarily exist[a]
Orderly	Intermediate	Does not ordinarily exist[b]
Orderly	High	Crystalline solid

[a] Exception: solid helium near 0°K.
[b] Exception: "liquid crystals"—liquids that have some optical properties typical of crystals.

as gels, rubber, and plastics, are not at all well classified by this system. Attempts to apply this classification are made, nonetheless; a textile chemist who examines cellulose fibers by X-ray diffraction patterns (page 146) will say that a well-ordered (repetitive) pattern indicates that the cellulose has a high degree of "crystallinity."

THE PURIFICATION OF MATERIALS / 1.4

We have defined a pure substance as one with a fixed composition that is associated with its properties. We usually think of it as being entirely composed of like molecules. However, the most reliable way to detect the presence of "foreign" molecules in a sample is to separate the foreign matter, and to note that the properties of the new samples differ from those of the old. If attempts at separation fail (which is to say that no further change in properties is detected), we may *assume* that the sample is pure. As far as laboratory operations are concerned, there is no way of showing that the assumption is false except by developing a more effective method to separate the "pure" substance into new components. In laboratory practice, then, the criterion of purity of a substance is simply *the inability of the experimenter to isolate or otherwise detect foreign material*.

Purification involves physical separation, which occurs as a consequence of the differences in properties among the components to be separated. For example, bricks and grains of sand differ from each other in size; a sieve will retain the bricks and allow the sand to fall through.

Materials may differ from each other in many attributes other than particle size—electrical, magnetic, and solubility properties are among those which frequently serve as bases for purification procedures. Finally, the presence of the purified component(s) must be detected in some way so that we know that separation has occurred.

Various traditional purification methods, such as distillation, crystallization, and sublimation, which have been used since the beginnings of chemistry, are described in undergraduate laboratory manuals and will not be discussed here. Instead, we will outline a more recent set of methods to illustrate how chemists purify materials.

In 1906 the botanist Mikhail Tswett showed that colored plant pigments, such as chlorophylls and carotenoids,* could be separated into components of different hues by passing a solution of the pigment through a column of a finely divided solid such as precipitated chalk (calcium carbonate). In his pioneer experiments Tswett used an extract of green leaves in petroleum ether (a solvent similar to gasoline). This solution, on trickling through his chalk column, produced sharply differentiated colored bands, comprising various yellows and greens. This process is called CHROMATOGRAPHY (Fig. 1.2). Today, the meaning of this term has broadened far beyond its initial connotation of a visible separation of colored components. It now refers to any separation process in which the motion of the components is effected by a moving stream of gas or liquid. The fluid and the material to be separated move together through a tube. To effect the separation, there must be some medium

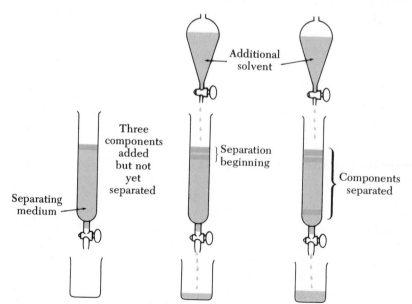

Fig. 1.2. *Column chromatography.*

Additional solvent

Three components added but not yet separated

} Separation beginning

Components separated

Separating medium

* Carotenoids are yellow-red pigments that impart characteristic color to carrots, butter, and other natural organic products.

(like Tswett's chalk) in the tube that retards some components more than others, and allows them to emerge from the tube at different times. Finally, we must be able to detect the separated components.

The development of a great variety of highly selective retention media, and of several very sensitive methods of detection, have made chromatographic methods the most widely used and effective means of analytic separation of materials. For example, the components of marijuana have been separated and identified by this technique. Generally, the sample analyzed is about 10^{-3} g and the amount of component detected may be as small as 10^{-12} g.

PROBLEMS

1. Definitions. Define the following terms: (a) substance, (b) mixture, (c) ion, (d) decomposition, (e) homogeneous, (f) heterogeneous, (g) phase.

2. Definitions. Discuss the inadequacies and limitations of the following definitions. Supply better ones. (a) A molecule is the smallest particle of matter that retains the specific properties of the matter. (b) A solid is matter that retains its shape. (c) A liquid is matter that takes the shape of its container. (Is this true of a liquid in a vehicle in interplanetary space?) (d) An element is a substance that cannot be decomposed by chemical means. (e) A compound is a substance that contains more than one element.

3. Development of chemistry. Imagine that you visited another planet inhabited by rational beings and learned their language. You then examined their chemistry textbooks, chemical laboratories, and chemical manufacturing plants. Which of these three would you expect to be most recognizable by resemblance to their counterparts on Earth? Defend your answer.

4. Chemical and physical change. Identify the chemical and the physical changes in the following sequences: (a) Carrots are scraped, cooked, mashed, sieved, swallowed, digested, and utilized metabolically for coenzyme manufacture. (b) Oxygen is liquefied, poured into rocket storage tanks, used as a fuel component during flight, used to pressurize the capsule, inhaled by the astronauts, and converted during respiration to CO_2 and H_2O. (c) A lump of sugar is ground to a powder and then heated in air. It melts, then darkens, and finally bursts into flame and burns. (d) Gasoline is sprayed into the carburetor, mixed with air, converted to vapor, burned, and the combustion products expand in the cylinder.

5. Phases of matter. A sealed glass bulb is half filled with water, on which some ice and wood are floating. The remainder of the bulb is filled with air. How many phases does this system contain? Identify them.

6. States of matter. Identify the state of aggregation of each of the samples of matter described: (a) A lump of rigid matter cannot be deformed by hand pressure. When struck with a hammer, however, it breaks into fragments that have randomly curved surfaces. (b) A lump of rigid matter cannot be deformed by hand pressure. Several adjoining planar faces of the sample form angles of 120° with each other. When struck with a hammer, the sample breaks into fragments some of which have new sets of faces that form 120° angles. (c) A colored transparent material is in a closed glass bottle. When the bottle is opened, the color of the material becomes lighter, first near the top, then gradually throughout the sample, until it disappears entirely from the bottle.

7. States of matter. Suppose we classified the orderliness of arrangements of molecules in matter as (1) none, (2) intermediate, (3) high, and the degree of cohesiveness as (a) low, unable to maintain a rigid shape under the stress of gravity, (b) high, able to maintain a rigid shape under the stress of gravity. Outline the states of aggregation of matter based on this classification and, where feasible, assign the appropriate common name of a state of matter to each category.

8. Purity of matter. In 1875 a sample of distilled water resisted further attempts at purification and was called "pure." Techniques available in 1975 made it possible to separate the "pure" sample into several components, including protium deuterium oxide and dideuterium oxide. Has the purity of the unchanged sample deteriorated because of advance in technique? Was the sample pure in 1875? Is it possible for future advances in methods to show that components separated today are "impure"?

9. Scientific method. Does a scientific investigation start with accumulation of facts, formation of hypotheses, both, or neither? Defend your answer.

2

GASES

THE PROPERTIES OF GASES / 2.1

Gaseous materials are well known in our experience, and, under ordinary conditions, we have no trouble in differentiating them from solids, liquids, or glasses. A qualitative summary of the characteristic properties of gases under ordinary conditions is given in Table 2.1.

Quantitative experimental investigations of gases were begun in the seventeenth century. We shall discuss the findings of some of these experiments and then consider a general theory of gas behavior.

Table 2.1
Characteristic Properties of Gases

Color	All gases are transparent; almost all are colorless; some exceptions: fluorine, chlorine (both green–yellow), bromine (red–brown), iodine (violet), nitrogen dioxide, dinitrogen trioxide (both red–brown)
Mobility	Compared with other forms of matter, gases show little resistance to flow; gases disperse rapidly in space and effuse through porous barriers
Response to changes of pressure, temperature	Compared with other forms of matter, gases expand greatly on heating or on reduction of applied pressure

BOYLE'S LAW / 2.2

Robert Boyle (1627–1691) investigated pressure–volume relationships of air by pouring successive quantities of mercury into the open arm of a J-shaped tube as shown in Fig. 2.1. After each addition,

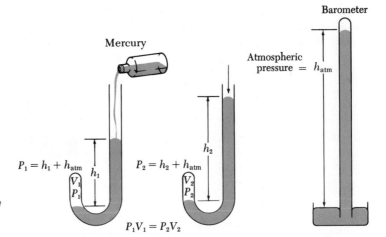

Fig. 2.1. *Boyle's apparatus. The liquid in the tubes is mercury; the pressure is measured in terms of the heights of the columns. The addition of the barometer reading takes into account the atmospheric pressure on the right arms of the J-tubes.*

Mercury

Atmospheric pressure $= h_{atm}$

Barometer

$P_1 = h_1 + h_{atm}$ h_1 V_1 P_1

$P_2 = h_2 + h_{atm}$ h_2 V_2 P_2

$P_1 V_1 = P_2 V_2$

he measured (a) the volume of air in the closed section of the tube and (b) the difference between the heights of the mercury columns in the two arms. The pressure on the trapped air in the lower section is the *sum* of the pressures exerted by (1) the difference in the mercury levels plus (2) the atmospheric pressure. Gas pressures have traditionally been expressed in terms of heights of columns of mercury. (Refer to Appendix I.10 for a more complete discussion.) Thus, the pressure of the trapped gas in V_1 is expressed as the sum of the height of the mercury column that confines it, h_1, plus the height of mercury in a barometer, h_{atm}. We shall express pressure in TORR units, where 1 torr is the pressure exerted by 1 mm of mercury at 0°C and standard gravity.

Boyle was probably aware that minor fluctuations in room temperature would not affect his results significantly. His PV values were constant to within about 1%. Thus, Boyle's law states that *the volume of a gas at constant temperature is inversely proportional to the pressure,* $V \propto 1/P$, or $PV = constant$.

Problems dealing with P–V relationships among gases can be solved as follows: Let P_1, V_1 refer to one set of conditions and P_2, V_2 to another set of conditions of the same quantity of the same gas at the same temperature. Then $P_1 V_1$ and $P_2 V_2$ both equal the same constant and therefore equal each other:

$$P_2 V_2 = P_1 V_1 \tag{1}$$

The effect on gas volume of a change in pressure can be calculated by substituting values in Equation (1).

Alternatively, we can convert the original volume to the new volume by multiplying by the ratio of the pressures. The form of this correction factor is established by our "common sense" knowledge of gas behavior. Thus, an increase of pressure will make the volume less; therefore, the larger of the two pressures goes in the denominator. A decrease of pressure increases the volume; therefore, the larger of the two pressures then goes in the numerator.

STANDARD PRESSURE, referring to gases, means 1 atmosphere, or 760 torr.

Example 1 The pressure on 10.0 liters of gas at 760 torr is reduced to 700 torr at constant temperature. What will the new volume be?

Answer $V_1 = 10.0$ liters, $P_1 = 760$ torr, $P_2 = 700$ torr.

From Equation (1)	*Common sense method*
$$V_2 = V_1 \times \frac{P_1}{P_2}$$	New volume = original volume $\times$ pressure correction (the gas will expand because of pressure reduction, so correction factor is greater than unity and larger pressure goes into numerator)
$$= 10.0 \text{ liters} \times \frac{760 \text{ torr}}{700 \text{ torr}}$$	$$= 10.0 \text{ liters} \times \frac{760 \text{ torr}}{700 \text{ torr}}$$

Therefore $V_2 = 10.9$ liters.

THE LAW OF CHARLES AND GAY-LUSSAC; ABSOLUTE TEMPERATURE / 2.3

Jacques Charles measured the relationship between temperature* and volume of gases around 1787, but his data are not available. His work was confirmed in 1802 by Joseph Gay-Lussac.

The results of such a study show that the volume of a given sample of gas at constant pressure increases linearly with increasing temperature. If 273 ml of any gas is warmed from 0 to 1°C, it expands to 274 ml. This *same sample* of gas will expand by 1 ml during *any* 1-deg rise in temperature, or will contract by 1 ml for any 1-deg fall in temperature. This statement implies that if the gas is cooled to −273°C, it will shrink by 273 ml, to nothing. Leaving aside the question of whether such a disappearance can really occur, a graph of this relationship becomes a straight-line plot of gas volume versus temperature, such as appears in Fig. 2.2.

The lowest possible temperature of a gas corresponds to a theoretical total loss of volume. This temperature is called **ABSOLUTE ZERO**, because it is absurd to suppose that any substance might possess a negative volume. The best current value for absolute zero is −273.15°C.

---

* Thermometers had been developed in Europe starting around 1600. Earliest Western concepts of gas expansion by heating probably originated in the mechanical toys of the Hellenistic period, such as Hero's engine.

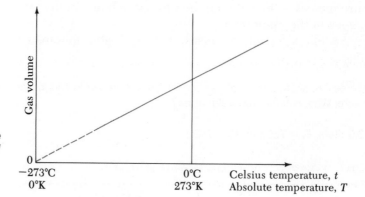

Fig. 2.2. Volume–temperature relationship for a gas obtained from experiments at low constant pressure (Charles' law).

This temperature is taken as the zero point of the **KELVIN* SCALE OF ABSOLUTE TEMPERATURE**. Temperatures on this scale are written as, for example, 150K or 150°K (Appendix I.6). An unspecified absolute temperature is represented by T, while Celsius temperatures are represented by t. Then,

$$T(°K) = t(°C) + 273.15 \tag{2}$$

Now, we affix the Kelvin scale to the horizontal axis of Fig. 2.2, so that zero gas volume corresponds to 0°K. Then the graph corresponds to Charles' law, which is

gas volume $\propto$ absolute temperature

or

$$V = constant \times T$$

Charles' law can be used to solve problems concerning V–T relationships of gases at constant pressure. If V_1, T_1 refer to one set of conditions and V_2, T_2 refer to another set of conditions of the same quantity of the same gas at the same pressure, then

$$\frac{V_2}{V_1} = \frac{T_2}{T_1} \tag{3}$$

The "common sense" application of correction factors can also be used (see below).

The term "**STANDARD TEMPERATURE**," as used in calculations involving gases, means 0°C. The term "**STANDARD CONDITIONS**" (sc) means standard temperature and pressure, 0°C and 760 torr.

Example 2 10.0 ml of gas at 20.0°C is cooled at constant pressure to 0°C. What will the new volume be?

---

* Named after Lord Kelvin (1824–1907), who explained the fundamental significance of this scale.

Answer $V_1 = 10.0$ ml, $T_1 = (20 + 273)°K$, $T_2 = (0 + 273)°K$.

New volume = original volume × temperature correction
(because cooling reduces gas volume,
correction factor must be less than unity)

$$V_2 = 10.0 \text{ ml} \times \frac{273°\cancel{K}}{293°\cancel{K}} = 9.32 \text{ ml}$$

AVOGADRO'S LAW / 2.4

In 1738 Daniel Bernoulli proposed a concept of gas behavior based on the assumption that gases are composed of invisibly small, separated particles called MOLECULES. The idea that the volume of a gas depends on the number of molecules it contains was vaguely conceived by scientists as early as about 1760. The first clear statement of this relationship was made by Amedeo Avogadro in 1811, but it was not fully accepted until the later work of Cannizzaro, about 1860. In modern terms, it says, "*equal volumes of all gases at the same temperature and pressure contain the same number of molecules.*" Expressed as an equation, where N is the number of molecules, Avogadro's law is

$$N \propto V \quad \text{or} \quad N/V = constant$$

The concept is highly useful since it teaches us that we may select equal numbers of molecules simply by selecting equal volumes of different gases at the same temperature and pressure.

Example 3   One gram of radium emits α particles at the rate of 1.16×10^{18}/year. Each α particle becomes a molecule of helium gas. The total volume of the 1.16×10^{18} molecules is 0.0430 ml at standard conditions. How many molecules are there in 1 liter of helium under standard conditions? In 1 liter of oxygen? In 1 liter of any gas?

Answer

$$\frac{N}{V} = \frac{1.16 \times 10^{18} \text{ molecules}}{0.043 \cancel{\text{ml}}} \times 1000 \frac{\cancel{\text{ml}}}{\text{liter}}$$

$$= 2.70 \times 10^{22} \frac{\text{molecules}}{\text{liter}}$$

Since N/V is a constant at constant temperature and pressure, regardless of the nature of the gas, the ratio applies to helium, oxygen, or any other gas.

THE GAS LAW; THE MOLE / 2.5

Boyle's, Charles', and Avogadro's laws, taken together, are expressed by the relationship $PV \propto NT$, or $PV = constant \times NT$. This relationship is called the IDEAL GAS LAW. It is found, within experimental precision, that the proportionality constant is the same for all gases. To evaluate the constant, we must establish values for the four variables, P, V, N, and T. We can readily measure three of these: pressure, volume, and temperature, but it is rather more difficult to count molecules. Let us then arbitrarily select a standard number of molecules, which we call the AVOGADRO NUMBER, N_A. We define N_A as the number of molecules in 31.9988 g of oxygen. (This value is selected because it is the molecular weight of oxygen; page 34.) The volume occupied by this mass of oxygen at 0°C and 1 atm pressure is 22.4 liters. Avogadro's law teaches us that this is also the volume occupied by N_A molecules of *any* gas at the same conditions, independent of their mass.

The quantity of any substance containing an Avogadro number of molecules is called a MOLE. The reader should recognize that a mole is simply a standard number, like a dozen. To illustrate the similarity:

$$n \text{ (moles)} \times N_A \left(\frac{\text{molecules}}{\text{mole}} \right) = N \text{ (molecules)} \tag{4}$$

$$(\text{dozens}) \times 12 \left(\frac{\text{eggs}}{\text{dozen}} \right) = (\text{eggs})$$

where N_A = number of molecules in 22.4 liters of gas at 0°C and 1 atm (analogous to a dozen eggs in an egg box).

Example 4 Using the data of Example 3, calculate the Avogadro number.

Answer The Avogadro number is the number of molecules in 1 mole of a gas (22.4 liters at sc). From Example 3, this is

$$2.70 \times 10^{22} \, \frac{\text{molecules}}{\cancel{\text{liter}}} \times 22.4 \, \frac{\cancel{\text{liters}}}{\text{mole}} = 6.05 \times 10^{23} \text{ molecules/mole}$$

The Avogadro number has been determined by several methods in addition to the one described by Example 3, and the best currently accepted value is $(6.0225 \pm 0.0003) \times 10^{23}$ molecules/mole.

The ideal gas equation may now be expressed in terms of moles; the constant is then represented by R, and is known as the GAS CONSTANT.

$$PV = nRT \tag{5}$$

R may then be evaluated from the following data:

$n = 1.00$ mole
$V = 22.4$ liters
$\left. \begin{array}{l} P = 1.00 \text{ atm} \\ T = 273°\text{K} \end{array} \right\}$ "standard conditions" (sc)
$R = PV/nT$

$$= \frac{1.00 \text{ atm} \times 22.4 \text{ liters}}{1.00 \text{ mole} \times 273°\text{K}}$$

$$= 0.0821 \text{ liter atm/°K mole}$$

Equation (5) is especially useful for solving problems involving a single set of conditions, rather than ratios between different sets. In using this equation, it is very important to use the correct units. Most errors involving this equation result from inconsistency in the units in which the five quantities are expressed.

Example 5 To what temperature must 1.00×10^{-4} mole of a gas be heated in a 25.0-liter container to obtain a pressure of 1.00×10^{-1} torr?

Answer

$$T = \frac{PV}{nR}$$

$$= \frac{1.00 \times 10^{-1} \, torr \times 25.0 \, liters}{1.00 \times 10^{-4} \, mole \times 0.0821 \, \dfrac{liter \, atm}{mole \, °K} \times 760 \, \dfrac{torr}{atm}}$$

$$= 401°K, \text{ or } 128°C$$

DALTON'S LAW OF PARTIAL PRESSURES / 2.6

The pressure of a gas in a rigid container at constant temperature depends only on the number of moles:

$$P = \left(\frac{RT}{V}\right)n$$

where (RT/V) is constant. Now, what if the gas is a mixture of gaseous substances, like air? Remember, R applies to all gases; each gas is at the same temperature as the mixture of gases; and, of course, *each gas occupies the same total volume*—the volume of the container. Therefore (RT/V) is also constant for a *mixture* of gases in a rigid container at constant temperature. Let us then consider the total pressure, P_{tot}, of a mixture of n_1 moles of gas$_1$ plus n_2 moles of gas$_2$. Then,

$$P_{tot} = (RT/V) \times (n_1 + n_2) \tag{6}$$

Let us now separate the two gases into individual containers,* each having the original volume. The pressures exerted in the new containers are called the **PARTIAL PRESSURES**, p_1 and p_2. The partial pressure is defined simply as the pressure that one component of a gas mixture would exert if it occupied the same volume alone. Then,

$$p_1 = (RT/V) \times n_1 \quad \text{and} \quad p_2 = (RT/V) \times n_2$$

and, by addition,

$$p_1 + p_2 = (RT/V) \times (n_1 + n_2) \tag{7}$$

---

* This experiment need not be performed. It is sufficient just to think of it. Werner Heisenberg called such efforts "*Gedanken*" (imaginary) experiments.

By combining Equations (6) and (7) we have $P_{tot} = p_1 + p_2$ (at constant temperature and constant volume). Similar reasoning can be applied to any number of components, so that

$$P_{tot} = p_1 + p_2 + \cdots \qquad (8)$$

Equation (8) is Dalton's law of partial pressures (published in 1803 by John Dalton, of atomic theory fame): *In a mixture of gases, the total pressure equals the sum of the partial pressures.*

This relationship is important in measurements of gases confined over volatile liquids. The evaporation of the liquid adds molecules of a new kind to the gas and makes up a part of its total pressure in accordance with Dalton's law. Fortunately for measurement purposes, the pressure contribution from such evaporation at constant temperature reaches a constant value called the VAPOR PRESSURE OF THE LIQUID. In laboratory practice, the confining liquid is often water.

Example 6 The composition of dry air "by volume" is 78.1% nitrogen, 20.9% oxygen, and 1.0% of other gases. Calculate the partial pressures, in atmospheres, in a tank of dry air compressed to 10.0 atm.

Answer "Composition by volume" is a poor expression, although it is universally used. As we stated earlier, each component of a gaseous mixture occupies the *entire volume,* not a percentage of it. What the expression really means is "composition by number of molecules." Thus, 78.1% of the molecules are nitrogen, 20.9% are oxygen, etc. Since the pressure is proportional to the number of molecules, the so-called proportion by volume is also the proportion by pressure, or

$$\text{proportion by volume} = \frac{\text{partial pressure}}{\text{total pressure}}$$

Then

$$\text{partial pressure} = \text{proportion by volume} \times \text{total pressure}$$

By Dalton's law, the sum of the partial pressures is the total pressure. Therefore,

$$p_{nitrogen} = \frac{78.1}{100} \times 10.0 \text{ atm} = 7.81 \text{ atm}$$

$$p_{oxygen} = \frac{20.9}{100} \times 10.0 \text{ atm} = 2.09 \text{ atm}$$

$$p_{other\ gases} = \frac{1.00}{100} \times 10.0 \text{ atm} = 0.10 \text{ atm}$$

Example 7 10.0 ml of oxygen is collected over water at 20°C and 770.0 torr. What is the volume of the dry gas at standard conditions? The vapor pressure of water at 20°C is 17.5 torr (Appendix V).

Answer $V_1 = 10.0$ ml, $T_1 = 293°$K. To find P_1 use the partial pressure law, $P_{tot} = p_{oxygen} + p_{water}$. Therefore,

$$P_1 = p_{oxygen} = P_{tot} - p_{water} = 770.0 \text{ torr} - 17.5 \text{ torr} = 752.5 \text{ torr}$$

P_2 and T_2 are the standard conditions (sc), 760 torr and 273°K. Then (common sense)

$$V_2 = 10.0 \text{ ml} \times \frac{273°\text{K}}{293°\text{K}} \times \frac{752.5 \text{ torr}}{760 \text{ torr}}$$

$$= 9.23 \text{ ml}$$

IDEAL GASES; THE KINETIC MOLECULAR HYPOTHESIS / 2.7

Most scientists are dissatisfied with purely mathematical descriptions of natural phenomena. They find that their insights are improved if they conceive of phenomena in terms of physical models. A model is a representation of reality, and a good model is a successful compromise between simplicity and accuracy. The observed behavior of gases is fairly well explained by a model that we call the IDEAL GAS.

The ideal gas consists of molecules that have mass and velocity, but no volume, and that exhibit no attractive or repulsive forces among themselves or with other matter. The ideal gas, when unconfined, will rapidly disperse. In a rigid container, however, the molecules will bounce off the inside walls without loss of energy.

The pressure of the gas is the force that it exerts per unit area of its container. This force is applied by the collisions of the molecules with the walls. Consider a molecule of mass m which approaches the wall at velocity $-u$ (one direction) and rebounds at equal speed with velocity $+u$ (reverse direction). The change in velocity is therefore $2u$ (Fig. 2.3). The force exerted by this collision is

$$\frac{\text{force}}{\text{collision}} = \text{mass} \times \text{acceleration (Appendix I.9)}$$

$$= \text{mass} \times \frac{\text{change in velocity}}{\text{time between collisions}} = m \times \frac{2u}{\text{time}}$$

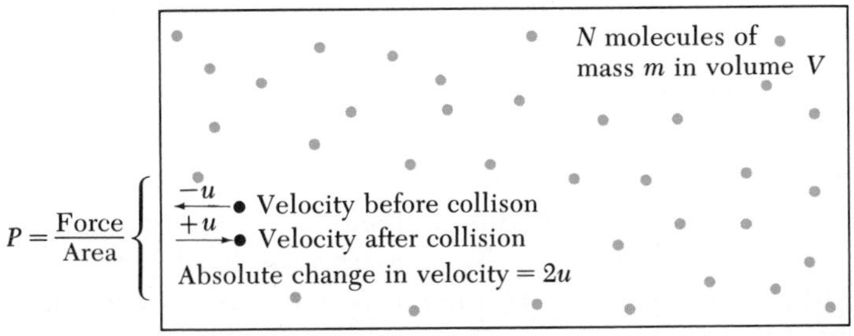

$$P = \frac{\text{Force}}{\text{Area}} \begin{cases} \end{cases}$$

N molecules of mass m in volume V

$-u$ • Velocity before collison
$+u$ • Velocity after collision
Absolute change in velocity $= 2u$

Fig. 2.3. Gas molecules confined in a container.

We assume that collisions occur so frequently that the pressure remains practically constant. Then,

$$\text{pressure } (P) = \frac{\text{force}}{\text{collision}} \times \frac{\text{collisions}}{\text{area}}$$

$$= \frac{2mu}{\text{time}} \times \frac{\text{collisions}}{\text{area}} = 2mu \times \frac{\text{collisions}}{\text{time} \times \text{area}}$$

Letting

$$Z = \frac{\text{collisions}}{\text{time} \times \text{area}}$$

we have

$$P = 2muZ \tag{9}$$

Since Z is proportional to the frequency of collisions, it follows that Z is proportional to Nu/V. This proportionality can be understood from the following relationships:

Z is proportional to N/V, the number of molecules per unit volume, because the more molecules in a unit of volume the more frequent will be the collisions with the walls.

Z is proportional to the molecular speed, u, because the faster the molecules move, the more frequent will be the collisions with the walls. Thus,

$$Z = constant \times \frac{Nu}{V}$$

If the container were a cube, and all the molecules approached the walls perpendicularly, then only $\frac{1}{6}$ of the molecules would be approaching a given wall, because a cube has six sides. Of course, not all molecules move along perpendicular lines, and not all containers are cubes, but a rigorous proof (which will not be presented here) confirms that the constant $\frac{1}{6}$ applies to all cases. Then,

$$Z = Nu/6V$$

Substituting this value for Z in Equation (9), we have,

$$P = 2mu \times \frac{Nu}{6V} = \frac{1}{3}\frac{Nmu^2}{V}, \quad \text{or} \quad PV = \frac{1}{3}Nmu^2 \tag{10}$$

We now ask two questions about Equation (10).

First, what about the molecules of various masses, such as N_1 molecules of mass m_1, N_2 of mass m_2, etc? Second, what about molecules of various speeds, such as N_1 molecules of speed u_1, N_2 molecules of speed u_2, etc?

To accommodate these variations, we may rewrite Equation (10) as follows:

$$PV = \tfrac{1}{3}N\overline{mu^2} \qquad (11)$$

The term $\overline{mu^2}$ is the average, for all the molecules, of the product of the mass and the square of the speed.

The kinetic energy of a moving molecule (or of any other particle, Appendix I.11) is $\tfrac{1}{2}mu^2$. It will therefore be helpful to rewrite Equation (10) by multiplying by $\tfrac{2}{2}$, obtaining:

$$PV = \tfrac{2}{3}N(\tfrac{1}{2}\overline{mu^2})$$

and, substituting the average kinetic energy $\overline{E}$ of the molecules for $(\tfrac{1}{2}\overline{mu^2})$, we have

$$PV = \tfrac{2}{3}N\overline{E} \qquad (12)$$

Equation (12) tells us that for any given number of molecules N of a gas, the PV product is proportional to the average kinetic energy $\overline{E}$ of all the molecules. Experiment (Boyle's and Charles' laws) tells us that PV is proportional to the absolute temperature T. Therefore, the temperature of a gas must be a manifestation of the average kinetic energy of its molecules. The relationship can be obtained by combining Equations (5) (page 14) and (12),

$$PV = \tfrac{2}{3}N\overline{E} = nRT \qquad (13)$$

from which

$$\overline{E} = \frac{3}{2}\frac{n}{N}RT$$

and, since [from Equation (4), page 14] $n/N = 1/N_A$,

$$\overline{E} = \frac{3}{2}\frac{RT}{N_A}$$

The ratio R/N_A is called BOLTZMANN'S CONSTANT, k, and has the value 1.38×10^{-16} ergs/°K molecule. Then,

$$\overline{E} = \tfrac{3}{2}kT$$

We see from Equation (13) that the model we have called the "ideal gas" is consistent with the ideal gas law, Equation (5).

THE DISTRIBUTION OF MOLECULAR SPEEDS / 2.8

The molecules of an ideal gas, being volumeless points, would never hit each other. Real molecules, however, occupy volume, and therefore collide with each other frequently. On colliding, some molecules speed up while others slow down. As a result, not all the molecules in a given sample have the same speed at any one time, even at a constant temperature. Even if all the molecules in a container had the same initial speed, the effects of collisions would soon disrupt such uniformity. To visualize this effect, consider two molecules at the same speed approaching each other at an angle, and hitting each other with a glancing blow. Upon collision one molecule may speed up, the other may slow down. An individual molecule in a gas sample will, therefore, have different speeds at different times, depending on the kind of collision it underwent last. However, a measurable sample of gas has so many molecules (for example, 10^{20} molecules) that at any one time the fraction of molecules having a speed close to any particular value will be constant. A very small fraction of the molecules will have speeds close to zero; another small fraction will have abnormally high speeds; most of the molecules will have speeds close to some average value. The shape of the distribution curve agrees with theory and experiment.

The predicted shape is shown in Fig. 2.4, in which the relative number of molecules is plotted against their speeds. The curve is not symmetrical because it is limited to zero at the left end, but it is limitless at the right. Therefore, it flattens out to account for the few speedy molecules. The most probable speed of a molecule, u_{mp}, is represented by the curve's peak, but the average speed $\bar{u}$ is somewhat greater than this value because the curve is skewed to the right. The *area* under the curve *represents the total number of molecules in the sample.* A typical average molecular speed is about 400 m/second for nitrogen at 0°C. Some students find the meaning of this curve hard to visualize. Look at it this way: Take a large number of people. Find out how old each one is (to the nearest year). For each person, place a little square just above his year as shown in the graph. Now the *area* (the number of squares) under the curve between, say, age 20 and age 40 gives the number of people in that range of ages. The *total area* under the curve gives the total number of people because everyone has some age. Now imagine that the squares are molecules and the "age" is their speed and you have the distribution of molecular speeds.

The experiment illustrated in Fig. 2.5 is an adaptation of a method of determining molecular speeds originally devised by Otto Stern. A narrow beam of molecules is aimed at a rapidly rotating solid band that is split in one position. Once every rotation some molecules enter the slit and hit the opposite inside surface of the band. They do not all hit at the same spot, however, because they are traveling at different speeds. The faster ones arrive first and deposit on the leading surface of the band. Then, most of the molecules with about average speeds arrive, hitting the band to the left of the faster molecules. Finally, the slower molecules arrive last, hitting further to the left, on the trailing surface of the band. The location of any given molecule along the band is, therefore, related to its speed, and a plot of the density of deposition is, therefore, the distribution of molecular speeds.

If a gas sample is heated from T_1 to a higher temperature T_2 the

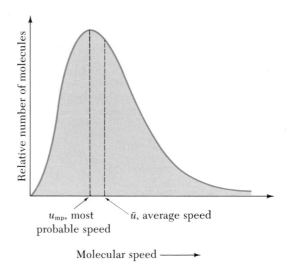

u_{mp}, most probable speed

$\bar{u}$, average speed

Molecular speed ⟶

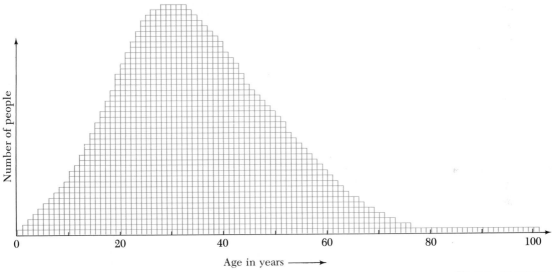

Age in years ⟶

Fig. 2.4. *Distributions of molecular speeds and ages of people. (The age distribution refers to an arbitrary group—not a typical population.)*

average speed of the molecules must increase, but of course the number of molecules remains constant and the lowest possible speed is still zero. The result is as if one tugged at the right tail of the curve while the left was fixed to the origin; the shape becomes flatter and more skewed to the right (Fig. 2.6). The important consequence for chemistry is not so much that the average speed has increased somewhat, but that the proportion of really speedy molecules has increased considerably. Let us say that only molecules faster than u_1 will undergo a given chemical reaction. Figure 2.6 shows that the reaction would hardly occur at all at temperature T_1, but would occur much more rapidly

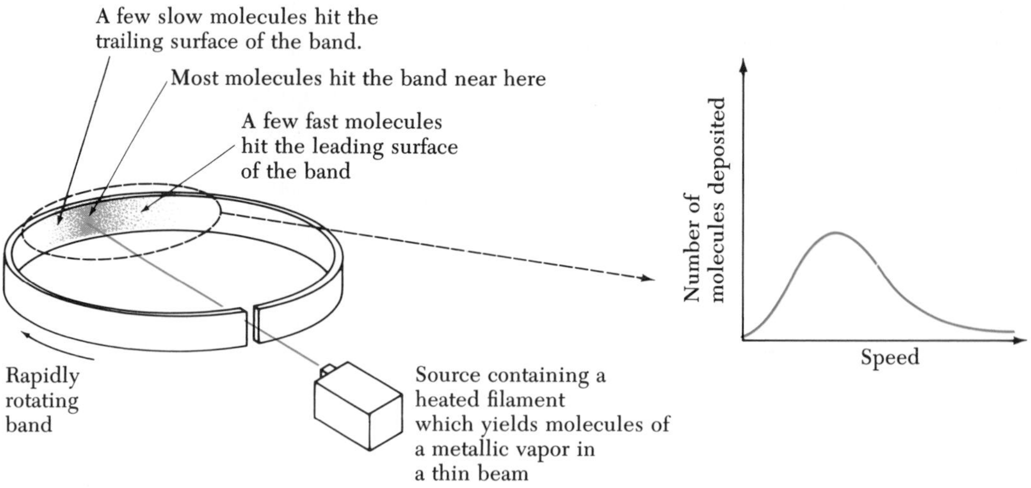

A few slow molecules hit the
trailing surface of the band.

Most molecules hit the band near here

A few fast molecules
hit the leading surface
of the band

Rapidly
rotating
band

Source containing a
heated filament
which yields molecules of
a metallic vapor in
a thin beam

Number of
molecules deposited

Speed

Fig. 2.5. *Experimental
verification of the distribution
of molecular speeds.*

at T_2. The reason is that the colored area under the T_2 curve to the right
of u_1 (representing the molecules speedy enough to react) is about 70
times greater than that under the T_1 curve.

DEVIATIONS FROM IDEAL BEHAVIOR / 2.9

Real gases do not behave in an ideal way. The important differences
between ideal and real gases, and the consequences of these differences, are the following:

(a) **The volume occupied by molecules.** Real molecules have volume.

The pressure exerted by a given sample of gas in a container at a
given temperature, as predicted by the ideal gas law, is

$$P_{\text{ideal}} = \frac{nRT}{V} \qquad \text{or} \qquad \frac{P_{\text{ideal}}V}{nRT} = 1$$

in which V *is the volume of the container.*

However, in a real gas the void space is *less* than the volume of the
container V, because *some of the space is occupied by molecules.*
Therefore, the molecules act as if they were in a smaller container, and
they hit a wall more frequently, thus exerting a greater pressure. Let us
designate the measured or observed pressure P_{obs}. Then

$$P_{\text{obs}} > P_{\text{ideal}}$$

Consequently,

$$\frac{P_{\text{obs}}V}{nRT} > 1$$

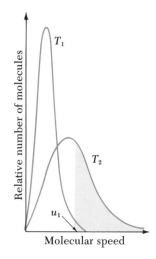

Relative number of molecules

T_1

T_2

u_1

Molecular speed

Fig. 2.6. *Distribution of
molecular speeds at T_1 (lower
temperature) and T_2 (higher
temperature).*

(b) The attraction of molecules for each other. Real molecules attract each other. Mutually attracting molecules, when they approach each other, may actually remain together for a short time (a sticky collision), or they may merely deflect each other's paths somewhat. If they stick together, then the number of independent molecules in the sample at any one time is decreased. As a result, the gas pressure will be reduced, and $P_{obs} < P_{ideal}$, or

$$\frac{P_{obs}V}{nRT} < 1$$

If the molecules do not actually stick together, but just deflect each other, then their paths become curved instead of straight, and the molecules must travel farther to hit the wall. As a result, they hit it less frequently and therefore exert less pressure. Again, the result is that $P_{obs} < P_{ideal}$, which is the same as if the molecules had actually stuck together.

Deviations from ideal behavior are shown in Fig. 2.7. For hydrogen at 0°C, the molecular attractive forces are low and do not produce a predominant effect; instead, the volume of the molecules yields PV/nRT values exceeding the ideal value, 1, at all pressures shown on the graph. For nitrogen, the intermolecular attractive forces are great enough to yield negative deviations ($PV/nRT < 1$) up to about 150 atm. Intermolecular attractions in carbon dioxide, even at 40°C, are obviously much more important than those in nitrogen at 0°C. For either nitrogen or carbon dioxide a pressure exists (150 or 600 atm, respec-

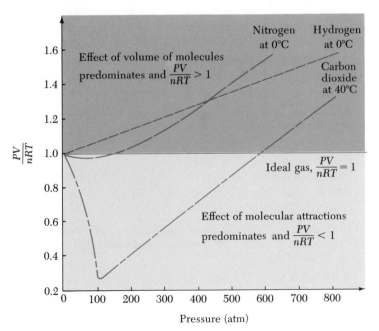

Fig. 2.7. *Deviations from ideal gas behavior.*

tively) at which the two effects cancel each other and the PV/nRT product is ideal, that is, it equals 1.

At low pressures, the volume of a given quantity of gas becomes large, and the fraction of the volume occupied by the molecules becomes relatively small. Since the molecules become more widely separated at low pressures, their mutual attractions and tendencies to aggregate lessen. At high temperatures, the molecules have greater kinetic energy and their tendencies to aggregate lessen. Thus, high temperatures and low pressures both favor ideal gas behavior. At room temperature, and 1 atm pressure, deviations from ideality are of the order of 1% for most gases.

Chemists have devised several equations relating P, V, n, and T that give better approximations of the behavior of real gases than the ideal gas law provides. These equations are more complicated than $PV = nRT$ because they include several empirical constants to account for the nonideal effects described above. One such relationship, the van der Waals equation, is given in Problem 25, page 27.

PROBLEMS

1. Properties of gases. Suggest explanations for the facts that gases (a) are transparent, (b) do not settle to the bottom of a container, (c) diffuse through all the space available to them.

2. Temperature scales. Convert the following to the Kelvin scale: (a) 100°C, (b) 273°C, (c) −273°C, (d) −78.2°C. Convert the following to the Celsius scale: (e) 0.5°K, (f) 1.1×10^5 K, (g) 273°K.

3. Boyle's law. A sample of helium occupies 500 ml at 2.00 atm. Assuming that the temperature remains constant, what volume will the helium occupy at (a) 4.00 atm; (b) 760 torr; (c) 1.8×10^{-3} torr?

4. Charles' law. The volume of a gas at 25.0°C is 10.5 liters. (a) If the gas temperature rises to 250°C at constant pressure, would you expect the volume to increase by a factor of 10 to 105 liters? Explain. (b) Calculate the new volume.

5. Charles' law. The following data are obtained for a given sample of gas at constant pressure:

T, °K	V, ml
250	5.0
300	6.0
350	7.0

(a) Evaluate the Charles' law constant for the conditions of this experiment. (b) Will the constant depend on the number of molecules of gas in the sample? On the pressure of the gas?

6. Dalton's law. A mixture of nitrogen and oxygen at 1.00 atm pressure is stored in a 3.71-liter iron container at constant temperature. The iron eventually reacts with all the oxygen, converting it to a solid oxide of negligible volume.

The final pressure is 450 torr. (a) What is the final volume of the nitrogen? (b) What are the initial and final partial pressures of N_2 and O_2?

7. Avogadro's law. A 100-ml container holds 5.0×10^{10} molecules of an insect-attracting gas at a given temperature and pressure. (a) How many molecules would remain if the container volume were decreased to 5.0 ml? (b) Assume that a given insect can be stimulated by as few as 10,000 of these molecules. What volume would this number of molecules occupy at the original temperature and pressure?

8. Ideal gas law. A sample of gas occupies 350 ml at 25.0°C and 600 torr. What will its volume be at 600°C and 25.0 torr?

9. Ideal gas law. 65.0 ml of hydrogen is collected over water at 20.0°C and 765 torr. Calculate the volume of the dry gas at standard conditions. (See Appendix V.)

10. Deviations from ideal behavior. (a) Referring to Fig. 2.7 (page 23), compare, qualitatively, the molecular attractive forces among nitrogen, hydrogen, and carbon dioxide at 100 atm pressure. (b) At 600 atm and 0°C, the PV product for nitrogen exceeds that for hydrogen. Does this mean that the intermolecular forces in hydrogen are the greater at this pressure? Why or why not?

ADDITIONAL PROBLEMS

11. Properties of gases. Suppose you were asked to supply a stated mass of a specific gas in a given rigid container at a specified pressure and temperature. Would it be likely that you could fulfill the request? Explain.

12. Properties of gases. State whether each of the following samples of matter is or is not a gas, or whether the information is insufficient for deciding: (a) A material is in a steel tank under 100 atm pressure. When the tank is opened to the atmosphere, the material suddenly expands, increasing its volume by 10%. (b) A material, on being emitted from an industrial smokestack, rises about 25 ft into the air; viewed against a clear sky, it has a dense, white appearance. (c) 1.0 ml of material weighs 8.2 g. (d) When a material is released from a point 30 ft (equivalent in pressure to about 76 cm of mercury) below the level of a lake at sea level, it rises rapidly to the surface, at the same time doubling its volume. (e) A material is transparent and pale green in color. (f) One cubic foot of a material contains as many molecules as 1 ft³ of air at the same temperature and pressure.

13. Boyle's law. Assume that the value of k in $PV = k$ is 12. Plot the graph of P (x axis) vs V (y axis) for the values $V = 1, 2, 3, 4, 6,$ and 12. What is the shape of the curve? Plot the graph of $1/P$ (x axis) versus V (y axis) for the same values of V. What plot is obtained?

14. Charles' law. The device shown in Fig. 2.8 is a GAS THERMOMETER. (a) At the ice point, the gas volume is 1.25 liters. What would be the new volume if the gas temperature were raised from the ice point to 35°C? (b) Assume the cross-sectional area of the graduated arm is 1.0 cm². What would be the difference in height if the gas temperature went up from 0 to 35°C? (c) What modifications may be made to increase the sensitivity of the thermometer?

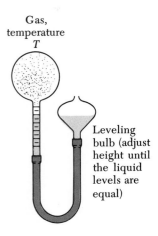

Gas, temperature T

Leveling bulb (adjust height until the liquid levels are equal)

Fig. 2.8. *Gas thermometer.*

15. Absolute zero. One gram of dry air occupies 1060 ml at 100°C and 776 ml at 0.00°C. Construct a graph like that of Fig. 2.2 and plot a straight line corresponding to these data. From your graph, estimate the value of "absolute zero" in °C.

16. Dalton's law. A volume of 200 ml of oxygen gas is collected over water at 25°C and 750 torr. The oxygen is separated from the water and dried at constant temperature and 750 torr. (a) What is the new volume of the dry oxygen? (b) If the water vapor that was removed from the oxygen were stored at the same initial temperature and pressure (25°C and 750 torr), what volume would it occupy? (c) What is the sum of your answers to parts (a) and (b)?

17. Gas density. Starting with the ideal gas law, prove that, for any gas, $P/\rho T$ is a constant, where ρ is the density of the gas.

18. Gas density. 10.0 g of a gas occupies 45.0 liters at 20.0°C and 800 torr. What is the density of the gas at standard conditions?

19. Gas density. At standard conditions 22.4 liters of carbon monoxide weigh 28.0 g. What is the density of carbon monoxide in grams per liter at 20°C and 600 torr?

20. Ideal gas law. A sample of gas occupies 14.3 liters at 19°C and 790 torr. What will its volume be at 190°C and 79.0 torr?

21. Ideal gas law. The gas pressure on the moon's surface is about 10^{-10} torr. Assuming a temperature of 100°K, what volume of the lunar "atmosphere" would have to be sampled at this pressure to obtain (a) 1×10^6 molecules of gas; (b) 1 millimole (10^{-3} mole) of gas?

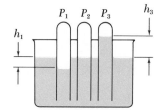

P_1 P_2 P_3 h_3

h_1

Fig. 2.9. *Gas confined by a liquid.*

22. Pressure of a gas confined by a liquid. Consider gases collected over a liquid, as shown in Fig. 2.9. (a) If the liquid is mercury at 0°C, the barometric pressure is P_{atm}, the distances h_1 and h_3 are measured in millimeters, and the ratio (torr/millimeters of mercury) is used as conversion factor, express P_1, P_2, and P_3 in terms of P_{atm} and the height or depth of the mercury columns. (Refer to Appendix I.10.) (b) Assume that the gases are confined not by mercury, but by another liquid of density d g/ml. The density of mercury at 0°C is 13.595 g/ml. Express P_1, P_2, and P_3 in terms of P_{atm}, the height or depth of the liquid columns, and d.

23. Values of the gas constant. Evaluate the gas constant in the ideal gas equation when the other factors have the following values:

P	V	N	T,°K	Constant
(a) 760 torr	22.4 liters	1 mole	273	R'
(b) 1 atm	22.4 liters	6.02×10^{23} molecules	273	R''
(c) $PV = 542$ cal		1 mole	273	R'''

24. Graham's law. This law states that the rates of EFFUSION (the escape of molecules through orifices) of gases are inversely proportional to the square roots of their densities, at constant temperature. Derive this law from the following assumptions: (*i*) The temperature is directly proportional to the average kinetic energy of the molecules, $T \propto \frac{1}{2}\overline{mu^2}$, where m is mass and u is speed. (*ii*) The rate of effusion of a gas is directly proportional to the ROOT MEAN SQUARE SPEED, $\sqrt{\overline{u^2}}$, of its molecules. (*iii*) The density of a gas (at constant temperature and pressure) is directly proportional to the mass of its molecules.

25. van der Waals' equation. This is a modification of the ideal gas law. It incorporates a pressure correction that accounts for intermolecular attractive forces, an^2/V^2, that is added to the measured pressure, and a volume correction that accounts for the volume of the molecules, nb, that is subtracted from the measured volume; a and b are constants for a given gas, and the other symbols have the same meanings as in the ideal gas law. The equation is

$$\left(P + \frac{an^2}{V^2}\right)(V - nb) = nRT$$

Does the behavior of a gas become more ideal ($PV/nRT \rightarrow 1$) or less ideal if: (a) the gas is compressed to a smaller volume at constant temperature; (b) more gas is forced into the same volume at constant temperature; (c) the temperature of the gas is raised at constant volume? (Suggestion: Multiply out the left side and solve for PV/nRT.)

3

ATOMIC AND MOLECULAR WEIGHTS

The atomic theory was first proposed on a philosophical basis about the fifth century B.C. by Leucippus and Democritus. In 1803, John Dalton reaffirmed the theory and put it on a quantitative basis by assigning relative masses to atoms, thereby accounting for the mass relationships observed in chemical reactions. The kinetic molecular theory of gases also evolved during the eighteenth and nineteenth centuries. The great advances in chemistry during the nineteenth century, such as the structural theory of molecules from which organic chemistry (Chapter 23) is derived and the concept of the periodic table (Chapter 7) which systematizes the properties of chemical elements, were based on molecular and atomic theories. All this might suggest that by the start of the twentieth century almost everyone believed in atoms.* Not so. The struggle over the acceptance of the existence of atoms is apparent from the words of Ira Remsen in 1902, taken from his presidential address to the American Chemical Society:

> The doctrine of atoms is still alive although it came into being about a hundred years ago. It has proved to be illogical just as the ether that fills all space has been shown to be incapable of existence. Properties must be ascribed to the atom that it cannot possess and the same is true of the ether. What are we to do? Throw over the atom and the ether? Although both have been convicted of being illogical, I do not think it would be logical to give them up, for they are helpful in spite of their shortcomings, and in some way they suggest great truths. They are symbolic.

Other scientists, especially those who studied spectra of elements (Chapter 6), strongly opposed the nineteenth century criticisms of the atomic theory. The theory proved indeed so "helpful" that it has outlived all the skeptics.

---

* A-*tomos* means "without cutting."

Measurements show that, within the experimental error, the mass (weight)* of substances in a sealed container does not change during a chemical change. For example, suppose that hydrogen, H_2, and air containing oxygen, O_2, are enclosed in a tube so that no matter can enter or escape. The tube and contents are weighed. Passage of an electric spark across the tube causes hydrogen and oxygen gases to disappear, but liquid water and heat are produced. After restoring the original temperature, the tube and contents are reweighed. The observed masses, before and after the chemical change, are identical within the precision of the balance. Such observations are summarized in the LAW OF CONSERVATION OF MATTER: *the mass of a chemically reacting system remains constant.* If matter is created or destroyed (page 69), the quantity is less than can be detected with the best available balance.

THE LAW OF DEFINITE PROPORTIONS / 3.2

Analyses of compounds show that when elements form a given compound, they always combine in the same mass ratio. For example, independently of the source or method of formation, the percent composition of silicon dioxide, SiO_2, is 46.7% by mass of silicon and 53.3% of oxygen. This knowledge is summarized in the LAW OF DEFINITE PROPORTIONS: *the mass composition of a given compound is constant.*

Example 1 10.0 g of silicon dust, Si, is exploded with 100.0 g of oxygen, O_2, forming silicon dioxide, SiO_2. How many grams of SiO_2 are formed and how many grams of O_2 remain uncombined?

Answer From the percent composition of SiO_2, we know that to make 100 g of SiO_2, 46.7 g of Si must combine with 53.3 g of O_2. The mass ratio between the oxygen and the silicon is

$$\frac{53.3 \text{ g } O_2}{46.7 \text{ g Si}}$$

Therefore, for 10.0 g of Si, the quantity of O_2 required is

$$10.0 \text{ g Si} \times \frac{53.3 \text{ g } O_2}{46.7 \text{ g Si}} = 11.4 \text{ g } O_2$$

The mass of SiO_2 formed is 10.0 g Si + 11.4 g O_2 = 21.4 g SiO_2 and the mass of uncombined O_2 is 100.0 g O_2 − 11.4 g O_2 = 88.6 g O_2.

Alternatively, we can write the mass ratio between silicon and silicon dioxide as

$$\frac{100 \text{ g } SiO_2}{46.7 \text{ g Si}}$$

* "Weight" and "mass" will be used synonymously (Appendix I).

Then, for 10.0 g of Si, the quantity of SiO_2 formed is

$$10.0 \text{ g Si} \times \frac{100 \text{ g } SiO_2}{46.7 \text{ g Si}} = 21.4 \text{ g } SiO_2$$

The quantity of O_2 consumed is 21.4 g SiO_2 − 10.0 g Si = 11.4 g O_2 and the mass of uncombined O_2 is 100.0 g O_2 − 11.4 g O_2 = 88.6 g O_2.

THE ATOMIC THEORY / 3.3

The mass relationships of substances participating in chemical reactions are clearly explained in terms of the atomic theory. Although John Dalton (1803) is generally recognized as the originator of the theory, he was anticipated by other scientists, particularly William Higgins (1789). The novel point of Dalton's activities was the determination of the relative masses of atoms, also called ATOMIC WEIGHTS.

The assumptions of the atomic theory were

(i) *The elements are composed of indivisible particles called atoms.*

(ii) *All the atoms of a given element possess identical properties. For example, all atoms of a given element are assumed to have the same mass.*[*]

(iii) *The atoms of different elements differ in properties.*

(iv) *These atoms are the units of chemical changes; chemical changes merely involve the combination, separation, or rearrangement of atoms; atoms are not destroyed, created, or changed.*

(v) *When atoms combine, they combine in fixed ratios of whole numbers forming particles known as molecules.*[†] *Since atoms are indivisible, a fraction of an atom cannot combine.*

This theory differs from previous theories of matter in that it endows atoms with definite properties, particularly with definite masses, and limits the number of different kinds of atoms to the number of elements known.

The theory offers an acceptable explanation of the laws of chemical change:

(a) *The conservation of matter:* Since atoms merely combine or rearrange during a chemical change, the number of atoms of each element in the PRODUCTS of the reaction is the same as the original number in the REACTANTS or starting materials. Since it is further assumed that the mass of an atom does not change, the mass of a chemically reacting system remains constant. Thus, treat 100×10^{10} carbon atoms, C, and 65×10^{10} oxygen atoms, O, in a tube at 1000°C so that 20×10^{10} C atoms combine with 20×10^{10} atoms of O, forming carbon monoxide, CO. The

---

[*] Present knowledge recognizes that the atoms of a given element do not possess uniform masses. Silver, for example, occurs as atoms of two different masses (ISOTOPES, page 38).

[†] The term MOLECULE, meaning a particle composed of similar or dissimilar atoms, was introduced by Avogadro in 1811.

tube now contains 20×10^{10} C atoms and 20×10^{10} O atoms combined in CO plus 80×10^{10} C atoms and 45×10^{10} O atoms uncombined. These add up to the original number of atoms and, since the mass of each atom is constant, the total mass remains constant during the chemical change.

(b) *The law of definite proportions:* Three concepts are involved: (i) A given compound contains only one kind of molecule. (ii) A given molecule always contains the same whole number of atoms of each element. (iii) Each kind of atom has a fixed mass. It follows that the mass ratio of the elements in a given compound must be constant. For example, 20×10^{10} C atoms combine with 20×10^{10} O atoms forming 20×10^{10} CO molecules. Carbon monoxide consists of only one kind of molecule, CO. Each CO molecule contains 1 C atom and 1 O atom. Every C atom has the same mass and every O atom has the same mass. Since each molecule contains the same number of C and O atoms, and since the mass of each of these atoms is also constant, the mass ratio of carbon to oxygen must be constant in any given quantity of carbon monoxide.

Example 2 As most people know, although few understand how the chemist arrived at the conclusion (page 39), water vapor consists of molecules composed of 1 oxygen atom to 2 hydrogen atoms, H_2O. Assume the mass of a hydrogen atom is 1.7×10^{-24} g and that of an oxygen atom is 2.7×10^{-23} g.

(a) Calculate the mass of one molecule. (b) Calculate the mass percent of hydrogen in one molecule. (c) Calculate the mass of 10^3 water molecules. (d) Calculate the mass percent of hydrogen in 10^3 molecules.

Answer
(a) The mass of one molecule of H_2O is given by

$$2 \text{ atoms H} \times (1.7 \times 10^{-24}) \frac{g}{\text{atom H}} + 1 \text{ atom O} \times (2.7 \times 10^{-23}) \frac{g}{\text{atom O}}$$

$$= 3.04 \times 10^{-23} \text{ g}$$

(b) The mass percent of hydrogen is given by

$$\frac{2 \times (1.7 \times 10^{-24}) \text{ g H}}{3.04 \times 10^{-23} \text{ g } H_2O} \times 100\% = 11.2\% \text{ hydrogen}$$

(c) Since 10^3 molecules contain 2×10^3 hydrogen atoms and 10^3 oxygen atoms, the mass of 10^3 water molecules is given by

$$2 \times 10^3 \text{ atoms H} \times (1.7 \times 10^{-24}) \frac{g}{\text{atom H}} + 10^3 \text{ atoms O} \times (2.7 \times 10^{-23}) \frac{g}{\text{atom O}}$$

$$= (3.04 \times 10^{-23}) \times 10^3 \text{ g} = 3.04 \times 10^{-20} \text{ g}$$

(d) The mass percent of hydrogen is given by

$$\frac{2 \times 10^3 \times (1.7 \times 10^{-24}) \text{ g H}}{3.04 \times 10^{-20} \text{ g } H_2O} \times 100\% = 11.2\% \text{ hydrogen}$$

The atomic theory restricts combinations to whole numbers of atoms; fractional numbers are excluded. Further, the number ratio of atoms is fixed for any given combination. The theory, however, does not restrict the number of possible combinations between atoms. The same elements, for example carbon and oxygen, may react under different conditions to form different oxides, such as carbon monoxide, CO, carbon dioxide, CO_2, and carbon suboxide, C_3O_2.

THE LAW OF COMBINING VOLUMES; THE AVOGADRO LAW / 3.4

As a result of his experiments on the volumes of gases involved in chemical changes, Joseph Louis Gay-Lussac concluded (1808) that *when gases react, the volumes consumed and produced, measured at the same temperature and pressure, are in ratios of small whole numbers.* For example, at the same temperature and pressure,

10.6 ml hydrogen gas combines with 10.6 ml chlorine gas to produce 21.2 ml hydrogen chloride gas (*ratios, 1:1:2*)

20.4 ml hydrogen gas combines with 10.2 ml oxygen gas to produce 20.4 ml water vapor (*ratios, 2:1:2*)

12 ml hydrogen gas combines with 4.0 ml nitrogen gas to produce 8.0 ml ammonia gas (*ratios, 3:1:2*)

6.3 ml water vapor combines with solid carbon to produce 6.3 ml hydrogen and 6.3 ml carbon monoxide gas (*ratios, 1:1:1*)

Atoms combine in ratios of small whole numbers. Volumes of gases react and are produced in ratios of small whole numbers. These results suggest strongly that gas volumes have something to do with whole numbers of atoms or molecules.

In 1811 Amedeo Avogadro, whose ideas were ignored for half a century, explained these results by assuming that molecules of elements are composed of similar atoms, that molecules of compounds are composed of dissimilar atoms, and, further, that *equal volumes of all gases at the same temperature and pressure contain the same number of molecules* (page 13). Thus, let N be the number of molecules in a given volume of gas; then, according to the Avogadro hypothesis, the experimental observation

$$\text{1 volume of hydrogen} + \text{1 volume of chlorine} \longrightarrow \text{2 volumes of hydrogen chloride}$$

means that

$$N \text{ molecules of hydrogen} + N \text{ molecules of chlorine} \longrightarrow 2N \text{ molecules of hydrogen chloride}$$

and dividing by N yields

$$\underset{\text{of hydrogen}}{\text{1 molecule}} + \underset{\text{of chlorine}}{\text{1 molecule}} \longrightarrow \underset{\text{hydrogen chloride}}{\text{2 molecules of}}$$

The measured relative volumes give us the relative numbers of molecules reacting and produced.

Since a molecule of hydrogen chloride must contain at least 1 atom of hydrogen and 1 atom of chlorine, the symbol HCl may be used to represent a hydrogen chloride molecule. But there are 2 hydrogen chloride molecules formed for every 1 hydrogen molecule and 1 chlorine molecule used. Hence, to satisfy the law of conservation of matter, 1 hydrogen molecule must contain at least 2 hydrogen atoms and 1 chlorine molecule must contain at least 2 chlorine atoms:

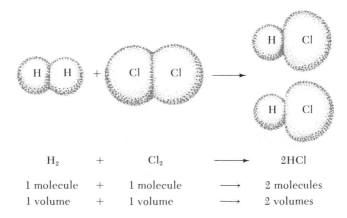

H_2	+	Cl_2	$\longrightarrow$	$2HCl$
1 molecule	+	1 molecule	$\longrightarrow$	2 molecules
1 volume	+	1 volume	$\longrightarrow$	2 volumes

This explanation does not prove that hydrogen and chlorine molecules in the gaseous state are each composed of 2 atoms (DIATOMIC). Nor does it prove that a hydrogen chloride molecule is composed of 1 atom of each constituent element. If we assume H_4, Cl_4, and H_2Cl_2 to be the molecules, we can write

$$H_4 \quad + \quad Cl_4 \quad \longrightarrow 2H_2Cl_2$$
$$\text{1 molecule} + \text{1 molecule} \longrightarrow \text{2 molecules}$$
$$\text{1 volume} \quad + \text{1 volume} \quad \longrightarrow \text{2 volumes}$$

This assumption also explains the measured relative volumes, $1:1:2$. We have only proved that the gaseous molecules of hydrogen and chlorine contain an even number of atoms, e.g., H_2 or H_4, Cl_2 or Cl_4, and not an odd number, e.g., H or H_3, Cl or Cl_3.

Conversely, suppose that we know the equation $H_2 + Cl_2 \rightarrow 2HCl$. Then, according to the Avogadro law,* the measured relative volumes of gases can be predicted from the relative numbers of molecules reacting and produced.

* The expression "Avogadro hypothesis," still in use, is left over from the time when most people did not accept it.

Although we cannot weigh a single molecule by any method, it should be possible to determine their relative masses (weights) by weighing *large* but *identical* numbers of molecules. We illustrate with the fictitious molecules AB and CD:

Number weighed	Total mass, g
2.00×10^{22} molecules of AB	33.0
2.00×10^{22} molecules of CD	6.0
4.00×10^{22} molecules of AB	66.0
4.00×10^{22} molecules of CD	12.0

The relative masses of the molecules of AB and CD are then 33.0 to 6.0, the same as 66.0 to 12.0. Observe that identical relative masses are obtained when *equal numbers* of molecules are weighed, irrespective of that number. Observe that relative masses of the molecules are obtainable without knowing the number being weighed, provided the number weighed is identical. The masses of x molecules of AB and x molecules of CD will be in the ratio of 33.0 to 6.0, the same as the relative mass of 1 molecule of AB to 1 molecule of CD.

Thus, when a chemist says, "The molecular weight of water is 18, the molecular weight of carbon dioxide is 44, and the molecular weight of sulfur dioxide is 64," he means there are as many molecules in 18 g of water as in 44 g of carbon dioxide, or in 64 g of sulfur dioxide. Also, since these masses contain the same number of molecules, the values 18, 44, and 64 may be used to represent the relative masses of 1 molecule of each of these substances.

The question now is, "How do we know when we are weighing equal numbers of molecules?" Stanislao Cannizzaro showed in 1858 that the Avogadro law may be used to determine molecular weights of gases: by common consent,[*] we take as a standard quantity 31.9988 g of oxygen, using air as the source of oxygen. This quantity contains the **AVOGADRO NUMBER** of molecules, N_A, page 14. Next, we determine that the volume occupied by this quantity of oxygen at standard conditions (sc), 0°C and 760 torr, is 22.4136 liters (sc).[†]

It follows, then, from the Avogadro law that 22.4 liters (sc) of any other gas contains N_A molecules. Hence, the mass in grams of 22.4

---

[*] By a majority vote of chemists taken in 1905 through the International Union of Pure and Applied Chemistry, 32.0000 g of oxygen was chosen to avoid a number smaller than 1 for the atomic weight of hydrogen. In 1961, this was changed to 31.9988 g with the adoption of the atomic weight scale based on the isotope of carbon, ^{12}C = exactly 12 (further treated on page 38).

[†] This is the volume that oxygen would have at standard conditions if it were an ideal gas.

liters at sc of a gas will be its MOLECULAR WEIGHT because it contains N_A molecules. The amount of substance which contains N_A molecules, or atoms, or in general, N_A particles is called a MOLE (page 14); N_A is $(6.0225 \pm 0.0003) \times 10^{23}$ particles. Thus, 1 mole of hydrogen atoms, H, contains 6.02×10^{23} H atoms; 1 mole of hydrogen molecules, H_2, contains 6.02×10^{23} H_2 molecules; 1 mole of hydrogen ions (H^+ or H^-) contains 6.02×10^{23} H^+ or H^- ions; 1 mole of sulfate ions contains 6.02×10^{23} SO_4^{2-} ions. Then, the quantity of matter in a mole may be expressed as 6.02×10^{23} particles per mole and 22.4 liters or 22.4×10^3 ml (sc) per mole of gas. The volume (sc) occupied by 1 mole of a gas is known as the MOLAR VOLUME. The dimensional units of molecular weight are grams per mole. We can then write

$$\text{number of moles, } n = \frac{\text{mass in grams}}{\text{molecular weight, g/mole}}$$

For example, the number of moles in 8.00 g of O_2, molecular weight 32.0 g/mole, is

$$n = \frac{8.00 \text{ g}}{32.0 \frac{\text{g}}{\text{mole}}} = 0.250 \text{ mole}$$

and the volume (sc) occupied by this quantity is

$$0.250 \text{ mole} \times 22.4 \frac{\text{liters (sc)}}{\text{mole}} = 5.60 \text{ liters (sc)}$$

The equation for an ideal gas (page 14) can be used to calculate the molecular weight of a gas. Using the equation $PV = nRT$, we can calculate the number of moles of a gas from its pressure, volume, and temperature:

$$n = \frac{PV}{RT} = \text{number of moles of gas}$$

Observe that the number of moles of gas is the same for all gases occupying the same volume at the same temperature and pressure, (merely a statement of the Avogadro law). But the mass of gas in this volume differs for different gases. The number of moles, n, is given by

$$n = \frac{w}{M}$$

in which w is the mass of the gas, and M is its molecular weight. Hence

$$PV = \frac{w}{M} RT \qquad \text{or} \qquad M = \frac{wRT}{PV}$$

Example 3 0.482 g of pentane occupies 204 ml as a vapor at 102°C and 767 torr. Calculate its molecular weight.

Answer Using $M = wRT/PV$ and the given data but being careful to express all quantities in the proper units,[*]

$$T = 102°C + 273 = 375°K;$$

$$P = \frac{767 \text{ torr}}{760 \frac{\text{torr}}{\text{atm}}} = 1.009 \text{ atm};$$

$$V = 0.204 \text{ liter}$$

$$M = \frac{0.482 \text{ g} \times 0.0821 \frac{\text{liter atm}}{\text{mole deg}} \times 375 \text{ deg}}{1.009 \text{ atm} \times 0.204 \text{ liter}}$$

$$= 72.1 \frac{\text{g}}{\text{mole}}$$

Example 4 The molecular weight of the hydrocarbon butane is 58.1. Calculate the number of moles, the number of molecules, and the volume in liters (sc) corresponding to 12.0 g of butane.

Answer One mole has a mass of 58.1 g; the number of moles is given by

$$\frac{12.0 \text{ g}}{58.1 \frac{\text{g}}{\text{mole}}} = 0.207 \text{ mole}$$

and since there are 6.02×10^{23} molecules in 1 mole,

$$0.207 \text{ mole} \times 6.02 \times 10^{23} \frac{\text{molecules}}{\text{mole}} = 1.25 \times 10^{23} \text{ molecules}$$

and since 1 mole occupies 22.4 liters (sc),

$$0.207 \text{ mole} \times 22.4 \frac{\text{liters (sc)}}{\text{mole}} = 4.64 \text{ liters (sc)}$$

ATOMIC WEIGHTS; MASS SPECTROSCOPY / 3.6

The first set of accepted atomic weights was deduced from the molecular weights of substances by Cannizzaro. Atomic weight is analogous to molecular weight. The atomic weight of an element is not the mass of one atom; rather, it is the mass of one atom of the element relative to the mass of an atom of another element. It is also the mass of N_A, 6.0225×10^{23}, atoms of the element. Thus when a chemist says, "The atomic weight of mercury, Hg, is 200.6, the atomic weight of oxygen, O, is 16.00, and the atomic weight of lead, Pb, is 207.2," he means that 6.0225×10^{23} mercury atoms weigh 200.6 g, the same number of oxygen atoms weigh 16.00 g, and the same number of lead atoms weigh

---

[*] The use of dimensional units is strongly urged to verify your calculations. Multiplication of incorrect units will be obvious.

207.2 g. The dimensional units of an atomic weight are grams per mole of atoms.*

Cannizzaro's method, though clever and intellectually satisfying (Problem 25, page 44), is now only of historical interest. Very precise atomic weights are now obtained by **MASS SPECTROSCOPY**. The principle of mass spectroscopy is based on the fact that when a charged particle enters a magnetic field, the particle *moves in a circular path*. For a given magnetic field strength H, the *radius of the path, r, is proportional to the mass* of the particle m and its velocity v, but *inversely proportional to the charge Q on the particle*.

Although it appears possible to measure the mass of a single kind of charged atom from its motion in a magnetic field, it is more practical to measure relative masses. An apparatus, Fig. 3.1, is set up so that H, Q, and v remain constant, making the mass of the ion proportional to the radius of its path:

$$\frac{m_1}{m_2} = \frac{r_1}{r_2}$$

The lighter the ion, the smaller is the radius. The *relative masses, m_1/m_2*, of ions may now be determined directly from the radii described by the particles in a magnetic field. The ions are detected by a metal plate connected to a galvanometer in a mass spectrometer or by a photographic plate in a mass spectroscope.

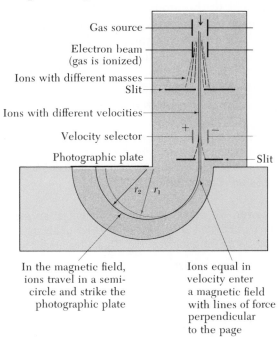

Gas source

Electron beam (gas is ionized)

Ions with different masses

Slit

Ions with different velocities

Velocity selector

Photographic plate

Slit

r_2 r_1

In the magnetic field, ions travel in a semi-circle and strike the photographic plate

Ions equal in velocity enter a magnetic field with lines of force perpendicular to the page

Fig. 3.1. *Schematic representation of the Bainbridge mass spectrograph. The apparatus is vacuum sealed. The velocity selector consists of a magnetic field perpendicular to an electric field. The radius r is measured in centimeters.*

* "Grams per gram atomic weight" and "grams per gram-atom" are also in use.

Example 5 Atoms of the element nitrogen (with identical positive charges and velocities) produce two circular paths in a magnetic field; the radius ratio is 1.0712. What are the relative masses of the atoms of the element nitrogen?

Answer Since the masses of the ions are directly proportional to the radii of the circular paths, the relative masses of the ions are in the ratio of 1.0712, or 1 atom of nitrogen is 1.0712 times heavier than the other atom.

The data in Example 5 tell us that there are two kinds of nitrogen atoms differing in mass. The atoms of the same element having different masses are called ISOTOPES.* Oxygen possesses three isotopes. To compare the masses of the atoms of the elements, a standard must be established; the presently accepted standard is the isotope of carbon, carbon-12, $^{12}C =$ exactly 12. 12.0000 g of ^{12}C contains as many atoms as there are atoms of oxygen in 15.9994 g of oxygen obtained from air; it contains the Avogadro number of atoms.

From the radii described by ^{12}C and ^{16}O, the atomic weight of the isotope ^{16}O is found to be 15.9949. The atomic weights of the other isotopes of oxygen are 16.9991 and 17.9992, abbreviated as ^{17}O and ^{18}O. The atomic weights of the isotopes of nitrogen are 14.003 and 15.000, abbreviated as ^{14}N and ^{15}N. In these symbols, the superscript, the nearest whole number to the atomic weight of an isotope, is called the MASS NUMBER. To convert the atomic weights of the *isotopes* of an element to the atomic weight of the *element,* the relative number of atoms of each isotope present must also be known. In mass spectroscopy, the relative abundance of isotopes can be obtained from the relative darkness of the lines developed on the photographic plate or from the relative currents produced by the ions of the isotopes. Figure 3.2 illustrates the MASS SPECTRUM† of an element, a graph showing the relative abundance plotted against mass number.

Example 6 Calculate the atomic weight of oxygen found in air and of naturally occurring nitrogen from the previous and the following data. Atmospheric oxygen contains 99.756% ^{16}O, 0.039% ^{17}O, and 0.205% ^{18}O. Atmospheric nitrogen contains 99.64% ^{14}N and 0.36% ^{15}N.

Answer To obtain the atomic weight of an element from mass spectroscopic data, calculate the average relative weight from the relative weight and the relative abundance of each isotope. Thus, the atomic weight of atmospheric oxygen is

$$(15.9949 \times 0.99756) + (16.9991 \times 0.00039) + (17.9992 \times 0.00205) = 15.9994$$

---

* Named, from the Greek word meaning *"same place,"* by Frederick Soddy (1911) to indicate that these atoms occupy the same position in the periodic system of elements (pages 78, 106).

† A spectrum (Latin, *image*) is a separated group of components arranged in some sequence.

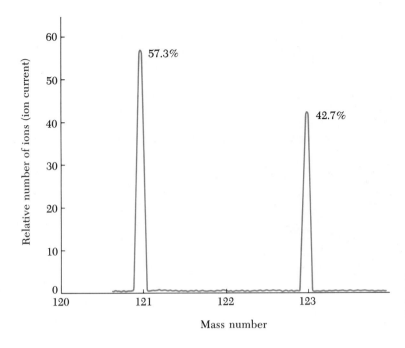

Fig. 3.2. *The mass spectrum of antimony, a plot showing the relative abundance (peaks) of the two naturally occurring isotopes of antimony.*

and the atomic weight of nitrogen is

$$(14.003 \times 0.9964) + (15.000 \times 0.0036) = 14.007$$

The International Commission on Atomic Weights was established in 1900 by the International Union of Pure and Applied Chemistry and given the duty of issuing periodically a table of atomic weights after the consideration of all papers dealing with the subject. The values chosen by the Commission (see back cover) become the ACCEPTED ATOMIC WEIGHTS.

MOLECULAR FORMULAS / 3.7

A chemical symbol such as C is not only used to represent the element C but also stands for 1 mole of carbon atoms or 12.0 g. A molecular formula such as C_6H_6, benzene, tells you that there are 6 carbon atoms and 6 hydrogen atoms in 1 molecule of benzene. It also tells you that there are 6 moles of carbon atoms and 6 moles of hydrogen atoms in 1 mole of benzene. Thus, the molecular formula gives us the actual composition in number of atoms per molecule or in the number of moles of each kind of atom per mole of compound. It also represents a definite quantity of the substance, 1 mole or 78 g, the sum of the atomic weights of its constituent elements, $6 \times 12.0 + 6 \times 1.0$.

Example 7 Analysis of the insecticide hexachlorocyclohexane, molecular weight of 291 g/mole, shows 24.7% carbon, 2.06% hydrogen, and 73.2% chlorine by mass. Using 12.0, 1.0, and 35.5 as the atomic weights, respectively, of these elements, calculate the molecular formula of the hexachlorocyclohexane.

Answer Since a molecular formula represents the number of moles of each element per mole of compound, first calculate the mass of each element in 1 mole of compound from its molecular weight and composition. (Recall that mass percentage of an element is its mass in 100 g of the compound.)

$$291 \; \frac{\text{g compd}}{\text{mole compd}} \times \frac{24.7 \text{ g carbon}}{100 \text{ g compd}} = 71.9 \; \frac{\text{g carbon}}{\text{mole compd}}$$

$$291 \; \frac{\text{g compd}}{\text{mole compd}} \times \frac{2.06 \text{ g hydrogen}}{100 \text{ g compd}} = 5.99 \; \frac{\text{g hydrogen}}{\text{mole compd}}$$

$$291 \; \frac{\text{g compd}}{\text{mole compd}} \times \frac{73.2 \text{ g chlorine}}{100 \text{ g compd}} = 213 \; \frac{\text{g chlorine}}{\text{mole compd}}$$

Then convert these masses to the corresponding numbers of moles of atoms:

$$\frac{71.9 \; \frac{\text{g carbon}}{\text{mole compd}}}{12.0 \; \frac{\text{g carbon}}{\text{mole C}}} = 5.99 \; \frac{\text{moles C}}{\text{mole compd}} = 6.00^* \; \frac{\text{moles C}}{\text{mole compd}}$$

$$= 6.00 \; \frac{\text{atoms}}{\text{molecule}}$$

$$\frac{5.99 \; \frac{\text{g hydrogen}}{\text{mole compd}}}{1.0 \; \frac{\text{g hydrogen}}{\text{mole H}}} = 6.0 \; \frac{\text{moles H}}{\text{mole compd}} = 6.0 \; \frac{\text{atoms}}{\text{molecule}}$$

$$\frac{213 \; \frac{\text{g chlorine}}{\text{mole compd}}}{35.5 \; \frac{\text{g chlorine}}{\text{mole Cl}}} = 6.00 \; \frac{\text{moles Cl}}{\text{mole compd}} = 6.00 \; \frac{\text{atoms}}{\text{molecule}}$$

Therefore, the molecular formula is $C_6H_6Cl_6$.

EMPIRICAL FORMULAS / 3.8

It is evident that if the molecular weight of a substance is unknown, then its molecular formula cannot be calculated. For such substances, it is, however, possible to calculate the EMPIRICAL FORMULA—*the simplest integral ratios in which the atoms combine*. For example, the empirical formula corresponding to $C_6H_6Cl_6$ is CHCl. The weight composition expresses the relative masses of the constituent elements in a particular compound. If we divide each relative mass by the atomic weight of the corresponding element, the relative number of moles of atoms of each constituent element is obtained.

* The difference between 5.99 and 6.00 is within the experimental error.

Example 8 SAP ("sodium acid pyrophosphate") is added to frankfurters and other sausages to accelerate development of a rosy-red color. Its mass composition is 20.7% sodium, 0.910% hydrogen, 27.9% phosphorus, and 50.5% oxygen. Find its simplest (empirical) formula.

Answer Dividing the relative mass of each element by its atomic weight gives the relative number of moles of atoms of the element:

$$\text{Sodium: } \frac{20.7 \, \cancel{g}}{23.0 \, \frac{\cancel{g}}{\text{mole}}} = 0.900 \text{ mole Na}$$

$$\text{Hydrogen: } \frac{0.910 \, \cancel{g}}{1.01 \, \frac{\cancel{g}}{\text{mole}}} = 0.901 \text{ mole H}$$

$$\text{Phosphorus: } \frac{27.9 \, \cancel{g}}{31.0 \, \frac{\cancel{g}}{\text{mole}}} = 0.900 \text{ mole P}$$

$$\text{Oxygen: } \frac{50.5 \, \cancel{g}}{16.0 \, \frac{\cancel{g}}{\text{mole}}} = 3.16 \text{ mole O}$$

The ratios of the numbers of moles of atoms of the elements are

$$Na_{0.900}H_{0.901}P_{0.900}O_{3.16}$$

To find the relative whole number ratios, divide by the smallest number, 0.900 in this case:

$$Na_1H_1P_1O_{3.5}$$

Such a division often yields whole numbers but in this case multiplication by 2 is required, yielding the empirical formula of the compound

$$Na_2H_2P_2O_7$$

whose name is disodium dihydrogen diphosphate (old name: sodium acid pyrophosphate).

The sum of the atomic weights of all the atoms indicated by either an empirical formula or a molecular formula is called a FORMULA WEIGHT; the molecular weight is, of course, related to the formula weight by a whole number. By general usage, 1 mole of a substance means 1 molecular weight or 1 formula weight expressed in grams; the dimensional units of a formula weight are then grams per mole.

Example 9 What is the formula weight of disodium dihydrogen diphosphate, $Na_2H_2P_2O_7$?

Answer The formula weight $= 2 \times$ atomic weight of sodium $+ 2 \times$ atomic weight of hydrogen $+ 2 \times$ atomic weight of phosphorus $+ 7 \times$ atomic weight of oxygen $=$ 222. The formula weight is 222 or 222 g/mole.

Formulas derived from molecular weights of substances in the gaseous state are valid only for the gaseous state. We are not justified in assuming that the particles of liquids and solids must be identical with the molecules of the corresponding gaseous state. For example, the molecular weight of sodium chloride vapor is about 59, determined at 1970°C. This indicates that the vapor consists essentially of NaCl molecules; NaCl is thus the molecular formula for sodium chloride in the vapor state at about 2000°C. But *solid sodium chloride* is a typical ionic compound (page 152). 1 mole of NaCl has a mass of 58.5 g and 1 mole of $CaCl_2$ has a mass of 111 g. It is impossible to write a molecular formula for an ionic solid. An empirical formula* is, however, acceptable. Nevertheless, we do refer to the gram formula weight of an ionic substance as 1 mole, so that the dimensional units of a formula weight of an ionic substance are grams per mole.

1. Conservation. A tube containing mercury(II) oxide, HgO, and potassium chlorate, $KClO_3$, was sealed, weighed, and then heated to decompose an unknown amount of the solids. After the original temperature was restored, the tube with its contents was reweighed:

weight of tube + contents before decomposition = 10.96545 ± 0.00002 g
weight of tube + contents after decomposition = 10.96544 ± 0.00002 g

Is matter destroyed in this experiment? Explain.

2. Mass composition. 1.00 g of magnesium reacts completely with chlorine forming 3.92 g of magnesium chloride. Calculate the mass percent of magnesium in the chloride and the mass ratio, Mg to Cl.

3. Definite proportions. What is the mass percentage of nitrogen in nitrosomethylurea, a potent carcinogen, $N_3O_2C_2H_5$? Find the mass of nitrogen in 50.0 g of $N_3O_2C_2H_5$.

4. Combining volumes. 1.00 liter of oxygen is consumed in the *Kel-Chlor* process for converting HCl to Cl_2,

$$2NOCl\ (g) + O_2\ (g) + 4HCl\ (g) \longrightarrow 2NO\ (g) + 3Cl_2\ (g) + 2H_2O\ (l)$$

How many (a) liters of HCl are used, (b) liters of Cl_2 are produced at constant T and P? (c) Can you predict the volume of water produced by inspection of the chemical equation? (See table on page 47.)

5. Avogadro law. Use the measured volume ratios, 1 volume of nitrogen + 1 volume of oxygen → 2 volumes of nitrogen oxide, to show that the molecules of nitrogen and oxygen can be diatomic but cannot be monatomic. Can both gases be tetraatomic? If so, write the balanced equation.

6. Atomic theory. 10.0 g of element X combines with 40 g of element Y,

---

* Ionic solids such as sodium chloride might be more clearly represented by ionic formulas, Na^+Cl^-, but common practice dictates the use of empirical formulas, NaCl.

forming a compound X_2Y. What are the relative weights of the atoms of X and Y?

7. Empirical formula. A platinum(II) compound which has antitumor activity contains 65.0% Pt, 23.6% Cl, 9.33% N, and 2.02% H by mass. Calculate its empirical formula.

8. Mole. (a) One mole of saccharin has a mass of 183 g. Calculate the mass of 0.300 mole and the number of molecules in 91.5 g of saccharin. (b) Calculate the number of moles and number of molecules in 32.0 lb of oxygen.

9. Molecular weight. An instrument (*molecular weight analyzer*) records the pressure produced on evaporation of a weighed sample injected into a flask of fixed volume (3000 ml) at a fixed temperature (150°C). The instrument recorded 3.80 torr when 99.7 mg of dimethylmercury was injected. (a) Calculate the molecular weight of dimethylmercury. (b) The flesh of a salmon (1-kg mass) is found to contain 0.100 mg of dimethylmercury. Find the number of molecules of dimethylmercury in the flesh.

10. Molecular formula. (a) Caffeine found in tea, coffee, and added to many soft drinks is a white solid, 194.1 g/mole, containing 49.5% carbon, 5.20% hydrogen, 28.9% nitrogen, and 16.5% oxygen by mass. Calculate its molecular formula. (b) If a scientist synthesized a nitrogen compound containing 7.0 g of nitrogen/mole, would this affect the molecular weight or the calculated formula for caffeine? Explain and give the "new" formula for caffeine.

11. Atomic weight. Naturally occurring carbon is a mixture of 98.89% ^{12}C; 1.11% ^{13}C, 13.003; and about 10^{-10}% ^{14}C, 14.008. Calculate the atomic weight of natural carbon.

ADDITIONAL PROBLEMS

12. Conservation. A mixture containing 10.0 g of solid sodium carbonate and 10.0 g of solid sodium hydrogen carbonate is heated, driving off 3.7 g of water vapor and carbon dioxide. What mass of solid remains?

13. Mass composition. Hydrogen gas removes 4.422 g of oxygen from a solid oxide in forming 4.976 g of water. Calculate (a) the mass of hydrogen used to produce the water; (b) the mass of hydrogen that combines with 16.0 g of oxygen; (c) the mass percent of oxygen in water.

14. Definite proportions. Calculate the mass percentage of Cl in (a) 2,4,5-T, trichlorophenoxy acetic acid, $C_6H_2Cl_3OCH_2COOH$, a weed killer and defoliant; (b) DDT, $(C_6H_4Cl)_2CHCCl_3$.

15. Definite proportions. A molecule recently discovered in outer space contains 3.05% hydrogen and 96.95% sulfur. Calculate (a) the mass of compound required to produce 8.0 g of sulfur, (b) the empirical formula of the compound.

16. Atomic theory. Given the following oxides: NO, N_2O, NO_3, NO_2, N_2O_3, and N_2O_5. (a) Show that for a given mass of nitrogen, the masses of oxygen in these combinations are in ratios of whole numbers. (b) The generalization of this relationship to include any series of compounds formed from the same two elements is called the LAW OF MULTIPLE PROPORTIONS. Suggest a statement of this law.

17. Combining volumes. What volume of chlorine will combine with 5.0 liters of nitrogen at the same temperature and pressure, to form what volume of NCl_3?

18. Avogadro law. Given the following gaseous reactions at the same T and P, involving the elements E, F and the compounds G, H:

1 volume of E + 3 volumes of F $\longrightarrow$ 1 volume of G
1 volume of E + 4 volumes of F $\longrightarrow$ 2 volumes of H

What are the molecular formulas of E, F, G, and H? Are your answers the only possible answers? If they are not, give another set of molecular formulas consistent with the experimental facts.

19. Mole. Calculate the number of moles and atoms in 50.0-g samples of each of the following elements: (a) neon, Ne; (b) silicon, Si; (c) osmium, Os.

20. Mole. Express the following quantities (a) of oxygen in moles of O atoms: (i) 25.0 g of O_2 molecules; (ii) 25.0 moles of O_2 molecules; (iii) 25.0×10^{23} atoms; (iv) 25.0×10^{23} O_2 molecules; (b) of chlorine in moles of Cl_2: (i) 50.0 g of Cl; (ii) 50.0 moles of Cl; (iii) 50×10^{23} Cl atoms; (iv) 50×10^{23} Cl_2 molecules.

21. Mole. Express in your own words and illustrate the meaning of the following definition adopted by IUPAC: "The mole is the amount of substance of a system which contains as many elementary entities as there are carbon atoms in 0.012 kilogramme of carbon-12."

22. Mole. The accepted proton and electron masses are, respectively, 1.67252×10^{-24} g and 9.1091×10^{-28} g. Calculate the mass of one mole of ^{1}H atoms.

23. Mole. Assume the ideal gas law. As a result of a nuclear explosion, 10^3 g of radioactive ^{90}Sr is dispersed uniformly throughout the atmosphere. Calculate the number of ^{90}Sr atoms per liter of air at 300°K and 1.0 atm air pressure. Required information: air, 29 g/mole; 1.0 atm equals 10^3 g/cm²; surface area of earth, 5.1×10^{18} cm².

24. Atomic weights. The compound A_2B contains 40% A and 60% B by mass. Calculate the relative mass of atoms of A to B.

25. Cannizzaro method. Cannizzaro (1858) deduced atomic weights from the molecular weights of gases. Since the mass of a molecule is the sum of the masses of all the atoms in the molecule, the mass of an element found in 1 mole of its compounds is the atomic weight of the element or is related to the atomic weight by a whole number. The following data on chlorine and platinum–chlorine compounds are given:

Compound number	Molecular weight (g/mole)	Mass % composition	
		Pt	Cl
1	337	57.9	42.1
2	231	84.5	15.4
3	302	64.6	35.4
4	266	73.3	26.7
chlorine	71.0		100

(a) Calculate the mass of each element in 1 mole of each compound. Do you notice anything about the answers for Pt? For Cl? (b) Calculate the atomic weight of platinum and chlorine and the molecular formula of each of the substances. What data would you have to discover to compel a change in the atomic weight of platinum? Would you undertake such a research project?

26. Atomic weight scale. Before 1961, the atomic weight scale was based on oxygen (from air) = 16 (exactly). On the present scale, based on $^{12}C = 12$ (exactly), the atomic weight of oxygen is 15.9994 and the atomic weight of natural carbon is 12.011. If the standard were changed to natural carbon = 12(exactly), state whether each of the following would be increased, decreased, or unchanged: (a) the mass of an atom of carbon; (b) the atomic weight of ^{12}C; (c) the atomic weight of natural carbon; (d) the atomic weight of oxygen; (e) the Avogadro number.

27. Atomic weight. Natural boron consists of 19.8% ^{10}B, atomic weight 10.0129, and 80.2% ^{11}B, atomic weight 11.0093. Calculate the atomic weight of natural boron.

28. Empirical formula. The relative composition of sodium cyclamate (201 g/mole) is 18 g C, 3.0 g H, 0.75 mole of O, and the same number, 1.51×10^{23}, of Na, S, and N atoms. Find its empirical and molecular formulas.

29. Empirical formula. (a) The decomposition of a given mass of a xenon–fluorine compound yields 3.73 ml (sc) of Xe and 11.2 ml (sc) of F_2. Find the empirical formula of the compound. (b) The mass composition of a compound is 84.7% C and 14.3% H. Find its empirical formula. If combustion of one mole of the compound produces two moles of CO_2, what is the molecular formula of the compound?

30. Molecular weight. 0.910 mg methylmercury dicyandiamide, a poisonous antifungal agent, occupies 100 ml as a gas at 127°C and 1.00×10^{-3} atm. Calculate its molecular weight.

31. Molecular formula. Cholesterol, 386 g/mole, is composed of 84.0% C, 12.0% H, and 4.15% O by mass. Find its molecular and empirical formulas.

32. Mass spectroscopy. Two compounds, $C_{15}H_{10}O_3$ and $C_{14}H_{10}O_2N_2$, have the same molecular weight to four significant figures. A mass spectroscope shows a separation of these compounds corresponding to 238.06299 and 238.07422. Check these values using the isotopic weights $^{1}H = 1.007825$, $^{12}C = 12$ exactly, $^{14}N = 14.003074$, $^{16}O = 15.994915$.

33. Mass spectroscopy. The isotopes ^{79}Br and ^{81}Br are nearly equiabundant. The mass spectrum of $HBrO_x$ shows two equiabundant ions with molecular weights 144 and 146. Is the acid bromic ($HBrO_3$) or perbromic ($HBrO_4$)?

4

CHEMICAL EQUATIONS AND
CHEMICAL ARITHMETIC

The formulas of such compounds as HCl, H_2O, NH_3, NaCl, $CaCl_2$, H_2SO_4, and H_3PO_4 reveal that the COMBINING CAPACITY of atoms is not uniform. The combining capacity of an atom is commonly referred to as its VALENCE (Latin, "capacity"). The formula HCl shows that the atoms of hydrogen and chlorine have the same combining capacity and, therefore, the same valence. The formula CaO shows that calcium and oxygen also have the same valence. The formula H_2O shows that the valence of oxygen is twice that of hydrogen. If, for convenience, we assign a valence of 1 to the hydrogen atom, then the valence of the chlorine atom is 1 and that of the oxygen atom 2. It also follows from the formula $CaCl_2$ that the valence of Ca is twice that of Cl; the valence of Ca is therefore 2.

Certain combinations of elements, known as GROUPS, remain combined and behave as units during many chemical reactions. The sulfate group is such a combination. From the formula for sulfuric acid, H_2SO_4, a valence of 2 may be assigned to the sulfate, (SO_4), group.

The knowledge of the valences of the atoms and groups furnishes information for writing the empirical formulas of many compounds, especially those composed of 2 atoms or groups. The valence (see Table II.1, Appendix II) of sodium is 1 and the valence of the phosphate group is 3. This means that the combining capacity of the phosphate group is 3 times greater than that of the sodium atom; thus, if we take 3 sodium atoms to 1 phosphate group, the total combining capacity of sodium atoms and of the phosphate group becomes identical. The empirical formula for sodium phosphate is then Na_3PO_4. Similarly, to make the total combining capacity of aluminum atoms and of sulfate groups identical, we take these in the ratio of 2:3; the simplest formula for aluminum sulfate is then $Al_2(SO_4)_3$, in which the subscript outside refs to all atoms within the parentheses. This formula assigns a total combining capacity of 6 to the 2 Al atoms and likewise a total capacity of 6 to the 3 sulfate groups.

The fact that we can write a formula for a compound from the valences of the constituent atoms does not mean the compound must exist. We can write AgOH for the compound "silver hydroxide," but no compound of this definite composition has yet been isolated. Conversely, many compounds exist whose formulas cannot be deduced by this scheme; examples are benzene, C_6H_6, and carbon suboxide, C_3O_2.

The valence concept described here is meaningful only for compounds made of two elements or groups. In Chapter 8, we shall define the *covalency* and the *oxidation number* of an atom — refinements of valence capable of describing and classifying more complicated cases.

The nomenclature of inorganic compounds (Appendix II) should be reviewed.

CHEMICAL EQUATIONS / 4.2

Chemical equations tell us what substances react and what substances are produced. The physical states of the substances may also be indicated, according to the following usage:

Symbol	Meaning
(g)	Gas
(l)	Liquid
(c)	Crystalline solid
(amorph)	Amorphous solid
(aq)	Aqueous solution, relatively large amount of water present
(sol)	Solution other than aqueous

The products of a reaction are determined by experimentation. Thus, propane, a typical hydrocarbon, burns in the presence of oxygen to form carbon dioxide and water. First, since the formulas for these substances are known, we may write

$$C_3H_8(g) + O_2(g) \longrightarrow CO_2(g) + H_2O(g) \quad \text{(not balanced)}$$

Second, since atoms are not created or destroyed but merely rearranged in chemical changes, we BALANCE THE EQUATION by making the number of atoms of each element participating in the reaction the same as that appearing in the products. This is accomplished by placing the required number before each formula.* Thus, the 8 atoms of hydrogen in the C_3H_8 must form 4 molecules of H_2O, while the 3 atoms of carbon

---

* A systematic method is considered later (page 224) for more complicated reactions.

must form 3 molecules of CO_2; but the 3 molecules of CO_2 and the 4 molecules of H_2O require 10 atoms or 5 molecules of oxygen. The balanced equation is

$$C_3H_8(g) + 5O_2(g) \longrightarrow 3CO_2(g) + 4H_2O(g)$$

The number before each formula, such as the 5 before the O_2, is known as its COEFFICIENT. The coefficient is a multiplier for the entire formula, never for only a part of it. Thus, $3CaCl_2(H_2O)_6$ includes 3 Ca atoms, 6 Cl atoms, 36 H atoms, and 18 O atoms. It should be stressed that the *subscripts in the formulas must not be altered*. This error would alter the nature of the substances involved, thereby violating the experimental observation; H_2O_2 is the formula for hydrogen peroxide, different from water, H_2O. Also, the balanced equation is a statement of the relative numbers of moles of reactants and products involved in a chemical change. It does not, however, show the actual processes (Chapter 20) by which reactants are converted to products.

QUANTITATIVE INFORMATION FROM CHEMICAL EQUATIONS / 4.3

In this section, we learn about the information obtainable from a balanced chemical equation. The balanced equation, for example,

$$C_3H_8(g) + 5O_2(g) \longrightarrow 3CO_2(g) + 4H_2O(g)$$

gives the relative number of molecules of each substance reacting (reactants) and the relative number of molecules of each substance formed (products). Thus, for every 1 molecule of propane, 5 oxygen molecules react, producing 3 carbon dioxide molecules and 4 water molecules; or, in larger units, for every 1 mole of propane consumed, 5 moles of oxygen react, producing 3 moles of carbon dioxide and 4 moles of water. These statements may be abbreviated as

C_3H_8	+	$5O_2$	$\longrightarrow$	$3CO_2$	+	$4H_2O$
1 molecule		5 molecules		3 molecules		4 molecules
1 mole		5 moles		3 moles		4 moles
6.02×10^{23} molecules		$5 \times 6.02 \times 10^{23}$ molecules		$3 \times 6.02 \times 10^{23}$ molecules		$4 \times 6.02 \times 10^{23}$ molecules
22.4 liters (sc)		5×22.4 liters (sc)		3×22.4 liters (sc)		4×22.4 liters (sc)
44.1 g		5×32.0 g		3×44.0 g		4×18.0 g

As shown above, the formulas in chemical equations represent either relative molecular or molar quantities. The quantities (except the first) *under each formula* represent the same amount of substance. Thus, we can express the quantities of substances undergoing chemical change in terms of any molar quantities we choose. [Remember, however, that the molar volume, 22.4 liters (sc), applies only to gases.] For example,

$$1 \text{ mole } C_3H_8 + 5 \times 6.02 \times 10^{23} \text{ molecules } O_2 \longrightarrow 3 \times 22.4 \text{ liters (sc) } CO_2 + 4 \times 18.0 \text{ g } H_2O$$

Similarly, from the coefficients in the balanced equation,

$$3H_2(g) + Fe_2O_3(c) \longrightarrow 2Fe(c) + 3H_2O(g)$$

it follows that

$$3 \text{ moles } H_2 + 1 \text{ mole } Fe_2O_3 \longrightarrow 2 \text{ moles } Fe + 3 \text{ moles } H_2O$$

$$3 \times 22.4 \text{ liters (sc) } H_2 + 1 \text{ mole } Fe_2O_3 \longrightarrow 2 \times 6.02 \times 10^{23} \text{ atoms } Fe + 3 \times 18 \text{ g } H_2O$$

Note that each *quantity includes a unit*; refer to Appendix I for a review of dimensional numbers. The rigorous use of dimensional units is particularly important in chemical arithmetic.

Chemical equations thus make it possible to calculate quantities of materials required to produce a definite quantity of a desired product.

Example 1 Tetrachloroethylene, C_2Cl_4, used mainly as a dry cleaning fluid, also serves as a source of chlorine atoms (nuclei) for the detection of neutrinos (page 535). The manufacture of C_2Cl_4 from acetylene, C_2H_2, is summarized as

$$C_2H_2(g) + 3Cl_2(g) + Ca(OH)_2(c) \longrightarrow C_2Cl_4(l) + CaCl_2(c) + 2H_2O(l)$$

Find (a) the mass in grams of C_2Cl_4 produced from 50.0 g of Cl_2; (b) the volume of Cl_2 in liters at 27°C and 810 torr required to produce 100 g of C_2Cl_4; (c) the mass in grams of chlorine that will react with 75.0 liters of C_2H_2, measured at 24.0°C and 790 torr. Atomic weights: $C = 12.0$, $Cl = 35.5$.

Answer (a) From the balanced equation

3 moles Cl_2 yields 1 mole C_2Cl_4

or, since the molecular weight of Cl_2 is 71.0 and that of C_2Cl_4 is 166 (from $2 \times 12.0 + 4 \times 35.5 = 166$),

3×71.0 g Cl_2 yields 166 g C_2Cl_4

The ratio between the mass of C_2Cl_4 and the mass of chlorine is

$$\frac{166 \text{ g } C_2Cl_4}{3 \times 71.0 \text{ g } Cl_2}$$

Then the number of grams of C_2Cl_4 obtained from 50.0 g of Cl_2 is

$$50.0 \text{ g } Cl_2 \times \frac{166 \text{ g } C_2Cl_4}{3 \times 71.0 \text{ g } Cl_2} = 39.0 \text{ g } C_2Cl_4$$

(b) From the balanced equation

1 mole of C_2Cl_4 requires 3 moles Cl_2

or

166 g of C_2Cl_4 requires 3×22.4 liters (sc) Cl_2

The ratio between the volume of Cl_2 and the mass of C_2Cl_4 is

$$\frac{3 \times 22.4 \text{ liters (sc) } Cl_2}{166 \text{ g } C_2Cl_4}$$

Then the number of liters of Cl_2 (sc) required for 100 g of C_2Cl_4 is

$$100 \text{ g } C_2Cl_4 \times \frac{3 \times 22.4 \text{ liters (sc) } Cl_2}{166 \text{ g } C_2Cl_4} = 40.5 \text{ liters (sc) } Cl_2$$

The gas laws are used to calculate the volume corresponding to the conditions of the problem (cp),

$$V \text{ (cp)} = 40.5 \text{ liters (sc) } Cl_2 \times \frac{300°K \times 760 \text{ torr}}{273°K \times 810 \text{ torr}} = 41.8 \text{ liters (cp) } Cl_2$$

(c) From the balanced equation

3 moles Cl_2 reacts with 1 mole C_2H_2

or

3×71.0 g Cl_2 reacts with 22.4 liters (sc) C_2H_2

The ratio between the mass of Cl_2 and the volume of C_2H_2 is

$$\frac{3 \times 71.0 \text{ g } Cl_2}{22.4 \text{ liters (sc) } C_2H_2}$$

Then the number of grams of Cl_2 that reacts with 75.0 liters (cp) of C_2H_2 is

$$75.0 \text{ liters (cp) } C_2H_2 \times \frac{3 \times 71.0 \text{ g } Cl_2}{22.4 \text{ liters (sc) } C_2H_2}$$

But this is incomplete because we cannot cancel liters (cp) and liters (sc); consequently, the 75 liters (cp) must be converted to the corresponding volume at sc,

$$V \text{ (sc)} = 75.0 \text{ liters (cp) } C_2H_2 \times \frac{273°K \times 790 \text{ torr}}{297°K \times 760 \text{ torr}} = 71.7 \text{ liters (sc) } C_2H_2$$

Then the number of grams of Cl_2 that reacts with 75.0 liters (cp) or 71.7 liters (sc) of C_2H_2 is

$$71.7 \text{ liters (sc) } C_2H_2 \times \frac{3 \times 71.0 \text{ g } Cl_2}{22.4 \text{ liters (sc) } C_2H_2} = 682 \text{ g } Cl_2$$

Note that the *choice of units* is determined by the statement of the problem; also notice that the use of units verifies the method used for the solution of these problems because the correct unit is obtained for the answer (Appendix I). If we had incorrectly inverted the ratio in the last step of the previous problem

$$71.7 \text{ liters (sc) } C_2H_2 \times \frac{22.4 \text{ liters (sc) } C_2H_2}{3 \times 71.0 \text{ g } Cl_2} = 7.54 \frac{[\text{liters (sc) } C_2H_2]^2}{\text{g } Cl_2}$$

the error would be obvious since

$$\frac{[\text{liters (sc) C}_2\text{H}_2]^2}{\text{g Cl}_2}$$

does not correspond to the desired answer, g Cl_2. Note finally that the arithmetical procedure is similar to that used in everyday life. A typical problem is, "What is the cost in dollars of 12.2 gross of nails selling at 11.2 cents per hundred?" From the price, the ratio between the number of cents and the number of nails is

$$\frac{11.2 \text{ cents}}{100 \text{ nails}}$$

Then the number of cents required for 12.2 gross of nails is

$$12.2 \text{ gross} \times \frac{144 \text{ nails}}{\text{gross}} \times \frac{11.2 \text{ cents}}{100 \text{ nails}}$$

and, converted to dollars,

$$12.2 \text{ \sout{gross}} \times \frac{144 \text{ \sout{nails}}}{\text{\sout{gross}}} \times \frac{11.2 \text{ \sout{cents}}}{100 \text{ \sout{nails}}} \times \frac{1 \text{ dollar}}{100 \text{ \sout{cents}}} = 1.97 \text{ dollars}$$

Ratios such as

$$\frac{100 \text{ cents}}{1 \text{ dollar}} \quad \text{or} \quad \frac{12 \text{ inches}}{1 \text{ foot}} \quad \text{or} \quad \frac{454 \text{ g}}{1 \text{ pound}}$$

are known as *conversion factors* because they are used to convert a given unit to another. But the ratios used in Example 1,

$$\frac{166 \text{ g C}_2\text{Cl}_4}{3 \times 71 \text{ g Cl}_2}, \quad \frac{3 \times 22.4 \text{ liters (sc) Cl}_2}{166 \text{ g C}_2\text{Cl}_4}, \quad \frac{3 \times 71.0 \text{ g Cl}_2}{22.4 \text{ liters (sc) C}_2\text{H}_2}$$

are also "conversion factors." The first ratio, for example, "converted" the given quantity of Cl_2 to the quantity of C_2Cl_4 that it produces in this particular reaction.

Example 2 When an iodate is mixed with an excess of iodine in sulfuric acid, a solution is formed having the brown–red color of the ion I_3^+ formed by the reaction

$$HIO_3 + 7I_2 + 8H_2SO_4 \longrightarrow 5I_3^+ + 3H_3O^+ + 8HSO_4^-$$

Calculate the number of I_3^+ ions produced when 1.5 g of HIO_3 reacts.

Answer Essentially, we wish to "convert" 1.5 g of HIO_3 to a number of ions of I_3^+. From the given chemical equation, the ratio between these two quantities is

$$\frac{5 \text{ moles } I_3^+}{1 \text{ mole } HIO_3} \quad \text{or} \quad \frac{5 \times 6.0 \times 10^{23} \text{ ions } I_3^+}{176 \text{ g } HIO_3}$$

Then, following the same arithmetical procedure by which feet are converted to inches,

$$1.5 \text{ \sout{g HIO}}_3 \times \frac{5 \times 6.0 \times 10^{23} \text{ ions } I_3^+}{176 \text{ \sout{g HIO}}_3} = 2.6 \times 10^{22} \text{ ions } I_3^+$$

Reactants are seldom added in the exact relative amounts required by the balanced equation. Often, only one reactant is entirely consumed.* Its quantity fixes the quantity of product(s) obtained. The other reactants are present in excess.

Example 3 2.00 g of lead nitrate, $Pb(NO_3)_2$, is added to a solution containing 1.00 g of potassium iodide, KI; insoluble lead iodide forms,

$$Pb(NO_3)_2(aq) + 2KI(aq) \longrightarrow PbI_2(c) + 2KNO_3(aq)$$

Find the mass of PbI_2 formed. Which reactant is present in excess and by how many grams?

Answer First determine which reactant is present in excess by calculating the amount of one reactant required to react with the given amount of the second reactant. This calculation, *regardless of the choice* we make for the first reactant, tells us which reactant is in excess. Thus, suppose we calculate how many grams of $Pb(NO_3)_2$, 331 g/mole, are needed for the 1.00 g of KI, 166 g/mole,

$$1.00 \text{ g KI} \times \frac{331 \text{ g } Pb(NO_3)_2}{2 \times 166 \text{ g KI}} = 0.997 \text{ g } Pb(NO_3)_2 \qquad (1)$$

Since 2.00 g of $Pb(NO_3)_2$ is given, $Pb(NO_3)_2$ is in excess by 1.00 g.

If we had calculated how many grams of KI are needed for the 2.00 g of $Pb(NO_3)_2$,

$$2.00 \text{ g } Pb(NO_3)_2 \times \frac{2 \times 166 \text{ g KI}}{331 \text{ g } Pb(NO_3)_2} = 2.01 \text{ g KI}$$

we would have concluded that 1.00 g of KI is insufficient to react with 2.00 g of $Pb(NO_3)_2$. Then, to determine the quantity of $Pb(NO_3)_2$ in excess, we would have to proceed as above (Equation 1).

These calculations also show that the quantity of PbI_2, 461 g/mole, formed is determined by the 1.00 g of KI [or 0.997 g of $Pb(NO_3)_2$] consumed:

$$1.00 \text{ g KI} \times \frac{461 \text{ g } PbI_2}{2 \times 166 \text{ g KI}} = 1.39 \text{ g } PbI_2$$

or

$$0.997 \text{ g } Pb(NO_3)_2 \times \frac{461 \text{ g } PbI_2}{331 \text{ g } Pb(NO_3)_2} = 1.39 \text{ g } PbI_2$$

Equations with fractional coefficients, such as

$$H_2(g) + \tfrac{1}{2}O_2(g) \longrightarrow H_2O(l)$$

are frequently used. This balanced equation is read as: 1 mole of H_2 combines with $\frac{1}{2}$ mole of O_2 to form 1 mole H_2O and not as 1 molecule of H_2 combining with $\frac{1}{2}$ molecule of O_2. Unless oxygen atoms are actually used in the experiment, it is incorrect to rewrite the equation

---

* The conversion of reactants to products is frequently incomplete (Chapter 11).

as $H_2(g) + O(g) \rightarrow H_2O(l)$. This equation describes a different reaction and is read as: 1 mole of H_2 combines with 1 mole of O (oxygen atoms).

For a variety of reasons, reactants often yield quantities of products that are *less* than those calculated from the balanced chemical equation. The quantity calculated from the chemical equation is referred to as the THEORETICAL YIELD, and the PERCENT YIELD is given by

$$\% \text{ yield} = \frac{\text{actual yield}}{\text{theoretical yield}} \times 100\%$$

Example 4 Ethylene oxide, C_2H_4O, is manufactured by the oxidation of ethylene in air,

$$C_2H_4 + \tfrac{1}{2}O_2 \longrightarrow C_2H_4O$$

Undesirable events are (a) the failure of some ethylene to react; (b) the oxidation of some ethylene to formaldehyde (H_2CO), CO, and CO_2; (c) the decomposition of some ethylene to carbon (smoke) and other products. If 60 g of C_2H_4O is obtained from 42 g of C_2H_4, what is the percent yield?

Answer The theoretical yield, the quantity of C_2H_4O that could be obtained from 42 g of C_2H_4, is

$$42 \text{ g } C_2H_4 \times \frac{44 \text{ g } C_2H_4O}{28 \text{ g } C_2H_4} = 66 \text{ g } C_2H_4O$$

The actual yield obtained, however, is 60 g C_2H_4O. The percent yield is, therefore,

$$\frac{60 \text{ g } C_2H_4O}{66 \text{ g } C_2H_4O} \times 100\% = 91\%$$

PROBLEMS

1. Valence and formula. (a) What is the valence of the element combined with oxygen in each of the following oxides: CO, BrO_2, N_2O, and Bi_2O_5? (b) Write the simplest formula for each of the following compounds: calcium chlorate, sodium peroxide, magnesium phosphate, and aluminum acetate. (c) The formula for radium bromide is $RaBr_2$. Write the formula for each of the following compounds of radium: the sulfate, the nitrate, and the permanganate.

2. Nomenclature. (Review Appendix II.) (a) HNO_3 is the formula for nitric acid; write the formulas for nitrous acid and lead nitrate. (b) H_3PO_4 is the formula for phosphoric acid; name the acids $H_3PO_3(H_2PHO_3)$ and H_3PO_2 (HPH_2O_2). (c) Name $Mg(ClO_4)_2$. (d) $HBrO_3$ is the formula for bromic acid; write the formulas for perbromic and hypobromous acids.

3. Nomenclature. Name each of the following compounds: (a) AlN; (b) Cl_2O; (c) $FeCl_3$; (d) $LiBrO_3$; (e) K_2SO_3; (f) $Ca(H_2PO_4)_2$; (g) $PbCrO_4$. Write the formula for each of the following compounds: (a) disodium sulfide; (b) cobalt(II) chromate; (c) zinc arsenate; (d) tin(IV) sulfate; (e) nickel(II) acetate; (f) zinc cyanide; (g) strontium peroxide.

4. Chemical equations. Write the balanced equation for each reaction:
(a) $NaNO_3(c) \rightarrow NaNO_2(c) + O_2(g)$
(b) $Mg_3N_2(c) + H_2O \rightarrow Mg(OH)_2(c) + NH_3(g)$
(c) $BCl_3(l) + H_2O \rightarrow H_3BO_3(aq) + HCl(g)$
(d) $Ca(OH)_2(c) + H_3PO_4(l) \rightarrow Ca_3(PO_4)_2(c) + H_2O(l)$
(e) $K(l) + Al_2O_3(c) \rightarrow Al(l) + K_2O(c)$

5. Conversion factors. Write the conversion factor you would use for each of the following problems: (a) How many feet in 116 inches? (b) How many liters in 30.0 ml? (c) How many molecules of B_5H_9 are obtained when 10^2 mole of B_2H_6 reacts as follows? $5B_2H_6(g) \rightarrow 2B_5H_9(g) + 6H_2(g)$. (d) How many grams of iodine triperchlorate may be prepared by the reaction, $3AgClO_4 + 2I_2 \rightarrow 3AgI + I(ClO_4)_3$,
   (i)   if 4.2 moles of silver perchlorate react?
   (ii) if 10.0 ml (sc) of iodine gas reacts?
   (iii) if 16.5 g of silver iodide is produced?
   (iv) if 0.50 mole of I_2 reacts?
(e) How many grams of Br^- are in a sample that yields 0.5672 g of silver bromide when all the Br^- in the sample reacts as follows? $Ag^+(aq) + Br^-(aq) \rightarrow AgBr(c)$.

6. Chemical arithmetic. Find the mass of Cl_2 obtained by decomposition of 35 g of $FeCl_3$, $2\ FeCl_3(l) \rightarrow 2FeCl_2(c) + Cl_2(g)$.

7. Chemical arithmetic. A spaceship is propelled by hydrazine fuel. Calculate the number of liters of N_2O_4 at (a) standard conditions, (b) 27°C and 2.00 atm needed to burn 100 g N_2H_4; $2N_2H_4(l) + N_2O_4(g) \rightarrow 3N_2(g) + 4H_2O(g)$.

8. Chemical arithmetic. How many water molecules are produced when 100 liters of N_2O_4 at standard conditions is consumed? See Problem 7.

9. Excess reactant. 4.0 g of magnesium wire is ignited in 4.0 g of nitrogen, $3Mg(c) + N_2(g) \rightarrow Mg_3N_2(c)$. (a) Which reactant is present in excess? (b) Find the mass, in grams, of the excess. (c) How many grams of Mg_3N_2 are formed?

10. Pollution. Methanogenic bacteria in the mud bottom of lakes convert $HgCl_2$ to dimethylmercury, $HgCl_2(aq) + 2CH_4(g) \rightarrow (CH_3)_2Hg(aq) + 2HCl(aq)$. (a) Find the mass of $(CH_3)_2Hg$ formed when 2.0×10^{-9} g $HgCl_2$ reacts. (b) Find the volume of CH_4, in milliliters at standard conditions, that reacts with 2.5×10^{-10} g $HgCl_2$.

11. Yield. In the laboratory, oxygen may be prepared by heating potassium chlorate. During the decomposition, $2KClO_3(c) \rightarrow 2KCl(c) + 3O_2(g)$, a number of other reactions occur. 25.0 g of $KClO_3$ was heated and yielded 9.6 g O_2. Calculate the percent yield of O_2.

ADDITIONAL PROBLEMS

12. Nomenclature. (a) Write the formula for (i) iron(III) sulfate and iron(II) sulfate; (ii) dibismuth trioxide; (iii) phosphorus pentachloride. (b) Of what elements is the compound trichromium dicarbide composed?

13. Chemical equations. Write the balanced equation for each reaction:

(a) $C_6H_6(l) + O_2(g) \rightarrow$

(b) $SO_2(g) + NaNO_3(l) \rightarrow Na_2SO_4(c) + NO_2(g)$

(c) $NO_2(g) + HCl(g) \rightarrow NO(g) + Cl_2(g) + H_2O(l)$

(d) $Al(c) + H_2O(l) + NaOH(aq) \rightarrow Na_3Al(OH)_6(aq) + H_2(g)$

14. Chemical arithmetic. (a) Find the volume of nitrogen in liters (sc) obtainable from decomposition of 8.00 g ammonium nitrite, $NH_4NO_2(aq) \rightarrow N_2(g) + 2H_2O(l)$. (b) What is the volume at 127°C and 0.500 atm?

15. $LiNO_3$ decomposes at 750°K, $LiNO_3(l) \rightarrow LiNO_2(l) + \frac{1}{2}O_2(g)$. Find the mass (a) of $LiNO_3$ that will yield 50.0 g of O_2; (b) of $LiNO_2$ formed at the same time.

16. Given the reaction, $5O_3(g) + I_2(c) + H_2O \rightarrow 5O_2(g) + 2HIO_3(aq)$, (a) Find the mass of HIO_3 produced from 40.0 g of O_3. (b) What volume of O_2 in liters at 273°C and 100 torr is produced from 50.0 g of I_2? (c) Find the mass of O_3 consumed when 2.00 liters of O_2 at 127°C and 700 torr form.

17. (a) Find the mass of O_2 obtained from the decomposition of 15.0 g silver oxide, $2Ag_2O(c) \rightarrow 4Ag(c) + O_2(g)$. (b) A sample of silver oxide was heated until all the oxygen was driven off. The loss of mass of the sample was 4.00 g. Calculate the mass of the silver produced.

18. Isopropanol (rubbing alcohol) is combustible, $2C_3H_7OH(l) + 9O_2(g) \rightarrow 6CO_2(g) + 8H_2O(l)$. (a) What volume of oxygen (sc) would be required to burn 10.0 g? (b) How many grams of CO_2 would be produced in (a)?

19. Gravimetric analysis is a technique in which the mass of a substance in a mixture is determined. A mixture containing no fluorine compound except methyl fluoroacetate, FCH_2COOCH_3 (MFA), yields 12.1 mg CaF_2. Find the mass of MFA in the mixture.

20. Excess reagent. Methanol (wood alcohol), formerly produced by the distillation of wood, is produced by the *Lurgi process*, $CO(g) + 2H_2(g) \xrightarrow{Cu} CH_3OH(l)$. If equal masses of reactants are used, which one is present in excess?

21. Composition. The compound H_2O_3 is formed when slightly acidified water containing oxygen is irradiated with high-energy electrons. (a) What is the mass percentage of oxygen in H_2O_3? Is this value related to the mass percentage of oxygen in water by small whole numbers? (b) Find the mass of oxygen combined with 1.0 g of hydrogen in the two compounds; are these values related by small whole numbers? Explain.

22. Composition. Dalton erroneously assigned oxygen the atomic weight of 8.0 relative to hydrogen (1.0). How would Dalton have written the formulas of the following molecules: HO, H_2O, H_2O_2, H_2O_3, H_2O_4?

23. Yield. One approach to the control of SO_2 emission is absorption by molten carbonates $[SO_2(g) + Na_2CO_3(l) + \frac{1}{2}O_2(g) \rightarrow Na_2SO_4(l) + CO_2(g)]$ and regeneration $[Na_2SO_4(l) + 2CO(g) + 2H_2(g) \rightarrow Na_2CO_3(l) + H_2S(g) + CO_2(g) + H_2O(g)]$. Experiment shows that 15.0 mg SO_2 yields 30.0 mg Na_2SO_4. Calculate the percent yield of Na_2SO_4.

5

THE FIRST LAW OF THERMODYNAMICS

THERMODYNAMICS (*heat + work*) is the study of processes which involve the transfer of heat and the performance of work. The quantities of heat and work can be measured experimentally. Thermodynamics is *not* concerned with theories of the structure of matter; it is, therefore, independent of any theory of atoms or molecules.

The science of thermodynamics had its original applications in the design and use of steam and internal combustion engines, but its influence in modern physics, chemistry, biology, biochemistry, space science, and engineering cannot be overstressed. A living cell evolves heat, does work, and requires energy obtained by the enzymatic oxidation of food. An understanding of the principles of thermodynamics is therefore essential to the interpretation of biological processes in molecular terms.

The first law of thermodynamics is based on experimental results: for example, the interconvertibility of matter and energy (page 69) and the failure to invent a perpetual motion machine that can provide work without consuming fuel. Hess' law (page 64) teaches us that we cannot obtain more heat from a chemical reaction by changing the method of carrying out the reaction. Experience also teaches us that the energy of a system may be altered by changing its temperature or letting it do work. A SYSTEM is any particular piece of matter under consideration. It may be a confined gas or a mixture in a beaker or a very complicated apparatus. Increasing the temperature or doing work on the system increases the energy of the system. Decreasing the temperature or letting the system do work decreases the energy of the system. In a series of experiments (1845–1878) James P. Joule showed that the same amount of work, 4.18×10^7 ergs, *always* produces the same change in temperature as obtained from 1 calorie of heat. We may say that 4.18×10^7 ergs of work has the same effect as 1 cal of heat. Heat and work evolved historically as two independent concepts. We therefore have heat units such as calories and work units such as ergs. Since Joule's time, however, it has been recognized that heat and work are merely two forms of energy, and we now use heat and work units interchangeably. These facts are summarized in the first law of thermodynamics: *the energy of a system is fixed and independent of the method of preparation of the system or the method of attaining the energy.* This statement is also called the law of conservation of energy. To illustrate this law, let us take as our system 1.000 g of water at 14.50°C and 1.00 atm. The energy of our system is a fixed value; let us call it E_1. These specifications—mass, chemical composition, temperature, and pressure—fix the ENERGY STATE of the system, simply called the *state* of the system. We can then call E_1 the energy of the *initial state*. When our system undergoes a change in state—for example, the water is heated or decomposed into H_2 and O_2—its energy is changed. Let us add 1.000 cal so that the temperature of our system is raised to 15.50°C. However, the energy of the system in this second state is also a fixed value, E_2, independent of the method of preparation. It then follows that the difference in the energies of these two states, $E_2 - E_1$, must have a definite value regardless of the method of arriving at these two states: $E_2 - E_1 = \Delta E = 1.000$ cal, regardless of how state 1 is changed to state 2. No attempt is made to define absolute values of E; we are concerned only with the difference in the energies of two states, ΔE. If E_2 is greater than E_1, ΔE is positive; if E_2 is less than E_1, ΔE is negative.

In this chapter, we concern ourselves only with heat and mechanical work.[*] If we

(1) add heat to the system, its energy increases;
(2) remove heat from the system, its energy decreases;
(3) do work on the system, its energy increases;
(4) let the system do work, its energy decreases.

These relations may be summarized[†] as

$$E_2 - E_1 = \Delta E = q - w \tag{1}$$

where q is the heat absorbed from the surroundings, and w is the work done by the system upon the surroundings. The SURROUNDINGS refer to any matter that can interact with the system (e.g., add or remove heat; compress the system or allow it to expand). The energy of the system increases by the amount of heat added, and decreases by the amount of work done by the system.

We pause briefly to explain the sign conventions we shall use in this text. When a system *absorbs heat* from the surroundings, this heat is considered positive, that is, $q > 0$. When heat is *given out* by the

---

[*] Electrical work is considered in Chapter 17.

[†] This is another way of stating the first law.

system, the heat is considered negative: $q < 0$. Conventions like this save words. Data for processes, each involving a certain amount of heat, may then be expressed in either of the following forms:

Process	Heat
1	49 cal absorbed or $q = +49$ cal
2	17 cal absorbed or $q = +17$ cal
3	40 cal given out or $q = -40$ cal

In the case of work, the convention is that work is *positive* $(w > 0)$ when the system *does work* on the surroundings – for example, a gas expands against a pressure. Where work is *done on* the system – the gas is compressed – the work is *negative* $(w < 0)$:

Process	Work
1	35 cal done by system or $w = +35$ cal
2	24 cal done on system or $w = -24$ cal
3	68 cal done by system or $w = +68$ cal

When these conventions are adhered to, we can substitute the numbers into Equation (1) and get correct answers.

When the final and initial states of the system are fixed, ΔE is fixed, but the heat gained or lost by the system and the work done on or by the system are not fixed. For example, let us assign arbitrarily an energy value of 90 cal to a particular quantity of a gas in an initial state, and suppose that its energy in the final state is 10 cal more, or 100 cal; then

$$\Delta E = E_2 - E_1 = 100 \text{ cal} - 90 \text{ cal} = 10 \text{ cal}$$

This change may be effected by any of an infinite number of methods (paths), four of which are illustrated:

Path I. Add 10 cal of heat $(q = +10$ cal) to the gas, *no work* being done on or by the gas $(w = 0)$; then

$$\Delta E = q - w$$
$$= 10 \text{ cal} - 0 = 10 \text{ cal}$$

Path II. Do 10 cal of work *on* the gas $(w = -10$ cal) *without permitting the evolution or absorption of heat* $(q = 0)$; then

$$\Delta E = 0 \text{ cal} - (-10 \text{ cal})$$
$$= 0 \text{ cal} + 10 \text{ cal} = 10 \text{ cal}$$

Path III. Add 35 cal of heat $(q = +35$ cal) and let the system *do* 25 cal of *work* $(w = +25$ cal); then

$$\Delta E = 35 \text{ cal} - 25 \text{ cal} = 10 \text{ cal}$$

Path IV. Do 40 cal of work *on* the gas ($w = -40$ cal) and let the gas *evolve* 30 cal of heat ($q = -30$ cal); then

$$\Delta E = -30 \text{ cal} - (-40 \text{ cal})$$
$$= -30 \text{ cal} + 40 \text{ cal} = 10 \text{ cal}$$

One can imagine many other schemes by which this gas may pass from the initial to the final state. In words, we say ΔE *is independent of the path between the initial and the final state but q and w are path dependent.*

ENTHALPY / 5.2

The term "work" is used in physics and chemistry in a restricted sense; for example, it does not apply to "mental work." Work is done only when a force is exerted on a body so as to make it *move*. Thus, the effort of holding a 10-lb weight in your hand is not considered work since no motion is involved. Lift the weight and you do work. The WORK done by a body (solid, liquid, or gas) is the *product of the force, F, and the distance, d, through which the body moves* AGAINST *the force.* A gas does work by expanding *against* a pressure. The work done by a gas in expanding from V_1 to V_2 against a constant pressure P is

$$w = P(V_2 - V_1) = P \,\Delta V$$

(Appendix I.11). When P is expressed in atmospheres and V in liters, the units of w are liter atmosphere. One liter atm equals 24.2 cal.

When the *only* kind of work is $P \,\Delta V$, the first law may be written

$$\Delta E = q - P \,\Delta V$$

Let us now define

q_v = heat transferred to system at constant volume
q_p = heat transferred to system at constant pressure

For chemical reactions that occur in a container of *fixed volume* ($\Delta V = 0$), $P \,\Delta V = 0$ and

$$\Delta E = q_v \qquad\qquad\qquad\qquad\qquad (2)$$

In words, the increase (or decrease) in energy is equal to the heat absorbed (or evolved) when the process occurs at fixed volume.

However, chemical reactions occur more frequently in open vessels at constant atmospheric pressure; then

$$\Delta E = q_p - P \,\Delta V$$

or

$$q_p = \Delta E + P \,\Delta V$$

Since reactions are often conducted at constant pressure, we define a new quantity, called the ENTHALPY* H, by the equation

$$H = E + PV$$

It then follows that at constant pressure

$$\Delta H = \Delta E + P \, \Delta V = q_p$$

or

$$\Delta H = q_p \tag{3}$$

In words, the increase (or decrease) in enthalpy is equal to the heat absorbed (or evolved) when the process occurs at constant pressure and the only work is $P \, \Delta V$. As with E, no attempt is made to define absolute values of H for a system. Of interest, rather, is the change in H that occurs when the state of a system is changed at constant pressure.

Assuming ideal behavior and constant temperature, the relation between ΔH and ΔE can be derived from the equation (page 14)

$$PV = nRT \qquad \text{or} \qquad V = nRT/P$$

Let Δn be the total number of moles of *gaseous* products minus the total number of moles of *gaseous* reactants:

$$\Delta n = n_{\text{products}} - n_{\text{reactants}}$$

The ideal gas law can be written $V_{\text{products}} = n_{\text{products}} RT/P$ and $V_{\text{reactants}} = n_{\text{reactants}} RT/P$; then $\Delta V = (\Delta n)RT/P$, and

$$\Delta H = \Delta E + P\Delta V = \Delta E + (\Delta n)RT \tag{4}$$

The volumes of solids and liquids are so small that they do not contribute appreciably to ΔV.

The value of R is 1.99 cal/mole-deg. If expansion accompanies a chemical reaction at constant pressure, $n_{\text{products}} > n_{\text{reactants}}$, work is done *by* the system, and ΔH is greater than ΔE. If contraction accompanies a chemical change, $n_{\text{products}} < n_{\text{reactants}}$, work is done *on* the system, and ΔH is less than ΔE.

Thermochemical data are usually expressed in terms of enthalpy change, ΔH. The use of Equation (4) is illustrated in Section 5.3 in which we also study how the heat evolved or absorbed during a chemical change is measured.

The energy, E, of a system depends only on its state, and is independent of the path taken to reach that state. But the same is true of the product PV. Therefore, H, the sum of $E + PV$, depends only on the state of a system. Consequently, ΔH, like ΔE, depends only on the final and initial states.

* The term "enthalpy" is derived from the Greek word *enthalpein*, meaning "warming up." The more descriptive but misleading term, "heat content," is sometimes used synonymously with enthalpy.

Although the first law summarizes important experimental observations, it cannot tell us whether or not a given process will occur. It tells us that if carbon and hydrogen, state 1, change to methane gas, state 2, energy is conserved. However, it cannot tell us whether or not the change can occur. This subject is considered in Chapters 11, 14, and 17.

THERMOCHEMISTRY / 5.3

Chemical changes are nearly always accompanied by energy changes. The combustion of fossil fuels, such as coal, petroleum products, and natural gas, now constitutes man's major energy source. Chemical reactions giving off heat to the surroundings are called EXOTHERMIC; the surroundings become hot. If heat is absorbed from the surroundings, the reaction is said to be ENDOTHERMIC; the surroundings become cold. In an exothermic reaction, the original temperature can be restored only by removing heat from the system. In an endothermic reaction, the original temperature can be restored only by adding heat to the system.

The HEAT CAPACITY—or "the capacity for heat"—refers to the amount of heat (energy) required to raise the temperature of a given quantity of material 1 Celsius degree (units: cal/deg).

The SPECIFIC HEAT of a substance is the heat required to warm 1 g of the substance 1 Celsius degree (units: cal/g-deg). The relationship is

$$\text{specific heat} = \frac{\text{heat absorbed or released (cal)}}{\text{mass (g)} \times \text{temp rise or fall (C deg)}}$$

Specific heats vary markedly from substance to substance, and vary to some extent with temperature for any given substance. Some approximate values under ordinary conditions are: air (standard pressure), 0.25 cal/g-deg; water, 1.00; ice, 0.5; alcohol, 0.58; copper, 0.09. Heat measurements are made by mixing known amounts of reactants in a calorimeter (Fig. 5.1). The heat evolved by the reaction is equal to the heat absorbed by a known quantity of water, the metal bucket, the metal reaction chamber containing known quantities of reactants, the stirrer, and the thermometer. The heat capacity of the calorimeter is determined by putting in a known amount of energy and measuring the temperature rise.

Example 1 The combustion of 0.100 g of liquid benzene, C_6H_6, in a calorimeter whose heat capacity is 382.9 cal/deg produced a temperature rise of 2.609°C. Calculate the amount of heat evolved in the combustion of 1 mole of C_6H_6, the MOLAR HEAT OF COMBUSTION.

Answer The quantity of heat emitted is

$$382.9 \ \frac{\text{cal}}{\text{deg}} \times 2.609 \ \text{deg} = 999 \ \text{cal}$$

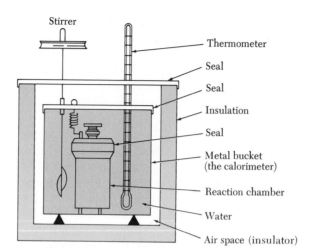

Fig. 5.1. *Representation of a calorimeter. The stirrer maintains uniform temperature.*

Stirrer

Thermometer

Seal

Seal

Insulation

Seal

Metal bucket (the calorimeter)

Reaction chamber

Water

Air space (insulator)

This is the amount of heat that would have to be removed to restore the calorimeter to its original temperature. Since 999 cal are emitted during the combustion of 0.100 g of C_6H_6, the heat emitted by 1 g of C_6H_6 is

$$\frac{999 \text{ cal}}{0.100 \text{ g}} = 9990 \; \frac{\text{cal}}{\text{g}}$$

Then the heat emitted by the combustion of 1 mole of C_6H_6, 78.1 g/mole, is

$$9990 \; \frac{\cancel{\text{cal}}}{\cancel{\text{g}}} \times 78.1 \; \frac{\cancel{\text{g}}}{\text{mole}} \times \frac{1 \text{ kcal}}{10^3 \; \cancel{\text{cal}}} = 780 \; \frac{\text{kcal}}{\text{mole}}$$

Calorimetric measurements are generally made in a container of *fixed volume*. From Equation (2), page 59, the value obtained in such an experiment, q_v, is equal to ΔE of the reaction. Thus, the **THERMOCHEMICAL** equation for the combustion of liquid C_6H_6,

$$C_6H_6(l) + 7\tfrac{1}{2}O_2(g) \longrightarrow 3H_2O(l) + 6CO_2(g) \qquad \Delta E = -780 \text{ kcal}$$
$$\text{(exothermic reaction)}$$

means that when 1 mole of liquid benzene reacts with $7\frac{1}{2}$ moles of oxygen, forming 3 moles of liquid water and 6 moles of carbon dioxide at constant volume and constant temperature, 780 kcal are evolved. The energy of 3 moles of $H_2O(l)$ and 6 moles of $CO_2(g)$ is *780 kcal less* than the energy of 1 mole of $C_6H_6(l)$ and $7\frac{1}{2}$ moles of $O_2(g)$. Since $\Delta E = E_{\text{products}} - E_{\text{reactants}}$, ΔE is negative for the combustion of $C_6H_6(l)$.

The thermochemical equation for the decomposition of nickel oxide is

$$NiO(c) \longrightarrow Ni(c) + \tfrac{1}{2}O_2(g) \qquad \Delta E = +57.0 \text{ kcal}$$
$$\text{(endothermic reaction)}$$

When 1 mole of solid NiO is decomposed to 1 mole of solid Ni and $\frac{1}{2}$ mole of O_2 (g) at *constant volume* and *constant temperature*, 57.0 kcal is absorbed from the surroundings so that the energy of the 1 mole of Ni and $\frac{1}{2}$ mole of O_2 is 57.0 $kcal$ $larger$ than the energy of 1 mole of NiO; ΔE for the reaction is therefore positive.

Since the energy of a substance depends on its state, the ΔE of a reaction also depends on the state of each reactant and product. The data given in this text are for substances in their stable states at $25.0°C$ and 1 atm pressure. Graphite is the stable state for carbon.

Example 2 For the combustion of liquid benzene at 25.0°C,

$$C_6H_6(l) + 7\tfrac{1}{2}O_2(g) \longrightarrow 3H_2O(l) + 6CO_2(g) \qquad \Delta E = -780 \text{ kcal}$$

Calculate ΔH.

Answer Six moles of gas are produced for 7.5 moles of gas consumed. From Equation (4), page 60,

$$\Delta n = 6.00 \text{ moles} - 7.50 \text{ moles} = -1.50 \text{ moles}$$

and

$$\Delta H = \Delta E + (\Delta n)RT$$

$$= -780 \text{ kcal} + (-1.50 \text{ moles}) \times 1.99 \, \frac{\text{cat}}{\text{mole deg}} \times \frac{\text{kcal}}{10^3 \text{ cat}} \times 298 \text{ deg}$$

$$= -781 \text{ kcal}$$

Note that ΔE and ΔH are almost equal. $(\Delta n)RT$ is usually a small correction and very different answers for ΔE and ΔH probably mean that you have confused calories and kilocalories.

The thermochemical equation for the combustion of C_6H_6 is

$$C_6H_6(l) + 7\tfrac{1}{2}O_2(g) \longrightarrow 3H_2O(l) + 6CO_2(g) \qquad \Delta H = -781 \text{ kcal}$$

This means that when 1 mole of liquid benzene reacts with $7\frac{1}{2}$ moles of oxygen, forming 3 moles of liquid water and 6 moles of carbon dioxide at *constant pressure* and *constant temperature*, 781 kcal are evolved; therefore, the enthalpy of the 3 moles of $H_2O(l)$ and 6 moles of $CO_2(g)$ is 781 kcal *less* than the enthalpy of 1 mole of $C_6H_6(l)$ and $7\frac{1}{2}$ moles of $O_2(g)$. Since $\Delta H = H_{products} - H_{reactants}$, ΔH is negative for the combustion of $C_6H_6(l)$. For the decomposition of nickel oxide at 25.0°C,

$$NiO(c) \longrightarrow Ni(c) + \tfrac{1}{2}O_2(g) \qquad \Delta H = +57.3 \text{ kcal}$$

This means that when 1 mole of solid NiO decomposes to 1 mole of solid Ni and $\frac{1}{2}$ mole of $O_2(g)$ at *constant temperature* and *constant pressure*, 57.3 kcal are absorbed from the surroundings. The enthalpy of the products is larger than the enthalpy of the reactants. ΔH for the reaction is positive.

When the reaction produces no volume change, $\Delta H = \Delta E$, as illustrated by

$$H_2(g) + Cl_2(g) = 2HCl(g) \qquad \Delta H = \Delta E = -44.12 \text{ kcal}$$

Thermochemical equations may be reversed, in which case the sign of ΔE or ΔH is changed. For example,

$$H_2(g) + \tfrac{1}{2}O_2(g) \longrightarrow H_2O(l) \qquad \Delta H = -68.3 \text{ kcal}$$
$$H_2O(l) \longrightarrow H_2(g) + \tfrac{1}{2}O_2(g) \qquad \Delta H = +68.3 \text{ kcal}$$

means the heat evolved in the formation of 1 mole of liquid water is equal to the heat required to decompose 1 mole of liquid water. The biochemically important role of *adenosine triphosphate, ATP,* is another illustration. ATP serves to store (absorb) and then evolve energy on command in many biochemical processes. The reaction is a decomposition to *adenosine diphosphate, ADP,* and phosphoric acid:

$$\text{adenosine}-\text{O}-\underset{\underset{\text{O}}{|}}{\overset{\overset{\text{OH}}{|}}{\text{P}}}-\text{O}-\underset{\underset{\text{O}}{|}}{\overset{\overset{\text{OH}}{|}}{\text{P}}}-\text{O}-\underset{\underset{\text{O}}{|}}{\overset{\overset{\text{OH}}{|}}{\text{P}}}-\text{OH}(aq) + H_2O \xrightarrow[\text{energy}]{\text{evolves}} \text{ADP}(aq) + H_3PO_4(aq)$$

$$\Delta H = -8 \text{ kcal}$$

$$\text{adenosine}-\text{O}-\underset{\underset{\text{O}}{|}}{\overset{\overset{\text{OH}}{|}}{\text{P}}}-\text{O}-\underset{\underset{\text{O}}{|}}{\overset{\overset{\text{OH}}{|}}{\text{P}}}-\text{OH}(aq) + H_3PO_4(aq) \xrightarrow[\text{energy}]{\text{absorbs}} \text{ATP}(aq) + H_2O \qquad \Delta H = +8 \text{ kcal}$$

The heat evolved is exactly equal to the heat absorbed per mole of ATP.

Thermochemical equations may be added and subtracted as ordinary algebraic equations. This generalization is known as **HESS' LAW.** For example, the addition of the two thermochemical equations

$$C_2H_4(g) + H_2(g) + 3\tfrac{1}{2}O_2(g) \longrightarrow 2CO_2(g) + 3H_2O(l) \qquad \Delta H = -405.5 \text{ kcal}$$
$$2CO_2(g) + 3H_2O(l) \longrightarrow C_2H_6(g) + 3\tfrac{1}{2}O_2(g) \qquad \Delta H = +372.8 \text{ kcal}$$

predicts that the heat of addition of H_2 to ethylene, C_2H_4, to form ethane, C_2H_6, is

$$C_2H_4(g) + H_2(g) \longrightarrow C_2H_6(g) \qquad \Delta H = -405.5 + 372.8 = -32.7 \text{ kcal}$$

The experimentally determined heat of reaction is

$$C_2H_4(g) + H_2(g) \longrightarrow C_2H_6(g) \qquad \Delta H = -32.6 \pm 0.1 \text{ kcal}$$

When a formula appears on the right of one equation and on the left of another, the substance will not appear in the final equation if it is produced and consumed in equal quantities. We must be careful to cancel out a formula only when the state (g, l, c) is the same on both sides of the equation. For example, the enthalpy of $H_2O(g)$ is greater than that of $H_2O(l)$ by 10.5 kcal/mole,

$$H_2O(l) \longrightarrow H_2O(g) \qquad \Delta H = +10.5 \text{ kcal}$$

Hess' law shows that the heat of a reaction, ΔE or ΔH, depends only on the final products and initial reactants, and is independent of the path (how the chemical change is carried out, in one or several steps).

Example 3 Use the thermochemical equations

$$
\begin{array}{lll}
\tfrac{1}{2}H_2(g) + \tfrac{1}{2}I_2(c) \longrightarrow HI(g) & \Delta H = +6.33 \text{ kcal} & (5) \\
\tfrac{1}{2}H_2(g) \longrightarrow H(g) & \Delta H = +52.1 \text{ kcal} & (6) \\
\tfrac{1}{2}I_2(g) \longrightarrow I(g) & \Delta H = +18.1 \text{ kcal} & (7) \\
I_2(c) \longrightarrow I_2(g) & \Delta H = +14.9 \text{ kcal} & (8)
\end{array}
$$

to calculate ΔH for the formation of gaseous hydrogen iodide, HI, from gaseous atomic hydrogen and gaseous atomic iodine,

$$H(g) + I(g) \longrightarrow HI(g)$$

Answer Manipulate the given chemical equations as you would algebraic equations so as to eliminate those substances not appearing in the final desired equation. Reverse Equations (6), (7), and (8) and also multiply Equation (8) by $\tfrac{1}{2}$ to give

$$
\begin{array}{lll}
\tfrac{1}{2}H_2(g) + \tfrac{1}{2}I_2(c) \longrightarrow HI(g) & \Delta H = +6.33 \text{ kcal} & (5) \\
H(g) \longrightarrow \tfrac{1}{2}H_2(g) & \Delta H = -52.1 \text{ kcal} & (6) \text{ reversed} \\
I(g) \longrightarrow \tfrac{1}{2}I_2(g) & \Delta H = -18.1 \text{ kcal} & (7) \text{ reversed} \\
\tfrac{1}{2}I_2(g) \longrightarrow \tfrac{1}{2}I_2(c) & \Delta H = -7.45 \text{ kcal} & (8) \text{ reversed} \times \tfrac{1}{2}
\end{array}
$$

which on addition yield

$$H(g) + I(g) \longrightarrow HI(g) \qquad \Delta H = +6.33 - 52.1 - 18.1 - 7.45 = -71.3 \text{ kcal}$$

Thermochemical data are recorded as HEATS OF FORMATION of substances, ΔH_f, defined as the change in enthalpy when 1 mole of the substance is formed from its elements. It is assumed that the elements are initially in their stable states at 25°C and 1 atm pressure.

Example 4 Given the following heats of formation (ΔH_f, kcal/mole): acetylene, $C_2H_2(g)$, +54.2; $CO_2(g)$, −94.1; and $H_2O(l)$, −68.3. Calculate the heat of combustion of acetylene.

Answer The thermochemical equations for the heats of formation are as follows:

$$
\begin{array}{ll}
2C(graphite) + H_2(g) \longrightarrow C_2H_2(g) & \Delta H = +54.2 \text{ kcal} \\
C(graphite) + O_2(g) \longrightarrow CO_2(g) & \Delta H = -94.1 \text{ kcal} \\
H_2(g) + \tfrac{1}{2}O_2(g) \longrightarrow H_2O(l) & \Delta H = -68.3 \text{ kcal}
\end{array}
$$

To obtain the heat of combustion, reverse the first equation, double the second, and then add all three. When the coefficients of an equation are doubled, ΔH is doubled:

$$
\begin{array}{lll}
C_2H_2(g) \longrightarrow 2C(graphite) + H_2(g) & \Delta H = - 54.2 \text{ kcal} \\
2C(graphite) + 2O_2(g) \longrightarrow 2CO_2(g) & \Delta H = -188.2 \text{ kcal} \\
\underline{H_2(g) + \tfrac{1}{2}O_2(g) \longrightarrow H_2O(l)} & \underline{\Delta H = - 68.3 \text{ kcal}} \\
C_2H_2(g) + 2\tfrac{1}{2}O_2(g) \longrightarrow 2CO_2(g) + H_2O(l) & \Delta H = -310.7 \text{ kcal}
\end{array}
$$

BOND ENERGY / 5.5

The formation of 1 mole of $HI(g)$ from gaseous atoms evolves 71.3 kcal, while the dissociation of 1 mole of $HI(g)$ into gaseous atoms requires 71.3 kcal. The quantity of energy required to break one bond in a molecule is called the **BOND DISSOCIATION ENERGY**, commonly called "**BOND ENERGY.**" Bond energy is the energy that the bond does *not* have, and will not get until some outside source of energy comes along and breaks the bond. Some bond energies are given in Table 5.1, expressed as ΔH in kilocalories per mole of bonds.

The bond energy depends to some extent on the molecule in which the bond is found. For example, the N—F bond energies in NF_3, NF_2, and NF are somewhat different:

$$
\begin{array}{lll}
NF_3(g) \longrightarrow NF_2(g) + F(g) & \Delta H = + 58 \text{ kcal} \\
NF_2(g) \longrightarrow NF(g) + F(g) & \Delta H = + 76 \text{ kcal} \\
\underline{NF(g) \longrightarrow N(g) + F(g)} & \underline{\Delta H = + 72 \text{ kcal}} \\
NF_3(g) \longrightarrow N(g) + 3F(g) & \Delta H = +206 \text{ kcal}
\end{array}
$$

The removal of any one of the three F atoms in NF_3 requires 58 kcal while the removal of either one of the two F atoms in NF_2 requires 76 kcal.

Now let us imagine that we did not have the ΔH values for the successive dissociations of each of the 3 F atoms from NF_3, but only the value for complete dissociation of NF_3, +206 kcal/mole. We could assume that equal energies are required to remove the fluorine atoms successively from the nitrogen atom. Then we may *assign* to each N—F bond an **AVERAGE BOND ENERGY** of 206/3 or 68.7 kcal/mole, called the *average bond energy* of the N—F bond. The decomposition of NF_3 into its constituent atoms will then require 3×68.7 or 206 kcal/mole.

Since no experimental data for the stepwise dissociation of most polyatomic molecules exist, the assignment of average bond energies to chemical bonds is common practice.

Example 5 Calculate the average bond energy for the C—H bond from the following experimental data:

$$C(graphite) + 2H_2(g) \longrightarrow H{-}\underset{\underset{\displaystyle H}{|}}{\overset{\overset{\displaystyle H}{|}}{C}}{-}H(g) \qquad \Delta H = -17.9 \text{ kcal}$$

methane

$$C(g) \longrightarrow C(graphite) \qquad \Delta H = -171.3 \text{ kcal}$$
$$2H(g) \longrightarrow H_2(g) \qquad \Delta H = -104.2 \text{ kcal}$$

Answer Multiply the third equation by 2 and add the three equations to give

$$C(g) + 4H(g) \longrightarrow CH_4(g) \qquad \Delta H = -397.6 \text{ kcal}$$

Notice that the only reaction now occurring is the formation of four C—H bonds; if we assume that an equal amount of energy is required in forming each C—H bond, it follows that in forming one bond

$$C(g) + H(g) \longrightarrow C{-}H(g) \qquad \Delta H = \frac{-397.6}{4} = -99.4 \text{ kcal}$$

or breaking one bond

$$C{-}H(g) \longrightarrow C(g) + H(g) \qquad \Delta H = +99.4 \text{ kcal}$$

The average bond energy for the C—H bond is, therefore, 99.4 kcal/mole.

  Average bond energies are usually called simply "bond energies"; Table 5.1 lists a number of bond energies. The energy required to decompose a gaseous molecule completely into isolated atoms must equal the sum of the bond energies for all the bonds within the molecule. Precise values may, of course, be calculated, as in Example 3 or 5, if thermochemical data are available. But suppose, as is frequently the situation, that experimental data are not available. If we *assume that the bond between a given pair of atoms is independent of the molecule in which it resides*, then the type of data given in Table 5.1 can be used to predict the energy required to dissociate any molecule into its constituent atoms.

Table 5.1
Bond Energy Values[a] for the Process: Bond AB
in a Gaseous Molecule $\longrightarrow$ A(g) + B(g)

C—H	99.4	C—Cl	78	H_2	104.2
C—C (ethane bond)	81	H—F	136	N_2	226.0
C=C (ethylenic bond)	141	H—Cl	103	F_2	37.8
C≡C (acetylenic bond)	194	H—Br	87.5	Cl_2	58.2
C—F	117	H—I	71.3	Br_2	46.1

  [a] ΔH in kilocalories per mole.

Example 6 Calculate the heat of the reaction

$$\text{H}-\overset{\displaystyle \text{H}}{\underset{\displaystyle \text{Cl}}{\underset{|}{\overset{|}{\text{C}}}}}-\text{H}(g) \longrightarrow C(g) + 3H(g) + Cl(g) \qquad \Delta H = ?$$

Answer This reaction involves breaking three C—H bonds, $\Delta H = +3 \times 99.4 = +298$ kcal, and breaking one C—Cl bond, $\Delta H = +78$ kcal. Note that bond breaking is endothermic, and ΔH is positive. ΔH for the reaction is, therefore, $298 + 78$ or $+376$ kcal/mole of CH_3Cl. The experimental value is 375.8 kcal/mole.

Example 7 Calculate the heat of the reaction

$$CH_4(g) + 4F_2(g) \longrightarrow CF_4(g) + 4HF(g)$$

Answer This reaction involves

(a) breaking four C—H bonds:

$$CH_4 \longrightarrow C + 4H \qquad \Delta H = +4 \times 99.4 \text{ kcal}$$

(b) breaking four F—F bonds:

$$4(F-F) \longrightarrow 8F \qquad \Delta H = +4 \times 37.8 \text{ kcal}$$

(c) forming four C—F bonds:

$$C + 4F \longrightarrow CF_4 \qquad \Delta H = -4 \times 117 \text{ kcal}$$

Note that bond forming is exothermic, and ΔH is negative.

(d) forming four H—F bonds:

$$4H + 4F \longrightarrow 4(H-F) \qquad \Delta H = -4 \times 136 \text{ kcal}$$

The calculated ΔH for the reaction is, therefore, $+4 \times 99.4 + 4 \times 37.8 - 4 \times 117 - 4 \times 136 = -463$ kcal/mole of CH_4. The measured value is -463 kcal/mole.

Although the agreement is usually good, for many reactions a significant difference occurs between the heat of the reaction as calculated in Examples 6 and 7 and the experimental heat of reaction. These discrepancies compel the reexamination of our assumption that the properties of a bond are independent of its molecular environment. This reexamination will lead us (page 384) to a better understanding of the nature of the bonds holding atoms together in molecules.

The larger the bond energy, the larger is the energy needed to break the bond, and the greater is the BOND STABILITY. Thus, H—H (104 kcal/mole) is a more stable bond than F—F (37.8 kcal/mole). And, of course, the formation of H_2 from 2H evolves more energy than the formation of F_2 from 2F.

The conversion of one form of matter to another involves the conversion of matter to energy if the reaction is exothermic; conversely, in an endothermic reaction, the heat energy absorbed is converted into matter. This means that matter and energy are not conserved separately; rather, the principle of conservation of energy is broadened to include matter as another form of energy. The quantity of energy, expressed in grams, liberated or absorbed is *exactly equal* to the quantity of matter, expressed in grams, destroyed or created.

The interconvertibility of matter and energy was predicted by Albert Einstein in 1905. It is described by the equation $\Delta E = (\Delta m)c^2$. ΔE is the energy in ergs when Δm, the quantity of matter converted, is in grams; the constant c, the speed of light, is 3.00×10^{10} cm/second. An erg has the dimensions of g cm^2/sec^2, and there are 4.18×10^7 ergs in 1 cal.

Example 8 In the combustion of about 400 g of gasoline, 4.9×10^6 cal are evolved,

$$\text{reactants} \longrightarrow \text{products} \qquad \Delta E = -4.9 \times 10^6 \text{ cal}$$

Calculate the decrease in mass in grams accompanying this reaction.

Answer The mass of the products will be less by the amount calculated from the Einstein equation ($\Delta E = \Delta mc^2$),

$$\Delta m = \frac{\Delta E}{c^2}$$

$$= \frac{-4.9 \times 10^6 \, \cancel{\text{cal}} \times 4.2 \times 10^7 \, \frac{\text{ergs}}{\cancel{\text{cal}}}}{(3.0 \times 10^{10})^2 \, \frac{\text{cm}^2}{\text{sec}^2}}$$

$$= -2.3 \times 10^{-7} \text{ ergs } \frac{\text{sec}^2}{\text{cm}^2}$$

and since an erg is 1 g $\frac{\text{cm}^2}{\text{sec}^2}$

$$\Delta m = -2.3 \times 10^{-7} \, \frac{\text{g} \, \cancel{\text{cm}^2} \, \cancel{\text{sec}^2}}{\cancel{\text{sec}^2} \, \cancel{\text{cm}^2}} = -2.3 \times 10^{-7} \text{ g}$$

2.3×10^{-7} g is the quantity of matter liberated as heat to the surroundings.

This quantity is so small that unless nuclear reactions (Chapter 25) are involved, we can say that the total mass of matter of a chemically reacting system is *practically* constant.

1. Definitions. Define and illustrate: (a) first law, (b) endothermic process, (c) exothermic process, (d) q_v, (e) q_p, (f) ΔE, (g) ΔH, (h) heat capacity, (i) specific heat, (j) Hess' law, (k) bond energy, (l) bond stability, (m) Einstein's law.

2. First law. (a) A system absorbs 300 cal while 4.00 liter atm of work is done *on* the system. Find ΔE for the system in calories. (b) Describe another path leading to the same ΔE.

3. First law. A closed vessel of constant volume contains 10.0 g of liquid water at 100°C. One gram of the liquid is converted to steam at 100°C and 1.00 atm requiring 499 cal. When the same process is done at constant pressure (1.00 atm) and constant temperature (100°C), the steam does 40 cal of work. Find (a) w, (b) q_p, (c) ΔE, (d) ΔH.

4. First law. When 2.00 g of water is boiled at constant pressure, 2.0 atm, the steam does 84.4 cal of work and the heat of vaporization is 525.4 cal/g. Find (a) ΔH, (b) q_p, (c) w, (d) ΔE in calories.

5. First law. The combination of NH_3 and BF_3 in a bomb calorimeter (volume fixed) evolves 41 kcal of heat per mole of product at 25°C, $NH_3(g) + BF_3(g) \rightarrow H_3NBF_3(c)$. During the course of the reaction at constant pressure, 1.0 kcal of work/mole of product is done *on* the system. Find in kilocalories per mole (a) q_v, (b) ΔE, (c) w, (d) ΔH, (e) q_p.

6. Specific heat. Calculate the heat in kilocalories required to raise the temperature of 50.0 g of liquid water from its freezing point, 0°C, to its boiling point, 100°C. Specific heat = 1.00 cal/g-deg.

7. Thermochemistry. (a) An exothermic reaction occurs very rapidly in air. Will the temperature of the air increase or decrease? (b) An endothermic reaction occurs very rapidly in air. Will the temperature of the air increase or decrease? (c) A person pours a liquid into the palm of his hand. As the liquid evaporates, his hand feels cold. Is the evaporation of this liquid an exothermic or an endothermic process? (d) As a solid dissolves, the temperature of the solution decreases. Is the process of solution exothermic or endothermic?

8. Thermochemistry. (a) 0.504 g of H_2 reacts with O_2 at constant pressure evolving 17.08 kcal. Calculate ΔH for the reaction $H_2(g) + \frac{1}{2}O_2(g) \rightarrow H_2O(l)$. (b) For the reaction $2N_2(g) + O_2(g) \rightarrow 2N_2O(g)$, $\Delta H = +39$ kcal. Calculate (*i*) the heat absorbed when 6.50 g of N_2O forms; (*ii*) the heat evolved when 3.00 g of N_2O decomposes.

9. ΔE, ΔH. Serum albumin, a protein, combines with 3.0 g Cu^{2+} in water to form a charged complex with the absorption of 0.14 kcal. Calculate ΔE and ΔH for the reaction: serum albumin$(aq) + Cu^{2+}(aq) \rightarrow$ serum albumin—$Cu^{2+}(aq)$. (Solids and liquids undergo relatively small changes in volume so that w is negligible.)

10. Hess' law. (a) Interconversions between hydroquinone and quinone are involved in many essential biochemical reactions. The reaction responsible for the defensive explosive discharge of the bombardier beetle is $C_6H_4(OH)_2(aq)$ *(hydroquinone)* $+ H_2O_2(aq) \rightarrow C_6H_4O_2(aq)$*(quinone)* $+ 2H_2O(l)$. Calculate ΔH for the reaction from

$$C_6H_4(OH)_2(aq) \longrightarrow C_6H_4O_2(aq) + H_2(g) \qquad \Delta H = +42.4 \text{ kcal}$$
$$O_2(g) + 2H_2O(l) \longrightarrow 2H_2O_2(aq) \qquad \Delta H = +45.2 \text{ kcal}$$
$$H_2O(l) \longrightarrow H_2(g) + \tfrac{1}{2}O_2(g) \qquad \Delta H = +68.3 \text{ kcal}$$

(b) Use your answer to calculate (*i*) the heat in calories evolved by 8×10^{-6} mole of hydroquinone and (*ii*) the temperature rise of 4.0 mg of water for this quantity of heat.

11. Hess' law. (a) Calculate ΔH for the reaction $2H(g) + O(g) \rightarrow H_2O(g)$ from the following data:

$$2H_2(g) + O_2(g) \longrightarrow 2H_2O(l) \qquad \Delta H = -136.6 \text{ kcal}$$
$$H_2O(l) \longrightarrow H_2O(g) \qquad \Delta H = +10.5 \text{ kcal}$$
$$\tfrac{1}{2}H_2(g) \longrightarrow H(g) \qquad \Delta H = +52.1 \text{ kcal}$$
$$O_2(g) \longrightarrow 2O(g) \qquad \Delta H = +119.1 \text{ kcal}$$

(b) How many O—H bonds are in one molecule of $H_2O(g)$? Assign a bond energy value to the O—H bond.

12. Bond energies. See Table 5.1, page 67. Calculate ΔH, kcal/mole, for the reactions:
  (a) $CH_4(g) + 2Cl_2(g) \rightarrow CH_2Cl_2(g) + 2HCl(g)$
  (b) $CH_2Cl_2(g) \rightarrow C(g) + 2H(g) + 2Cl(g)$
CH_2Cl_2 has 2 C—H bonds and 2 C—Cl bonds.

ADDITIONAL PROBLEMS

13. Bond energies. See Table 5.1, page 67. Predict ΔH for the reaction

ethylenic bond   difluoroethane

Note: use C=C (ethylenic bond) value and C—C value. The measured value is -127 kcal. Are bond energies independent of their molecular environment?

14. ΔH, ΔE. For each of the following reactions, explain whether the heat evolved at constant pressure is smaller than, larger than, or the same as the heat evolved at constant volume:
  (a) $CH_4(g) + \tfrac{3}{2}O_2(g) \rightarrow 2H_2O(g) + CO(g)$
  (b) $CH_4(g) + 2O_2(g) \rightarrow 2H_2O(l) + CO_2(g)$
  (c) $2C_3O_2(g) + O_2(g) \rightarrow 6CO(g)$

15. $\Delta E = \Delta mc^2$. Explain the statement, "If the heat of a reaction is considered as one of the products of the reaction and the heat is not removed, then the mass of the products *exactly* equals the mass of the reactants."

16. $\Delta E = \Delta mc^2$. The heat of formation of solid barium sulfate from its ions is given by the following equation:

$$Ba^{2+}(aq) + SO_4^{2-}(aq) \longrightarrow BaSO_4(c) \qquad \Delta H = -5 \text{ kcal}$$

Calculate the decrease in mass (after restoring the original temperature) that occurs when 0.10 mole of $BaSO_4$ forms. Do you think that balances with a capacity of about 100 g can be manufactured today with the precision necessary to detect this mass decrease?

17. Joule. (a) A gas is confined in a container of fixed volume. (b) A gas is so confined that it may be compressed or allowed to expand against an opposing pressure. For each case, is it possible to add heat to or remove heat from the gas without changing its temperature? Explain.

18. Joule. 10 g of H_2O can be heated from 25 to 30°C by putting in 50 cal of heat. Describe how the same change can be brought about by doing work on the water when it is in an insulated container. How many joules of work would have to be done?

19. First law. A gas undergoes a series of physical changes which return it to its initial state. During these changes, the gas evolves 100 cal and absorbs 150 cal of heat. Calculate (a) ΔE, (b) q, and (c) w for the series of changes. (d) Was work done by or on the gas?

20. Specific heat of gases. The specific heat of helium is 0.752 cal/g-deg at constant volume. Should the specific heat at constant pressure be the same as or smaller or larger than 0.752 cal/g-deg? Explain.

21. Calorie saver. The Food and Drug Administration limits intake of sodium cyclamate ($C_6H_{12}NSO_3Na$, 201 g/mole, heat of combustion 771 kcal/mole) to no more than 168 mg/day person. What is the saving in Calories* when the cyclamate (5 times sweeter than sucrose) is substituted for 840 mg sucrose (table sugar), $C_{12}H_{22}O_{11}$, 342 g/mole, heat of combustion 1350 kcal/mole?

22. Hess' law. Given:

$$H_2(g) + \tfrac{1}{2}O_2(g) \longrightarrow H_2O(l) \qquad \Delta H = -68.3 \text{ kcal}$$
$$C(graphite) + O_2(g) \longrightarrow CO_2(g) \qquad \Delta H = -94.1 \text{ kcal}$$
$$12C(graphite) + 11H_2(g) + 5\tfrac{1}{2}O_2(g) \longrightarrow C_{12}H_{22}O_{11}(c) \qquad \Delta H = -535 \text{ kcal}$$

Find ΔH for (a) the combustion of sucrose,

$$C_{12}H_{22}O_{11}(c) + 12O_2(g) \longrightarrow 12CO_2(g) + 11H_2O(l).$$

(b) the decomposition of sucrose into carbon and water,

$$C_{12}H_{22}O_{11}(c) \longrightarrow 12C(graphite) + 11H_2O(l)$$

23. Thermochemistry. When 2.000 g of magnesium is burned in fluorine to magnesium fluoride at constant pressure, 44.75 kcal are evolved. Calculate ΔH_f, the molar heat of formation of $MgF_2(c)$.

24. Rocket fuel. Given the following data:

$$H_2(g) + \tfrac{1}{2}O_2(g) \longrightarrow H_2O(l) \qquad \Delta H = -68.3 \text{ kcal}$$
$$C_{10}H_8(c) + 12O_2(g) \longrightarrow 10CO_2(g) + 4H_2O(l) \qquad \Delta H = -1228 \text{ kcal}$$

Pound for pound, which is the more energetic rocket fuel, hydrogen or naphthalene ($C_{10}H_8$)?

---

* 1 kcal = 1 Cal (capital C), the energy unit used in nutrition. The use of cyclamate is now banned by the FDA.

25. Bond energy. Calculate the Mg—Cl bond energy from the data

$$Mg(c) \longrightarrow Mg(g) \qquad \Delta H = +67 \text{ kcal}$$
$$Cl_2(g) \longrightarrow 2Cl(g) \qquad \Delta H = +58 \text{ kcal}$$
$$Mg(c) + \tfrac{1}{2}Cl_2(g) \longrightarrow MgCl(g) \qquad \Delta H = -10 \text{ kcal}$$

26. Bond energy. Use the data

$$AlF_3(g) \longrightarrow AlF_2(g) + F(g) \qquad \Delta H = +156 \text{ kcal}$$
$$AlF_2(g) \longrightarrow AlF(g) + F(g) \qquad \Delta H = +107 \text{ kcal}$$
$$AlF(g) \longrightarrow Al(g) + F(g) \qquad \Delta H = +160 \text{ kcal}$$

to assign a bond energy value to the Al—F bond.

27. $\Delta E = \Delta mc^2$. Energy is emitted by the sun at the rate of 2 ergs/g(of sun)-second (for comparison, the rate of heat energy generated during animal metabolism is about 250 ergs/g-second). The estimated mass of the sun is 2×10^{33} tons. Estimate the decrease in tons and percent decrease in the mass of the sun per day. (1 metric ton $= 10^6$ g.)

28. Calorimetry. (a) An insulated closed system contains 1.020 kg of an alloy (specific heat, 0.1550 cal/g-deg), 1.151 kg of water (1.000 cal/g-deg), a stirrer (1.75 cal/deg), and a temperature recording device (0.160 cal/deg). If the temperature increases 1.240°C, how many kilocalories are absorbed by the system? (b) How many kilograms of water would be raised 1.240°C by the same quantity of heat?

29. Calorimetry. The temperature of a calorimeter (including all component parts) is raised 1.105°C by passing an electric current of 1.950 A for 2.000 seconds through a 500-Ω resistor inside the calorimeter. Calculate the heat capacity of the calorimeter in kilocalories per degree. Energy(joules) = current²(A²) × resistance(Ω) × time(seconds).

30. Energy consumption. (a) The world production of fossil fuels (1967) is given with the heat evolved per gram. 1 metric ton $= 10^6$ g. Calculate the world's energy consumption in kilocalories for 1967.

Fuel	Production, millions of metric tons	kcal/g
Coal[a]	1750	7.3
Lignite[a]	2360	3.6
Oil	1630	10.2
Natural gas	655	18.9

[a] Coal composition varies considerably. Approximate carbon content of coal: ("hard") anthracite 70 to 98%, ("soft") bituminous 40 to 80%, lignite 30 to 40%.

(b) The earth's surface receives solar radiation at the rate of about 0.5 cal/cm²-min. Calculate the quantity of energy in kilocalories the earth receives in 1 year. 5.3×10^5 minutes = 1 year, surface area of earth $= 5.1 \times 10^{18}$ cm². Compare with (a) and draw a conclusion.

6

STRUCTURE OF ATOMS

In this chapter we consider some of the experiments that led to the conclusion that the atom is electrical in nature. Dalton's concept of a structureless atom provided no mechanism to explain these observations. These experiments, started over 150 years ago, also culminated in the discovery of X rays and radioactivity (Section 25.1). The spontaneous disintegration of naturally radioactive atoms into smaller particles contradicts the Daltonian hypothesis that atoms are unalterable. In turn, these discoveries inaugurated a more complete theory of the structure of atoms—the nuclear theory of the atom—and reaffirmed the atom as the unit of chemical changes.

QUANTIZATION OF ELECTRICITY / 6.1

When a property of matter exists in discrete amounts, we say the property is "quantized." For example, in accord with the atomic theory of matter, mass is quantized. Solid iron, liquid water, gaseous hydrogen are not continuous; they are composed of atoms or molecules and therefore their mass is quantized. Electricity in Dalton's time was regarded as a continuous fluid. However, this assumption could not survive when confronted by the new discoveries of the nineteenth century.

Electrolysis. When a direct current is passed through molten sodium chloride, chemical changes occur. Michael Faraday studied the relation between the quantity of electricity and the quantities of the products and discovered in 1833 that the mass of a substance produced at an electrode is proportional to the charge passed through the liquid. Also, he discovered that the same quantity of charge produces 1 mole of atoms of any univalent element (hydrogen, for example, in solution as hydrochloric acid, or silver in solution as silver nitrate). But to discharge 1 mole of atoms of a divalent element, copper in solution as copper(II) sulfate, exactly twice this quantity is required, and for a trivalent element, exactly three times this quantity is required.*

From these results, George Johnstone Stoney in 1874 concluded that electricity, like matter, consists of particles. He called the charge of a single particle of electricity the ELECTRON† and estimated 10^{-20} coulomb (C) for the electronic charge, later corrected to 1.6×10^{-19} C. Electricity is thus a stream of electrons, each electron carrying 1.6×10^{-19} C of electricity.

Discharge tubes. The passage of the electric current through a gas, called an electric discharge, was studied by Faraday, William Crookes, and many others. The results of these studies are best explained in terms of the electron particle suggested by Stoney.

A sealed glass tube with two electrodes attached to a high voltage source constitutes a discharge tube (Fig. 6.1). When the gas pressure in the tube is reduced to about 10 torr, the gas becomes a conductor, current flows, and the gas emits light as in a neon sign. When the pressure is further reduced, emission of light by the gas ceases. The current,

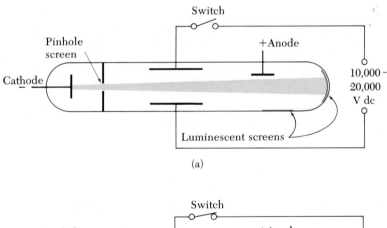

(a)

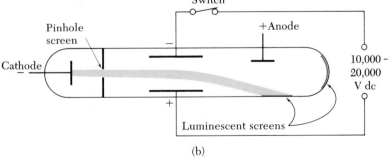

(b)

Fig. 6.1. *A discharge tube. Experiment showing that the cathode rays possess a negative charge: (a) position of ray in the absence of an electric field; (b) position of ray in an electric field. The pressure in the tube is about 10^{-3} torr.*

* Electrical conductance and electrolysis are covered in Chapter 12.

† *Elektron* (Greek), meaning amber; amber is *electrified* when rubbed with wool.

however, still flows between the electrodes, and the glass at the positive end of the tube glows. It thus appears that a radiation capable of causing the glow is emitted from the cathode (negative end); this radiation, because of its origin, is named the "CATHODE RAY."

These rays possess momentum, a characteristic property of matter in motion, and thus they possess mass. Also, they are deflected in electric fields in a manner predicted for negatively charged particles (see Fig. 6.1). These rays move in practically straight lines, and are capable of penetrating very thin metallic plates. They also darken a photographic plate. These experiments demonstrated the particle character of the cathode rays. Moreover, Joseph J. Thomson in 1897 showed the particles to be *electrons* by measuring in a primitive mass spectrograph the ratio of the charge of the particle, e, to its mass, m. This ratio is $e/m = 1.76 \times 10^8$ C/g. From other measurements the accepted mass of the particle is 9.11×10^{-28} g. Then, from the known value of e/m the charge of the particle is calculated to be 1.60×10^{-19} C. This value is in perfect agreement with the calculated electronic charge based on Faraday's discoveries.

The ratio e/m is independent of the nature of the gas in the discharge tube and of the materials of which the tube is composed. Also, electrons from other sources yield the same ratio. This, with the observation that the mass of the electron is about 2×10^3 times smaller than the mass of the hydrogen atom, the lightest atom known, indicates that the electron is a universal constituent of matter. The electron charge is, therefore, used as a unit of charge, written as -1. The more accurate name for cathode rays is "electron beams."

Wilhelm Röntgen, while performing experiments on the light produced by electron beams, discovered in 1895 a radiation that penetrates metal plates opaque to light and electron beams. This radiation, named the X RAY, is produced when high-speed electrons strike an object. X rays darken a photographic plate, exhibit the phenomena of diffraction, refraction, and reflection, but are not deflected in electric fields. They are, therefore, of the same nature as light, radio waves, and ultraviolet radiation. Their wavelengths are of the order of 1 Å, as compared to 4–7×10^3 Å for visible light. In addition, they ionize gases.

THE POSITIVE IONS / 6.2

In the discharge tube, neutral particles—atoms or molecules—are ionized into electrons and positive ions:

$$\text{Ne} \longrightarrow \text{Ne}^+ + e^- \quad (\textit{ionization})$$
$$\quad\textit{atom} \qquad \textit{ion} \quad \textit{electron}$$

At low pressures, positive ions hit the cathode. If a perforated cathode is used, some positive ions will pass through and emerge

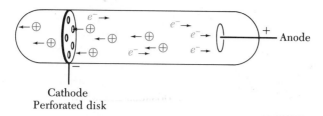

Fig. 6.2. *Discharge tube for producing positive ions.*

behind the cathode (Fig. 6.2). Eugen Goldstein performed such an experiment in 1886 and detected the positive ions ("rays"). Their behavior in electric fields conforms to that predicted for positively charged particles, but, unlike the electron beam, the properties of the positive ion are *characteristic* of the gas in the tube.

Since positive ions are produced by the removal of one or more electrons—particles of small mass—the mass of the remaining ion (page 37) is almost equal to that of the neutral particle. Natural hydrogen consists of three isotopes. Of all the positive ions whose masses have been determined, the *lightest ion found*, mass number 1, atomic weight 1.00728, is obtained from protium, the lightest isotope of hydrogen, atomic weight 1.007825. This ion, named the **PROTON**, H^+, carries a charge equal to that of an electron but opposite in sign. The ion obtained from deuterium, an isotope of hydrogen of mass number 2, is known as the **DEUTERON**, D^+. The removal of two electrons from a neutral helium atom produces an **ALPHA** (α) **PARTICLE**, He^{2+}.

THE RUTHERFORD THEORY OF THE ATOM / 6.3

The first recognized attempt to account for the presence of electrons in atoms was made by J. J. Thomson in 1904. He postulated that an atom is composed of a sphere of positive electricity in which are embedded a number of electrons sufficient to neutralize the positive charge. But the mass of an electron is exceedingly small compared to the mass of an atom. This means that nearly all the mass of an atom is associated with the positive charge. The model of a **THOMSON ATOM** containing 3 electrons in a uniform sphere of positive electricity with a total charge equal and opposite to 3 electrons is illustrated in Fig. 6.3. Let us now make a prediction based on this model: If a high velocity alpha particle, He^{2+}, strikes a "Thomson atom," it should not be deflected from its original path ("scattered") but should pass through the atom. Deflection in this case would result from the mutual repulsion of the positive charges on the α particle and the sphere of positive electricity. But the positive electricity would *uniformly* occupy the volume of the atom and, therefore, is not concentrated in any region. An alpha particle entering such an atom, Thomson reasoned, would be encircled by positive electricity and so be repelled almost equally on all sides. An atom, Thomson concluded, should produce no appreciable deflec-

Positive charge of
uniform density

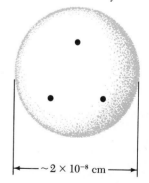

$\sim 2 \times 10^{-8}$ cm

Fig. 6.3. *The Thomson atom; a sphere of positive electricity in which are imbedded a number of electrons (●) so that the atom is electrically neutral.*

tion of an alpha particle. If, therefore, a beam of alpha particles strikes a metal film, practically all the particles should pass through the film undeflected (Fig. 6.4a); a few may be deflected through small angles.

Such an experiment (Fig. 6.5) was performed by Ernest Rutherford with Hans Geiger and Ernest Marsden in 1909. As predicted, practically all (about 99.9%) of the particles passed through the film without deviation or were only slightly deflected from the original path, but, much to their amazement, a small fraction of the particles was deflected through large angles; in fact, a minute fraction was almost completely reversed in direction. In Rutherford's own words, "It was almost as incredible as if you fired a 15-inch shell at a piece of tissue paper and it came back and hit you."

To explain these results, Rutherford, in 1911, assumed that the atom must be almost completely empty space; this would account for the passage of most of the particles in a straight line through the film. To account for the repulsive force required to produce the observed large deflections, he further assumed that the positive electricity of the atom must be concentrated in a very small volume, called the NUCLEUS OF THE ATOM. Therefore, nearly all the mass of the atom resides in the nucleus. When an alpha particle approaches a nucleus, it is repelled; the closer the approach to the nucleus, the greater is the deflection (Fig. 6.4b). Since the atom is electrically neutral, there must be electrons, sufficient in number to equal the positive charge of the nucleus, around the nucleus. The space they occupy determines the effective volume of the atom.

Rutherford concluded that the *number of unit positive charges on a nucleus is* equal to the ATOMIC NUMBER, the *ordinal number* of the atom in the periodic classification* (Chapter 7). He also calculated a value for the diameter of the nucleus, $\approx 10^{-13}$ cm.

Composition of the nucleus. A hydrogen atom, atomic number 1, consists of a proton, charge +1, as its nucleus, surrounded by 1 elec-

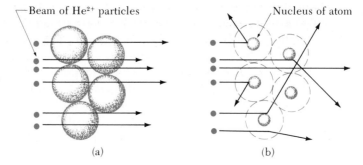

Fig. 6.4. *Representation of the deflection of alpha, He^{2+}, particles by a metal foil, as predicted (a) by the Thomson model of an atom, and (b) by the Rutherford nuclear model of an atom.*

—Beam of He^{2+} particles   Nucleus of atom

(a)   (b)

* The number of the space the element occupies in the periodic table (H 1, He 2, Li 3, Be 4, . . .).

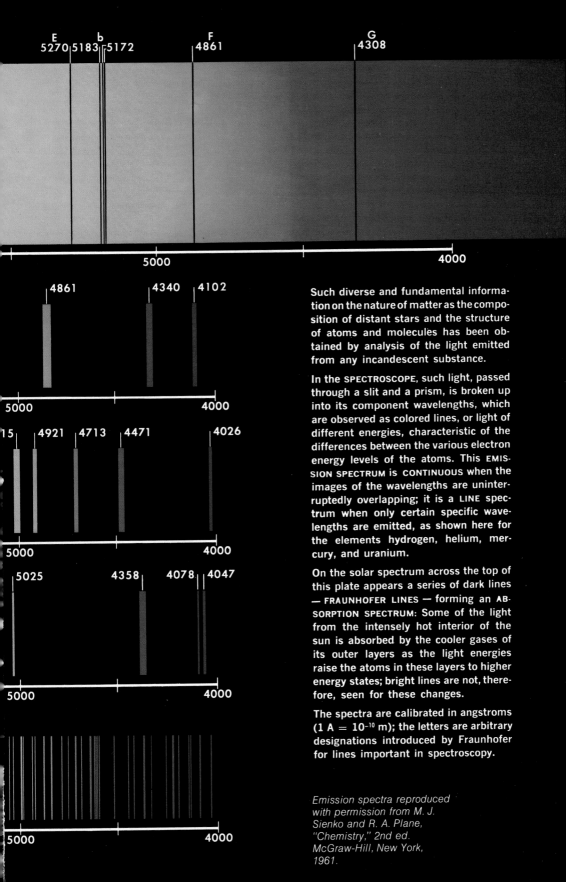

E
5270 5183 5172

b

F
4861

G
4308

5000   4000

4861   4340 4102

5000   4000

15 4921 4713 4471   4026

5000   4000

5025   4358   4078 4047

5000   4000

5000   4000

Such diverse and fundamental information on the nature of matter as the composition of distant stars and the structure of atoms and molecules has been obtained by analysis of the light emitted from any incandescent substance.

In the SPECTROSCOPE, such light, passed through a slit and a prism, is broken up into its component wavelengths, which are observed as colored lines, or light of different energies, characteristic of the differences between the various electron energy levels of the atoms. This EMISSION SPECTRUM is CONTINUOUS when the images of the wavelengths are uninterruptedly overlapping; it is a LINE spectrum when only certain specific wavelengths are emitted, as shown here for the elements hydrogen, helium, mercury, and uranium.

On the solar spectrum across the top of this plate appears a series of dark lines — FRAUNHOFER LINES — forming an ABSORPTION SPECTRUM: Some of the light from the intensely hot interior of the sun is absorbed by the cooler gases of its outer layers as the light energies raise the atoms in these layers to higher energy states; bright lines are not, therefore, seen for these changes.

The spectra are calibrated in angstroms (1 A = 10^{-10} m); the letters are arbitrary designations introduced by Fraunhofer for lines important in spectroscopy.

Fig. 6.7. *Electromagnetic spectrum:*

Gamma rays: Radiation from atomic nuclei
X rays: Radiation from electrons striking a target
Ultraviolet rays: Radiation from arcs and electric discharges in gases
Visible light: Radiation from stars, hot objects, hot gases, electric discharges
Infrared rays: Radiation from warm objects
Hertzian waves: Radiation from alternating electric currents: radio waves, television, microwaves, radar [10 Å = 1 nanometer (nm); 10^7 nm = 1 cm].

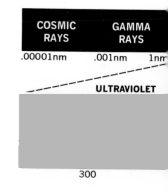

COSMIC RAYS	GAMMA RAYS	
.00001nm	.001nm	1nm

ULTRAVIOLET

300

Fig. 6.13. *Representation of a prism bending rays of white light. The effect of the prism is to bend shorter wavelengths more than longer wavelengths, separating them into distinctly identifiable lines of color.*

Figs. 6.7 and 6.13 reproduced with permission from "Light and Color" published by General Electric Company, Large Lamp Department, Nela Park, Cleveland, Ohio.

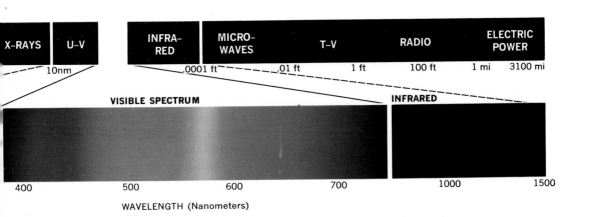

X-RAYS	U-V		INFRA-RED	MICRO-WAVES	T-V	RADIO	ELECTRIC POWER

10nm   .0001 ft   .01 ft   1 ft   100 ft   1 mi   3100 mi

VISIBLE SPECTRUM   **INFRARED**

400   500   600   700   1000   1500

WAVELENGTH (Nanometers)

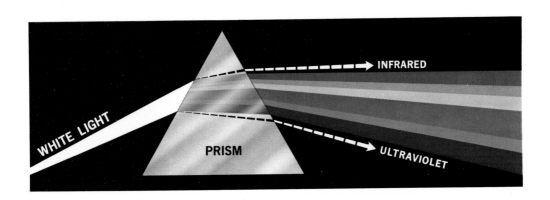

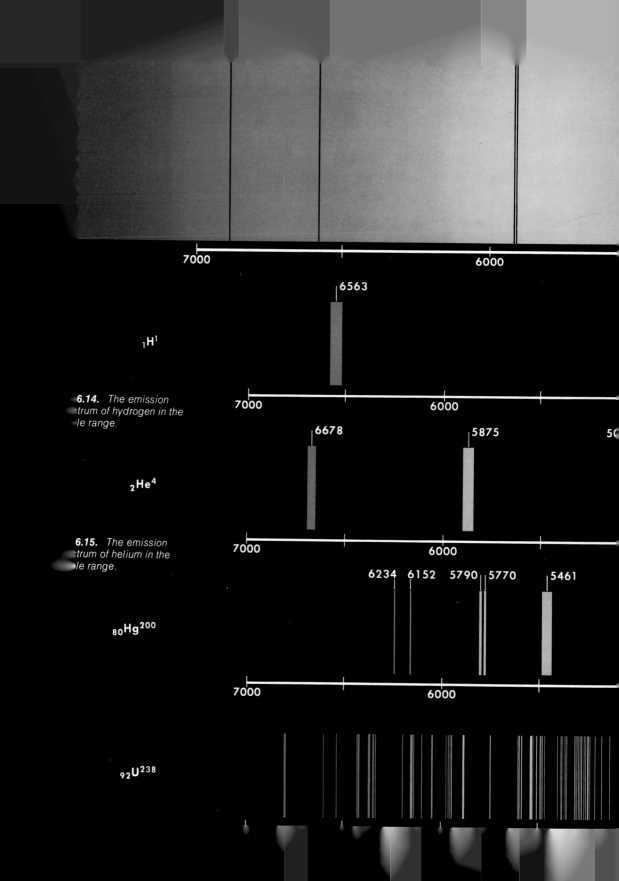

7000   6000

|6563

$_1H^1$

6.14. The emission
ctrum of hydrogen in the
le range.

7000   6000

|6678 |5875   5(

$_2He^4$

6.15. The emission
ctrum of helium in the
le range.

7000   6000

6234   6152   5790 5770 |5461

$_{80}Hg^{200}$

7000   6000

$_{92}U^{238}$

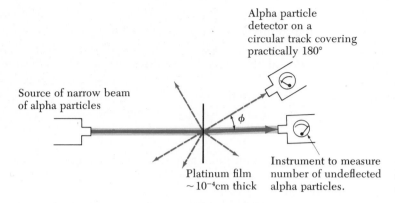

Source of narrow beam
of alpha particles

Alpha particle
detector on a
circular track covering
practically 180°

ϕ

Platinum film
~10^{-4}cm thick

Instrument to measure
number of undeflected
alpha particles.

Fig. 6.5. *Illustrating the principle of the alpha-particle deflection (scattering) experiments. Positively charged helium atoms are directed at a metal film, and its effect on the particles observed with detecting devices placed at various angles. The angle (ϕ) is a measure of the deflection.*

tron, charge -1. It is assumed that protons are a common constituent of atomic nuclei, sufficient in number to give a nucleus its characteristic positive charge. For example, the nucleus of any atom of neon, atomic number 10, contains 10 protons.

These protons account for only about one-half the mass of a neon atom. The remainder of the mass is attributed to the NEUTRONS in the nucleus. The neutron, an electrically neutral particle with mass number 1, atomic weight 1.00867, was discovered in 1932 by James Chadwick. The number of neutrons in a nucleus together with the number of protons in the nucleus make up the mass number of the atom:

mass number = number of neutrons + atomic number

Thus, the three isotopes of neon, atomic weights 19.99, 20.99, and 21.99, have the same atomic number (10) but different mass numbers. All the nuclei of neon have 10 protons, but the nuclei of the isotope of mass number 20 have 10 neutrons, those of mass number 21 have 11 neutrons, and those of mass number 22 have 12 neutrons. Symbolically, these nuclei are represented as $^{20}_{10}\text{Ne}$, $^{21}_{10}\text{Ne}$, $^{22}_{10}\text{Ne}$. The mass number is the *nearest integer* to the atomic weight of the isotope.

When it is unnecessary to distinguish between protons and neutrons, both are referred to as NUCLEONS. The mass number thus represents the total number of nucleons in the nucleus.

THE NATURE OF LIGHT / 6.4

The discussion of the electronic structure of an atom is so closely interwoven with an understanding of the nature of light that we shall include here a brief review on the nature of light.

James Clerk Maxwell in 1864 predicted that an alternating current in a circuit would radiate energy in the form of ELECTROMAGNETIC WAVES traveling through a vacuum with the speed of light. A wave, by

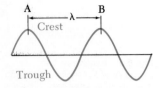

Fig. 6.6. *Illustrating the wavelength of the wave set up in a vibrating wire, the distance between two consecutive crests, AB.*

definition, *transmits energy without the transmission of matter.** This prediction led to the development of wireless telegraphy and the radio. The number of such waves passing a given point in 1 second is known as the **FREQUENCY** of the wave, *f*; the unit of frequency† is sec⁻¹. The distance between crests of two consecutive waves is called the **WAVELENGTH**, λ, illustrated in Fig. 6.6; the unit of wavelength is centimeter. The speed of the wave, *c*, is given by

$$c = \lambda f$$

The units of *c* are centimeters per second. Since the speed of electromagnetic waves is a constant (2.998×10^{10} cm/second), the shorter the wavelength, the greater is the frequency.

Light is an electromagnetic radiation that produces a visual effect. Light waves differ from radio waves in frequency; the light waves possess higher frequencies. The electromagnetic spectrum is given in Fig. 6.7 (*color foldout following page 78*). There are no upper or lower limits (except zero) to the wavelength.

Light waves, like sound and water waves, are **DIFFRACTED**; they spread out in passing through a hole. When light of a given frequency (**MONOCHROMATIC**) is passed through a hole, not too large compared to its wavelength, and allowed to fall upon a screen, the screen pattern consists of a series of dark and bright rings. It does not consist of a single spot of light. Such observations are consistent with a wave theory of light. See Fig. 9.1 (page 146) and Figs. 6.8 and 6.9. These diffraction patterns are the basis upon which the wavelength of light is *calculated*. Experimenters *do not measure wavelengths of light* from a wave shape.

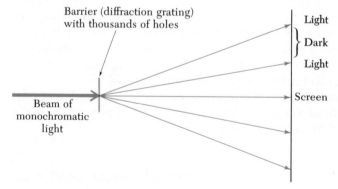

Fig. 6.8. *Representation of the diffraction of a light beam. The original beam is split upon passing through the holes. Actual patterns seen on the screen are illustrated in Fig. 6.9.*

---

* Although the word "wave" connotes a visible shape like that of a water wave or a vibrating wire (Fig. 6.6), no material form or shape or sequence of waves need be imagined. Unlike a water wave, matter is *not* required for the transmission of *electromagnetic waves*.

† Frequency is also represented by ν.

Fig. 6.9. *Diffraction pattern obtained with light.*

In spite of the success of the electromagnetic wave theory in explaining a vast range of observations, many phenomena can be explained only by attributing particle properties to light. For example,

(1) In the late nineteenth century, many scientists studied the distribution of the wavelengths and the energies of the light waves emitted by hot bodies, such as a hot tungsten filament. Science, at that time, recognized the quantization of matter and of electricity, but radiant energy was believed to be continuous. That is, the electromagnetic wave theory assumed that the energy of monochromatic radiation may have *any* value, from infinitely small to infinitely large. However, all arguments based on this assumption failed miserably to account for the energy distribution found experimentally. To obtain agreement with experiment, Max Planck, in 1900, found it necessary to invent what is now known as the *quantum theory:* radiant energy is composed of discrete units—"atoms of light"—called PHOTONS.* Like matter and electricity, *light is quantized*. The energy of a photon is given by

$$E = hf$$

---

* Named by the chemist Gilbert N. Lewis in 1926. Originally, the term *quantum* (plural, *quanta*) was used instead of photon.

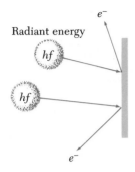

Radiant energy

Fig. 6.10. *The photoelectric effect.*

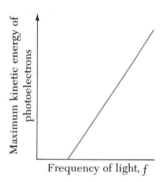

Maximum kinetic energy of photoelectrons

Frequency of light, *f*

Fig. 6.11. *A plot of the measured maximum kinetic energy of photoelectrons and the frequency of the incident light.*

where f is the frequency of the radiation and h is a constant, 6.6256×10^{-27} erg seconds/particle, known as Planck's constant. Only a whole number of photons may be absorbed or emitted by a body, in the same sense that any mass of an element can be only a whole number multiple of the mass of one atom of it.

In terms of the quantum theory, the intensity* of an electromagnetic radiation is determined by the number of photons striking a unit area in unit time.

(2) When light falls upon a metal, electrons (photoelectrons) are ejected from its surface. Such emission is known as the PHOTOELEC-TRIC EFFECT (Fig. 6.10). Experiments show that the maximum kinetic energy of the ejected electrons depends on the *frequency* of the mono-chromatic light. The greater the frequency, the greater is the kinetic energy of the ejected electron (Fig. 6.11). Increasing the intensity of the light increases only the number of electrons and *not the energy* of each electron. These results can only be explained (Albert Einstein, 1905) in terms of the quantum (photon) theory of radiation: The energies of all photons for a given frequency, f, are identical. Each photon has the energy hf. For each photon absorbed, one electron is ejected. Then, the energy required to remove the electron from the surface (a constant, characteristic of the nature of the surface) plus the kinetic energy of the ejected electron must equal the energy of the photon,

$$hf = \tfrac{1}{2}mv^2 + constant$$

m is the mass of an electron, and v is their measured maximum speed. This equation predicts that as the frequency of the light is increased, hf increases and the kinetic energy of the photoelectrons should increase. This prediction agrees with Fig. 6.11; h, Planck's constant, is the slope of the line. The modern theory of radiation, indeed, includes all the successes of Maxwell's electromagnetic wave theory and also explains the particle properties of light.

MATTER WAVES / 6.5

To this point, we have been content to accept the electron as a par-ticle, a bit of matter. However, Louis de Broglie in 1924 reasoned that since particle properties, as well as wave properties, are associated with light, it is likely that a wave character is associated with a particle, such as an electron, a proton, an atom, a molecule, or a piece of chalk. The wave associated with a particle is called a *matter wave*. He pre-dicted that the wavelength of a matter wave is

$$\lambda = \frac{h}{mv}, \qquad m = \text{mass of particle}, \qquad v = \text{its speed}$$

* Expressed as watts/cm².

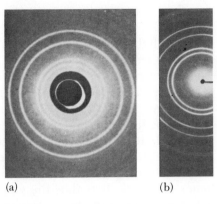

(a)   (b)

Fig. 6.12. *Comparison of electron and X-ray diffraction. Pattern produced by passing (a) X rays through gold foil; (b) an electron beam through gold foil.*

In words, an electron beam should exhibit wave properties; for example, like light, it should produce a diffraction pattern. This prediction has been verified by numerous experiments, Fig. 6.12. Thus, like electromagnetic radiations, particles of matter exhibit wave properties and particle properties. The electron microscope, now a common laboratory tool (Fig. 26.6, page 562), is an application of the de Broglie concept.

Every particle has a wavelength. A 100 g mass moving with a speed of 60 miles/hour possesses a wavelength of about 10^{-32} cm, evidently too small to be physically meaningful.*

SPECTRA OF ELEMENTS / 6.6

When a beam of parallel rays of light enters a triangular glass prism, it is dispersed into its component wavelengths, as illustrated in Fig. 6.13 (*see color plate following page 78*). This separation of light is the basis of the SPECTROSCOPE.

There are two kinds of spectra. When light *emitted* from a source is analyzed, the spectrum obtained is called an EMISSION SPECTRUM. Conversely, the spectrum obtained after light from some source *has passed through* a substance is known as an ABSORPTION SPECTRUM. Emission spectra usually have a few colored lines—the emitted frequencies—on a black background. Absorption spectra show all colors interspersed with black lines—the absorbed frequencies.

Solids, liquids, and dense gases become incandescent at high temperatures. If the emitted light is examined with a spectroscope, it will consist of a *continuous band of colors* as observed in a rainbow. An incandescent substance thus *emits* a CONTINUOUS SPECTRUM. This kind of spectrum is typical of matter in which the atoms are packed closely

---

* Matter waves possess wavelike characteristics; they are *not*, however, electromagnetic waves. They are *not* radiated into space or emitted by the particle; they are never dissociated from the particle. The speed of a matter wave is not the same as the speed of light, nor is it a constant.

together. Gases at *low* pressure, as in a neon sign, behave quite differently.

If the light from a neon sign is examined with a spectroscope, the resulting emission spectrum does not look like a rainbow. Instead, it consists of several isolated colored lines as illustrated in Figs. 6.14 and 6.15 (*see color foldout following page 78*). This is called a LINE SPECTRUM. If a volatile salt is sprayed into a flame, light is emitted. When this light is examined in a spectroscope, again, a line spectrum is observed. It is noteworthy that the line spectrum of an element is characteristic of that element regardless of its source. Sodium, for example, always yields the same line spectrum, differing from that of all other elements. It is thus the *atoms* of the element that emit the line spectrum. The study of the spectra of elements is, therefore, important in the determination of the structure of atoms.

The colored lines in the spectra of all elements are observed to come closer and closer to each other at the lower frequencies. This suggests that the frequencies of the lines may be given by a mathematical expression. The frequencies of the lines in the hydrogen atom spectrum, for example, can be represented by a relation *involving whole numbers*,

$$f = 3.288 \times 10^{15} \text{ sec}^{-1} \left(\frac{1}{n_1^2} - \frac{1}{n_2^2} \right) \tag{1}$$

where f is the frequency of the line in sec^{-1}, and $3.288 \times 10^{15} \text{ sec}^{-1}$ is a constant for hydrogen, known as the *Rydberg constant**; n_1 is a dimensionless integer and n_2 is another dimensionless integer greater than n_1. Thus, when $n_1 = 1$, n_2 may be any whole number greater than 1 and each value of n_2 yields a frequency, f, corresponding to a line in the spectrum of hydrogen. When $n_1 = 2$, n_2 is any integer greater than 2, etc. The relationship for atoms more complex than hydrogen is more complicated, but each frequency can still be represented as a difference of two terms involving integers.

The energy of a photon, E, is given by

$$E = hf = \frac{hc}{\lambda} \tag{2}$$

Each spectral line corresponds to photons possessing a definite energy. For example, the 4102-Å line in the spectrum of hydrogen atoms corresponds to 4.842×10^{-12} erg/atom,

$$E = h \frac{c}{\lambda} = 6.6256 \times 10^{-27} \frac{\text{erg } \cancel{\text{seconds}}}{\text{atom}} \times \frac{2.9979 \times 10^{10} \frac{\cancel{\text{cm}}}{\cancel{\text{second}}} \quad 10^8 \frac{\text{Å}}{\cancel{\text{cm}}}}{4102 \cancel{\text{Å}}}$$

$$= 4.842 \times 10^{-12} \frac{\text{erg}}{\text{atom}}$$

* Named for Johannes Robert Rydberg, who discovered this empirical relation in 1890 after a study of a large mass of data.

The atoms therefore emit or absorb only photons of definite characteristic energies, fixed by the whole numbers n_1 and n_2.

QUANTIZATION OF THE ENERGY OF AN ELECTRON ASSOCIATED WITH A NUCLEUS / 6.7

An atom consists of a positively charged nucleus surrounded by electrons. The arrangement of these electrons, we believe, determines the physical and chemical properties of the elements.

Theory and experiment agree that it is impossible to measure at the same time, and therefore futile to talk about, both the precise position and the precise energy of an electron in an atom.[*] This means that the more precisely we measure the energy of the electron in the hydrogen atom, the less we can know about its position. No one has yet devised an experiment capable of locating the electron, but if it were possible to do so, then we would know almost nothing about its energy. Since experiments give us the precise energy of the electron in a hydrogen atom (7 to 8 significant figures), we satisfy ourselves with this knowledge and must forgo knowing the exact position of the electron. On the other hand, theory permits us to talk about the *probability* of finding the electron at a given distance from the nucleus of the H atom.

The fundamental assumption of this theory is the *Schrödinger equation*, conceived by Erwin Schrödinger in 1926 from the wave and particle properties of matter. This equation, the basis of QUANTUM (WAVE) MECHANICS, is, in principle, applicable to the problems that arise when particles, such as electrons, nuclei, atoms, and molecules, are subject to a force. Many phenomena—for example, the stability of the covalent bond (Chapter 18) and the intensities of spectral lines—could not be explained prior to the application of quantum mechanics.

We have always *taken for granted* that the equations of motion developed by the great physicists of the past centuries are applicable to all bodies, regardless of size. Quantum mechanics, of necessity, contradicts this concept; *it does not depict the structure of the atom in terms that are familiar to us from the behavior of visible bodies.* Nevertheless, quantum mechanics is a part of the evolution of our knowledge, born from and retaining many characteristics of classical mechanics. As Paul Dirac wrote, "Quantum mechanics works so well that nobody can afford to disagree with it." For example, quantum mechanics predicts that when electrons or electromagnetic radiations are passed through narrow slits, they should spread and form the diffraction patterns as shown in Figs. 6.9 (page 81) and 6.12 (page 83). Quantum mechanics stands presently as the most embracing and successful description of nature.

The form of the Schrödinger equation (page 101) need not concern

---

[*] A version of the UNCERTAINTY PRINCIPLE, due to Werner Heisenberg (1927)

us. It is difficult to solve except in simple cases, but its results can be verbalized into understandable concepts. Applied to the H atom, the equation tells us that the *energy of the electron* (also referred to as the energy of the H atom) *is quantized*. Only certain fixed and definite energies are permitted. These energy values, referred to as the ENERGY LEVELS of the electron or as the energy levels of the hydrogen atom, are determined by four dimensionless numbers called QUANTUM NUMBERS. These quantum numbers, symbolized by the letters n, l, m_l, and m_s, are related as shown:

n can be any whole number 1, 2, 3, . . . but not 0 or fractional numbers such as 1.5 or 2.3.

l can be a whole number starting with 0 up to a maximum of $n-1$. For example, when $n=1$, $l=0$, but when $n=4$, $l=0$, 1, 2, or 3. Note that there are n values of l.

m_l can be a whole number starting with the value of $+l$ and descending to zero and then to $-l$. For example, when $l=0$, $m_l=0$ but when $l=2$, $m_l=2, 1, 0, -1, -2$. Note that there are $2l+1$ values of m_l.

m_s can have only values of $+\frac{1}{2}$ or $-\frac{1}{2}$.

To summarize, the following energy levels are available to an electron whose n value is 2:

$$n=2 \quad \begin{cases} l=1 \longrightarrow m_l = +1, 0, -1 \\ \\ l=0 \longrightarrow m_l = 0 \end{cases}$$

For each m_l value m_s can be $+\frac{1}{2}$ or $-\frac{1}{2}$.

$n=2$	$l=1$	$m_l=-1$	$m_s=-\frac{1}{2}$
$n=2$	$l=1$	$m_l=-1$	$m_s=+\frac{1}{2}$
$n=2$	$l=1$	$m_l=0$	$m_s=-\frac{1}{2}$
$n=2$	$l=1$	$m_l=0$	$m_s=+\frac{1}{2}$
$n=2$	$l=1$	$m_l=+1$	$m_s=-\frac{1}{2}$
$n=2$	$l=1$	$m_l=+1$	$m_s=+\frac{1}{2}$
$n=2$	$l=0$	$m_l=0$	$m_s=-\frac{1}{2}$
$n=2$	$l=0$	$m_l=0$	$m_s=+\frac{1}{2}$

In the simplest case of an isolated H atom, its energy is determined almost entirely by n. For this reason, n is known as the PRINCIPAL QUANTUM NUMBER.

An electron is commonly designated by a symbol that gives its n and

l values. The letters* s, p, d, f, g, etc. correspond, respectively, to l values of 0, 1, 2, 3, 4, etc. For example, a $3d$ electron has the values $n = 3$ and $l = 2$.

The energy of the electron in the H atom can take only the values given by

$$-\frac{21.79 \times 10^{-12} \text{ erg}}{n^2}$$

Here, the energy is expressed as the work required to remove the electron to infinity and *the energy of the electron at infinity is arbitrarily taken as zero.* At infinity, the electron is completely free from the influence of the proton. When the electron is attracted to a proton, energy is *evolved,* and therefore the energy of the electron *decreases* and becomes negative. When $n = 1$, the energy of the electron in the H atom is then -21.79×10^{-12} erg; the minus sign tells us that work must be done to move the electron away from the proton. The energy levels of the electron in the H atom are restricted to the values given in Table 6.1. The first energy level, when the principal quantum number of the

Table 6.1
Some Energy Levels of the Hydrogen Atom

Principal quantum number of the electron, n	Energy in erg per atom[a] $\dfrac{-21.79 \times 10^{-12}}{n^2}$		Number of the energy level
∞		Zero	Infinite
6	$\dfrac{-21.79 \times 10^{-12}}{36}$	$= -0.6053 \times 10^{-12}$	Sixth
5	$\dfrac{-21.79 \times 10^{-12}}{25}$	$= -0.8716 \times 10^{-12}$	Fifth
4	$\dfrac{-21.79 \times 10^{-12}}{16}$	$= -1.362 \times 10^{-12}$	Fourth
3	$\dfrac{-21.79 \times 10^{-12}}{9}$	$= -2.421 \times 10^{-12}$	Third
2	$\dfrac{-21.79 \times 10^{-12}}{4}$	$= -5.448 \times 10^{-12}$	Second
1	$\dfrac{-21.79 \times 10^{-12}}{1}$	$= -21.79 \times 10^{-12}$	First

(Left margin: Energy of electron increases ↑)

[a] The energy is expressed as the work necessary to separate the electron from the hydrogen atom against the attractive force of the nucleus. Zero energy then means that no work is required to remove the electron when the distance between the proton and the electron is infinite.

* Originate from words previously used to describe spectral lines: sharp, principal, diffuse, and fundamental.

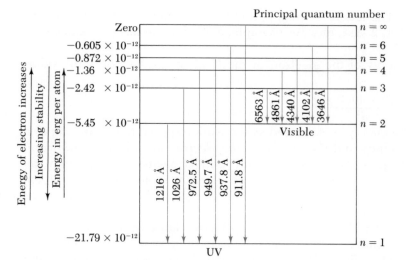

Fig. 6.16. *Energy-level diagram of the hydrogen atom. Wavelengths are in angstrom units. Compare the predicted lines in the visible range with the line emission spectrum of hydrogen, Fig. 6.14.*

electron is 1, is known as the **GROUND STATE** of the hydrogen atom; all other energy levels are known as **EXCITED STATES**. A convenient method of representing the ground and excited states of the hydrogen atom, Fig. 6.16, shows the energy levels as lines. Such a figure is known as an energy level diagram.

ORIGIN OF SPECTRAL LINES / 6.8

The stage is now set for the explanation of the observed spectrum of the hydrogen atom. Niels Bohr assumed (1913) that the energy of a photon emitted or absorbed by the H atom is the difference in energy of two energy levels between which the electron transition occurs. The *absorption* of radiation will correspond to a transition between any two of the possible energy levels, from a *lower* to a *higher* level. The *emission* of radiation will correspond to a transition between any two of the possible energy levels, but from a *higher* to a *lower* level. Thus, in absorption, the energy of the electron is increased and, in emission, the energy of the electron is decreased, but only in accord with the differences between any two energy levels. The energy lost appears as a photon. Since we have restricted (quantized) the energy of the electron, *the hydrogen atom can emit or absorb only photons whose energy is equal to the difference between the two energy levels.* A spectral line thus originates when the electron "jumps" or "drops" from one energy level to another:

$$E \text{ (energy of photon emitted)} = E_{\text{upper}} - E_{\text{lower}}$$

Example 1 The electron of an H atom is excited from its first ($n = 1$) to its sixth energy level ($n = 6$). This electron then "drops" to its second energy level ($n = 2$). What is the energy of the photon emitted?

Answer The energy of the photon emitted must be the difference between the higher ($n = 6$) and the lower ($n = 2$) energy levels of the electron; hence, from Table 6.1 or Fig. 6.16,

$$E \text{ (energy of the photon)} = E_6 - E_2$$

$$= (-0.605 \times 10^{-12}) \frac{\text{erg}}{\text{atom}} - (-5.45 \times 10^{-12}) \frac{\text{erg}}{\text{atom}}$$

$$= 4.84 \times 10^{-12} \frac{\text{erg}}{\text{atom}}$$

The emitted photon in Example 1 corresponds to the wavelength 4102 Å, in perfect agreement with one of the lines (Fig. 6.14) in the spectrum of hydrogen.

It is noteworthy that the lines in the ultraviolet range, given in Fig. 6.16, as well as other series of lines, were discovered after Bohr predicted their existence.

THE ELUSIVE ELECTRON / 6.9

An electron of definite energy cannot be assigned a definite position or a definite path relative to the nucleus; it can take any position from inside the nucleus to infinity. Instead, for each energy value (set of quantum numbers) the Schrödinger equation yields an expression, called "psi squared," ψ^2, that gives the distribution of electrons in an atom or molecule. Applied to the H atom, ψ^2 specifies the probability of finding the electron in a unit of volume at a given distance from the nucleus.

The electron may be visualized as a three-dimensional cloud, "an electron cloud," about the proton. The CLOUD (ELECTRON) DENSITY at any distance from the nucleus is related to the chance of finding the electron at that distance. For the 1s state in H (Fig. 6.17), the electron density is a maximum at the nucleus, decreases rapidly as r increases, and becomes practically zero at about 2×10^{-8} cm. Thus, the shorter the distance, the more effectively the proton holds the electron. The electron "likes" to be near the attractive nucleus, but is too "nervous" to stand still. The outer diameter of the electron cloud is not exactly definable; see Fig. 6.18. We may portray the electron cloud as possessing a shape and size depending upon the energy of the electron.

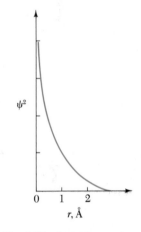

Fig. 6.17. *A plot in angstrom units of ψ^2 versus r, the distance between the proton and the electron of the hydrogen atom in the ground (1s) state.*

ATOMIC ORBITALS / 6.10

The electron distribution patterns (Fig. 6.18) identified by the three quantum numbers n, l, and m_l are called ATOMIC ORBITALS. An orbital is commonly designated by a symbol that gives its l value, or its n and l values. For example, a 3d orbital refers to an orbital for which $n = 3$ and $l = 2$. The possible orbitals for $n = 1$, 2, and 3 are given in Table 6.2.

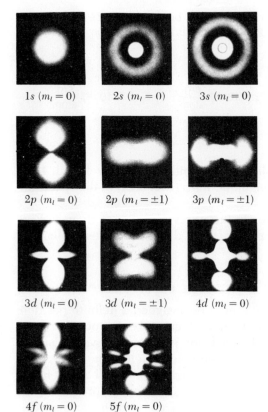

Fig. 6.18. *Electron-cloud diagrams of several states of the hydrogen atom according to the predictions of quantum mechanics for several values of n, l, and m_l. These should be visualized as three-dimensional clouds surrounding the nucleus of the hydrogen atom. The density of the shading (the cloud) is proportional to ψ^2. [H. E. White, Phys. Rev. 37, 1416 (1931)]. NOTE: These are photographs not of electrons but of mechanical models whose rotations correspond to the mathematical solutions of the Schrödinger equation.*

$1s\ (m_l = 0)$ $2s\ (m_l = 0)$ $3s\ (m_l = 0)$

$2p\ (m_l = 0)$ $2p\ (m_l = \pm 1)$ $3p\ (m_l = \pm 1)$

$3d\ (m_l = 0)$ $3d\ (m_l = \pm 1)$ $4d\ (m_l = 0)$

$4f\ (m_l = 0)$ $5f\ (m_l = 0)$

The orbitals possessing the same principal quantum number, n, are said to belong to a LEVEL or a PRINCIPAL QUANTUM LEVEL, called K when $n = 1$, L when $n = 2$, M when $n = 3$, etc. The set of orbitals possessing the same n and l values is called a SUBLEVEL.* See Table 6.2.

Atomic orbitals have characteristic shapes, as represented (crudely) in Fig. 6.18. These shapes correspond to the mathematical solutions of the Schrödinger equation for different quantum numbers. For example, the s orbitals have no angular dependence. The only solid figure that has no angular dependence is a sphere and, therefore, s orbitals are considered to be spherical. The shapes of p, d, and f orbitals have angular dependence.

Schematic representations of the s, p, and d orbitals are given in Fig. 6.19. These regions encompass about 90–95% of the electron density. Note that the three p orbitals differ in that one is oriented along the x axis (p_x), a second is oriented along the y axis (p_y), and the third is oriented along the z axis (p_z).

One of the five d orbitals (d_{z^2}) is p-like in appearance. The other four d

---

* Levels and sublevels are also known as shells and subshells.

Table 6.2
The Orbitals for n = 1, 2, and 3

Quantum numbers							
n	*l*	*m_l*	*Level*	*Orbital symbol*	*Sublevel designation*	*Number of sublevels*	*Number of orbitals per sublevel*
1	0	0	K	1s	1s	One for K	One s
2	0	0	L	2s	2s	Two for L	One s
2	1	1		2p			
2	1	0		2p	2p		Three p
2	1	−1		2p			
3	0	0	M	3s	3s	Three for M	One s
3	1	1		3p			
3	1	0		3p	3p		Three p
3	1	−1		3p			
3	2	2		3d			
3	2	1		3d			
3	2	0		3d	3d		Five d
3	2	−1		3d			
3	2	−2		3d			

orbitals resemble four-leaf clovers. Picture one "clover" ($d_{x^2-y^2}$) along the x and y axes; rotate the clover 45° without tilting and you produce the d_{xy} orbital. Picture another clover (d_{yz}) upright between the y and z axes. Rotate this clover 90° so that it now faces you; you are looking at the d_{xz} orbital.

In summary, we conceive of the atomic orbitals as being related to the n, l, and m_l quantum numbers as follows:

n is related to the size of the orbital;
l is related to the shape of the orbital;
m_l is related to the spatial orientation.

The shapes of orbitals (electron cloud diagrams) and Fig. 6.19, we emphasize, are *pictorial representations* of the mathematical solutions of the Schrödinger equation. *They do not represent reality;* the shapes are not pictures of electric charges or of matter.

DISTRIBUTION OF ELECTRONS IN ATOMS / 6.11

Any atom other than hydrogen has more than one electron. The presence of several electrons creates complex mathematical problems because electrons repel each other. This makes the energy of an electron dependent on all four quantum numbers. However, the distribution patterns are similar to those for an electron in an H atom. Therefore, orbitals with the same general shapes as the H atom orbitals may be used

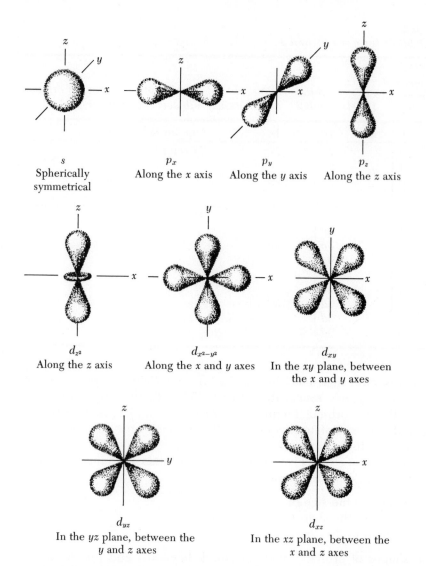

Fig. 6.19. (Schematic) The angular distribution* of the s, the three p, and the five d electrons (orbitals). The nucleus is at the origin of the axes.

s
Spherically symmetrical

p_x
Along the x axis

p_y
Along the y axis

p_z
Along the z axis

d_{z^2}
Along the z axis

$d_{x^2-y^2}$
Along the x and y axes

d_{xy}
In the xy plane, between the x and y axes

d_{yz}
In the yz plane, between the y and z axes

d_{xz}
In the xz plane, between the x and z axes

to describe the electronic structures of atoms with more than one electron.

In the absence of an external magnetic field, the energy of an orbital is determined by only two quantum numbers, n and l; the energy is practically independent of m_l and m_s. Therefore, all the orbitals in a given sublevel have the same energy. In neutral isolated atoms, the most stable orbital has the lowest energy; it is usually the orbital for which the sum of n and l is lowest. When two orbitals have the same $(n + l)$ value, the more stable orbital is the one with the lower n value. For example, the 4s sublevel $(n + l = 4 + 0 = 4)$ is more stable than the

* The representations are also referred to as boundary-surface diagrams.

4d sublevel ($n + l = 4 + 2 = 6$) and the 3d sublevel ($n + l = 3 + 2 = 5$). However, the 3p sublevel ($n + l = 3 + 1 = 4$) is more stable than the 4s sublevel. The orders of the energy and stability of the orbitals are given in Fig. 6.20.

Thus, the K level consists of one 1s orbital. The L level contains one 2s orbital and three 2p orbitals of *equal energy*, while the M level contains one 3s orbital, three 3p orbitals of *equal energy*, and five 3d orbitals of *equal energy*.

An atomic orbital is defined by three quantum numbers, n, l, and m_l. From a study of atomic spectra, Wolfgang Pauli concluded that *no two electrons in an atom can have the same four quantum numbers*.* In a given orbital, the fourth quantum number, m_s, can have only two possible values: $+\frac{1}{2}$ or $-\frac{1}{2}$. Therefore, *an atomic orbital can accommodate a maximum of 2 electrons*. It then follows:

An s orbital can accommodate 2 electrons.
Three p orbitals can accommodate a total of 6 electrons.
Five d orbitals can accommodate a total of 10 electrons.
Seven f orbitals can accommodate a total of 14 electrons.

The symbol 1s^2 represents 2 electrons for which $n = 1$ and $l = 0$; 1s^3 is an impossible case since the maximum number of electrons for which $l = 0$ is 2. The symbol 4p^3 represents 3 electrons for which $n = 4$ and $l = 1$; 4p^7 is an impossible case since the maximum number of electrons for which $l = 1$ is 6. The symbol 3d^4 represents 4 electrons for

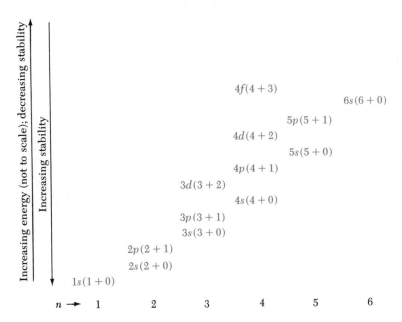

Fig. 6.20. ($n + l$) values. The order of the energies and stabilities of atomic orbitals in neutral isolated atoms; not applicable to positive ions.

* This law is known as the PAULI EXCLUSION PRINCIPLE.

which $n = 3$ and $l = 2$; $3d^{11}$ is an impossible case since the maximum number of electrons for which $l = 2$ is 10.

We are now ready to distribute the electrons in the ground states of the atoms. To obtain the ground state, the electrons must be distributed so as to make the energy of the atom as low as possible. The basis of the distribution is: (1) the number of electrons in an atom is equal to the atomic number; and (2) each added electron will enter the orbitals in the order of increasing energy, Fig. 6.20.

$_1$H – the 1 electron occupies the $1s$ orbital; the electron configuration of hydrogen is, therefore, $1s^1$.

$_2$He with 2 electrons – both can go into the $1s$ orbital; the configuration of helium is thus $1s^2$.

$_3$Li with 3 electrons – since the rules prohibit more than 2 electrons in $1s$, we put 2 electrons in $1s$ and the third one in the next available orbital, namely, $2s$. The configuration of lithium is, therefore, $1s^2 2s^1$.

$_4$Be with 4 electrons – since the $2s$ orbital can take 2 electrons, we put the fourth electron in it, yielding $1s^2 2s^2$ as the configuration of beryllium.

$_5$B with 5 electrons – the fifth is put in the next available sublevel, $2p$. The configuration of boron is, therefore, $1s^2 2s^2 2p^1$.

$_6$C with 6 electrons – since the $2p$ sublevel can take 6 electrons, we put the sixth electron in $2p$ and the configuration of carbon is $1s^2 2s^2 2p^2$. The same argument yields the following configurations:

$_7$N	$1s^2 2s^2 2p^3$	$_8$O	$1s^2 2s^2 2p^4$
$_9$F	$1s^2 2s^2 2p^5$	$_{10}$Ne	$1s^2 2s^2 2p^6$

The $2p$ sublevel is now "closed."

$_{11}$Na with 11 electrons – since the $2p$ sublevel is closed, the eleventh electron is put in the next available sublevel, $3s$, yielding $1s^2 2s^2 2p^6 3s^1$ for sodium.

$_{18}$Ar $1s^2 2s^2 2p^6 3s^2 3p^6$ closes the $3p$ sublevel.

For $_{19}$K and $_{20}$Ca, the next available sublevel, $4s$, is used, yielding the configurations

$_{19}$K $1s^2 2s^2 2p^6 3s^2 3p^6 4s^1$ $_{20}$Ca $1s^2 2s^2 2p^6 3s^2 3p^6 4s^2$

The next available sublevel is the $3d$, which can take 10 electrons. Hence, for the elements $_{21}$Sc, $_{25}$Mn, and $_{30}$Zn the configurations are

$_{21}$Sc	$1s^2 2s^2 2p^6 3s^2 3p^6 4s^2 3d^1$	$_{25}$Mn	$1s^2 2s^2 2p^6 3s^2 3p^6 4s^2 3d^5$
$_{30}$Zn	$1s^2 2s^2 2p^6 3s^2 3p^6 4s^2 3d^{10}$		

The letters $K, L, M, \ldots$ are used to represent filled levels; K means $1s^2$, L means $2s^2 2p^6$, etc. Thus, we could write

$_{30}$Zn $KLM4s^2$

Electronic configurations may be assigned to the remaining elements by similar reasoning. Although orbital energies are a function of the atomic number, the order given in Fig. 6.20 closely parallels the order of stability obtained from empirical properties of the elements. However, several of the predicted electron configurations do not agree with the empirically assigned configurations (Appendix III).

It is frequently convenient to express the ground state of an atom in terms of the orbitals. A dash is used to represent an orbital, an arrow to represent an electron, arbitrarily pointing down ($\downarrow$) to represent the m_s quantum number of $-\frac{1}{2}$ and up ($\uparrow$) to represent $m_s = +\frac{1}{2}$. The electron configuration of beryllium is represented as $_4$Be $\frac{\downarrow\uparrow}{1s}\frac{\downarrow\uparrow}{2s}$. But for carbon, the order of occupancy is significant: Do the $2p^2$ electrons occupy one or two orbitals? The question is answered by the *Hund rule*,* applicable to atoms, ions, and molecules: *electrons stay unpaired in orbitals of the same energy until each such orbital has at least one electron in it.* Then, the order of occupancy of three $2p$ orbitals is

$_5$B $\frac{\downarrow\uparrow}{1s}$ $\frac{\downarrow\uparrow}{2s}$ $\frac{\downarrow}{}\frac{}{2p}\frac{}{}$ $_6$C $\frac{\downarrow\uparrow}{1s}$ $\frac{\downarrow\uparrow}{2s}$ $\frac{\downarrow}{}\frac{\downarrow}{2p}\frac{}{}$

$_7$N $\frac{\downarrow\uparrow}{1s}$ $\frac{\downarrow\uparrow}{2s}$ $\frac{\downarrow}{}\frac{\downarrow}{2p}\frac{\downarrow}{}$ $_8$O $\frac{\downarrow\uparrow}{1s}$ $\frac{\downarrow\uparrow}{2s}$ $\frac{\downarrow\uparrow}{}\frac{\downarrow}{2p}\frac{\downarrow}{}$

$_9$F $\frac{\downarrow\uparrow}{1s}$ $\frac{\downarrow\uparrow}{2s}$ $\frac{\downarrow\uparrow}{}\frac{\downarrow\uparrow}{2p}\frac{\downarrow}{}$ $_{10}$Ne $\frac{\downarrow\uparrow}{1s}$ $\frac{\downarrow\uparrow}{2s}$ $\frac{\downarrow\uparrow}{}\frac{\downarrow\uparrow}{2p}\frac{\downarrow\uparrow}{}$

For $_{25}$Mn, the order of occupancy is

$_{25}$Mn KL $\frac{\downarrow\uparrow}{3s}$ $\frac{\downarrow\uparrow}{}\frac{\downarrow\uparrow}{3p}\frac{\downarrow\uparrow}{}$ $\frac{\downarrow}{}\frac{\downarrow}{}\frac{\downarrow}{3d}\frac{\downarrow}{}\frac{\downarrow}{}$ $\frac{\downarrow\uparrow}{4s}$

These electron configurations agree with the magnetic properties of atoms.

The Hund rule is most likely associated with repulsion effects between electrons. In carbon atoms, for example, repulsion is greater when the $2p$ electrons are in the same orbital compared to the arrangement when they are in separate orbitals. An unusual stability is generally associated with a half-filled sublevel† as in $_7$N. However, filled sublevels, characteristic of the noble gases, are of greater stability.

* Named in honor of Friedrich Hund for his work with atomic and molecular spectra.

† The s sublevels excepted.

In quantum mechanics, although the trajectory of the electron is not described, it is useful for educational purposes and for some theoretical calculations to associate the m_s quantum number with some kind of definite movement of the electron in an atom. We assume that the electron spins like a top about its own axis. The values $+\frac{1}{2}$ and $-\frac{1}{2}$ are interpreted to mean that the spin is oriented in one of two possible ways, clockwise or counterclockwise.*

Atomic sodium, atomic hydrogen, molecular oxygen, and iron are attracted into magnetic fields. They are typical **PARAMAGNETIC** substances; they *possess permanent magnetic moments.*

An electric current (electrons) flowing through a wire produces a magnetic field around the wire. Magnetic fields are thus produced by the motion of charged particles. Then, a single spinning electron *should act as if it were a small bar magnet with a permanent magnetic moment.* The magnetism of an isolated atom thus results from the spin of the electron around its axis, illustrated for the hydrogen atom in Fig. 6.21. The electron-magnetic bar analogy in Fig. 6.22 may help to visualize the magnetic property of an electron. However, *only two spin orientations* (relative to the magnetic field) are permitted for 1 electron, cor-

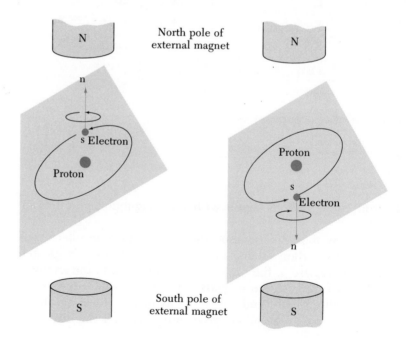

Fig. 6.21. *Two spin orientations of the electron, represented as $\overset{n}{\underset{s}{\uparrow}}$ and $\overset{s}{\underset{n}{\downarrow}}$, may occur in a magnetic field.*

North pole of external magnet

South pole of external magnet

* The "electron-spin" hypothesis was formulated in 1925 by George Uhlenbeck and Samuel Goudsmit to account for the splitting of spectral lines in a magnetic field. The "spinning" of the electron is not known to be a real physical phenomenon.

responding to the m_s quantum numbers $-\tfrac{1}{2}$ or $+\tfrac{1}{2}$. These orientations are arbitrarily "pictured" as

$$\begin{array}{ccc} s & & n \\ \downarrow & \text{or} & \uparrow \\ n & & s \end{array}$$

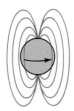

Paramagnetism, a property of substances with permanent magnetic moments, is associated with UNPAIRED ELECTRONS *in an atom, ion, or molecule.* These permanent micromagnets are attracted into a magnetic field and align with the field, similarly to the magnetic needle of a compass. Experimentally, the magnetic moment of isolated atoms is related to the number of unpaired electron spins.*

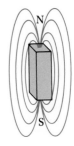

The observation that paired electrons are not paramagnetic is explained by the restriction placed on the spin of an electron. When two electrons occupy the same orbital, they spin in opposite directions so

that the paramagnetism of one electron $\begin{array}{c} n \\ \uparrow \\ s \end{array}$ is canceled by the paramag-

netism of the second electron $\begin{array}{c} s \\ \downarrow \\ n \end{array}$.

Fig. 6.22. *The electron in motion sets up a magnetic field whose properties are identical to the field set up by a bar magnet or by the earth. (The magnetic moments, of course, differ.)*

PROBLEMS

1. Quantization. Is it possible to select a weight that cannot be balanced (within the errors of observation) by a quantity of pennies? Of hydrogen? Of water? Explain.

2. Atomic structure. (a) In 1895, P. Lenard detected electrons outside a thin glass tube in which they were produced, and concluded that the atoms in the glass must have a very open structure. Explain these results in terms of the Rutherford atom. (b) Describe the fundamental differences among the Dalton, Thomson, Rutherford, and Schrödinger models of the atom.

3. Line spectra. (a) Is it possible to detect a line in the spectrum of atomic hydrogen corresponding to a transition $n = 4.4 \to n = 3.4$? (b) Use Equation (1), page 84, to calculate the energy in ergs per photon of the hydrogen spectral line corresponding to the transition $n = 5 \to n = 2$. Check your answer against Table 6.1 (page 87).

4. Nuclear atom. (a) For a given set of conditions, how should the number of α particles deflected through a given angle (*i*) vary (increase or decrease) with the kinetic energy of the particle, and (*ii*) be related (directly or inversely) to the number of unit positive charges on the nucleus of the atom? (b) How should the angle of deflection vary (increase or decrease) with the closeness of the α particle to the nucleus?

5. Nuclear composition. What is the nuclear composition of each of the isotopes of boron, atomic isotopic weights: 9.0133, 10.0129, and 11.0093.

---

* The contribution of the angular momentum about the nucleus to the magnetic moment can usually be ignored.

6. Nuclear composition. Complete the following table by substituting a numerical value when an x appears, and a symbol where an E appears:

Isotope	Mass number	Atomic number	Number of neutrons
$^{54}_{x}$Mn	x	x	x
$^{x}_{x}$E	87	x	51
$^{x}_{x}$Ce	135	x	x
$^{x}_{x}$E	x	48	60
$^{x}_{x}$E	205	86	x
$^{x}_{x}$S	x	x	14
$^{x}_{x}$Al	28	x	x

7. Nuclear atom. Calculate the energy of the hydrogen atom when the principal quantum number of the electron is 100. What does the minus sign signify?

8. Ionization. Find the work in kilocalories per mole of H required to remove the electron in the H atom from the energy level $n = 5$ to infinity.

9. Spectrum. Energy is transferred to an H atom causing the transition $n = 1$ to $n = 6$; then the transitions $n = 6$ to $n = 4$, $n = 4$ to $n = 2$, and $n = 2$ to ground state occur. How many photons are emitted by the atom and what is the energy of each photon? Use Table 6.1, page 87.

10. Spectrum. H atoms in the ground state are subjected to collisions with electrons having a maximum kinetic energy of 17.3×10^{-12} erg per electron. What is the number of lines in the spectrum emitted by the H atoms? Use Table 6.1, page 87.

11. n and l. (a) Give the n and l value for each of the following orbitals: $2p$, $4s$, $3d$, and $6s$. (b) Use the $(n + l)$ rule to predict the electron configuration of $_{40}$Zr and $_{41}$Nb in the ground state. Check your answers in Appendix III. (c) What are the total numbers of s, p, and d electrons in $_{40}$Zr?

12. Paramagnetism. (a) Pick the atoms that should be attracted into a magnetic field: (*i*) Ag, (*ii*) Na, (*iii*) Ne, (*iv*) Mg. (b) Pick the ion having five unpaired electrons: (*i*) $_{24}$Cr^{2+}, (*ii*) $_{27}$Co^{4+}, (*iii*) $_{29}$Cu^{3+}. Assume that the $3d$ sublevel is more stable than the $4s$ in ions.

13. Paramagnetism. Which of the following particles possess paramagnetic properties?
(a) $_{29}$Cu KL $3s^{2}3p^{6}3d^{10}4s^{1}$, (b) $_{29}$Cu^{+} KL $3s^{2}3p^{6}3d^{10}$, (c) $_{29}$Cu^{2+} KL $3s^{2}3p^{6}3d^{9}$, (d) $_{29}$Cu^{18+} KL $3s^{1}$.

14. Wave mechanics. (a) Does H^{+} have a $4s$ orbital? (b) Is the electron density for the H atom in the $1s$ state zero at 4 Å from the nucleus? Explain your answer. (c) Is the electron density of a $2p_{z}$ electron at the point on the x axis 0.5 Å from the nucleus high, low, or zero compared to the same distance on the z axis?

15. Electronic charge. Stoney calculated the magnitude of the electronic charge from the following ideas: during electrolysis, a univalent silver ion picks up 1 electron becoming a silver atom. Then, to liberate N_A atoms (1 mole), N_A electrons are required. Let e be the charge associated with 1 electron; then the quantity of electricity required to liberate 108 g of silver is $N_A e$. This quantity of electricity is easily measurable, 9.7×10^4 C. Calculate "the natural unit of electricity."

16. Electronic charge. *Millikan oil-drop experiment.* In an apparatus (Fig. 6.23) the velocity of rise of a positively charged oil drop in an electric field is measured. With the electric field turned off, its velocity of fall under the action of gravity is measured. From these data, the charge on the drop may be calculated. Some typical results are

Oil drop number	Measured charge on drop (C)
1	16.0×10^{-19}
2	1.60×10^{-19}
3	12.9×10^{-19}
4	1.59×10^{-19}
5	24.0×10^{-19}

(a) What is the electronic charge, and (b) the charge on each drop in units of the electronic charge? (c) What charge on a drop would have to be measured to compel a change in the accepted value of the electronic charge?

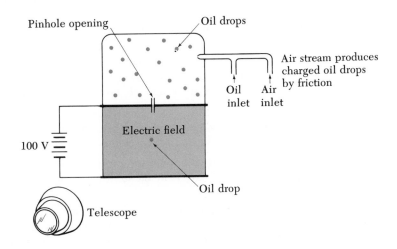

Fig. 6.23. *The Millikan apparatus for determining the electronic charge. Oil drops pass through the pinhole into the electric field. A telescope containing a length scale is used to measure the velocity of rise and fall.*

17. Electronic mass. From mass spectrographic measurements, the ratio of the mass of a hydrogen atom to that of an electron is 1.84×10^3. Calculate (a) the atomic weight of the electron, (b) the mass of an electron.

18. Nuclear atom. A stream of alpha particles impinges upon a gold film of thickness 4.0×10^{-4} mm; 100 alpha particles are detected in the region between $\phi = 180°$ and $92°$; see Fig. 6.5 (page 79). When the experiment is repeated with identical conditions except that the thickness is increased to 5.0×10^{-4} mm, 124 alpha particles are detected. Use these results to argue against the hypothesis that these deflections are caused by a surface reflection (similar to the action of a wall upon a handball).

19. Quantum theory. Predict the minimum energy that an electron must possess so that, upon collision with a mercury atom, it causes the appearance of the mercury line, $\lambda = 2537 \times 10^{-8}$ cm.[*]

20. Quantum theory. The measured frequency of photons emitted by the planet Jupiter is 1.9×10^7 sec^{-1}. Calculate the photon energy.

21. Spectrum. A particle "collides" with an H atom causing its electron to "jump" to the energy level $n = 4$. What is the energy and wavelength of the photon emitted for the transition $n = 4$ to $n = 1$?

22. Bohr atom. A rocket-born photometer has detected ultraviolet night glow radiations among which the 1026-Å line (Fig. 6.16, page 88) is identified. Check this value by calculating the λ of the line corresponding to the transition $n = 3 \rightarrow n = 1$.

23. Stability of atom. Given the charged spheres at rest at the distances shown:

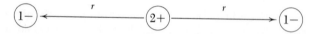

Calculate (a) the repulsion force between the 2 negative spheres, (b) the force of attraction of the positive sphere for each negative sphere. Which force is greater? What statement can you make about the stabilization of the Rutherford atom by electrostatic forces only?

24. 4s versus 3d. The electron configuration of $_{20}$Ca is $KL\ 3s^23p^64s^23d^0$. Three protons are added to the nucleus; the $3d$ sublevel becomes more stable than the $4s$. Write the chemical symbol and electron configuration of the ion produced.

25. 4s versus 3d. The magnetic moment of $_{24}$Cr in the gaseous state corresponds to six unpaired electrons. Write the electron configuration of Cr.

26. Paramagnetism. Calculate the number of unpaired electrons in (a) $_{29}Cu^{6+}$, (b) $_{30}Zn^+$, (c) $_{30}Zn^{2+}$, (d) $_{27}Co^{5+}$. The $3d$ sublevel is more stable than the $4s$ in these ions.

27. Quantum mechanics. (a) Did anyone prove that particles of atomic and subatomic dimensions obey Newton's laws of motion? (b) Has anyone proved that an electron is without structure? (c) Has anyone ever observed light waves

[*] This prediction was verified in 1913 by the experiment of James Franck and Gustav Herz, in which electrons of precisely known energy were passed through mercury vapor at low pressure.

with or without the aid of instruments? (d) Pick the correct statement: orbital shapes (patterns of electron distribution) are (i) established by experiment, (ii) calculated from theory. (e) In studying the trajectory of an ordinary moving pendulum, is it possible to observe the exact position and the frequency of the oscillation at any instant? (f) Do electrons travel around the nucleus in definite orbits? Justify your answer.

28. Quantum mechanics. Use the following results for the hydrogen atom in the 2s state to plot ψ_{2s}^2 (y axis) versus r (x axis):

r, cm	ψ_{2s}		ψ_{2s}^2, cm^{-3}	
0.0	5.20	$\times 10^{11}$	27.04	$\times 10^{22}$
0.5×10^{-8}	1.70	$\times 10^{11}$	2.89	$\times 10^{22}$
1.0×10^{-8}	0.10	$\times 10^{11}$	0.010	$\times 10^{22}$
1.06×10^{-8}	0.00		0.00	
1.5×10^{-8}	-0.52	$\times 10^{11}$	0.27	$\times 10^{22}$
2.0×10^{-8}	-0.65	$\times 10^{11}$	0.42	$\times 10^{22}$
3.0×10^{-8}	-0.58	$\times 10^{11}$	0.34	$\times 10^{22}$
5.0×10^{-8}	-0.18	$\times 10^{11}$	0.032	$\times 10^{22}$
10.0×10^{-8}	-0.0037	$\times 10^{11}$	0.00001	$\times 10^{22}$
∞	0.0		0.00	

Correlate your plot with the electron density diagram for the 2s electron, Fig. 6.18 (page 90).

29. Schrödinger equation. (*This problem is intended only for students with experience in differential calculus.*) The Schrödinger equation in one dimension (x) is written

$$\frac{\partial^2 \psi}{\partial x^2} + \frac{8\pi^2 m}{h^2}(E - P)\psi = 0 \tag{3}$$

where $h =$ Planck's constant, $m =$ the mass of the particle located in space by the coordinate x, $E =$ the total energy of the particle, the sum of its kinetic and potential energy, $P =$ the potential energy of the particle. Potential energy refers to the *work done against a force* in changing the position of a particle. ψ, called a WAVE FUNCTION, is expressed in terms of the coordinates of the particle. For an electron confined to a fixed distance l in one dimension, undergoing an oscillatory translational motion, and not subjected to a force (except when it collides with the barriers at $x = 0$ and $x = l$), the wave equation becomes

$$\frac{d^2\psi}{dx^2} + \frac{8\pi^2 m}{h^2} E\psi = 0 \tag{4}$$

and ψ is subject to the boundary conditions

$$\psi(0) = 0, \ \psi(l) = 0 \tag{5}$$

Let

$$\beta^2 = \frac{8\pi^2 m}{h^2} E \tag{6}$$

Then

$$\frac{d^2\psi}{dx^2} + \beta^2\psi = 0 \tag{7}$$

Show that Equations (5) and (7) are satisfied when

$$\psi = k \sin \beta x \quad \text{and} \quad \beta = \frac{a\pi}{l} \quad \text{where } a = 1, 2, 3, \ldots$$

Show that if the only possible values of β are

$$\beta = \frac{a\pi}{l}, \quad a = 1, 2, 3 \ldots$$

then the kinetic energy of the electron can only have values given by

$$E = \frac{a^2h^2}{8ml^2}$$

30. Experiment versus theory. Which of the following are observable? (a) Position of electron in H atom; (b) frequency of radiation emitted by H atoms; (c) path of electron in H atom; (d) wave motion of electrons; (e) diffraction patterns produced by electrons; (f) energy required to remove electrons from H atoms; (g) a light wave; (h) a water wave.

7

CHEMICAL PERIODICITY

Elements can be grouped by properties into various classes or families, somewhat as living things are classified by biologists. For example, the metals iron, copper, and lead differ as a class from the nonmetals sulfur, phosphorus, and carbon. Within the metallic class, copper, silver, and gold (COINAGE METALS) differ as a group from calcium, strontium, and barium (ALKALINE EARTH METALS). These facts suggest that atoms are themselves made up of other fundamental particles, because the similarity of the properties of, say, the alkaline earth metals may be the result of a common structural feature of their atoms. This notion foreshadowed more refined concepts of atomic structure.

Hence, the effort of chemists to classify the elements was one of the necessary first steps leading to our current understanding of atomic structure. The first classifications were based on relating chemical properties (especially valence) to the atomic weights of the elements; such systems, therefore, could not be formulated until accepted methods for determining valences and atomic weights became available in the middle of the nineteenth century.

In 1864, John Newlands noted that when the elements were arranged in the increasing order of their atomic weights, similar properties recurred at periodic intervals. He called this relationship, by a rather inapt analogy to the musical scale, the LAW OF OCTAVES. Thus, lithium (thought then to be the second element) and sodium (thought to be the ninth) are an "octave" apart.[*] This arrangement was the beginning of our idea of "atomic number"—a concept vital to our present expressions of chemical periodicity and the electronic structure of atoms. Newlands' contribution, ridiculed at first, was later given recognition (Davy Medal, 1887).

---

[*] Helium and neon had not yet been discovered.

In 1869 Dmitrii Ivanovich Mendeleev and several months later Julius Lothar Meyer published independent versions of a periodic system of the elements. Credit for the "periodic law" is given to these two scientists (especially Mendeleev) not because they were the first to describe periodic classifications of the elements, but more particularly for the following reasons:

(a) The significance of periodicity was more fully appreciated, the system of classification was more elaborate, and attention was given to a broader range of physical and chemical properties than was examined by earlier investigators. Table 7.1 is an adaptation of Mendeleev's periodic table of 1871.

(b) Mendeleev recognized that, when the atomic weight order produced a dislocation of elements in the periodic system, adherence to atomic weight order must be sacrificed. For example, if iodine (atomic weight in 1869: 127) is placed before tellurium (atomic weight in 1869: 128), iodine appears in a group with sulfur and selenium, and tellurium finds itself in the company of chlorine and bromine. However, iodine is chemically similar to chlorine and bromine, not to sulfur and selenium; tellurium is similar to sulfur and selenium, not to chlorine and bromine. This unreasonable arrangement must therefore mean that the

Table 7.1
Periodic Table of the Elements (Adapted from Mendeleev, 1871)

Groups:	I		II		III		IV		V		VI		VII		VIII
Type formulas:	R_2O RCl		RO RCl_2		R_2O_3 RCl_3		RO_2 RH_4		R_2O_5 RH_3		RO_3 RH_2		R_2O_7 RH		RO_4
Subgroup or family:	A	B	A	B	A	B	A	B	A	B	A	B	A	B	
1		H													
2	Li		Be		B		C		N		O		F		
3		Na		Mg		Al		Si		P		S		Cl	
4	K		Ca		—		Ti		V		Cr		Mn		Fe, Co, Ni, (Cu)
5		(Cu)		Zn		—		—		As		Se		Br	
6	Rb		Sr		Yt?		Zr		Nb		Mo		—		Ru, Rh, Pd, (Ag)
7		(Ag)		Cd		In		Sn		Sb		Te		I	
8	Cs		Ba		Di?		Ce?		—		—		—		— — — —
9		—		—		—		—		—		—		—	
10	—		—		Er?		La?		Ta		W		—		Os, Ir, Pt, (Au)
11		(Au)		Hg		Tl		Pb		Bi					
12	—		—		—		Th		—		U		—		— — — —

atomic weights either are wrong or are not the fundamental basis of chemical periodicity. Mendeleev imagined that the weights were in error; actually, they were not far from today's accepted values.

(c) Mendeleev avoided what would otherwise be additional chemical inconsistencies by boldly leaving gaps in his table, predicting that elements would be discovered to fill the gaps, and describing the properties that the to-be-discovered elements would have. For example, in 1871, the element after zinc was arsenic. But arsenic does not belong in the same chemical group with aluminum or with silicon (see Table 7.1, groups III and IV). Arsenic is chemically like phosphorus, and therefore belongs in group V. Succeeding elements (selenium, bromine, and so on) also fall into reasonable locations if they follow arsenic located in group V. This means that two elements, one like aluminum and one like silicon, are missing.

For testing a theory, there is nothing like a prediction. The more spectacular the confrontation between predicted and observed data, the better the test. Table 7.2 shows the degree to which such predictions

Table 7.2
Prediction of the Properties of Germanium

| Property | "Eka-silicon" (Es) predicted in 1871 by Mendeleev | Germanium (Ge) | |
		Reported in 1886 by Clemens Alexander Winkler	Currently accepted
Atomic weight	72	72.32	72.59
Density (g/ml)	5.5	5.47	5.35
Melting point	High	–	947°C
Specific heat (cal/g-deg)	0.073	0.076	0.074
Molar volume[a]	13	13.22	13.5
Color	Dark gray	Grayish white	Grayish white
Valence	4	4	4
Reaction with acids and bases	Es will be slightly attacked by such acids as HCl, but will resist attack by NaOH	Ge is dissolved by neither HCl nor dilute NaOH, but is dissolved by concentrated NaOH	Ge is dissolved by neither HCl nor dilute NaOH, but is dissolved by concentrated NaOH
Boiling point of the tetraethyl derivative	160°C	160°C	185°–187°C
Density of the dioxide (g/ml)	4.7	4.703	4.228
Density of the tetrachloride (g/ml)	1.9	1.887	1.8443
Boiling point of the tetrachloride	100°C	86°C	84°C

[a] The volume occupied by 1 mole of atoms in the solid state, cm³/mole.

were successful for *eka*-silicon* (germanium). *Eka*-aluminum (gallium) and *eka*-boron (scandium) are other examples of Mendeleev's successful predictions.

THE PERIODIC LAW AND PERIODIC TABLE; TYPES OF ELEMENTS / 7.3

The properties of the elements are periodic† functions of their atomic numbers. This "law" is a broad, imprecise statement that still functions as a much-used framework on which comparisons and generalizations of chemical behavior are based. Chemists everywhere agree that the periodic law expresses deep-seated relationships in chemical properties and in the structure of atoms, but no two chemists are likely to offer the same detailed interpretation of the law.

Starting with Mendeleev and Meyer, and continuing to the present, enough periodic tables have been published to fill a book. We shall examine the Mendeleev table of 1871 (Table 7.1) because some of its usages and conventions still persist. We shall then look at modern periodic tables.

Mendeleev's table. Mendeleev divided the elements into eight vertical columns, called GROUPS. The horizontal rows are called PERIODS. The most important characteristic common to elements in the same group is their VALENCE—the numerical capacity of atoms of an element to combine with other atoms. If the other atoms are taken to be H, Cl, or O, it is possible to write "type formulas" that characterize the valences of elements within a group. Such formulas appear in Table 7.1, R denoting any element in the group. It was apparent to Mendeleev, however, that the elements within a given valence group exhibited not one, but two more or less distinct sets of properties. In group I, for example, Li, K, Rb, and Cs differ markedly from Cu, Ag, and Au. Accordingly, he divided each group into two SUBGROUPS or FAMILIES, to correspond to this chemical duality. Some of his assignments, however, were uncertain.

A modern table: the separated form; types of elements. Modern periodic tables attempt to depict both chemical and electronic relationships. One such form is shown as Table 7.3. Note the following features:

(a) The upper section of Table 7.3 comprises the elements that represent the regular progression of chemical properties on which the original periodic systems were based. They are therefore called REPRESENTATIVE ELEMENTS, and are designated as R groups. The concept of "rep-

---

* *Eka* (Greek) means "first after" or "first beyond."

† "Periodic" is used here in the loose sense of "more or less regularly recurring."

Table 7 3
Periodic Table, Separated Form.

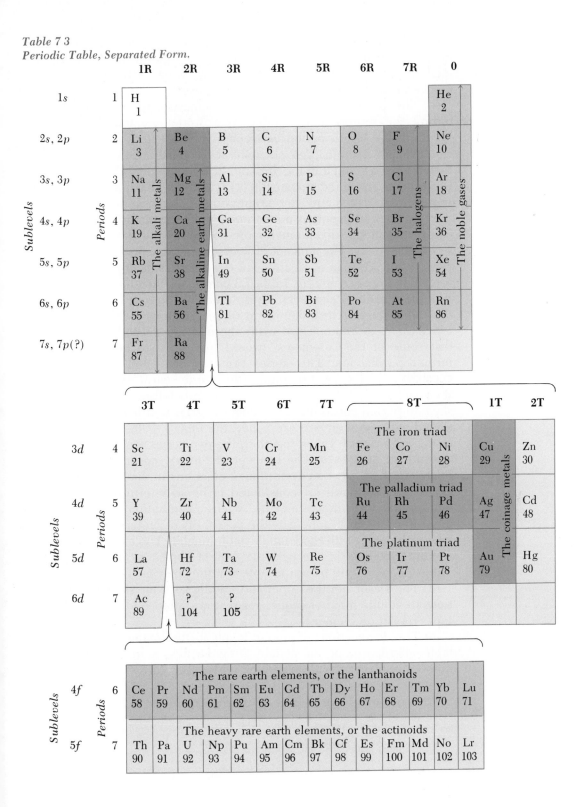

		1R	2R	3R	4R	5R	6R	7R	0
1s	1	H 1							He 2
2s, 2p	2	Li 3	Be 4	B 5	C 6	N 7	O 8	F 9	Ne 10
3s, 3p	3	Na 11	Mg 12	Al 13	Si 14	P 15	S 16	Cl 17	Ar 18
4s, 4p	4	K 19	Ca 20	Ga 31	Ge 32	As 33	Se 34	Br 35	Kr 36
5s, 5p	5	Rb 37	Sr 38	In 49	Sn 50	Sb 51	Te 52	I 53	Xe 54
6s, 6p	6	Cs 55	Ba 56	Tl 81	Pb 82	Bi 83	Po 84	At 85	Rn 86
7s, 7p(?)	7	Fr 87	Ra 88						

Sublevels / *Periods*

The alkali metals · The alkaline earth metals · The halogens · The noble gases

		3T	4T	5T	6T	7T	8T			1T	2T
							The iron triad			The coinage metals	
3d	4	Sc 21	Ti 22	V 23	Cr 24	Mn 25	Fe 26	Co 27	Ni 28	Cu 29	Zn 30
							The palladium triad				
4d	5	Y 39	Zr 40	Nb 41	Mo 42	Tc 43	Ru 44	Rh 45	Pd 46	Ag 47	Cd 48
							The platinum triad				
5d	6	La 57	Hf 72	Ta 73	W 74	Re 75	Os 76	Ir 77	Pt 78	Au 79	Hg 80
6d	7	Ac 89	? 104	? 105							

Sublevels / *Periods*

		The rare earth elements, or the lanthanoids												
4f	6	Ce 58	Pr 59	Nd 60	Pm 61	Sm 62	Eu 63	Gd 64	Tb 65	Dy 66	Ho 67	Er 68	Tm 69	Yb 70	Lu 71
		The heavy rare earth elements, or the actinoids												
5f	7	Th 90	Pa 91	U 92	Np 93	Pu 94	Am 95	Cm 96	Bk 97	Cf 98	Es 99	Fm 100	Md 101	No 102	Lr 103

Sublevels / *Periods*

resentative elements" is related to the progressive addition of electrons in the s and p sublevels of the atoms. The "group 0" elements, also included here, are the noble gases, which were once thought to have no chemical reactivity at all, hence "zero valence."

(b) The middle section of Table 7.3 comprises the elements that interrupt the representative series. Interruption by ten elements first occurs in the fourth period between groups 2 and 3, that is, between calcium and gallium. Ten elements also constitute the transition between groups 2 and 3 in the fifth and in the sixth periods. These groups of elements are called TRANSITION (T) ELEMENTS. In comparison with the representative elements, the progression of chemical properties among the transition elements is not so marked. The transition elements thus constitute, chemically, a much more homogeneous group than the representative ones. The concept of "transition elements" is related to the progressive addition of electrons in the d sublevels of the atoms.*

(c) The lower section of Table 7.3 comprises the elements that interrupt the transition elements. These elements were frustrating because they all seemed to be alike; it was difficult to separate them, to characterize them, to find out how many there were, and to be sure, indeed, that some of the "elements" would not prove to be mixtures after further purification. This series came to be called the RARE-EARTH ELEMENTS, or the LANTHANOIDS† (the word "earth" was used in the sense of "oxide"). That the fourteen elements following actinium in the seventh period constitute an analogous series was learned much later; accordingly, its members are called the HEAVY RARE-EARTH ELEMENTS, or the ACTINOIDS.† Taken together, the two series are called the INNER TRANSITION ELEMENTS. The concept of "inner transition elements" is related to the progressive addition of electrons in the f sublevels of the atoms. The actinoids are considered to be an interruption in the seventh-period transition elements, whose resumption starts with element 104. Chemists anticipate the possibility of synthesis of a few more heavy elements with moderately stable nuclei. The expected positions of new elements in the periodic system appear as dashed boxes in Table 7.4. Most will be so unstable that they may never be synthesized.

(d) The slashes in the upper and middle sections show the interruptions in atomic number sequences that are filled by the transition and inner transition elements.

(e) The designations of energy levels in the listing to the left of the table afford, except for certain irregularities, a quick count of the number of electrons in the various sublevels.

---

* The International Union of Pure and Applied Chemistry (IUPAC) defines transition elements as those whose atoms or ions have an incomplete d subshell. This definition excludes Zn, Cd, and Hg. The problem is discussed in more detail in Chapter 21.

† Old names: lanthanides and actinides.

Example 1 How many 5f electrons and how many 6d electrons does $_{93}$Np have?

Answer Table 7.3 shows that Np is the fourth element in the 5f series; therefore, it is $5f^4$. This entire period is taken from the position just after $_{89}$Ac; therefore it is $6d^1$. It must be recognized that the convenience of this method does not remove the uncertainty of some of these electronic assignments (Appendix III).

Another modern periodic table: the long period form. This form, which appears as Table 7.4, has eighteen vertical columns and separates the R subgroups from the T subgroups. In the second and third periods, where no subgroup distinction exists, this wide separation, occurring between groups 2 and 3, has no significance.

Placing the lanthanoid and actinoid elements into the body of the table would impose strains on the dimensions of charts and books, and is therefore not usually attempted.

Table 7.4
Periodic Table, Long Form[a]

Periods	1R	2R	3T	4T	5T	6T	7T	8T			1T	2T	3R	4R	5R	6R	7R	0
1	H 1																	He 2
2	Li 3	Be 4					*Transition elements*						B 5	C 6	N 7	O 8	F 9	Ne 10
3	Na 11	Mg 12											Al 13	Si 14	P 15	S 16	Cl 17	Ar 18
4	K 19	Ca 20	Sc 21	Ti 22	V 23	Cr 24	Mn 25	Fe 26	Co 27	Ni 28	Cu 29	Zn 30	Ga 31	Ge 32	As 33	Se 34	Br 35	Kr 36
5	Rb 37	Sr 38	Y 39	Zr 40	Nb 41	Mo 42	Tc 43	Ru 44	Rh 45	Pd 46	Ag 47	Cd 48	In 49	Sn 50	Sb 51	Te 52	I 53	Xe 54
6	Cs 55	Ba 56	57–71	Hf 72	Ta 73	W 74	Re 75	Os 76	Ir 77	Pt 78	Au 79	Hg 80	Tl 81	Pb 82	Bi 83	Po 84	At 85	Rn 86
7	Fr 87	Ra 88	89–103	? 104	? 105	106	107	108	109	110	111	112	113	114	115	116	117	118
8(?)	119	120	121															

	La 57	Ce 58	Pr 59	Nd 60	Pm 61	Sm 62	Eu 63	Gd 64	Tb 65	Dy 66	Ho 67	Er 68	Tm 69	Yb 70	Lu 71
Lanthanoids															
Actinoids	Ac 89	Th 90	Pa 91	U 92	Np 93	Pu 94	Am 95	Cm 96	Bk 97	Cf 98	Es 99	Fm 100	Md 101	No 102	Lr 103

122	123	124	125	126

[a] The heavy line approximately separates the metallic from the nonmetallic elements.

The most striking and significant periodic variation among the elements is that of valence. Valence periodicity is exhibited most consistently among the representative elements; it is somewhat of a strain to apply the system to the transition elements, and hopeless for the inner transition elements.

Valences of the representative elements. If we designate the group number as G, we may write

$$\text{valence of representative elements} = G \quad \text{or} \quad (8 - G) \tag{1}$$

To illustrate Equation (1), it is best to consider the valence of an element to be the number of atoms of H, Cl, or F per atom of the element in a binary (two-element) compound. For example,

Element	Group number, G	Valence, G or (8-G)	Compound
Na	1	1	NaCl
O	6	2	H_2O
C	4	4	CH_4, CCl_4, CF_4
P	5	5 or 3	PCl_5, PCl_3
S	6	6 or 2	SF_6, H_2S

Equation (1) does not always predict correctly which compounds exist and which do not exist. Thus, there are no substances corresponding to the formula $NaCl_7$ or H_6O, nor does it apply to the noble gases

Table 7.5
Some Compounds of the Representative Elements

		Group					
	1R	2R	3R	4R	5R	6R	7R
Some compounds whose formulas correspond to Equation (1)	H_2O	$BeCl_2$	BF_3	CH_4	NH_3	H_2O	HF
	Li_2O	Be_2C	Al_2O_3	CO_2	N_2O_5	H_2S	HCl
	NaH	CaH_2	$InCl_3$	$SiHF_3$	PCl_3	SO_3	Cl_2O_7
	KCl	SrO	TlF_3	$GeBr_4$	PCl_5	SF_6	BrCl
	$RbNO_3$	BaO		SnS_2	AsI_3	SeO_3	Br_2O
	CsF	$Ra(NO_3)_2$		PbO_2	AsP	TeO_3	KI
					Sb_2O_3		BaI_2
					BiOCl		
Some compounds whose formulas do not correspond to Equation (1)			$B_{10}H_{14}$	CO	N_2O	SO_2	ClO_2
			TlF	$C_{10}H_{20}$	AsI_2	SeO_2	BrF_5
				PbO		TeF_4	ICl_3

(Group 0). However, all the representative elements do form some compounds whose formulas correspond to Equation (1). Table 7.5 separates various compounds of several representative elements into two groups, those whose valences correspond to Equation (1) and those whose valences do not. On this basis, and from considerations of electronic structure, it is reasonable to designate as "typical" those valences of the representative elements which correspond to Equation (1).

SIZES OF ATOMS AND IONS / 7.5

The distance between the nuclei of two bonded atoms or ions, called the BOND DISTANCE, can be measured by X-ray or electron diffraction methods. The actual radius of an individual atom or ion, on the other hand, is *not* a definite, physically measurable value, because the positions of electrons are indefinite. However, the *sum* of the radii of two atoms or ions that are bonded to each other is assumed to be equal to the bond distance. Then, if enough assumptions are made, a radius can be assigned to each atom or ion, whose addition for a pair of atoms or ions would equal the measured internuclear distance.

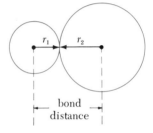

The radius of an atom or an ion is a function of its environment. An extreme example is the hydrogen atom, which must be assigned one radius when it combines with itself, a second radius when it combines with elements of the second period, a third radius when it combines with elements of the third period, and so on. Most ionic radii, however, may be treated as fairly constant. It is, therefore, advantageous for many purposes to assign a radius to each atom, called the COVALENT RADIUS, and to each ion, called the IONIC or CRYSTAL RADIUS.

The variation in the size of atoms and ions within a period and a group is illustrated for various elements in Table 7.6.

Within a given period, the atomic radius decreases with increasing atomic number owing to the effect of increasing the positive nuclear charge while adding electrons to the same level. The irregularity at nitrogen may be associated with the stability of the half-filled p orbitals (page 95). Within a given group the atomic radius increases with atomic number because of the addition of another level. Several notable exceptions occur in which the atomic radius within a group increases only very slightly or actually decreases with increasing atomic number; for example, $_{13}$Al, 1.25 Å and $_{31}$Ga, 1.26 Å; $_{46}$Pd, 1.28Å and $_{78}$Pt, 1.29 Å; $_{40}$Zr, 1.45 Å and $_{72}$Hf, 1.44 Å. This results from the presence of the intervening transition elements and, in particular, the inner transition elements, in which the added electrons generally enter an inner level. We would then expect the added electrons virtually to nullify the electrostatic attraction effect of the increased nuclear charge, maintaining, thereby, an approximately constant atomic size for each transition series. Actually, however, small but significant decreases

Table 7.6
Covalent[a] and Ionic (Crystal) Radii in Angstrom Units

Period	1R	2R	3R	4R	5R	6R	7R
1	H 0.28 to 0.38						
	H^- 2.08						
2	Li 1.33	Be 0.90	B 0.80	C 0.77 / C^{4+} 0.16	N 0.73	O 0.74	F 0.71
	Li^+ 0.60	Be^{2+} 0.31	B^{3+} 0.23	C^{4-} 2.60	N^{3-} 1.71	O^{2-} 1.40	F^- 1.36
3	Na 1.54	Mg 1.36	Al 1.25	Si 1.15	P 1.1	S 1.02	Cl 0.99
	Na^+ 0.95	Mg^{2+} 0.65	Al^{3+} 0.50	Si^{4+} 0.42	P^{3-} 2.12	S^{2-} 1.84	Cl^- 1.81
4	K 1.96	Ca 1.74	Ga 1.26	Ge 1.22	As 1.19	Se 1.16	Br 1.14
	K^+ 1.33	Ca^{2+} 0.99	Ga^{3+} 0.62	Ge^{4+} 0.53	As^{3-} 2.22	Se^{2-} 1.98	Br^- 1.95
5	Rb 2.16	Sr 1.92	In 1.44	Sn 1.41	Sb 1.38	Te 1.35	I 1.33
	Rb^+ 1.48	Sr^{2+} 1.13	In^{3+} 0.81	Sn^{4+} 0.71	Sb^{3-} 2.45	Te^{2-} 2.21	I^- 2.16
6	Cs 2.35	Ba 1.98	Tl 1.48	Pb 1.47	Bi 1.46	Po	At
	Cs^+ 1.69	Ba^{2+} 1.35			Bi^{3+} 0.96		
7	Fr	Ra					

Transition elements

Period	3T	4T	5T	6T	7T	8T	8T	8T	1T	2T
4	Sc 1.44	Ti 1.32 / Ti^{2+} 0.90	V 1.22 / V^{3+} 0.74	Cr 1.18 / Cr^{3+} 0.65	Mn 1.17 / Mn^{2+} 0.80	Fe 1.17 / Fe^{2+} 0.75	Co 1.16 / Co^{2+} 0.82	Ni 1.15 / Ni^{2+} 0.78	Cu 1.17 / Cu^+ 0.96	Zn 1.25
	Sc^{3+} 0.81	Ti^{4+} 0.68	V^{5+} 0.59	Cr^{6+} 0.52	Mn^{7+} 0.46	Fe^{3+} 0.67	Co^{3+} 0.29	Ni^{3+} 0.35	Cu^{2+} 0.72	Zn^{2+} 0.74
5	Y 1.62	Zr 1.45	Nb 1.34	Mo 1.30	Tc 1.27	Ru 1.25	Rh 1.25	Pd 1.28	Ag 1.34	Cd 1.48
	Y^{3+} 0.93	Zr^{4+} 0.80							Ag^+ 1.26	Cd^{2+} 0.97
6	La 1.69	Hf 1.44	Ta 1.34	W 1.30	Re 1.28	Os 1.26	Ir 1.27	Pt 1.30	Au 1.34	Hg 1.49
									Au^+ 1.37	Hg^{2+} 1.10
7	Ac									

[a] Covalent radii (taken mostly from *Table of Interatomic Distances*, London, Chem. Soc., Special Publ., 1958 and from the publications of R. T. Sanderson) are applicable only to single-bonded atoms in mainly covalent molecules. Crystal (ionic) radii are from the publications of Linus Pauling.

occur in each of these series. Hence, the size increase characteristic of families generally does not appear after these transition series. Thus:

$_{13}$Al KL $3s^23p^1$

$_{31}$Ga KL $3s^23p^63d^{10}$ $4s^24p^1$

Most of the 18-electron difference between $_{13}$Al and $_{31}$Ga is in the third level, just about canceling the 18-proton difference in the nuclear charge. The radii, which are determined by the outer electrons, thus remain nearly constant.

The ionic radius increases downward in a group. Down a group, the maximum value of n (the principal quantum number) and the nuclear charge increase. However, in going, for example, from $_3$Li to $_{55}$Cs, the effect of the added electrons more than cancels the increase in charge, and the radius increases.

The ionic radius decreases across a period for ISOELECTRONIC IONS, ions possessing the same electron configuration, illustrated by Li^+ and Be^{2+} (both $1s^2$), and by N^{3-}, O^{2-}, F^-, Na^+, Mg^{2+}, and Al^{3+} (all $1s^22s^22p^6$).

IONIZATION ENERGY / 7.6

The minimum energy required to remove an electron (e^-) from a gaseous atom in its ground state is called the first IONIZATION ENERGY; the second ionization energy refers to the removal of a second electron from the positive ion.

Ionization process	Ionization energy
$He(g) \longrightarrow He^+(g) + e^-$	+567 kcal/mole
$He^+(g) \longrightarrow He^{2+}(g) + e^-$	+1254 kcal/mole

The ionization energy is measured in a discharge tube containing the gas. Initially, the current flow through the tube is practically zero. The voltage is gradually increased until ionization occurs. This event is signaled by a sudden very large increase in the current flow through the tube. The voltage at which this occurs is proportional to the ionization energy. With hydrogen atoms, for example, the sharp increase in current occurs at 13.60 V; the ionization energy of the hydrogen atom is then said to be 13.60 eV/atom. One eV/atom equals 23.1 kcal/mole (Appendix I.11); the ionization energy of hydrogen is therefore 314 kcal/mole.

Note the following trends as illustrated by the data in Table 7.7:

1. Among the representative elements, the first ionization energy increases irregularly across a period. There are two abnormally high values, one coinciding with the completion of an s sublevel (Be, $2s^1$), the other with a half-filled p sublevel (N, $2s^22p^3$).

Table 7.7
First Ionization Energies, kcal/mole

1	H $1s^1$ 314							He $1s^2$ 567
2	Li $2s^1$ 124	Be $2s^2$ 215	B $2s^22p^1$ 191	C $2s^22p^2$ 260	N $2s^22p^3$ 336	O $2s^22p^4$ 314	F $2s^22p^5$ 402	Ne $2s^22p^6$ 497
3	Na 119		Al 138					
4	K 100		Ga 138					
5	Rb 96.3		In 133					
6	Cs 89.8		Tl 141					

	$_{21}$Sc	$_{22}$Ti	$_{23}$V	$_{24}$Cr	$_{25}$Mn	$_{26}$Fe	$_{27}$Co	$_{28}$Ni	$_{29}$Cu
4	154	157	155	156	171	182	181	176	178

2. Within a group, the first ionization energy decreases downward, with some exceptions. The transition elements do not conform fully to these patterns. We shall, therefore, content ourselves with the broad generalization that *the first ionization energy increases across a period and decreases down a group.*

3. Among the representative elements, metallic character—malleability, good conduction of heat and electricity, tendency to produce positive ions—decreases across a period and increases downward within a group. This gradation of metallic properties is related to the trends in ionization energy; *the smaller the ionization energy, the greater the metallic character.*

The chemical properties of the elements are related not only to ionization energies but also to excitation energies—the energy needed to promote an electron to the next available orbital. For example,

Be, $1s^22s^2$	He, $1s^2$
Excitation possible within occupied level → $1s^22s^12p^1$. This requires relatively little energy and therefore Be is reactive.	Excitation requires promotion to a higher level → $1s^12s^1$. Requires ~10^4 times as much energy as for Be promotion. Therefore He is inert.

Electronic structure of positive ions. The success of the $(n + l)$ rule of orbital stability (page 92) in predicting electron configurations is restricted to neutral atoms in the ground state. The rule is not necessarily applicable to positive ions. For example, the configuration $_{21}Sc$ KL $3s^23p^63d^14s^2$ is correct but, since the $4s(4 + 0)$ is calculated to be more stable than the $3d(3 + 2)$, the rule predicts the configuration KL $3s^23p^63d^04s^2$ for Sc^+; hence, Sc^+ should not be paramagnetic. But it is; its magnetic moment corresponds to two unpaired electrons. This means that the ionization of Sc yields Sc^+ KL $3s^23p^63d^14s^1$; thus for the Sc^+ ion, the $3d^14s^1$ state is more stable (lower energy) than the $3d^04s^2$ state. The configuration of Sc^{2+} is KL $3s^23p^63d^1$. Thus, for the *ions* of the first transition elements, the $3d$ subshell is usually more stable than the $4s$. As another example, the ionization of $_{27}Co$ KL $3s^23p^63d^74s^2$ yields $_{27}Co^+$ with a $3d^84s^0$ configuration:

$$_{27}Co \quad \underline{2}\ \underline{2}\ \underline{1}\ \underline{1}\ \underline{1}\ \ \underline{2} \atop \quad\qquad 3d \qquad\qquad 4s$$

$$_{27}Co^+ \quad \underline{2}\ \underline{2}\ \underline{2}\ \underline{1}\ \underline{1}\ \ \underline{} \atop \qquad\quad 3d \qquad\qquad 4s$$

However, the formation of $_{39}Y^+$ KLM $4s^24p^64d^05s^2$ shows that the $5s$ sublevel is more stable than the $4d$ for this ion. This means that, for any atom except hydrogen, the energy of one atomic orbital is actually dependent in a complicated manner on the number of electrons in all the other atomic orbitals. It is therefore unreliable to predict the electron configurations of the positive ions of the transition elements.

ELECTRON AFFINITY / 7.7

Neutral atoms generally have some attraction for electrons. The energy change accompanying the addition of an electron to a gaseous atom in its ground state is called the **ELECTRON AFFINITY**, exemplified by the equations

$$\begin{aligned} Cl(g) + e^- &\longrightarrow Cl^-(g) & \Delta H &= -83.32 \text{ kcal/mole} \\ O(g) + e^- &\longrightarrow O^-(g) & \Delta H &= -53.8 \text{ kcal/mole} \\ O^-(g) + e^- &\longrightarrow O^{2-}(g) & \Delta H &= +210 \ \text{ kcal/mole} \end{aligned}$$

We should expect the attraction to be greatest for the smallest atom within a given group because its ionization energy is greatest. However, the electron affinity is difficult to measure, and the meager available data indicate that it is unwise to generalize on the relationship between size and electron affinity of atoms. Some values are given in Table 7.8.

Table 7.8
Electron Affinities, kcal/mole[a]

H −17.2							
Li −14			C −48±20	N +16	O −53.8	F −79.52	
Na −19						Cl −83.32	
K −16						Br −81.6	
						I −70.64	

[a] Electron affinities have been derived from studies of the equilibrium constant for the reaction $X(g) + e^- \rightleftharpoons X^-(g)$ as a function of temperature. The − sign convention in this table is consistent with our previous usage for ΔH values; thus the capture of an electron by a chlorine atom is exothermic. However, the opposite convention, in which the energy released is given a + sign, is also in use. Sources of the data are: H. O. Pritchard, *Chemical Reviews* **52**, 554 (1953); R. S. Berry and C. W. Reimann, *Journal of Chemical Physics* **38**, 1540 (1963); G. Klopman, *Journal of the American Chemical Society* **86**, 4550 (1964).

PROBLEMS

1. Periodic table. Define the following terms related to the periodic table: (a) period, (b) group, (c) type formula, (d) representative element, (e) transition element, (f) inner transition element.

2. Periodic table. Locate the following classes of elements on Table 7.4, page 109: (a) the alkali metals, (b) the halogens, (c) the noble gases, (d) the alkaline earth metals, (e) the iron triad, (f) the rare earth elements, (g) the coinage metals.

3. Valence. Ag, Zn, Zr, Cr, and Os each form an oxide in which the valence of the metal equals its periodic group number. Write the formulas for these oxides.

4. Ionic radii. Negative ions are always larger than positive ions in the same period. Explain.

5. Size. (a) Pick from the following list the element with (*i*) the largest atom, (*ii*), the smallest atom, (*iii*) the highest first ionization energy: $_4\text{Be}$, $_5\text{B}$, $_7\text{N}$, $_{12}\text{Mg}$; (b) pick the ion with (*i*) the largest radius, (*ii*) the smallest radius: $_7\text{N}^{3-}$, $_8\text{O}^{2-}$, $_{13}\text{Al}^{3+}$, $_{15}\text{P}^{3-}$.

6. Ionization energy. Which factor, the radius or the nuclear charge, is more significant in determining the ionization energy of the elements in a group? Explain.

7. Promotion energy. Pick the promotion requiring the larger energy: B, $1s^2 2s^2 2p^1 \rightarrow$ B, $1s^2 2s^1 2p^2$; or Ne, $1s^2 2s^2 2p^6 \rightarrow$ Ne, $1s^2 2s^2 2p^5 3s^1$.

8. Size and ionization energy. For each of the following pairs, state which one has the larger radius, and which has the greater ionization energy. (a) I, I$^-$; (b) Br, K; (c) S, Te; (d) Sr, Ra; (e) Li, Li$^+$; (f) Br, I.

ADDITIONAL PROBLEMS

9. Mendeleev's table. Write the type formulas for the hydrides of elements of groups I, II, and III, and the chlorides of elements of groups IV, V, VI, and VII. Criticize the use of such type formulas in predicting the compositions of real compounds.

10. Periodic table. It has been postulated that a hitherto undiscovered element, *eka*-platinum, may be stable enough to exist in minute traces on earth. (a) Predict the atomic number and electron configuration of *eka*-platinum. (b) In minerals of what elements would you look for it?

11. Periodic table. Where on Table 7.3 (page 107) would elements 105 through 125 be located?

12. Periodic table. Use Table 7.3 (page 107) to predict the electronic configurations of (a) $_{82}$Pb; (b) $_{77}$Ir; (c) $_{67}$Ho; (d) $_{99}$Es. Compare your predictions with the information given in Appendix III.

13. Periodicity of valence. For each of the compounds in the upper section of Table 7.5 (page 110) state whether the valence of the element that corresponds to the listed group number is G or $8 - G$.

14. Periodicity of valence. For each of the following compounds, state whether the valence of the underlined representative element corresponds to G, to $8 - G$, or to neither one. G is the periodic group number. Assume for valences of elements not underlined that H, F, Cl, Br, I, (NO$_3$) = 1; O, S, Se, Te, (SO$_4$) = 2; N, P, (PO$_4$) = 3. (a) $\underline{\text{Al}}$PO$_4$; (b) $\underline{\text{Ga}}_2$S$_3$; (c) $\underline{\text{Ge}}$H$_4$; (d) $\underline{\text{Ge}}$Cl$_2$; (e) $\underline{\text{P}}_2$I$_4$; (f) Pb$\underline{\text{N}}_6$; (g) $\underline{\text{B}}_2$H$_6$; (h) $\underline{\text{I}}_2$O$_5$; (i) $\underline{\text{Fr}}$F; (j) $\underline{\text{Sb}}_2$Te$_3$; (k) $\underline{\text{Si}}$C; (l) $\underline{\text{Rb}}_2$S; (m) $\underline{\text{Al}}$N; (n) Be$_3$($\underline{\text{P}}$O$_4$)$_2$; (o) $\underline{\text{B}}_4$H$_{10}$; (p) Na$\underline{\text{H}}$S; (q) $\underline{\text{Mg}}$S; (r) $\underline{\text{As}}$F$_5$; (s) $\underline{\text{Pb}}_3$O$_4$; (t) $\underline{\text{Rb}}$NO$_3$; (u) $\underline{\text{Tl}}$NO$_3$.

15. Size. From the ionic radii (Table 7.6, page 112) calculate the K$^+$—Br$^-$ distance in a potassium bromide crystal, and the Mg^{2+}—O^{2-} distance in a magnesium oxide crystal. The accepted experimental values are 3.293 and 2.104 Å, respectively.

16. Radii. Explain the following data: the decrease in the radius for the elements $_{11}$Na to $_{17}$Cl is 0.55 Å while the decrease for $_{21}$Sc to $_{30}$Zn is only 0.19 Å.

17. Radii. Note the covalent and ionic radii of the atoms from $_{15}$P to $_{21}$Sc as given in Table 7.6 (page 112). (a) With what neutral atom are all these ions isoelectronic? (b) Explain the variation in atomic size, and the relation between the size of each ion and of its parent atom.

18. Ionic radii. Account for the very large difference between the ionic radii of Sb and Bi as shown in Table 7.6 (page 112).

19. Isoelectronic particles. What would have to be the charge and electronic configuration of a sodium ion for it to be isoelectronic with Li^+? What can you say about the size of this hypothetical sodium ion compared to Li^+ and to Na?

20. Ionization energy. (a) Plot the ionization energy (y axis) for the alkali metals as a function of the period number (x axis). (b) Each transition and inner transition series may be considered a family in the sense we consider, for example, the halogens to be a family. What prediction can you make about the shape of the graph of ionization energy versus atomic number for the transition and inner transition series?

21. Ionization energy. Plot the first ionization energy (y axis) versus atomic number (x axis) for period 2 and for the transition elements Sc to Cu on the same scale. See Table 7.7, page 114, for data. What conclusion(s) could you draw from these graphs?

22. Ionization energy. Should you use the electron in the energy level $n = 1$, $n = 2$, or $n = 3$ (page 87) to predict the ionization energy of hydrogen? Calculate the ionization energy, and compare with the experimental value given in Table 7.7, page 114.

23. Ionization energy. Predict a value for the ionization energy of francium.

24. Promotion energy. Helium atoms can be promoted to an excited state whose energy is 19.6 eV/atom. This excited helium can ionize other gases by transferring its excess energy to them. Which element(s) shown in Table 7.7 *cannot* be ionized by excited helium?

8

TYPES OF CHEMICAL BONDS

Under terrestrial conditions most elements rarely exist as isolated atoms. The atoms of most known elements are chemically bonded to other atoms. For example, oxygen, nitrogen, hydrogen, and the halogens are diatomic molecules. Yellow sulfur and white phosphorus exist as molecules whose formulas are S_8 and P_4, respectively. The molecules of diamond or graphite (both forms of carbon) and of red phosphorus consist of many millions of atoms. Metallic elements, too, such as copper and potassium, are composed of bonded atoms, generally in a crystalline form.

Of course, some elements, notably the noble gases such as helium and argon, exist as unbonded atoms. To break the bonds and free the atoms of other substances, however, requires large expenditures of energy. At temperatures above 5000°C, most matter is in the gaseous, monatomic state.

How do atoms combine and what are the forces that bind them? These questions are fundamental in the study of chemistry since chemical change is essentially an alteration of chemical bonds. Of the three types of attractive force—gravitational, magnetic, and electrostatic—only the last is strong enough to account for observed bond energies. An important clue to the understanding of the driving force for chemical bonding was the discovery of the noble gases and their apparently inert chemical behavior. The relationship between the bonding of atoms and the electronic configuration of noble gases was proposed in 1916 independently by Walther Kossel and Gilbert N. Lewis, and extended in 1919 by Irving Langmuir. It was suggested that *atoms interact by changing the number of their electrons so as to acquire the electronic structure of a noble gas.* With the exception of helium, which has a $1s^2$ electronic configuration, each noble gas has 8 electrons with an s^2p^6 distribution in its highest principal energy level (page 94). The need for 8 electrons gives the name OCTET RULE to this concept. There are, however, *many exceptions to the octet rule* — even compounds of the noble gases have been synthesized.

LEWIS SYMBOLS / 8.2

In the following discussion, Lewis symbols are used to represent the atoms. Such symbols comprise the letter designation of the element, and dots to represent electrons in the highest principal energy level. They are the electrons usually involved in bonding. For the atoms of the representative elements, the number of electrons equals the periodic group number. Lewis symbols for the first ten elements and some of their ions are shown. Note that O^{2-} and F^- have the same electronic structure as Ne; these species are said to be ISOELECTRONIC.

1	2	3	4	5	6	7	8
H· H⁺ H:⁻							He:
Li· Li⁺	·Be· Be²⁺	·B·	·C·	·N· :N:³⁻	:O· :O:²⁻	:F· :F:⁻	:Ne:

This symbolism is very useful for the representative elements, but difficulties arise when Lewis symbols are used for the transition elements. Bonding of the transition elements can involve electrons in more than one principal energy level.

IONIC BOND / 8.3

Let us now consider the kind of bond that results when potassium bromide is formed from its elements. Were a potassium atom, K· (atomic number 19), to lose 1 electron thereby becoming a potassium ion, K⁺, it would have the same electron configuration as argon (atomic number 18), a noble gas. Were a bromine atom, :Br· (atomic number 35), to gain 1 electron thereby becoming a bromide ion, :Br:⁻, it would have the same electron configuration as krypton (atomic number 36), a noble gas. The potassium bromide, K⁺Br⁻, formed by such transfer of electrons, is said to have an IONIC BOND.

When atoms react by electron transfer, the number of electrons gained and lost must be equal; the resulting compound is neutral. Illustrations of this principle are given below:

$$\overset{\cdot}{Ca}\cdot\ +\ 2H\cdot\ \longrightarrow\ Ca^{2+}\ +\ 2H\!:^{-} \qquad (CaH_2)\ \text{calcium hydride}$$

$$2Li\cdot\ +\ \cdot\overset{\cdot\cdot}{O}\!:\ \longrightarrow\ 2Li^{+}\ +\ :\overset{\cdot\cdot}{O}\!:^{2-} \qquad (Li_2O)\ \text{lithium oxide}$$

$$\cdot\overset{\cdot}{Al}\cdot\ +\ 3\cdot\overset{\cdot\cdot}{F}\!:\ \longrightarrow\ Al^{3+}\ +\ 3:\overset{\cdot\cdot}{F}\!:^{-} \qquad (AlF_3)\ \text{aluminum fluoride}$$

$$3\overset{\cdot}{Mg}\cdot\ +\ 2\cdot\overset{\cdot}{N}\!:\ \longrightarrow\ 3Mg^{2+}\ +\ 2:\overset{\cdot\cdot}{N}\!:^{3-} \qquad (Mg_3N_2)\ \text{magnesium nitride}$$

The number of electrons lost or gained by an atom in forming an ionic bond is equal to its VALENCE. Atoms that readily lose electrons are said to be ELECTROPOSITIVE, those that readily gain electrons to be ELECTRONEGATIVE. The loss of electrons is called OXIDATION,* and the gain of electrons is called REDUCTION. By definition, the formation of an ionic bond from elements must involve an oxidation–reduction (REDOX) reaction. The more electropositive element is oxidized, and the more electronegative one is reduced.

Formation of an ionic bond most often occurs when a metallic element with a low ionization energy (page 113) reacts with a nonmetallic element of high electron affinity (page 115).

FORMATION OF IONIC SOLIDS / 8.4

The formation of a solid ionic compound such as KBr from its elements is exothermic. Does this mean that the transfer of an electron from K to Br is exothermic? Electron transfer can be represented by two steps:

$$K\cdot\ (g) \longrightarrow K^{+}(g) + e^{-} \qquad \Delta H = +100.0\ \text{kcal/mole (ionization energy)}$$
$$:\overset{\cdot\cdot}{Br}\!\cdot (g) + e^{-} \longrightarrow :\overset{\cdot\cdot}{Br}\!:^{-}(g) \qquad \Delta H = -81.6\ \text{kcal/mole (electron affinity)}$$

$$K\cdot\ (g) + :\overset{\cdot\cdot}{Br}\!\cdot(g) \longrightarrow K^{+}(g) + :\overset{\cdot\cdot}{Br}\!:^{-}(g) \qquad \Delta H = +18.4\ \text{kcal/mole}$$

The gain of an electron by a Br atom is exothermic. However, the electron loss by K is endothermic. The sum of the two steps is endothermic, not exothermic, as we might have expected. The reason is that we have not represented the reaction completely. Potassium and bromine consist of bonded rather than individual atoms; also potassium bromide is not a collection of isolated ions. The actual reaction is

$$K(c) + \tfrac{1}{2}Br_2(l) \longrightarrow KBr(c) \qquad \Delta H = -93.7\ \text{kcal/mole (heat of formation)}$$

Somewhere in this reaction the solid potassium crystal lattice must be destroyed and the bromine molecules must be dissociated and a

---

* Oxidation originally referred to the addition of oxygen to elements. The broader concept of electron transfer was first used by Wilhelm Ostwald in 1903.

solid, crystalline KBr salt must be produced. To understand why this reaction is so exothermic we need not know how or in what sequence these events occur because the heat of formation is independent of the sequence. We just assume reasonable steps with measurable ΔH values, and according to the Hess' law, these individual ΔH values, when totaled, should equal the heat of formation.

		ΔH in kcal
Vaporization of liquid Br_2	$\frac{1}{2}Br_2(l) \longrightarrow \frac{1}{2}Br_2(g)$	$+$ 3.6 ($\frac{1}{2}$ heat of vaporization)
Dissociation of $Br_2(g)$	$\frac{1}{2}Br_2(g) \longrightarrow Br(g)$	$+$ 23.1 ($\frac{1}{2}$ bond energy)
Conversion of $K(c)$ to individual atoms	$K(c) \longrightarrow K(g)$	$+$ 21.5 (heat of sublimation)
Ionization of K atoms	$K(g) \longrightarrow K^+(g) + e^-$	$+100.0$ (ionization energy)
Formation of Br^-	$Br(g) + e^- \longrightarrow Br^-(g)$	$-$ 81.6 (electron affinity)

The sum of these ΔH values, which equals ΔH for the conversion of liquid bromine and crystalline potassium, is even more positive:

$$K(c) + \tfrac{1}{2}Br_2(l) \longrightarrow K^+(g) + Br^-(g) \qquad \Delta H = +66.6 \text{ kcal}$$

However, to this net equation we must add yet one more equation, the one for the combination of the individual gaseous K^+ and Br^- ions to form the solid crystalline KBr:

$$K^+(g) + Br^-(g) \longrightarrow KBr(c) \qquad \Delta H = -159.9 \text{ kcal}$$

$$\text{(crystal lattice energy)}$$

When this last addition is made, we get the overall equation for the reaction and the ΔH, calculated by Hess' law, is $+66.6 - 159.9 = -93.3$, which is close to the observed value of -93.7 kcal. This thermochemical analysis makes clear that the exothermicity of the overall reaction comes from the formation of crystalline KBr. The energy released in this last reaction, called the CRYSTAL LATTICE ENERGY, arises because the ions in the solid are closely arranged so that there is a strong net electrostatic attraction between ions of opposite charge.

COVALENT BOND; COVALENCY / 8.5

We have just seen that electron transfer leading to an ionic bond most often occurs when one atom gains and the other loses electrons, and both resulting ions are isoelectronic with noble gases.

What happens when two atoms, both of which require electrons to become isoelectronic with noble gases, react? The simplest example of such a situation is the combination of 2 H· atoms to form an H_2 molecule. Each H· atom needs 1 electron to become isoelectronic with He:, so that a transfer of electrons cannot satisfy the requirements of both. Instead, the hydrogen atoms *mutually share their electrons*. The shared pair is said to "belong" to both; each hydrogen atom can be considered to have gained an electron, and to have acquired the helium structure:

$$H\cdot + H\cdot \longrightarrow H\!:\!H \qquad \Delta H = -104 \text{ kcal}$$

Such a sharing of a pair of electrons results in a COVALENT BOND, the presence of which is usually depicted by a dash, H—H, or a pair of dots, H:H. Once the covalent bond has formed, the two bonding electrons are attracted by *two nuclei instead of one*, and the bonded state is therefore more stable than the nonbonded state.

The number of covalent bonds formed by an atom is termed its COVALENCY. The covalency of an atom is equal to the number of electrons the atom needs to become isoelectronic with a noble gas. Some of the more common elements have the following covalencies when they follow the octet rule and also have *no charge*: hydrogen and the halogens, 1; oxygen and sulfur, 2; nitrogen and phosphorus, 3; carbon and silicon, 4. LEWIS ELECTRONIC FORMULAS,* in which bonds and unshared electrons are shown, are given for a few typical compounds of these elements:

ammonia	phosphorus trichloride	silicon tetrafluoride	hydrogen sulfide

Remembering the covalencies of these common elements greatly simplifies the writing of a structural or an electronic formula.

Example 1 Write the electronic formula for CH_4O which satisfies the covalency requirements and the octet rule (duet rule for hydrogen).

Answer The covalency requirements of the atoms in this molecule are $C = 4$, $O = 2$, and $H = 1$. The correct structure is

$$H\!-\!\overset{\displaystyle H}{\underset{\displaystyle H}{\overset{|}{\underset{|}{C}}}}\!-\!\ddot{\overset{..}{O}}\!-\!H$$

methyl alcohol

* The student need not be concerned here with angles between atoms.

Below are given three incorrect electronic formulas for CH_4O, and the errors are indicated:

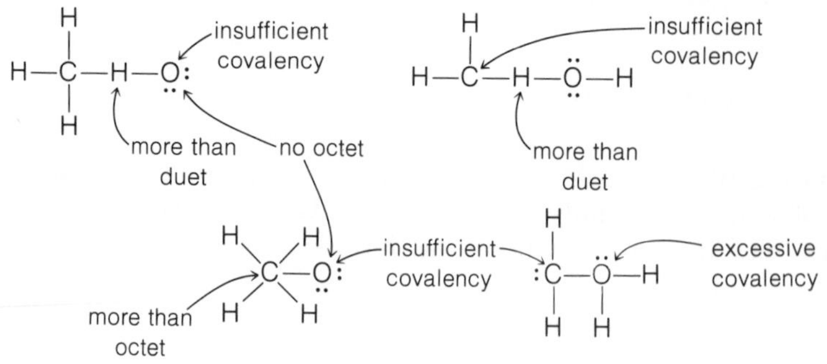

MULTIPLE BONDS / 8.6

To satisfy the octet rule and their covalency requirements, two atoms must often share more than one pair of electrons. This leads to the assignment of MULTIPLE BONDS. The sharing of two pairs of electrons leads to a DOUBLE BOND; the sharing of three pairs of electrons leads to a TRIPLE BOND. Examples of multiple-bonded molecules are phosgene, $COCl_2$, which contains a double bond; N_2 and HCN, each of which contains a triple bond; and CO_2, in which carbon has two double bonds:

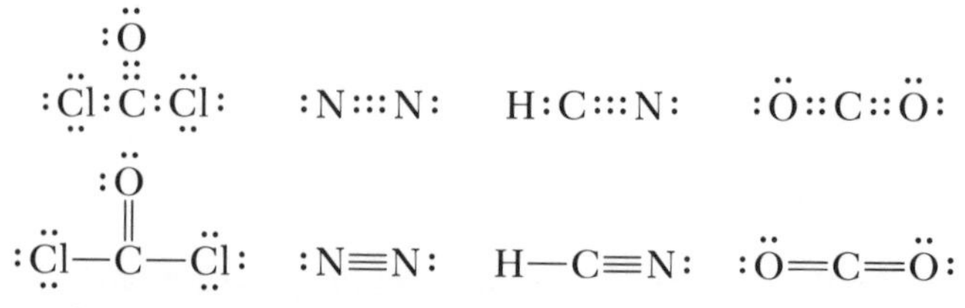

| phosgene | nitrogen | hydrogen cyanide | carbon dioxide |

In general, within a given group in the periodic table, the ability to form multiple bonds, *while maintaining an octet of electrons,* diminishes with increasing size of the atom. With few exceptions, such as sulfur, only atoms in the second period, especially C, N, and O, can have a multiple bond and an octet of electrons; for example,

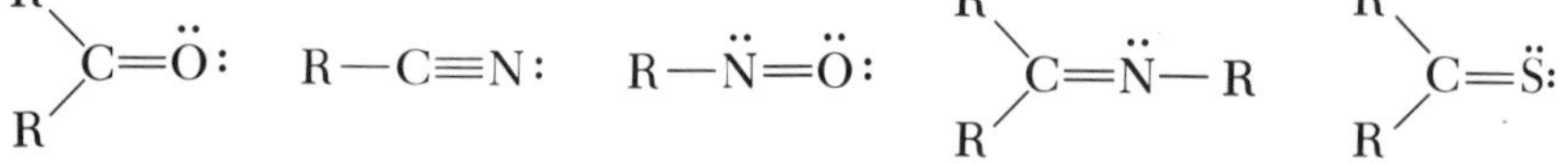

R = another atom or group

The number of compounds in which sulfur forms multiple bonds are few, and the bonds are much weaker than their oxygen analogs.

Let us consider what effects on the structure of molecules may result from the tendency or the disinclination of atoms to undergo multiple bonding. Compare the oxides of empirical formulas CO_2 and SiO_2. Both C and Si have a covalency of 4, but C forms multiple bonds in CO_2 and Si is singly bonded in SiO_2. For CO_2, we can write :Ö=C=Ö:, corresponding to the molecular formula CO_2, identical with the empirical formula. Now consider SiO_2. In the absence of multiple bonds, the covalency of 4 for Si is satisfied by writing

$$
\begin{array}{c}
| \\
O \\
| \\
-O-Si-O- \\
| \\
O \\
|
\end{array}
$$

But the covalency of 2 for O is satisfied by writing

$$
\begin{array}{c}
| \qquad | \\
-Si-O-Si- \\
| \qquad |
\end{array}
$$

If we start to write a structural formula to satisfy both these covalency requirements, we write indefinitely; there is no end to the "molecule." Silicon dioxide, a component of sand and rocks, is then actually a three-dimensional network of bonded atoms with no identifiable individual molecules. "SiO_2" is an empirical formula, not a molecular one. Figure 8.1 (page 128) shows a model of the SiO_2 lattice.

We also find that while molecular nitrogen possesses a triple bond, each phosphorus atom in a molecule of white phosphorus, P_4 (Fig. 21.2), has an unshared pair of electrons and forms single covalent bonds with the three other P atoms.

EXCEPTIONS TO THE OCTET RULE / 8.7

Although the octet rule is a useful generalization, it is not wise to be too dogmatic in its application. Some molecules and ions have atoms with less than an octet, while others have atoms with more than an octet.

(a) *Atoms with less than an octet of electrons.* When an atom with fewer than 4 electrons shares them to form covalent bonds, it may have less than an octet of electrons. Boron, ·Ḃ· , the first member of group 3R, and beryllium, ·Be· , the first member of group 2R, are typical examples. Boron and fluorine form the covalent compound boron tri-fluoride, BF_3 (b.p. $= -101°C$),

$$\cdot \overset{.}{B} \cdot + 3 : \overset{..}{\underset{..}{F}} \cdot \longrightarrow : \overset{..}{\underset{..}{F}} : \overset{:\overset{..}{F}:}{\underset{}{B}} : \overset{..}{\underset{..}{F}} :$$

in which boron is surrounded by only 6 electrons.

When beryllium reacts with chlorine it forms a covalent compound, $BeCl_2$, in which Be has only 4 electrons:

$$\cdot Be \cdot + 2 : \overset{..}{\underset{..}{Cl}} \cdot \longrightarrow : \overset{..}{\underset{..}{Cl}} : Be : \overset{..}{\underset{..}{Cl}} :$$

(b) *Free radicals.* Nitric oxide, NO, is another example of a molecule possessing an atom with less than an octet. A possible electronic structure of the molecule

$$: \overset{.}{N} :: \overset{..}{\underset{..}{O}} :$$

shows the oxygen atom surrounded by 8 and the nitrogen atom surrounded by 7 electrons. Molecules in which 1 or more electrons are unpaired are known as FREE RADICALS. Free radicals are characterized by two properties: they are paramagnetic and often colored. Thus, nitrogen dioxide

$$: \overset{..}{\underset{..}{O}} : \overset{.}{N} :: \overset{..}{\underset{..}{O}} :$$

is a brown gas, and chlorine dioxide

$$: \overset{..}{\underset{..}{O}} : \overset{.}{\underset{..}{Cl}} : \overset{..}{\underset{..}{O}} :$$

is a yellow gas; NO is colorless in the gaseous state, but is blue in the liquid state.

Although an electronic formula consistent with the octet rule can be written for O_2, experiment shows that O_2 has a permanent magnetic moment corresponding to 2 unpaired electrons. Consequently, molecular oxygen must be a free radical with an unpaired electron on each atom,

$$: \overset{..}{\underset{..}{O}} :: \overset{..}{\underset{..}{O}} : \qquad\qquad : \overset{..}{\underset{.}{O}} : \overset{..}{\underset{.}{O}} :$$

a possible structure   a free radical structure
for O_2 for O_2

We shall see (page 366) that neither of these structures accurately represents the O_2 molecule.

Two free radical molecules may form a covalent bond by pairing the unpaired electrons, a process known as DIMERIZATION:

$$H_3C\cdot + \cdot CH_3 \longrightarrow H_3C:CH_3$$
dimerization of free radicals

Most free radicals tend to dimerize rather than remain in the odd electron state; NO, NO_2, O_2, and ClO_2 are among the few exceptions. At low temperatures, even some of these dimerize as shown for NO_2,

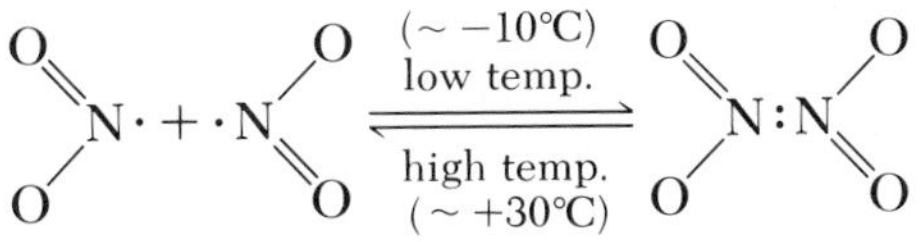

nitrogen dioxide
(brown)

dinitrogen
tetroxide
(colorless)

The dimerization is reversed when the temperature is increased.

(c) *Atoms with more than an octet. The electron octet is not exceeded by atoms in the second period of the periodic table.* No stable molecules are yet known in which the atoms Li, Be, B, C, N, O, and F are surrounded by more than 8 electrons. It is possible, however, for certain atoms in the higher periods to be surrounded by more than 8 electrons. Stable molecules and ions such as SiF_6^{2-}, PCl_5, SF_6, and ICl_3 are known. In these molecules the P and I atoms are surrounded by 10 electrons and the Si and S atoms by 12 electrons:

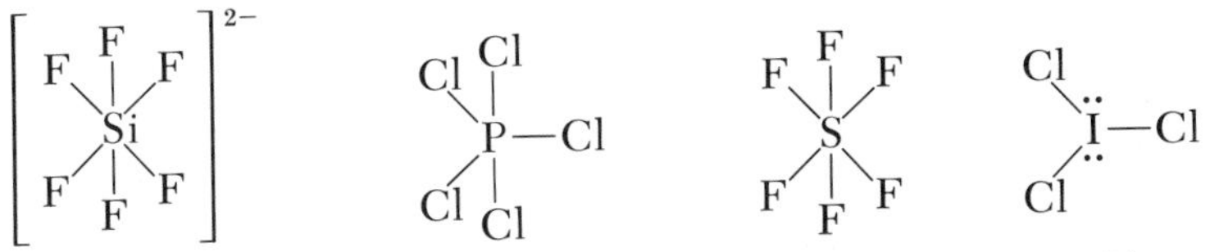

SiF_6^{2-}

silicon hexafluoride
anion

PCl_5

phosphorus
pentachloride

SF_6

sulfur
hexafluoride

ICl_3

iodine
trichloride

(d) *Ions of transition elements and ions of certain representative elements.* The ions of many transition elements are not isoelectronic with noble gases. Table 8.1 lists some typical transition elements and their ions, and the configuration of the nearest noble gas. Scandium (Sc), the first of the transition elements in period 4, loses 3 electrons to give the ion Sc^{3+}, which is isoelectronic with argon. The other elements, however, form stable ions not isoelectronic with argon or with krypton.

Table 8.1
Electronic Configuration of Typical Transition Elements and Ions

Element	Ion	Nearest noble gas
Sc (Ne core, $3s^23p^63d^14s^2$)	Sc^{3+} (Ne core, $3s^23p^6$)	Ar (Ne core, $3s^23p^6$)
Cr (Ne core, $3s^23p^63d^54s^1$)	Cr^{3+} (Ne core, $3s^23p^63d^3$)	Ar (Ne core, $3s^23p^6$)
Mn (Ne core, $3s^23p^63d^54s^2$)	Mn^{2+} (Ne core, $3s^23p^63d^5$)	Ar (Ne core, $3s^23p^6$)
Fe (Ne core, $3s^23p^63d^64s^2$)	Fe^{3+} (Ne core, $3s^23p^63d^5$)	Ar (Ne core, $3s^23p^6$)
Cu (Ne core, $3s^23p^63d^{10}4s^1$)	Cu^{1+} (Ne core, $3s^23p^63d^{10}$)	Kr (Ar core, $3d^{10}4s^24p^6$)
Zn (Ne core, $3s^23p^63d^{10}4s^2$)	Zn^{2+} (Ne core, $3s^23p^63d^{10}$)	Kr (Ar core, $3d^{10}4s^24p^6$)

Several of the heavy atoms in groups 3R to 5R of the periodic table form ions that do not have a noble gas configuration. A few examples are shown:

Element	Group	Ion	Typical salt
Tin (KLM $4s^24p^64d^{10}5s^25p^2$)	4R	Sn^{2+} (KLM $4s^24p^64d^{10}5s^2$)	SnF_2
Antimony (KLM $4s^24p^64d^{10}5s^25p^3$)	5R	Sb^{3+} (isoelectronic with Sn^{2+})	$Sb_2(SO_4)_3$
Thallium ($KLMN$ $5s^25p^65d^{10}6s^26p^1$)	3R	Tl^+ ($KLMN$ $5s^25p^65d^{10}6s^2$)	TlF
Lead ($KLMN$ $5s^25p^65d^{10}6s^26p^2$)	4R	Pb^{2+} (isoelectronic with Tl^+)	PbF_2
Bismuth ($KLMN$ $5s^25p^65d^{10}6s^26p^3$)	5R	Bi^{3+} (isoelectronic with Tl^+)	BiF_3

PROPERTIES OF IONIC AND COVALENT COMPOUNDS / 8.8

By what criteria are compounds classified as ionic or covalent? The distinction is often made on the basis of physical properties. In order to discuss the physical properties of *covalent* substances, it is necessary to classify them into two types:

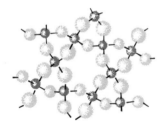

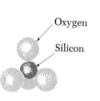

Oxygen

Silicon

Fig. 8.1. *Structure of silicon dioxide. Si = dark blue; O = light blue.*

(a) The **MOLECULAR COVALENT** type, exemplified by CO_2, I_2, P_4, and Cl_2, consists of distinguishable molecules. The atoms in the molecules are held together by strong covalent bonds, but *the forces between the molecules themselves are very weak*. As a result of weak intermolecular forces, the molecules are easily separated from each other. Hence, molecular covalent compounds are often gases, liquids, or solids that melt, boil, or vaporize at comparatively low temperatures. It is rare to find a molecular compound that melts above 300° or boils above 600°C.

(b) The **NETWORK-TYPE** species is exemplified by silicon dioxide (Fig. 8.1) and diamond (Fig. 8.2). These substances are extended, three-dimensional aggregates of covalently bonded atoms. The attractive forces between the atoms are very strong. Network compounds are, therefore, invariably solids with very high melting and boiling points. For example, silicon dioxide (quartz) melts at about 1710° and boils at 2230°C, and diamond melts at 3500° and boils at 4200°C.

The property that most clearly identifies an IONIC compound is the *ability to conduct an electric current only when liquefied.* The compound must be molten or dissolved so that the ions are free to move and carry the charge. Ionic compounds usually have high melting points and boiling points, both above about 500°C. These properties are a result of the large amount of energy that must be supplied to overcome the strong interionic attractive force. However, the melting and boiling points of ionic compounds are not so high as those of network type covalent solids. For example, NaCl melts at 800.4° and boils at 1413°C, and NaOH melts at 318.4° and boils at 1390°C. Table 8.2 summarizes the property differences. For comparison, the metallic type (page 367) is included.

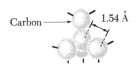

Carbon   1.54 Å

Fig. 8.2. *Structure of diamond.*

Table 8.2
A Comparison of Distinguishable Properties of the Main Types of Substances

	Ionic	Covalent	Network type	Metallic
Electrical conductivity				
Solid	No[a]	No	No	Yes (high)
Liquid	Yes	No[a]	No	Yes (high)
Approximate range of boiling points	700° to 3500°C	−250° to +600°C	2000° to 6000°C	650° to 6000°C

[a] A few exceptions exist.

POLAR COVALENT BONDS; ELECTRONEGATIVITY / 8.9

Equal sharing of a pair of electrons occurs in molecules of diatomic elements such as H:H, and between identical atoms with identical neighbors, as illustrated by the two C atoms in ethane, $H_3C—CH_3$. But if the two bonded atoms are dissimilar, as in H:Cl, or are identical but not in identical surroundings, as the two C atoms in $H_3C—CCl_3$, the sharing is unequal. One atom is likely to attract electrons more strongly than the other. The atom which more strongly attracts electrons develops some negative charge; the other atom develops some positive charge. These charges have *less* than unit +1 or −1 values and are designated as *partial charges* δ^+ and δ^-. For example, a Cl atom is more electron attracting than an H atom; hydrogen chloride is depicted, therefore, as

$$\overset{\delta^+}{H}—\overset{\delta^-}{Cl}$$

Such covalent bonds are said to be POLAR, as distinguished from the Cl—Cl or H—H bonds, which are called NONPOLAR. The most extreme case of unequal sharing of a pair of electrons is the ionic bond in such compounds as CsF, NaCl, or CaF_2. According to our simplified picture of ionic bonding, one atom has usurped the pair, and the partial charges approach integral values of electronic charge. Actually, even these bonds are intermediate between ionic and covalent. Thus, the *nonpolar covalent bond and the ionic bond are extremes for the distribution of a pair of electrons between two nuclei.* Between these extremes are the many polar covalent bonds.

The relative tendency of a bonded atom in a molecule to attract electrons is expressed by the term ELECTRONEGATIVITY. This word does not mean the actual content of negative charge, just the tendency to acquire it. Thus, F is highly electronegative; F^- is not. Electronegativity values are not measured directly; they are derived from other data and from various sets of assumptions on which there is not unanimous agreement among chemists. The values given in Table 8.3 are derived from bond-energy data. Note that *electronegativities increase from left to right in a period* and, with a number of exceptions, *from the bottom to the top of a group.* Note also that fluorine and oxygen are the most electronegative elements.

Table 8.3
Electronegativities of the Representative Elements[a]

$_1$H 2.1							
$_3$Li 1.0	$_4$Be 1.5	$_5$B 2.0	$_6$C 2.5	$_7$N 3.0	$_8$O 3.5	$_9$F 4.0	
$_{11}$Na 0.9	$_{12}$Mg 1.2	$_{13}$Al 1.5	$_{14}$Si 1.9	$_{15}$P 2.1	$_{16}$S 2.5	$_{17}$Cl 3.0	
$_{19}$K 0.8	$_{20}$Ca 1.0	$_{31}$Ga 1.8	$_{32}$Ge 2.0	$_{33}$As 2.0	$_{34}$Se 2.4	$_{35}$Br 2.8	
$_{37}$Rb 0.8	$_{38}$Sr 1.0	$_{49}$In 1.5	$_{50}$Sn 1.7	$_{51}$Sb 1.9	$_{52}$Te 2.1	$_{53}$I 2.4	
$_{55}$Cs 0.7	$_{56}$Ba 0.9	$_{81}$Tl 1.4	$_{82}$Pb 1.6	$_{83}$Bi 1.9	$_{84}$Po	$_{85}$At	
$_{87}$Fr	$_{88}$Ra						

[a] The values for noble gases are too uncertain and therefore are not included.

We have seen that certain covalent bonds are polar. Because of the polarity of individual bonds, the entire molecule *may* have separated centers of positive and negative charge. Such a molecule constitutes a dipole. A crude illustration of a polar molecule is

The dipole is symbolized by ↦, where the arrow points toward the negative pole. Polar molecules possess a dipole moment, μ, which is the product of the magnitude of partial charge, q, and the distance, d, between the centers of opposite charge (Appendix I.12):

$$\mu = q \times d$$

The unit for μ is the *debye* (D) named in honor of Peter Debye, who pioneered in this field of study.

As a consequence of the dipole moment, polar molecules tend to be oriented in an electric field with the positive ends directed toward the negative electric plate and the negative ends toward the positive plate (Fig. 8.3). The orientations are far from perfect because of the randomness imposed on them by the kinetic energy of the molecules.

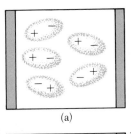

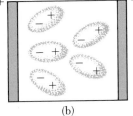

Fig. 8.3. *Orientation of polar molecules in an electric field; (a) field off, (b) field on.*

FORMAL CHARGE AND OXIDATION NUMBER / 8.11

In this section we discuss two conventions for assigning charges to atoms. Although these charges are artificial, they are useful in understanding the reactivity of covalent molecules. These conventions are based on how we view the sharing of the pair of electrons in covalent bonds.

Formal charge. For formal charge we assume all electrons are equally shared regardless of bond polarity, and we assign one of a pair of shared electrons to each atom. Formal charge is then defined as

$$\left(\begin{array}{c}\text{number of outer electrons}\\\text{in lone atom}\end{array}\right) - \left(\begin{array}{c}\text{number of assigned}\\\text{outer electrons}\end{array}\right) = \text{formal charge}$$

In this and in all following definitions and discussions we are concerned only with electrons in the highest principal energy level; that is, those appearing in the Lewis symbol. Consider the following sequence:

$$\cdot\ddot{\text{N}}\cdot \quad \xrightarrow{+3\text{H}\cdot} \quad \text{H}:\overset{\displaystyle \text{H}}{\underset{\displaystyle \ \ }{\ddot{\text{N}}}}:\text{H} \quad \xrightarrow{+\text{H}^+} \quad \left[\ \text{H}:\overset{\displaystyle \text{H}}{\underset{\displaystyle \text{H}}{\ddot{\text{N}}}}:\text{H}\ \right]^+$$

lone	lone	ammonia	hydrogen ion	ammonium ion
N atom	H atoms	molecule	(a proton)	

In $:NH_3$, the N atom has an octet of electrons, but 6 of these are shared with H atoms. We assign 3 of these 6 to the N atom which now has 5 electrons since we also count the unshared pair of electrons:

$$2 \text{ unshared electrons} + [\tfrac{1}{2}(6 \text{ shared electrons})] = 5 \text{ assigned electrons.}$$

Five is the same number of electrons as is present in the lone N atom. The formal charge on N in NH_3 is $5 - 5 = 0$.

The same calculation for N in NH_4^+ leads to:

$$(\text{number of assigned electrons}) = \tfrac{1}{2}(8 \text{ shared electrons}) = 4$$

The formal charge is $5 - 4 = +1$. Since one-half the number of shared electrons equals the number of covalent bonds we can summarize as follows:

$$\text{formal charge} = \begin{bmatrix} \text{number of electrons} \\ \text{in lone atom } (G) \end{bmatrix} - \begin{bmatrix} \text{unshared} \\ \text{electrons } (U) \end{bmatrix} + \begin{bmatrix} \text{number of} \\ \text{covalent bonds } (C) \end{bmatrix}$$

Note also that for representative elements the number of electrons in the lone atom is the periodic group number of the atom.

Example 2 Calculate the formal charges of the atoms in nitric acid.

Answer

$$G - (U + C)$$
formal charge of $\;$ H $= 1 - (0 + 1) = 0$
formal charge of αO $= 6 - (4 + 2) = 0$
formal charge of βO $= 6 - (6 + 1) = -1$
formal charge of γO $= 6 - (4 + 2) = 0$
formal charge of $\;$ N $= 5 - (0 + 4) = +1$
$$\text{Sum} = \overline{0}$$

The N atom in nitric acid has four covalent bonds, thereby exceeding its normal covalency of three. The N atom also bears a formal charge. These circumstances are related. Invariably, *an atom with an octet of electrons will have a formal charge when not exhibiting its normal covalency.* (See Problem 8, page 143.)

The sum of the formal charges equals zero for a molecule, or the ionic charge for an ion.

Example 3 Which of the following structures are ions? Determine their charges.

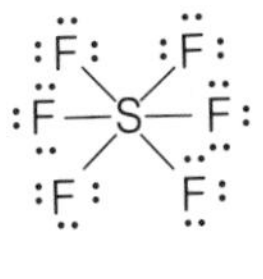

(a)

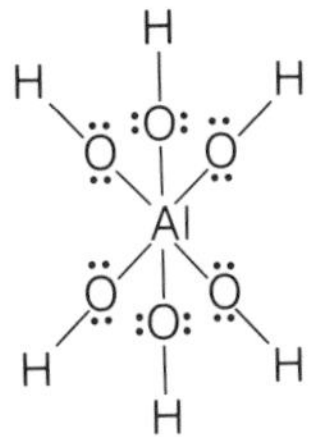

(b)

Answer

(a)

$$G - (U + C)$$

formal charge of $S = 6 - (0 + 6) = 0$
formal charge of $F = 7 - (6 + 1) = 0$; SF_6 is a molecule.

(b)

formal charge of $Al = 3 - (0 + 6) = -3$
formal charge of $O = 6 - (4 + 2) = 0$
formal charge of $H = 1 - (0 + 1) = 0$; $Al(OH)_6$ is an ion; charge $= -3$.

Oxidation number. Consider the combination

$$H \cdot + \cdot \ddot{\underset{..}{Cl}} : \longrightarrow H : \ddot{\underset{..}{Cl}} :$$

The assignment of formal charges (zero for each atom), which derives from the assumption that the bonding electrons are equally shared, *ignores the polar quality of the bond.* If we arbitrarily *assign both electrons to the more electronegative atom*, Cl, then we consider Cl to have 8 electrons and hydrogen none, as though HCl consisted of separate ions. Charges assigned on such a basis are called **OXIDATION NUMBERS**, and are said to represent the **OXIDATION STATES** of the element. They are calculated as follows:

$$\begin{Bmatrix} \text{oxidation} \\ \text{number} \end{Bmatrix} = \begin{Bmatrix} \text{number of} \\ \text{electrons in} \\ \text{lone atom } (G) \end{Bmatrix} - \begin{Bmatrix} \text{number of} \\ \text{electrons assigned to} \\ \text{atom on the basis} \\ \text{that all the bonded} \\ \text{electrons belong to} \\ \text{the more electroneg-} \\ \text{ative atom } (A) \end{Bmatrix} + \begin{Bmatrix} \text{unshared} \\ \text{electrons } (U) \end{Bmatrix} \quad (4)$$

Example 4 Calculate the oxidation numbers of the atoms in nitric acid (see Example 2).

Answer Since O is more electronegative than N or H, we have

$$G - (A + U)$$

oxidation number of $\,$ H $= 1 - (0 + 0) = +1$
oxidation number of αO $= 6 - (4 + 4) = -2$
oxidation number of βO $= 6 - (2 + 6) = -2$
oxidation number of γO $= 6 - (4 + 4) = -2$
oxidation number of $\,$ N $= 5 - (0 + 0) = \underline{+5}$
$$\text{Sum} = \;\; 0$$

Example 5 Calculate the oxidation number of each atom in the thiosulfate ion, $S_2O_3{}^{2-}$, whose electronic formula is

Answer Oxygen is more electronegative than sulfur, so all electrons shared by these atoms are assigned to oxygen. Since the S atoms have the same electronegativity, the electrons in the S—S bond are presumed to be equally shared, and 1 electron is assigned to each S atom. Thus, we have

$$G - (A + U)$$

oxidation number of central S $= 6 - (1 + 0) = +5$
oxidation number of outer S $\;= 6 - (1 + 6) = -1$
oxidation number of each O $\;= 6 - (2 + 6) = -2$

Note that the sum of the oxidation numbers of all the atoms equals the charge on the ion, $3(-2) + 5 + (-1) = -2$.

This method of calculating oxidation numbers requires a prior knowledge of Lewis electronic formulas. Such formulas may be difficult to write for molecules and ions containing transition elements. Fortunately, rules derived from the method just discussed can be used to get oxidation numbers directly from molecular or empirical formulas:

Rule (1). Whenever an atom is found in its elementary state, its oxidation number is zero, as for N_2, Cl_2, C, S_8, and P_4.

Rule (2). The oxidation number of a monatomic ion equals the charge of the ion; Na^+, Cu^{2+}.

Rule (3). The oxidation number of some common atoms are: oxygen usually -2, covalently bonded hydrogen $+1$, and halogen, when bonded to less electronegative atoms, -1.

Rule (4). The sum of the oxidation numbers of all atoms in a molecule is zero; in an ion, the sum is the charge of the ion.

Example 6 Calculate the oxidation number of the S atoms in $S_2O_3^{2-}$.

Answer

($2 \times$ oxid. number of S) + ($3 \times$ oxid. number of O) = charge on ion.

Let x = oxidation number of S. Then

$$2x + 3(-2) = -2$$
$$x = +2$$

Note that the value obtained for the S atoms in thiosulfate ion, $S_2O_3^{2-}$, is the average of the values obtained for the individual S atoms by the method of allotting electrons, as done in Example 5.

In summary, the assignment of electrons for calculating the oxidation number recognizes the ionic character of the bond, whereas the method for assigning formal charge recognizes the covalent character of the bond. Since most bonds are somewhere between purely covalent and purely ionic, neither assigned charge is real.

Oxidation numbers are useful in identifying oxidation–reduction reactions involving covalent species. The terms oxidation and reduction have been defined in terms of transfer of electrons (page 121). However, in reactions of covalent substances the concept of oxidation–reduction is less clear. For example, does oxidation–reduction occur in the following reaction (oxidation numbers are shown)?

$$\underset{-4\;+1}{CH_4} + \underset{0}{Cl_2} \longrightarrow \underset{-2\;+1}{CH_3Cl} + \underset{+1\;-1}{HCl}$$

The answer is "yes" since there is a change in oxidation number of certain atoms as shown. For reactions involving alterations in covalent bonds, we define *reduction* as *a decrease in oxidation number,* and *oxidation* as *an increase in oxidation number.* Thus, in the above reaction methane, CH_4, is oxidized and chlorine is reduced.

WRITING LEWIS FORMULAS / 8.12

The previous discussion stresses the necessity of being able to write Lewis formulas. Although it is impossible to give foolproof rules, some helpful guides can be offered.

The formulas of many molecules and ions take the form AB_x or AB_xD_y. In most cases, the sequential arrangement of atoms, the so-called SKELETON, has x atoms of B and y atoms of D attached to the CENTRAL ATOM, A, as in the sulfate ion, SO_4^{2-}, and sulfuryl chloride, $SOCl_2$:

$$\left[\begin{array}{c} O \\ | \\ O-S-O \\ | \\ O \end{array} \right]^{2-} \qquad \begin{array}{c} O \\ \| \\ Cl-S-Cl \\ \| \\ O \end{array}$$

sulfuryl chloride

In the structure AB_x, A *will be the central atom if it has a higher covalency or is capable of having a more positive oxidation number than* B. For example, Cl in ClO_4^- has an oxidation number of $+7$, S in SO_4^{2-} is $+6$, while O has an oxidation number of -2 in these ions. Otherwise, we may have an exceptional structure in which one of the B atoms is central. For example, since oxygen has a smaller covalency than nitrogen, the skeleton for nitrous oxide, N_2O, is N—N—O rather than N—O—N.

Structures of the type A_xB_y pose a problem. One can make a judicious guess by writing a *symmetrical skeleton*. Thus, for N_2O_3, N_2O_4, and N_2O_5, we have the following symmetrical skeletons (shown with the correct bonding):

$$O{=}N{-}O{-}N{=}O$$

N_2O_3	N_2O_4	N_2O_5
dinitrogen trioxide	dinitrogen tetroxide	dinitrogen pentoxide

However, this type of reasoning can occasionally lead to incorrect structures, as evidenced by the unsymmetrical structure of the thiosulfate ion, $S_2O_3^{2-}$ (page 134).

The number of electrons (dots) showing in the Lewis formulas is the sum of the electrons in the Lewis formula of each atom (group number). If an ion is negative (an ANION), we add its charge. If an ion is positive (a CATION), we subtract its charge. For example,

$$\left[\begin{array}{c} :\overset{\cdot\cdot}{O}: \\ | \\ :\overset{\cdot\cdot}{O}-S-\overset{\cdot\cdot}{O}: \\ | \\ :\overset{\cdot\cdot}{O}: \end{array} \right]^{2-} \qquad \left[\begin{array}{c} H \\ | \\ H-N-H \\ | \\ H \end{array} \right]^{+}$$

sulfate ion   ammonium ion

Group number: S = O = 6; charge -2 N = 5, H = 1; charge $+1$
Total number of electrons: $6 + (4 \times 6) + 2 = 32$ $5 + (4 \times 1) - 1 = 8$

When the skeleton and the number of electrons are established, the dots are placed around each atom so as to satisfy the octet rule (duet rule for H) and when possible the covalency requirements. In some cases more than one pair of electrons may be shared by two atoms. For example, carbon dioxide, CO_2, has 16 electrons and the skeleton is O—C—O. We could place 8 electrons around each O atom as shown below.

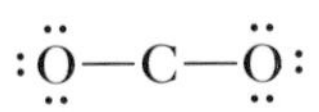

$$:\overset{..}{\underset{..}{O}}-C-\overset{..}{\underset{..}{O}}:$$

This structural formula is incorrect because the C atom does not have an octet of electrons. We could improve the structural formula by showing each O atom sharing two pairs of electrons with the carbon. Thus, the correct structural formula would show two double bonds:

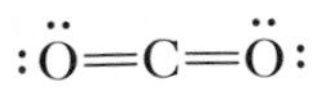

$$:\overset{..}{O}=C=\overset{..}{O}:$$

Example 7 Draw a Lewis structural formula for (a) COS; (b) ClF_3; (c) $(NH_4)_2(PHO_3)$.

Answer (a) Since C has the higher covalency, it is the central atom. The skeleton is OCS. There are 16 electrons, $C(4) + O(6) + S(6)$, and 2 double bonds:

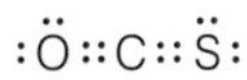

$$:\overset{..}{O}::C::\overset{..}{S}:$$

(b) Cl is the central atom and there are a total of 28 electrons, $Cl(7) + [3 \times F(7)]$ as shown:

$$:\overset{..}{\underset{..}{F}}:\overset{..}{\underset{..}{Cl}}:\overset{..}{\underset{..}{F}}:$$
$$:\overset{..}{\underset{..}{F}}:$$

To form this compound, more than 8 electrons must surround the Cl atom.

(c) This is an ionic compound (a salt). Each ion is shown independently.

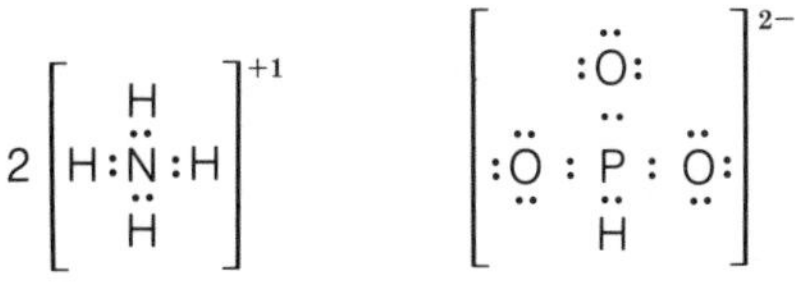

$$2\left[\begin{matrix} H \\ H:N:H \\ H \end{matrix} \right]^{+1} \qquad \left[\begin{matrix} :\overset{..}{O}: \\ :\overset{..}{O}:P:\overset{..}{O}: \\ H \end{matrix} \right]^{2-}$$

PERIODICITY OF BONDING, VALENCE, AND OXIDATION NUMBER / 8.13

(a) **Formation of anions of representative elements.** When an atom of a representative element forms a *monatomic anion* (transition elements do not form anions) that is isoelectronic with the noble gas of next higher atomic number, the charge on the anion is $G - 8$, where G is the periodic group number. Thus, sulfur (group 6R) would become $S^{2-} (6 - 8 = 2)$, and nitrogen (group 5R) would become N^{3-}. The smaller the size and the lesser the charge on an anion the more stable it is and the easier it is to form. These differences in stability can be summarized in certain periodic trends.

(*i*) *Within a period, the ability of an element to participate in ionic bonding by forming an anion increases as the group number increases* (*proceed across the period from left to right*). Thus, there are more ionic fluorides $(G = 7)$ than nitrides $(G = 5)$. This trend is a consequence of the fact that fluorine must gain only 1 electron to form F^-, whereas nitrogen must gain 3 electrons to form N^{3-}. It becomes increasingly difficult to acquire successive numbers of electrons.

(*ii*) *The ability of elements, within a given group, to form anions decreases with increasing atomic weight.* Thus, there are more ionic fluorides than chlorides, bromides, or iodides.

(b) **Formation of cations of representative elements.** When an atom forms a *cation* that is isoelectronic with the noble gas of the nearest lower atomic number, the charge on the cation must be numerically equal to the group number (G), for example, Li^+, Mg^{2+}, and Al^{3+}. The ionic bonds with the least amount of covalent character occur in the anhydrous salts (free from water) of cations with polyatomic anions, such as perchlorate, ClO_4^-; nitrate, NO_3^-; sulfate, SO_4^{2-}; carbonate, CO_3^{2-}; and phosphate, PO_4^{3-}. If an element does not form a salt even with these anions, then the element cannot sustain a cationic state, and it will *share* rather than *lose* electrons. The larger the radius and the smaller the positive charge of a cation, the greater is the chance that the cation will form. These factors can be summarized in terms of trends in the periodic table.

(*i*) *On proceeding down a group, the ability to form a cation increases.* Thus, there are more ionic compounds of thallium (Tl) than of aluminum (Al).

(*ii*) *Among representative elements, the increase in the ability to form cations becomes most pronounced on proceeding from the second to the third period.* For this reason the chemistry of the elements in the second period is atypical of their group. Thus, we find that although all the alkali metals (group 1R) form cations, a tendency toward covalent bonding is sometimes observed for lithium. For example, when combined with carbon, as in ethyllithium, CH_3CH_2Li, lithium forms bonds with some covalent character. Also unlike the other members of their respective groups, beryllium (group 2R) rarely forms Be^{2+}, and boron (group 3R) never forms B^{3+}.

(iii) *Metals are more likely to form ionic compounds in their lower oxidation states than in their higher ones.* Thus, $PbCl_2$ (m.p. 501°C) is ionic and $PbCl_4$ (m.p. − 15°C) is covalent.

INTERMOLECULAR FORCES / 8.14

The preceding sections were concerned with forces of attraction between atoms in a molecule. We now turn our attention to forces of attraction *between molecules.* These INTERMOLECULAR forces determine whether a substance is found in a gaseous, liquid, or solid state. They also affect many chemical reactions which involve interactions among molecules.

(a) **Dipole interactions.** When polar molecules attract each other, they so orient themselves that the positive pole of one molecule is close to the negative pole of another molecule, as shown in Fig. 8.4. Molecules in the gaseous state under normal pressures are far apart, and their dipole–dipole attractions are very small. As the pressure of the gas increases, the molecules approach each other more closely, and, as the temperature decreases, the kinetic energy also decreases. The combination of these effects permits the dipole–dipole interaction to cause liquefaction and eventually solidification.

The extent of dipole–dipole interaction is one of the factors that determine the melting and boiling points of polar substances. Other factors such as molecular weight and molecular shape being equal, a substance with zero dipole moment will have a lower boiling point and melting point than a polar molecule. Thus, the nonpolar molecules N_2 and O_2 have boiling points of −196°C and −183°C, respectively, whereas the somewhat polar NO ($\mu = 0.070$ D) boils at −151°C.

(b) **Hydrogen bonding.** The attractive forces between certain molecules are greater than can be accounted for by ordinary dipole–dipole interactions. Such unusual intermolecular attractions are evidenced by the abnormally high boiling points of H_2O, NH_3, and HF as compared with other covalent hydrides, as shown in Fig. 8.5. The group 4R hydrides, CH_4, SiH_4, GeH_4, and SnH_4, are nonpolar. As the atomic weight of the central atom decreases, there is a regular decrease in boiling point. In the absence of dipole–dipole interaction and major differences in shape, this direct relationship between boiling point and molecular weight is typical. Even in the case of the polar hydrides, those of groups 5R, 6R, and 7R, this direct relationship prevails except for the hydrides of the first members of these groups, NH_3, H_2O, and HF. These molecules are more polar than the other hydrides in their respective groups because N, O, and F are the most electronegative elements. Nevertheless, their startlingly high boiling points indicate

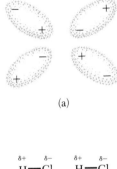

(a)

$$\overset{\delta+}{H}—\overset{\delta-}{Cl} \qquad \overset{\delta+}{H}—\overset{\delta-}{Cl}$$

$$\overset{\delta-}{Cl}—\overset{\delta+}{H} \qquad \overset{\delta-}{Cl}—\overset{\delta+}{H}$$

(b)

Fig. 8.4. (a) Dipole–dipole attraction; (b) exemplified by HCl.

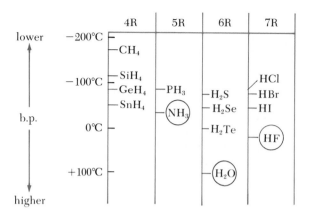

Fig. 8.5. Boiling points of hydrides of Groups 4R, 5R, 6R, and 7R. The hydrides are, positioned according to their boiling points on the vertical scale at left. Note that as the molecular weight increases (going down a group) so does the boiling point. The exceptions are the encircled hydrides, NH_3, H_2O, and HF, which are strongly hydrogen bonded.

the presence of a force stronger than a dipole–dipole interaction. H_2O, HF, and NH_3 have one structural feature in common: Each has at least 1 hydrogen atom covalently bonded to a highly electronegative atom with at least one unshared pair of electrons. Covalently bonded O, F, and N atoms attract the somewhat positive H atom of another molecule,

$$\overset{\delta-}{:X}\!-\!\overset{\delta+}{H}\text{-}\text{-}\text{-}\overset{\delta-}{:X}\!-\!\overset{\delta+}{H}$$

covalent bond

hydrogen bond

The hydrogen atom increases the intermolecular attraction by bridging the two molecules. Since the very small hydrogen atom is the bridge between the 2 electronegative atoms, the molecules can approach sufficiently close to each other to produce an attraction strong enough to be considered a *bond*, rather than just another dipole–dipole attraction. This unique type of bond is called the **HYDROGEN BOND**. The energy[*] of the hydrogen bond, which is much less than that of a typical covalent bond, varies from 3 to 10 kcal/mole.

In a typical H bond, —A—H---A— or —A—H---B—,

(*i*) A and B are usually the very electronegative atoms, F, O, N, or Cl. The strength of the H bond depends partly on the electronegativities of A and B and usually decreases in the order F > O > N > Cl.

(*ii*) The strongest bond is achieved when the three atoms are in a straight line.

(*iii*) Usually, the H atom is nearer to one atom than the other. It is covalently bonded to the nearer one, H bonded to the other. An exception is hydrogen difluoride ion, HF_2^-,

[*] Recall (page 66) that the "energy of a bond" means the energy required to break the bond.

$$HF + F^- \longrightarrow [F \cdots H \cdots F]^- \text{ or } [F^-H^+F^-]$$

where the H is midway between the F atoms. This ion, which has the strongest known H bond, may be regarded as two F^- ions held together by an H^+. Another strong hydrogen bond is observed in hydrogen fluoride, whose formula should be written $(HF)_x$ since it exists as linear or cyclic aggregates, as shown in Figs. 8.6a and 8.6b.

If it is true that the hydrogen bond in $(HF)_x$ is stronger than the hydrogen bond in water, how can we account for the observation that water (b.p. 100°C, molecular weight 18) has a much higher boiling point than hydrogen fluoride (b.p. 19.4°C, molecular weight 20)? The answer comes from considering the geometry of the two hydrogen-bonded systems. In the case of hydrogen fluoride, any one F atom can be surrounded by only 2 hydrogen atoms, and so can participate in only one hydrogen bond. In ice, the oxygen atom is surrounded by 4 hydrogen atoms, and so participates in two hydrogen bonds (Fig. 8.6c). The presence of two hydrogen bonds per H_2O molecule increases the attraction between the individual H_2O molecules. Anything with this kind of network structure holds together firmly because a molecule could get free only if several bonds broke at the same time. The unusually high melting point of 0°C for a molecule with a molecular weight of 18 g/mole is accounted for by the cross-linked nature of the hydrogen bonds. Since cross-linked hydrogen bonds persist to some extent in the liquid state, water also has an inordinately high boiling point. As ice melts, some H bonds break causing a partial collapse of the network structure. Below 4°C, the breakdown of the structure brings the H_2O molecules closer together. The density of water increases, therefore, from 0° to a maximum at 4°C. Above 4°C, the increase in kinetic energy of the molecules is sufficient to cause the molecules to begin to disperse, and the density steadily decreases with increasing temperature. In the vapor phase, water exists as individual H_2O molecules.

(c) **London forces.** All gases, including the elements with nonpolar molecules such as O_2, N_2, and F_2, and even the noble monatomic gases like Ne and He, can be liquefied. Apparently there is some attractive force even among these nonpolar molecules. Since these substances have very low boiling points, the attractive forces are relatively weak, certainly weaker in most cases even than dipole–dipole forces. These weak forces are called LONDON FORCES after Fritz London, who developed a theoretical explanation of them in 1928.

The electrons in a molecule are in constant motion. The electron density of an He atom, for example, is spherically symmetrical *on the average*, but at any instant, there may be an imbalance of charge distribution in the atom. Thus, an atom or nonpolar molecule may be momentarily *self-polarized* because of the unbalanced charge distribu-

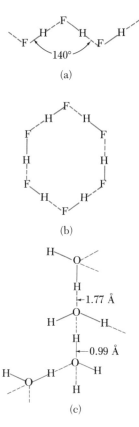

Fig. 8.6. *H bonded structure for (a) HF (linear); (b) $(HF)_6$ (cyclic) and (c) ice (cross-linked); broken lines indicate H bond.*

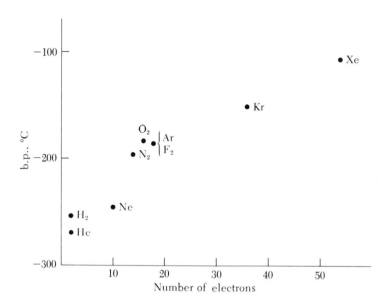

Fig. 8.7. *Boiling points of nonpolar gases relative to number of electrons per molecule.*

tion. This polarized molecule induces transient dipole moments in neighboring molecules. These induced dipoles then cause the nonpolar molecules to be mutually attracted. London forces are the weakest of all known intermolecular forces. They are present between all molecules but are generally overshadowed when stronger forces are present.

The data in Fig. 8.7 for some noble gases and nonpolar molecules reveal correlation between the boiling point and the number of electrons. The boiling point increases as the number of electrons increases. The London forces apparently are roughly proportional in strength to the number of electrons per molecule. It is also true that for more complicated series of nonpolar molecules the boiling point increases with increasing molecular weight, assuming that the molecular structures are comparable.

The correlation of boiling point with molecular weight is also considerably influenced by the shape of the molecule. A comparison of the boiling points of *n*-pentane and neopentane, which have the same molecular formula, C_5H_{12}, reveals this dependence. Both molecules have the same kind and number of atoms and therefore the same number of electrons, yet the boiling points differ by 27°. A molecule of *n*-pentane may be regarded as a zig-zag chain (Fig. 8.8a), a molecule of neopentane almost as a sphere (Fig. 8.8b). For *n*-pentane, the approach between the two molecules can occur over the entire length of the chain, whereas for neopentane the approach can occur only at a point. The lateral, side-to-side approach of two *n*-pentane molecules involves more contact, which increases the London forces, and therefore the boiling point of *n*-pentane is higher.

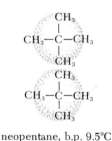

n-pentane, b.p. 36.2°C

(a)

neopentane, b.p. 9.5°C

(b)

Fig. 8.8. *Representation of molecular shape of (a) n-pentane; (b) neopentane.*

1. Definitions. Define and illustrate the following terms: (a) octet rule, (b) Lewis symbol, (c) ionic bond, (d) covalent bond, (e) crystal lattice energy, (f) free radical, (g) network covalent compound, (h) electronegativity, (i) polar molecule, (j) dipole moment, (k) formal charge, (l) oxidation number, (m) hydrogen bond, (n) dipole–dipole attraction, (o) London forces.

2. Structural formulas (single bonds only). Write Lewis formulas consistent with the octet rule and covalency requirements for the following: (a) CH_4S, (b) CH_2ClBr, (c) S_2Cl_2, (d) NH_2Cl, (e) CH_5N, (f) C_2H_6S (two structural possibilities), (g) H_3NO, (h) H_4N_2, (i) Si_2H_5F.

3. Structural formulas (with multiple bonds). Write Lewis formulas consistent with the octet rule and covalency requirements for (a) CS_2; (b) C_2H_4; (c) $ClNO$; (d) C_2H_2; (e) H_2CO; (f) $ClCN$; (g) C_2N_2; (h) C_3H_6; (i) $HNCO$; (j) N_2F_2; (k) H_2C_2O; (l) H_2CO_2.

4. Ionic and covalent compounds. State whether each of the following substances has a molecular covalent, network covalent, or ionic structure. Give your reason for each answer. (a) HBr, m.p. $-88.5°C$, b.p. $-67°C$; (b) BN, covalency of each atom is 3, has no multiple bonds; (c) $MgCl_2$, m.p. $708°C$, b.p. $1412°C$, the molten substance conducts electricity; (d) $AsCl_3$, m.p. $-8.5°C$, b.p. $130.2°C$; (e) CS_2, has double bonds.

5. Electronegativity. Indicate the partial charges, if any, on the atoms in each of the following bonds (see Table 8.3, page 130): (a) F—O, (b) N—Cl, (c) C—H, (d) O—N, (e) S—H, (f) P—H.

6. Free radical. Cl forms the oxides Cl_2O, ClO_2, ClO_4, Cl_2O_8, and Cl_2O_7. Which must be free radicals? Make a generalized statement about the formula of the oxide and its free radical character.

7. Oxidation number. Evaluate the oxidation number of the underlined atom in (a) $\underline{C}O_2$, (b) $\underline{S}OCl_2$, (c) $Na_2\underline{S}$, (d) $K_3\underline{P}O_4$, (e) $K_2\underline{P}HO_3$, (f) $K\underline{P}H_2O_2$, (g) $K_4\underline{P}_2O_7$, (h) $K(\underline{V}O_2)SO_4$, (i) $K_2\underline{S}_2O_7$, (j) $K_2\underline{Mn}O_4$, (k) $H\underline{C}O_2H$, (l) $\underline{N}_2H_4$.

8. Formal charge and oxidation number. For the following species (a) locate all formal charges that are not zero, (b) determine the oxidation number of each atom, (c) compare the covalency of each atom having a formal charge with the normal value. Make a useful generalization based on your comparison.

| sulfurous acid (i) | chloric acid (ii) | phosphorus oxychloride (iii) | amide ion (iv) | ammonium ion (v) |

9. Dipole moment. Account for the following order of dipole moments observed for the hydrogen halides: HF, 1.9 D; HCl, 1.03 D; HBr, 0.74 D; HI, 0.38 D.

10. Hydrogen bonding. Suggest a structure for H_3F_3, a component of liquid hydrogen fluoride.

11. Hydrogen bonding. Account for the following facts. (a) Although ethyl alcohol, C_2H_5OH (b.p. 80°C), has a larger molecular weight than water (b.p. 100°C), it has a lower boiling point. (b) Salts of the HCl_2^- anion are known. (c) Mixing 50 ml each of water and ethyl alcohol gives a solution whose volume is less than 100 ml.

12. Intermolecular attraction. Select the substance in each of the following pairs which you think has the higher boiling point: (a) O_2, oxygen, or H_2S, hydrogen sulfide; (b) Ar, argon, or Xe, xenon; (c) CH_3CH_2OH, ethanol, or CH_3OCH_3, dimethyl ether; (d) $CH_3CH_2CH_2CH_2OH$, n-butyl alcohol, or $(CH_3)_3COH$, t-butyl alcohol; (e) $(CH_3)_3N$, trimethylamine, or $CH_3CH_2CH_2NH_2$, n-propyl amine. Explain your choice. Check your results in a handbook.

13. Lewis structural formulas. Write a suitable Lewis formula for each of the following species and state in which cases the octet is exceeded: (a) NO_2Cl, (b) H_2NCN, (c) XeF_2, (d) ICl_4^-, (e) $COCl_2$, (f) SF_4, (g) BiF_3.

ADDITIONAL PROBLEMS

14. Thermodynamics. Calculate the heat of formation of MgO, $Mg(c) + \frac{1}{2}O_2(g) \rightarrow MgO(c)$, from the following data:

		ΔH, kcal/mole
Electron affinity	$O(g) + 2e^- \longrightarrow O^{2-}(g)$	+156
Bond dissociation energy	$O_2(g) \longrightarrow 2O(g)$	+119.1
Heat of sublimation	$Mg(c) \longrightarrow Mg(g)$	+ 36.5
Ionization energy	$Mg(g) \longrightarrow Mg^{2+}(g) + 2e^-$	+523
Lattice energy	$Mg^{2+}(g) + O^{2-}(g) \longrightarrow MgO(c)$	−919

15. Electronegativity. (a) List three plausible covalent bonds between dissimilar atoms which would show practically no polarity. (b) List three ionic bonds which you think exhibit extreme ionic character. (c) List three covalent bonds you think are highly polar. (See Table 8.3, page 130.)

16. Hydrogen bonding. Acetic acid, $H_3CC\overset{\displaystyle \|}{\underset{\displaystyle O}{}}$—O—H, has a strong tendency to form dimers (structures made up of two individual molecules). Draw a likely structural formula for the dimer of acetic acid.

17. Free radicals. For each of the following, write a Lewis formula and indicate whether or not it is a free radical: (a) O^{2-}, (b) OH^-, (c) H_2O_2, (d) HO_2^-, (e) O_2^{2-}, (f) O_2^-.

18. Free radicals. Tetramethyl lead, $(CH_3)_4Pb$, is a liquid, b.p. 110°C. When heated, it decomposes, leaving lead and a very reactive substance that rapidly becomes converted to ethane, $H_3C:CH_3$. (a) Write the structural formula for $(CH_3)_4Pb$. (b) Write the Lewis formula for the reactive substance. (c) Write the equation for the conversion of the reactive substance to ethane.

19. Bond type. (a) Select A and B from among the elements of atomic number 1–9. Write electronic formulas and indicate the type of bond formed in each compound: (*i*) 3 compounds of formula AB_3 or A_3B; (*ii*) 5 compounds of formula AB_2 or A_2B; (*iii*) 4 compounds of formula AB; (*iv*) 2 compounds of formula AB_4 or A_4B; (*v*) 2 compounds of formula A_2B_3 or A_3B_2. (b) Indicate which compounds do not fit the octet rule. (c) Which compounds are likely to have a network covalent structure? (d) Which are likely to have a multiple bond?

20. Formal charge and oxidation number. (a) Write a Lewis formula for hydrazoic acid, HN_3, whose skeleton is H—N—N—N. Make certain the structure obeys the octet rule. (b) Locate all formal charges. (c) Calculate the oxidation number for each atom. (d) Account for the fact that, for the middle N atom, the oxidation number and formal charge are identical.

21. Mixed bonding types. Write Lewis formulas consistent with the octet rule for the following ionic compounds: (a) calcium carbide, CaC_2; (b) sodium peroxide, Na_2O_2; (c) magnesium cyanide, $Mg(CN)_2$; (d) sodium formate, $NaHCO_2$.

22. Thermodynamics. Calculate the lattice energy of NaCl from the following data:

		ΔH, *kcal/mole*
Heat of sublimation	$Na(c) \longrightarrow Na(g)$	+26.0
Ionization energy	$Na(g) \longrightarrow Na^+(g) + e^-$	+118.0
Bond energy	$Cl_2(g) \longrightarrow 2Cl(g)$	+58.2
Electron affinity	$Cl(g) + e^- \longrightarrow Cl^-(g)$	−88.2
Heat of formation	$Na(c) + \frac{1}{2}Cl_2(g) \longrightarrow NaCl(c)$	−98.2

23. Periodicity of chemical bonding. Account for each of the following facts. (a) There are fewer ionic nitrides than oxides. (b) There are fewer ionic tellurides than sulfides. (c) There are fewer ionic compounds of germanium than of lead. (d) SbF_3 is ionic; SbF_5 is covalent.

24. Lewis formulas. Write Lewis formulas for (a) sulfuryl chloride, SO_2Cl_2; (b) hypophosphorous acid, HPH_2O_2; (c) nitrous anhydride, N_2O_3; (d) nitrous oxide, N_2O; (e) phosphorus oxychloride, $POCl_3$; (f) perchloric anhydride, Cl_2O_7; (g) diazomethane, H_2CN_2.

9

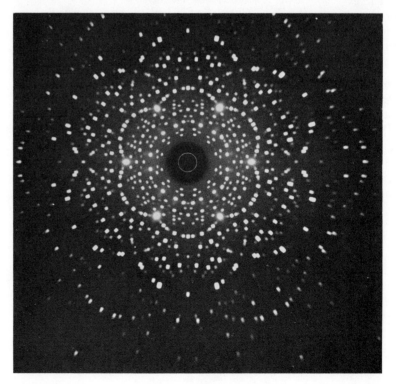

CONDENSED STATES OF MATTER

Crystals are characterized by the orderly, cohesive arrangement of their atoms, ions, or molecules (Table 1.1, page 5). As a result of this internal orderliness, a crystal assumes a recognizable external shape. The angles at which its surfaces (faces) meet each other are a characteristic and reproducible property of the crystal. These geometrical patterns have been of interest to man since earliest civilization, and more

Fig. 9.1. *X-ray diffraction pattern of beryl,* $Be_3Al_2Si_6O_{18}$. *(Courtesy Eastman Kodak Laboratories.)*

especially since the invention of the microscope. The study of such patterns is the science of CRYSTALLOGRAPHY. The internal structures of crystals can be studied more effectively, however, by measuring the degree to which they reflect or scatter radiation. Since the distances between atoms in matter are of the same order of magnitude as X-ray wavelengths, X-ray patterns (Fig. 9.1) may be expected to give useful information about the arrangement of atoms in crystals. This idea was experimentally verified by Max von Laue in 1912. In 1913, William Henry Bragg and William Lawrence Bragg (father and son) calculated the spacing between layers of atoms by measuring the intensities of X rays reflected from crystals at different angles. X-ray analysis is a very powerful, if complicated, method of examining orderly arrangements in crystals, crystalline powders, and large molecules like those of proteins, rubber, and textile fibers.

THE CRYSTAL LATTICE; THE UNIT CELL / 9.2

The internal structure of a crystal is characterized by regularity in three dimensions. Let us consider an elemental crystal, in which all the atoms are alike. If we represent the center of each atom by a point corresponding to its position in space,* the arrangement of such points is called a CRYSTAL LATTICE, as shown, for example, in Fig. 9.2. Note that the straight lines connecting the lattice points outline a block of unit shapes, called cells. All such cells are identical, and any one of them is called a UNIT CELL. The crystal lattice may be considered to consist of an indefinite number of unit cells, each one having sides in common with those of its nearest neighbors, and all similarly oriented in space.

The simplest type of unit cell is a cube. There are three cubic

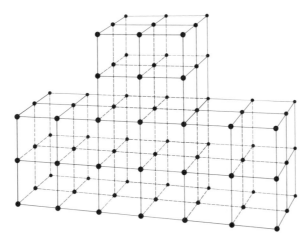

Fig. 9.2. Space lattice.

* Since the atoms vibrate in the crystal because of their thermal energy, each designated position is actually an average position, or a "center of vibration."

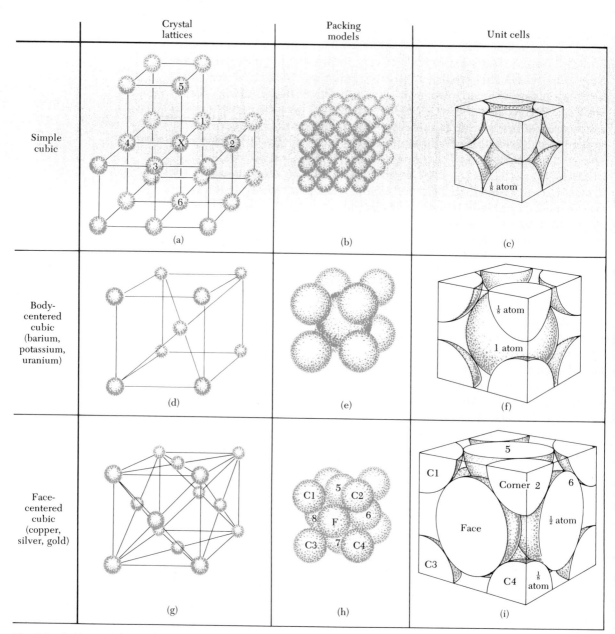

	Crystal lattices	Packing models	Unit cells
Simple cubic	(a)	(b)	(c)
Body-centered cubic (barium, potassium, uranium)	(d)	(e)	(f)
Face-centered cubic (copper, silver, gold)	(g)	(h)	(i)

Fig. 9.3. *Cubic crystal systems; C = corner atom, F = face atom.*

systems, whose names refer to the locations of atoms in the unit cell: SIMPLE CUBIC, BODY-CENTERED CUBIC, and FACE-CENTERED CUBIC. Figures 9.2 and 9.3a show a simple cubic lattice; note that the centers of the atoms are at the corners of the cubes. The lines between the atoms are not real, of course; they are drawn to illustrate the cubic arrangement. If we think, more realistically, of the atoms as being packed close together, a better illustration is the simple cubic packing model of Fig. 9.3b. The simple cubic unit cell is shown in Fig. 9.3c. The body-centered cubic lattice, containing one atom in the center of the cube, appears in Fig. 9.3d, the packing model in Fig. 9.3e, and the unit cell in Fig. 9.3f. Finally, Figs. 9.3g, 9.3h, and 9.3i represent the face-centered cubic lattice, the packing model, and the unit cell, respectively.

We now ask two questions: (a) How many atoms exist within the space defined by the cubic unit cell, and (b) how many atoms "touch" (or are the nearest neighbors to) a given atom?

Assignment of atoms to the cubic unit cell is made on the following basis:

A **CORNER ATOM** (Fig. 9.3c, 9.3f, or 9.3i) is shared equally by all the unit cells that touch the same point. In two dimensions, four squares can touch a given point. In three dimensions, eight cubes can touch a given point. Therefore, only $\frac{1}{8}$ of a corner atom is assigned to one unit cell.

A **FACE ATOM** (Fig. 9.3i) is shared equally between the two cells that face each other. Therefore, $\frac{1}{2}$ of a face atom is assigned to one unit cell.

An **EDGE ATOM** (Fig. 9.4) is shared equally among the 4 unit cells that have a common edge. Therefore, $\frac{1}{4}$ of an edge atom is assigned to one unit cell.

A **BODY-CENTERED ATOM** (Fig. 9.3f) belongs entirely to its unit cell.

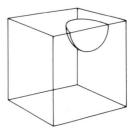

Fig. 9.4. *Edge atom ($\frac{1}{4}$ inside cube).*

In summary:

Location of atom	Portion in cubic unit cell
Corner	$\frac{1}{8}$
Face	$\frac{1}{2}$
Edge	$\frac{1}{4}$
Center	1

Example 1 How many atoms are assigned to the unit cell of the body-centered cube?

Answer Eight corners = $8 \times \frac{1}{8} = 1$; 1 center = 1; total = 2.

The **COORDINATION NUMBER** of an atom is its number of nearest-neighboring atoms. For a given crystal lattice, the coordination number is established by inspection of the model. In the **SIMPLE CUBE**, each atom touches six adjacent atoms. Note that in Fig. 9.3a atom X is closest to the six atoms numbered 1–6.

In the **BODY-CENTERED CUBE**, each atom may be considered to be in the center of a cube and touching the eight corner atoms (Figs. 9.3d–9.3f and Fig. 9.5).

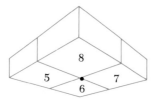

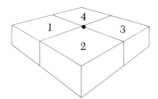

In the **FACE-CENTERED CUBE** the coordination number can be visualized from Fig. 9.3h or 9.3i by noting how many atoms touch a face atom. There are four adjacent corner atoms (marked C1 to C4 in Fig. 9.3h) lying in the same plane as the face atom F. Four others (5, 6, 7, 8) behind the face plane also touch the face atom. Finally, there is room for four more touching atoms (not shown in the figure) in front of the face plane directly forward of atoms 5, 6, 7, 8. This arrangement, which gives a coordination number of 12, is called **CUBIC CLOSEST PACKING**. It is impossible to pack more uniform spheres into a given space in any other way.* In summary:

Lattice of uniform spheres	Coordination number
Simple cubic	6
Body-centered cubic	8
Face-centered cubic	12

Fig. 9.5. *Eight cubes can touch a given point if four cubes are laid exactly on four other cubes.*

Example 2 (a) Argon is an element in which each molecule is a single atom. Argon freezes at about −189°C to form a crystalline solid with a face-centered cubic lattice. How many atoms are there in the unit cell? (b) The density of solid argon is 1.7 g/cm³. The edge of the unit cell is 5.4 Å in length. Calculate the number of atoms in 1 mole (40 g) of argon (the Avogadro number).

Answer (a) Each face-centered cube contains 8 corner atoms (=1) plus 6 face atoms (=3), or a total of 4 atoms/unit cell. (b) We first calculate how many unit cells make 40 g, and then, from the number of cells, we can calculate the number of atoms in the 40 g:

$$\text{volume of one unit cell} = \left(\frac{5.4 \text{ Å}}{10^8 \text{ Å/cm}}\right)^3 = 157 \times 10^{-24} \text{ cm}^3$$

From the density, we find that

$$\text{volume of 1 mole of solid argon} = \frac{40 \text{ g/mole}}{1.7 \text{ g/cm}^3} = 23.5 \frac{\text{cm}^3}{\text{mole}}$$

$$\text{number of unit cells per mole} = \frac{23.5 \text{ cm}^3/\text{mole}}{157 \times 10^{-24} \text{ cm}^3/\text{unit cell}}$$

$$= 1.5 \times 10^{23} \frac{\text{unit cells}}{\text{mole}}$$

$$\text{number of atoms per unit cell} = 4 \text{ atoms [from part (a) above]}$$

$$\text{total number of atoms in 40 g (1 mole) of argon} = 1.5 \times 10^{23} \frac{\text{unit cells}}{\text{mole}} \times \frac{4 \text{ atoms (exactly)}}{\text{unit cell}} = 6.0 \times 10^{23} \frac{\text{atoms}}{\text{mole}}$$

* There is another equally closely packed arrangement based on a hexagonal crystal type called **HEXAGONAL CLOSEST PACKING**.

A compound lattice contains more than one kind of atom (or ion); for example, both sodium and chloride ions make up the sodium chloride lattice. The different kinds of atoms may differ in both number and size. A representation of a compound lattice must identify as well as locate the atoms in the unit cell. A compound lattice may also be described in terms of separate unit cells for each type of atom, and of the spatial relationships among the different unit cells. It is easier to visualize this first in two dimensions, as shown in Fig. 9.6a, then in three dimensions, Fig. 9.6b.

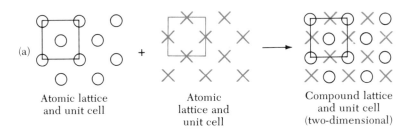

(a)

Atomic lattice and unit cell

Atomic lattice and unit cell

Compound lattice and unit cell (two-dimensional)

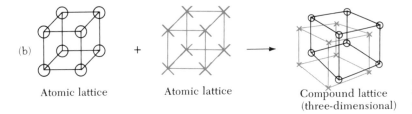

(b)

Atomic lattice

Atomic lattice

Compound lattice (three-dimensional)

Fig. 9.6. *Two- and three-dimensional compound crystal lattices.*

Figure 9.7 shows the lattices, packing models, and unit cells of both sodium chloride and cesium chloride crystals. Sodium chloride crystallizes as a face-centered cubic compound lattice with alternating sodium and chloride ions. (However, see Problems 12 and 13, page 164.) The coordination number of each ion is 6; the nearest neighbors of any one ion are 6 ions of the other kind. Cesium chloride is a body-centered cubic lattice in which the central ion's nearest neighbors are the eight ions of the other kind.

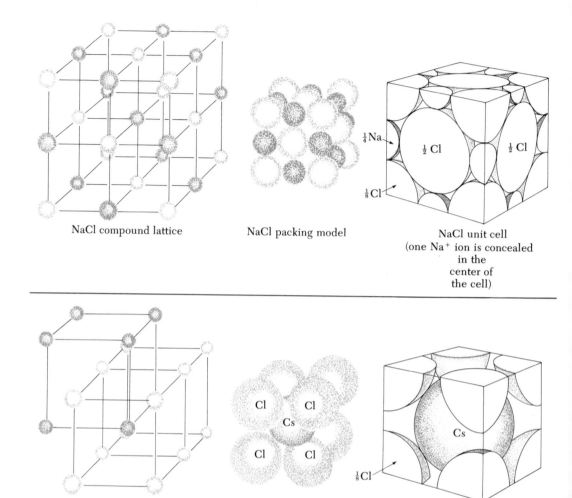

NaCl compound lattice

NaCl packing model

NaCl unit cell
(one Na$^+$ ion is concealed
in the
center of
the cell)

CsCl compound lattice

CsCl packing model

CsCl unit cell

Fig. 9.7. *Compound lattices and unit cells of sodium chloride and cesium chloride.*

REAL CRYSTALS / 9.4

Like different buildings constructed of identically shaped bricks, real crystals do not necessarily look like the crystal lattices or unit cells of which they are composed. The shape of real crystals is determined in large part by their relative rates of growth in different directions. For example, a face-centered cubic structure may grow to form a cube, an octahedron, or a cube with its corners cut off. These forms are shown in Fig. 9.8.

The shape in which a crystal usually grows is called its HABIT. The perfect order depicted in the foregoing sections, like the complete randomness discussed for gases, is an ideal picture. Real crystals are very

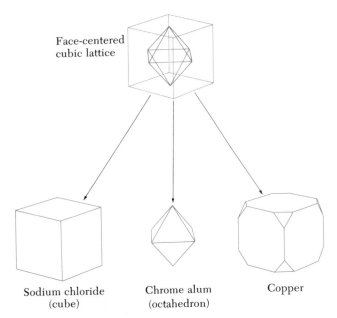

Face-centered
cubic lattice

Sodium chloride
(cube)

Chrome alum
(octahedron)

Copper

Fig. 9.8. *Crystal habits from a face-centered cubic lattice.*

likely to be imperfect. When a crystal breaks, the fracture usually occurs along the defect lines or planes.

Another effect of lattice imperfection is solid-state diffusion. It is found, for example, that gold plated on lead will gradually diffuse into it, forming a lead–gold alloy. It is postulated that lattice vacancies in the lead allow the almost equal-sized gold atoms to occupy the empty sites. The random shifting of both lead and gold atoms into these holes promotes diffusion. Unequally sized atoms may exhibit solid-state diffusion if the smaller atoms can slip among the larger ones (for example, carbon in iron), like a cyclist in stalled truck traffic.

LIQUIDS; GLASSES / 9.5

The liquid state cannot yet be described by any concept that has the simplicity of the theories of the gaseous or crystalline states. Nonetheless, important advances have been made by Henry Eyring, J. D. Bernal, and others.

Is there any order at all in liquid structure? X-ray examination shows some limited regularity—but much less than the sharp patterns obtained from crystalline solids. One way to regard the liquid state is as a crystallinelike mass interspersed with molecular-sized spaces or "holes" whose arrangements are entirely random (Fig. 9.9). Molecules may enter or leave these "holes"; this interchange produces overall disorder and fluidity.

"holes"

Fig. 9.9. *Two-dimensional model of the liquid state, showing long-range disorder.*

If a liquid is cooled until the decrease in molecular energy robs it of its fluidity, and if crystallization does not occur, then the substance becomes a GLASS. A glass has the overall randomness of a liquid, but flows so slowly that it is almost indistinguishable from a rigid body. (Not entirely, however. Windows in old cathedrals are somewhat thicker toward the bottom because of downward flow of the glass.) When a glass is heated, fluid properties increase gradually over a range of temperature, since no sudden breakup of an orderly structure is required.

Examples of liquids which, on cooling, are likely to produce glasses rather than crystalline solids include molten tar, melted sugar, and fused metallic silicates. The last of these yields the material called "glass," which has given its name to the state of matter that it typifies.

The various states of matter are generally interconvertible. As the energy of matter is increased, the arrangement of its constituent particles becomes more disordered; as the energy is decreased, the opportunities for orderly arrangements increase. There are limitations, however, in both directions. Increase of energy may destroy particles

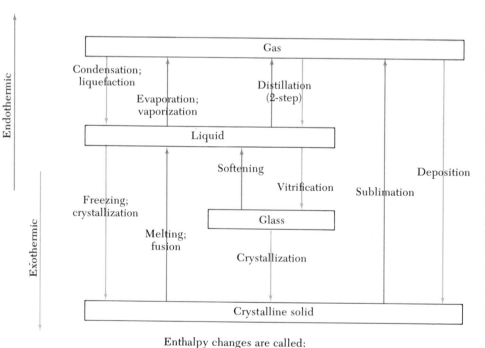

Fig. 9.10. *Vocabulary of changes of state.*

Enthalpy changes are called:

Heat of fusion   (solid $\leftrightarrows$ liquid)
Heat of vaporization (liquid $\leftrightarrows$ gas)
Heat of sublimation   (solid $\leftrightarrows$ gas)

besides producing disorder. For example, crystalline insulin (a protein) decomposes before it melts; less energy is required to destroy the molecules than to make them exhibit mobile flow. In the reverse direction, loss of energy may afford a better chance for an orderly pattern, but it does not assure it. Instead (as discussed earlier), a glass may form.

It will be helpful to learn the vocabulary of changes of state, as illustrated in Fig. 9.10. A change of state involves gain or loss of heat, but does not involve a temperature change. Thus the **HEAT OF FUSION**, ΔH_{fus}, of a substance is the amount of heat necessary to melt 1 g of the solid to a liquid *at the same temperature*. The heat of fusion, as illustrated in Fig. 9.10, is absorbed by the substance from the environment; the opposite process (freezing) involves the same quantity of heat per gram of substance, but the heat is released by the substance to the environment. Analogous expressions are used for the other changes of state.

Any of the processes in Fig. 9.10 that proceed in the direction solid → liquid → gas produces a twofold change in the affected matter:

(1) The energy of the matter is increased because work is being done to overcome attractive forces.

(2) The orderliness of the molecular arrangement is decreased.

Conversely, in any of the processes that proceed in the direction gas → liquid → solid, the energy is decreased and the order is increased.

Table 9.1 presents data on heats of fusion and vaporization.

Table 9.1
Heats of Fusion and Vaporization

Substance	Melting point, °C	Boiling point, °C	Heat (cal/g) of	
			Fusion	Vaporization
Aluminum	658	2057	94	2.5×10^3
Beeswax	62	—	42.3	—
Carbon tetrachloride	−24	76.8	4.2	46.4
Ethyl alcohol	−114	78.3	24.9	204
Mercury	−39	357	2.8	70.6
TNT	79	—	22.3	—
Water	0	100	79.7	539.6

SPONTANEOUS CHANGE; ENTROPY / 9.7

Which set of processes shown in Fig. 9.10, the absorption or the liberation of heat, would be expected to occur naturally? The direction of heat flow will be influenced by the relative temperatures of the system and the surroundings; if the surroundings are colder, then heat will tend to flow out of the matter, and the gas → solid processes will be favored. The reverse will be true in a hot environment.

There is also, however, a second factor—*a natural tendency for disorder*. This deceptively simple idea has far-reaching consequences. For changes of state, it means there is *always* a drive that favors the solid → gas processes, regardless of the outside temperature. Thus, it is not necessary to place solids or liquids in a warm environment for them to evaporate; evaporation will occur even when they are cooling. Examples: snow sublimes even on a cold winter day; moth-repellent crystals sublime in a closet, even though the temperature may be falling.

The greater the disorder or randomness of a system, the higher its ENTROPY. The relationship between the entropy and the properties of a system can be arrived at either from statistical or from thermodynamic (heat/work) considerations. We will reserve the thermodynamic discussion for Chapter 17.

Statistically, we may argue that disordered conditions are more probable than ordered ones because there are many more of the former. To understand this, we must first think about what we mean by order and disorder. If we walked into a library and saw books on the floor, tables, chairs, and shelves, many of the books open (but not to the same page number), and the books oriented in various directions in space, we would call the arrangement disorderly. However, if all the books were on the shelves, in the sequence of their call numbers as given in the catalog file, we would consider that the books were orderly. Why? Because *order is characterized by repetition*. The books on the shelves occur *repeatedly* in the same orientation (vertical and parallel); their arrangement by title *repeats* the arrangement in the catalog file. In fact, the file arrangement itself is orderly because its sequence *repeats* the alphabet, or some other known succession of symbols. The books lying around in various ways are disorderly because *very little repeats itself*, not the page numbers to which the books lie open nor their orientation in space, nor does the sequence of books in any direction match the sequence of call numbers, or any other sequence familiar to us.

Now, we have said that when many arrangements are possible, the disorderly ones far outnumber the orderly ones. We realize this fact intuitively. To return to the library, if we see that the order of books on the shelves matches the call numbers, we assume that they were so arranged by the conscious act of the librarian. We are confident that order is no accident because we realize that if the books were arranged by chance, a very much larger number of disorderly sequences would be possible. Thus, disorder is favored, or, in other words, *the entropy of the system tends to increase*.

Now we turn from books in a library to gaseous molecules in a box. Let us divide the box into two compartments (Fig. 9.11), allow the

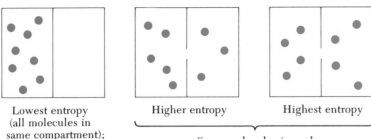

Lowest entropy
(all molecules in
same compartment);
1 chance in 128

Higher entropy

Highest entropy

Some molecules in each
compartment; 127 chances in 128

Fig. 9.11. *Eight molecules in two compartments.*

molecules to be anywhere in either compartment, and ask, "What is the chance at any one instant that all the molecules will be in one of the compartments?" (Note that we are assuming that the system is isolated from all outside influences so that there is no bias which would favor any arrangements over any others.) Such an event, if it occurred, would *decrease* the entropy of the system, because all the molecules would be found *repeatedly* in the same compartment. The chance of having all the molecules in one compartment is small compared with other arrangements, and becomes smaller and smaller as the number of molecules increases.* These considerations illustrate the statistical basis for our conclusion that *spontaneous decreases in entropy are impossible in isolated systems.* Obviously, this conclusion does not apply to nonisolated systems that may be biased by some outside influence. Thus, a human being may develop from a single cell, an event that would be quite unlikely by chance alone. We will see later, however (Chapter 17), that the decrease of entropy in one system must be accompanied by an increase in entropy somewhere else.

LIQUID–GAS INTERCONVERSION; VAPOR PRESSURE / 9.8

Molecules of a liquid, like those of a gas (see Fig. 2.4, page 21), exhibit a distribution of energies. The molecules that are most energetic and close to the surface have the best chance to escape (vaporize). The escaped molecules may be recaptured if they collide with and are held by the attractive forces of the liquid (condensation). Any condensable gas is, in fact, called a **VAPOR**.

A high *net* rate of vaporization (evaporation) is therefore favored by (1) high temperature of the liquid, (2) small attractive forces in the

---

* The general relationship that expresses the probability (P) that all the molecules (N) will be on one side is $P = 2^{1-N}$. If the box contains $\frac{1}{2}$ mole (3×10^{23} molecules) then the probability that all the molecules appear on one side is one chance in $10^{(10^{23})}$. If we were to write out all the zeros in this number, they would extend a distance of some 100,000 light years! Of course, results that are so highly improbable do not occur in our experience.

liquid, (3) a large surface area, (4) low atmospheric pressure above the liquid—to decrease the number of collisions with other molecules and thereby minimize return of molecules to the liquid—and (5) a motion of the atmosphere (breeze) above the liquid to carry away vaporized molecules and minimize reentry. Some of these factors are highly favorable when wet laundry is spread on a clothesline on a warm, dry, windy day.

If the enclosed space above a liquid is small enough, and if the temperature is held constant, a condition will be reached in which the rates of vaporization and of condensation are equal. For every molecule that enters the vapor, another one returns to the liquid. The system is now said to be in DYNAMIC EQUILIBRIUM, or simply in EQUILIBRIUM.

A system at equilibrium is defined as one in which two opposing processes are going on at the same time and at the same rate. The effect is no net change.

The VAPOR PRESSURE of a liquid is the pressure exerted by the vapor that is in equilibrium with the liquid at a definite temperature. The vapor pressure depends on the temperature and on the nature of the liquid; it is, within the limits of the ideal gas law, independent of the presence of other gases. Solids, too, exert vapor pressures, but they are usually lower than those of liquids.

Vapor pressure may be measured in a number of ways. A direct method is illustrated schematically in Fig. 9.12.

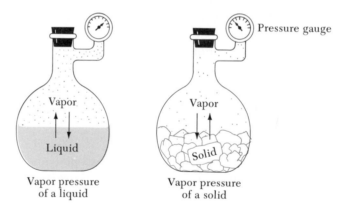

Pressure gauge

Vapor

Liquid

Vapor

Solid

Fig. 9.12. *Measurement of vapor pressure* (schematic). *Temperature is constant.*

Vapor pressure of a liquid

Vapor pressure of a solid

CRITICAL CONSTANTS / 9.9

For every gas, there is a CRITICAL TEMPERATURE above which its molecules have enough kinetic energy to overcome the attractive forces that would cause liquefaction. The critical temperature is therefore *the highest temperature at which a vapor can be liquefied.* The CRITICAL PRESSURE is *the minimal pressure needed to liquefy a gas at the critical temperature.*

The critical temperature of a substance is related to the magnitude of the attractive forces among its molecules. If the attractive forces are great, liquefaction can be achieved in spite of molecular motion at elevated temperatures. The critical temperature is, therefore, high. If attractive forces are small, liquefaction is easily opposed by thermal agitation, and the critical temperature is low. Refer to Table 9.2 for data.

Table 9.2
Critical Constants

Substance	Temperature, °C	Pressure, atm
Air	−140.6	37.2
Carbon dioxide	31.0	72.9
Ethyl alcohol	243	63.0
Helium	−267.9	2.26
Hydrogen	−239.9	12.8
Mercury	900	180
Nitrogen	−147	33.5
Oxygen	−118.8	49.7
Water	374.2	218.3

BOILING, MELTING, AND FREEZING POINTS / 9.10

There are two cases of phase equilibrium that are worthy of note. One is BOILING, the process of vaporization of a liquid that is usually accompanied by the formation of bubbles of vapor in the interior of the liquid. This occurs when the vapor pressure is equal to the pressure of the atmosphere in contact with the liquid. Thus, the BOILING POINT of a liquid is *the temperature at which the liquid is in equilibrium with its vapor at the same pressure as the surrounding atmosphere.*

The boiling point of a liquid, therefore, varies with changes in the atmospheric pressure. The NORMAL BOILING POINT is that which occurs at standard atmospheric pressure (1 atm or 760 torr).

The bubbling action in ordinary boiling is called EBULLITION. A pure liquid, heated carefully in a smooth, clean vessel, often reaches a temperature considerably above its boiling point (up to about 20 Celsius degrees higher) without bubbling. This phenomenon is called SUPERHEATING, and is explained as follows: Evaporation occurs from the surface of a liquid. Any tiny bubbles, such as air bubbles, or bits of gas adhering to dust particles or to the wall of the container, provide internal space into which the liquid can evaporate. At the boiling point, the vaporization of the liquid expands the little bubbles, resulting in ebullition. In the absence of such little "nuclei," ebullition does not get started, and superheating occurs instead.

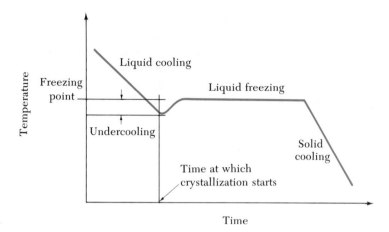

Fig. 9.13. *Cooling curve of a pure liquid, with undercooling.*

Liquid cooling

Freezing point

Liquid freezing

Undercooling

Solid cooling

Time at which crystallization starts

Temperature

Time

The MELTING POINT and FREEZING POINT of a substance are equivalent; they are the temperatures at which liquid and solid are in equilibrium at standard atmospheric pressure. The melting point is reached by warming the solid, the freezing point by cooling the liquid. The effect of change of atmospheric pressure is usually so small that it need not be considered.

If heat is added to a mixture of solid and liquid at equilibrium, some solid melts without change of temperature. If heat is removed, some liquid freezes, also at constant temperature. If heat is slowly and continuously extracted from a pure liquid substance, the liquid will cool until it starts to freeze, then the liquid will freeze at constant temperature until it is all solid, and finally the solid will start to cool. The graphical representation of this sequence, called a COOLING CURVE, is shown in Fig. 9.13. Note that the liquid may cool *below* its freezing point before crystallization occurs and solid–liquid equilibrium is established. This phenomenon is called SUPERCOOLING, or, more aptly, UNDERCOOLING. The time direction of this curve may be reversed by warming the solid. Such a graph is called a WARMING CURVE. Superheating does not occur during warming of a crystal.

Cooling curves obtained in practice do not show such sharp changes of direction because it is difficult to keep the temperature of the entire sample uniform, especially when much of it is solid.

THE PHASE DIAGRAM / 9.11

The temperature–pressure relationships among the solid, liquid, and gaseous states of a substance may be illustrated in a single PHASE DIAGRAM, which is determined experimentally. In Fig. 9.14, referring to water, an area represents one phase, a line two phases, and a point (the TRIPLE POINT) three phases. If we specify a combination of temperature and pressure, the location of these coordinates on the diagram will in-

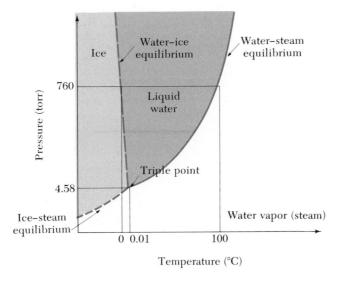

Fig. 9.14. *Phase diagram for water (not to scale).*

dicate what the *most stable condition* for water will be. Thus it is seen, for example, that the melting point of ice (water–ice equilibrium line) decreases as the pressure increases.

The reader should verify that there are seven possibilities: three single phases (ice,* water, and steam), three two-phase equilibria (water–ice, water–steam, and steam–ice), and one three-phase equilibrium (steam–water–ice at the triple point, 0.01°C and 4.58 torr).

The small effect of pressure variation on the freezing point is illustrated by the steepness of the solid–liquid equilibrium line. Its negative slope (leaning to the left) is not typical of most substances.

The physical states of a substance are by no means limited to three phases. The coexistence of more than one solid phase is quite common (for example, the diamond and graphite forms of carbon, or the rhombic and monoclinic forms of sulfur, page 435). Any three-phase equilibrium, such as two solid phases plus one liquid phase, may be represented by a triple point.

COLLOIDS; ADSORBENTS / 9.12

The forces that attract molecules in the solid state take on special significance at the surface of the solid where the molecules are bound but not surrounded by the solid lattice. These surface molecules, therefore, exert residual forces that attract any foreign molecules happening by.

When the surface area of any body is very large compared with its bulk volume, the special properties of the surface molecules influence the overall properties of the body to a significant degree. Large surface-

---

* At high pressures, several different ice phases are known to exist.

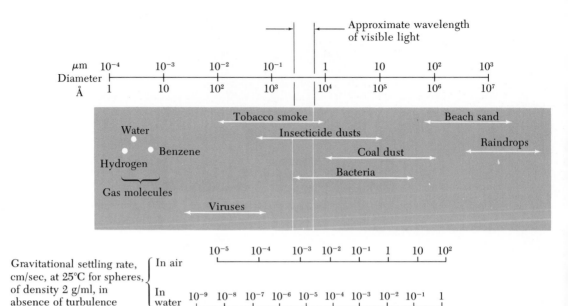

Fig. 9.15. *Small particles.* (∘)

to-volume ratios exist when either the particles of the substance are very small, or the body is interlaced with small pores throughout its bulk volume.

Particles so small that surface effects assume great relative importance in determining their properties or behavior are called COLLOIDS. Colloidal particles dispersed in air are called AEROSOLS. It will be helpful to refer to Fig. 9.15 to note the particle sizes of some common aerosols in comparison with coarse matter (such as beach sand) and with gas molecules. Particles smaller than the wavelength range of visible light cannot be seen with the ordinary microscope. The lower limit of microscopic visibility, about 0.2 μm, is frequently taken to be the upper limit of colloidal particle sizes. The lower size limit of the colloid range is taken to be the size of such large molecules as those of protein or starch and of some small viruses, roughly 50 Å.

Colloidal particles have the ability to scatter visible light. A thin beam of light passing through a colloid in a gas or a liquid can be seen at right angles because of the scattering. This phenomenon, called the TYNDALL EFFECT, was discovered in 1857, and constitutes a simple (and still reliable) criterion of the colloidal state of matter. Colloids are characterized by very slow settling rates. Note (Fig. 9.15) that at 0.2 μm, the rate of settling of particles in air is about 5×10^{-4} cm/sec and in water is several hundredfold slower.

As a result of the surface forces mentioned above, any gas, vapor, or liquid adheres to some degree to any surface of a body. This phenomenon is called ADSORPTION, and solid bodies having large surface areas compared to their bulk volumes are called ADSORBENTS. Colloids, because of their small size, are excellent adsorbents. Even large bodies, however, may have large surface-to-volume ratios if a network of pores within them provides extensive surface area. Matter of this type can be produced from coconut shells, for example, by charring and then heating them in a steam atmosphere to about 1500°C to drive out all material which can be gasified at that temperature. This process is called ACTIVATION. The matter left behind, consisting mainly of carbon and retaining the many pores created by the activation process, is called ACTIVATED CARBON. The adsorbing capacity of such material is startlingly high. For example, activated carbon at room temperature can adsorb up to three-fourths its own weight of carbon tetrachloride or bromine from air that is saturated with the corresponding vapor. Other effective adsorbents may be made from metallic oxides, silicates, and metals.

PROBLEMS

1. Definitions. Define and illustrate the following terms: (a) crystal lattice, (b) unit cell, (c) coordination number, (d) compound lattice, (e) vapor pressure, (f) critical temperature, (g) critical pressure, (h) colloid, (i) adsorbent.

2. Unit cell. Designate a unit cell in (a) a uniform sheet of postage stamps, (b) black and white squares arranged in a striped pattern, (c) a checkerboard.

3. Space lattice. Lithium crystallizes in a cubic lattice whose unit cell edge is 3.50 Å long at 20°C. The density of lithium at 20°C is 0.534 g/cm³. (a) How many atoms of lithium are assigned to one of its unit cells? (b) To which cubic crystal system does this number correspond?

4. Physical states. A zinc metal rod and a glass rod are heated in the absence of air to a temperature somewhat above 419°C. The zinc rod remains rigid, but liquid zinc begins to drip from it. The glass rod sags, but does not drip. Account for these phenomena.

5. Changes of state. Complete the following sentences: The boiling point of "isooctane" (the 100-octane gasoline standard) is 99°C, and its freezing point is −107°C. It is, therefore, a _____ in its stable condition at 100°C and 1 atm pressure, and a _____ under the ordinary conditions of man's environment. If the _____ is distilled, it is converted first to a _____ and then to a _____. When the liquid is cooled sufficiently, the isooctane will _____ and become a _____ if the molecules arrange themselves in an orderly lattice, but if molecular disorder is maintained, the isooctane will _____ and become a _____. If helium gas is swept over isooctane crystals at −110°C, the crystals will gradually disappear by the process of _____, being converted to a _____.

6. Critical temperature. The intermolecular attractive forces in liquid methane are greater than those in liquid argon. Which substance has the higher critical temperature? Explain.

7. Cooling curve. Draw a complete time (x axis)–temperature (y axis) cooling curve for water at a pressure of 1 atm, showing the cooling of steam, its condensation at 100°C, the cooling, undercooling, and freezing of water, and the cooling of ice. The specific heat of steam and of ice are both about 0.5 cal/g °C. Label all parts.

8. Phase diagram. Solid carbon dioxide ("dry ice"), when heated at standard atmospheric pressure, vaporizes completely without melting. What is characteristic of the phase diagrams of substances that behave in this way?

9. Adsorption. It is observed that when air containing gasoline vapor passes through a bed of activated carbon, the gasoline vapor is removed from the air and the temperature of the carbon bed rises. Account for these phenomena.

10. Unit cell. What fraction of a circle in each of the following locations is located within a two-dimensional square unit cell: (a) corner, (b) side, (c) center?

11. Unit cell. Draw a diagram showing how the diameter of a sphere may be part of an edge of a cube. If each edge contains one such sphere, what is the total number of spheres assigned to the space within the cube?

12. Crystal lattice. Refer to the diagram of the sodium chloride lattice (Fig. 9.7, page 152). What is the coordination number (a) of the sodium; (b) of the chlorine? (c) Tabulate the number of corner, edge, face, or center particles of sodium atoms (smaller spheres) and of chlorine atoms (larger spheres) in the unit cell of sodium chloride. How many atoms of each are assigned to the unit cell? (d) If all the sodium ions were removed from the lattice, what would the lattice structure of the chloride ions be? If all the chloride ions were removed from the lattice, what would the lattice structure of the sodium ions be?

13. Crystal lattice. Refer to the diagram of the cesium chloride lattice (Fig. 9.7, page 152). What is the coordination number (a) of the cesium; (b) of the chlorine? (c) How many atoms of each are assigned to the unit cell? (d) If all the cesium ions were removed, what would the lattice structure of the chloride ions be? (e) Describe another unit cell from which the cesium chloride lattice could be generated.

14. Crystal lattice. Gold, Au, crystallizes as a face-centered cubic lattice whose unit cell edge is 4.08 Å at 20°C. (a) How many atoms are there in the unit cell? (b) Predict the density of gold. (c) Assuming that the Au atoms are packed as closely as the face-centered cube will allow, calculate the radius of the Au atom. (*Hint:* The atoms "touch" each other along the diagonals of the faces of the cube.)

15. Space lattice. Two empty swimming pools of the same size are filled with uniform spheres of ice packed as closely as possible. The spheres in the first pool are as large as grains of sand; those in the second pool are as large as oranges. The ice in both pools melts. In which pool, if either, will the water level be higher?

16. Unit cell. Tantalum, Ta, crystallizes in a body-centered cubic lattice. The density of Ta is 16.7 g/cm³ at 20°C. (a) How many Ta atoms are assigned to one unit cell? (b) What is the coordination number of Ta in the lattice? (c) What volume does one mole of atoms occupy? (d) What is the volume of 1 unit cell of

Ta, in cubic angstrom units? (e) Assume that the Ta atoms are packed as closely as the body-centered cube will allow. Calculate the radius of the Ta atom in Å units. (*Hint:* The atoms "touch" each other along the body diagonal of the cube. This diagonal is $\sqrt{3}$ times as long as the edge of the cube.)

17. Coordination number. Consider a crystalline substance that consists of two elements A and B, whose formula is $A_?B_?$. The coordination numbers of A and B are a and b, respectively. Write a formula for the substance in which the subscripts are expressed in terms of the coordination numbers in place of the question marks.

18. Lattice defects. A recent report in a dental journal states that "silver" tooth fillings, which consist of an alloy of silver and mercury, gradually shrink and leave voids from which new cavities may develop. The alloy shrinks because, at the temperature of hot foods, the lattice gradually changes from simple cubic to close-packed hexagonal (coordination number = 12). (a) Does the fact that the lattice changes without melting imply that it is imperfect? Explain your answer. (b) Complete the sentence: If a crystal structure changes so that the coordination number of its atoms increases, the density of the substance will _____ and the volume of a given sample will, therefore, _____.

19. Liquid and glassy states. A man wearing spiked shoes finds that he sinks slowly when he stands on a tar road or on an ice rink. Ice is crystalline; tar is not. Account for both phenomena.

20. Liquid crystals. Some soapy solids (for example, potassium oleate), on heating, melt sharply to form homogeneous, turbid liquids that have some of the optical properties of crystalline solids, and that give much sharper X-ray diffraction patterns than ordinary liquids give. On further heating, an abrupt change to a clear liquid occurs. Such turbid liquids are sometimes called LIQUID CRYSTALS and their state of aggregation is called the "mesomorphic state." Suggest a theory of molecular arrangement to account for the mesomorphic state.

21. Critical constants. Large quantities of liquid oxygen are transferred to and kept at steel mills in containers that are vented to the atmosphere. Liquid carbon dioxide, however, is handled and shipped in closed steel cylinders. Explain.

22. Phase diagram. Copy or trace Fig. 9.14 on your own paper, and draw straight lines on the figure to represent each of the following processes: (a) The melting of ice by compressing it at constant temperature; (b) the melting of ice, followed by the warming and boiling of water, all at sea level; (c) the melting of ice, followed by the warming and boiling of water on a mountain top; (d) the sublimation of ice (as in "freeze-drying"), at constant temperature.

10

SOLUTIONS

A **SOLUTION** *is a mixture of two or more substances dispersed as molecules, atoms, or ions* rather than as larger aggregates. If we mix sand and water, the sand grains are dispersed in the water; since the grains are much larger than molecules, we call this mixture a *suspension*, not a solution. After a while the sand will settle to the bottom by gravity. Imagine doing this experiment with finer and finer grains. When the grains are small enough, they will not sink to the bottom, no matter how long you wait. We now have a *colloidal dispersion* (page 161). Though we cannot see the individual grains, the mixture appears cloudy in a strong beam of light (Tyndall effect). If, however, we stir sugar with water, the grains disappear and the result is a liquid that does not scatter light any more than water itself. This is a true solution, with individual sugar molecules dispersed among the water molecules.

This chapter omits two important classes of solutions: gaseous mixtures, such as air, and, for the most part, ionic solutions, in which some of the constituent particles are electrically charged. Because of the special properties and great importance of ionic solutions, Chapter 12 has been reserved for them.

LIQUID SOLUTIONS / 10.2

The word "solution" suggests a liquid to most of us, and liquid solutions are indeed the most important and interesting ones. The molecules in a liquid are in intimate contact, and thus the properties of each component in a liquid solution are considerably influenced by the presence of the others. For example, the energy of a molecule is affected by the nature of the molecules surrounding it; as a result, heat is usually absorbed or emitted when a solution is formed. In some cases, there is actually a chemical reaction between the components, so that the molecules present in solution are quite different from the molecules of the pure components.

The SOLVENT is the component which is visualized as dissolving the other component, the SOLUTE. This distinction, though arbitrary, is often useful. For a liquid solution, if one component is a liquid when pure and the other is a solid or a gas, the liquid is considered the solvent: for example, water dissolving sugar or ammonia. When both are liquids, the more abundant component is the solvent: in vinegar, water is the solvent and acetic acid the solute; in acetic acid slightly contaminated with water, it is the other way around.

A DILUTE SOLUTION is one which contains only a small quantity of solute (or solutes) relative to the quantity of solvent. A CONCENTRATED SOLUTION contains a large proportion of solute. These terms are no more precise than the words "large" and "small"—as many a student has learned when he used "dilute hydrochloric acid" and found that it was 20 or 30 times more concentrated than the acid he was supposed to use.

MEASURES OF COMPOSITION FOR SOLUTIONS / 10.3

In discussing solutions, we must be able to specify their compositions—that is, the relative amounts of the several components. Composition is expressed in a number of ways. Let us assume that we have a solution of two components, A (solvent) and B (solute). We adopt the following notation:

w_A, w_B = mass ("weight") in grams of A or B in the solution
n_A, n_B = number of moles of A or B
V = total volume of the solution, in liters

The most important measures of composition for liquid solutions are the following:

(a) WEIGHT PERCENTAGE (more properly "mass percentage") of B is

$$\frac{w_B}{w_A + w_B} \times 100\%$$

If we have a solution containing 40 g of water (A) and 10 g of sucrose (B), the percentage of sucrose is

$$\frac{10 \text{ g}}{40 \text{ g} + 10 \text{ g}} \times 100\% = 20\%$$

This is one of the simplest and most useful measures; nothing need be known about the components except their masses, which are easily determined experimentally, and which always add up to the mass of the solution.

(b) **MOLE FRACTION** of B, denoted by x_B, is the ratio of the number of moles of B to the total number of moles:

$$x_B = \frac{n_B}{n_A + n_B}$$

A mole is a fixed number (Avogadro's number) of molecules; therefore, a mole fraction is a number fraction—the mole fraction of B is the fraction of all the molecules in the solution that are B molecules. It follows from the definition that $x_A + x_B = 1$ or, for more than two components, $x_A + x_B + x_C + \cdots = 1$. If we have 3.42 g of sucrose ($C_{12}H_{22}O_{11}$, 342 g/mole) dissolved in 18.0 g of water (18.0 g/mole), then we have 0.0100 mole sucrose and 1.00 mole water. The mole fractions are

$$x_{sucrose} = \frac{0.0100}{1.00 + 0.0100} = 0.0099$$

$$x_{water} = \frac{1.00}{1.00 + 0.0100} = 0.99$$

MOLE PERCENTAGE is 100 times mole fraction: this solution is 0.99 mole % sucrose and 99 mole % water.

(c) **MOLALITY** of B (m_B) is the number of moles of B dissolved in 1 kg (1000 g) of A. The mass of solvent in grams is w_A; the mass in kilograms is $w_A/1000$. Then

$$m_B = \frac{n_B}{(w_A/1000)} = \frac{1000\ n_B}{w_A}$$

or

$$m_B = \frac{\text{moles B}}{\text{kg A}}$$

The solution in the last paragraph had 0.0100 mole sucrose dissolved in 18.0 g = 0.0180 kg of water. The molality is

$$m_{sucrose} = \frac{0.0100 \text{ mole sucrose}}{0.0180 \text{ kg water}}$$

$$= 0.556 \text{ mole sucrose/kg water}$$

(d) **MOLARITY, CONCENTRATION,** or **MOLAR CONCENTRATION** of B is the number of moles of the solute B per liter of solution: n_B/V, with V in liters. If we dissolve our 0.0100 mole sucrose in water and then add water to make the volume of the solution 100 ml = 0.100 liter, the molarity is

$$[C_{12}H_{22}O_{11}] = \frac{0.0100 \text{ mole sucrose}}{0.100 \text{ liter}} = 0.100 \text{ mole/liter}$$

Note that we did not use 100 ml of water. We used whatever amount was needed to make the volume of the *solution* equal to 100 ml.

The molarity of B is represented by [B], M_B, or c_B. When we know the molarity of a solution, we can measure a certain volume of it and calculate the number of moles of B in that volume. Against this virtue must be set the disadvantage, not shared by molality or mole fraction, that the molarity changes on a mere change of temperature because of the thermal expansion or contraction of the solution.

The following statements are equivalent: the molarity of B is 0.1; the molarity of the solution with respect to B is 0.1; the solution is 0.1 M (0.1 molar) with respect to B. The same forms are used in specifying molality.

It is customary to refer to the molality or molarity of a solution with respect to the solute (B), not the solvent. These measures of composition are most useful for dilute solutions.

Example 1   A solution of 20.0% ethanol, C_2H_5OH, and 80.0% water, by mass (weight), has density 0.966 g/ml at 25°C. Find (a) the mole fraction, (b) the molality, (c) the molarity of ethanol in this solution.

Answer   The molecular weights are 18.0 g/mole for water and 46.1 g/mole for ethanol. We fix our attention on a definite but arbitrary quantity of solution, say 100 g. In this 100 g of solution there are 20.0 g ethanol and 80.0 g water; the numbers of moles of the components are

$$n_{water} = \frac{80.0 \text{ g}}{18.0 \text{ g/mole}} = 4.44 \text{ moles}$$

$$n_{ethanol} = \frac{20.0 \text{ g}}{46.1 \text{ g/mole}} = 0.434 \text{ mole}$$

(a) The mole fraction of ethanol is

$$X_{ethanol} = \frac{0.434}{4.44 + 0.434} = \frac{0.434}{4.87} = 0.0891$$

(b) The molality of the solution with respect to ethanol is

$$m_{ethanol} = \frac{0.434 \text{ mole } C_2H_5OH}{0.0800 \text{ kg } H_2O} = 5.43 \frac{\text{mole}}{\text{kg}}$$

(c) From the given density, we can calculate that the volume of the 100 g of solution is

$$\frac{100.0 \text{ g}}{0.966 \text{ g/ml}} = 103.5 \text{ ml} = 0.1035 \text{ liter}$$

and the molarity of the solution with respect to ethanol is

$$[C_2H_5OH] = \frac{0.434 \text{ mole}}{0.1035 \text{ liter}} = 4.19 \frac{\text{mole}}{\text{liter}}$$

A solution of known molarity can be prepared without knowing its density or the mass of solvent used, with the aid of a VOLUMETRIC FLASK

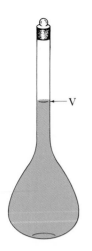

Fig. 10.1. *Volumetric flask. The capacity, up to the mark V, is accurately known.*

(Fig. 10.1). An accurately measured mass of solute is dissolved in the solvent, and the solution is transferred completely to the flask. Solvent is then added carefully to the mark and the solution well mixed.

Example 2 (a) A solution was prepared by dissolving 86.53 g of sodium carbonate, Na_2CO_3, in water in a 1000-ml volumetric flask at 20°C, adding water to the mark, and mixing. Find the molarity of the solution. (b) The density of the solution at 20°C is 1.0816 g/ml. Find its molality.

Answer (a) The molecular weight of Na_2CO_3 is 105.99 g/mole. The number of moles used is

$$\frac{86.53 \ \cancel{g}}{105.99 \ \cancel{g}/mole} = 0.8164 \ mole$$

The volume of the solution is 1.000 liter; the molarity is, therefore, 0.8164 mole/liter. (b) The mass of 1 liter of solution is

$$1.0816 \ \frac{g}{\cancel{ml}} \times 1000 \ \frac{\cancel{ml}}{liter} = 1081.6 \ \frac{g}{liter}$$

We know that this quantity of solution contains 86.53 g Na_2CO_3, and it must therefore contain 995.1 g H_2O:

$$1081.6 \ g \ total - 86.53 \ g \ Na_2CO_3 = 995.1 \ g \ H_2O$$

There is 0.8164 mole Na_2CO_3 in 995.1 g H_2O; in 1000 g H_2O, there is

$$0.8164 \ mole \times \frac{1000}{995.1} = 0.8204 \ mole$$

The molality is 0.8204 mole/kg.

In calculations involving molarity it is often convenient to switch from grams, liters, and moles to MILLIGRAMS (mg), MILLILITERS (ml), and MILLIMOLES (mmole): $1 \ g = 10^3 \ mg$, $1 \ liter = 10^3 \ ml$, $1 \ mole = 10^3$ mmole, etc. (To avoid confusion, convert either *everything* or *nothing* to milliunits.) The number of millimoles per milliliter is the same as the number of moles per liter (*i.e.*, equal to the molarity):

$$0.40 \ \frac{\cancel{mole}}{\cancel{liter}} \times 10^3 \ \frac{mmole}{\cancel{mole}} \times \frac{1 \ \cancel{liter}}{10^3 \ ml} = 0.40 \ \frac{mmole}{ml}$$

The molecular weight is the same number whether it is expressed in g/mole or in mg/mmole:

$$105.99 \ \frac{\cancel{g}}{\cancel{mole}} \times 10^3 \ \frac{mg}{\cancel{g}} \times \frac{1 \ \cancel{mole}}{10^3 \ mmole} = 105.99 \ \frac{mg}{mmole}$$

Since volumes measured in the laboratory are usually in the range of milliliters rather than liters, the milliquantities have the advantage of not requiring very small numbers. However, masses are usually given in grams, and one must not forget to convert them to milligrams.

When a solid or a gas is added to a liquid solvent at a given temperature, the dissolving process gradually slows down as the solution becomes more concentrated. When dissolving stops—in spite of the presence of some undissolved solute—the solution is said to be **SATURATED**. The maximum quantity of solute (relative to the quantity of solvent) that will dissolve is the **SOLUBILITY**. It may be expressed in any of the measures of composition described in Section 10.3: weight percent, mole fraction, molarity, etc.

Why does a solution become saturated? Consider a gas being dissolved in a liquid, with the entire gas–liquid system enclosed (Fig. 10.2a). The gas molecules are moving about in all directions. Molecules frequently strike the surface of the liquid; some merely bounce off, but others are captured and diffuse into the body of the liquid. Conversely, a molecule already in the liquid may happen to reach the surface with enough kinetic energy to escape and become part of the gas again. At a fixed temperature (and thus fixed average kinetic energy), the rate at which gas molecules enter the liquid depends on the number of collisions with the liquid surface, and thus on the pressure of the gas; the rate at which gas molecules leave the liquid depends on the number already dissolved. When the rate of escape of gas from the solution equals the rate of dissolving, the composition of the solution remains constant; no more of the gas dissolves. We say that **EQUILIBRIUM** has been attained. The processes of capture and escape are still taking place, but their rates are equal, and there is no further net change in the pressure of the gas or the composition of the solution. The solution is then saturated. The idea of equilibrium here is the same one discussed in connection with the vapor pressure of a pure liquid (page 158).

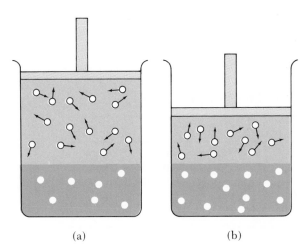

(a)   (b)

Fig. 10.2. *Dissolving of a gas in a liquid at (a) low pressure; (b) higher pressure.*

If the pressure of the gas is increased (Fig. 10.2b), the number of collisions with the surface is increased, and the rate of capture of gas molecules by the liquid therefore increases. The quantity of dissolved gas thus increases until the rate of escape has built up to equality with the new rate of capture. Thus, the *solubility of a gas in a liquid increases with rising pressure.* A carbonated beverage must be kept under pressure; when the cap or cork is removed, the pressure drops, the solubility of the gas decreases, and the beverage fizzes.

Raising the temperature decreases the solubility of a gas in a liquid. Although higher temperature results in more frequent collisions of gas molecules with the liquid surface, the principal effect of raising the temperature is that a larger fraction of the dissolved molecules have the kinetic energy needed in order to escape from the liquid.

Table 10.1 gives the solubilities of some common gases in water. The very large solubility of SO_2 is noteworthy; this is one of the cases in which dissolving is accompanied by a chemical reaction,

$$SO_2 + H_2O \longrightarrow H_2SO_3 \quad \text{(sulfurous acid)}$$

A solid dissolves in a liquid by essentially the same process by which a gas dissolves, except that we should picture the solute molecules not as colliding with the solvent, but as diffusing away from the surface of the solid and becoming surrounded by—and to some extent, attached to—the solvent molecules. When some of the solute has gone into solution, some solute molecules will be recaptured by the solid from the solution, and the rate of this redeposition will increase as the solution becomes more concentrated. If enough solute is present, the solution will attain a composition at which there is a molecule coming out of solution for every molecule that goes into solution. When this condition is attained, the solution is saturated.

When the solution is saturated, the rate of dissolving (molecules going into solution per second) is equal to the rate of deposition (molecules coming out of solution per second). Imagine that we double the surface area of the solid (cut it into smaller pieces). The rate of dissolving is doubled; the rate of deposition is doubled; the two rates are

Table 10.1
Solubilities of Gases in Water[a]

	0°C	25°C	50°C
N_2	0.00294	0.00175	0.00122
O_2	0.00694	0.00393	0.00266
H_2	0.000192	0.000154	0.000129
CO_2	0.335	0.145	0.076
SO_2	22.83	9.41	

[a] In g gas/100 g H_2O, total pressure 1 atm.

still equal, and the solution is still saturated. Therefore, the solubility is independent of the surface area.* At this point, we should note the distinction between *solubility* and *rate of dissolving*. A finely divided solid dissolves much more rapidly than large lumps—the solution becomes saturated sooner—but that does not mean that the total amount which can be dissolved is greater. Similarly, stirring accelerates the dissolving process, but, once the solution is saturated, you can stir your arm off without dissolving another milligram.

Many pairs of liquids are completely MISCIBLE—that is, they dissolve in each other in all proportions. Examples are water and acetic acid, water and glycerol, benzene and toluene. In such cases, there is no such thing as a saturated solution. Other pairs of liquids, however, are only partially miscible: Each dissolves in the other to some extent, giving two saturated solutions. Even so-called immiscible liquids—water and carbon tetrachloride, for example—usually have a measurable solubility in each other. A familiar example of partial miscibility is water and diethyl ether, $(C_2H_5)_2O$; when these two liquids are shaken together at 25°C, two layers are obtained: 94.1% water and 5.9% ether (by mass) in the bottom layer, 1.3% water and 98.7% ether in the top layer.

SOLUBILITY AND TEMPERATURE / 10.5

We observed in the last section that, at constant pressure, the solubility of a gas in a liquid decreases with increasing temperature. The solubility of a solid in a liquid may change with temperature in either direction, as illustrated by the following solubilities in water, in grams of solute per 100 g H_2O:

	Temperature, °C			
	20	*40*	*60*	*80*
Sucrose ($C_{12}H_{22}O_{11}$)	204	238	287	362
Lithium carbonate (Li_2CO_3)	1.33	1.17	1.01	0.85

The typical behavior is that of sucrose: The solubility of a solid in a liquid usually increases with increasing temperature.

Two things happen when a solid dissolves: (a) Solute molecules are pulled apart, as in melting (or even as in evaporation), which requires an input of heat. (b) Solute molecules become associated with solvent molecules, with a resultant output of heat. This attachment of solvent molecules to solute molecules is known as SOLVATION or, when the solvent is water, HYDRATION. For most solutions of *solids* in liquids, the

---

* When the particles become very small, however, solubility increases. This effect begins to be important for particles with diameters less than about 10^{-5} cm—a few hundred molecules across.

"pulling apart" effect predominates. When a *gas* dissolves in a liquid, only the solvation effect is present, and heat is given out. The change of solubility with temperature can be predicted if we know whether heat is given out or absorbed when the solution is formed. When 1 mole of Li_2CO_3 is dissolved in 220 moles of water, 3060 cal of heat is given out; when 1 mole of SO_2 is dissolved in 2000 moles of water, 8554 cal of heat is given out. In both these cases, the solubility decreases with increasing temperature. However, when 1 mole of sucrose is dissolved in a large quantity of water, 1320 cal of heat is absorbed. In this case, the solubility increases with increasing temperature. The rule is general: *If heat is given out on forming the solution, the solubility decreases with increasing temperature; if heat is absorbed, the solubility increases with increasing temperature.*

Example 3 Given the following enthalpy changes at 25°C:

$$HgCl_2(c) \longrightarrow HgCl_2(g) \qquad \Delta H = +18.8 \text{ kcal}$$
$$HgCl_2(g) + aq \longrightarrow HgCl_2(aq) \qquad \Delta H = -15.5 \text{ kcal}$$

(*aq* represents a large quantity of water used as a solvent.) Predict whether the solubility of $HgCl_2(c)$ decreases or increases with increasing temperature.

Answer For the process of dissolving $HgCl_2$, the equation is

$$HgCl_2(c) + aq \longrightarrow HgCl_2(aq)$$

The Hess law (page 64) gives $\Delta H = +18.8 - 15.5 = +3.3$ kcal for this process. Positive ΔH corresponds to absorption of heat. Therefore, the solubility should increase with increasing temperature. Experimentally, the solubility increases by a factor of 7 from 20° to 100°C.

In contrast to the effect of pressure on gases, the solubilities of solids and liquids are affected very little by changes of pressure. For example, the solubility of sodium chloride, NaCl, in water is multiplied by only 1.025 when the pressure is increased from 1 to 1000 atm.

SUPERSATURATION / 10.6

Suppose that we prepare a solution of a solid in a liquid at a high temperature, and then cool the solution. In most cases, the solubility decreases on cooling, and the solution may, at some temperature, become saturated. If we expect a crystal to appear at the temperature of saturation, we may be disappointed. It is often possible to cool the solution far below this temperature without the appearance of solid. Such a solution is more concentrated than a saturated solution at the same temperature; it is said to be SUPERSATURATED.

A supersaturated solution is analogous to a book standing on end. Each may remain that way indefinitely, but each could exist in a more stable condition: saturated solution plus excess solute, or the book lying flat. Each may require a little shove before it can reach its more stable condition.

Supersaturated solutions are common because it is difficult to initiate the growth of a crystal. The embryonic crystal must be formed by several molecules or ions that come together in the correct configuration and stay that way long enough for other particles to deposit on them. Such an event remains improbable until the solution is considerably supersaturated. An incipient crystal of only a few molecules is likely to be disrupted by collisions before it can grow by addition of other molecules. It is only when the crystal has grown past a critical size that further growth is assured. This critical size may not be achieved for a long time, if ever, and some supersaturated solutions can persist indefinitely in that condition—especially viscous solutions, like sugar syrups, in which the molecules cannot easily migrate over long distances. Supersaturation can be relieved by deposition of solute molecules on a dust particle, the container wall, or any other solid present. The best crystallization nucleus is a fragment of the solute itself. When such a "seed" is introduced, crystals usually form very rapidly, leaving a saturated solution.

Supersaturated solutions of gases in liquids are also common. A bubble, like a crystal, has a poor chance of survival until it has grown past a critical size.

SOLUBILITY AND MOLECULAR STRUCTURE / 10.7

When a solid or a liquid dissolves in a liquid, the molecules of each kind must lose some of their like neighbors, obtaining in exchange neighbors of the other kind. The stronger the attractive forces between *unlike* molecules, the greater the solubility. The stronger are the forces *between molecules of the solute*, however, the *less* is its solubility. Strong forces between solute molecules inhibit solubility unless the molecules of the solvent can exert a comparable attraction. Similarly, strong forces between solvent molecules prevent solute molecules from elbowing in, unless there is strong solute–solvent attraction.

It is a time-honored, though ambiguous, maxim among chemists that "like dissolves like." Forces between *similar* molecules are comparable to forces between *identical* molecules. When the solute and solvent molecules are similar, solute molecules can easily replace solvent molecules. In Chapter 8, the forces between molecules were classified into several kinds, including London forces, dipole–dipole forces, and hydrogen bonds (in ascending order of strength). Liquids and solids can be classified, in a rough-and-ready way, according to which kind of force is mainly responsible for holding their molecules together. Two substances may be expected to mix more easily when their intermolecular attractions fall into the same category (both hydrogen bonding, say, or both merely London forces), but less easily when they belong to different categories. Thus, water is a good example of a hydrogen-bonding substance. Water molecules hang firmly together (page 141). If another molecule is to break into this happy arrangement, it must have positive and negative charges to attract the opposite charges in the water molecule. Still better, it should be able to participate in hydrogen bonding: It should have hydrogen atoms attached to highly electronegative atoms (especially N, O, F). At least it should have the electronegative atoms, which can attract the H atoms of H_2O. Examples of substances which are very soluble in water are

$$H-\overset{\cdot\cdot}{\underset{\overset{|}{H}}{N}}-H \qquad H_3C\overset{\overset{H}{|}}{\underset{\overset{|}{H}}{C}}\overset{\cdot\cdot}{\underset{}{O}}{-}H \qquad H_3C\overset{}{\underset{\overset{\|}{\overset{}{:O:}}}{C}}CH_3$$

| ammonia | ethanol | acetone |
| (gas) | (liquid) | (liquid) |

In contrast to these polar molecules, 2,2,4-trimethylpentane, a constituent of gasoline, is almost nonpolar:

$$H_3C-\overset{\overset{\displaystyle CH_3}{|}}{\underset{\underset{\displaystyle CH_3}{|}}{C}}-\overset{\overset{\displaystyle H}{|}}{\underset{\underset{\displaystyle H}{|}}{C}}-\overset{\overset{\displaystyle CH_3}{|}}{\underset{\underset{\displaystyle H}{|}}{C}}-CH_3$$

It should not surprise us that water and gasoline are not appreciably miscible. But gasoline is good solvent for oils, greases, and waxes, which also have nearly nonpolar molecules; neither substance has very strong intermolecular forces in these cases.

DETERGENCY / 10.8

A single molecule can have a part that is polar, or even charged—and therefore has a strong affinity for water—and another part that consists of a long hydrocarbon* chain. Such a molecule or ion—the most common examples are ions—should be somewhat soluble both in oil and in water. More important, it can facilitate the dispersion of oil into water, or of water into oil. A salt containing such ions may be useful as a DETERGENT, or cleansing agent, because of its ability to bring the oily, greasy components of dirt into colloidal dispersion in water.

* A *hydrocarbon* is a compound of C and H only (page 479), for example, 2,2,4-trimethylpentane.

Molecules at the surface of a phase have higher energy than molecules deeper within the phase (page 161). When two immiscible liquids, or a liquid and a solid that does not dissolve in it, are in contact, the area of contact counts as liquid surface. Thus, the surface area of water is increased when droplets of oil are suspended in it. Work must be done to create this additional surface, just as work must be done (by the lungs) to blow up a soap bubble. Work is necessary because molecules that were formerly in the interior of the liquid must be raised to the higher-energy state of being in the surface. The work required, per unit increase in area, is called the INTERFACIAL TENSION. When one phase is a gas, the interfacial tension becomes the SURFACE TENSION of the liquid.

Detergent ions find an especially congenial environment at the interface between oil and water, for they can have their hydrocarbon ends in the oil and their ionic ends in the water—the best of both worlds (Fig. 10.3). The oil and water molecules are likewise attracted to the corresponding ends of the detergent ions. The interfacial tension is therefore much less when a detergent solution is used than it would be for oil and plain water. It thus becomes possible to detach oil or grease from a soiled surface and bring it into a finely divided state, dispersed throughout the water, with the surface of each globule protected by a layer of detergent ions.

The most common detergents have either —COO$^-$ or —OSO$_3^-$ as their ionic parts. Two examples are shown below:

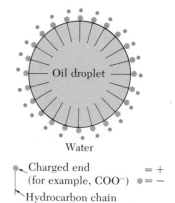

Charged end (for example, COO$^-$) $\bullet$ = +
$\bullet$ = −
Hydrocarbon chain

Fig. 10.3. Detergent ions at an oil–water interface.

H—C—C—C—C—C—C—C—C—C—C—C—C—C—C—C—C—C—C(=O)O$^-$ Na$^+$
sodium stearate

H—C—C—C—C—C—C—C—C—C—C—C—C—O—S(=O)(=O)—O$^-$ Na$^+$
sodium lauryl sulfate

The word "soap" is used technically to refer to any salt in which the negative ion contains the carboxylate group, —COO$^-$, and a long hydrocarbon chain. Thus, sodium stearate is a soap, but sodium lauryl sulfate is not; it is a nonsoap detergent or synthetic detergent. Carboxylate salts of metals other than the alkali metals are usually only slightly soluble in water, and therefore not useful as detergents, but they are nevertheless called "soaps." Some of them find industrial application as lubricants. However, insoluble soaps are more often nuisances. Natural waters usually contain multiply charged positive ions, especially Ca^{2+}, Mg^{2+}, and Fe^{2+}. Such water is described as "hard," and these positive ions react with, for example, the stearate ion, forming nearly insoluble soaps. The result is that soap is wasted—a portion is consumed in reacting with the troublesome metal ions—and the insoluble soap clings to the fabric or the sides of the bathtub. The greatest advantage of the synthetic detergents is that their negative ions form relatively soluble salts with Ca^{2+}, Mg^{2+}, etc.; they are therefore much better than soaps for use with hard water.

SOLID SOLUTIONS / 10.9

A solid solution is usually prepared by mixing the liquids and freezing the liquid solution. Although solid solutions are not at all uncommon, the number of pairs of substances that show appreciable miscibility as solids is much less than the number of pairs that mix in the liquid state. The regular arrangement of molecules in a crystal will tolerate only those replacements that resemble the solvent molecules in their type of intermolecular forces, and have nearly the correct size and shape to fit in without causing serious distortion of the lattice. Solid solutions are most frequently found among metals; all metals are made of spherical ions and free electrons, with essentially the same type of bond (page 367) holding them together. Therefore, ions of similar size should be able to replace each other. Thus, copper and nickel form crystals with the same structure, and the radii of the atoms (half the internuclear distances) are not too different: 1.278 and 1.246 Å, respectively. These two metals are completely miscible in the solid state. The same is true of silver and gold, with radii of 1.444 and 1.442 Å, respectively. Copper and silver, however, are only partially miscible.

VAPOR PRESSURES OF SOLUTIONS

GENERAL REMARKS / 10.10

We saw in Chapter 9 that a solid or liquid, placed in an enclosed space, evaporates to some extent. As the pressure of the vapor increases, the rate of return of vapor molecules to the solid or liquid increases, until finally the rate of evaporation equals the rate of condensation, and equilibrium is attained. The pressure of the vapor at equilibrium is the **VAPOR PRESSURE** of the substance. The vapor pressure increases steeply with temperature, but is independent of the size of the container, as long as some of the solid or liquid is present. To a good approximation, it is independent of the presence of foreign gases in the space above the liquid.

Each component of a solution has a certain vapor pressure, which is *less than the vapor pressure of the pure substance* at the same temperature. In order to obtain quantitative information about the vapor pressures of solutions, one may prepare a series of solutions of two components, measure the total vapor pressure of each solution at a fixed temperature, and then analyze samples of the vapor to obtain the partial pressure (page 15) of each component. The vapor pressures are then plotted against some measure of composition, usually mole fraction (x). Examples of such graphs are given in Figs. 10.4 and 10.5. When the mole fraction of component A is zero, its vapor pressure is, of course, zero; when the mole fraction is 1, its vapor pressure p_A is that of the pure substance (p_A^0). In between, the vapor pressure of each component rises with increasing mole fraction. In Fig. 10.4, one of the dashed straight lines is a plot of $x_{CCl_4} p_{CCl_4}^0$ versus mole fraction. This plot shows the value that p_{CCl_4} would have if it were directly proportional at all compositions to the mole fraction (x_{CCl_4}) of CCl_4. The solid curve represents the actual values of p_{CCl_4} measured over the various solutions. We note that the solid and dashed curves are, in this case, quite close together. Similarly, the solid (experimental) curve for C_6H_6 is close to the dashed line for C_6H_6, which is a plot of $x_{C_6H_6} p_{C_6H_6}^0$.

There are many liquid solutions in which *the partial vapor pressure*

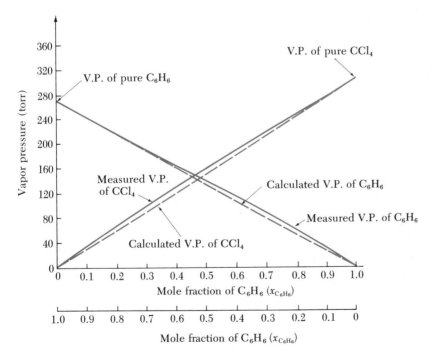

Fig. 10.4. Vapor pressures (V.P.) of benzene (C_6H_6)–carbon tetrachloride (CCl_4) solutions at 49.99°C. The solid curves represent the experimental vapor pressures; the dashed lines represent the vapor pressures predicted by Raoult's law.

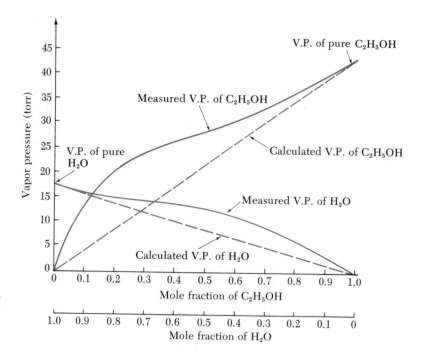

Fig. 10.5. *Vapor pressures (V.P.) of water–ethanol solutions at 20°C. The solid curves represent the experimental vapor pressures; the dashed lines represent the vapor pressures predicted by Raoult's law.*

of each component is nearly equal to the mole fraction of that component times the vapor pressure of the pure component:

$$p_A = x_A p_A^0$$

This relation was first pointed out by François Marie Raoult in 1886, and is called **RAOULT'S LAW**. It is a good approximation for solutions in which the different molecules are very similar in size and polarity. Imagine a liquid (A) in equilibrium with its vapor (Fig. 10.6). For simplicity, we assume that there are only 15 A molecules in the container. In a unit time interval, each molecule in the liquid has a certain probability of escaping, and each gas molecule has a certain probability of returning to the liquid. In Fig. 10.6a (pure A), 100% of the molecules in the liquid are A molecules. In Fig. 10.6b (solution), only $\frac{10}{12} = \frac{5}{6} = 83\%$ of the molecules in the liquid are A molecules. When an A molecule escapes from the liquid, it must come out of the surface layer. But now, some of the A molecules in the surface have been replaced by B molecules. The probability that we will see, in unit time, an A molecule escaping from the solution has been decreased. We may suppose that the probability that some A molecule will escape from the liquid per unit time is cut down by the factor $\frac{5}{6}$. This reasoning is based on the assumption that each individual A molecule has the same chance of escaping whether it is surrounded by all A's or by some A's and some B's—true only if A and B are very much alike. However, the rate of return of A molecules from gas to liquid is unaffected by the presence of B; a molecule has about the same chance of sticking to the liquid whether it collides with another A or with the similar B. To restore equality between the rate of escape and the rate of return, we must cut down the number of A molecules in the vapor by the same factor, $\frac{5}{6}$. The vapor pressure will thus be only $\frac{5}{6}$ of what it was before. That is what Raoult's law tells us: The vapor pressure of A is proportional to the mole fraction of A in the liquid.

A solution in which the components have vapor pressures as given by Raoult's law is called an IDEAL SOLUTION. (An "ideal" solution has nothing to do with an "ideal" gas, except that each is described by an especially simple law.) Probably there are no exactly ideal solutions, but many solutions are nearly ideal. C_6H_6 and CCl_4 illustrate a case in which two liquids form a slightly but measurably nonideal solution.

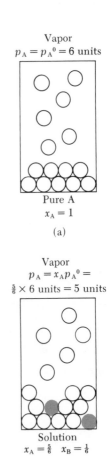

Fig. 10.6. A simplified illustration of the change of vapor pressure with mole fraction of solvent (Raoult's law). (Open circle) A molecule; (filled circle) B molecule.

DEVIATIONS FROM RAOULT'S LAW / 10.12

When the attraction between different molecules is weaker than between identical molecules, the molecules of each kind can escape from the solution more easily than from the pure substance, and the vapor pressures are, therefore, greater than one would predict from Raoult's law; the solution is said to exhibit POSITIVE DEVIATION from Raoult's law. When the attraction is stronger between different than between identical molecules—a less common situation—the molecules are held more firmly in the solution than in the pure substances, the vapor pressures are lower than predicted, and the solution exhibits NEGATIVE DEVIATION from Raoult's law. Figures 10.4 and 10.5 illustrate, respectively, slight positive deviation and strong positive deviation.

The measured ratio of the vapor pressure of dissolved A to that of pure A, $p_A/p_A{}^0$, is called the ACTIVITY of A in the solution and is represented by a_A. For each component of an ideal solution, $p_A/p_A{}^0 = x_A =$ the mole fraction of A. In this case, activity is the same as mole fraction: $a_A = x_A$. The ACTIVITY COEFFICIENT, γ_A, is defined by

$$\gamma_A = \frac{a_A}{x_A} = \frac{p_A}{x_A p_A{}^0}$$

The activity coefficient is the ratio of the actual vapor pressure (p_A) to the vapor pressure calculated from Raoult's law ($x_A p_A{}^0$). It is also the correction factor that converts mole fraction to activity. Finally, it gives a direct measure of the deviation from Raoult's law. If the solution is ideal, $\gamma_A = 1$. When $\gamma_A > 1$, the deviations from Raoult's law are positive; when $\gamma_A < 1$, the deviations are negative.

An important generalization is well illustrated by water in Fig. 10.5. In the solutions containing only a little ethanol, the vapor pressure of water is close to the Raoult's law line, although the solution is far from ideal. This behavior is general: *The vapor pressure of the* SOLVENT *in any dilute solution is given approximately by Raoult's law.* The activity coefficient of the solvent in a dilute solution is therefore close to 1. How dilute the solution must be depends on the identities of the solvent and solute, and on how good an approximation we demand; no general rule can be laid down.

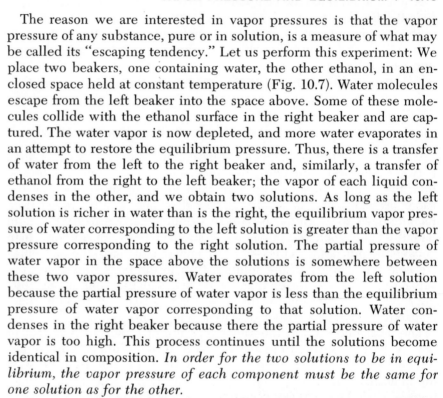

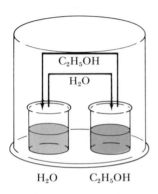

Fig. 10.7. *Transfer of vapor between liquid solutions.*

The reason we are interested in vapor pressures is that the vapor pressure of any substance, pure or in solution, is a measure of what may be called its "escaping tendency." Let us perform this experiment: We place two beakers, one containing water, the other ethanol, in an enclosed space held at constant temperature (Fig. 10.7). Water molecules escape from the left beaker into the space above. Some of these molecules collide with the ethanol surface in the right beaker and are captured. The water vapor is now depleted, and more water evaporates in an attempt to restore the equilibrium pressure. Thus, there is a transfer of water from the left to the right beaker and, similarly, a transfer of ethanol from the right to the left beaker; the vapor of each liquid condenses in the other, and we obtain two solutions. As long as the left solution is richer in water than is the right, the equilibrium vapor pressure of water corresponding to the left solution is greater than the vapor pressure corresponding to the right solution. The partial pressure of water vapor in the space above the solutions is somewhere between these two vapor pressures. Water evaporates from the left solution because the partial pressure of water vapor is less than the equilibrium pressure of water vapor corresponding to that solution. Water condenses in the right beaker because there the partial pressure of water vapor is too high. This process continues until the solutions become identical in composition. *In order for the two solutions to be in equilibrium, the vapor pressure of each component must be the same for one solution as for the other.*

The same principle applies to any two phases: They are in equilibrium if, and only if, the vapor pressure of each component common to the two phases is the same for the two phases. A substance always *tends* to go from the place where its vapor pressure is *higher* to the place where it is *lower*, regardless of whether the transfer takes place through the vapor phase or in some other way, regardless of whether or not vapor is actually present, and regardless of whether or not the vapor pressure is high enough to be measured. However, a process may "tend" to occur, and yet be immeasurably slow. If one beaker in Fig. 10.7 had contained sodium chloride, this compound would have stayed right there, because its vapor pressure is practically zero at room temperature, and its transfer by way of the vapor is too slow to be observed.

COLLIGATIVE PROPERTIES OF SOLUTIONS

We have seen that, in any dilute solution, the vapor pressure of the solvent A is approximately equal to the mole fraction of A times the vapor pressure of pure A (Raoult's law)*: $p_A \approx x_A p_A{}^0$. Let us assume that there is only one solute, B. Then

$$x_A = 1 - x_B$$

and Raoult's law may be written

$$p_A \approx (1 - x_B) p_A{}^0$$

or, after rearrangement,

$$p_A{}^0 - p_A \approx x_B p_A{}^0 = \frac{n_B}{n_A + n_B} p_A{}^0$$

where n_A and n_B are the numbers of moles of the two components. If the solution is dilute, n_B is much less than n_A, and we can simplify this equation to

$$p_A{}^0 - p_A \approx \frac{n_B}{n_A} p_A{}^0$$

The quantity $p_A{}^0 - p_A$ is the VAPOR PRESSURE DEPRESSION caused by the addition of the solute to the solvent. Any property of a solution that is approximately proportional to the number of moles (or molecules) of solute per unit quantity of solvent, independently of the identity of the solute, is called a COLLIGATIVE PROPERTY. Vapor pressure depression is an example of such a property.

BOILING-POINT ELEVATION AND

Two other colligative properties can be understood with the aid of Fig. 10.8. The solid lines reproduce the pressure–temperature phase diagram for a typical substance (call it A), as shown in Fig. 9.14 (page 161). The curve VS gives the vapor pressure of the solid plotted against temperature; SL gives the vapor pressure of the liquid. MS gives the melting point of the solid plotted against pressure. F and B represent the freezing and boiling points of pure A at 1 atm pressure. Now, we add to the solvent A a solute C that has a negligible vapor pressure. The vapor pressure of the liquid is lowered by the presence of the solute. The dashed curve $S'L'$ represents the vapor pressure of A in equilibrium with the solution. The temperature, $t_b{}'$, at which this

---

* The wiggly equal sign ($\approx$) means "approximately equal to."

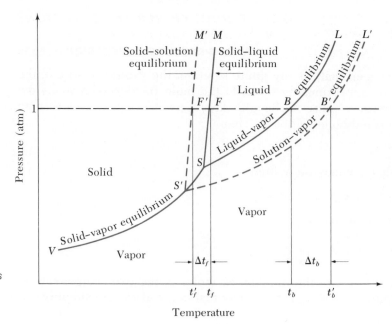

Fig. 10.8. *Phase diagram for pure solvent and solution. Δt_f is the freezing-point depression; Δt_b is the boiling-point elevation.*

vapor pressure becomes equal to 1 atm is the boiling point of the solution (point B'). As we see, t_b' is greater than t_b, the boiling point of pure A. The boiling-point elevation is approximately proportional to the mole fraction of C,

$$t_b' - t_b \approx K x_C = K \frac{n_C}{n_A + n_C} \approx K \frac{n_C}{n_A}$$

The last step assumes that the solution is dilute enough so that n_C is much less than n_A, and therefore $n_A + n_C \approx n_A$.

Similarly, the triple point S has been moved to S', and the solid–liquid equilibrium curve now extends up from S' instead of S. The solid–vapor equilibrium curve is unaffected because the solid that freezes from the solution is, in most cases, pure solvent. The freezing point (t_f') is lower than it was (t_f) for the pure solvent. The depression of the freezing point $(t_f - t_f')$ is, like the boiling-point elevation, a colligative property.

By convention, these two properties are calculated from the molality, rather than the mole fraction:

$$\Delta t_b = t_b' - t_b \approx K_b m \tag{1}$$
$$\Delta t_f = t_f - t_f' \approx K_f m \tag{2}$$

The proportionality constants K_b and K_f, called the boiling-point elevation constant and the freezing-point depression constant, respectively, are characteristic of the solvent. Constants for some common solvents are listed in Table 10.2.

Table 10.2
Boiling-Point Elevation and Freezing-Point Depression Constants

Solvent	Boiling point (°C) at 1 atm	K_b, deg kg mole	Freezing point (°C)	K_f, deg kg mole
Acetic acid (CH₃COOH)	118.1	3.07	16.6	3.90
Benzene (C₆H₆)	80.1	2.53	5.51	4.90
Carbon tetrachloride (CCl₄)	76.8	5.03	−22.8	31.8
Ethanol (C₂H₅OH)	78.5	1.22	−117.3	1.99
Water	100	0.512	0	1.86

Example 4 Find the freezing and boiling points of a solution containing 2.00 g sucrose ($C_{12}H_{22}O_{11}$) dissolved in 100 g water.

Answer We first need to calculate the molality of the solution. The molecular weight of sucrose is 342.3 g/mole. The number of moles of sucrose is

$$\frac{2.00\ g}{342.3\ g/mole} = 0.00584\ mole$$

This amount of sucrose is dissolved in 100 g = 0.100 kg H_2O. The molality is

$$m = \frac{0.00584\ mole}{0.100\ kg} = 0.0584\ \frac{mole}{kg}$$

K_f and K_b for water are found in Table 10.2. Then the freezing-point depression is

$$t_f - t_f' \approx K_f m = 1.86\ \frac{deg\ kg}{mole} \times 0.0584\ \frac{mole}{kg} = 0.109°C$$

and the freezing point is

$$0°C - 0.109°C = -0.109°C$$

The boiling-point elevation is

$$t_b' - t_b \approx K_b m = 0.512\ \frac{deg\ kg}{mole} \times 0.0584\ \frac{mole}{kg} = 0.0299°C$$

and the boiling point is

$$100°C + 0.0299°C = 100.03°C$$

When the solution contains two or more solutes, the molality to be used in calculating the colligative properties is the *total molality of all solutes*:

$$m = m_C + m_D + \cdots$$

Equations (1) and (2) are applicable only under certain conditions:

(a) They are approximations good only for dilute solutions.

(b) Equation (1) is only for the case of a *nonvolatile solute*—one with vapor pressure nearly equal to zero. The boiling points in Fig. 10.8 are the temperatures at which the *solvent* vapor pressure becomes 1 atm; if the solute gets into the act, everything is more complicated.

(c) Equation (2) assumes that the solid which appears at the freezing point is pure *solvent*—not the solute, a compound of solvent and solute, or a solid solution.

OSMOTIC PRESSURE / 10.16

Some membranes—for example, cellophane, parchment paper, the walls of living cells—are SEMIPERMEABLE. Such a membrane allows some molecules, but not others, to pass through. In some cases, the membrane seems to act simply as a sieve, passing small molecules but not large ones.

Suppose we have a membrane that is permeable to the solvent but not to the solute in a certain solution. An example would be a cellophane membrane with an aqueous solution of sugar. Let us separate a sugar solution from pure water by means of this membrane (Fig. 10.9). A simplified explanation of what happens is this: Water molecules are continually colliding with the membrane from both sides. However, the rate of collision is somewhat less on the solution side (right) than on the pure water side (left) because the concentration of water molecules in a solution is less than their concentration in pure water. The membrane is permeable to water, which means that some of the colliding molecules manage to get through. Since there are more collisions from the left than from the right, more water molecules pass through from

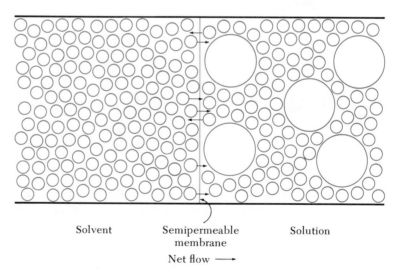

Solvent   Semipermeable   Solution
              membrane

Net flow ⟶

Fig. 10.9. *Osmotic flow from pure solvent to solution.*

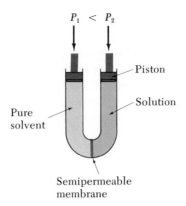

$P_1 < P_2$

Piston

Solution

Pure solvent

Semipermeable membrane

Fig. 10.10. *An apparatus to illustrate osmotic pressure. To maintain equilibrium between solvent and solution, the pressure P_2 on the right piston must be greater than the pressure P_1 on the left piston. The difference, $\pi = P_2 - P_1$, is the osmotic pressure of the solution.*

left to right than from right to left. This flow from solvent to solution, or from one solution to another, through a semipermeable membrane is known as OSMOSIS (Greek, "push"). Sugar molecules are also colliding with the membrane from the right, but they are just butting their heads against a wall.

Water can also be made to flow through the membrane by applying a greater pressure to one side than to the other. Figure 10.10 illustrates an apparatus that can be used in this way. The application of a greater pressure (P_2) on the right piston than on the left (P_1) pushes the water from right to left. The left-to-right osmotic flow can be stopped by just the correct excess pressure on the right piston. The excess pressure required to stop the flow and keep the solution in equilibrium with the pure solvent is the OSMOTIC PRESSURE (π) of the solution: $\pi = P_2 - P_1$. For dilute solutions, π is given by the equation

$$\pi \approx [B]RT$$

where [B] is the molarity of the solute (moles per liter), R is the gas constant (0.08206 liter-atm/deg-mole), and T is the absolute temperature. Since π is approximately proportional to the concentration of the solute but independent of the identity of the solute, it is a colligative property.

Osmosis is a process of overwhelming importance in living organisms. Cell membranes are generally permeable to water and to some solutes, but more or less impermeable to others. Thus, different concentrations of various dissolved substances can be maintained inside and outside the cell. Nutrients pass through the intestinal walls into the blood and lymph, while wastes pass through the kidney membrane into the urine. The normal osmotic pressure of blood and other body fluids in man, measured with a membrane permeable only to water, is equal to that of a 0.16 M NaCl solution. If one ingests a solution, such as sea water, with a higher osmotic pressure than blood, the flow of water will be from the blood to the intestinal contents — the opposite of the effect of drinking fresh water.

When the excess pressure $P_2 - P_1$ is made *greater* than the osmotic pressure, there is a flow of solvent from solution to pure solvent. This REVERSE OSMOSIS can be used to purify polluted water.

The solvent will flow from a solution with a lower to a solution with a higher osmotic pressure. Like vapor pressure, osmotic pressure is a measure of the escaping tendency of the solvent, but in reverse: The *greater* the osmotic pressure, the *less* the escaping tendency.

DETERMINATION OF MOLECULAR WEIGHTS / 10.17

The strongest reason for interest in colligative* properties is that they provide a means of measuring the number of moles of solute, n, present; if we also know the mass of the solute, w, we can calculate its molecular weight, w/n. Freezing-point depression is the easiest of the properties to measure experimentally, except with solutes of very high molecular weight (Chapter 26), for which osmotic pressure is more useful.

Example 5 The freezing point of pure benzene was found to be 5.49°C. With the same benzene and equipment, the freezing point of a solution containing 0.100 g naphthalene in 10.0 g benzene was found to be 5.10°C. Find the molecular weight of naphthalene.

Answer The freezing-point depression is

$$\Delta t_f = t_f - t_f' = 5.49° - 5.10° = 0.39°C$$

The freezing-point depression constant is

$$K_f = 4.90 \; \frac{\text{deg kg}}{\text{mole}} \qquad \text{(Table 10.2)}$$

From the equation $\Delta t_f = K_f m$, we can calculate the molality of naphthalene:

$$m = \frac{\Delta t_f}{K_f} = \frac{0.39 \; \text{deg}}{4.90 \; \text{deg kg/mole}} = 0.080 \; \frac{\text{mole}}{\text{kg}}$$

The mass of naphthalene per kilogram of benzene is

$$w = \frac{0.100 \; \text{g naphthalene}}{10.0 \; \text{g benzene}} \times 1000 \; \frac{\text{g}}{\text{kg}} = 10.0 \; \frac{\text{g naphthalene}}{\text{kg benzene}}$$

The molecular weight is, therefore,

$$\frac{w}{m} = \frac{10.0 \; \text{g/kg}}{0.080 \; \text{mole/kg}} = 125 \; \frac{\text{g}}{\text{mole}} \qquad \left(\text{actual value, } 128 \; \frac{\text{g}}{\text{mole}} \right)$$

---

* The word "colligative" means, literally, "pertaining to binding together." The reference is to the fact that the four colligative properties are interrelated, not to the binding together of atoms or molecules.

Example 6 An aqueous solution containing 1.00 g bovine insulin (a protein) per liter has osmotic pressure 3.1 torr at 25°C. Find the molecular weight of bovine insulin.

Answer The osmotic pressure is related to the concentration by $\pi = cRT$. We first calculate the concentration:

$$c = \frac{\pi}{RT}$$

$$= \left(\frac{3.1 \text{ torr}}{760 \text{ torr/atm}}\right) \times \frac{1}{(0.08206 \text{ liter atm/°K mole})(298°K)}$$

$$= 1.7 \times 10^{-4} \text{ mole/liter}$$

We are told that there is 1.00 g insulin/liter. Then the molecular weight is

$$\frac{1.00 \text{ g/liter}}{1.7 \times 10^{-4} \text{ mole/liter}} = 6.0 \times 10^3 \text{ g/mole} \qquad \text{(actual value, 5733 g/mole)}$$

Solutes of high molecular weight, like insulin, give osmotic pressures that are easily measured, but imperceptibly small freezing-point depressions. Solutes of low molecular weight, like naphthalene, give convenient freezing-point depressions, but osmotic pressures so high that the membrane is likely to break before equilibrium is attained—if, indeed, one can find a suitable semipermeable membrane. The two methods thus complement each other nicely: freezing-point depression when the molecular weight is low, osmotic pressure when it is high.

PROBLEMS

1. Definitions. Define briefly: (a) mole fraction, (b) molality, (c) molarity, (d) solubility, (e) saturated solution, (f) supersaturated solution, (g) ideal solution, (h) activity, (i) colligative property, (j) freezing-point depression, (k) osmotic pressure.

2. Molarity. Solutions were prepared by dissolving 50.00 g of each of the following solutes in water, and adding water to the mark in a 500-ml volumetric flask. Find the molarity of each solution: (a) KBr; (b) Na_2HPO_4; (c) fructose, $C_6H_{12}O_6$.

3. Molarity. We need 500 ml of a 0.100 M solution of HCl. A 3.00 M solution of HCl is available in the laboratory, as is a 500-ml volumetric flask. (a) How many moles of HCl do we need? (b) What volume of the 3.00 M solution should be measured out and diluted with water in the flask?

4. Composition. A bottle of concentrated hydrobromic acid has the following information on the label: "Assay 48% HBr. Sp. gr. 1.5." The assay is the percentage by mass; the specific gravity is numerically equal to the density in grams per milliliter. Calculate (a) the mole fraction and (b) the molarity of HBr in this solution.

5. Milliunits. A solution of HCl is labeled "0.200 M." (a) How many millimoles of HCl are in 50.0 ml of this solution? (b) How many milligrams of

HCl are in 50.0 ml of the solution? (c) What volume of this solution would you take in order to prepare 500 ml of a 0.0500 M solution? How would you then prepare the desired solution?

6. Solubility. The solubility of $MgCl_2$ in water at 20°C is 54 g $MgCl_2$/100 g H_2O. Calculate the solubility in (a) weight percent; (b) molarity. The density of the saturated solution is 1.32 g/ml.

7. Solubility and temperature. Given the following enthalpy changes:

$$N_2(g) + 2H_2(g) + C(graphite) + \tfrac{1}{2}O_2(g) \longrightarrow (NH_2)_2CO(c) \quad \Delta H = -79.63 \text{ kcal}$$
$$N_2(g) + 2H_2(g) + C(graphite) + \tfrac{1}{2}O_2(g) + aq \longrightarrow (NH_2)_2CO(aq) \quad \Delta H = -76.30 \text{ kcal}$$

Does the solubility of urea, $(NH_2)_2CO$, in water increase or decrease as the temperature increases?

8. Raoult's law. At 25°C, the vapor pressure of liquid benzene, C_6H_6, is 95 torr; the vapor pressure of liquid toluene, C_7H_8, is 28 torr. Assume that solutions of these two compounds are ideal. Find the total vapor pressure at 25°C of a liquid solution containing 30% benzene and 70% toluene by mass.

9. Colligative properties. A solution contains 2.00% ethylene glycol, $C_2H_4(OH)_2$, and 98.00% water by mass. The density of the solution is 1.00 g/ml. Find (a) the freezing point of the solution; (b) the osmotic pressure of the solution at 25°C, with a membrane permeable to water but not to ethylene glycol.

10. Molecular weight. An aqueous solution contains 0.20 g of an unknown solute and 100.0 g H_2O. It freezes at −0.050°C. Find the molecular weight of the solute.

11. Osmotic pressure. An aqueous solution containing 5.0 g starch/liter gave an osmotic pressure of 12 torr at 25°C. (a) Find the molecular weight of the starch. (Because not all the starch molecules are identical, the result will be an average.) (b) What is the freezing point of the solution? Would it be easy to find the molecular weight of the starch by measuring the freezing-point depression? (Assume that molality and molarity are equal.)

ADDITIONAL PROBLEMS

12. Molarity. What mass of each of the following compounds should be weighed out in order to prepare a 0.1000 M solution with the aid of a 500.0-ml volumetric flask? (a) Sodium nitrate, $NaNO_3$; (b) iron(II) ammonium sulfate hexahydrate ("Mohr's salt"), $Fe(NH_4)_2(SO_4)_2(H_2O)_6$; (c) potassium hydrogen phthalate, $KHC_8H_4O_4$.

13. Composition. The molarity of NH_3 in an aqueous solution is 10.06 moles/liter. The density of the solution is 0.930 g/ml. Find (a) the weight percentage and (b) the mole percentage of NH_3 in this solution. (c) The answers to (a) and (b) are nearly the same. Explain.

14. Molarity and temperature. The coefficient of expansion of H_2O near 20°C is 2.1×10^{-4}/deg; this means that the volume of a sample of H_2O is multiplied by 1.00021 when the temperature is raised 1°C. A certain aqueous solution is 0.1000 M at 25°C. Find its molarity at 20°C, assuming that the coefficient of expansion of this solution is the same as that of pure water.

15. Composition measures. A bottle of concentrated ammonia solution ("ammonium hydroxide") has the following information on the label: "Assay 30% NH_4OH. Sp. gr. 0.90." (See Problem 4.) Find (a) the percentage by mass of NH_3 in this solution (note that "NH_4OH" is $NH_3 + H_2O$); (b) the molarity of NH_3; (c) the volume of the solution that must be used to prepare 100 ml of 2.0 M solution by dilution with water.

16. Solubility and temperature. The solubility of ammonium bromide, NH_4Br, in water is 60.5 g/100 g H_2O at 0°C and 126.2 g/100 g H_2O at 80°C. A solution was prepared by dissolving NH_4Br in 200 g H_2O until no more would dissolve at 80°C, and the solution was then cooled to 0°C. How many grams of NH_4Br precipitated? Assume that no water evaporated and that there was no supersaturation.

17. Solubility and temperature. Saturated solutions of potassium nitrate, KNO_3, in water contain 17.2% KNO_3 (by mass) at 10°C and 62.8% KNO_3 at 80°C. A solution prepared from 50.0 g KNO_3 and 50.0 g H_2O was cooled from 80° to 10°C. What mass of KNO_3 precipitated? Make the same assumptions as in the preceding problem.

18. Solubility and molecular structure. Draw a structural formula for a hydrogen-bonded combination of H_2O with (a) NH_3; (b) C_2H_5OH; (c) $(CH_3)_2CO$. (See page 176.) The H atoms attached to C do not participate in hydrogen bonding.

19. Solubility and molecular structure. In each of the following pairs, select the compound that you would expect to be more soluble in water; explain.
(a) C_2H_5F and C_2H_5I
(b) C_3H_7OH and C_3H_7SH
(c)

(The circles are explained on page 489.)
(d) NO and H_2NNH_2
(e) NH_3 and AsH_3

20. Detergents. Which of the following compounds is more promising as a detergent? Why?

21. Density. At 20°C, the density of benzene, C_6H_6, is 0.879 g/ml; the density of carbon tetrachloride, CCl_4, is 1.595 g/ml. Assuming no volume change on mixing, find the density of a solution prepared by mixing 40.0 g benzene and 60.0 g carbon tetrachloride.

22. Volume change on mixing. At 20°C, the density of ethanol, C_2H_5OH, is 0.7893 g/ml; the density of water is 0.9982 g/ml; the density of a 50% (by mass) solution of ethanol in water is 0.9131 g/ml. A solution is prepared by mixing 50.0 g ethanol and 50.0 g water at 20°C. (a) What is the volume of the ethanol? (b) What is the volume of the water? (c) What is the volume of the solution? (d) Is the volume of the solution more or less than the sum of the volumes of the components? How much more or less?

23. Raoult's law. At 27°C, the vapor pressure of carbon disulfide, CS_2, is 390 torr; of ethyl acetate, $CH_3COOC_2H_5$, 100 torr. Find the total vapor pressure at 27°C of a solution prepared by mixing 20.0 g CS_2 and 30.0 g $CH_3COOC_2H_5$. Assume that Raoult's law is applicable. (b) A solution of CS_2 and $CH_3COOC_2H_5$ has a total vapor pressure of 300 torr at 27°C. Calculate the mole fraction of CS_2 in this solution.

24. Molecular weight. A solution of 5.00 g acetic acid in 100 g benzene freezes at 3.38°C. A solution of 5.00 g acetic acid in 100 g water freezes at −1.49°C. Find the molecular weight of acetic acid from each of these data. What can you conclude about the state of the acetic acid molecules dissolved in each of these solvents?

25. Freezing-point depression. The two most commonly used antifreezes are methanol, CH_3OH, and ethylene glycol, $(CH_2OH)_2$. (a) Estimate the mass of each of these solutes that must be added to 1 kg of water in order to prevent the water from freezing at −10°C. Why are your answers only estimates? (b) A certain solution of water and methanol has a freezing point of −10°C. This means that pure ice deposits when the solution is cooled to −10°C. An automobile radiator is filled with this solution, tightly capped, and kept at −11°C for an indefinitely long time. Will the liquid in the radiator eventually freeze solid? Explain.

26. Molecular weight. A compound of C and H contains 4.03% H by mass. A solution containing 1.00 g of this compound and 50.0 g benzene freezes 0.327°C lower than the freezing point of pure benzene. Find (a) the molecular weight and (b) the molecular formula of the compound.

27. Osmotic pressure. (a) An aqueous solution containing 1.00 g hemoglobin (a protein) per liter shows an osmotic pressure of 0.290 torr at 27°C. Find the molecular weight of hemoglobin. (b) Estimate the freezing-point depression of this solution. (c) Which method, osmotic pressure or freezing-point depression, would you recommend for determining molecular weights of proteins?

11

CHEMICAL EQUILIBRIUM

A gaseous mixture of hydrogen and iodine, in a closed vessel, will react to form hydrogen iodide:

$$H_2(g) + I_2(g) \longrightarrow 2HI(g) \tag{1}$$

If we had started with pure HI at the same temperature, it would have decomposed to give hydrogen and iodine:

$$2HI(g) \longrightarrow H_2(g) + I_2(g) \tag{2}$$

Now, the HI that is formed in reaction (1) acts like any other HI. As soon as some of it has been formed, the decomposition [reaction (2)] sets in. The H_2 and I_2 formed in reaction (2) will react to give HI, as in reaction (1). Thus, in any mixture of the three gases H_2, I_2, and HI, reactions (1) and (2) will both take place. This fact is expressed by saying that reaction (1) or (2) is REVERSIBLE, and writing the equation with a double arrow:

$$H_2 + I_2 \rightleftharpoons 2HI$$

or

$$2HI \rightleftharpoons H_2 + I_2$$

The rate of a reaction usually increases as the concentrations of reactants increase, and decreases as they decrease. If reaction (1) is faster than reaction (2), there is a net formation of HI. Its concentration increases with time, while the concentrations of H_2 and I_2 decrease. As the concentrations of H_2 and I_2 decrease, reaction (1) slows down, while the rising concentration of HI speeds up reaction (2). Conversely, if reaction (2) is faster, the loss of HI will slow it down, and the gain of H_2 and I_2 will speed up reaction (1). A state of equilibrium will eventually

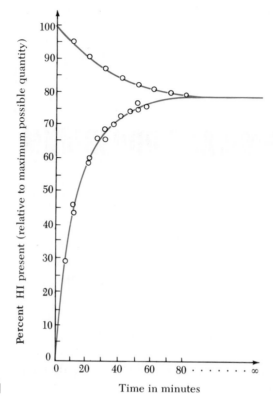

Fig. 11.1. *Experimental data on the reaction*
$H_2(g) + I_2(g) \rightleftharpoons 2HI(g)$
at 445°C. [*After M. Bodenstein, Z. Phys. Chem.* **13**, 111 (1894).]

Percent HI present (relative to maximum possible quantity)

Time in minutes

be reached when the forward and reverse reactions are occurring at the same rate. From then on, there will be *no further net change* in the composition of the reaction mixture, unless some external disturbance is imposed. The concept of equilibrium is the same here as for a saturated solution, in which the two processes of dissolving and coming out of solution are equal in rate at equilibrium.

Figure 11.1 shows experimental data which illustrate how the reaction $H_2 + I_2 \rightleftharpoons 2HI$ approaches equilibrium. The horizontal axis represents the time elapsed from the beginning of the reaction. The vertical axis represents the composition of the reaction mixture, from $H_2 + I_2$ at the bottom to pure HI at the top. If we start at the bottom with $H_2 + I_2$, the percentage converted to HI increases rapidly at first, then more slowly, approaching a limit around 78%. If we start at the top with pure HI, the same final state is attained. In this particular case, the mixture reaches equilibrium in an hour or two; in other reactions, it may take a tiny fraction of a second, and there are many reactions that would be nowhere close to equilibrium after a million years. Some reactions settle down to equilibrium after formation of only minute amounts of the products; in other reactions, only minute amounts of the reactants remain unreacted at equilibrium. Sometimes, as in the HI reaction, appreciable quantities of both reactants and products are present in the equilibrium mixture.

Whatever appears on the left-hand side of an equation is called a "reactant"; whatever appears on the right-hand side is called a "product." This terminology results from our habit of reading from left to right. The reaction can go either way, depending on what we start with. A reaction is said to go "forward" or "as written" when it goes from left to right, "backwards" or "in reverse" when it goes from right to left. When we write "$\Delta H = -100$ kcal" for a reaction with a double arrow, it is understood that -100 kcal is the enthalpy change in the *left-to-right* reaction.

The changes in reaction rates that lead to equilibrium are associated with changes in the concentrations in moles per liter (page 168) of reactants and products. The same information can also be given by other measures of composition. For gases, the partial pressure is related to the concentration in a simple way (page 15):

$$p_A = \frac{n_A RT}{V} = [A]RT$$

where $[A] = n_A/V$ is the concentration of A.

THE LAW OF CHEMICAL EQUILIBRIUM / 11.2

Suppose that the substances A, B, C, and D undergo the reaction

$$aA + bB \rightleftharpoons cC + dD$$

A and B are the "reactants"; C and D are the "products." The small letters are the coefficients. We define the REACTION QUOTIENT for this reaction:

$$Q = \frac{[C]^c[D]^d}{[A]^a[B]^b}$$

For example, if the reaction is

$$2Cl_2(g) + 2H_2O(g) \rightleftharpoons 4HCl(g) + O_2(g)$$

then $a = 2$, $b = 2$, $c = 4$, and $d = 1$; the reaction quotient is

$$Q = \frac{[HCl]^4[O_2]}{[Cl_2]^2[H_2O]^2}$$

Remember: *Products over reactants;* each concentration is raised to a power equal to the corresponding coefficient in the chemical equation; concentrations are multiplied, never added.

The LAW OF CHEMICAL EQUILIBRIUM says that equilibrium is established when Q reaches a certain value, called the EQUILIBRIUM CONSTANT K:

$$Q = K \qquad\qquad (3)$$

For the previous examples,

$$\frac{[C]^c[D]^d}{[A]^a[B]^b} = K$$

$$\frac{[HCl]^4[O_2]}{[Cl_2]^2[H_2O]^2} = K$$

This equation is the EQUILIBRIUM CONDITION for the reaction. The value of K depends on what the reaction is, on how the equation is written (page 197), and on the temperature. It does not, however, depend on the composition of a particular mixture; that is what makes it a "constant." *The concentrations will adjust themselves until Equation (3) is satisfied.*

When $Q < K$, the concentrations of the products are smaller than they would be at equilibrium, and the concentrations of the reactants are larger. The reaction can then proceed only in the *forward* direction, reactants → products, so that Q approaches K. Conversely, when $Q > K$, the concentrations of the products are greater than they would be at equilibrium, and the concentrations of the reactants are smaller. The reaction can then proceed only in the *reverse* direction, products → reactants, and again Q approaches K. In summary,

When $Q < K$	When $Q > K$
Numerator (product concentrations) is too small	Numerator (product concentrations) is too large
Denominator (reactant concentrations) is too large	Denominator (reactant concentrations) is too small
Reaction can proceed ⟶	Reaction can proceed ⟵

The reaction *can* proceed in the direction making Q *more nearly* equal to K. We have not said that the reaction *will* take place in that direction; there may be no reaction at all, or there may be an immeasurably slow reaction. All we know is that the reaction will *not* go in the other direction.

The law of chemical equilibrium is exact for ideal gases (page 14). It is a good approximation for real gases at moderate pressures. It is a not so good, but still useful, approximation for liquid solutions. We will first consider gases.

An example of a reversible gas reaction is

$$2SO_2(g) + O_2(g) \rightleftharpoons 2SO_3(g)$$

This reaction is the crucial step in the usual process for making sulfuric acid, H_2SO_4 (page 444). The reaction is extremely slow under ordinary conditions, but proceeds quite rapidly above 500°C in the presence of a catalyst (page 420), such as divanadium pentoxide (V_2O_5) or platinum. Table 11.1 shows data from experiments (Max Bodenstein,

Table 11.1
Determination of the Equilibrium Constant of $2SO_2(g) + O_2(g) \rightleftharpoons 2SO_3(g)$ *at 1000°K*

Initial concentrations (mole/liter)		Concentrations at equilibrium (mole/liter)			$\dfrac{[SO_3]^2}{[SO_2]^2[O_2]}$
$[SO_2]$	$[O_2]$	$[SO_2]$	$[O_2]$	$[SO_3]$	
0.00675	0.00544	0.00377	0.00430	0.00412	278
0.00847	0.00372	0.00557	0.00219	0.00444	290
0.00864	0.00354	0.00590	0.00201	0.00435	270
0.00939	0.00280	0.00681	0.00123	0.00403	285
0.00433	0.00165	0.00302	0.000975	0.00156	274
				Average	279
			Relative average deviation[a]		2.3%

[a] Average deviation = 6.4 (page 309); r.a.d. = (6.4/279) × 100% = 2.3%.

1905) in which the equilibrium constant of this reaction was measured. A mixture of SO_2 and O_2, in the concentrations shown, was passed through a tube containing platinum sponge heated to 1000°K. The outcoming gas was analyzed for SO_2, SO_3, and O_2. The equilibrium condition in this case is

$$\frac{[SO_3]^2}{[SO_2]^2[O_2]} = K$$

The reaction quotient is approximately constant, as it should be at equilibrium.

CHANGE OF K WITH FORM OF EQUATION / 11.3

Instead of writing

$$2SO_2 + O_2 \rightleftharpoons 2SO_3$$

with the equilibrium condition

$$\frac{[SO_3]^2}{[SO_2]^2[O_2]} = K$$

we could double all the coefficients and write the equally correct equation

$$4SO_2 + 2O_2 \rightleftharpoons 4SO_3$$

The equilibrium condition then becomes

$$\frac{[SO_3]^4}{[SO]^4[O_2]^2} = K'$$

Obviously, $K' = K^2$. When the chemical equation is multiplied through by a constant n, the equilibrium constant is raised to the nth power. If the reaction is written backwards,

$$2SO_3 \rightleftharpoons 2SO_2 + O_2$$

the equilibrium condition becomes

$$\frac{[SO_2]^2[O_2]}{[SO_3]^2} = K'' = \frac{1}{K}$$

Reversing the chemical equation changes the equilibrium constant to its reciprocal. In summary,

If the coefficients in the chemical equation are	Then K is
Doubled	Squared
Reversed in sign (left and right sides interchanged)	Inverted (the -1 power)
Halved	Replaced by its square root (the $\frac{1}{2}$ power)
Multiplied by n	Raised to the nth power

Thus, to speak of "the equilibrium constant for the formation of SO_3 from SO_2 and O_2" is ambiguous — one should write the chemical equation.

COMBINATION OF EQUILIBRIA / 11.4

It is possible to find the equilibrium constant for a reaction when the constants for certain other reactions are known. One may be interested, for example, in the reaction

$$SO_2(g) + CO_2(g) \rightleftharpoons SO_3(g) + CO(g) \tag{4}$$

but may find that data on it are not readily available. However, the equilibrium constants for the reactions

$$SO_2(g) + \tfrac{1}{2}O_2(g) \rightleftharpoons SO_3(g) \tag{5}$$

and

$$CO_2(g) \rightleftharpoons CO(g) + \tfrac{1}{2}O_2(g) \tag{6}$$

are well known. We can write for them the equilibrium conditions

$$\frac{[SO_3]}{[SO_2][O_2]^{1/2}} = K_5$$

$$\frac{[CO][O_2]^{1/2}}{[CO_2]} = K_6$$

respectively. (The constants are labeled by the same numbers as the equations.) In a mixture of all five substances — SO_2, SO_3, CO, CO_2, and O_2 — these two conditions must be satisfied simultaneously at equilibrium. If we multiply the left-hand side of one equation by the left-hand side of the other, and similarly multiply the right-hand sides together, we obtain another equation:

$$\frac{[SO_3]}{[SO_2][O_2]^{1/2}} \times \frac{[CO][O_2]^{1/2}}{[CO_2]} = K_5 K_6$$

or

$$\frac{[SO_3][CO]}{[SO_2][CO_2]} = K_5 K_6 \qquad (7)$$

But the equilibrium condition for Equation (4) is

$$\frac{[SO_3][CO]}{[SO_2][CO_2]} = K_4 \qquad (8)$$

On comparing Equations (7) and (8), we see that

$$K_4 = K_5 K_6$$

Of course, K_4 is the same whether or not O_2 is present.

The procedure can be simplified if we observe that, when two reactions are added to obtain a third reaction, their equilibrium constants are multiplied to give the equilibrium constant of the third reaction. Thus, we may add reactions (5) and (6):

$$SO_2(g) + \tfrac{1}{2}O_2(g) \rightleftharpoons SO_3(g) \qquad (5)$$
$$\underline{CO_2(g) \rightleftharpoons CO(g) + \tfrac{1}{2}O_2(g)} \qquad (6)$$
$$SO_2(g) + CO_2(g) \rightleftharpoons SO_3(g) + CO(g) \qquad (4)$$

Without writing out the equilibrium conditions, we can see, from the way the reaction quotient is constructed, that $K_5 K_6 = K_4$. In general, when a number of reactions are *added*, their equilibrium constants are *multiplied* together to give the equilibrium constant of the overall reaction. The process is similar to the calculation of a heat of reaction with the aid of Hess' law (page 64), except that equilibrium constants are multiplied, while heats of reaction are added.

PRINCIPLE OF LE CHATELIER / 11.5

The composition of a mixture at equilibrium depends, in general, on the quantities of the reacting substances that are initially brought together, on the volume of the container in which they are placed, and on the temperature. Only the temperature affects the actual value of the equilibrium constant. However, a change in volume, or the addition or removal of one of the substances involved in the reaction, may require a change in the composition of the mixture so that the equilibrium condition is still satisfied ($Q = K$). In this section, we shall develop some useful rules enabling us to predict qualitatively the direction in which the composition of an equilibrium mixture will change.

(a) **Change of quantity.** If a mixture of SO_2, O_2, and SO_3 is in equilibrium and more O_2 is added without changing the volume, the initial effect is to make $[SO_3]^2/[SO_2]^2[O_2]$ $(= Q)$ smaller than K because the denominator has become larger. If equilibrium is to be restored, the numerator must increase, or the denominator must decrease, or both. Actually, both will change: SO_3 can be produced only by consuming SO_2 and O_2. When this happens, the reaction $2SO_2 + O_2 \leftrightarrows 2SO_3$ is said to *shift to the right*. Similarly, adding SO_2 causes a shift to the right; adding SO_3 causes a shift to the left (formation of more SO_2 and O_2). The general rule is that *increasing the concentration of one substance in an equilibrium mixture displaces the equilibrium in that direction which consumes some of the material added*. Conversely, decreasing the concentration of a substance causes the production of more of that substance.

(b) **Change of volume.** The concentrations of gases are changed, and an equilibrium mixture is thus disturbed, if the volume of the container is changed. For the reaction $2SO_2 + O_2 \leftrightarrows 2SO_3$,

$$Q = \frac{[SO_3]^2}{[SO_2]^2[O_2]}$$

Suppose we have 1.00 liter of the equilibrium mixture described in the first line of Table 11.1, with $[SO_3] = 0.00412$ mole/liter, $[SO_2] = 0.00377$ mole liter, and $[O_2] = 0.00430$ mole/liter. Then $Q = 278 = K$. Let us assume that the reaction is slow enough so that we can disturb the mixture and then study it before it has changed toward its new equilibrium. Now we double the volume of the container to 2.00 liters at constant temperature. Each concentration is thereby divided by 2: instead of

$$[SO_3] = \frac{0.00412 \text{ mole}}{1.00 \text{ liter}} = 0.00412 \frac{\text{mole}}{\text{liter}}$$

we now have

$$[SO_3] = \frac{0.00412 \text{ mole}}{2.00 \text{ liters}} = 0.00206 \frac{\text{mole}}{\text{liter}}$$

Similarly,

$$[SO_2] = \frac{0.00377 \text{ mole}}{2.00 \text{ liters}} = 0.001885 \frac{\text{mole}}{\text{liter}}$$

and

$$[O_2] = \frac{0.00430 \text{ mole}}{2.00 \text{ liters}} = 0.00215 \frac{\text{mole}}{\text{liter}}$$

Now the reaction quotient is

$$Q = \frac{(0.00206)^2}{(0.001885)^2 \times 0.00215} = 555$$

Since $Q > K$, the mixture is no longer in equilibrium. To restore equilibrium, more SO_2 and O_2 must form at the expense of SO_3. Decreasing the volume, conversely, results in the formation of more SO_3. The rule is that *increasing the volume of the container favors that reaction which produces more moles of gas.* $2SO_2 + O_2$ is 3 moles of gas, $2SO_3$ is 2 moles of gas, and the equilibrium $2SO_2 + O_2 \rightleftharpoons 2SO_3$ is therefore shifted to the left when the volume increases, to the right when it decreases.

A few other examples of the effect of a volume change are the following:

	Increase volume	*Decrease volume*
$CO(g) + Cl_2(g) \rightleftharpoons COCl_2(g)$ 2 moles of gas 1 mole of gas	Shifts $\longleftarrow$	Shifts $\longrightarrow$
$N_2(g) + O_2(g) \rightleftharpoons 2NO(g)$ 2 moles of gas 2 moles of gas	No effect	No effect
$H_2S(g) \rightleftharpoons H_2(g) + \frac{1}{2}S_2(g)$ 1 mole of gas $1\frac{1}{2}$ mole of gas	Shifts $\longrightarrow$	Shifts $\longleftarrow$

In general, when we decrease the volume available to a mixture, we induce the molecules to react in a way that results in fewer molecules. Conversely, increasing the available volume favors the process which increases the number of molecules.

(c) **Change of temperature.** An equilibrium "constant" is constant only as long as the temperature does not change. A change in temperature results in a change, usually large, in the equilibrium constant. When K increases, the equilibrium must shift to the right; when K decreases, the equilibrium must shift to the left. *Increasing the temperature causes reaction to occur in that direction which results in absorption of heat; decreasing the temperature causes reaction to occur in that direction which results in emission of heat.* In the formation of SO_3, heat is given out:

$$2SO_2 + O_2 \longrightarrow 2SO_3 \qquad \Delta H = -47 \text{ kcal}$$

Therefore, raising the temperature shifts the equilibrium to the left, and K must have decreased; lowering the temperature increases K.[*]

The rules given in this section are summarized by **LE CHATELIER'S**

---

[*] The *rate* of attainment of equilibrium is *increased by raising the temperature,* regardless of the effect that temperature has on the final state of equilibrium. An increase in temperature speeds up both the formation and the decomposition of SO_3, but speeds up the decomposition more than the formation, and thus less SO_3 is present when the two rates become equal.

PRINCIPLE*: *When a system in equilibrium is disturbed, the equilibrium shifts so as to undo, in part, the effect of the disturbance.* Adding a reactant results in partial consumption of that reactant. Rasing the temperature results in an endothermic reaction, which uses up some of the added heat and makes the rise in temperature less than it would be if the same quantity of heat were added and no reaction occurred. The volume change can be fitted into this pattern with a touch of the shoehorn. When the volume of the container is *decreased,* the total pressure is *increased.* The shift in the equilibrium now decreases the number of moles of gas and makes the increase of pressure less than it would be if the equilibrium did not shift. The pressure increase, not the volume decrease, is the effect that is partly undone. Thus, if the volume is halved, the pressure is not doubled (as Boyle's law would predict); the pressure might be multiplied only by 1.8 or 1.9 — something between 1 and 2.

If the pressure is increased by adding a gas that is not involved in the reaction, the equilibrium is *not* shifted. That is why we emphasize the volume change rather than the pressure change. If helium, for example, is added to an equilibrium mixture of SO_2, O_2, and SO_3 (with the volume kept constant), these three kinds of molecules collide with, but otherwise ignore, the He atoms. Their reactions with each other are unaffected.

Another change that does not cause a shift in equilibrium is the addition or removal of a catalyst (page 420). A catalyst changes the rate of approach to equilibrium, but must affect the rates of the forward and reverse reactions in the same way, so that the composition of the equilibrium mixture is unchanged by its presence.

HETEROGENEOUS EQUILIBRIUM / 11.6

When chemical equilibrium is established in one phase — a mixture of gases, a liquid solution — we have a case of HOMOGENEOUS EQUILIBRIUM. An equilibrium involving more than one phase — gas and solid, for example, or liquid and solid — is said to be HETEROGENEOUS. A saturated solution in equilibrium with solute is a familiar instance of heterogeneous equilibrium. For the present we shall concern ourselves with gas–solid reactions.

We have studied the formation of HI from H_2 and I_2 at a temperature at which I_2 is a gas. At room temperature, however, I_2 is a solid. The composition of the equilibrium mixture of H_2 and HI in the reaction

$$H_2(g) + I_2(c) \rightleftharpoons 2HI(g) \tag{9}$$

* After Henri-Louis Le Chatelier, who formulated the principle in 1888.

is independent of the amount of solid I_2 present, as long as some of it is present. The equilibrium condition for reaction (9) is

$$\frac{[HI]^2}{[H_2]} = K$$

$[I_2(c)]$ does not appear in the reaction quotient. Why?

The rates of reactions, and thus the conditions at which the rates are equal, are determined by the concentrations, partial pressures, mole fractions, etc. of the reactants and products. The concentration of a solid is the number of moles of solid *per liter of solid*. At 25°C, the density of $I_2(c)$ is 4.93 g/ml; its concentration is

$$[I_2(c)] = \frac{4.93 \text{ g/ml}}{254 \text{ g/mole}} \times \frac{1000 \text{ ml}}{\text{liter}} = 19.4 \text{ moles/liter}$$

a constant, independent of the volume of the container and of the quantity of solid. The partial pressure of $I_2(g)$, in equilibrium with $I_2(c)$ at 25°C, is 0.307 torr—the vapor pressure of I_2—also independent of volume and quantity. The mole fraction of I_2 in pure $I_2(c)$ is 1.00. All the properties of $I_2(c)$ that are relevant to equilibrium are constant at constant temperature.

The *rate* of a reaction between two phases depends on the area of the surface where the phases meet—in this case, the surface area of the solid I_2. However, any change in this area has the same effect on the rates of forward and reverse reactions. If equilibrium is established, the two rates are equal and remain equal after the change in surface area.

A pure liquid is omitted from the equilibrium condition for the same reasons. For the reaction

$$2Hg(l) + O_2(g) \rightleftharpoons 2HgO(c)$$

the equilibrium condition at 25°C is

$$\frac{1}{[O_2]} = K = 7.9 \times 10^{21}$$

Le Chatelier's principle applies to a heterogeneous equilibrium, but with a few new twists which follow from the preceding discussion:

(1) Adding or subtracting a solid or liquid has no effect, as long as some of each substance involved in the reaction is still present.

(2) Changing the concentration of a gas shifts the equilibrium in the direction that counteracts the change, just as in an all-gas reaction.

(3) To decide what the effect of changing the volume of the container is, we count only the moles of *gas*. Thus, the equilibrium

$$H_2(g) + I_2(c) \rightleftharpoons 2HI(g)$$
1 mole of gas   2 moles of gas

is shifted to the left when the volume of the container is decreased, to the right when it is increased; there are fewer moles of gas on the left than on the right. The solid does not count. The equilibrium

$$2Hg(l) + O_2(g) \rightleftharpoons 2HgO(c)$$
$$\text{1 mole of gas} \qquad \text{0 mole of gas}$$

is shifted to the right when the volume is decreased. The equilibrium

$$4H_2O(g) + 3Fe(c) \rightleftharpoons 4H_2(g) + Fe_3O_4(c)$$
$$\text{4 moles of gas} \qquad\qquad \text{4 moles of gas}$$

is unaffected by a change of volume. (The temperature is high enough so that no liquid water is present.)

EQUILIBRIUM IN SOLUTIONS / 11.7

A solute in a liquid solution, like a gas, has a concentration that may change when a reaction occurs. We shall confine our attention to dilute solutions, in which one component—the solvent—is much more abundant than all the other components together. For example, suppose that the reaction

$$C_2H_5OH(aq) + CH_3COOH(aq) \rightleftharpoons CH_3COOC_2H_5(aq) + H_2O(l) \quad (10)$$
$$\text{ethanol} \qquad\qquad \text{acetic acid} \qquad\qquad \text{ethyl acetate}$$
$$\text{(an alcohol)} \qquad\qquad\qquad\qquad\qquad\qquad \text{(an ester)}$$

occurs in a dilute aqueous solution (mostly water). The equilibrium condition for this reaction is

$$\frac{[CH_3COOC_2H_5]}{[C_2H_5OH][CH_3COOH]} = K \tag{11}$$

The solvent (H_2O in this case) *is omitted*. The reasons are similar to the reasons for omitting a pure solid or liquid. As long as the solution is dilute, the concentration, vapor pressure, and mole fraction of the solvent are practically the same as for pure solvent. We could allow for the fact that the solution is really not pure solvent by including the mole fraction of the solvent (x_{H_2O}) as a correction factor in the equilibrium condition:

$$\frac{[CH_3COOC_2H_5]x_{H_2O}}{[C_2H_5OH][CH_3COOH]} = K \tag{12}$$

However, in a dilute solution, solvent molecules far outnumber all the others: $x_{H_2O} \approx 1$. That makes Equation (12) practically the same as Equation (11).

If the solvent does not appear in the chemical equation, the question of including or omitting its concentration or mole fraction does not arise. However, the equilibrium constant for any reaction *depends on the identity of the solvent.* For example, molecules of acetic acid behave quite differently when surrounded by water molecules and when surrounded by benzene molecules (Problem 24, page 192), and still differently when far apart in the gas phase. If we were to add a large quantity of, say, acetone or sugar to the solution in which reaction (10) occurs, we would not find the same equilibrium constant as when the solution was mostly water.

The identity of the solvent is especially important for ionic reactions. An example is the reaction

$$2AgCN(c) + Br^- \rightleftharpoons Ag(CN)_2^- + AgBr(c)$$

Some values of K (at 25°C) are 8.6 in H_2O, 42.6 in $C_3H_5(OH)_3$(glycerol), and very large (too large to measure) in CH_3OH (methanol). The application of the law of chemical equilibrium to ionic reactions will be considered in more detail in Chapters 14 and 17.

As before, undissolved solids are omitted from the equilibrium condition, and the concentrations of gases appear. The equilibrium condition for the reaction

$$AgCl(c) \rightleftharpoons Ag^+(aq) + Cl^-(aq)$$

is

$$[Ag^+][Cl^-] = K$$

and for the reaction

$$2H_2O_2(aq) \rightleftharpoons 2H_2O(l) + O_2(g)$$

is

$$\frac{[O_2]}{[H_2O_2]^2} = K$$

The principle of Le Chatelier is the same for solution reactions as for the other cases we have studied, but two points should be noted: (1) If one of the reacting substances is a pure solid (or, less commonly, a pure liquid that does not mix with the solution), its quantity does not matter, as long as some of it is present. (2) Increasing the volume by adding solvent shifts the equilibrium in the direction that produces a larger number of *dissolved* particles (molecules or ions). Thus, the equilibrium

$$AgCl(c) \rightleftharpoons Ag^+ + Cl^-$$

is shifted to the right by adding solvent (water); there are 2 moles of solute on the right and zero on the left. The equilibrium

$$2AgCN(c) + Br^- \rightleftharpoons Ag(CN)_2^- + AgBr(c)$$

is unaffected by addition or removal of solvent; each side of the equation shows 1 mole of solute. In each case, the solids are ignored.

When the values at equilibrium of all the concentrations are known, the calculation of the equilibrium constant involves merely a substitution of the numbers into the reaction quotient. More commonly, however, what we know is the initial quantity of each reactant or product, and the final quantity of only one substance.

In the reaction

$$2SO_2 + O_2 \rightleftharpoons 2SO_3$$

the equation tells us that, for every mole of O_2 consumed, 2 moles of SO_2 must also be consumed, and 2 moles of SO_3 must be produced. Conversely, if the reaction goes from right to left, 2 moles of SO_2 and 1 mole of O_2 must be produced for every 2 moles of SO_3 consumed. Contrary to a common misconception, the equation does *not* tell us that SO_2 and O_2 must be mixed in the ratio of 2 moles SO_2 for each 1 mole O_2. We can start with any amounts; let us introduce 1.00 mole SO_2 and 1.00 mole O_2, and no SO_3, into a 1.00-liter container. We cannot tell from this information how much of the SO_2 and O_2 will have reacted at equilibrium, or how much SO_3 will have been formed, but some conclusions can be drawn from the chemical equation alone. Let y be the number of moles of O_2 consumed at equilibrium. Then the number of moles of SO_3 formed is $2y$, and the number of moles of SO_2 consumed is also $2y$. We can display these numbers in tabular form:

	$2SO_2$	$+$	O_2	$2SO_3$
Initial moles	1.00		1.00	0
Change	$-2y$		$-y$	$+2y$
Moles at equilibrium	$1.00 - 2y$		$1.00 - y$	$0 + 2y$

We now need to measure the final quantity of any one of the three gases. This will determine y, and thus the other numbers of moles. From these numbers, the equilibrium constant can be calculated.

Example 1 1.00 mole SO_2 and 1.00 mole O_2 were confined at 1000°K in a 1.00-liter container. At equilibrium, 0.925 mole of SO_3 had been formed. Calculate K for the reaction $2SO_2(g) + O_2(g) \rightleftharpoons 2SO_3(g)$ at 1000°K.

Answer Let y be the number of moles of O_2 consumed at equilibrium. Then, as shown above, $2y$ moles of SO_3 must be formed, and $1.00 - 2y$ mole of SO_2 must remain. At equilibrium,

$n_{SO_3} = 0.925$ mole $= 2y$
$n_{SO_2} = 1.00 - 2y = 0.075$ mole

Then

$y = 0.463$ mole
$n_{O_2} = 1.00 - y = 0.537$ mole

The equilibrium concentrations are the same as the numbers of moles, since the volume is 1.00 liter: $[SO_3] = 0.925$ mole/liter, $[SO_2] = 0.075$ mole/liter, $[SO_2] = 0.537$ mole/liter. Then

$$K = \frac{[SO_3]^2}{[SO_2]^2[O_2]} = \frac{0.925^2}{0.075^2 \times 0.537} = 283^*$$

Example 2 3.00 moles of pure SO_3 were introduced into an 8.00-liter container at 1105°K. At equilibrium, 0.95 mole of O_2 had been formed. Calculate K for the reaction $2SO_2(g) + O_2(g) \leftrightharpoons 2SO_3(g)$ at 1105°K.

Answer Let y be the number of moles of O_2 formed:

	$2SO_2$	$+ O_2$	$\leftrightharpoons$ $2SO_3$
Initial moles	0	0	3.00
Change	$+ 2y$	$+ y$	$- 2y$
Moles at equilibrium	$0 + 2y$	$0 + y$	$3.00 - 2y$

At equilibrium,

$$y = n_{O_2} = 0.95 \text{ mole}$$
$$n_{SO_2} = 2y = 1.90 \text{ moles}$$
$$n_{SO_3} = 3.00 - 2y = 1.10 \text{ moles}$$
$$[O_2] = \frac{0.95 \text{ mole}}{8.00 \text{ liters}} = 0.119 \frac{\text{mole}}{\text{liter}}$$
$$[SO_2] = \frac{1.90 \text{ moles}}{8.00 \text{ liters}} = 0.238 \frac{\text{mole}}{\text{liter}}$$
$$[SO_3] = \frac{1.10 \text{ moles}}{8.00 \text{ liters}} = 0.138 \frac{\text{mole}}{\text{liter}}$$
$$K = \frac{[SO_3]^2}{[SO_2]^2[O_2]} = \frac{0.138^2}{0.238^2 \times 0.119}$$
$$= 2.83$$

The given numbers of moles can be converted to concentrations, and calculations can be made thereafter in terms of concentrations. The table in the preceding example would then have the form

	$2SO_2$	$+ O_2$	$\leftrightharpoons$ $2SO_3$
Initial concentration (mole/liter)	0	0	$3.00/8.00 = 0.375$
Change	$+ 2z$	$+ z$	$- 2z$
Concentration at equilibrium	$2z$	z	$0.375 - 2z$

$$z = \frac{0.95 \text{ mole}}{8.00 \text{ liter}} = 0.119 \text{ mole/liter}$$

The reader can verify that the same answer is obtained for K.

---

* One might be inclined to write $K = (0.925$ mole/liter$)^2/(0.075$ mole/liter$)^2(0.537$ mole/liter$) = 283$ liters/mole. The units thus obtained depend on the form of the chemical equation. There are advantages in attaching units to equilibrium constants, but there are also various objections to it. We shall follow the common practice of writing equilibrium constants as dimensionless numbers.

Example 3 1.00 mole ethanol and 1.00 mole acetic acid were dissolved in water and kept at 100°C. The volume of the solution was 250 ml. At equilibrium, 25% of the acetic acid had been consumed. Calculate the equilibrium constant at 100°C for the reaction

$$C_2H_5OH(aq) + CH_3COOH(aq) \rightleftharpoons CH_3COOC_2H_5(aq) + H_2O(l)$$

Answer At equilibrium, 25% of the acid, or 0.25 mole, had been consumed; therefore, 0.25 mole of ethanol must also have been consumed, while 0.25 mole of ethyl acetate had been formed:

	C_2H_5OH	$+$	CH_3COOH	$\rightleftharpoons$	$CH_3COOC_2H_5 + H_2O$
Initial moles	1.00		1.00		0
Change	−0.25		−0.25		+0.25
Moles at equilibrium	0.75		0.75		0.25
Concentration at	$\dfrac{0.75 \text{ mole}}{0.250 \text{ liter}}$		$\dfrac{0.75 \text{ mole}}{0.250 \text{ liter}}$		$\dfrac{0.25 \text{ mole}}{0.250 \text{ liter}}$
equilibrium	3.0		3.0		1.0 mole/liter

$$K = \frac{[CH_3COOC_2H_5]}{[C_2H_5OH][CH_3COOH]}$$

$$= \frac{1.0}{(3.0)(3.0)} = 0.11$$

CALCULATIONS FROM THE EQUILIBRIUM CONSTANT / 11.9

A type of problem more common than those of the preceding section is this: Given the equilibrium constant and the initial numbers of moles or concentrations, find the quantities present at equilibrium. These problems sometimes present a certain amount of computational difficulty because of the need for solving quadratic, cubic, and worse equations.

Example 4 At 1557.8°K, the equilibrium constant of the reaction

$$Br_2(g) \rightleftharpoons 2Br(g)$$

is $1.04 \times 10^{-3} = K$. 0.10 mole of Br_2 is confined in a 1.0-liter container and heated to 1557.8°K. Calculate (a) the concentration of Br atoms present at equilibrium, and (b) the fraction of the initial Br_2 that is dissociated into atoms.

Answer The initial concentration of Br_2, if there were no dissociation, would be

$$[Br_2] = \frac{0.10 \text{ mole}}{1.0 \text{ liter}} = 0.10 \frac{\text{mole}}{\text{liter}}$$

	Br_2	$\rightleftharpoons$	$2Br$
Initial concentration (mole/liter)	0.10		0
Change	$-y$		$+2y$
Concentration at equilibrium	$0.10 - y$		$2y$

$$\frac{[Br]^2}{[Br_2]} = K$$

$$\frac{(2y)^2}{0.10 - y} = 1.04 \times 10^{-3}$$

$$\frac{y^2}{0.10 - y} = 2.6 \times 10^{-4}$$

$$y^2 + 2.6 \times 10^{-4}y - 2.6 \times 10^{-5} = 0$$

The quadratic formula (Appendix I.16) gives

$$y = \frac{-2.6 \times 10^{-4} \pm \sqrt{2.6^2 \times 10^{-8} + 4 \times 2.6 \times 10^{-5}}}{2}$$

$$= 5.0 \times 10^{-3}$$

(The negative root is physically meaningless and is therefore rejected.)

(a) At equilibrium

$$[Br] = 2y = 1.0 \times 10^{-2} \text{ mole/liter}$$

(b) Fraction dissociated equals

$$\frac{y}{0.10} = \frac{5.0 \times 10^{-3}}{0.10} = 0.050 \quad \text{or} \quad 5.0\%$$

Example 5 The reaction

$$N_2(g) + O_2(g) \rightleftharpoons 2 \, NO(g)$$

contributes to air pollution whenever a fuel is burned in air at a high temperature, as in a gasoline engine. At 1500°K, $K = 1.0 \times 10^{-5}$. A sample of air is heated to 1500°K and compressed. Before any reaction, $[N_2] = 0.80$ mole/liter and $[O_2] = 0.20$ mole/liter. Calculate the concentration of NO at equilibrium.

Answer

	N_2	$+$	O_2	$\rightleftharpoons$	$2NO$
Initial concentration (mole/liter)	0.80		0.20		0
Change	$-y$		$-y$		$+2y$
Concentration at equilibrium	$0.80 - y$		$0.20 - y$		$2y$

$$\frac{[NO]^2}{[N_2][O_2]} = K$$

$$\frac{(2y)^2}{(0.80 - y)(0.20 - y)} = 1.0 \times 10^{-5}$$

Since the right-hand side of this equation is small, we try the approximation (Appendix 1.16) $y \ll$ ("much less than") 0.20. Then $0.80 - y \approx 0.80$, $0.20 - y \approx 0.20$.

$$\frac{(2y)^2}{0.80 \times 0.20} = 1.0 \times 10^{-5}$$

$$y = 6.3 \times 10^{-4}$$

y is indeed much smaller than 0.20:

0.20
−0.00063
0.20

The approximation is therefore justified.

$$[NO] = 2y = 1.3 \times 10^{-3} \text{ mole/liter}$$

In most equilibrium calculations, a quantity *y* may be neglected in comparison to another quantity *z* if *y* is less than 5% of *z*.

Example 6 A solution was made by dissolving 0.10 mole $C_6H_{10}O$ in CH_3OH to make 250 ml of solution. For the reaction

$$C_6H_{10}O + 2CH_3OH \rightleftharpoons C_6H_{10}(OCH_3)_2 + H_2O$$

cyclohexanone   methanol   1,1-dimethoxycyclohexane

at 25°C in the solvent CH_3OH, $K = 3.7$. Calculate the concentration of H_2O in the equilibrium solution.

Answer The initial concentration of $C_6H_{10}O$ is

$$\frac{0.10 \text{ mole}}{0.250 \text{ liter}} = 0.40 \frac{\text{mole}}{\text{liter}}$$

Since CH_3OH is the solvent, it will not appear in the equilibrium condition.

$$2CH_3OH + C_6H_{10}O \rightleftharpoons C_6H_{10}(OCH_3)_2 + H_2O$$

Initial concentration (mole/liter)	0.40	0	0
Change	−y	+y	+y
Concentration at equilibrium	0.40 − y	y	y

$$\frac{[C_6H_{10}(OCH_3)_2][H_2O]}{[C_6H_{10}O]} = K$$

$$\frac{y^2}{0.40 - y} = 3.7$$

$$y^2 + 3.7y - 1.48 = 0$$

$$[H_2O] = y = 0.36 \text{ mole/liter}$$

(Note that H_2O is a *solute*.)

PROBLEMS

1. Definitions. Define briefly: (a) equilibrium, (b) equilibrium constant, (c) reaction quotient, (d) equilibrium condition, (e) heterogeneous equilibrium, (f) Le Chatelier principle. Choose a reaction and use it to illustrate each of these concepts.

2. Equilibrium condition. Write the equilibrium condition for each of the following reactions:
(a) $CO(g) + \frac{1}{2}O_2(g) \rightleftharpoons CO_2(g)$
(b) $N_2(g) + 3H_2(g) \rightleftharpoons 2NH_3(g)$

(c) $C(c) + CO_2(g) \leftrightharpoons 2CO(g)$

(d) $NH_4HS(c) \leftrightharpoons NH_3(g) + H_2S(g)$

(e) $C_6H_5COOH(solute) + C_2H_5OH(solvent) \leftrightharpoons C_6H_5COOC_2H_5(solute) + H_2O(solute)$

(f) $H_2C_4H_2O_4(aq) + H_2O(l) \leftrightharpoons H_2C_4H_4O_5(aq)$

(g) $Li_2CO_3(c) + CO_2(l, solvent) + H_2O(solute) \leftrightharpoons 2LiHCO_3(solute)$

3. Le Chatelier principle. Consider the reaction

$$2Cl_2(g) + 2H_2O(g) \rightleftharpoons 4HCl(g) + O_2(g) \qquad \Delta H = +27 \text{ kcal}$$

The four gases Cl_2, H_2O, HCl, and O_2 are mixed and the reaction is allowed to come to equilibrium. State and explain the effect (increase, decrease, no change) of the operation in the left column (below) on the equilibrium value of the quantity in the right column. Each operation is to be considered separately. Temperature and volume are constant except when the contrary is stated.

(a) Increasing the volume of the container	Number of moles of H_2O
(b) Adding O_2	Number of moles of H_2O
(c) Adding O_2	Number of moles of O_2
(d) Adding O_2	Number of moles of HCl
(e) Decreasing the volume of the container	Number of moles of Cl_2
(f) Decreasing the volume of the container	Partial pressure of Cl_2
(g) Decreasing the volume of the container	K
(h) Raising the temperature	K
(i) Raising the temperature	Concentration of HCl
(j) Adding He	Number of moles of HCl
(k) Adding catalyst	Number of moles of HCl

4. Combination of equilibria. The following equilibrium constants at 700°C are given:

$$H_2(g) + CO_2(g) \rightleftharpoons H_2O(g) + CO(g) \qquad K = 0.62$$
$$FeO(c) + H_2(g) \rightleftharpoons Fe(c) + H_2O(g) \qquad K = 0.42$$

Calculate K for the reaction

$$FeO(c) + CO(g) \rightleftharpoons Fe(c) + CO_2(g)$$

5. Equilibrium constant. NO(g) and $O_2(g)$ were mixed in a container of fixed volume kept at 1000°K. Their initial concentrations were 0.0200 mole/liter for NO and 0.0300 mole/liter for O_2. When the reaction

$$2NO(g) + O_2(g) \rightleftharpoons 2NO_2(g)$$

had come to equilibrium, the concentration of NO_2 was 2.2×10^{-3} mole/liter. Calculate (a) the concentration of NO at equilibrium; (b) the concentration of O_2 at equilibrium; (c) the equilibrium constant for the reaction.

6. Equilibrium constant. The following equilibrium is established in the presence of a certain enzyme:

$$\underset{\text{fumaric acid}}{H_2C_4H_2O_4(aq)} + H_2O \rightleftharpoons \underset{\text{malic acid}}{H_2C_4H_4O_5(aq)}$$

One liter of solution was prepared from water and 0.100 mole of fumaric acid. At equilibrium, the solution contained 0.078 mole of malic acid. Find the equilibrium constant for the reaction.

7. Gas equilibrium. For the reaction

$$PCl_3(g) + Cl_2(g) \Longleftrightarrow PCl_5(g)$$

$K = 20$ at 240°C. When PCl_3 and PCl_5 have equal concentrations at equilibrium, what is the concentration of Cl_2?

8. Gas equilibrium. At 25°C, the equilibrium constant for the reaction

$$N_2O_4(g) \Longleftrightarrow 2NO_2(g)$$

is $K = 5.85 \times 10^{-3}$. 10.0 g N_2O_4 is confined in a 10.0-liter flask at 25°C. Calculate (a) the number of moles of NO_2 present at equilibrium; (b) the percentage of the original N_2O_4 that is dissociated.

9. Solution equilibrium. A solution contains 1.00 mole benzoic acid, C_6H_5COOH, 1.00 mole methanol, CH_3OH, and water in 1.00 liter of solution. The solution is kept at 200°C (under pressure) until the reaction

$$C_6H_5COOH(aq) + CH_3OH(aq) \Longleftrightarrow C_6H_5COOCH_3(aq) + H_2O(l)$$

has come to equilibrium. At equilibrium, 5.5% of the benzoic acid has reacted. (a) Calculate the equilibrium constant for the reaction. (b) Calculate the percentage of the acid that reacts when 1.00 mole benzoic acid and 5.00 moles methanol are dissolved in water to make 1.00 liter and allowed to come to equilibrium at 200°C.

10. Solid–gas equilibrium. At 21.8°C, the equilibrium constant for the reaction

$$NH_4HS(c) \Longleftrightarrow NH_3(g) + H_2S(g)$$

is 1.2×10^{-4}. (a) Solid NH_4HS was introduced into an empty container and allowed to come to equilibrium at 21.8°C. Find the total concentration and the total pressure of the gas mixture at equilibrium. (b) Find the equilibrium concentration of H_2S when NH_4HS is introduced into a container which also contains NH_3 at an initial concentration of 0.010 mole/liter.

ADDITIONAL PROBLEMS

11. Equilibrium condition. (a) For each of the five equilibrium mixtures in Table 11.1, calculate the expression $[SO_3]/[SO_2][O_2]$. Calculate the average and relative average deviation. Is this expression as nearly constant as the correct reaction quotient? (b) Which of the following expressions should be constant at equilibrium? (i) $[SO_3]/[SO_2][O_2]^{1/2}$; (ii) $[SO_3]^{1/2}/[SO_2]^{1/2}[O_2]$; (iii) $[SO_2]^2[O_2]/[SO_3]$.

12. Combination of equilibria. Given the following equilibrium constants at 1362°K:

$$2H_2(g) + S_2(g) \Longleftrightarrow 2H_2S(g) \qquad K = 72.3$$
$$H_2(g) + Br_2(g) \Longleftrightarrow 2HBr(g) \qquad K = 7.2 \times 10^{-4}$$

Find K for the reaction

$$Br_2(g) + H_2S(g) \rightleftharpoons 2HBr(g) + \tfrac{1}{2}S_2(g)$$

at 1362°K.

13. Equilibrium constant. CO_2 and H_2 were mixed in a container kept at 959°K. Their initial concentrations were 0.0200 mole/liter for CO_2 and 0.0100 mole/liter for H_2. The reaction

$$CO_2(g) + H_2(g) \rightleftharpoons CO(g) + H_2O(g)$$

then occurred. At equilibrium, the concentration of H_2O was found to be 8.5×10^{-3} mole/liter. Calculate the concentrations of CO_2, H_2, and CO at equilibrium, and K for the reaction.

14. Le Chatelier principle. $K = 1.00$ for the reaction $X_2(g) \rightleftharpoons 2X(g)$ at 366°K $(RT = 30.0$ liter atm/mole). 1.00 mole X_2 is placed in a 1.00-liter flask and allowed to come to equilibrium. (a) Calculate $[X_2]$, $[X]$, and the total pressure P_1 at equilibrium. (b) The volume of the container is decreased to 0.50 liter at constant temperature. The new total pressure is P_2. What is the ratio P_2/P_1 if the equilibrium does not shift? (c) The reaction now comes to equilibrium under the new conditions. Does the Le Chatelier principle predict that P_2/P_1 will be less than, greater than, or equal to your answer to (b)? (d) Calculate $[X_2]$, $[X]$, P_2, and P_2/P_1 at equilibrium.

15. Equilibrium constant. CO_2 was passed over graphite at 1000°K. The emerging gas stream consisted of 28 mole % CO_2 and 72 mole % CO. The total pressure was 2.00 atm throughout. Assume that equilibrium was attained. (a) Calculate the concentrations of CO_2 and CO. (b) Find K for the reaction

$$C(graphite) + CO_2(g) \rightleftharpoons 2CO(g)$$

16. Equilibrium constant. 0.0200 mole NH_4Cl and 0.0100 mole NH_3 were placed in a closed 1.00-liter container and heated to 603°K. At this temperature, all the NH_4Cl vaporized. When the reaction

$$NH_4Cl(g) \rightleftharpoons NH_3(g) + HCl(g)$$

had come to equilibrium, 7.7×10^{-3} mole HCl was present. Calculate K for this reaction at 603°K.

17. Gas equilibrium. The equilibrium constant of the reaction

$$H_2(g) + Br_2(g) \rightleftharpoons 2HBr(g)$$

is 1.6×10^5 at 1297°K and 3.5×10^4 at 1495°K. (a) Is ΔH for this reaction positive or negative? (b) Find K for the reaction

$$\tfrac{1}{2}H_2(g) + \tfrac{1}{2}Br_2(g) \rightleftharpoons HBr(g)$$

at 1297°K. (c) Pure HBr is placed in a container of constant volume and heated to 1297°K. What percentage of the HBr is decomposed to H_2 and Br_2 at equilibrium? Is any superfluous information given in (c)?

18. Gas equilibrium. 1.00 mole H_2O was placed in a 50-liter container and heated to 1705°K. The equilibrium constant for the reaction

$$2H_2O(g) \rightleftharpoons 2H_2(g) + O_2(g)$$

is $K = 1.35 \times 10^{-11}$. Find the number of moles of O_2 present at equilibrium. Make reasonable approximations to avoid having to solve a cubic equation.

19. Gas equilibrium. When acetic acid is a gas, some of it exists as dimers (double molecules):

$$2CH_3COOH(g) \rightleftharpoons (CH_3COOH)_2(g)$$

For this reaction, $K = 3.18 \times 10^4$ at 25°C. (a) The vapor pressure of acetic acid at 25°C is 15.4 torr. Calculate the total concentration of acetic acid vapor, including both monomers (single molecules) and dimers, in equilibrium with the liquid at 25°C. (b) Let y be the concentration of monomer. What is the concentration of dimer, in terms of y? Calculate the concentrations of monomer and dimer and the percentage of the vapor molecules that are dimers.

20. Heterogeneous equilibrium. The equilibrium constant for the reaction

$$H_2(g) + Br_2(l) \rightleftharpoons 2HBr(g)$$

is $K = 1.84 \times 10^{17}$ at 25°C. The vapor pressure of liquid Br_2 at this temperature is 0.28 atm. (a) Find the concentration of $Br_2(g)$ in equilibrium with $Br_2(l)$ at 25°C. (b) Find K at 25°C for the reaction.

$$H_2(g) + Br_2(g) \rightleftharpoons 2HBr(g)$$

(c) How will the equilibrium in (a) be shifted by a decrease in the volume of the container if (*i*) liquid Br_2 is absent, (*ii*) liquid Br_2 is present? Explain why the effect is different in the two cases.

21. Heterogeneous equilibrium. At −10°C, $Br_2(H_2O)_{10}(c)$ is in equilibrium with $Br_2(g)$, $H_2O(g)$, and $H_2O(c)$. The equilibrium concentrations of the two gases are 0.014 mole/liter for Br_2 and 1.21×10^{-4} mole/liter for water vapor. Find the equilibrium constant for the reaction

(a) $Br_2(H_2O)_{10}(c) \rightleftharpoons Br_2(g) + 10H_2O(g)$

and for the reaction

(b) $Br_2(H_2O)_{10}(c) \rightleftharpoons Br_2(g) + 10H_2O(c)$

22. Heterogeneous equilibrium. For the reaction

$$\tfrac{1}{8}S_8(c,\ rhombic) + 2HI(g) \rightleftharpoons I_2(c) + H_2S(g)$$

$K = 3.13 \times 10^7$ at 300°K. H_2S was admitted to a rigid container until its concentration was 2.00×10^{-3} mole/liter. Solid S_8 and solid I_2 were then introduced and the system was allowed to come to equilibrium at 300°K. Calculate the concentration of HI at equilibrium.

23. Heterogeneous equilibrium. In the distant future, when hydrogen may be cheaper than coal (page 353), steel mills may make iron by the reaction

$$Fe_2O_3(c) + 3H_2(g) \rightleftharpoons 2Fe(c) + 3H_2O(g)$$

For this reaction, $\Delta H = 23$ kcal and $K = 0.35$ at 400°C. (a) What percentage of the H_2 will remain unreacted after the reaction has come to equilibrium at 400°C? (Suggestion: The cube root of 0.35 is 0.705.) (b) Would this percentage be greater or less if the temperature were increased above 400°C?

24. Solution equilibrium. (a) 25 ml of a solution was prepared by dissolving 1.99×10^{-3} mole acetone, $(CH_3)_2CO$, in the solvent methanol, CH_3OH. When the reaction

$$(CH_3)_2CO + 2CH_3OH \rightleftharpoons (CH_3)_2C(OCH_3)_2 + H_2O$$

had come to equilibrium at 25°C, 61.5% of the acetone was still present. Calculate K for this reaction. (b) Another solution is prepared with the initial concentration of the acetone equal to 0.010 mole/liter. What percentage of the acetone will remain unreacted at equilibrium?

25. Gas–solution equilibrium. Hemoglobin (Hb) can form a complex with either O_2 or CO. For the reaction

$$O_2Hb(aq) + CO(g) \rightleftharpoons COHb(aq) + O_2(g)$$

at body temperature, $K \approx 200$. If the ratio $[COHb]/[O_2Hb]$ comes close to 1, death is probable. What concentration of CO in the air is likely to be fatal? (Assume that the partial pressure of O_2 is 0.2 atm and the air temperature is 25°C.)

12

IONIC SOLUTIONS

As long ago as the eighteenth century, certain aqueous solutions were found to have a special property not shared by the pure solutes, by water, or by most other liquids: These solutions easily conduct electricity. Solutes that show this property are of the kinds commonly known as acids, bases, and salts. The first rough measurements of the conductivities of such solutions were made by Henry Cavendish in 1777: He compared the lengths of different solutions through which the discharge of a battery of Leyden jars gave equal shocks. More recent data, obtained by more accurate and less heroic methods, on the conductivities (page 230) of solutions, in comparison to other conductors and insulators, are shown in Table 12.1.

The solutions are far better conductors than the insulators, though hardly in the same class as the metals. There is a striking difference between the solutions and the metals in their behavior as conductors. When a current is passed through a copper wire, the copper is unaf-

Table 12.1
Conductivities at Room Temperature (in ohm^{-1} cm^{-1})

Metals		Aqueous solutions		Insulators	
Ag	6.1×10^5	0.1 M HCl	3.5×10^{-2}	H_2O (pure)	6×10^{-8}
Cu	5.8×10^5	0.1 M KCl	1.1×10^{-2}	Maple wood	3×10^{-11}
Fe	1.0×10^5	0.1 M NaCl	9.2×10^{-3}	Glass	10^{-14}
Hg	1.0×10^4	0.1 M $HC_2H_3O_2$	4.7×10^{-4}	Hard rubber	10^{-15}
C[a]	1.2×10^3	0.1 M NH_3	3.1×10^{-4}		

[a] Graphite. Carbon is not a metal, but the electrical properties of graphite are similar to those of metals.

fected by this process. However, when a direct current is passed through a solution of, say, NaCl, chemical reactions occur at the electrodes (the metallic conductors in contact with the solution), shown in this case by gas bubbles. There is a reaction at each electrode; the pair of reactions is called ELECTROLYSIS. A solute that dissolves in a suitable solvent (usually water) to give an electrically conducting solution is called an ELECTROLYTE. Many electrolytes conduct not only when dissolved but also when melted.

COLLIGATIVE PROPERTIES OF SOLUTIONS OF ELECTROLYTES / 12.2

Another peculiarity of solutions of electrolytes was discovered in 1884 by François Marie Raoult, of vapor-pressure fame. He observed that the freezing-point depression in an aqueous solution of NaCl is almost twice what one would expect from the molality of the solution. If we dissolve 5.85 g of NaCl in 1000 g of water, the solution is 0.100 molal, and we expect it to freeze approximately 0.186°C below the freezing point of pure water (page 184). The actual freezing point is −0.348°C, which gives a depression nearly twice the calculated depression of 0.186°C. In the case of zinc chloride, $ZnCl_2$, the freezing-point depression is nearly three times as great as we would calculate from its formula weight. The more dilute the solution, the more nearly is the measured depression equal to an integer times the calculated depression. The ratio of the observed to the calculated value of the freezing-point depression is called the VAN'T HOFF FACTOR, and is represented by i. Table 12.2 illustrates how i depends on molality for several typical electrolytes. Similar results are found for the other colligative properties: Boiling-point elevation, vapor-pressure depression, and osmotic pressure are multiplied by the same integral factor in dilute solutions of a given electrolyte.

Table 12.2
van't Hoff Factors of Aqueous Solutions of Electrolytes

Molality (m), moles/kg H_2O	Calculated freezing-point depression (°C) (1.86m)	van't Hoff factor $i = \dfrac{\text{measured freezing-point depression}}{\text{calculated freezing-point depression}}$		
		NaCl	$ZnCl_2$	$MgSO_4$
0.100	0.186	1.87	2.66	1.21
0.0100	0.0186	1.94	2.77	1.53
0.0025	0.00465	—	2.91	—
0.00100	0.00186	1.97	—	1.82

The conclusion is inescapable: What is present in the solution is not NaCl molecules, or $ZnCl_2$ molecules, but individual Na and Cl particles, or Zn and Cl particles. In a $MgSO_4$ solution, however, the calculated colligative properties are multiplied by approximately 2, not 6, so that this substance must dissociate not into Mg, S, and four O's, but into two fragments—probably Mg and SO_4, since the SO_4 group turns up in many compounds and is undoubtedly a stable entity.

ELECTRONIC AND IONIC CONDUCTION / 12.3

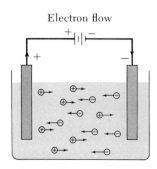

Electron flow

Fig. 12.1. *The motion of ions in an electric field.*

An electric current in a metal is a flow of electrons. A few outer electrons from each atom are free to move through the metal under the influence of an electric field. The nuclei do not migrate in this process. Every electron leaving the metal at one end is replaced by an identical electron entering at the other end. This type of conduction is called **ELECTRONIC** or **METALLIC CONDUCTION**, and it produces no observable change in the metal.

The remarkable properties of electrolytes can be understood if we assume that the solute is present in the form of charged particles, that is, *ions* (page 4). When an electric field is applied to a liquid containing ions, the positive ions will move toward the negative electrode and the negative ions will move toward the positive electrode (Fig. 12.1). In a metal, the only mobile charged particles are negative electrons, but, in the solution, both positive and negative ions move. Positive ions are also called **CATIONS** ("cat-ions," not "cay-shuns"); negative ions are **ANIONS**.

In electrolysis, electrons flow into the negative electrode, while an equal number of electrons are removed from the positive electrode. When a positive ion reaches the negative electrode, the journey ends. It may accept one or more electrons from the electrode, or it may just hang around while some other ion or molecule accepts the electrons. Similarly, a negative ion may give electrons to the positive electrode, or it may remain unchanged while something else gives up electrons. Every measurable volume of the solution remains electrically neutral during conduction; the total charge of the positive ions equals the total charge of the negative ions. Ion migration and the electrode reactions work together to maintain this overall neutrality.

When, for example, a solution of $CuBr_2$ is electrolyzed by a direct current between unreactive electrodes such as platinum, copper is deposited on one electrode, while bromine, as a red gas or liquid, appears at the other. These processes are represented, respectively, as

$$Cu^{2+} + 2e^- \longrightarrow Cu$$

and

$$2Br^- \longrightarrow Br_2 + 2e^-$$

An electron is represented by e^-. If the solution is NaBr instead of $CuBr_2$, the Na^+ ions are attracted to the negative electrode but do not react there; instead, the reaction is

$$2H_2O + 2e^- \longrightarrow H_2(g) + 2OH^-$$

The Br^- ions and some of the new OH^- ions move away from the negative electrode. The solution near this electrode gradually changes from a solution of NaBr to a solution of NaOH.

A transfer of electrons between chemical species is an oxidation–reduction reaction (page 121). Oxidation occurs at one electrode, reduction at the other.

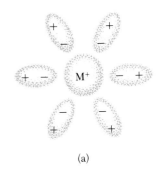

(a)

SOLVATION OF IONS / 12.4

When Svante Arrhenius proposed in 1887 that electrolytes exist in solution as ions, most chemists greeted his theory with less than enthusiasm. The principal objection was that forces between charged particles are very strong, and it seemed implausible that, in solution, these particles would be separated from each other and be free to move independently. An explanation is suggested by the fact that the solvents in which electrolytic conduction is observed—such as water, liquid NH_3, and liquid SO_2—consist of molecules with large dipole moments (Appendix I.12). An ion attracts the oppositely charged ends of such solvent molecules almost as strongly as it attracts another ion. The ions are said to be **SOLVATED**, or, when the solvent is water, **HYDRATED** (Fig. 12.2). Solvation often involves the formation of covalent bonds between the ion and the adjacent solvent molecules, in addition to simple electrostatic attraction (page 465).

A negative ion usually has unshared electron pairs. Its negative charge greatly enhances the ability of these electrons to form hydrogen bonds (page 139) to water molecules, even when the champion hydrogen bonders (N, O, F) are not present (Fig. 12.3a). One common positive ion, the ammonium ion (NH_4^+), can also form hydrogen bonds (Fig. 12.3b); a result is the very large solubility of ammonium salts in water. Solvents like water, having at least one hydrogen atom capable of hydrogen bonding, are called **PROTIC** solvents.

There are also polar solvents that do not possess hydrogen atoms capable of hydrogen bonding. An example of such an **APROTIC** solvent is dimethyl sulfoxide,

$$\overset{\displaystyle \overset{..}{O}:}{\underset{\displaystyle H}{\overset{\displaystyle \|}{H_3C-\underset{..}{\overset{..}{S}}-CH_3}}}$$

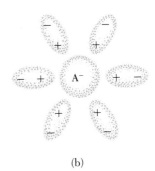

(b)

Fig. 12.2. *Ion-dipole attraction: (a) positive ion; (b) negative ion.*

$:\overset{..}{\underset{..}{Cl}}: \text{----} H \diagdown \underset{O}{} \diagup H$

(a)

$H-\overset{\displaystyle H}{\underset{\displaystyle H}{N^+}}-H\text{----}O\diagdown^H_H$

(b)

Fig. 12.3. *Hydrogen bonding between ions and water. (a) Chloride ion; (b) ammonium ion.*

Any positive ion dissolved in this solvent is strongly attracted to the oxygen atom. However, since hydrogen bonding cannot occur, and the sulfur atom is only slightly attracted to the negative ion, the negative ion is left in a relatively "free" state, unsolvated by solvent molecules. Because of this "free" state, negative ions are much more reactive in aprotic solvents.

An ionic crystal represents a state of low energy because the oppositely charged ions are close together (page 122). Therefore, the dissociation of a crystal is very endothermic ($\Delta H > 0$), which means that the separated ions have much higher energy than the crystal:

$$K^+Cl^-(c) \longrightarrow K^+(g) + Cl^-(g) \qquad \Delta H_1 = \text{lattice energy} = 165 \text{ kcal}$$

However, when the crystal dissolves in a solvent like water, the energy of the hydrated ions is much lower than the energy of the free ions because energy is released when the ions attract the solvent molecules[*]:

$$K^+(g) + Cl^-(g) + aq \longrightarrow K^+(aq) + Cl^-(aq)$$
$$\Delta H_2 = \text{hydration energy} = -161 \text{ kcal}$$

The hydration energy is negative ($\Delta H_2 < 0$) and the process of dissolving the free ions is thus exothermic. The actual process of dissolving a crystal is the sum of these two imaginary steps, vaporizing to ions and dissolving the separated ions:

$$K^+Cl^-(c) + aq \longrightarrow K^+(aq) + Cl^-(aq)$$
$$\Delta H_3 = \text{heat of solution} = \Delta H_1 + \Delta H_2$$
$$= \text{lattice energy} + \text{hydration energy}$$
$$= +165 - 161 = +4 \text{ kcal}$$

The heat of solution may be either positive (KCl, $AgNO_3$) or negative (LiCl, NaI), depending on the relative magnitudes of the lattice energy and the hydration energy.

The number of water molecules closely associated with a given ion can be measured only by indirect and ingenious methods. Typical numbers range from 4 to 6, increasing with the size and charge of the ion. We usually write Na^+, Cl^-, etc., rather than $Na(H_2O)_x^+$, $Cl(H_2O)_y^-$, for simplicity and because x and y are not accurately known or even precisely defined. However, when there is strong covalent bonding between the ion and water (as with many of the transition metals), it is reasonable to write a definite formula showing the water molecules, for example, $Fe(H_2O)_6^{3+}$.

DIELECTRIC CONSTANT / 12.5

When two charged bodies are immersed in some other substance (called the "medium"), the force between them is less than it would be

---

[*] An indefinitely large amount of water used as solvent is represented by aq.

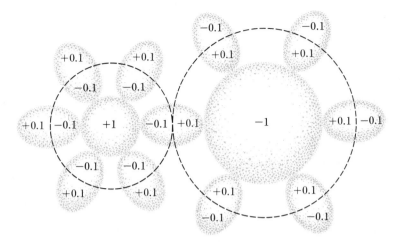

Fig. 12.4. *Two ions (+1 and −1 charges) immersed in a polar solvent.*

in a vacuum by a factor $1/D$, where D is the DIELECTRIC CONSTANT of the medium (Appendix I.12):

$$\text{force in medium} = \frac{\text{force in vacuum}}{D}$$

If the molecules of the medium are polar and are free to rotate, they are predominantly aligned with their negative ends adjacent to a positively charged body. The result is a layer of negative charge just outside the positive body; the effect is to cancel part of the charge on the body. Similarly, there is a layer of positive charge just outside a negatively charged body. Figure 12.4 shows, in a simplified way, how this cancellation occurs. Each solvent molecule is assumed to have a charge of $+0.1$ unit at one end and -0.1 unit at the other end. The net charge within the dotted circle on the left is $1 - 6(0.1) = 0.4$ unit; in the dotted circle on the right, -0.4 unit. The attraction is smaller than it would be if the ions were at the same distance in a vacuum by an estimated factor $(+0.4)\,(-0.4)/(+1)(-1) = 0.16$. It is said that the charged bodies are *shielded* from each other by the medium. We expect, therefore, that liquids with high dielectric constants—the same ones that have the most polar molecules—will be effective in decreasing the forces between separated ions and will thus be good solvents for ionic crystals. The best solvent is water, which has the very high dielectric constant 78.5 at 25°C; other ionizing solvents, with their dielectric constants for the liquid state, are NH_3, 22; HCN, 116; CH_3OH, 33; SO_2, 15.6. Liquids composed of nonpolar (and slightly polar) molecules, however, have much lower dielectric constants, e.g., benzene, C_6H_6, 2.27; CCl_4, 2.24; CS_2, 2.65. Here, the molecules have no tendency to orient themselves around an ion. Ionic salts seldom have appreciable solubility in these solvents.

Ionic crystals are not the only solutes that yield ions in solution. Many covalent molecules react with the solvent to give ions, usually by the transfer of a proton from one molecule to the other. For example, hydrogen chloride is a gas under ordinary conditions and consists of covalently bonded molecules. When it is dissolved in water, the reaction

$$HCl + H_2O \longrightarrow H_3O^+ + Cl^-$$

produces the HYDRONIUM or HYDROGEN ION, H_3O^+, and Cl^-. Ammonia reacts by taking a proton away from water:

$$NH_3 + H_2O \rightleftharpoons NH_4^+ + OH^-$$

A proton donor, as HCl, is called an ACID; a proton acceptor, as NH_3, is a BASE. Acids and bases will be discussed in detail in Chapters 13 and 14.

Most acids and bases are WEAK ELECTROLYTES. This means that a reaction such as

$$NH_3 + H_2O \xleftharpoons{\hspace{1em}} NH_4^+ + OH^-$$

or

$$HC_2H_3O_2 + H_2O \xleftharpoons{\hspace{1em}} H_3O^+ + C_2H_3O_2^-$$
acetic acid   acetate ion

reaches equilibrium when only a small fraction of the molecules are converted to ions (as suggested by the length of the arrows). However, some acids (such as HCl and HNO_3) and most salts* (which are already ionic in the crystalline state) are STRONG ELECTROLYTES—they exist in dilute solution mostly as ions (pages 233 and 235).

Our usual practice will be to write the hydronium ion as H^+. All ions are hydrated in aqueous solution; for most purposes, there is no need to show the water of hydration in the formula. As we usually write Na^+ instead of $Na(H_2O)_x^+$, or Fe^{3+} instead of $Fe(H_2O)_6^{3+}$, so we may write H^+ with the understanding that this represents not a free proton, but a proton attached to one or more water molecules.

NET IONIC EQUATIONS / 12.7

As soon as we recognize that ions exist in solution as more or less independent chemical species, it becomes natural to write chemical equations in terms of ions. In order to decide when and how to write an equation in ionic form, we must know which substances exist primarily as ions. The custom is to represent each substance involved in a reaction by the formula corresponding to the form in which most of it exists.

° The category of "salts" includes ionic bases like NaOH.

Thus, we write the reaction between hydrochloric acid and ammonia, in aqueous solution, as

$$H^+ + Cl^- + NH_3 \longrightarrow NH_4^+ + Cl^- \qquad (1)$$

The strong acid HCl is represented as $H^+ + Cl^-$ since it is almost completely ionized. Ammonium chloride, NH_4Cl, is a strong electrolyte, and is therefore written as the two distinct species, $NH_4^+ + Cl^-$. Ammonia, however, is a weak electrolyte and is represented by its molecular formula, NH_3. If the acid were acetic acid, which is weak, the equation would be

$$HC_2H_3O_2 + NH_3 \longrightarrow NH_4^+ + C_2H_3O_2^-$$

The weak acid is represented by its molecular formula $HC_2H_3O_2$; the product, ammonium acetate, is a typical strong electrolyte and is written in terms of ions.

Inspection of Equation (1) shows that Cl^- appears on both sides, which means that it was present initially and is not changed by the reaction. Any species remaining unchanged can be omitted from a chemical equation. Equation (1) is therefore written as

$$H^+ + NH_3 \longrightarrow NH_4^+ + H_2O \qquad (2)$$

Equation (2) is called a **NET IONIC EQUATION** because nonreacting species have been omitted, leaving only those that actually participate in the reaction.* We know that negative ions must be present to keep the solution electrically neutral, but Equation (2) does not tell us what they are, nor do we necessarily care. The reaction is the same whether the acid is hydrochloric, hydrobromic, or sulfuric, as long as it is a strong acid. Only when the solution is evaporated, to obtain a solid product, does the identity of the acid make a difference; the solid product obtained will then be NH_4Cl, NH_4Br, or $(NH_4)_2SO_4$, depending on which acid is used.

Undissolved solids, including ionic crystals, and gases are represented by their molecular (or empirical) formulas. When solutions of the strong electrolytes KCl and $AgNO_3$ are mixed, a white precipitate of silver chloride, AgCl, is formed. We write the formulas of the soluble salts in ionic form, but silver chloride, which is present almost entirely as the solid, is represented as AgCl:

$$K^+ + Cl^- + Ag^+ + NO_3^- \longrightarrow AgCl(c) + K^+ + NO_3^-$$

K^+ and NO_3^- appear on both sides and may be omitted to give the net equation

$$Cl^- + Ag^+ \longrightarrow AgCl(c)$$

This form of the equation emphasizes that the actual chemical change is the combination of Ag^+ and Cl^- ions.

---

* Ions which are present but do not participate may be called "spectator ions."

Equations for oxidation–reduction (redox) reactions are often difficult to balance. The method to be described in this section is especially appropriate for reactions involving ions in solution. The method resolves the reaction into two parts, or **PARTIAL REACTIONS**: a loss of electrons (**OXIDATION**) and a gain of electrons (**REDUCTION**). In reality, most oxidation–reduction reactions probably do not proceed in this way.[*] The method is essentially a bookkeeping device enabling us to deal with the oxidation and with the reduction separately, and finally to combine the results into a balanced equation. This artificial recipe, while giving the correct answer for the overall reaction, *should not be interpreted as an explanation of the reaction.*

We illustrate the method with the reaction

$$Cu(c) + HNO_3(aq) \longrightarrow Cu(NO_3)_2(aq) + H_2O + NO(g)$$

which occurs when copper is added to dilute nitric acid. The first step is to write the equation in ionic form. HNO_3 and $Cu(NO_3)_2$ are strong electrolytes and are represented as ions:

$$Cu + H^+ + NO_3^- \longrightarrow Cu^{2+} + 2NO_3^- + H_2O + NO$$

Next, we observe which species are involved in the oxidation and in the reduction (although it is not yet necessary to decide which process is which). We write a skeleton equation for oxidation and a skeleton equation for reduction. Here, it is evident that Cu becomes Cu^{2+},

$$Cu \longrightarrow Cu^{2+} \quad \text{(unbalanced)} \tag{3}$$

and that NO_3^- somehow becomes NO,

$$NO_3^- \longrightarrow NO \quad \text{(unbalanced)} \tag{4}$$

Equation (3) is almost balanced, but not quite. An equation must have not only the same number of atoms of each kind on each side but also the same total electric charge on each side. Equation (3) has a charge of 0 on the left and +2 on the right. This situation is rectified by adding 2 electrons to the right-hand side:

$$Cu \longrightarrow Cu^{2+} + 2e^- \quad \text{(oxidation)} \tag{3'}$$

We have now achieved a balanced **PARTIAL EQUATION** for the oxidation of Cu.

Equation (4) is more difficult, for it is unbalanced both electrically and atomically. We first balance it atomically. There are 3 O atoms on the left and 1 on the right. Where did the other two go? They did not

---

[*] In an electrolytic or galvanic cell (Chapter 16), electrons must be lost at one electrode and gained at the other. For a reaction occurring in such a cell, therefore, the partial reactions have physical reality.

become O_2 gas; O_2 is not a product of the reaction. The only other fate possible for the O atoms is to become part of the water molecules:

$$NO_3^- \longrightarrow NO + 2H_2O$$

But now the equation is unbalanced with respect to H. H_2 gas is not a reactant and our source of the needed H must be the H^+ ion:

$$NO_3^- + 4H^+ \longrightarrow NO + 2H_2O$$

The equation is now balanced atomically, but not electrically. The net charge is $-1 + 4 = +3$ on the left and 0 on the right; we add 3 electrons to the left to balance it electrically*:

$$NO_3^- + 4H^+ + 3e^- \longrightarrow NO + 2H_2O \qquad \text{(reduction)} \qquad (4')$$

It remains to combine Equations (3′) and (4′) to obtain a balanced equation for the overall reaction. We cannot simply add them, for there would then be one more electron on the left-hand side than on the right-hand side. *The number of electrons given out in the oxidation must be equal to the number of electrons taken up in the reduction.* For every 3 Cu atoms oxidized, two NO_3^- ions must be reduced, so that 6 electrons are lost and 6 gained. We multiply Equation (3′) by 3, Equation (4′) by 2, and add them:

$$3Cu \longrightarrow 3Cu^{2+} + 6e^-$$
$$\underline{2NO_3^- + 8H^+ + 6e^- \longrightarrow 2NO + 4H_2O}$$
$$3Cu + 2NO_3^- + 8H^+ \longrightarrow 3Cu^{2+} + 2NO + 4H_2O$$

In checking the balancing of an equation like this, one must verify that the net charge is the same on both sides—in addition, of course, to counting atoms.

When the equation to be balanced was first given, it was not necessary to show H_2O or H^+ in the equation. If we are told that the reaction occurs in an acidic† solution (understood to be aqueous), we know that H_2O is available as a source of O and H^+ as a source of H. The problem might have been posed as: Balance the equation $Cu + NO_3^- \rightarrow Cu^{2+} + NO$ in acidic solution. We would then be justified in adding H_2O and H^+ as needed to balance the partial equations.

If the solution is basic, it contains very little H^+, but does contain a more or less high concentration of OH^-. We may balance the equation in the same way as for an acidic solution, but at the end we must return

---

* We should observe that the number of electrons appearing in the partial equation is the same as the change in oxidation number (page 133) of the atom undergoing a change in oxidation number. In the oxidation, $Cu \rightarrow Cu^{2+} + 2e^-$, the oxidation number of Cu changes from 0 to +2, corresponding to a *loss* of 2 electrons. In the reduction, the oxidation number of N changes from +5 (in NO_3^-) to +2 (in NO), a change of −3; accordingly, the partial equation shows that NO_3^- *gains* 3 electrons.

† An acidic solution is a solution containing H^+; a basic solution is a solution containing OH^- (page 248).

to reality by neutralizing the H^+ with OH^-. The whole procedure, in any case, is just bookkeeping; the steps we go through do not matter as long as the end product is correct.

As an example of a reaction in basic solution, we consider the reaction

$$MnO_4^- + SO_3^{2-} \longrightarrow MnO_2(c) + SO_4^{2-}$$

The unbalanced partial equations are

$$MnO_4^- \longrightarrow MnO_2 \tag{5}$$

and

$$SO_3^{2-} \longrightarrow SO_4^{2-} \tag{6}$$

Equation (5) is short by two O on the right; therefore, we add $2H_2O$ to the right, requiring $4H^+$ on the left:

$$MnO_4^- + 4H^+ \longrightarrow MnO_2 + 2H_2O$$

Since the net charge is $+3$ on the left and zero on the right, we add $3e^-$ to the left:

$$MnO_4^- + 4H^+ + 3e^- \longrightarrow MnO_2 + 2H_2O \tag{7}$$

Equation (6) is balanced similarly. We add H_2O to the left, leaving $2H^+$ on the right:

$$SO_3^{2-} + H_2O \longrightarrow SO_4^{2-} + 2H^+$$

For electrical balancing, we add $2e^-$ to the right:

$$SO_3^{2-} + H_2O \longrightarrow SO_4^{2-} + 2H^+ + 2e^- \tag{8}$$

Before adding the partial equations, we must multiply Equation (7) by 2 and Equation (8) by 3:

$$2MnO_4^- + 8H^+ + 6e^- \longrightarrow 2MnO_2 + 4H_2O$$
$$3SO_3^{2-} + 3H_2O \longrightarrow 3SO_4^{2-} + 6H^+ + 6e^-$$
$$\overline{2MnO_4^- + 8H^+ + 3SO_3^{2-} + 3H_2O \longrightarrow 2MnO_2 + 4H_2O + 3SO_4^{2-} + 6H^+}$$

Since H^+ and H_2O appear on both sides, they should be canceled in part:

$$2MnO_4^- + 2H^+ + 3SO_3^{2-} \longrightarrow 2MnO_2 + H_2O + 3SO_4^{2-}$$

Finally, we account for the basicity of the solution by adding $2\,OH^-$ to each side of the equation:

$$2MnO_4^- + 2H^+ + 2OH^- + 3SO_3^{2-} \longrightarrow 2MnO_2 + H_2O + 3SO_4^{2-} + 2OH^-$$

We now eliminate H^+ by combining each H^+ with OH^- to give H_2O:

$$2MnO_4^- + 2H_2O + 3SO_3^{2-} \longrightarrow 2MnO_2 + H_2O + 3SO_4^{2-} + 2OH^-$$

or

$$2MnO_4^- + H_2O + 3SO_3^{2-} \longrightarrow 2MnO_2 + 3SO_4^{2-} + 2OH^-$$

Example 1 Balance the equation

$$I^- + H_2O_2(aq) \longrightarrow I_2(aq) + H_2O$$

in acidic solution.

Answer

(a) Write unbalanced partial equations:

$$I^- \longrightarrow I_2 \qquad\qquad\qquad H_2O_2 \longrightarrow H_2O$$

(b) Balance with respect to atoms:

$$2I^- \longrightarrow I_2 \qquad\qquad H_2O_2 \longrightarrow 2H_2O \qquad \text{(balanced for O)}$$
$$H_2O_2 + 2H^+ \longrightarrow 2H_2O \qquad \text{(balanced for H and O)}$$

(c) Balance electrically:

$$2I^- \longrightarrow I_2 + 2e^- \qquad H_2O_2 + 2H^+ + 2e^- \longrightarrow 2H_2O$$

(In this case, the number of electrons lost happens to equal the number gained.)

(d) Add partial equations:

$$H_2O_2 + 2H^+ + 2I^- \longrightarrow I_2 + 2H_2O$$

Example 2 Balance the equation $CrO_4^{2-} + I^- \rightarrow Cr^{3+} + IO_3^-$ in basic solution.

Answer

(a) $\ CrO_4^{2-} \longrightarrow Cr^{3+}$ $\qquad\qquad\qquad\qquad I^- \longrightarrow IO_3^-$

(b) $\ CrO_4^{2-} + 8H^+ \longrightarrow Cr^{3+} + 4H_2O \qquad I^- + 3H_2O \longrightarrow IO_3^- + 6H^+$

(c) $\ CrO_4^{2-} + 8H^+ + 3e^- \longrightarrow Cr^{3+} + 4H_2O \quad I^- + 3H_2O \longrightarrow IO_3^- + 6H^+ + 6e^-$

(d) $\ 2CrO_4^{2-} + 16H^+ + 6e^- \longrightarrow 2Cr^{3+} + 8H_2O \quad I^- + 3H_2O \longrightarrow IO_3^- + 6H^+ + 6e^-$

(e) $\ 2CrO_4^{2-} + 16H^+ + I^- + 3H_2O \longrightarrow 2Cr^{3+} + 8H_2O + IO_3^- + 6H^+$
$\qquad 2CrO_4^{2-} + 10H^+ + I^- \longrightarrow 2Cr^{3+} + 5H_2O + IO_3^-$

(f) $\ 2CrO_4^{2-} + (10H^+ + 10OH^-) + I^- \longrightarrow 2Cr^{3+} + 5H_2O + IO_3^- + 10OH^-$
$\qquad 2CrO_4^{2-} + 10H_2O + I^- \longrightarrow 2Cr^{3+} + 5H_2O + IO_3^- + 10OH^-$
$\qquad 2CrO_4^{2-} + 5H_2O + I^- \longrightarrow 2Cr^{3+} + IO_3^- + 10OH^-$

ELECTRODE REACTIONS / 12.9

In Section 12.8, we learned how to write partial equations for oxidation and for reduction. These equations are just what we need to describe what happens when an electric current passes through a solution. At one electrode, called the ANODE, oxidation takes place. At the other electrode, the CATHODE, reduction takes place. The number of electrons required to produce or consume a given amount of material can be seen from the partial equation. For example, suppose that we electrolyze water, containing some sulfuric acid, H_2SO_4 ($H^+ + HSO_4^-$), to make it a conductor. The partial equations are

$$2H_2O \longrightarrow O_2(g) + 4H^+ + 4e^- \qquad \text{(oxidation, at anode)} \qquad (9)$$
$$\frac{4H^+ + 4e^- \longrightarrow 2H_2(g)}{2H_2O \longrightarrow 2H_2(g) + O_2(g)} \qquad \text{(reduction, at cathode)} \qquad (10)$$

Equation (9) says that 4 electrons must be removed from water molecules at the anode in order to produce one molecule of O_2. At the same time, 4 other electrons enter at the cathode to produce 2 molecules of H_2 [Equation (10)]. If you, in your imagination, sit anywhere along the wire watching the electrons go by, you will count 4 electrons passing you while one O_2 molecule and two H_2 molecules are being produced, and two water molecules are being decomposed.

Because we usually deal with quantities larger than a few molecules, we may say that 4 *moles** of electrons pass through the circuit for every mole of O_2 and every 2 moles of H_2 that are produced. A mole of electrons is given a special name:

1 FARADAY = N_A electrons = 1 mole of electrons

The charge of N_A electrons is represented by $\mathscr{F}$ and is called FARADAY'S CONSTANT†:

$$\mathscr{F} = 9.6487 \times 10^4 \text{ coulombs/faraday}$$

The faraday is a unit of charge, like the coulomb but bigger. When 4 faradays pass through the circuit, 2 moles H_2O are decomposed, 2 moles H_2 are produced, and 1 mole O_2 is produced. Or, if we prefer, when 1 faraday passes through the circuit, $\frac{1}{2}$ mole H_2O is decomposed, $\frac{1}{2}$ mole H_2 is produced, and $\frac{1}{4}$ mole O_2 is produced.

The relation between the charge and the quantity of matter in an electrode reaction is given by FARADAY'S LAWS. When Michael Faraday discovered these laws in 1832–1833, he did not know about partial reactions, or electrons, or a unit named after him. Rather, his laws led to these concepts. All he had were chemicals, a balance, and electricity (from a battery, not from his local utility). He did know something about atomic and molecular weights. His first law seems (in hindsight) obvious:

The mass of product produced (or reactant consumed) is proportional

---

* Reminder: 1 mole = N_A molecules, atoms, ions, electrons, etc.; N_A = Avogadro's number = 6.023×10^{23}.

† Charge is measured in *coulombs* (C); current is measured in *amperes* (amp or A). One ampere is one coulomb per second. See Appendix I.12.

to the quantity of charge (current × time) that has passed through the circuit.

The second law is the interesting one:

The mass (in grams) of a product (or reactant) produced (or consumed), when 1 faraday passes through the circuit, is equal to the atomic or molecular weight of the substance divided by a small integer. The small integer is the number of faradays required to produce or consume 1 mole of the substance; its reciprocal $(1, \frac{1}{2}, \frac{1}{3}, \ldots)$ is the fraction of a mole corresponding to 1 faraday.

For example, passing 1 faraday through appropriate solutions will produce

$\frac{1}{4}$ mole (8.00 g) of O_2
$\frac{1}{2}$ mole (1.008 g) of H_2
1 mole (107.87 g) of Ag
$\frac{1}{2}$ mole (31.77 g) of Cu
$\frac{1}{3}$ mole (8.99 g) of Al
$\frac{1}{2}$ mole (35.453 g) of Cl_2

In each case,

1 mole requires 1, 2, 3, . . . faradays

$\left.\begin{array}{l} \text{1 mole} \\ N_A \text{ molecules} \end{array}\right\}$ requires 1, 2, 3, . . . × $\mathscr{F}$ coulombs

1 molecule requires 1, 2, 3, . . . × $(\mathscr{F}/N_A)$ coulombs

The fact that 1 molecule always requires an integral number of these $\mathscr{F}/N_A$ units is highly suggestive. It suggests that $(\mathscr{F}/N_A)$ coulombs is the charge of an "atom of electricity," now called an *electron*. If electrons could be subdivided, we would expect to find, sooner or later, a reaction in which one molecule was produced by $\frac{1}{2}(\mathscr{F}/N_A)$ coulombs, or $\frac{2}{3}(\mathscr{F}/N_A)$ coulombs, or something like that. No one has ever obtained such a result. Faraday's experiments were thus the first indication that electricity, like matter, consists of indivisible particles (page 75).

Example 3 An aqueous solution of gold(III) nitrate, $Au(NO_3)_3$, was electrolyzed with a current of 0.500 A until 1.200 g of Au had been deposited on the cathode. At the other electrode, the reaction was the evolution of O_2. Find (a) the number of moles, (b) the volume at standard conditions, and (c) the mass of O_2 liberated; (d) the number of coulombs passed through the circuit; and (e) the duration of the electrolysis.

Answer The atomic weight of Au is 197.0 g/mole. The cathode reaction is

$Au^{3+} + 3e^- \longrightarrow Au$

Three faradays pass through the circuit for each mole of Au deposited. The number of faradays is therefore

$$\frac{1.200 \text{ g}}{197.0 \text{ g/mole}} \times \frac{3 \text{ faraday}}{\text{mole}} = 0.01827 \text{ faraday}$$

(a) From the partial equation

$$2H_2O \longrightarrow O_2 + 4H^+ + 4e^-$$

we see that 1 mole of O_2 requires 4 faradays, so that

$$\frac{0.01827 \text{ faraday}}{4 \text{ faraday/mole}} = 0.00457 \text{ mole of } O_2 \text{ is liberated.}$$

(b) $0.00457 \text{ mole} \times 22.4 \dfrac{\text{liters}}{\text{mole}} = 0.102 \text{ liter (sc)}$

(c) $0.00457 \text{ mole} \times 32.00 \dfrac{\text{g}}{\text{mole}} = 0.146 \text{ g}$

(d) $0.01827 \text{ faraday} \times 9.65 \times 10^4 \dfrac{\text{C}}{\text{faraday}} = 1.76 \times 10^3 \text{ C}$

(e) The charge (Q) in coulombs is the current (I) in amperes times the time (t) in seconds:

$$Q = It$$

Then

$$t = \frac{Q}{I} = \frac{1.76 \times 10^3 \text{ C}}{0.500 \text{ C/second}} = 3.52 \times 10^3 \text{ seconds}$$

CONDUCTIVITY AND EQUIVALENT CONDUCTANCE / 12.10

We noticed at the beginning of the chapter that a solution of an electrolyte is characterized by its ability to conduct electricity. We will now look at this ability in a more quantitative way. We will thus see the experimental basis for some of the assumptions we have been making, especially the distinction between strong and weak electrolytes.

To specify how *poor* a conductor is, we give its RESISTANCE, R (Appendix I.12). If we want to specify how *good* a conductor is, we give its CONDUCTANCE, $K = 1/R$. The unit of resistance is the OHM (Ω); the unit of conductance is the RECIPROCAL OHM (Ω^{-1}), or MHO. Electrolytic solutions are customarily discussed with reference to their conductances.

Conductors differ in conductance for several reasons, some trivial and some interesting. The following paragraphs identify the trivial reasons and show how they can be put aside, to enable us to focus our attention on the interesting reasons.

(1) *The size and shape of the conductor.* A short, fat conductor is a better conductor than a long, thin conductor. We choose a standard shape for comparison: a cube 1 cm on an edge. The conductance of this cube is the CONDUCTIVITY or SPECIFIC CONDUCTANCE of the material, represented by k. The units of k are $\Omega^{-1} \text{ cm}^{-1}$.

(2) *The concentration of the solution.* A more concentrated solution is a better conductor than a less concentrated solution, simply because it has more ions in a cubic centimeter. We expect that, as a rough

approximation, the conductivity should be proportional to the number of ions in the cube. When two quantities are roughly proportional, it is helpful to consider their ratio—for the ratio should be roughly constant, and any deviations from constancy are then an interesting topic of discussion. We will therefore be concerned not so much with k as with the ratio k/c, where c is the concentration in moles/liter.

(3) *The charge of the ions.* Some ions have a built-in advantage as charge carriers: they have charges of 2, 3, or more units. What determines the conductivity is not the *number* of ions per unit volume, but the *total charge* of these ions. *One* Mg^{2+} ion transports as much charge as *two* Na^+ ions moving the same distance. We can take account of these differences by considering, instead of k/c, the ratio k/nc, where n is the number of charges on the ions of each sign in the formula. Another way of saying it is that n is the number of faradays of positive charge (the same as the number of faradays of negative charge) on the ions obtained when 1 mole of the electrolyte is dissolved. The units of n are thus faradays per mole. For NaCl, $n = 1$, because 1 mole dissolves to give Na^+ ions with a charge of $+1$ faraday and Cl^- ions with a charge of -1 faraday. For $MgSO_4$, 1 mole dissolves to give Mg^{2+} ions with a charge of $+2$ faradays and SO_4^{2-} ions with a charge of -2 faradays; therefore, $n = 2$. For $CaCl_2$, $n = 2$; for $Al_2(SO_4)_3$, $n = 6$.

The EQUIVALENT CONDUCTANCE, Λ,* is defined by

$$\Lambda = \frac{k}{n(c/1000)} = \frac{1000\ k}{nc}$$

The factor 1000 cm³/liter is introduced to change moles/liter (the units of c) to moles/cm³ (the units of $c/1000$). The units of Λ are

$$\frac{(1000\ cm^3\ liter^{-1}) \times (k\Omega^{-1}\ cm^{-1})}{(n\ faraday\ mole^{-1}) \times (c\ mole\ liter^{-1})} = \frac{1000k}{nc}\ \Omega^{-1}\ cm^2\ faraday^{-1}$$

The equivalent conductance is the quantity in terms of which we can compare different solutions. All the trivial factors—size and shape, concentration, ionic charge—have been corrected for, and we have a real measure of the current-carrying ability of the electrolyte.

For a weak electrolyte, n is the number of faradays on the ions of each sign that *would* be produced by 1 mole *if* it were completely ionized. Thus, for acetic acid, $n = 1$ faraday/mole, as indicated by the equation

$$HC_2H_3O_2 \Longleftrightarrow H^+ + C_2H_3O_2^-$$

For oxalic acid, $n = 2$ faradays/mole, for 1 mole of oxalic acid would yield, on complete ionization, 2 moles of H^+ ions and 1 mole of $C_2O_4^{2-}$ ions:

$$H_2C_2O_4 \Longleftrightarrow 2H^+ + C_2O_4^{2-}$$

---

* Λ, capital lambda, for the German *Leitfähigkeit*, "conductivity."

The *conductivity* (k) of an ionic solution *increases* with increasing concentration. We guessed that the *equivalent conductance* (Λ) should be independent of concentration. However, this guess is not quite correct. Instead of being constant, Λ normally *decreases* with increasing concentration. As the solution becomes more concentrated, k and c both increase, but k does not increase so rapidly as c; the ratio $\Lambda = 1000 \, k/nc$ therefore decreases. The following data for KCl in water at 25°C illustrate this behavior:

c, mole/liter	k, $\Omega^{-1} \, cm^{-1}$	Λ, $\Omega^{-1} \, cm^2 \, faraday^{-1}$
0.01000	0.001409	140.9
0.1000	0.01286	128.6
1.000	0.1113	111.3

This distinction between the increase in k and the decrease in Λ should be noted. There is a common misconception that a solution becomes a poorer conductor as its concentration increases. The opposite is true—it becomes a *better* conductor. However, the improvement in its conductivity does not keep pace with the increase in its concentration; that is what we mean by saying that Λ decreases. The variation of Λ with c, for a few typical electrolytes, is shown in Fig. 12.5. We plot Λ against $\sqrt{c}$ rather than against c because a more nearly straight line is obtained in this way.

How can we account for this change of Λ with concentration? Arrhenius' explanation was that an equilibrium is established between ions and undissociated molecules; for example, $KCl(aq) \leftrightharpoons K^+ + Cl^-$. As the concentration increases, the equilibrium should be shifted toward neutral molecules,* with a consequent loss in the percentage available for conduction. Qualitatively, of course, this assumption is in agreement with the facts. It fails, however, to predict the variation of equivalent conductance even approximately. If the equilibrium is established, the equilibrium condition

$$\frac{[K^+][Cl^-]}{[KCl]} = K$$

should be satisfied. An *apparent* equilibrium constant can be calculated from conductance data, but is far from being constant—at 18°C, for example, $K = 2.3$ for a 1 M solution and 15.2 for a 0.01 M solution. Similar

---

* This is an application of Le Chatelier's principle: increasing the concentration favors the reaction which decreases the number of dissolved particles (page 205).

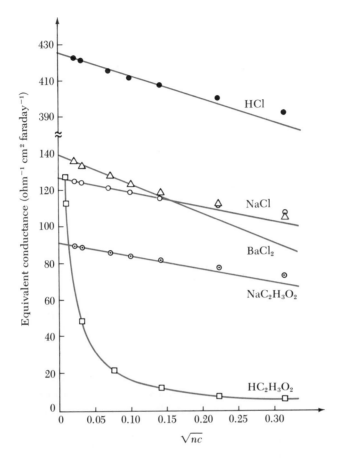

Fig. 12.5. *Equivalent conductances (Λ) of several electrolytes at 25°C, plotted against the square root of the concentration. In dilute solutions of strong electrolytes, the experimental points are well fitted by straight lines.*

variations are found in the values of K calculated from colligative properties.

STRONG ELECTROLYTES / 12.12

There is a large class of electrolytes which have certain common features:

(1) They have relatively large equivalent conductances.

(2) Λ is a linear function of $\sqrt{c}$ at low concentrations.

(3) The variation of equivalent conductance with concentration cannot be accounted for on the basis of equilibrium between ions and molecules.

(4) Most, but not all, of these electrolytes exist as ionic crystals in the solid state.

These are the strong electrolytes, and are believed to be present mostly as ions in dilute solutions.

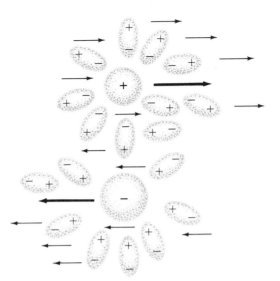

Fig. 12.6. *Two oppositely charged ions moving in opposite directions through a polar solvent.*

The equivalent conductance depends on the speed at which the ions move in a given electric field. The faster the ions, the better the conductor. Even if the electrolyte is present entirely as ions, it is possible for its equivalent conductance to decrease as the concentration increases, if the speed of the ions decreases as concentration increases. This is indeed what happens. Each ion has neighbors predominantly of the opposite charge; it is attracted by these ions moving in the opposite direction, and its progress is thus hindered. What is equally bad is that the oppositely charged ions nearby are dragging solvent molecules with them in the wrong direction—each ion is attempting to keep its sheath of solvent molecules with it—increasing the resistance that any moving particle must overcome (Fig. 12.6). These effects are, of course, greater when the concentration is greater. They are greater for ions with multiple charge, which attract each other and the solvent molecules more strongly than singly charged ions. Figure 12.5 shows that the variation of Λ with c is greater for $BaCl_2$ than for $NaCl$.

Because the graph of Λ versus $\sqrt{c}$ (Fig. 12.5) becomes nearly a straight line at low concentrations, it can be extrapolated with confidence to $c = 0$. The limiting value of Λ obtained by this extrapolation, Λ_0, is called the **EQUIVALENT CONDUCTANCE AT INFINITE DILUTION**; it would be the equivalent conductance of the electrolyte if the ions were so far apart as to have no effect on each other.

Not only the conductance but also the colligative properties of strong electrolytes show deviations from the values to be expected on the basis of complete ionization. The freezing-point depression of, for example, a $NaCl$ solution is less than we would expect for 2 moles of ions per mole of $NaCl$; the van't Hoff factor approaches 2 only in very dilute solutions (Table 12.2, page 217). These diminutions in the colligative properties are also accounted for by the concept of interionic attraction. In general, the effect of attraction is to make the ions behave as if they were not all there as individual ions, but were partly combined into pairs. Indeed, in more concentrated solutions, and especially in solvents with lower dielectric constants than water, a substantial fraction of the ions are actually associated into *ion pairs*. This physical pairing is in addition to the apparent pairing resulting from attraction.

In addition to the ionic salts (including the soluble hydroxides, as NaOH), some acids behave as strong electrolytes in aqueous solution: HCl, HBr, HI, H_2SO_4 (sulfuric acid), HNO_3 (nitric acid), and $HClO_4$ (perchloric acid) are the most common of them. These acids, which give H^+ (actually H_3O^+) in solution, have especially large equivalent conductances, as illustrated for HCl in Fig. 12.5. The explanation is that the positive charge of an H_3O^+ ion can travel in two ways: The ion, like any other, can push its way through the solution ("ferryboat conduction"); or the ion can pass its extra proton to the next H_2O molecule, which thus becomes an H_3O^+ ion and can pass a proton to the next H_2O molecule, which thus becomes an H_3O^+ ion, etc. ("bucket-brigade conduction"). The second process is much faster than the first and accounts for the large conductance of solutions containing H_3O^+. OH^- also shows bucket-brigade conduction, though to a smaller extent than H_3O^+. This accounts for the large equivalent conductance of ionic hydroxides (for example, $\Lambda_0 = 248 \ \Omega^{-1} \ cm^2 \ faraday^{-1}$ for NaOH).

WEAK ELECTROLYTES / 12.13

Acetic acid, on being dissolved in water, yields hydronium ions and acetate ions by the reaction

$$HC_2H_3O_2 \rightleftharpoons H^+ + C_2H_3O_2^-$$

Λ_0 for acetic acid cannot be obtained by extrapolation of Λ to $c = 0$, but it can be calculated from other data (specifically, from the Λ_0's of $NaC_2H_3O_2$, HCl, and NaCl). The result of this calculation is $\Lambda_0 = 390.7 \ \Omega^{-1} \ cm^2 \ faraday^{-1}$. If the acid were completely ionized, then its equivalent conductance would be close to Λ_0, as for the strong electrolytes in Fig. 12.5. As we can see in the figure, however, the equivalent conductance of acetic acid is nowhere near this value, even in the most dilute solutions for which measurements have been made. We also note that Λ decreases, as c increases, much more rapidly for acetic acid than for strong electrolytes, with no semblance of a linear dependence on $\sqrt{c}$. Even so, the *conductivity* (k) increases as the concentration increases,[*] as for strong electrolytes. The following data for acetic acid at

---

[*] Pure acetic acid, like pure water, is practically a nonconductor. Therefore, as the concentration increases, the conductivity of acetic acid solutions must eventually reach a maximum and decline. The maximum occurs at $c = 2.7$ mole/liter (16% acetic acid by mass).

25°C will illustrate the distinction:

c, moles/liter	k, $\Omega^{-1}\,cm^{-1}$	Λ, $\Omega^{-1}\,cm^2\,faraday^{-1}$
2.801×10^{-5}	0.589×10^{-5}	210
1.028×10^{-3}	$4.96\ \times 10^{-5}$	48.2
5.912×10^{-3}	$12.4\ \times 10^{-5}$	21.0
5.000×10^{-2}	$36.0\ \times 10^{-5}$	7.20

The most likely explanation for this great change of equivalent conductance with concentration is that we have in this case a partial dissociation into ions, just as Arrhenius postulated for all electrolytes. An electrolyte which dissociates only to a small extent into ions at equilibrium is known as a WEAK ELECTROLYTE. Most acids are of this kind. Weak bases are ammonia, NH_3, and many compounds related to it by the replacement of one or more hydrogen atoms by other atoms or groups of atoms; examples are methylamine, H_3CNH_2, and hydrazine, H_2NNH_2, an important rocket fuel. There are a relatively few weakly ionized salts which exist predominantly as undissociated molecules even in very dilute solutions: $CdCl_2$, $Pb(C_2H_3O_2)_2$, $HgCl_2$, for example.

The maximum equivalent conductance of completely ionized acetic acid would be 390.7 Ω^{-1} cm^2 faraday^{-1}. The observed equivalent conductance at 25°C of a 0.050 M solution of acetic acid is 7.20 Ω^{-1} cm^2 faraday^{-1}. The ratio of these numbers is

$$\frac{\Lambda}{\Lambda_0} = \frac{7.20}{390.7} = 0.0184 = 1.84\%$$

The solution is only 1.84% as good a conductor as it would be if the acid were completely ionized. We conclude that 1.84% of the acetic acid molecules have reacted with water to become H^+ and $C_2H_3O_2^-$ ions, and 98.16% of the acetic acid is still present as uncharged molecules. The fraction (α) of a weak electrolyte that is present as ions is called the DEGREE OF IONIZATION of the electrolyte; it depends, of course, on the concentration of the solution, the identity of the solvent, and the temperature. To a good approximation, $\alpha = \Lambda/\Lambda_0$.

In contrast to the situation with strong electrolytes, the degree of ionization of weak electrolytes is in good agreement with the law of chemical equilibrium. For the reaction

$$HC_2H_3O_2 \leftrightharpoons H^+ + C_2H_3O_2^-$$

the equilibrium constant at 25°C calculated from conductance data is $K = 1.77 \times 10^{-5}$ for a 2.80×10^{-5} M solution and $K = 1.81 \times 10^{-5}$ for a 0.231 M solution. We can conclude from data like these that partial ionization is the major cause of the variation of Λ with concentration.

1. Definitions. Define: (a) theory of electrolytic dissociation, (b) lattice energy, (c) hydration energy, (d) Faraday's laws, (e) faraday (unit), (f) conductivity, (g) equivalent conductance, (h) strong electrolyte, (i) weak electrolyte.

2. Freezing-point depression. A 0.00200 molal aqueous solution of $KCo(NH_3)_2(NO_2)_4$ freezes at $-0.00736°C$. How many moles of ions are present in a dilute solution containing 1 mole of this compound?

3. Balancing equations. Balance each of the following equations by the ion–electron method. Show the balanced partial equations for oxidation and reduction, and the ionic equation for the overall reaction:

(a) $Pb(c) + PbO_2(c) \longrightarrow Pb^{2+}$ (in acidic solution)

(b) $Cl_2(g) + CN^- \longrightarrow OCN^- + Cl^-$ (in basic solution)

cyanate ion

(c) $Cr_2O_7^{2-} + H_2S(aq) \longrightarrow Cr^{3+} + S_8(c)$ (in acidic solution)

(d) $SO_3^{2-} + MnO_4^- \longrightarrow SO_4^{2-} + MnO_4^{2-}$ (in strongly basic solution)

manganate ion

(e) $I_2(c) \longrightarrow I^- + IO_3^-$ (in basic solution)

(f) $Bi(OH)_3(c) + Sn(OH)_4^{2-} \longrightarrow Bi(c) + Sn(OH)_6^{2-}$ (in basic solution)

(g) $HgCl_4^{2-} + Sn^{2+} \longrightarrow Hg_2Cl_2(c) + SnCl_6^{2-}$ (in acidic solution)

(h) $H_2CO + Ag(NH_3)_2^+ \longrightarrow CHO_2^- + Ag(c) + NH_3$ (in basic solution)

formaldehyde   formate ion

4. Faraday's laws. An aqueous solution of $CuSO_4$ was electrolyzed between inert electrodes for 60.0 minutes. The cathode reaction was the deposition of 0.368 g of Cu. The anode reaction was the formation of O_2. How many (a) moles, (b) grams, (c) liters (sc) of O_2 were evolved at the anode? (d) How many coulombs passed through the solution? (e) What was the average current in amperes?

5. Faraday's laws. Rhodium was plated from an aqueous solution containing Rh^{y+} for 2.00 hours with a current of 0.0650 A. The rhodium deposit on the cathode weighed 0.1664 g. What is the charge of the rhodium ion?

6. Equivalent conductance. The conductivity of a 0.100 M solution of $Ca(NO_3)_2$ is 0.01932 Ω^{-1} cm^{-1} at 18°C. Find its equivalent conductance.

7. Degree of ionization. The equivalent conductance of benzoic acid, $HC_7H_5O_2$, at infinite dilution at 25°C is 382.1 Ω^{-1} cm^2 faraday^{-1}. The equivalent conductance of 0.10 M benzoic acid at 25°C is 8.8 Ω^{-1} cm^2 faraday^{-1}. Find the degree of ionization of the acid in this solution.

ADDITIONAL PROBLEMS

8. Freezing-point depression. Estimate the freezing point of a 0.01000 molal aqueous solution of $KClO_3$, an ionic salt.

9. Freezing-point depression. The freezing point of a 0.10 molal aqueous solution of sulfuric acid, H_2SO_4, is $-0.40°C$. (a) What is the van't Hoff factor for H_2SO_4 at this molality? (b) Is the principal ionization reaction $H_2SO_4 \rightarrow H^+ + HSO_4^-$ or $H_2SO_4 \rightarrow 2H^+ + SO_4^{2-}$? Do both reactions occur to some extent?

10. Osmotic pressure. Estimate the osmotic pressure of sea water at $25°C$. Assume that it contains 36 g NaCl (a strong electrolyte) per liter. What is the minimum pressure needed to purify sea water by reverse osmosis (page 188)?

11. Balancing equations. Balance each of the following equations by the ion–electron method. Show the balanced partial equations, and the overall equation in ionic form. (Electrolytes are strong unless otherwise indicated.)

(a) $Zn(c) + HNO_3(aq) \rightarrow Zn(NO_3)_2(aq) + NH_4NO_3(aq)$ (in acidic solution)

(b) $H_2C_2O_4(aq,$ weak acid$) + KMnO_4(aq) + H_2SO_4(aq) \rightarrow 2CO_2(g) + MnSO_4(aq) + K_2SO_4(aq)$ (in acidic solution) (Note that H_2SO_4 is not oxidized or reduced.)

(c) $CoSO_4(aq) + Na_2O_2(c) \rightarrow Co(OH)_3(c) + Na_2SO_4(aq)$ (in basic solution)

(d) $I_2(aq) + Na_2S_2O_3(aq) \rightarrow Na_2S_4O_6(aq) + NaI(aq)$ (in acidic or neutral solution)

(e) $Cu(NH_3)_4^{2+} + CN^- \rightarrow Cu(CN)_2^- + OCN^- + NH_3$ (in basic solution)

12. Partial equations. Chemists have long wanted to synthesize "squaric acid,"

both as an interesting compound and as a starting point for useful products. It was made in 1972 by electrochemical reduction of CO under pressure in a nonaqueous solvent containing solvated H^+ ions. Write a partial equation for this reduction.

13. Faraday's laws. Three electrolytic cells are connected so that the same current passes through all 3 cells. In the first Ag^+ is reduced to Ag. In the second, Zn is oxidized to Zn^{2+}. In the third, Hg_2^{2+} is oxidized to Hg^{2+}. If 1.00 g of Ag is deposited in the first cell, find the mass of Zn that is removed from the anode in the second cell, and the mass of $Hg(NO_3)_2$ that can be recovered from the solution in the third cell.

14. Faraday's laws. A current of 0.500 A is used for the decomposition of water. (a) How long (in hours) will it take to produce 1.00 liter (sc) of O_2? (b) What mass of H_2 will be produced in the same time?

15. Faraday's laws. Iridium was plated from a solution containing $IrCl_6^y$ for 2.00 hours with a current of 0.0750 A. The Ir deposit on the cathode weighed 0.359 g. (a) What is the oxidation number of Ir in $IrCl_6^y$? (b) What is the charge (y) on this ion?

16. Faraday's laws. In a chemical plant, fused NaCl is electrolyzed with a current of 100 A. (a) How many kilograms of Na are produced every 24 hours? (b) How many liters (sc) of Cl_2 are produced?

17. Electrode reactions. Malodorous materials can be removed from industrial stack gases by oxidation to harmless products. One possible process is oxidation in acid solution by $CeOH^{3+}$ (which is reduced to Ce^{3+}), followed by electrolytic regeneration of $CeOH^{3+}$ with H_2 as a byproduct. (a) Write four partial equations for these processes, using as the pollutant acrolein, C_3H_4O, which is oxidized to CO_2 and H^+. (b) Write the equation for the overall process. (c) How many coulombs are needed to dispose of 1.00 kg of acrolein?

18. Electrolysis. Solid polymer electrolytes have been developed for use in large-scale electrolysis of water. They are sheets of ion-exchange resin (page 561) about 0.3 mm thick, containing mobile H^+ ions and fixed $-SO_3^-$ groups. (a) In 1974, current densities of 10^4 A/m^2 had been achieved. How many kilograms of H_2 could be produced per hour per square meter of electrolyte? (b) What advantages would this system have over electrolysis in an ordinary liquid solution? over a polymer with fixed positive ions and mobile OH^- ions?

19. Faraday's laws. In the electrolysis of a $CuSO_4$ solution, there were two reactions at the anode:

$$Cu \longrightarrow Cu^{2+} + 2e^- \qquad\qquad (i)$$
$$2H_2O \longrightarrow O_2 + 4H^+ + 4e^- \qquad\qquad (ii)$$

A current of 1.00 A passed for 1.00 hour. The loss in mass of the Cu anode was 0.600 g. (a) If all the current had been used for reaction (i), how much Cu would have reacted? (b) What percentage of the current was used to produce O_2? (c) How many moles of O_2 were liberated?

20. Units. The scale of atomic weights was changed in 1961. On the old (chemical) scale, the atomic weight of natural oxygen was exactly 16. On the new (unified) scale, the atomic weight of natural oxygen is 15.9994. Did the number of coulombs per faraday change in 1961? If so, what is the ratio of the new to the old faraday?

21. Practical electrochemistry. Electrochemical machining is a technique of shaping metal by controlled oxidation to ions. The metal to be shaped is the anode and a complementary shape—with elevations where depressions are to be made, etc.—is the cathode. The anode and cathode are very close together but must not touch. High current densities are needed to make the process reasonably fast. (a) Why does an elevation in the cathode produce a depression in the anode? (Suggestion: What determines where the greatest current flows?) (b) Why must the electrodes be close together (besides minimizing resistance)? (c) Solution is continuously pumped through the gap between the electrodes. Why? What would happen if it were not pumped?

22. Equivalent conductance. In pure water at 25°C, $[H^+] = [OH^-] = 1.0 \times 10^{-7}$ mole/liter. (Is this close to "infinite dilution"?) For the imaginary electrolyte H^+OH^-, $\Lambda_0 = 548$ Ω^{-1} cm^2 faraday^{-1}. Find the conductivity of perfectly pure water at 25°C.

23. Equivalent conductance. The conductivity of a saturated solution of AgCl at 25°C is 1.8×10^{-6} Ω^{-1} cm^{-1}, after subtraction of the conductivity of the water. Λ_0 for AgCl is 138 Ω^{-1} cm^2 faraday^{-1}. AgCl is so slightly soluble that its saturated solution may be considered "infinitely dilute." Find the solubility, in moles per liter, of AgCl in H$_2$O at 25°C.

24. Strong electrolytes. The equivalent conductance of an aqueous solution of HCl at 25°C is 425.13 Ω^{-1} cm^2 faraday^{-1} when the concentration is 2.841×10^{-5} mole/liter, and 418.10 when the concentration is 2.994×10^{-3} mole/liter. Λ_0 for HCl is 426.16 Ω^{-1} cm^2 faraday^{-1}. (a) From these data, calculate the apparent degree of ionization of HCl in each of the two solutions. (Suggestion: Calculate $1 - \alpha$, then α.) (b) Calculate the apparent values of the equilibrium constant for the reaction HCl $\leftrightharpoons$ H$^+$ + Cl$^-$. (c) What conclusions can you draw from the results of these calculations?

13

ACIDS AND BASES

Acidic and basic solutions were first defined in terms of several easily recognizable chemical and physical properties. Solutions were called "acidic" if they had a sour taste, changed the color of indicators (for example turned blue litmus red), and reacted with active metals such as zinc or magnesium to release hydrogen. Solutions were called "basic" if they felt slippery, tasted bitter, and changed red litmus to blue. Acidic and basic solutions were said to neutralize each other to produce water and a salt. In the late 1800's, chemists began to consider structural features that might account for the observed behavior. Svante Arrhenius, in 1887, offered a definition of acids and bases based on the knowledge that

(1) pure molecular acids, such as anhydrous sulfuric acid and 100% (glacial)* acetic acid, are nonelectrolytes, yet their aqueous solutions conduct a current;

(2) substances recognized as acids possess at least one hydrogen atom per molecule.

Arrhenius reasoned that a pure acid, such as hydrogen chloride, HCl, is a neutral molecule that dissolves in water to give $H^+(aq)$ and $Cl^-(aq)$ ions; the acidic solution is called *hydrochloric acid*. The abundance of the H^+ ion was thought to impart the familiar acidic properties to the solution.

Since all bases at the time of Arrhenius were known to possess the hydroxyl group, OH, and to conduct a current when dissolved in water, he assumed that a base is a substance that dissociates in water to give an OH^- ion and a cation: $NaOH(c) + aq \rightarrow Na^+(aq) + OH^-(aq)$. The

---

* One hundred percent acetic acid (freezing point, 16.6°C) is called "glacial" from the days when it froze in cold laboratories.

characteristic behavior of a base was attributed to the high concentration of the OH^- ion.

Equilibria were thought to exist between the molecules and the ions in solution. For a strong acid such as hydrochloric acid, and a strong base such as sodium hydroxide, the equilibrium constants are large. Consequently, for dilute solutions, ionization is practically complete in the direction indicated by the relative lengths of the arrows in the equations

$$HCl \rightleftharpoons H^+ + Cl^- \qquad K > 10^2$$
$$NaOH \rightleftharpoons Na^+ + OH^- \qquad K > 10^2$$

For a weak acid such as acetic acid, $HC_2H_3O_2$, and a weak base such as ammonia, the equilibrium constants are small and ionization is very incomplete:

$$HC_2H_3O_2 \rightleftharpoons H^+ + C_2H_3O_2^- \qquad K_a \approx 10^{-5}$$
$$NH_3 \cdot H_2O \rightleftharpoons NH_4^+ + OH^- \qquad K_b \approx 10^{-5}$$

In terms of the Arrhenius theory, we would define an acid–base reaction as one occurring specifically between H^+ and OH^- to form water, coincident with the formation of a salt:

$$\text{in water:} \quad \underbrace{H^+ + Cl^-}_{\text{acid}} + \underbrace{Na^+ + OH^-}_{\text{base}} \longrightarrow H_2O + \underbrace{Na^+ + Cl^-}_{\text{salt}}$$

BRÖNSTED–LOWRY CONCEPT OF ACID–BASE REACTIONS / 13.2

One limitation of Arrhenius' definition is its failure to deal with acid–base reactions that occur in the absence of water. Individually, hydrogen chloride and ammonia do not ionize in benzene, yet together they react to form ammonium chloride, the same salt obtained from an aqueous solution. Also, one would not expect an H^+ ion (a proton) to maintain its identity in water because it would be strongly attracted to any molecule or ion having a free pair of electrons. It could exist in an isolated state, as in an electric discharge tube, but certainly not in the regions of high electron density surrounding the oxygen atoms of water molecules.

A major improvement on Arrhenius' definition of acids and bases came in 1923 when Johannes N. Brönsted and Thomas M. Lowry independently proposed a more general theory. Their theory states that *an acid–base reaction involves a proton transfer; the acid is the proton donor and the base is the proton acceptor.* This theory emphasizes the interdependence of an acid and a base; one cannot react without the other. The stronger the acid, the more readily it donates a proton (large K_a); the stronger the base, the more readily it accepts a proton (large K_b).

(a) Conjugate acid–base pairs. In aqueous solution an acid, HA, ionizes by transferring its proton to water, the base*:

$$\begin{array}{c} H \\ \diagdown \\ :O: + H:A \rightleftharpoons \left[\begin{array}{c} H \\ \diagdown \\ :O:H \\ \diagup \\ H \end{array}\right]^+ + :A^- \\ \diagup \\ H \end{array} \tag{1}$$

base₁   acid₂   acid₁   base₂

The hydronium ion is further hydrated, but for simplicity it will be written as H_3O^+. In some equations it will be further simplified by writing H^+. In the reverse reaction, H_3O^+ is the acid that donates a proton to A^-, the base. The reaction of an acid and a base *always* leads to the formation of a new acid and a new base. The product acid, H_3O^+, arises when the reactant base, H_2O, accepts a proton; H_3O^+ and H_2O are said to constitute a **CONJUGATE ACID–BASE PAIR**. Similarly, the product base, A^-, results when the reactant acid, HA, loses a proton; HA and A^- also make up a conjugate acid–base pair. One conjugate acid–base pair in the reaction is designated by the subscript 1 and the other pair by the subscript 2, as demonstrated in Equations (1) and (2),

$$\begin{array}{c} H \qquad\qquad H \\ \diagdown \qquad\qquad \diagup \\ :O: \quad :N{-}H \rightleftharpoons :O: + \left[\begin{array}{c} H \\ \diagup \\ H{-}N{-}H \\ \diagdown \\ H \end{array}\right]^+ \\ \diagup \qquad \diagdown \qquad \diagup \\ H \qquad\quad H \qquad H \end{array} \tag{2}$$

acid₁   base₂   base₁   acid₂

In the reaction shown in Equation (2), NH_3 is the base and H_2O is the acid. The ammonium ion, NH_4^+, is the conjugate acid of NH_3, and OH^- is the conjugate base of H_2O. When NH_3 reacts with HCl in benzene solution, the HCl (acid) donates a proton to the NH_3 (base). The products are Cl^-, the conjugate base of HCl, and NH_4^+, the conjugate acid of NH_3, which form a salt (NH_4Cl) insoluble in benzene.

$$H_3N: + HCl \xrightleftharpoons{\text{in benzene}} [H_3N:H]^+Cl^- \quad (c)$$

base₁   acid₂   acid₁   base₂

Electronic formulas have been used in the preceding equations to reveal an important structural feature: *The base in each case has an unshared pair of electrons that can form a covalent bond with the*

---

* Curved arrows are used to indicate the fate of the reactants. In H_3O^+ all bonds to H atoms are alike; the formula is drawn this way to emphasize the change.

transferred proton. Because it acquires a proton (H^+), the base *gains* in positive charge when it is converted to its conjugate acid, and the acid *loses* positive charge or gains negative charge on being converted into its conjugate base. For example:

$$Base \xrightarrow{+H^+} Conjugate\ acid \qquad Acid \xrightarrow{-H^+} Conjugate\ base$$

$$NH_3 \xrightarrow{+H^+} NH_4^+ \qquad\qquad NH_4^+ \xrightarrow{-H^+} NH_3$$

$$OH^- \xrightarrow{+H^+} H_2O \qquad\qquad HNO_3 \xrightarrow{-H^+} NO_3^-$$

$$O^{2-} \xrightarrow{+H^+} OH^- \qquad\qquad HSO_4^- \xrightarrow{-H^+} SO_4^{2-}$$

$$HPO_4^{2-} \xrightarrow{-H^+} PO_4^{3-}$$

There is an interdependent relationship between the strengths of a given conjugate pair. In general, *the stronger an acid, the weaker its conjugate base; the stronger a base, the weaker its conjugate acid.* Conversely, *the weaker an acid, the stronger its conjugate base; the weaker a base, the stronger its conjugate acid.* Table 13.1 lists some common conjugate acid–base pairs, showing the interdependent relationship of their strengths.

We see from Table 13.1 that most uncharged acids have the acidic proton bonded to oxygen, a halogen, or sulfur. Occasionally an acid may have H bonded to N as in hydrazoic acid (H—N=N=N), or to C, as in hydrogen cyanide (H—C≡N). In most molecules an H atom bonded to C, N, P, Si, or B is not acidic. When writing the molecular formula of a compound with acidic and nonacidic H atoms, the acidic H atoms are shown first (see Appendix II).

From observing the relative positions of the equilibria of Equations (3) and (4),

$$
\begin{array}{cccccccc}
HCl & + & H_2O & \rightleftharpoons & H_3O^+ & + & Cl^- & \qquad K_a \approx 10^7 \quad (3)\\
acid_1 & & base_2 & & acid_2 & & base_1 \\
(stronger & & (stronger & & (weaker & & (weaker \\
than\ H_3O^+) & & than\ Cl^-) & & than\ HCl) & & than\ H_2O)
\end{array}
$$

$$
\begin{array}{cccccccc}
HC_2H_3O_2 & + & H_2O & \rightleftharpoons & H_3O^+ & + & C_2H_3O_2^- & \qquad K_a \approx 10^{-5} \quad (4)\\
acid_1 & & base_2 & & acid_2 & & base_1 \\
(weaker & & (weaker & & (stronger & & (stronger \\
than\ H_3O^+) & & than\ C_2H_3O_2^-) & & than\ HC_2H_3O_2) & & than\ H_2O)
\end{array}
$$

Table 13.1
Common Conjugate Acid–Base Pairs

Acid		Conjugate base	
Name	Formula	Formula	Name
Perchloric acid	$HClO_4$	ClO_4^-	Perchlorate ion
Sulfuric acid	H_2SO_4	HSO_4^-	Hydrogen sulfate ion
Hydriodic acid	HI	I^-	Iodide ion
Hydrobromic acid	HBr	Br^-	Bromide ion
Hydrochloric acid	HCl	Cl^-	Chloride ion
Nitric acid	HNO_3	NO_3^-	Nitrate ion
Hydronium ion	$\mathbf{H_3O^+}$	$\mathbf{H_2O}$	**Water**
Hydrogen sulfate ion	HSO_4^-	SO_4^{2-}	Sulfate ion
Phosphoric acid	H_3PO_4	$H_2PO_4^-$	Dihydrogen phosphate ion
Nitrous acid	HNO_2	NO_2^-	Nitrite ion
Acetic acid	$HC_2H_3O_2$	$C_2H_3O_2^-$	Acetate ion
Carbonic acid	$CO_2 + H_2O\ [H_2CO_3]$	HCO_3^-	Bicarbonate ion
Hydrogen sulfide	H_2S	HS^-	Hydrosulfide ion
Ammonium ion	NH_4^+	NH_3	Ammonia
Hydrogen cyanide	HCN	CN^-	Cyanide ion
Water	H_2O	OH^-	Hydroxide ion
Ammonia	NH_3	NH_2^-	Amide ion

(left margin, bottom-to-top arrow) Increasing strengths as an acid

(right margin, top-to-bottom arrow) Increasing strength as a base

it is evident that more molecules of HCl give up protons to water than do molecules of $HC_2H_3O_2$; hydrochloric acid is a stronger acid than acetic acid. From the relative extent of the reverse reactions, it is evident that acetate ion, $C_2H_3O_2^-$, is a stronger base than Cl^-, because of its greater tendency to accept a proton from H_3O^+.

In every acid–base reaction, the two existing bases—as H_2O and Cl^- in Equation (3)—compete for the proton. The stronger base prevails and is thereby converted to its conjugate acid. The weaker base fails to acquire the proton and remains unchanged. Consequently, *the equilibrium favors the presence of the weaker base.* Thus, in Equation (3), Cl^- is a weaker base than H_2O, which must also mean that HCl is a stronger acid than H_3O^+. Also, in Equation (4), $C_2H_3O_2^-$ is a stronger base than H_2O which means that between the two conjugate acids, H_3O^+ is a stronger acid than $HC_2H_3O_2$. We see that acid–base reactions "run downhill" *to favor the formation of the weaker base and weaker acid.*

(b) **Relative acidity and basicity.** *Strengths of acids must be compared with respect to the same base.* Dilute aqueous solutions of strong acids, those with $K > 2$, ionize practically completely so that the only acid present is H_3O^+:

$$HCl \quad + H_2O \longrightarrow H_3O^+ + \quad Cl^-$$
hydrochloric acid

$$HNO_3 \quad + H_2O \longrightarrow H_3O^+ + NO_3^-$$
nitric acid

$$HClO_4 \quad + H_2O \longrightarrow H_3O^+ + ClO_4^-$$
perchloric acid

strong acids of different
strength $(K > 2)$; all
stronger than H_3O^+

only acid
present

H_3O^+ is, thus, the strongest acid that can exist in water. Consequently, even though these acids are not all equally strong, water is a strong enough base to "level out" or obscure differences in acid strength. This behavior is called the LEVELING EFFECT, because it makes strong acids like $HClO_4$ and HCl appear as if they had equal acid strength. It is analogous to comparing the strengths of two men by asking each of them to lift a 5-lb weight.

Since weak acids ionize incompletely, equilibrium is established and an equilibrium constant, K_a, can be calculated (page 269),

$$HA + H_2O \rightleftharpoons H_3O^+ + A^-$$
$$K_a = \frac{[H_3O^+][A^-]}{[HA]}$$

The magnitude of K_a, the ionization constant, is a measure of acidity or acid strength—the larger the value of K_a, the stronger is the acid. Thus, acetic acid, $K_a = 1.8 \times 10^{-5}$, is more acidic than hydrocyanic acid, $K_a = 4.9 \times 10^{-10}$.

To avoid the use of exponential expressions, a new term, pK_a, is defined, such that $pK_a = -\log K_a$.

Example 1 Find pK_a for (a) acetic acid, and (b) hydrocyanic acid. (See Appendix I.15.)

Answer

(a) For acetic acid:
$$pK_a = -\log (1.8 \times 10^{-5})$$
$$= -(\log 1.8 + \log 10^{-5})$$
$$= -(0.25 - 5)$$
$$= 4.75$$

(b) For hydrocyanic acid:
$$pK_a = -\log (4.9 \times 10^{-10})$$
$$= -(\log 4.9 + \log 10^{-10})$$
$$= -(0.69 - 10.00)$$
$$= 9.31$$

The weaker acid thus has the larger pK_a value.

The same principles apply for comparing basicities. The bases B: and :ÖH⁻ compete for the proton,

$$B: + \underset{\underset{H}{\diagup} \overset{\ddot{O}}{} \underset{\diagdown}{}{H}}{} \rightleftharpoons B:H^+ + :\ddot{O}H^-$$

It happens that all neutral (uncharged) molecules are weaker bases than OH^-. Hence, the equilibrium does *not* favor the formation of OH^- ions. The weak ionization of the neutral base, ammonia, illustrates this point:

$$:NH_3 + \underset{\underset{H}{\diagup} \overset{\ddot{O}}{} \underset{\diagdown}{}{H}}{} \rightleftharpoons H:NH_3^+ + :\ddot{O}H^-$$

base₁	acid₂	acid₁	base₂
(weaker than OH^-)	(weaker than NH_4^+)	(stronger than H_2O)	(stronger than NH_3)

The equilibrium condition is

$$\frac{[NH_4^+][OH^-]}{[NH_3]} = K_b = 1.8 \times 10^{-5}$$

MECHANISM OF PROTON TRANSFER / 13.3

How is the proton transferred? Does it leave the acid and wriggle around until it forms a bond with the base? The answer is "NO!" First the base must come close enough to the H atom of the acid so that bond changes begin to occur, as shown:

$$\underset{\underset{H}{\diagup} \overset{H}{\diagdown}}{:\ddot{O}:} + H—\ddot{C}l: \rightleftharpoons \underset{\underset{H}{\diagup} \overset{H}{\diagdown}}{:\ddot{O}:}\cdots H—\ddot{C}l: \rightleftharpoons \underset{\underset{H}{\diagup} \overset{H}{\diagdown}}{:\ddot{O}}^{\delta+}—H\cdots^{\delta-}:\ddot{C}l: \rightleftharpoons \underset{\underset{H}{\diagup} \overset{H}{\diagdown}}{:\ddot{O}}{}^+—H + :\ddot{C}l:^-$$

acid forms H bond to basic water solvent	bond to base forms and acid bond breaks; incipient charges form	proton transfer is complete; ions form, each solvated by water

The sequence of bond changes that reactants undergo until the products emerge is called a **REACTION MECHANISM**. Knowing a mechanism helps focus attention on the factors influencing the reaction. For example, from this mechanism we can expect that the acidity of HCl in a basic solvent other than H_2O would depend on the basicity of the solvent and also on the ability of the solvent to solvate the ions formed. The poorer the ion-solvating ability of the solvent, the less stable are the ions in solution, the more the equilibrium lies towards the undis-sociated acid, and the weaker is the acid.

Water behaves either as an acid or as a base:

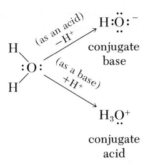

It is an example of a group of substances that are said to be AMPHIPROTIC or AMPHOTERIC. A substance with an H atom attached to a strongly electronegative atom with an unshared pair of electrons such as F, O, N, and Cl may be amphoteric; examples are $:NH_3$, $C_2H_5\ddot{O}H$, and $H\ddot{F}:$. Since water behaves as an acid and as a base, it reacts with it-self (self-ionization):

$$H\underset{\displaystyle \overset{\displaystyle\ddot{O}}{}}{\diagdown}H + :\overset{\displaystyle H}{\underset{\displaystyle H}{\diagup}}\ddot{O}: \;\rightleftharpoons\; \left[H\underset{\displaystyle \ddot{O}:}{\diagdown}\right]^- + \left[H-\overset{\displaystyle H}{\underset{\displaystyle H}{\diagup}}\ddot{O}:\right]^+$$

$$\text{acid}_1 \qquad \text{base}_2 \qquad\qquad \text{base}_1 \qquad\qquad \text{acid}_2$$

If

$[OH^-] = [H_3O^+]$, the solution is NEUTRAL*
$[OH^-] > [H_3O^+]$, the solution is BASIC (or ALKALINE)
$[OH^-] < [H_3O^+]$, the solution is ACIDIC

In aqueous solution a substance that produces an excess of OH^- is a BASE. A substance that produces an excess of H_3O^+ is an ACID. A substance that has no effect on either concentration is NEUTRAL.

Self-ionization is exhibited by many pure liquids; for example, in liquid ammonia,

$$\overset{\displaystyle H}{\underset{}{:\ddot{N}H_2}} + :\ddot{N}H_3 \;\rightleftharpoons\; :\ddot{N}H_2^- \;+\; [H:NH_3]^+$$

$$\text{acid}_1 \qquad \text{base}_2 \qquad \text{base}_1 \qquad\qquad \text{acid}_2$$
$$\qquad\qquad\qquad\text{(amide ion)} \quad \text{(ammonium ion)}$$

* Note that NEUTRAL is used in two different senses in chemistry: (a) in this case to mean neither acidic nor basic and (b) to mean neither positive nor negative.

We can generalize by saying that when a basic or acidic liquid undergoes self-ionization, *its conjugate acid and its conjugate base are formed* [Equation (5), below].

The equilibrium constant for the self-ionization of water is

$$K_w = [H_3O^+][OH^-]$$
$$= 1.0 \times 10^{-14} \text{ (at 25°C)}$$
$$pK_w = 14.0 \text{ (at 25°C)}$$

TYPES OF ACID–BASE REACTIONS / 13.5

Before the introduction of the Brönsted–Lowry theory, many reactions actually involving transfer of a proton were thought to be unique, and were given specific names. Several typical acid–base reactions are listed in Table 13.2 to emphasize their similarity, not to encourage their commitment to memory.

Table 13.2
Types of Acid–Base Reactions

Type	$Acid_1$	$+ Base_2$	$\longrightarrow Acid_2$	$+ Base_1$
(1) Ionization of uncharged acid	HCN	$+ H_2O$	$\longrightarrow H_3O^+$	$+ CN^-$
(2) Ionization of uncharged base	H_2O	$+ NH_3$	$\longrightarrow NH_4^+$	$+ OH^-$
(3) Self-ionization of water	H_2O	$+ H_2O$	$\longrightarrow H_3O^+$	$+ OH^-$
(4) Neutralization in water	H_3O^+	$+ OH^-$	$\longrightarrow H_2O$	$+ H_2O$
(4') Neutralization in liquid ammonia	NH_4^+	$+ NH_2^-$	$\longrightarrow NH_3$	$+ NH_3$
(5) Second ionization of a diprotic acid	HS^-	$+ H_2O$	$\longrightarrow H_3O^+$	$+ S^{2-}$
(6) Second ionization of a diprotic base	H_2O	$+ H_2NNH_3^+$	$\longrightarrow H_3NNH_3^{2+}$	$+ OH^-$
(7) Hydrolysis of an anion	H_2O	$+ CN^-$	$\longrightarrow HCN$	$+ OH^-$
(8) Hydrolysis of a cation	NH_4^+	$+ H_2O$	$\longrightarrow H_3O^+$	$+ NH_3$
(9) Hydrolysis of a hydrated cation	$Fe(H_2O)_6^{3+}$	$+ H_2O$	$\longrightarrow H_3O^+$	$+ Fe(OH)(H_2O)_5^{2+}$

(a) **Neutralization.** The reaction between an acid and a base was called **NEUTRALIZATION** even before the days of Arrhenius. In dilute aqueous solutions of strong acids, the only acid present to react with the OH^- ion is H_3O^+. Hence, in terms of the Brönsted–Lowry theory, neutralization is the reaction between H_3O^+ and OH^-, as shown in reaction (4) in Table 13.2. The term neutralization has been extended to include all reactions between the conjugate acid and conjugate base of the same substance. Hence reaction (4'), Table 13.2, represents a neutralization reaction between NH_4^+, the conjugate acid of NH_3, and amide ion, NH_2^-, the conjugate base of the same molecule.

$$\underset{\substack{\text{conjugate acid} \\ \text{of HX}}}{H_2X^+} + \underset{\substack{\text{conjugate base} \\ \text{of HX}}}{X^-} \underset{\text{self-ionization}}{\overset{\text{neutralization}}{\rightleftharpoons}} 2HX \qquad (5)$$

(b) Ionization of polyprotic acids and polyprotic bases. Some acids, such as sulfuric acid, hydrogen sulfide (H_2S), and phosphoric acid (H_3PO_4), have more than one proton that can be donated to a base. When such **POLYPROTIC ACIDS**, as exemplified by the **DIPROTIC** acid H_2S, dissolve in water, they release their protons in sequence:

$$H_2S + H_2O \rightleftharpoons H_3O^+ + HS^- \qquad K_{a_1} = 6.3 \times 10^{-8}$$
$$HS^- + H_2O \rightleftharpoons H_3O^+ + S^{2-} \qquad K_{a_2} = 10^{-14}$$

The values for the ionization constants reveal that HS^- is a much weaker acid than H_2S. This is understandable since the hydrosulfide ion, HS^-, has a negative charge and so is more likely to resist losing another proton to form the sulfide ion, S^{2-}. Conversely, the sulfide ion has a double negative charge and, therefore, is more likely to attract a proton than is the singly charged base, HS^-; the sulfide ion is a much stronger base than HS^-.

There are also basic compounds that can accept more than one proton in a stepwise process. For example, each nitrogen atom in hydrazine, H_2NNH_2, has an unshared pair of electrons available for bonding with a proton. As is typical of all such **POLYPROTIC BASES**, the ionization constant for the second step is much smaller than that for the first step:

$$H_2\ddot{N}-\ddot{N}H_2 + H_2O \rightleftharpoons \left[H_2\ddot{N}-\overset{H}{\ddot{N}}H_2 \right]^+ + OH^- \qquad pK_{b_1} = 6.04$$

$$\left[H_2\ddot{N}-\overset{H}{\ddot{N}}H_2 \right]^+ + H_2O \rightleftharpoons \left[H_2\overset{H}{\ddot{N}}-\overset{H}{\ddot{N}}H_2 \right]^{2+} + OH^- \qquad pK_{b_2} = 15$$

This is to be expected because $H_2N-NH_3^+$ already has a positive charge and resists accepting a second proton.

(c) Reactions of anions with water (hydrolysis). An aqueous solution containing the conjugate base of a weak acid is basic, as exemplified by a solution of sodium cyanide (Na^+CN^-). A solution of NaCN is basic because its $[OH^-]$ is greater than its $[H_3O^+]$. The greater $[OH^-]$ is a consequence of the fact that CN^- is sufficiently basic to remove a proton from water. The reaction of ions with water to give solutions that are not neutral is called **HYDROLYSIS**. More generally speaking, any reaction in which a water molecule is broken apart (lysed), usually into H and OH, is called hydrolysis.

$$\underset{\text{base}_1}{CN^-} + \underset{\text{acid}_2}{H_2O} \rightleftharpoons \underset{\text{acid}_1}{HCN} + \underset{\text{base}_2}{OH^-} \qquad \text{(hydrolysis of } CN^-\text{)}$$

The Na^+ plays no role in the reaction.

The equilibrium condition for this reaction with water is

$$K_b = \frac{[HCN][OH^-]}{[CN^-]} \tag{6}$$

The strength of the basic anion as measured by the magnitude of K_b^* is related to the strength of its conjugate acid and, therefore, to K_a. We illustrate the general relationship by considering HCN and CN⁻. The equation for the ionization of HCN is

$$HCN + H_2O \rightleftharpoons H_3O^+ + CN^-$$

and the equilibrium condition for this reaction is

$$K_a = \frac{[H_3O^+][CN^-]}{[HCN]} \tag{7}$$

If we multiply Equations (6) and (7), we get

$$K_a \cdot K_b = \frac{[H_3O^+][\cancel{CN^-}]}{[\cancel{HCN}]} \cdot \frac{[\cancel{HCN}][OH^-]}{[\cancel{CN^-}]} = [H_3O^+][OH^-]$$

> **SAFETY NOTE:** HCN is an extremely poisonous gas with a not very strong odor of bitter almonds. Its aqueous solution, hydrocyanic acid, is very weakly acidic. When an acid stronger than HCN is mixed with a cyanide (like NaCN), the relatively strong base CN⁻ reacts with the acid to produce HCN: HA + CN⁻ → HCN(g) + A⁻. *Never* add cyanide to a solution without first testing to be sure it is basic.

Since $[H_3O^+][OH^-] = K_w$,

$$K_a \cdot K_b = K_w$$

At 25°C, $K_w = 1.0 \times 10^{-14}$; therefore, at this temperature,

$$K_a \cdot K_b = 1.0 \times 10^{-14} \tag{8}$$

or in pK units

$$pK_a + pK_b = pK_w = 14$$

Equation (8) refers *only to an acid and its conjugate base.*

Although the equation $K_a \cdot K_b = K_w$ permits quantitative predictions to be made concerning the extent of reaction of anions with water, we will concern ourselves only with the following qualitative conclusions:

(1) *The conjugate base of an acid stronger than H_3O^+ (see Table 13.1, page 245) does not hydrolyze.* For example, ClO_4^-, Cl^-, and NO_3^- leave water neutral.

---

* These constants are also referred to as "hydrolysis constants," K_h.

(2) *The conjugate base of an acid that is moderately stronger than water undergoes some hydrolysis until a state of equilibrium is reached.* These ions make water slightly basic. Examples are nitrite ion, NO_2^- (pK_a of $HNO_2 = 3.35$; $NO_2^- + H_2O \rightleftharpoons HNO_2 + OH^-$), and acetate ion, $C_2H_3O_2^-$ (pK_a of $HC_2H_3O_2 = 4.8$).

(3) *The conjugate base of an acid that is approximately equal in strength to water hydrolyzes almost completely.* Examples are phosphate, PO_4^{3-} (pK_a of $HPO_4^{2-} = 12$; $PO_4^{3-} + H_2O \rightleftharpoons HPO_4^{2-} + OH^-$), and sulfide, S^{2-} (pK_a of $HS^- = 14.0$). These ions make water distinctly basic.

(4) *The conjugate base of an acid much weaker than water hydrolyzes completely.* Another way of stating this is to say that any base stronger than OH^- reacts with water so that OH^- remains as the only base. Therefore, water cannot be used to distinguish the basicities of very strong bases. This is another manifestation of the "leveling effect." Since amide, NH_2^-, hydride, H^-, and methide, CH_3^-, are stronger bases than OH^-, they react completely with water, forming their conjugate acids and OH^-,

$$:\ddot{N}H_2^- \ + \ H_2O \longrightarrow NH_3 \ + \ OH^- \quad \text{(as from sodium amide, } NaNH_2\text{)}$$

$$:H^- \ + \ H_2O \longrightarrow H_2 \ + \ OH^- \quad \text{(as from calcium hydride, } CaH_2\text{)}$$

$$H_3C:^- \ + \ H_2O \longrightarrow CH_4 \ + \ OH^- \quad \text{(as from sodium methide, } NaCH_3\text{)}$$

base₁   acid₂   acid₁   base₂

(strong bases of different strength; all stronger than OH^-)   (only base present)

The conjugate base of a weak diprotic acid is capable of behaving either as an acid or as a base, as exemplified by the amphoteric ion HS^-,

$$HS^- + H_2O \rightleftharpoons H_2S + OH^-$$
base   acid

$$HS^- + H_2O \rightleftharpoons S^{2-} + H_3O^+$$
acid   base

(d) Reactions of cations with water. Conjugate acids of weak bases, such as ammonium ion, NH_4^+, undergo hydrolysis. They are acidic enough to donate a proton to water:

$$NH_4^+ + H_2O \rightleftharpoons NH_3 + H_3O^+$$

and

$$K_a = \frac{[H_3O^+][NH_3]}{[NH_4^+]}$$

Since

$$K_a = \frac{K_w}{K_b} \quad \text{and} \quad K_b = 1.8 \times 10^{-5}$$

then

$$K_a = \frac{10^{-14}}{1.8 \times 10^{-5}} = 5.6 \times 10^{-10}$$

Salts of *trivalent* cations, such as aluminum sulfate, $Al_2(SO_4)_3$, ferric chloride, $FeCl_3$, and chromic nitrate, $Cr(NO_3)_3$, give acidic water solutions. Because of their high positive charge, Al^{3+}, Fe^{3+}, and Cr^{3+} strongly bind water molecules to themselves; the usual number is six:

$$M^{3+} + 6H_2O \longrightarrow \left[\begin{array}{c} OH_2 \\ H_2O \,\diagdown \; \Big| \; \diagup\, OH_2 \\ M \\ H_2O \,\diagup \; \Big| \; \diagdown\, OH_2 \\ OH_2 \end{array} \right]^{3+} \quad \text{(where M = for example, Al, Cr, Fe)} \quad (9)$$

These hydrated ions act as acids. $Al(H_2O)_6{}^{3+}$ is a triprotic acid; the end product is hydrated aluminum hydroxide, which slowly forms as a gelatinous precipitate when an aqueous solution of aluminum sulfate is allowed to stand:

$$Al(H_2O)_6{}^{3+} + H_2O \rightleftharpoons H_3O^+ + Al(OH)(H_2O)_5{}^{2+}$$
$$Al(OH)(H_2O)_5{}^{2+} + H_2O \rightleftharpoons H_3O^+ + Al(OH)_2(H_2O)_4{}^{+}$$
$$Al(OH)_2(H_2O)_4{}^{+} + H_2O \rightleftharpoons H_3O^+ + Al(OH)_3(H_2O)_3(c) \quad \text{(hydrated aluminum hydroxide)}$$

$$\text{acid}_1 \qquad\qquad \text{base}_2 \qquad \text{acid}_2 \qquad\qquad \text{base}_1$$

This precipitate has good adsorptive properties, and therefore Al(III) salts are used for water purification.

Some divalent cations may also form acidic hydrated cations. For example, an aqueous solution of $Zn(NO_3)_2$ is acidic because of the presence of $Zn(H_2O)_4{}^{2+}$. On the other hand, an aqueous solution of $Ca(NO_3)_2$ is neutral. In *general*, except for Be^{2+}, the divalent cations of Group 2R (since they possess a noble gas configuration) and all univalent cations, such as Na^+ and Ag^+, are not acidic in water solution.

When a strong base such as OH^- is added to hydrated aluminum hydroxide, the precipitate dissolves. Since each of the three water molecules is capable of donating a proton, the addition of hydroxide ion, OH^- (as NaOH), converts aluminum hydroxide to the aluminate ion, $Al(OH)_4(H_2O)_2{}^{-}$,

$$Al(OH)_3(H_2O)_3 + OH^- \longrightarrow Al(OH)_4(H_2O)_2{}^- + H_2O$$

Since $Al(OH)_3(H_2O)_3$ also dissolves in acid,

$$Al(OH)_3(H_2O)_3 + 3H_3O^+ \longrightarrow Al(H_2O)_6{}^{3+} + 3H_2O$$

it is an example of an amphoteric substance. It is noteworthy that $Fe(H_2O)_6{}^{3+}$ is a triprotic acid; in water it forms ferric hydroxide, $Fe(OH)_3(H_2O)_3$. Yet it does not dissolve in alkali hydroxides. Thus, not all acidic hydrated cations form amphoteric hydroxides.

LEWIS ACID–BASE CONCEPT / 13.6

Every base has an unshared pair of electrons to share with a proton. The proton is an electron-deficient particle during the process of bonding to the base. However, a proton is not the only species to form bonds with bases; other electron-deficient substances do likewise, as exemplified by the reaction of boron trifluoride, BF_3, with NH_3:

$$
\underset{\substack{\text{acid}\\\text{(electrophile)}}}{\underset{F}{\overset{F}{\underset{\diagdown}{B}}}\diagup F}
\;+\;
\underset{\substack{\text{base}\\\text{(nucleophile)}}}{:NH_3}
\;\longrightarrow\;
\underset{F}{\overset{F}{\underset{\diagdown}{\overset{\diagup}{F-B}}}}:NH_3
$$

Gilbert N. Lewis recognized the similarity in behavior of boron trifluoride and a transferred proton toward a base, and in 1923 enunciated a definition of acid–base reaction in terms of sharing of an electron pair—*a base donates an electron pair* in covalent bonding *and an acid accepts the pair*. The acid is called an ELECTROPHILE, and the base is called a NUCLEOPHILE. In the base, the atom with the unshared pair of electrons is an electron-rich site, and, in the acid, the atom that accepts the pair of electrons to form a covalent bond is an electron-deficient site. The Lewis theory focuses attention on the electron pair rather than on the proton, and in so doing broadens the concept of acidity. The transferred proton of a so-called Brönsted acid is a special case of a Lewis acid.

Lewis bases, like Brönsted bases, are necessarily substances with unshared pairs of electrons. They can be neutral molecules, such as $:NH_3$, or anions, such as cyanide ion, CN^-. Lewis acids appear in many guises.

EXAMPLES OF LEWIS ACIDS / 13.7

(a) **Molecules possessing an atom with less than an octet of electrons.** BF_3 falls into this category. A Lewis acid frequently used in organic chemistry is aluminum chloride, $AlCl_3$. The Al atom is electron-deficient, and can form a covalent bond with the electron-rich atom of an ether,

$$\underset{\underset{\text{Cl}}{|}}{\overset{\overset{\text{Cl}}{|}}{\text{Cl—Al}}} + :\overset{..}{\underset{..}{\text{O}}}\overset{\text{CH}_3}{\underset{\text{CH}_3}{<}} \longrightarrow \underset{\underset{\text{Cl}}{|}}{\overset{\overset{\text{Cl}}{|}}{\text{Cl—Al}}}{:}\overset{+}{\overset{..}{\text{O}}}\overset{\text{CH}_3}{\underset{\text{CH}_3}{<}}$$

dimethyl
ether

The product is shown with formal charges.

(b) **Cations.** Theoretically, all cations are potential Lewis acids. However, this property is negligible for univalent and divalent cations having a noble gas configuration. Only Li^+, Be^{2+}, and Mg^{2+} in this category show any significant tendency to behave as Lewis acids. The smaller is the cation, the greater is this tendency. M^{3+}-type ions are Lewis acids, as exemplified by the reaction of Al^{3+} and H_2O [Equation (9)]. Most cations, regardless of charge, that do not have a noble gas configuration, react with Lewis bases. The following reactions are typical:

$$Ag^+ + 2H_3N: \longrightarrow [H_3N:Ag:NH_3]^+$$

$$\underset{\substack{\text{Lewis} \\ \text{acid}}}{Cd^{2+}} + \underset{\substack{\text{Lewis} \\ \text{base}^\circ}}{4:\overset{..}{\underset{..}{\text{Cl}}}:^-} \longrightarrow \left[\begin{array}{c} :\overset{..}{\text{Cl}}: \\ :\overset{..}{\underset{..}{\text{Cl}}}:Cd:\overset{..}{\underset{..}{\text{Cl}}}: \\ :\overset{..}{\underset{..}{\text{Cl}}}: \end{array} \right]^{2-}$$

a complex

(c) **Molecules in which the central atom has available d orbitals and may acquire more than an octet of electrons.** Silicon tetrafluoride and tin(IV) chloride are typical examples:

$$SiF_4 + 2:\overset{..}{\underset{..}{\text{F}}}:^- \longrightarrow SiF_6{}^{2-}$$

$$SnCl_4 + 2:\overset{..}{\underset{..}{\text{Cl}}}:^- \longrightarrow SnCl_6{}^{2-}$$

$$\underset{\substack{\text{Lewis} \\ \text{acid}}}{} \quad \underset{\substack{\text{Lewis} \\ \text{base}}}{} \quad \underset{\text{complex}}{}$$

(d) **Molecules with a multiple bond between atoms of dissimilar electronegativities.** An example of such a molecule is carbon dioxide, $O{=}C{=}O$. The O atoms are more electronegative than the C atom. Hence, the C atom is electron deficient and is apt to form a bond with a

° Usually, the number of reacting Lewis bases is twice the charge on the cation.

Examples of Lewis Acids / 13.7

Lewis base such as OH⁻,

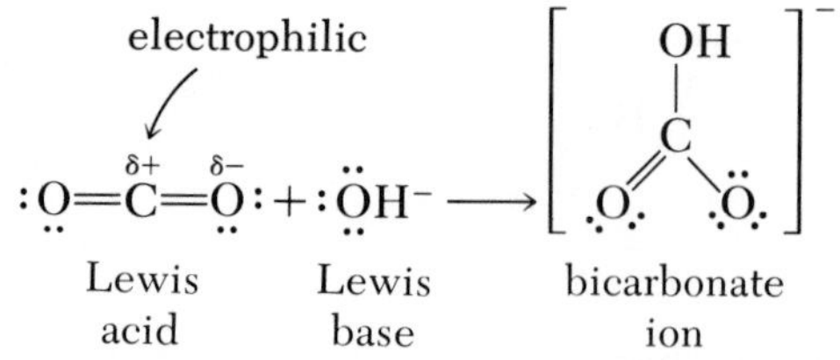

electrophilic

$$:O=C=O: + :\overset{..}{O}H^- \longrightarrow \left[\begin{array}{c} OH \\ | \\ C \\ \diagup \; \diagdown \\ .O. \quad :O. \end{array} \right]^-$$

| Lewis acid | Lewis base | bicarbonate ion |

Sulfur dioxide, SO_2, reacts like CO_2 toward OH⁻,

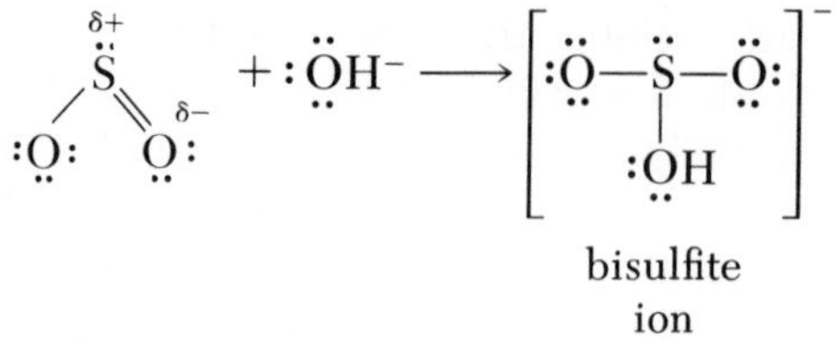

$$\underset{:O: \quad O:}{\overset{\overset{\delta+}{\ddot{S}}}{\diagup \diagdown}} + :\overset{..}{O}H^- \longrightarrow \left[:\overset{..}{O}-\overset{..}{S}-\overset{..}{O}: \atop \quad\; | \atop \quad :\overset{..}{O}H \right]^-$$

bisulfite
ion

NUCLEOPHILIC DISPLACEMENT / 13.8

The Brönsted acid–base reaction is a typical example of a very common type of reaction in which an atom or group of atoms in a molecule is displaced by another atom or group of atoms. This is called a DISPLACEMENT (substitution) reaction.

$$\widehat{D} + A \, \widehat{L} \longrightarrow DA + L$$

D is the displacing group, L is the leaving group, and the displacement is said to occur on atom A. The curved arrows indicate the course of events in the reaction. In one of the most common types of displacement reactions, the displacing group *has an unshared pair of electrons and behaves as a nucleophile.* The leaving group is also a nucleophile. The reaction is called a NUCLEOPHILIC DISPLACEMENT. The generalized equation is

$$\widehat{D:} + A:\widehat{L} \longrightarrow D:A \; + \; :L \tag{10}$$

| displacing group | | nucleophilic leaving group |

Proton transfer is a typical nucleophilic displacement occurring on an H atom.

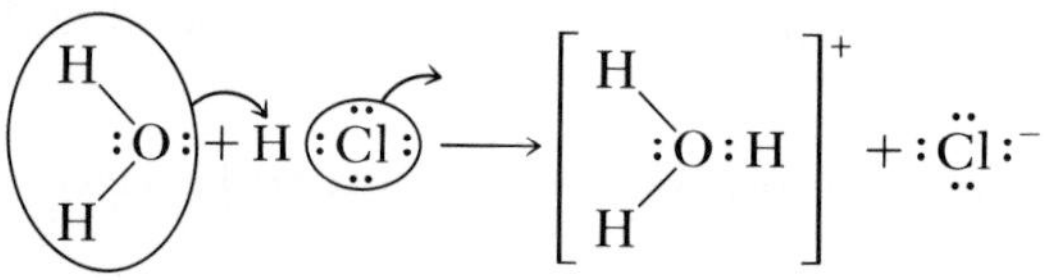

$$\left(\begin{array}{c} H \\ \diagdown \\ :O: \\ \diagup \\ H \end{array} \right) + H \, (:\overset{..}{\underset{..}{Cl}}:) \longrightarrow \left[\begin{array}{c} H \\ \diagdown \\ :O:H \\ \diagup \\ H \end{array} \right]^+ + :\overset{..}{\underset{..}{Cl}}:^-$$

In many cases, the transfer of electrons in an oxidation–reduction reaction occurs as a result of nucleophilic displacements. Consider the self-redox of chlorine in water to give hypochlorous and hydrochloric acids,

$$2H_2O + Cl_2 \longrightarrow HOCl + Cl^- + H_3O^+$$

oxidation number: $\qquad$ 0 $\qquad$ +1 $\quad$ −1

This reaction can be viewed as two nucleophilic displacements. In step (1), water is the nucleophile which displaces Cl^- from a Cl atom of Cl_2,

step (1)

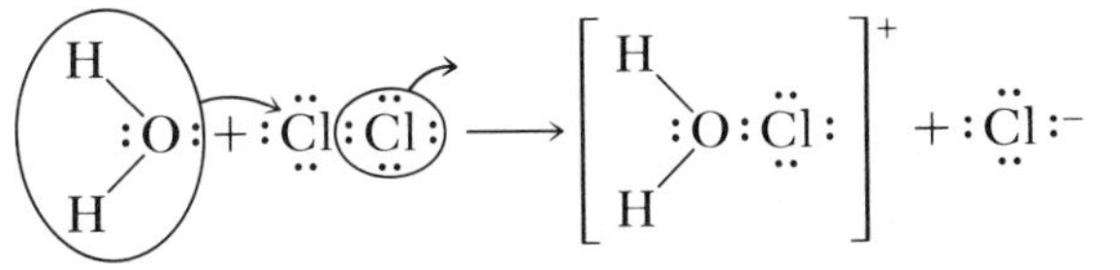

In the second step, water again acting as a nucleophile attacks an H atom of H_2OCl^+, and hypochlorous acid becomes the leaving group,

step (2)

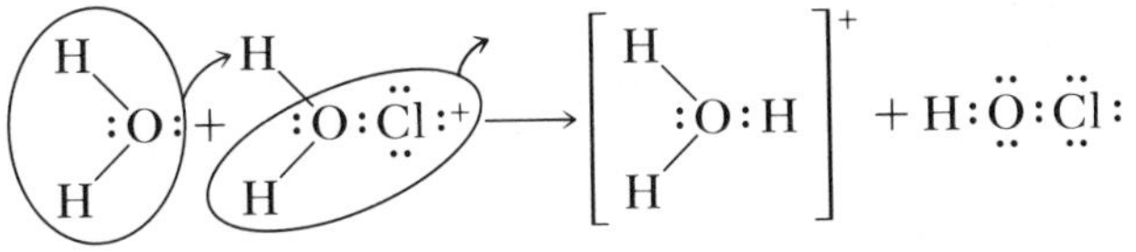

The oxidation of iodide ion by chlorine can also be formulated as a two-step nucleophilic displacement:

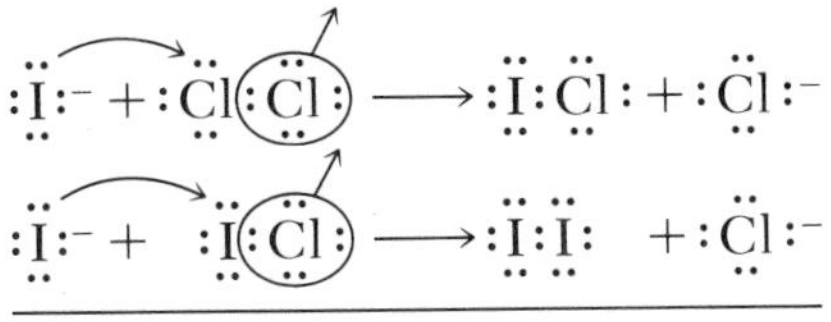

$$2I^- + \quad Cl_2 \longrightarrow I_2 \quad + 2Cl^-$$

Many redox reactions involving oxyanions and molecular oxides seem to involve a transfer of oxygen atoms as well as a transfer of electrons. A typical example is the oxidation of sulfite ion by chlorate ion,

$$SO_3^{2-} + ClO_3^- \longrightarrow SO_4^{2-} + ClO_2^-$$

Although this equation is balanced, let us use the ion–electron method (page 224) to find the partial equations:

$$
\begin{aligned}
SO_3^{2-} + H_2O &\longrightarrow SO_4^{2-} + 2H^+ + 2e^- \quad \text{(oxidation)} \\
ClO_3^- + 2H^+ + 2e^- &\longrightarrow ClO_2^- + H_2O \quad\quad\;\; \text{(reduction)} \\
\hline
SO_3^{2-} + ClO_3^- &\longrightarrow SO_4^{2-} + ClO_2^-
\end{aligned}
$$

It appears from the equation for the oxidation that the fourth oxygen atom needed to convert SO_3^{2-} to SO_4^{2-} comes from a water molecule. It also *appears* that in changing to ClO_2^-, the ClO_3^- ion loses its oxygen atom to two protons to form water. Does this really happen? To determine the fate of the various oxygen atoms, the chemist tags his materials with ^{18}O, an isotope of oxygen with atomic weight 18. When the reaction between $[S^{16}O_3]^{2-}$ and $[Cl^{16}O_3]^-$ is carried out in $H_2^{18}O$, practically no ^{18}O appears in the resulting sulfate ions. However, when $[Cl^{18}O_3]^-$ is mixed with $[S^{16}O_3]^{2-}$ in $H_2^{16}O$, the resulting sulfate ion is enriched with ^{18}O. These results indicate that the reaction proceeds mainly by direct transfer of an oxygen atom from chlorine to sulfur without the intervention of water:

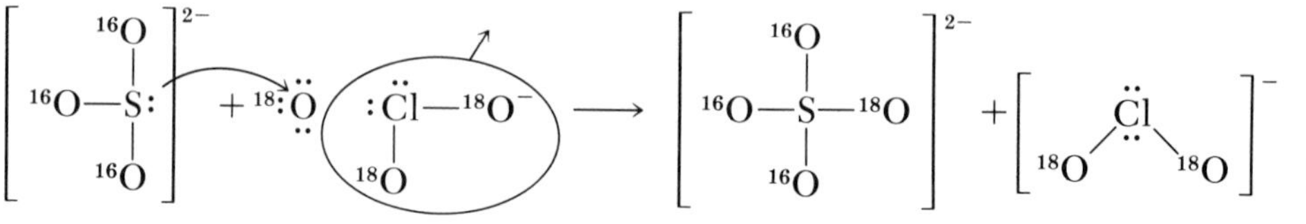

This process is a typical nucleophilic displacement. The nucleophile is SO_3^{2-}, attack occurs on an oxygen atom of the ClO_3^- ion, and the chlorite ion, ClO_2^-, is the leaving group. Thus, we should not construe the partial equations as representing a mechanism for the redox reaction.

Hydrolysis of the halides of nonmetals can be classified as a nucleophilic displacement in which a water molecule acts as the nucleophile. The reactions of silicon tetrachloride and boron trichloride are typical.

$$SiCl_4 + 8H_2O \longrightarrow Si(OH)_4 + 4H_3O^+ + 4Cl^-$$
$$(H_4SiO_4)$$
"silicic acid"

$$BCl_3 + 6H_2O \longrightarrow B(OH)_3 + 3H_3O^+ + 3Cl^-$$
$$(H_3BO_3)$$
boric acid

ELECTROPHILIC DISPLACEMENT / 13.9

Some displacements have an electrophile displacing another electrophile. The general equation for such an **ELECTROPHILIC DISPLACEMENT** is

electrophilic   electrophilic
displacing group   leaving group

The solution of aluminum chloride, $AlCl_3$, in phosgene, $Cl_2C=O$, exemplifies the reaction. Although neither substance alone conducts an electric current, the solution conducts. The presence of ions is explained by the following reaction:

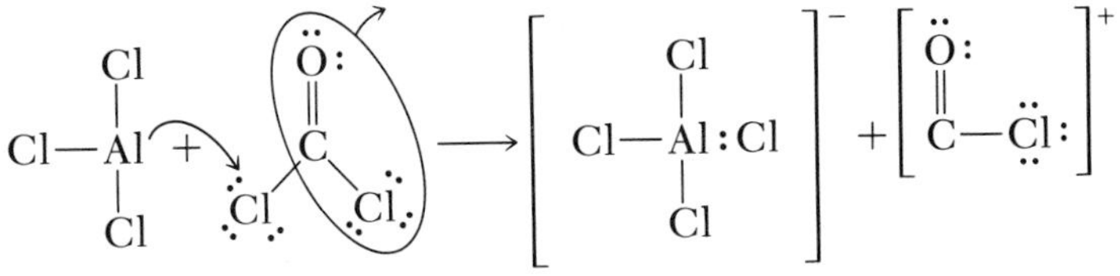

electrophile   displacement
                on Cl

electrophilic
leaving group

SOFT AND HARD ACIDS AND BASES / 13.10

The typical Lewis acid–base reaction can be generalized as

$$A \quad + \quad :B \quad \longrightarrow A:B$$

Lewis acid   Lewis base
(acceptor)   (donor)

The bond in AB can be ionic, polar, or nonpolar. It would be very beneficial if the chemist could predict the stability of A:B in terms of the natures of A and B. A concept designed to aid in this prediction is called the **PRINCIPLE OF "SOFT AND HARD" ACIDS AND BASES (SHAB)**, as proposed by Ralph G. Pearson (1963). He classified bases and acids as being soft or hard as follows: In a *soft base, the nucleophilic atom has a low electronegativity*. Molecules or ions whose nucleophilic atoms participate in multiple bonding, such as $:C\equiv O:$ and $:C\equiv N:^-$, are also soft. In a *hard base the nucleophilic atom has a high electronegativity*. In Table 13.3 (to be used as reference, not to be memorized) some typical bases are classified as hard, intermediate, or soft. Within a group in the periodic table, softness increases with increasing size of the donor

Table 13.3
Classification of Lewis Bases

Increasing softness	H_2O, OH^-, F^-, $C_2H_3O_2^-$ PO_4^{3-}, Cl^-, CO_3^{2-} ClO_4^-, NO_3^-	Hard	Increasing hardness
	NH_3		
	Br^-, NO_2^-, SO_3^{2-}	Intermediate	
	H^-		
	CN^-, CO	Soft	
	SH^-, I^-, $S_2O_3^{2-}$, S^{2-}		

Table 13.4
Classification of Lewis Acids

Increasing softness	$\left.\begin{array}{l} H^+, Li^+, Na^+, K^+ \\ Mg^{2+}, Ca^{2+} \\ Al^{3+} \\ Cr^{3+}, Co^{3+}, Fe^{3+} \\ BF_3, AlH_3 \\ SO_3, CO_2 \end{array}\right\}$	Hard	Increasing hardness
	$\left.\begin{array}{l} Fe^{2+}, Co^{2+}, Ni^{2+}, Cu^{2+} \\ Zn^{2+}, Pb^{2+}, Sn^{2+} \\ Sb^{3+}, Bi^{3+} \end{array}\right\}$	Intermediate	
	$\left.\begin{array}{l} Cu^+, Ag^+ \\ Cd^{2+}, Pt^{2+}, Hg^{2+} \\ I^+, Br^+, \\ I_2, Br_2 \\ \text{Some metal atoms} \\ \quad \text{such as Fe, Ni} \end{array}\right\}$	Soft	

atom. Thus, of the halide ions, F^- is the hardest and I^- is the softest. A *cationic hard acid*, such as Al^{3+}, generally will have *a small size, a high positive charge, and a noble gas electron configuration.* Shown in Table 13.4 among the hard acids are the cations of the representative elements whose charge is the same as the group number, and a few highly charged cations of transition metals such as Cr^{3+}, Fe^{3+}, and Co^{3+}. Soft acids, such as Ag^+, I^+, and Ni, have a large size, a low positive or zero charge, and do not have a noble gas configuration. Many transition metal cations fall into this category.

We now apply the idea of "soft" and "hard" acids and bases to the question of the stability of A:B. AB is *most stable when A and B are either both soft or both hard. It is least stable when one of the reactants is very hard and the other one is very soft.*

As an application of this principle, we can explain the fact that mercuric hydroxide, $Hg(OH)_2$, dissolves readily in an acidic aqueous solution, whereas mercuric sulfide, HgS, does not. Mercuric sulfide can be considered to be formed by a combination of Hg^{2+} and S^{2-}, a soft acid and a soft base. However, mercuric hydroxide is formed by a combination of Hg^{2+}, a soft acid, and OH^-, a hard base. Therefore, HgS (both soft) is more stable than $Hg(OH)_2$ (one soft and one hard).

The kinds of metal ores found in the earth's crust can be rationalized by using this principle. Thus, hard acids such as Ca^{2+}, Mg^{2+}, and Al^{3+} appear as $CaCO_3$, $MgCO_3$, and Al_2O_3, respectively. The anions CO_3^{2-} and O^{2-} are also hard. These three hard-acid cations are never found as their sulfides since S^{2-} is a soft base. However, soft acids such as Cu^+, Ag^+, and Hg^{2+} are found combined with the soft base S^{2-} as sulfides.

The intermediate acids such as Ni^{2+}, Pb^{2+}, and Cu^{2+} are found as both sulfides and carbonates.

1. Definitions. Define and illustrate the following terms: (a) Brönsted acid, (b) Brönsted base, (c) conjugate acid–base pair, (d) leveling effect, (e) amphoterism, (f) neutralization, (g) polyprotic acid, (h) hydrolysis, (i) Lewis acid (electrophile), (j) Lewis base (nucleophile), (k) soft and hard acid, (l) soft and hard base, (m) nucleophilic displacement.

2. Conjugate acid–base pairs. (a) List the conjugate acids of H_2O, Cl^-, PO_4^{3-}, NH_3, and HF. (b) List the conjugate bases of H_2O, HBr, $CH_3NH_3^+$, HSO_4^-, and OH^-.

3. Acid–base reactions. Write ionic equations for the reactions between the following pairs of acids and bases. Omit the "spectator" ions (page 223) and indicate all conjugate acid–base pairs: (a) $NaOH + HCl$, (b) $NH_3 + HNO_3$, (c) $NaCl + H_2SO_4$, (d) $NaHSO_3 + HCl$, (e) $NaHSO_3 + KOH$, (f) $NaCN + HBr$, (g) $NaHSO_3 + LiOH$.

4. Self-ionization. Write equations for the self-ionization of (a) HF, (b) H_2SO_4, (c) $HOCH_3$.

5. Amphoterism. Write balanced equations to illustrate the reaction of each of the following substances with an acid, HA, and with a base, B^-. Write formulas for products and designate acid–base pairs. (a) HCO_3^-, (b) $H_2PO_4^-$, (c) CH_3OH, (d) $Zn(OH)_2(H_2O)_2$, (e) NH_3.

6. Reaction of ions with water (hydrolysis). State whether an aqueous solution containing one of the following ions is acidic, basic, or neutral, assuming that no other solute is exerting any influence: (a) Br^-, (b) CN^-, (c) NH_4^+, (d) PO_4^{3-}, (e) HPO_4^{2-}, (f) Ca^{2+}, (g) Bi^{3+}, (h) Fe^{3+}, (i) CO_3^{2-}, (j) HCO_3^-.

7. Neutralization. Write an equation for a neutralization reaction occurring in each of the following liquids: (a) formic acid, $HCHO_2$; (b) NH_3; (c) HCl.

8. Polyprotic acids and bases. Write equations (designate conjugate pairs) for the stepwise reactions in water of (a) sulfuric acid, H_2SO_4; (b) phosphorous acid, H_2PHO_3; (c) CO_3^{2-}; (d) H_3AsO_4; (e) ethylene diammonium ion, $(H_3NCH_2CH_2NH_3)^{2+}$.

9. Lewis concept. Classify the following species as Lewis acids, bases, both, or neither of these. (A bonded Cl atom is not a good donor site.) All molecules are covalent. (a) Anhydrous $CdCl_2$, (b) $BeCl_2$, (c) CH_4, (d) I^-, (e) $GeCl_4$, (f) SO_2, (g) $H_2C{=}O$, (h) H_2, (i) AlH_3, (j) $(CH_3)_3C^+$, (k) $H_3C:^-$, (l) Ba^{2+}, (m) Zn^{2+}.

10. Nonaqueous solvents. Write equations and account for the fact that (a) HCN is a strong acid in liquid NH_3; (b) HCl is a weak acid in pure ("glacial") acetic acid; (c) sulfuric acid is a base in pure perchloric acid; (d) formaldehyde, $H_2C{=}O$, is a base in sulfuric acid; (e) HNO_3 is a base in H_2SO_4.

11. Nucleophilic displacement. Give the steps, each of which involves a nucleophilic displacement, for the following oxidation–reduction reactions:

(a) $Cl_2 + 2OH^- \rightarrow H_2O + Cl^- + ClO^-$ (2 steps)

(b) $Cl_2 + 2 :C\equiv N:^- \rightarrow N\equiv C-C\equiv N: + 2Cl^-$ (2 steps)

(c) $2H_2O + \overset{O}{\underset{O}{\overset{\diagdown\diagup}{N}}} - \overset{O}{\underset{O}{\overset{\diagup\diagdown}{N}}} \rightarrow HNO_2 + NO_3^- + H_3O^+$ (3 steps)

12. General. Discuss the validity of the following statements: (a) The acidity of an acid in a given solvent does not depend *only* on the proton-accepting ability of the basic solvent; (b) neutralization does not refer only to reactions between hydroxides (M^+OH^-) and acids (HA); (c) all acids are uncharged molecules; (d) it is not unreasonable to smell NH_3 and acetic acid, $HC_2H_3O_2$, when opening a bottle of the salt ammonium acetate, $NH_4C_2H_3O_2$.

ADDITIONAL PROBLEMS

13. Acid–base reactions. See problem 3 for instructions. (a) $HOCH_3 + NaH$, (b) $NH_4Cl + KCN$, (c) $NaH_2PO_4 + K_2CO_3$, (d) $NaH_2PO_4 + KHCO_3$, (e) $Na_2S + D_2O$ (D, deuterium, is an isotope of H), (f) $NaNH_2 + HCCl_3$.

14. Carbonate hard water. Water containing Ca^{2+} and bicarbonate ion, HCO_3^-, is called "carbonate hard water." On being heated, the Ca^{2+} which causes the hardness precipitates as $CaCO_3$. (a) Write an ionic equation for the formation of CO_3^{2-} resulting from a self-ionization of HCO_3^-. (b) Ascribe a role to the heat. (c) Ammonia and sodium phosphate, Na_3PO_4, are used in household cleansers to soften carbonate hard water by causing precipitation of $CaCO_3$. Write equations for the acid–base reactions resulting in the formation of CO_3^{2-} from HCO_3^-. Indicate the conjugate acid–base pairs.

15. Solvay process. (a) The Solvay process for the manufacture of sodium bicarbonate, $NaHCO_3$ (baking soda), utilizes CO_2, H_2O, and NH_3 to generate the bicarbonate ion, HCO_3^-. (*i*) Write the equation for the formation of carbonic acid from carbon dioxide and water. (*ii*) Write an ionic equation for the formation of HCO_3^- from carbonic acid and ammonia, indicating the conjugate acid–base pairs. (*iii*) In terms of chemical equilibrium, explain why CO_2 alone is not used to furnish the required concentration of HCO_3^-. (*iv*) Why cannot a base such as OH^- be used instead of NH_3 to produce the HCO_3^-? (b) Addition of NaCl to the concentrated solution of NH_4HCO_3 causes the precipitation of $NaHCO_3$, leaving ammonium chloride in solution. NH_3 is regenerated by heating the $NH_4Cl(aq)$ with calcium oxide, CaO. (The decomposition of $CaCO_3$ serves as the source of the CaO and CO_2.) (*i*) Write an ionic equation for the regeneration of the NH_3 from NH_4Cl and CaO solution. (*ii*) What substance is the net byproduct of the entire Solvay process? (Add up all equations.) (c) Sodium carbonate, Na_2CO_3 [washing soda is the hydrate $Na_2CO_3(H_2O)_{10}$], is made by heating $NaHCO_3$. The other products are H_2O and CO_2. Write an ionic equation for the reaction and indicate the type of Brönsted acid–base reaction it is. (See Table 13.2.)

16. H–D exchange. (a) Write the equation for the equilibrium established when (i) $HC_2H_3O_2$ is dissolved in D_2O; (ii) $DC_2D_3O_2$ is dissolved in H_2O. (b) How would you prepare DCN? (c) Can you prepare ND_3 by passing NH_3 into D_2O? Show with the aid of ionic equations. (D, deuterium, is an isotope of H.)

17. Relative acidity. To prepare sodium methoxide, $Na^+CH_3O^-$, sodium is added to pure methyl alcohol, CH_3OH, resulting in the generation of H_2. (a) Write an equation for the reaction. (b) Suggest a reason why sodium methoxide cannot be prepared by adding NaOH to CH_3OH.

18. Mechanism. In terms of the mechanism of proton transfer (a) account for the fact that acetic acid is a weaker acid in ethanol, C_2H_5OH (dielectric constant $= 24.2$), than in water (dielectric constant $= 79$); (b) account for the fact that NH_4Cl has about the same acidity in both solvents.

19. Nonaqueous solvents. If there were chemists on the planet Jupiter, which has an atmosphere rich in NH_3, they would probably classify both acetic and sulfuric acids as strong acids. Explain, with the aid of ionic equations.

20. Lewis concept. For each reaction write the structural formulas for reactants and products. Specify the Lewis acid and base.
(a) $BH_3 + Na^+H^- \rightarrow$
(b) $Mg^{2+}O^{2-} + CO_2 \rightarrow$
(c) $Fe + 5CO \rightarrow$
(d) $BF_3 + CH_3OH \rightarrow$
(e) $BeF_2 + 2F^- \rightarrow$
(f) $S^{2-} + SO_3 \rightarrow$

21. Soft and hard acids and bases. Explain the following: (a) Ga^{3+} and In^{3+} are much harder acids than Sb^{3+} and Bi^{3+}. (Consider the electronic structures.) (b) SH^- is a softer base than OH^-; (c) the order of softness is AsH_3 (softest) $> PH_3 > NH_3$ (hardest); (d) I_2 reacts with I^- to form I_3^-, but does not form I_2F^- with F^-; (e) Fe^{3+} does not form a complex with CO, whereas Fe does; (f) SiF_4 and $SiCl_4$ are stable compounds, but SiI_4 is not.

23. Leveling effect. (a) Elaborate on the statement, "The strongest base and acid that can exist in liquid NH_3 are NH_2^- and NH_4^+, respectively." (b) What are the strongest base and acid, respectively, that can exist in pure CH_3CH_2OH (ethyl alcohol)? (c) Generalize about the strongest acid and base that can exist in any self-ionizing solvent.

24. Practical acid–base reactions. Write an equation for each of the following acid–base reactions:

(a) $NaHCO_3(aq) + H_2SO_4$ ("wet" fire extinguisher)
(b) $NaHCO_3(aq) + K_2Al_2(SO_4)_4(H_2O)_{24}(aq)$ (baking powder reaction)
(c) Calcium hydride + water   (filling weather balloons)
(d) Calcium carbide (Ca^{2+} $:C\equiv C:^{2-}$) + water   (preparation of acetylene)
(e) $Na_3PO_4 + Ca(HCO_3)_2$ (softening of carbonate hard water)

25. Neutralization. ΔH for the reaction at 25°C of NaOH and HCl, or KOH and HNO_3, is -13.82 kcal/mole. That for $HC_2H_3O_2$ and NaOH is -13.52 kcal/mole. Is the dissociation of $HC_2H_3O_2$ exothermic or endothermic? Explain.

26. Nucleophilic displacement. See Problem 11 for instructions.

(a) $2F_2 + 2OH^- \rightarrow FOF + H_2O + 2F^-$ (3 steps)

(b) $2O_3S\!-\!S^{2-} + I_2 \rightarrow O_3S\!-\!S\!-\!S\!-\!SO_3^{2-} + 2I^-$ (2 steps)

(c) $Cl_2 + 4NH_3 \rightarrow H_2NNH_2(\text{hydrazine}) + 2NH_4^+ + 2Cl^-$ (4 steps)

27. Amphoterism. Mg liberates H_2 from an acidic aqueous medium faster than does Al, but Al liberates H_2 much faster from an aqueous solution of NaOH. (Al and NaOH are used in drain cleaners.) With the aid of chemical equations and bearing in mind the difference in behavior of $Mg(OH)_2$ and $Al(OH)_3$ towards OH^-, explain these observations.

28. Preparation of acids. (a) HCl is one of the few binary acids synthesized from its elements. Write a chemical equation for the reaction. (b) A volatile acid, such as nitric acid, HNO_3, is prepared by adding concentrated sulfuric acid, H_2SO_4, to a salt of its conjugate base, such as $NaNO_3$, commonly called "Chile saltpeter." Write an equation for the reaction.

29. K_a–K_b. Which of the following relationships are correct? (a) $K_{a(H_2S)} \times K_{b(HS^-)} = K_w$; (b) $K_{a(H_2S)} \times K_{b(S^{2-})} = K_w$; (c) $K_{a(HS^-)} \times K_{b(HS^-)} = K_w$; (d) $K_{a(HS^-)} \times K_{b(S^{2-})} = K_w$; (e) $K_{a(HS^-)} \times K_{b(H_2S)} = K_w$.

30. Electrophilic displacement. Account for the fact that nitrosyl chloride, $ClN\!=\!O$, and antimony pentachloride, $SbCl_5$, both of which are covalent molecules, react with each other to form a salt.

IONIC EQUILIBRIUM CALCULATIONS

Because of the great abundance and importance of water and its remarkable effectiveness as a solvent, many familiar chemical reactions occur in aqueous solutions. Most of these reactions involve ions, often along with neutral molecules or undissolved solids. Ionic reactions usually come rapidly to equilibrium, and the properties of the solution depend on the concentrations of the species present at equilibrium. For all these reasons, ionic equilibria deserve careful study.

We shall make some rather drastic simplifying assumptions throughout this chapter. We shall assume that strong electrolytes exist as free ions, that all activity coefficients are equal to 1, and that the solvent (water) behaves like pure water. These assumptions are accurate only in very dilute solutions. We shall also ignore various side reactions that would, if taken into account, complicate the calculations. As a result, many of our answers will be correct only to an order of magnitude (power of 10). More advanced treatises describe the revisions that must be made in the calculations to take account of the effects neglected here.

ACID–BASE EQUILIBRIA

THE IONIZATION OF WATER / 14.2

Even the purest water has some ability to conduct an electric current. This conductivity is attributed to the presence of H^+ and OH^- ions produced by the self-ionization reaction (page 248)

$$H_2O \rightleftharpoons H^+ + OH^-$$

The equilibrium constant of this reaction is known as the ION PRODUCT

OF WATER, and is represented by K_w:

$$[H^+][OH^-] = K_w = 1.00 \times 10^{-14} \text{ at } 25°C$$

At $0°C$, $K_w = 1.14 \times 10^{-15}$; at $50°C$, $K_w = 5.35 \times 10^{-14}$.

The concentrations of H^+ and OH^- must be equal in pure water, for two reasons:

(1) In the reaction $H_2O \rightleftharpoons H^+ + OH^-$, one H^+ ion is formed for every OH^- ion formed.

(2) Any solution must be electrically neutral, which means that the total positive charge equals the total negative charge; when H^+ and OH^- are the only ions present, this requires that $[H^+] = [OH^-]$.

The equilibrium concentrations of H^+ and OH^- in pure water at $25°C$ are found by letting $y = [H^+] = [OH^-]$, and substituting into the equilibrium condition:

$$[H^+][OH^-] = K_w$$
$$y^2 = 1.0 \times 10^{-14}$$
$$y = 1.0 \times 10^{-7}$$

At $50°C$, however, the concentrations are higher:

$$y^2 = 5.35 \times 10^{-14}$$
$$y = 2.3 \times 10^{-7}$$

If the water contains some source of H^+ or OH^-, other than H_2O itself, we can no longer expect that these two concentrations will be equal. For example, we might make $[H^+] = 0.10$ mole/liter by adding a suitable quantity of acid. Then, at $25°C$,

$$0.10 \times [OH^-] = 1.0 \times 10^{-14}$$
$$[OH^-] = 1.0 \times 10^{-13} \text{ mole/liter}$$

pH / 14.3

In many aqueous solutions, we are concerned with very small concentrations of H^+ and OH^-. In Chapter 13, the convenient notation $pK = -\log K$ was introduced. In the same way, we write[*]

$$pH = -\log [H^+]$$

or

$$pH = \log \left(\frac{1}{[H^+]} \right)$$

[*] Historically, pH preceded pK. The notation "pH" was introduced by Søren P. L. Sørensen in 1909, as an abbreviation of "power of hydrogen."

or

$$[H^+] = \text{antilog}(-pH)$$

or

$$[H^+] = 10^{-pH}$$

Values for pH are usually between 0 and 14, but not necessarily; if $[H^+] = 2$ moles/liter, $pH = -\log 2 = -0.3$, and if $[H^+] = 10^{-15}$ mole/liter, $pH = 15$. If $[H^+] = 1$ mole/liter, $pH = 0$.

Example 1 Find the pH of a solution in which $[H^+] = 6.38 \times 10^{-6}$ mole/liter.

Answer

$$\begin{aligned} pH &= -\log(6.38 \times 10^{-6}) \\ &= -\log 6.38 - \log 10^{-6} \\ &= -\log 6.38 - (-6) \\ &= -0.805 + 6 = 5.195 \quad \text{(see Appendix I.15)} \end{aligned}$$

Note the fact, surprising at first, that the pH in this case is between 5 and 6, not between 6 and 7. The reason is that 6.38×10^{-6} is between 10^{-5} and 10^{-6}, not between 10^{-6} and 10^{-7}.

Example 2 Calculate $[H^+]$ for a solution of pH 8.37.

Answer

$$[H^+] = 10^{-pH} = 10^{-8.37} = 10^{-9+0.63} = 10^{0.63} \times 10^{-9}$$
$$= 4.3 \times 10^{-9}$$

Example 3 A certain solution has pH 3.89 at 0°C. Find $[OH^-]$.

Answer

$$[H^+] = 10^{-pH} = 10^{-3.89} = 1.3 \times 10^{-4}$$
$$[H^+][OH^-] = K_w = 1.14 \times 10^{-15}$$

$$[OH^-] = \frac{K_w}{[H^+]} = \frac{1.14 \times 10^{-15}}{1.3 \times 10^{-4}} = 8.8 \times 10^{-12} \text{ mole/liter}$$

At 25°C, $pH = 7.00$ for a neutral solution; when $pH < 7$, the solution is acidic, and when $pH > 7$, the solution is basic. The number 7 may be remembered as the dividing line between acidic and basic solutions, but one should also remember that it refers only to 25°C.

Example 4 What is the pH of a neutral solution at 50°C?

Answer Let $y = [H^+] = [OH^-]$

$$[H^+][OH^-] = K_w$$
$$y^2 = 5.35 \times 10^{-14}$$
$$y = 2.31 \times 10^{-7}$$
$$pH = -\log y = 6.636$$

Throughout the remainder of this chapter, you may assume that the temperature is that for which the data are available, usually 25°C.

STRONG ACIDS AND BASES / 14.4

To calculate the pH of a dilute solution of a strong acid (page 235), such as HCl, or of an ionic hydroxide, such as NaOH, we assume that the reaction

$$HCl \longrightarrow H^+ + Cl^-$$

or

$$NaOH \longrightarrow Na^+ + OH^-$$

goes to completion. Then $[H^+]$ or $[OH^-]$ is equal to the concentration of the solute times the appropriate integer (1 for NaOH, 2 for $Ca(OH)_2$, etc.).

Example 5 Calculate the pH of (a) 1.0×10^{-2} M HCl; (b) 5.0×10^{-3} M KOH.

Answer

(a) $[H^+] = 1.0 \times 10^{-2}$ mole/liter
$$pH = -\log(1.0 \times 10^{-2}) = 2.0$$

(b) $[OH^-] = 5.0 \times 10^{-3}$ mole/liter

$$[H^+] = \frac{K_w}{[OH^-]}$$

$$= \frac{1.00 \times 10^{-14}}{5.0 \times 10^{-3}} = 2.0 \times 10^{-12} \text{ mole/liter}$$

$$pH = -\log 2.0 + 12 = 11.7$$

If the solution is extremely dilute, however, this simple method fails. Suppose we have a 1.0×10^{-8} M solution of HCl. We might say that $[H^+] = 10^{-8}$, pH = 8. This is obviously wrong because pH 8 corresponds to a basic solution, and this solution must be slightly acidic. We have overlooked the fact that the self-ionization of water also contributes something to $[H^+]$. Its contribution is negligible in 10^{-2} M HCl, but not in 10^{-8} M HCl. To find the pH, we first recall that the solution must be electrically neutral. The total number of moles of positive ions must equal the total number of moles of negative ions:

$$[H^+] = [OH^-] + [Cl^-]$$

In 1.0×10^{-8} M HCl, $[Cl^-] = 1.0 \times 10^{-8}$ mole/liter. Let $[H^+] = y$; then

$$[OH^-] = [H^+] - [Cl^-]$$
$$= y - 1.0 \times 10^{-8}$$
$$[H^+][OH^-] = K_w$$
$$y(y - 1.0 \times 10^{-8}) = 1.00 \times 10^{-14}$$
$$y^2 - 1.0 \times 10^{-8}\, y - 1.00 \times 10^{-14} = 0$$

By the quadratic formula (Appendix I.16),

$$y = 1.05 \times 10^{-7} \quad \text{or} \quad -9.5 \times 10^{-8} \text{ mole/liter}$$

Since a concentration cannot be negative, the negative root is rejected. Then

$$\text{pH} = -\log y = -\log(1.05 \times 10^{-7}) = 6.98$$

In pure water at 25°C, $[H^+] = [OH^-] = 10^{-7}$ mole/liter. If the concentration of the acid (or base) is much greater than 10^{-7}, we can neglect the contribution of self-ionization relative to the H^+ from the acid (or the OH^- from the base). When the concentration is around 10^{-7} mole/liter or less, we can no longer make this approximation.

IONIZATION OF WEAK ACIDS / 14.5

We represent a general acid by HA; for example, it might be acetic acid, $HC_2H_3O_2$,* or hydrofluoric acid, HF. The equilibrium constant for the reaction

$$HA \rightleftharpoons H^+ + A^-$$

is the ionization constant of HA, as in Chapter 13, and is represented by K_a or K_{HA}. Table 14.1 lists K_a for a number of common acids.

Most ionic equilibrium problems are solved in the same way as any other equilibrium problem (Chapter 11). Some cases involved added complications, which we shall consider later (page 278).

Example 6 Referring to Table 14.1, find the pH of a 0.20 M solution of formic acid.

---

* Formulas for carboxylic acids (containing the —COOH group) are written in several ways, depending on how much we want to tell about the structure of the molecule. In this chapter, we write them in the simplest way: the acidic hydrogen(s) at the beginning of the formula, then the remaining atoms (including other H atoms that are not donated to bases) merely counted.

Table 14.1
Ionization Constants of Acids and Bases in Water at 25°C

Name	Formula[a] Monoprotic acids	Constant K_a		
Acetic acid	$HC_2H_3O_2$, CH_3COOH	1.75×10^{-5}		
Benzoic acid	$HC_7H_5O_2$, C_6H_5COOH	6.46×10^{-5}		
Bromoacetic acid	$HC_2H_2O_2Br$, $CH_2BrCOOH$	2.05×10^{-3}		
Chloroacetic acid	$HC_2H_2O_2Cl$, $CH_2ClCOOH$	1.40×10^{-3}		
Dichloroacetic acid	$HC_2HO_2Cl_2$, $CHCl_2COOH$	3.32×10^{-2}		
Formic acid	$HCHO_2$, $HCOOH$	1.76×10^{-4}		
Hydrocyanic acid	HCN	4.93×10^{-10}		
Hydrofluoric acid	HF	6.4×10^{-4}		
Nitrous acid	HNO_2	4.5×10^{-4}		
Phenol (carbolic acid)	HC_6H_5O, C_6H_5OH	1.3×10^{-10}		
Propionic acid	$HC_3H_5O_2$, CH_3CH_2COOH	1.34×10^{-5}		
Trichloroacetic acid	$HC_2O_2Cl_3$, CCl_3COOH	2×10^{-1}		
	Polyprotic acids	K_1	K_2	K_3
Carbonic acid[b]	H_2CO_3	4.30×10^{-7}	5.62×10^{-11}	—
Hydrogen sulfide	H_2S	6.3×10^{-8}	10^{-14}	—
Malonic acid	$H_2C_3H_2O_4$, $CH_2(COOH)_2$	1.49×10^{-3}	2.03×10^{-6}	—
Oxalic acid	$H_2C_2O_4$, $(COOH)_2$	5.90×10^{-2}	6.40×10^{-5}	—
Phosphoric acid	H_3PO_4	7.52×10^{-3}	6.22×10^{-8}	4.8×10^{-13}
Phthalic acid	$H_2C_8H_4O_4$, $C_6H_4(COOH)_2$	1.26×10^{-3}	3.9×10^{-6}	—
Succinic acid	$H_2C_4H_4O_4$, $(CH_2COOH)_2$	6.89×10^{-5}	2.47×10^{-6}	—
Sulfuric acid	H_2SO_4	$\sim 10^{+11}$	1.2×10^{-2}	—
Sulfurous acid	H_2SO_3	1.7×10^{-2}	6.24×10^{-8}	—
Tartaric acid	$H_2C_4H_4O_6$, $(CHOHCOOH)_2$	1.04×10^{-3}	4.55×10^{-5}	—
	Monoprotic bases	K_b		
Ammonia	NH_3	1.77×10^{-5}		
Aniline	$C_6H_5NH_2$	4.27×10^{-10}		
Dimethylamine	$(CH_3)_2NH$	5.41×10^{-4}		
Ethylamine	$CH_3CH_2NH_2$	4.71×10^{-4}		
Hydroxylamine	$HONH_2$	1.1×10^{-8}		
Methylamine	CH_3NH_2	3.70×10^{-4}		
Pyridine	C_5H_5N	1.78×10^{-9}		
Trimethylamine	$(CH_3)_3N$	6.45×10^{-5}		
	Diprotic base	K_1	K_2	
Hydrazine	H_2NNH_2	9.1×10^{-7}	10^{-15}	

[a] See footnote on page 269.
[b] When CO_2 dissolves in water, a small fraction — less than 1% — reacts to form H_2CO_3: $CO_2 + H_2O \rightleftharpoons H_2CO_3$. Most of the dissolved CO_2 is present as CO_2. However, it is common practice to represent dissolved CO_2 by the formula H_2CO_3; $[H_2CO_3]$, therefore, means the total concentration of CO_2 and H_2CO_3.

Answer We construct a table as on page 206:

$$HCHO_2 \rightleftharpoons H^+ + CHO_2^-$$

Initial concentration	0.20	0	0
Change	$-y$	$+y$	$+y$
Concentration at equilibrium	$0.20 - y$	y	y

y represents the number of moles of acid (per liter) that ionize. The equilibrium condition is

$$\frac{[H^+][CHO_2^-]}{[HCHO_2]} = K_a$$

$$\frac{y^2}{0.20 - y} = 1.76 \times 10^{-4} \qquad (1)$$

This quadratic equation can be solved without undue labor, but a simplifying approximation (Appendix I.16) is often applicable. The acid is weak, meaning that only a small fraction of it ionizes. We thus expect that y, representing the number of moles of acid that has ionized per liter, will be much less than the initial concentration of acid. If this approximation is valid, $0.20 - y$ can be replaced by 0.20 without serious error, and Equation (1) becomes

$$\frac{y^2}{0.20} = 1.76 \times 10^{-4} \qquad (2)$$

$$y^2 = 35.2 \times 10^{-6}$$
$$y = 5.9 \times 10^{-3} \text{ mole/liter}$$
$$pH = -\log y = 3 - \log 5.9 = 2.23$$

We must now check whether or not the assumption that only a negligible amount of acid ionizes ($0.20 - y \approx 0.20$) is valid. Our result for y makes $0.20 - y = 0.20 - 6 \times 10^{-3} = 0.19$. For most calculations involving ionic equilibria, the discrepancy between 0.20 and 0.19 is not serious. In that case, the exact Equation (1) and the simplified Equation (2) are essentially the same equation and we have found the solution for both. [More accurate solution of Equation (1) gives $y = 5.8 \times 10^{-3}$, pH = 2.24.*]

The ionization of a weak acid is commonly described by giving the **DEGREE OF IONIZATION** (represented by α), which is the fraction of acid initially present that ionizes. α is the ratio of the number of moles per liter that ionize to the initial concentration (c) of the acid:

$$\alpha = \frac{[A^-]}{c}$$

---

* In the calculations of this chapter, it is seldom necessary to achieve an accuracy better than ±5% of a given concentration, or better than ±0.02 in a pH. In this example, $0.20 - 0.19 = 0.01$, which is 5% of 0.20. The approximate and accurate answers for y differ by $(5.9 - 5.8) \times 10^{-3} = 10^{-4}$, which is 2% of 5.8×10^{-3}. The pH's differ by $2.24 - 2.23 = 0.01$. These errors are all within the guidelines.

For the preceding example,

$$\alpha = \frac{5.8 \times 10^{-3}}{0.20} = 2.9 \times 10^{-2} \quad \text{or} \quad 2.9\%$$

Example 7 Find the degree of ionization of acetic acid in a 0.50 M solution, and the pH of the solution.

Answer From Table 14.1, $K_a = 1.75 \times 10^{-5}$. Then, for the reaction

$$HC_2H_3O_2 \rightleftharpoons H^+ + C_2H_3O_2^-$$

the equilibrium condition is

$$\frac{[H^+][C_2H_3O_2^-]}{[HC_2H_3O_2]} = K_a$$

By the same reasoning as in Example 6,

$$\frac{y^2}{0.50 - y} = 1.75 \times 10^{-5}$$

where y is the number of moles of acid per liter that ionize: $y = [H^+] = [C_2H_3O_2^-]$. If we assume that y is much less than 0.50,

$$\frac{y^2}{0.50} = 1.75 \times 10^{-5}$$

$$y^2 = 8.75 \times 10^{-6}$$
$$y = 3.0 \times 10^{-3} \text{ mole/liter} = [H^+]$$

which is, for our purposes, negligible in comparison to 0.50. Then

$$\alpha = \frac{y}{0.50} = \frac{3.0 \times 10^{-3}}{0.50} = 6.0 \times 10^{-3}$$

$$pH = -\log [H^+] = -\log 3.0 + 3 = 2.52$$

The following example illustrates a case in which we cannot avoid solving a quadratic equation.

Example 8 Find $[H^+]$ in a 0.50 M solution of trichloroacetic acid.

Answer From Table 14.1, $K_a = 0.2$ for the reaction

$$HC_2O_2Cl_3 \rightleftharpoons H^+ + C_2O_2Cl_3^-$$
$$\frac{[H^+][C_2O_2Cl_3^-]}{[HC_2O_2Cl_3]} = 0.2$$

$$\frac{y^2}{0.50 - y} = 0.2 \tag{3}$$

Let us assume that $y \ll 0.50$. This assumption gives

$$\frac{y^2}{0.50} = 0.2$$

$$y = 0.32$$

The assumption was a bad one, for 0.32 is clearly not negligible in comparison to 0.50. As a result, $y = 0.32$ is not a solution of Equation (3). We must solve Equation (3) as a quadratic equation (Appendix I.16). Equation (3) may be rewritten

$$y^2 + 0.2\,y - 0.1 = 0$$
$$y = +0.23 \quad \text{or} \quad -0.43$$

A concentration cannot be negative, and the answer -0.43 must therefore be rejected; then

$$[H^+] = y = 0.2 \text{ mole/liter}$$

(Why is only one significant figure retained?)

IONIZATION OF WEAK BASES / 14.6

The ionization constant of the base B is the equilibrium constant of the reaction

$$B + H_2O \rightleftharpoons BH^+ + OH^-$$

for which the equilibrium condition is

$$\frac{[BH^+][OH^-]}{[B]} = K_b \tag{4}$$

Table 14.1 gives the ionization constants of some common bases.

Example 9 Find the pH of a 0.100 M solution of ammonia, NH_3.

Answer The procedure is essentially the same as for weak acids:

	$NH_3 + H_2O \rightleftharpoons NH_4^+ + OH^-$		
Initial concentration	0.100	0	0
Change	$-y$	$+y$	$+y$
Concentration at equilibrium	$0.100 - y$	y	y mole/liter

The only difference from the acid case is that y now represents $[OH^-]$ instead of $[H^+]$. Equation (4) is, in this case,

$$\frac{[NH_4^+][OH^-]}{[NH_3]} = K_b = 1.77 \times 10^{-5}$$
$$y = [NH_4^+] = [OH^-]$$

Then

$$\frac{y^2}{0.100 - y} = 1.77 \times 10^{-5}$$

If $y \ll 0.100$,

$$\frac{y^2}{0.100} = 1.77 \times 10^{-5}$$
$$y^2 = 1.77 \times 10^{-6}$$
$$y = 1.33 \times 10^{-3} \text{ mole/liter} = [OH^-]$$

y is less than 5% of 0.100 (5×10^{-3}), and thus meets our requirement for being negligible. Since $[H^+][OH^-] = K_w = 1.00 \times 10^{-14}$,

$$[H^+] = \frac{K_w}{[OH^-]}$$

$$= \frac{1.00 \times 10^{-14}}{1.33 \times 10^{-3}} = 7.5 \times 10^{-12}$$

$$pH = -\log(7.5 \times 10^{-12}) = 12 - \log 7.5 = 11.12$$

CONJUGATE ACID–BASE PAIRS / 14.7

In Chapter 13, conjugate acid–base pairs were defined: HA is the conjugate acid of A^-, and A^- is the conjugate base of HA. A simple relation between the acidic and basic ionization constants in a conjugate pair was established (page 251): $K_a K_b = K_w$. Thus, if we know the ionization constant for an acid or base, we can find the ionization constant for its conjugate base or acid.

Example 10 Find K_b and pK_b for the acetate ion, $C_2H_3O_2{}^-$.

Answer The conjugate acid of $C_2H_3O_2{}^-$ is $HC_2H_3O_2$. The ionization constant of $HC_2H_3O_2$ is $K_a = 1.75 \times 10^{-5}$; $K_w = 1.00 \times 10^{-14}$. Then

$$K_b = \frac{K_w}{K_a}$$

$$= \frac{1.00 \times 10^{-14}}{1.75 \times 10^{-5}} = 5.71 \times 10^{-10}$$

$$pK_b = 10 - \log 5.71 = 9.243$$

Example 11 Find the degree of ionization of pyridinium ion, $C_5H_5NH^+$, in a 0.100 M solution of pyridinium chloride ("pyridine hydrochloride"), $(C_5H_5NH)^+Cl^-$. Assume that $(C_5H_5NH)^+Cl^-$ is a strong electrolyte, present in solution as $C_5H_5NH^+$ and Cl^- ions.

Answer The acidic ionization constant of $C_5H_5NH^+$, corresponding to the reaction

$$C_5H_5NH^+ \rightleftharpoons C_5H_5N + H^+$$

is obtained from K_b for the conjugate base C_5H_5N,

$$K_a = \frac{K_w}{K_b}$$

$$= \frac{1.00 \times 10^{-14}}{1.78 \times 10^{-9}} = 5.62 \times 10^{-6}$$

The equilibrium condition is

$$\frac{[C_5H_5N][H^+]}{[C_5H_5NH^+]} = 5.62 \times 10^{-6}$$

$$C_5H_5NH^+ \rightleftharpoons H^+ + C_5H_5N$$
$$0.100 - y \qquad y \qquad y \text{ mole/liter at equilibrium}$$

Then

$$\frac{y^2}{0.100 - y} = 5.62 \times 10^{-6}$$

just as with an uncharged acid. If $y \ll 0.100$,

$$\frac{y^2}{0.100} = 5.62 \times 10^{-6}$$

$$y^2 = 0.562 \times 10^{-6}$$
$$= 7.50 \times 10^{-4} \text{ mole/liter} = [H^+]$$

$$\alpha = \frac{y}{0.100} = 7.50 \times 10^{-3}$$

Example 12 Find the pH of a 0.30 M solution of NaF.

Answer We leave most of the details to the reader.

$$F^- + H_2O \rightleftharpoons HF + OH^-$$

$$K_b = \frac{K_w}{K_a} = 1.56 \times 10^{-11}$$

$$[OH^-] = [HF] = y$$

$$\frac{y^2}{0.30 - y} = 1.56 \times 10^{-11}$$

$$y^2 = 0.30 \times 1.56 \times 10^{-11} = 4.68 \times 10^{-12}$$
$$y = 2.16 \times 10^{-6} \text{ mole/liter} = [OH^-]$$

$$[H^+] = \frac{1.00 \times 10^{-14}}{2.16 \times 10^{-6}} = 4.62 \times 10^{-9} \text{ mole/liter}$$

$$pH = 8.34$$

POLYPROTIC ACIDS AND BASES / 14.8

Let H_2A represent a diprotic acid. The equilibrium constants for the two reactions

$$H_2A \rightleftharpoons H^+ + HA^- \tag{5}$$
$$HA^- \rightleftharpoons H^+ + A^{2-} \tag{6}$$

are represented by K_1 and K_2, respectively, and are called the first and second ionization constants of H_2A. For a triprotic acid, there is also a third constant K_3. We note that HA^- can act either as an acid, in reaction (6), or as a base, in the reaction

$$HA^- + H_2O \rightleftharpoons H_2A + OH^- \tag{7}$$

It is amphoteric (page 248). Likewise, A^{2-} is a base:

$$A^{2-} + H_2O \rightleftharpoons HA^- + OH^- \tag{8}$$

The equilibrium constants for reactions (7) and (8) are obtained from K_1 and K_2 by the usual relation between the ionization constants of an acid and of its conjugate base:

$$HA^- + H_2O \rightleftharpoons H_2A + OH^- \qquad K_{b(HA^-)} = \frac{K_w}{K_1}$$

$$A^{2-} + H_2O \rightleftharpoons HA^- + OH^- \qquad K_{b(A^{2-})} = \frac{K_w}{K_2}$$

Be careful to pick the correct conjugate acid for a given base, and conversely the correct conjugate base for a given acid. For example, the conjugate acid of A^{2-} is HA^-, not H_2A. Likewise, for a diprotic base B, there are two ionization constants:

$$B + H_2O \rightleftharpoons BH^+ + OH^- \qquad K_1$$
$$BH^+ + H_2O \rightleftharpoons BH_2^{2+} + OH^- \qquad K_2$$

and, correspondingly, one can calculate from these the acidic ionization constants of BH^+ and BH_2^{2+}. We observe that A^{2-} is a diprotic base, and BH_2^{2+} is a diprotic acid.

Example 13 Calculate the basic ionization constant of bicarbonate (hydrogen carbonate) ion, HCO_3^-.

Answer The conjugate acid of HCO_3^- is H_2CO_3:

$$K_{b(HCO_3^-)} \frac{K_w}{K_{a(H_2CO_3)}} = \frac{1.00 \times 10^{-14}}{4.30 \times 10^{-7}} = 2.33 \times 10^{-8}$$

Calculations involving the ionization of polyprotic acids and bases present no great difficulties, provided that one makes a simplifying assumption. Because the second ionization constant is usually much less than the first, we can assume that *the concentration of* A^{2-}, *which is produced in the second ionization, is much less than the concentration of* HA^-, *which is produced in the first ionization.* Another way of looking at it is that the first ionization accounts for most of the H^+, only a little more being added by the second ionization. Calculations involving the first ionization can then be performed as if the acid were monoprotic, and the results of this calculation can then be used to find $[A^{2-}]$. Corresponding remarks apply to ionization of polyprotic bases.

Example 14 Find $[H^+]$, $[HC_2H_2O_4^-]$, and $[C_3H_2O_4^{2-}]$ in a 0.700 M solution of malonic acid, $H_2C_3H_2O_4$.

Answer Let the acid and its ions be represented by H_2A, HA^-, and A^{2-}. We can find $[H^+]$ and $[HA^-]$ by considering only the first ionization:

$$H_2A \rightleftharpoons H^+ + HA^-$$
$$0.700 - y \qquad y \qquad y \quad \text{mole/liter}$$

The equilibrium condition is

$$\frac{[H^+][HA^-]}{[H_2A]} = K_1 = 1.49 \times 10^{-3}$$

or

$$\frac{y^2}{0.700 - y} = 1.49 \times 10^{-3}$$

If $y \ll 0.700$ ($y < 5\%$ of 0.700 or 0.035),

$$\frac{y^2}{0.700} = 1.49 \times 10^{-3}$$

$$y = 3.23 \times 10^{-2} \text{ mole/liter} = [H^+] = [HA^-]$$

y passes our test for "much less." For comparison, we solve the quadratic equation and get $y = 3.16 \times 10^{-2}$. We have now found $[H^+]$ and $[HA^-]$. To calculate $[A^{2-}]$, we need the equilibrium condition for the second ionization:

$$HA^- \rightleftharpoons H^+ + A^{2-}$$

$$\frac{[H^+][A^{2-}]}{[HA^-]} = K_2 = 2.03 \times 10^{-6}$$

We assume that $[H^+]$ and $[HA^-]$ are little changed by this reaction, and we use our previous answers for them:

$$\frac{3.16 \times 10^{-2}[A^{2-}]}{3.16 \times 10^{-2}} = 2.03 \times 10^{-6}$$

$$[A^{2-}] = 2.03 \times 10^{-6} \text{ mole/liter}$$

The result $[A^{2-}] = K_2$ will be true in any case in which HA^- is a weak acid and the solution is prepared by dissolving *only* H_2A in water.

There is a source of confusion that should be pointed out here. The misconception sometimes arises that one concentration of H^+ is produced by the reaction

$$H_2A \rightleftharpoons H^+ + HA^-$$

and another concentration of H^+ is produced by the reaction

$$HA^- \rightleftharpoons H^+ + A^{2-}$$

This is *not* the case. In a given solution, a given ion or molecule has only one concentration. The solution has only *one* value of $[H^+]$, to be used in all the equations in which this concentration appears. After all, an ion has no memory—it cannot recall by what reaction it was produced.

Example 15 Find the pH of a 0.25 M solution of Na_2CO_3, a strong electrolyte.

Answer The solution contains the diprotic base CO_3^{2-},

$$CO_3^{2-} + H_2O \rightleftharpoons HCO_3^- + OH^-$$

$$\frac{[HCO_3^-][OH^-]}{[CO_3^{2-}]} = K_{b(CO_3^{2-})} \qquad (9)$$

$$HCO_3^- + H_2O \rightleftharpoons H_2CO_3 + OH^-$$

$$\frac{[H_2CO_3][OH^-]}{[HCO_3^-]} = K_{b(HCO_3^-)}$$

The basic ionization constant of HCO_3^- was calculated in Example 13:

$$K_{b(HCO_3^-)} = 2.33 \times 10^{-8}$$

Similarly, the basic ionization constant of CO_3^{2-} is related to the acidic ionization constant of HCO_3^-, which is the second ionization constant of H_2CO_3:

$$K_{b(CO_3^{2-})} = \frac{K_w}{K_{a(HCO_3^-)}} = \frac{1.00 \times 10^{-14}}{5.62 \times 10^{-11}} = 1.78 \times 10^{-4}$$

Just as the second ionization constant of a diprotic acid is much less than the first, so here, for the diprotic base CO_3^{2-}, $K_{b(HCO_3^-)} \ll K_{b(CO_3^{2-})}$. We expect, then, that $[H_2CO_3]$ will be much less than $[HCO_3^-]$. As for the diprotic acid in Example 14, so also for the diprotic base CO_3^{2-} we can confine our attention to the first ionization,

$$\begin{array}{cccc} CO_3^{2-} + H_2O & \rightleftharpoons & HCO_3^- + & OH^- \\ 0.25 - y & & y & y \qquad \text{mole/liter} \end{array}$$

Equation (9) becomes

$$\frac{y^2}{0.25 - y} = 1.78 \times 10^{-4}$$

If $y \ll 0.25$, then

$$\frac{y^2}{0.25} = 1.78 \times 10^{-4}$$

$$y = 6.7 \times 10^{-3} \text{ mole/liter} = [OH^-]$$

$$[H^+] = \frac{K_w}{[OH^-]} = \frac{1.00 \times 10^{-14}}{6.7 \times 10^{-3}} = 1.49 \times 10^{-12}$$

$$pH = -\log[H^+] = 11.83$$

GENERAL TREATMENT OF SIMULTANEOUS EQUILIBRIA / 14.9

In our work thus far, we have swept something under the rug. Most of the time it can stay there, but some problems require that we do not overlook this troublesome matter. We will study in this section methods of attacking these more difficult problems.

The difficulties appear whenever *two or more different reactions contribute comparable concentrations of the same species*. We have already seen an example in a very dilute solution of a strong acid or base (page 268). H^+ comes not only from the reaction

$$HCl \longrightarrow H^+ + Cl^-$$

but also from the reaction

$$H_2O \rightleftharpoons H^+ + OH^- \tag{10}$$

In a 10^{-2} M solution of HCl, the first reaction contributes much more H^+ than the second and the second can be ignored. That is no longer true in a 10^{-8} M solution.

The same problem rears its head more often in connection with solutions of weak acids. We have been assuming that only one reaction occurs in the solution:

$$HA \rightleftharpoons H^+ + A^- \tag{11}$$

We have overlooked reaction (10), except when we used it to calculate $[OH^-]$. For these two reactions, there are two equilibrium conditions:

$$[H^+]\,[OH^-] = K_w$$
$$\frac{[H^+]\,[A^-]}{[HA]} = K_a$$

There are four concentrations to be determined: $[H^+]$, $[OH^-]$, $[HA]$, and $[A^-]$. There should be two additional conditions which they satisfy. Let us see what these conditions are.

(a) Some of the HA molecules in the solution lose protons and become A^-. We do not know in advance what percentage of the molecules becomes A^- and what percentage remains HA. However, we may assume that every HA molecule that was put into the solution ends up as one or the other; if it is not there as HA, it is there as A^-. We dissolved c moles of HA per liter of solution. The total number of moles of HA and A^- present (per liter) at equilibrium must then be equal to the number of moles dissolved per liter:

$$[HA] + [A^-] = c \tag{12}$$

For example, if we have a $c = 0.20$ M solution of $HCHO_2$, then $[CHO_2^-] = 5.8 \times 10^{-3}$ mole/liter and $[HCHO_2] = 0.19$ mole/liter; $[HCHO_2] + [CHO_2^-] = 0.19 + 5.8 \times 10^{-3} = 0.20$ mole/liter. This equation is called the CONSERVATION CONDITION.

(b) The solution must be *electrically neutral*. This means that the total number of positive charges is equal to the total number of negative charges. In the present case, there is only one kind of positive ion, H^+; there are two kinds of negative ion, A^- and OH^-. In each liter of solution, then, the number of H^+ ions must equal the total number of A^- and OH^- ions. Likewise, the number of *moles* of H^+ in a liter must equal the total number of moles of A^- and OH^- in a liter:

$$[H^+] = [A^-] + [OH^-] \tag{13}$$

This equation is the ELECTRICAL NEUTRALITY CONDITION for the solution.

If the acid is not extremely weak, the solution will be decidedly acidic, which means that $[OH^-] \ll [H^+]$, and Equation (13) can be written

$$[H^+] = [A^-] \tag{13$'$}$$

We can now let

$$y = [H^+] = [A^-] \tag{13$''$}$$

and Equation (12) gives us

$$[HA] = c - [A^-] = c - y \tag{14}$$

The equilibrium conditions are

$$\frac{[H^+][A^-]}{[HA]} = K_a \tag{15}$$

$$[H^+][OH^-] = K_w \tag{16}$$

Substitution of (13$''$) and (14) into (15) gives

$$\frac{y^2}{c - y} = K_a$$

which is the same equation that we obtained earlier by considering only reaction (11). The reason for this agreement may be understood thus: The concentrations of H^+ and of OH^- contributed by reaction (10) are equal. When we neglected reaction (10), we were neglecting the contribution of this reaction to $[H^+]$, and that is the same as neglecting $[OH^-]$ relative to $[H^+]$. Thus, the simple way we solved the problem on page 271, by considering only reaction (11), is applicable whenever $[OH^-] \ll [H^+]$—that is to say, whenever the solution is distinctly acidic. The same ideas apply to bases, but with OH^- and H^+ interchanged.

Example 16 (a) Write the equilibrium, conservation, and electrical neutrality conditions for a 0.20 M solution of NH_3 in water at 25°C. (b) Find the concentrations of OH^- and H^+ in this solution.

Answer

(a) $NH_3 + H_2O \rightleftharpoons NH_4^+ + OH^-$

Equilibrium condition:

$$\frac{[NH_4^+][OH^-]}{[NH_3]} = K_b = 1.77 \times 10^{-5} \tag{17}$$

$$H_2O \rightleftharpoons H^+ + OH^-$$

Equilibrium condition:

$$[H^+][OH^-] = K_w = 1.00 \times 10^{-14} \tag{18}$$

Conservation condition:

$$[NH_3] + [NH_4^+] = 0.20 \qquad (19)$$

Electrical neutrality condition:

$$[NH_4^+] + [H^+] = [OH^-] \qquad (20)$$

(b) If the solution is distinctly basic, $[H^+] \ll [OH^-]$. Then (20) becomes

$$[NH_4^+] = [OH^-] = y$$

and (19) is

$$[NH_3] = 0.20 - [NH_4^+]$$
$$= 0.20 - y$$

Substitution in (17) gives the familiar equation

$$\frac{y^2}{0.20 - y} = 1.77 \times 10^{-5}$$
$$y = 1.9 \times 10^{-3} \text{ mole/liter} = [OH^-]$$

From (18),

$$[H^+] = \frac{1.00 \times 10^{-14}}{1.9 \times 10^{-3}} = 5.3 \times 10^{-12} \text{ mole/liter}$$

We see that $[H^+] \ll [OH^-]$, as assumed.

Note that a conservation condition includes the concentrations of only those species which contain the atom or group being conserved. Equation (19) says that the total concentration of NH_3 in solution, whether in the form of NH_3 or of NH_4^+, is 0.20 mole/liter. Other concentrations have no business in this equation: $[NH_3] + [NH_4^+] + [H^+] + [OH^-] = 0.20$ is incorrect. Note also that the electrical neutrality condition involves only ions, not uncharged molecules.

In order to recognize when the more sophisticated approach is required, we must be aware of which quantities we are neglecting in the simple method of considering only one reaction at a time. Thus, in solving a weak-acid ionization problem one should *always verify* at the end that $[OH^-] \ll [H^+]$, for this assumption was implicit in using the same letter to represent both $[H^+]$ and $[A^-]$. With a diprotic acid, one should verify that $[A^{2-}] \ll [HA^-]$. The next example shows a problem where the simple method fails.

Example 17 Find the pH of a 1.0×10^{-5} M solution of hydrogen cyanide, HCN.

Answer

$$HCN \rightleftharpoons H^+ + CN^-, \qquad \frac{[H^+][CN^-]}{[HCN]} = 4.93 \times 10^{-10} \qquad (21)$$

$$H_2O \rightleftharpoons H^+ + OH^-, \qquad [H^+][OH^-] = K_w = 1.00 \times 10^{-14} \qquad (22)$$

Conservation: $\qquad$ $[HCN] + [CN^-] = 1.0 \times 10^{-5}$ $\qquad$ (23)

Electrical neutrality: $\qquad$ $[H^+] = [CN^-] + [OH^-]$ $\qquad$ (24)

Let $y = [CN^-]$ and $z = [OH^-]$. Then Equation (24) says that

$$[H^+] = y + z$$

and Equation (23) gives

$$[HCN] = 1.0 \times 10^{-5} - y$$

Since the acid is very weak, we assume $y \ll 10^{-5}$; $[HCN] = 1.0 \times 10^{-5}$. Substitution into Equation (21) yields

$$\frac{(y + z)y}{1.0 \times 10^{-5}} = 4.93 \times 10^{-10}$$

$$y + z = \frac{4.93 \times 10^{-15}}{y} \qquad (25)$$

We also have, from Equation (22),

$$(y + z)z = 1.00 \times 10^{-14}$$

or

$$y + z = \frac{1.00 \times 10^{-14}}{z} \qquad (26)$$

Comparing Equations (25) and (26), we have

$$\frac{1.00 \times 10^{-14}}{z} = \frac{4.93 \times 10^{-15}}{y}$$

and therefore

$$y = 0.493z$$
$$y + z = 0.493z + z = 1.493z$$

Then Equation (26) becomes

$$1.493z = 1.00 \times 10^{-14}/z$$
$$z^2 = 6.70 \times 10^{-15}$$
$$z = 8.18 \times 10^{-8} \text{ mole/liter} = [OH^-]$$
$$y = 0.493z = 4.03 \times 10^{-8} \text{ mole/liter} = [CN^-]$$
$$y \ll 10^{-5}, \text{ as assumed}$$
$$[H^+] = y + z = 1.22 \times 10^{-7} \text{ mole/liter}$$
$$pH = 7 - \log 1.22 = 6.91$$

SAFETY NOTE: See page 251.

COMMON-ION EFFECT / 14.10

Le Chatelier's principle (page 200) predicts that an equilibrium is shifted to the left when the concentration of a product of the reaction (appearing on the right-hand side of the equation) is increased. Thus, the equilibrium

$$HA \rightleftharpoons H^+ + A^-$$

is shifted to the left when H^+ (in the form of a strong acid, such as HCl) or A^- (in the form of a salt, such as Na^+A^-) is added to the solution. This effect is referred to as the COMMON-ION EFFECT because it appears when an electrolyte having an ion in common with HA is added. If HA is acetic acid, then the shift can be produced by HCl (common ion $= H^+$) or by sodium acetate (common ion $= A^-$), but not by NaCl (no ion in common with the ions yielded by HA in solution).

A shift to the left means that the degree of ionization of HA is decreased. If H^+ is added, the concentration of A^- is decreased. If A^- is added, the concentration of H^+ is decreased, and the solution thus becomes less acidic.

If a weak acid is present in a solution that also contains a strong acid, the concentration of H^+ is determined almost entirely by the concentration of the strong acid. A corresponding statement is true for bases.

Example 18 Find the concentration of $C_2H_3O_2^-$ and the degree of ionization of $HC_2H_3O_2$ in a solution 0.10 M with respect to HCl and 0.20 M with respect to $HC_2H_3O_2$.

Answer

$$HC_2H_3O_2 \rightleftharpoons H^+ + C_2H_3O_2^-$$
$$0.20 - y \qquad 0.10 + y \qquad y \qquad \text{mole/liter at equilibrium}$$

$[H^+] \approx 0.10$ mole/liter, since practically all of the H^+ is contributed by the HCl.

$$\frac{[H^+][C_2H_3O_2^-]}{[HC_2H_3O_2]} = 1.75 \times 10^{-5}$$

$$\frac{0.10y}{0.20 - y} = 1.75 \times 10^{-5}$$

$$y = 3.5 \times 10^{-5} \text{ mole/liter} = [C_2H_3O_2^-]$$

The degree of ionization is

$$\alpha = \frac{y}{0.20} = 1.75 \times 10^{-4}$$

It should be noted that the degree of ionization of the weak acid is much decreased by the presence of the strong acid; if the HCl were absent, α would be 9.3×10^{-3} instead of 1.75×10^{-4}.

Example 19 Find the concentration of S^{2-} in a solution 0.10 M with respect to H_2S and 0.30 M with respect to HCl.

Answer For the reactions

$$H_2S \rightleftharpoons H^+ + HS^-$$
$$HS^- \rightleftharpoons H^+ + S^{2-}$$

we have the equilibrium conditions

$$\frac{[H^+][HS^-]}{[H_2S]} = K_1 = 6.3 \times 10^{-8}$$

$$\frac{[H^+][S^{2-}]}{[HS^-]} = K_2 = 10^{-14}$$

Since $[HS^-]$ does not interest us in this problem, we are free to eliminate it by multiplying the two equilibrium conditions together:

$$\frac{[H^+]^2[S^{2-}]}{[H_2S]} = K_1K_2$$

$$[S^{2-}] = \frac{K_1K_2[H_2S]}{[H^+]^2}$$

$$= \frac{6.3 \times 10^{-8} \times 10^{-14} \times 0.10}{(0.30)^2}$$

$$= 7 \times 10^{-22} \text{ mole/liter}$$

Note that $[S^{2-}]$ is not equal to K_2. That result (page 277) applied only to the case in which the diprotic acid was the only solute. Here, the solution also contains HCl.

Example 20 Calculate the pH of a solution 0.20 M with respect to formic acid, $HCHO_2$, and 0.10 M with respect to sodium formate, $NaCHO_2$.

Answer

$$\begin{array}{cccc} HCHO_2 & \rightleftharpoons & H^+ + & CHO_2^- \\ 0.20 - y & & y & 0.10 + y \end{array} \quad \text{mole/liter at equilibrium}$$

The acid is weak, and its ionization is further repressed by the common-ion effect. We expect, therefore, that y will be very small, and we can replace $0.20 - y$ and $0.10 + y$ by 0.20 and 0.10, respectively. The equilibrium condition then is

$$\frac{[H^+][CHO_2^-]}{[HCHO_2]} = K_a$$

$$\frac{y(0.10)}{0.20} = 1.76 \times 10^{-4}$$

$$y = 3.52 \times 10^{-4} \text{ mole/liter} = [H^+]$$
$$pH = -\log y = -0.547 + 4 = 3.45$$

A comparison of this result with that of Example 6 (page 269) shows the effect of the common ion (formate) on the pH. The presence of sodium formate has increased the pH from 2.24 to 3.45. The solution is still acidic, but less so than the solution of formic acid alone. Another way of looking at it is that we have added a base, CHO_2^-, to an acidic solution, and this naturally makes the solution less acidic.

The pH of human blood is normally 7.40. If the pH were to fall below 7.0 or rise above 7.8, death would be likely, for the effectiveness of enzymes is very sensitive to pH. How can the pH be kept so constant during ingestion of sauerkraut (acid) followed by baking soda (base)? A solution with the property that the addition of acids or bases causes only a relatively small change in pH is known as a BUFFER SOLUTION. The way to make a buffer solution is *to dissolve an acid and its conjugate base, both moderately weak.* Buffer solutions are an application of the common-ion effect as illustrated in Example 20.

The pH of pure water, or of a solution that is not a buffer, is very sensitive to the addition of small quantities of acid or base. Suppose that we want a solution with pH 4.76. Such a solution can be prepared by dissolving 1.75×10^{-5} mole HCl in water to make a liter:

$$[H^+] = 1.75 \times 10^{-5}$$
$$pH = 4.76$$

Now, perhaps by accident, 0.10 mole HCl is added per liter of this solution:

$$[H^+] = 0.10 + 1.75 \times 10^{-5}$$
$$= 0.10 \text{ mole/liter (the second term is negligible)}$$
$$pH = 1.0$$

The pH has been changed drastically. Conversely, suppose that we add 0.10 mole NaOH/liter. Of the OH$^-$ added, 1.75×10^{-5} mole/liter is neutralized by the H$^+$ in the solution, leaving $[OH^-] = 0.10 - 1.75 \times 10^{-5} = 0.10$ mole/liter. Then $[H^+] = 10^{-13}$ mole/liter, $pH = 13.0$. Again there has been a large change in pH.

However, when a strong acid is added to a buffer solution, it reacts with the base (A$^-$) present in the solution:

$$H^+ + A^- \longrightarrow HA$$

Any added base reacts with the acid (HA) in the solution:

$$OH^- + HA \longrightarrow A^- + H_2O$$

The solution contains a reservoir of base that consumes the added acid, and a reservoir of acid to consume added base. In each case, the acid–base ratio is changed, but with only a moderate effect on the pH of the solution. Very little of the H$^+$ or OH$^-$ added will remain as such. The following example illustrates the quantitative treatment of buffer action.

Example 21 Calculate the pH of (a) a solution 0.50 M with respect to $HC_2H_3O_2$ and 0.50 M with respect to $NaC_2H_3O_2$; (b) the same solution after 0.10 mole HCl/liter has been added to it. Assume that the volume is unchanged.

Answer

(a) $HC_2H_3O_2 \rightleftharpoons H^+ + C_2H_3O_2^-$ (27)
             $0.50 - y$ y $0.50 + y$ mole/liter at equilibrium

$$\frac{[H^+][C_2H_3O_2^-]}{[HC_2H_3O_2]} = 1.75 \times 10^{-5}$$

If y is small enough so that $0.50 + y$ and $0.50 - y$ are both ≈ 0.50, then

$$\frac{[H^+](0.50)}{0.50} = 1.75 \times 10^{-5}$$

$$[H^+] = 1.75 \times 10^{-5}$$
$$pH = 4.76 \text{ (same as the HCl solution above)}$$

(b) We have added H^+, but that does not mean that $[H^+]$ has increased by 0.10 mole/liter. On the contrary, most of the H^+ added reacts with the base $C_2H_3O_2^-$ to form $HC_2H_3O_2$. The effect of adding the HCl has been to convert 0.10 mole/liter of $C_2H_3O_2^-$ to 0.10 mole/liter of $HC_2H_3O_2$. $[C_2H_3O_2^-]$ has gone down from 0.50 to 0.40 mole/liter; $[HC_2H_3O_2]$ has gone up from 0.50 to 0.60 mole/liter:

                      $C_2H_3O_2^- + H^+ \rightleftharpoons HC_2H_3O_2$

Initial concentration	0.50	small	0.50 mole/liter
Change	−0.10	small	+0.10
Final concentration	0.40	small	0.60

The equilibrium condition for reaction (27) is

$$\frac{[H^+][C_2H_3O_2^-]}{[HC_2H_3O_2]} = 1.75 \times 10^{-5}$$

$$\frac{[H^+](0.40)}{0.60} = 1.75 \times 10^{-5}$$

$$[H^+] = \frac{0.60}{0.40} \times 1.75 \times 10^{-5}$$

$$= 2.63 \times 10^{-5} \text{ mole/liter}$$
$$pH = 4.58$$

The decrease in pH is only 0.18, to be contrasted with a decrease of 3.76 for the unbuffered HCl solution. If we had added NaOH instead of HCl, the effect would have been to convert $HC_2H_3O_2$ to $C_2H_3O_2^-$.

A buffer solution retains its buffering action as long as the quantity of acid or base added is much less than the quantities of the weak acid and its conjugate base in the buffer solution. If the number of moles of H^+ added exceeds the number of moles of A^- in the solution, or if the number of moles of OH^- added exceeds the number of moles of HA in the solution, the buffer is destroyed and there is a large change in pH. With a given total concentration of acid and base, the maximum buffering capacity and most nearly constant pH are attained when the concentrations of the acid and base are initially equal. As we saw in Example 21a, $[H^+] = K_a$, or $pH = pK_a$, in this case. Similarly, with a base B and its conjugate acid BH^+, the equation

$$\frac{[BH^+][OH^-]}{[B]} = K_b$$

shows that when $[BH^+] = [B]$, then $[OH^-] = K_b$. If, then, we are told to prepare a buffer solution of given pH, we should seek an acid with $pK_a \approx pH$, or a base with $pK_b \approx pK_w - pH$. We cannot expect to find an available acid for which $pK_a = pH$ exactly, but we choose the one that is closest and adjust the acid–base ratio to make the pH come out right.

Example 22 (a) Choose from Table 14.1 a conjugate acid–base pair suitable for making a buffer solution of pH 7.4 (the pH of blood). (b) Calculate the acid–base ratio required in this buffer.

Answer (a) We wish to have $pK_a \approx 7.4$, or $K_a \approx 4 \times 10^{-8}$. The nearest is K_2 for H_3PO_4 (K_a for $H_2PO_4^-$), corresponding to the pair $H_2PO_4^- + HPO_4^{2-}$ (actually present in blood). Other possibilities are H_2CO_3 (also present), H_2S (toxic), and HSO_3^- (toxic).

(b) $\qquad H_2PO_4^- \rightleftharpoons H^+ + HPO_4^{2-}$

$$\frac{[H^+][HPO_4^{2-}]}{[H_2PO_4^-]} = K_a$$

$$\frac{[H_2PO_4^-]}{[HPO_4^{2-}]} = \frac{[H^+]}{K_a}$$

$$= \frac{10^{-7.4}}{6.22 \times 10^{-8}} = 0.64$$

SLIGHTLY SOLUBLE SALTS

SOLUBILITY PRODUCT / 14.12

Silver chloride is an insoluble salt. Like other "insoluble" salts, it has a slight but measurable solubility in water. The dissolving process is represented by the equation

$$AgCl(c) \rightleftharpoons Ag^+ + Cl^-$$

The equilibrium condition for this reaction is

$$[Ag^+][Cl^-] = K$$

The solid is omitted from the reaction quotient (page 205). K is called the SOLUBILITY PRODUCT of AgCl. It is also represented by K_{sp}. When this equilibrium is established, the solution is saturated, and the solubility product thus gives the product of the Ag^+ and Cl^- concentrations in a saturated solution. The individual concentrations may be changed—by adding NaCl or $AgNO_3$, for instance—but *the product, at any one temperature, remains constant.**

When the salt dissolves to give unequal numbers of positive and negative ions, each concentration must be raised to a power equal to the coefficient of that ion in the equation. For example:

Salt	Reaction	Equilibrium condition
CaF_2	$CaF_2(c) \rightleftharpoons Ca^{2+} + 2F^-$	$[Ca^{2+}][F^-]^2 = K$
Hg_2Cl_2	$Hg_2Cl_2(c) \rightleftharpoons Hg_2^{2+} + 2Cl^-$ (not $2Hg^+$)	$[Hg_2^{2+}][Cl^-]^2 = K$
$Al(OH)_3$	$Al(OH)_3(c) \rightleftharpoons Al^{3+} + 3OH^-$	$[Al^{3+}][OH^-]^3 = K$
$Ca_3(PO_4)_2$	$Ca_3(PO_4)_2(c) \rightleftharpoons 3Ca^{2+} + 2PO_4^{3-}$	$[Ca^{2+}]^3[PO_4^{3-}]^2 = K$

Solubility products of some slightly soluble salts are given in Table 14.2. We will see later (page 342) how these constants can be calculated from the emf's of galvanic cells.

Solubility and *solubility product* must not be confused with each other. The solubility of a salt is the quantity present in a unit amount of a saturated solution, expressed in moles per liter, grams per 100 ml, or other units. The solubility depends on what else is in the solution. The solubility *product*, being an equilibrium constant, depends only on temperature. There is a connection, of course, between the solubility and the solubility product; if one is known, the other can be calculated.

Example 23 Find the solubility (in moles/liter) of AgCl at 25°C in (a) pure water; (b) a 0.010 M solution of NaCl.

Answer (a) For every mole of AgCl that dissolves, we get 1 mole Ag^+ and 1 mole Cl^- in solution. Therefore, the concentration of either of these ions is equal to the solubility of AgCl in moles/liter. In the absence of any source of Ag^+ or Cl^- other than AgCl, these two concentrations must be equal. Let s represent either of them:

$$AgCl(c) \rightleftharpoons Ag^+ + Cl^-$$
$$[Ag^+][Cl^-] = K$$
$$s \cdot s = 1.74 \times 10^{-10}$$
$$s = 1.32 \times 10^{-5} \text{ mole/liter}$$

(b) The number of moles of AgCl that dissolve per liter (the new solubility, s') is equal to the number of moles of Ag^+ in the solution: $s' = [Ag^+]$. In a 0.010 M NaCl solution, there is already 0.010 mole Cl^-/liter. The dissolving of s' mole AgCl/liter raises $[Cl^-]$ to $0.010 + s'$:

$$AgCl(c) \rightleftharpoons Ag^+ + Cl^-$$
$$ s' \quad 0.010 + s'$$

---

* Solubility products are used only for slightly soluble salts because, at high concentrations, the disparity between concentration and activity is so great, and so dependent on the composition of the solution, as to make solubility products useless for calculation.

Table 14.2
Solubility Products in Water

Salt	Temperature, °C	K
AgBr	25	7.7×10^{-13}
AgCl	25	1.74×10^{-10}
Ag_2CrO_4	25	1.1×10^{-12}
AgI	25	1.5×10^{-16}
Ag_2S	25	1.9×10^{-49}
$Al(OH)_3$	25	10^{-32}
$BaCO_3$	25	8.1×10^{-9}
$BaCrO_4$	25	2.2×10^{-10}
BaF_2	25.8	1.73×10^{-6}
$BaSO_4$	25	1.08×10^{-10}
$CaCO_3$	25	8.7×10^{-9}
CaF_2	26	3.95×10^{-11}
CdS	18	3.6×10^{-29}
CoS	18	3×10^{-26}
CuS	18	8.5×10^{-45}
FeS	25	1.3×10^{-17}
$Hg_2Cl_2{}^a$	25	2×10^{-18}
HgS	25	10^{-52}
MgC_2O_4	18	8.57×10^{-5}
MnS	25	7.9×10^{-13}
NiS	18	1.4×10^{-24}
$PbCrO_4$	18	1.77×10^{-14}
PbI_2	25	1.39×10^{-8}
PbS	25	2.3×10^{-27}
$PbSO_4$	18	1.06×10^{-8}
ZnS	25	2.5×10^{-24}

^{a} $Hg_2Cl_2(c) \rightleftharpoons Hg_2{}^{2+} + 2Cl^-$.

Then, substituting into the equilibrium condition,

$$[Ag^+][Cl^-] = K$$
$$s'(0.010 + s') = K$$

It is a very good approximation that $s' \ll 0.01$, or $0.010 + s' = 0.010$. In other words, practically all the Cl^- is contributed by the NaCl, making $[Cl^-] = 0.010$ mole/liter. Then

$$s'(0.010) = K = 1.74 \times 10^{-10}$$
$$s' = 1.74 \times 10^{-8} \text{ mole/liter}$$

We observe that the solubility (s') is much less in the NaCl solution than in pure water (s). This result could have been predicted from Le

Chatelier's principle: the equilibrium

$$AgCl(c) \rightleftharpoons Ag^+ + Cl^-$$

is shifted to the left by addition of Cl^- or Ag^+; that is, the solubility of AgCl is decreased. This effect—like the effect (page 282) of a strong acid in repressing the ionization of a weak acid—is known as the common-ion effect, because the decrease in solubility is caused by addition of a salt having an ion in common with the slightly soluble salt.

In an unsaturated or supersaturated solution of AgCl, there is no equilibrium between solid and ions, and we therefore cannot expect to have $[Ag^+][Cl^-] = K$. If the solution is

Unsaturated	$[Ag^+][Cl^-] < K$
Supersaturated	$[Ag^+][Cl^-] > K$
Saturated	$[Ag^+][Cl^-] = K$

One use of the solubility product is to determine whether or not precipitation is possible when the ions have certain concentrations. We calculate the appropriate product of concentrations (Q) and compare it with K. If $Q > K$, the solution is supersaturated, and precipitation is possible (but it may not occur if the solution remains supersaturated); if $Q < K$, the solution is unsaturated, and precipitation is impossible.

Example 24 A solution is prepared by mixing equal volumes of 0.010 M $MgCl_2$ and 0.020 M $Na_2C_2O_4$ solutions at 18°C. Is it possible for MgC_2O_4 to precipitate in the resulting solution?

Answer The reaction in question is

$$MgC_2O_4(c) \rightleftharpoons Mg^{2+} + C_2O_4^{2-}$$

We ask whether or not $Q > K$. When the solutions are mixed, the volume is doubled (to a very good approximation), and the concentrations of Mg^{2+} and $C_2O_4^{2-}$ are half what they were originally: $[Mg^{2+}] = 0.005$, $[C_2O_4^{2-}] = 0.010$ mole/liter. Then

$$Q = [Mg^{2+}][C_2O_4^{2-}] = 5 \times 10^{-5}$$

which is less than $K = 8.57 \times 10^{-5}$ (Table 14.2). It is impossible for MgC_2O_4 to precipitate.

Example 25 What is the minimum concentration of S^{2-} which can cause precipitation of CdS from a 0.10 M solution of Cd^{2+} at 18°C?

Answer The minimum concentration is that corresponding to a saturated solution. In a saturated solution,

$$CdS(c) \rightleftharpoons Cd^{2+} + S^{2-}$$
$$[Cd^{2+}][S^{2-}] = 3.6 \times 10^{-29}$$

Since $[Cd^{2+}] = 0.10$ mole/liter,

$$[S^{2-}] = \frac{3.6 \times 10^{-29}}{0.10} = 3.6 \times 10^{-28} \text{ mole/liter}$$

Example 26 We wish to dissolve enough $BaCrO_4$ in water so that $[Ba^{2+}] = 0.10$ mole/liter. This result can be accomplished by adding a reagent that reacts with CrO_4^{2-}. How low must the concentration of CrO_4^{2-} be in order for $[Ba^{2+}]$ to be 0.10 mole/liter?

Answer

$$BaCrO_4(c) \rightleftharpoons Ba^{2+} + CrO_4^{2-}$$
$$[Ba^{2+}][CrO_4^{2-}] = 2.2 \times 10^{-10}$$
$$(0.10)(y) = 2.2 \times 10^{-10}$$
$$y = 2.2 \times 10^{-9} \text{ mole/liter} = [CrO_4^{2-}]$$

This is the maximum permissible concentration of CrO_4^{2-}.

SOLUBILITY AND pH / 14.13

When an acid is weak, the concentration of its conjugate base is affected by the pH of the solution. For example, the reaction

$$HS^- \rightleftharpoons H^+ + S^{2-}$$

is shifted to the left by an increase in $[H^+]$. In turn, the solubility of a metallic sulfide is affected by the concentration of S^{2-}. When acid is added to a saturated solution of ZnS, the concentration of S^{2-} is decreased, and the reaction

$$ZnS(c) \rightleftharpoons Zn^{2+} + S^{2-}$$

is shifted to the right, meaning that the solubility of ZnS increases. In general, *the solubility of a salt containing the conjugate base of a weak acid is increased by addition of a strong acid to the solution.*

The effect is especially pronounced with sulfides because S^{2-} is the conjugate base of the very weak acid HS^-, and this case is of interest because of its importance in qualitative analysis. If any two of the three concentrations $[H^+]$, $[S^{2-}]$, and $[H_2S]$ are given, the third can be calculated from the equation (Example 19, page 284),

$$\frac{[H^+]^2[S^{2-}]}{[H_2S]} = K_1K_2 = 6.3 \times 10^{-8} \times 10^{-14} = 6 \times 10^{-22}$$

In a solution of H_2S, saturated at 1 atm pressure and 25°C, $[H_2S] = 0.10$ mole/liter; under these conditions,

$$[H^+]^2[S^{2-}] = 6 \times 10^{-23} \tag{28}$$

Example 27 Find $[Zn^{2+}]$ in a saturated solution of ZnS in which $[H^+] = 0.30$ and $[H_2S] = 0.10$ mole/liter at 25°C.

Answer First we must calculate $[S^{2-}]$:

$$[S^{2-}] = \frac{6 \times 10^{-23}}{[H^+]^2} = \frac{6 \times 10^{-23}}{(0.30)^2} = 7 \times 10^{-22} \text{ mole/liter}$$

Then

$$[Zn^{2+}][S^{2-}] = 2.5 \times 10^{-24} \quad \text{(Table 14.2)}$$

$$[Zn^{2+}] = \frac{2.5 \times 10^{-24}}{7 \times 10^{-22}} = 4 \times 10^{-3} \text{ mole/liter}$$

It should, perhaps, be pointed out again that in a given solution there is only one concentration of (for example) S^{2-}. The value of $[S^{2-}]$ must be the same in the H_2S equilibrium condition and in the ZnS equilibrium condition.

Example 28 What is the maximum concentration of H^+ in a solution from which PbS will precipitate, when $[Pb^{2+}] = 0.0050$ and $[H_2S] = 0.10$ mole/liter?

Answer In a saturated solution of PbS,

$$PbS(c) \rightleftharpoons Pb^{2+} + S^{2-}$$
$$[Pb^{2+}][S^{2-}] = 2.3 \times 10^{-27}$$

$$[S^{2-}] = \frac{2.3 \times 10^{-27}}{0.0050} = 4.6 \times 10^{-25} \text{ mole/liter}$$

Precipitation is possible if $[S^{2-}]$ is slightly larger than this value. The concentration of $[H^+]$ must be slightly less than that given by Equation (28):

$$[H^+]^2 = \frac{6 \times 10^{-23}}{4.6 \times 10^{-25}} = 1.3 \times 10^2$$

$$[H^+] = 11 \text{ moles/liter}$$

Thus, PbS will precipitate (or remain undissolved) in all but very concentrated strong acids.

Example 29 A solution has $[H^+] = 0.30$, $[H_2S] = 0.10$, $[Cd^{2+}] = 0.010$, $[Fe^{2+}] = 0.010$, $[Hg^{2+}] = 0.010$, and $[Mn^{2+}] = 0.010$ mole/liter. Which of the four sulfides, CdS, HgS, MnS, and ZnS, will precipitate? (Assume that K is about the same at 18° and 25°C and that $[H^+]$ and $[H_2S]$ are kept constant.)

Answer $[H^+]^2[S^{2-}] = 6 \times 10^{-23}$

$$[S^{2-}] = \frac{6 \times 10^{-23}}{(0.30)^2} = 7 \times 10^{-22} \text{ mole/liter}$$

For each of the metal ions (M^{2+}),

$$[M^{2+}] = 0.010 \text{ mole/liter}$$
$$[M^{2+}][S^{2-}] = 0.010 \times 7 \times 10^{-22} = 7 \times 10^{-24}$$

If, for a given sulfide, $7 \times 10^{-24} < K$, that sulfide will not precipitate; if $7 \times 10^{-24} > K$, there will be a precipitate. Inspection of Table 14.2 shows that CdS and HgS will precipitate, while FeS and MnS will not.

The preceding example illustrates an important step in qualitative analysis: the separation of metals according to the solubility of their sulfides.

Complex-ion equilibria are treated in Section 22.10 (page 470).

The answers given are based on the assumption that all activity coefficients are 1. When the temperature is not given, take it to be the temperature for which the necessary data are tabulated.

1. Definitions. Define and illustrate briefly: (a) pH, (b) acidic solution, (c) basic solution, (d) conservation condition, (e) electrical neutrality condition, (f) common-ion effect, (g) buffer solution, (h) solubility product.

2. Strong acid. Calculate the pH of the following solutions of $HClO_4$, a strong acid, at 25°C: (a) 5.0×10^{-4} M; (b) 5.0×10^{-8} M.

3. Ionization of water. What is the pH of pure water at 0°C?

4. Weak acid. A 5.00×10^{-3} M solution of butyric acid, $HC_4H_7O_2$, has pH 3.58. Calculate K_a and pK_a for this acid.

5. Weak acid. Find (a) the degree of ionization and (b) the pH of a 0.20 M solution of HNO_2.

6. Weak base. Calculate the pH of a 0.10 M solution of CH_3NH_2 in H_2O.

7. Basic anion. Find the pH of a 0.20 M solution of sodium phenoxide, $Na^+C_6H_5O^-$.

8. Acidic cation. Find the pH of a 0.050 M solution of anilinium chloride, $(C_6H_5NH_3)^+Cl^-$.

9. Polyprotic base. Find the basic ionization constant of the tartrate ion, $C_4H_4O_6^{2-}$.

10. Mixture of acids. A solution is 0.25 M with respect to HCl and 1.00 M with respect to $HCHO_2$. Calculate the concentration of CHO_2^- in this solution.

11. Buffer solutions. (a) Find the pH of a solution 0.80 M with respect to NH_3 and 1.20 M with respect to NH_4Cl. (b) Find the pH of this solution after 0.10 mole HCl/liter has been added to it. (c) A solution of pH 9.07 is prepared by adding NaOH to pure water. Find the pH of this solution after 0.10 mole HCl/liter has been added to it.

12. Solubility product. The solubility of lead fluoride, PbF_2, in water is 0.064 g/100 ml solution at 20°C. Calculate its solubility product.

13. Solubility from solubility product. Calculate the concentration of CrO_4^{2-} in a solution saturated with $BaCrO_4$ and containing (a) no other solute; (b) 0.20 mole $BaCl_2$/liter.

14. Precipitation. A 0.0005 M solution of $AgNO_3$ and a 0.0005 M solution of NaBr are mixed in equal volumes at 25°C. Is it possible for AgBr to precipitate?

15. Solubility and pH. Find the solubility, in moles per liter, of ZnS in a solution in which $[H_2S] = 0.10$ mole/liter and the pH is (a) 3.0, (b) 9.0.

16. Ionization of water. Fill in the blanks in the following table. Each line (a, b, c, d) refers to a different solution.

Solution	Temperature, °C	$[H^+]$	$[OH^-]$	pH
(a)	25	1.00×10^{-6}	—	—
(b)	25	—	—	4.52
(c)	0	—	—	7.00
(d)	25	—	5.0×10^{-4}	—

17. pD. The equilibrium constant of the reaction

$$2D_2O \rightleftharpoons D_3O^+ + OD^-$$

(where D is deuterium, ^2H) is 1.35×10^{-15} at 25°C. Calculate the pD of pure deuterium oxide (heavy water) at 25°C.

18. Weak acid. Calculate the pH of a 0.060 M solution of propionic acid.

19. Weak acid. Find the pH of a 0.50 M solution of HCN.

20. Weak acid. Calculate (a) the degree of ionization and (b) the pH of bromoacetic acid in a 0.020 M solution.

21. Weak acid. You are told to add acetic acid from a stock solution (0.1000 M) to 1.000 liter of water at 25°C until the pH of the resultant solution is 4.00. How much of the stock solution would you add?

22. Approximate calculations. Calculate $[H^+]$ in a 0.0100 M solution of nitrous acid, (a) making the assumption that avoids a quadratic equation; (b) solving the quadratic equation. Compare the two results.

23. Infinite dilution. It is often said that a weak electrolyte becomes completely ionized at infinite dilution. (a) In an aqueous solution, what is $[H^+]$ equal to at infinite dilution (in pure water)? (b) For a weak acid, HA, the degree of ionization is $\alpha = [A^-]/c$, where $c = [HA] + [A^-]$. Write the equilibrium condition for the ionization of HA in terms of α, c, and $[H^+]$. (c) Substitute the answer to (a) into the equation in (b). Does c drop out of the equation? Is the equation valid in the limit $c \rightarrow 0$? (d) Solve for α at infinite dilution in terms of K_a and K_w. (e) Calculate α at infinite dilution at 25°C for acetic acid and for hydrocyanic acid. (f) Is the statement at the beginning of this problem correct for weak acids? for weak bases? for other weak electrolytes, such as $HgCl_2$?

24. Basic anion. Calculate the pH of a 0.40 M solution of $NaCHO_2$.

25. Weak acids and bases. (a) For a certain weak acid, HA, $K_a = 10^{-6}$ at 25°C. What is K_b for its conjugate base A^-? Is A^- a strong base or a weak base? Is it true that (as is sometimes said) "The conjugate base of a weak acid is a strong base"? (b) What is the value of K_a such that K_a for an acid is equal to K_b for its conjugate base at 25°C?

26. Acidic cation. The equilibrium constant for the reaction $ZnOH^+ \rightleftharpoons Zn^{2+} + OH^-$ is 4.0×10^{-5} at 25°C. (a) Calculate the acidic ionization constant of Zn^{2+} ($Zn^{2+} + H_2O \rightleftharpoons ZnOH^+ + H^+$). (b) Find the pH of a 0.050 M solution of $ZnCl_2$.

27. Polyprotic acids. Calculate (a) the pH, and (b) the concentration of SO_3^{2-}, in a 0.20 M solution of H_2SO_3.

28. Polyprotic acids. Calculate (a) $[H^+]$, (b) $[H_2PO_4^-]$, (c) $[HPO_4^{2-}]$, and (d) $[PO_4^{3-}]$ in a 0.10 M solution of phosphoric acid, H_3PO_4.

29. Simultaneous equilibria. A solution is prepared by dissolving 5.00×10^{-5} mole phenol in H_2O and adding water until the volume is 1.00 liter. (a) Write four independent equations in which the variables are $[H^+]$, $[OH^-]$, $[C_6H_5O^-]$, and $[HC_6H_5O]$. (b) Calculate the pH of the solution.

30. Simultaneous equilibria. (a) Calculate the basic ionization constant at $25°C$ of formate ion, CHO_2^-. (b) For a 1.00×10^{-4} M solution of potassium formate, write five independent equations in which the variables are $[K^+]$, $[H^+]$, $[OH^-]$, $[CHO_2^-]$, and $[HCHO_2]$. (c) Calculate the pH of this solution at $25°C$.

31. Simultaneous equilibria. In problems involving basic anions, we usually consider the reaction

$$A^- + H_2O \rightleftharpoons HA + OH^- \qquad\qquad (i)$$

The equilibrium

$$HA + H_2O \rightleftharpoons H_3O^+ + A^- \qquad\qquad (ii)$$

is also established in the solution, and it should be possible to solve a problem such as Problem 7 by starting with either equilibrium condition. Do Problem 7 by using the equilibrium condition for (*ii*) instead of (*i*). Show that the *correct* application of either method leads to the same answer. Explain why it is simpler to start with (*i*).

32. Mixture of acids. Find the concentration of (a) hydrogen oxalate ion and (b) oxalate ion in a solution 1.00 M with respect to HBr and 0.100 M with respect to $H_2C_2O_4$.

33. Mixture of acids. Find the degree of ionization of 0.30 M acetic acid in a solution of pH 2.00.

34. Buffer solutions. Find the pH of a solution 0.50 M with respect to formic acid and 0.40 M with respect to sodium formate.

35. Buffer solutions. A buffer solution of pH 4.80 is to be prepared from propionic acid and sodium propionate. The concentration of sodium propionate must be 0.50 mole/liter. What should be the concentration of the acid?

36. Buffer solutions. Assume that all the monoprotic acids listed in Table 14.1, and their sodium salts, are available. (a) Which acid–salt pair should be chosen for a buffer solution of pH 3.65? (b) What should be the molar ratio of acid to salt in this buffer solution?

37. Buffer solutions. In human blood, $[H_2CO_3] = 1.25 \times 10^{-3}$ mole/liter (including CO_2) and $[HCO_3^-] = 2.5 \times 10^{-2}$ mole/liter. Calculate the pH of blood. Assume that the equilibrium constants are the same at body temperature as at $25°C$. Compare with the experimental value on page 285.

38. Buffer solutions. What volumes of 0.1000 M acetic acid and 0.1000 M NaOH stock solutions should be mixed to prepare 1.000 liter of a buffer solution of pH 6.058 at $25°C$?

39. Solubility from solubility product. Calculate the concentration of Pb^{2+} in a solution saturated with $PbSO_4$ and (a) containing no other solute; (b) 0.20 M with respect to $MgSO_4$.

40. Solubility product. The solubility of $Ca(IO_3)_2$ in water is 0.10 g/100 ml at

0°C. Calculate (a) its solubility product; (b) its solubility, in grams per 100 ml, in 0.050 M $Mg(IO_3)_2$ solution.

41. Precipitation. 30 ml of a 0.020 M solution of $BaCl_2$ and 50 ml of a 0.0050 M solution of NaF were mixed at 25.8°C. (a) Find $[Ba^{2+}]$ and $[F^-]$ in the mixed solution before any reaction occurs. (b) Is it possible for BaF_2 to precipitate?

42. Solubility and pH. Find the minimum pH at which CdS will precipitate from a solution 0.10 M with respect to H_2S and 0.0050 M with respect to Cd^{2+}. (Assume that K is the same at 18° and at 25°C.)

43. Solubility and pH. (a) Find the minimum pH at which FeS will precipitate from a solution 0.10 M with respect to H_2S and 0.0050 M with respect to Fe^{2+}. (b) In what range should the pH be in order to separate Fe^{2+} and Cd^{2+} by precipitation of one sulfide and not the other? (See the preceding problem.)

15

ANALYSIS OF IONS
IN SOLUTIONS

The determination of the quantities of substances in a sample is known as QUANTITATIVE ANALYSIS. In a gravimetric analysis (page 55) a known mass of sample is dissolved and a reagent is added to precipitate the substance whose quantity is being determined. The precipitate is then separated, dried, and weighed. VOLUMETRIC ANALYSIS involves the addition of that volume of a solution of known concentration that reacts completely with the substance that is being determined. A solution of known concentration is called a STANDARD SOLUTION. In COLORIMETRY, the concentration of a colored substance is determined by comparison with a standard solution.

TITRATIONS AND MOLARITY / 15.1

Volumetric analyses are based on the process known as TITRATION, the addition of a measured volume of one solution to another solution. The measuring instrument used is the buret (Fig. 15.1), a long tube with volume markings and a stopcock or a valve at the bottom.

The aim of a titration is the addition of a quantity of the standard solution CHEMICALLY EQUIVALENT *to the quantity of the unknown.* This means the MOLAR RATIO of the known added to the unknown must conform to the ratio in the chemical equation. Addition of an excess or insufficient quantity of the known solution results in an incorrect analysis. The reaction may be an acid–base reaction, an oxidation–reduction, the precipitation of a slightly soluble salt, or any other reaction which goes practically to completion, without complicating side reactions.

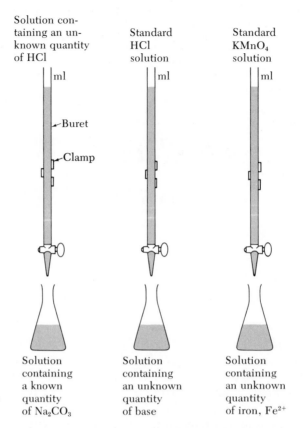

Solution containing an unknown quantity of HCl — ml — Buret — Clamp — Solution containing a known quantity of Na_2CO_3

Standard HCl solution — ml — Solution containing an unknown quantity of base

Standard $KMnO_4$ solution — ml — Solution containing an unknown quantity of iron, Fe^{2+}

Fig. 15.1. *Setups for some typical titrations.*

Example* 1 (a) Find the mass of sodium carbonate, Na_2CO_3, 106.0 g/mole, required to prepare 0.5000 liter of 0.1000 M solution. (b) 45.53 ml of this solution reacts completely (no excess) with 40.00 ml of HCl solution. Find the molarity of the acid. The reaction is

$$CO_3^{2-} + 2H^+ \longrightarrow CO_2 + H_2O \text{ (ionic equation)}$$
$$Na_2CO_3 + 2HCl \longrightarrow CO_2 + 2NaCl + H_2O \text{ (molecular equation)}$$

Here, 0.1000 M Na_2CO_3 is the standard solution while the HCl solution is the unknown.

Answer (a) The number of moles of Na_2CO_3 needed is

$$0.1000 \,\frac{\text{mole}}{\text{liter}} \times 0.5000 \,\text{liter} = 0.05000 \text{ mole}$$

The required mass is given by†

$$0.05000 \,\text{mole} \times 106.0 \,\frac{\text{g}}{\text{mole}} = 5.300 \text{ g}$$

* Calculations are done in the molar system. Equivalent weights and normality are treated on page 305.

† Use of a five-place log table is recommended in the absence of a calculating device.

(b) Since there are 10^3 millimoles (mmole) in 1 mole and 10^3 mg in 1 g, the unit mmole/ml is equivalent to mole/liter and mg/mmole is equivalent to g/mole (page 170). The number of mmoles of the known, Na_2CO_3, is

$$45.53 \text{ mt} \times 0.1000 \frac{\text{mmole}}{\text{mt}} = 4.553 \text{ mmole } Na_2CO_3$$

But 1 mmole of Na_2CO_3 reacts exactly with 2 mmoles of HCl; the number of mmoles of HCl must be

$$4.553 \text{ mmole } Na_2CO_3 \times \frac{2 \text{ mmole HCl}}{1 \text{ mmole } Na_2CO_3} = 9.106 \text{ mmole HCl}$$

and the molarity of the acid is

$$\frac{9.106 \text{ mmole}}{40.00 \text{ ml}} = 0.2277 \text{ M}$$

Example 2 Find the mass of Na_2CO_3 in a solution that reacts exactly with 26.75 ml of 0.2277 M HCl.

Answer The number of mmoles of HCl that reacts is

$$0.2277 \frac{\text{mmole}}{\text{mt}} \times 26.75 \text{ mt} = 6.091 \text{ mmole HCl}$$

From the balanced equation 1 mmole of Na_2CO_3 reacts with 2 mmoles of HCl. The mass of Na_2CO_3 is then given by

$$6.091 \text{ mmole HCl} \times \frac{106.0 \text{ mg } Na_2CO_3}{2 \text{ mmole HCl}} = 322.8 \text{ mg } Na_2CO_3$$

Example 3 28.22 ml of H_3PO_4 reacts exactly with 20.15 ml of 0.1512 M NaOH. The reaction is $H_3PO_4(aq) + 2NaOH(aq) \rightarrow Na_2HPO_4(aq) + 2H_2O$. Calculate the molarity of the phosphoric acid.

Answer The number of mmoles of NaOH is given by

$$20.15 \text{ mt} \times 0.1512 \frac{\text{mmole}}{\text{mt}} = 3.047 \text{ mmole NaOH}$$

The number of mmoles of H_3PO_4 is given by

$$3.047 \text{ mmole NaOH} \times \frac{1 \text{ mmole } H_3PO_4}{2 \text{ mmole NaOH}} = 1.524 \text{ mmole } H_3PO_4$$

so that its molarity is

$$\frac{1.524 \text{ mmole}}{28.22 \text{ ml}} = 0.05400 \text{ M}$$

The following example involves an oxidation–reduction reaction.

Example 4 (a) How many grams of sodium oxalate, $Na_2C_2O_4$ (134.0 g/mole), are required to prepare 0.5000 liter of 0.1500 M solution to be used as a reducing agent? (b) 25.00 ml of an acidified potassium permanganate, $KMnO_4$, solution reacts with 40.65 ml of the standard solution prepared in (a). Calculate the molarity of the $KMnO_4$ solution. The reaction is

$$2MnO_4^- + 5C_2O_4^{2-} + 16H^+ \longrightarrow 2Mn^{2+} + 10CO_2 + 8H_2O$$

or

$$2KMnO_4 + 5Na_2C_2O_4 + 8H_2SO_4 \longrightarrow 2MnSO_4 + 10CO_2 + K_2SO_4 + 5Na_2SO_4 + 8H_2O$$

Answer

(a) $0.1500 \dfrac{\cancel{mole}}{\cancel{liter}} \times 0.5000 \cancel{liter} \times 134.0 \dfrac{g}{\cancel{mole}} = 10.05 \text{ g}$

(b) The number of mmoles of $KMnO_4$ is obtained as follows:

$$40.65 \cancel{ml} \times 0.1500 \dfrac{mmole}{\cancel{ml}} = 6.098 \text{ mmole } Na_2C_2O_4$$

$$6.098 \cancel{mmole\ Na_2C_2O_4} \times \dfrac{2 \text{ mmole } KMnO_4}{5 \cancel{mmole\ Na_2C_2O_4}} = 2.439 \text{ mmole } KMnO_4$$

The molarity of the $KMnO_4$ solution is

$$\dfrac{2.439 \text{ mmole}}{25.00 \text{ ml}} = 0.09756 \text{ M}$$

Example 5 In an acidic solution, 25.00 ml of 0.02100 M $KMnO_4$ is used to oxidize Fe^{2+} to Fe^{3+}. Calculate the mass of Fe^{2+} (55.85 mg/mmole) oxidized. The reaction is

$$5Fe^{2+} + MnO_4^- + 8H^+ \longrightarrow Mn^{2+} + 5Fe^{3+} + 4H_2O$$

Answer The number of mmoles of MnO_4^- is

$$25.00 \cancel{ml} \times 0.02100 \dfrac{mmole}{\cancel{ml}} = 0.5250 \text{ mmole } MnO_4^-$$

From the balanced equation 5 mmoles of Fe^{2+} reacts with 1 mmole of MnO_4^- so that the mass of Fe^{2+} is

$$0.5250 \cancel{mmole\ MnO_4^-} \times \dfrac{5 \times 55.85 \text{ mg } Fe^{2+}}{1 \cancel{mmole\ MnO_4^-}} = 146.6 \text{ mg } Fe^{2+}$$

INDICATORS / 15.2

Since a common method of titration depends on the pH of the solution being titrated, we shall now discuss the use of indicators to determine the pH of a solution.

An INDICATOR is an acid (or base) having a different color from its conjugate base (or acid). Let us represent the acid form of the indicator by HIn and its conjugate base form by In⁻. An example is methyl orange, for which HIn is red and In⁻ is yellow. The equilibrium condition for

Table 15.1
Acid–Base Indicators[a]

| | Colors | | |
	Acid	Base	pH range
Indicator			
o-Cresol red	Red	Yellow	0–2
Methyl orange	Red	Yellow	3–4.4
Methyl red	Red	Yellow	4.4–6
Bromthymol blue	Yellow	Blue	6–8
Neutral red	Red	Yellow	7–8
Phenolphthalein	Colorless	Red	8–10
Thymolphthalein	Colorless	Blue	9.4–10.6
Alizarin yellow	Yellow	Violet	10–12
1,3,5-Trinitrobenzene	Colorless	Red	12–13.4

[a] Roger G. Bates, "Determination of pH, Theory and Practice," pp. 138–139. John Wiley & Sons, New York, 1964.

the reaction

$$HIn(aq) \rightleftharpoons H^+(aq) + In^-(aq)$$

is

$$\frac{[H^+][In^-]}{[HIn]} = K \qquad \text{or} \qquad \frac{[HIn]}{[In^-]} = \frac{[H^+]}{K}$$

Indicators are used in small quantities so that $[H^+]$ is determined by the other constitutents of the solution. The ratio $[HIn]/[In^-]$, which determines the color, depends on the pH of the solution. With methyl orange, for example, the solution is red if $[HIn] \gg [In^-]$, yellow if $[In^-] \gg [HIn]$, and varying shades of orange when $[HIn]$ and $[In^-]$ are comparable. Table 15.1 lists some common indicators, their colors, and the pH ranges in which the colors show perceptible gradations.

Indicators are fine for rough estimation of pH, or for detecting large changes in pH, but not for accurate measurements. The device now used is known as a pH METER. It is a galvanic cell whose emf depends on the pH of one of the solutions in it. Before use the instrument is calibrated with standard buffer solutions of known pH.

TITRATION CURVES / 15.3

We now consider an important question. "In a titration, how does the chemist know when to stop the addition?" In the type of titration involved in Example 4, this question has a comparatively simple answer. One drop, about 0.05 ml, of the permanganate solution added to as much as 500 ml of water suffices to impart a distinct pink color to the solution. Hence, with the addition of the permanganate solution to the colorless oxalate solution, the solution remains colorless as long as oxalate ions are in excess. When just sufficient permanganate is added so that exactly *equivalent** *quantities* have reacted, the solution is still colorless. The addition of one drop of the permanganate solution *in excess* then imparts a pink color to the solution. The stage at which equivalent quantities of the standard solution and the unknown have been mixed is called the EQUIVALENCE POINT OF THE TITRATION. The END POINT, signaled by a sudden change in the color of the solution, indicates that the titration is completed and the buret reading should be taken.

If the titration reactants do not produce a color change, an indicator may be added to the solution being titrated to signal the end point. A starch solution, for example, is used as the indicator in titrations with iodine solution; an excess of one drop of iodine solution imparts an unmistakable blue color to the resultant solution.

In titrations of acids and bases, advantage is taken of the observation that the *pH of the solution* being titrated *changes rapidly* as the equivalence point is approached. When an acid and a base have been mixed in equivalent quantities, they are said to have *neutralized* each other. This word is misleading, for the resulting solution may actually be neutral, acidic, or basic. Indicators change color within a characteristic pH range so that *the choice of an indicator is determined by the pH of the solution at the equivalence point.*

Let us say that an acid solution is in a flask; the basic solution, for example NaOH, is in the buret. As the base is added to the acid, the pH of the acid solution increases. The dependence of pH on the quantity of base added is shown in Fig. 15.2. If the acid is strong (Fig. 15.2a), the initial effect of adding the base is merely to decrease the concentration of the acid; if the acid is weak (Fig. 15.2b), the initial effect is to form a buffer solution containing the acid and its conjugate base (page 285). In either case, the pH rises slowly at first. *The curve becomes almost vertical when equivalent quantities of acid and base are present.* To provide the signal for the chemist to stop adding solution, the indicator must be so chosen that it changes color somewhere within the pH range spanned by the steeply rising portion of the curve. The difference between the end point and the equivalence point is not significant if the titration is done properly. Figure 15.2a shows that either phenolphthalein or methyl red is suitable for the HCl–NaOH titration, while Fig. 15.2b shows that phenolphthalein, but not methyl red, is suitable for the acetic acid–NaOH titration. Similarly, in a titration involving a strong acid and a weak base, the indicator would have to change color on the acid side.

---

* Equivalent quantities are the quantities whose molar ratios conform to the balanced equation for the reaction.

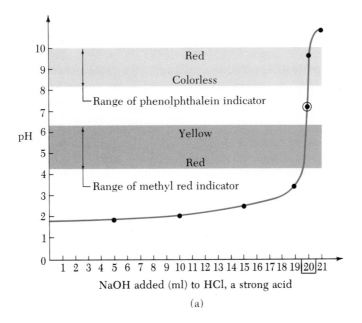

NaOH added (ml) to HCl, a strong acid

(a)

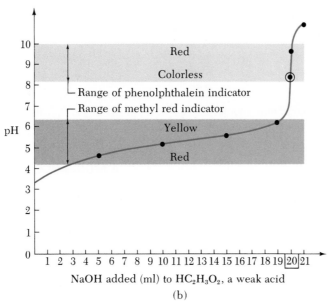

NaOH added (ml) to HC₂H₃O₂, a weak acid

(b)

Fig. 15.2. Titration curve; pH values plotted against the corresponding volume added from a buret. (a) 20.00 ml of 0.1000 M HCl diluted to 100.0 ml and titrated with 0.1000 M NaOH; (b) 20.00 ml of 0.1000 M HC₂H₃O₂ (acetic acid) diluted to 100.0 ml and titrated with 0.1000 M NaOH. Temperature is 25°C. Also shown is the pH range of the indicator, the range of pH values through which visible changes in color occur. The equivalence point is circled.

At the equivalence point, *the solution contains only the product* of neutralization of the acid and the base, and practically none of either the acid or the base used in the titration. Thus, if we mix 1 mmole of acetic acid and 1 mmole of NaOH, the reaction

$$HC_2H_3O_2(aq) + OH^-(aq) \longrightarrow H_2O + C_2H_3O_2^-(aq)$$

goes practically to completion (equilibrium constant for the reaction is very large) and *the only product* is an aqueous solution of sodium acetate, the same as if sodium acetate were added to water.

Example 6 At the equivalence point of a titration of acetic acid with NaOH, 2.500 mmoles of $NaC_2H_3O_2$ are in 125.0 ml of solution. For the acetate ion, $K_b = 5.71 \times 10^{-10}$. Select an indicator suitable for this titration.

Answer First calculate the pH of the solution at the equivalence point. An indicator that changes color at about this calculated pH value is then selected.

The solution at the equivalence point is identical to a 125.0 ml solution obtained on adding water to 2.500 mmoles $NaC_2H_3O_2$. The reaction of $C_2H_3O_2^-$ with H_2O,

$$C_2H_3O_2^- + H_2O \longrightarrow HC_2H_3O_2 + OH^-$$

then determines the pH of the solution (page 274),

$$[C_2H_3O_2^-] = \frac{2.500 \text{ mmoles}}{125.0 \text{ ml}} = 2.00 \times 10^{-2} \frac{\text{mmole}}{\text{ml}}$$

$$\frac{[HC_2H_3O_2][OH^-]}{[C_2H_3O_2^-]} = K_b = 5.71 \times 10^{-10}$$

$$[HC_2H_3O_2] = [OH^-] = y$$

$$\frac{y^2}{2.00 \times 10^{-2} - y} = \frac{y^2}{2.00 \times 10^{-2}} = 5.71 \times 10^{-10}$$

$$y = [OH^-] = 3.38 \times 10^{-6} \text{ mole/liter}$$

$$[H^+] = \frac{1.00 \times 10^{-14}}{3.38 \times 10^{-6}} = 2.96 \times 10^{-9}$$

$$pH = 8.53$$

The indicator should change color at about pH 9; Table 15.1 (page 301) shows that phenolphthalein is suitable.

ABSORBANCE; COLORIMETRY / 15.4

Photons may be absorbed by matter. Usually, only some of the photons are absorbed. The fraction of photons absorbed depends on

(a) The nature of the absorbing species;
(b) The concentration of the species—the higher the concentration, the more particles are present to absorb the photons;
(c) The length of the radiation path through the material—the longer the path, the larger the number of particles exposed and the greater the probability that a given photon will be absorbed.

ABSORBANCE is defined as

$$A = \log (I_0/I)$$

where I_0 is the intensity (watts/cm²) of the incident radiation and I is the intensity of the transmitted radiation. These relationships, illustrated in Fig. 15.3, are expressed by the equation

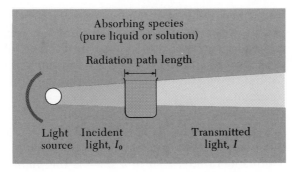

Fig. 15.3. *Representation of light absorption. Incident and transmitted light are compared for a given path length.*

absorbance $\propto$ concentration $\times$ path length

$$A = \epsilon c l \tag{1}$$

where A is absorbance, ϵ is a proportionality constant,* c is the concentration in mole per liter, and l is the path length in centimeters. The proportionality constant is a characteristic physical property of the absorbing species and varies only with the photon frequency. The greater the value of ϵ, the larger is the absorbance. From Equation (1), one can determine the concentrations of solutions of species absorbing photons. We first measure the absorbance, A_s, of a standard solution of known concentration, c_s:

$$A_s = \epsilon c_s l \quad \text{or} \quad \frac{A_s}{c_s} = \epsilon l \tag{2}$$

Then, *at the same wavelength and path length* (so that ϵ and l are constant) we measure the absorbance, A_u, of a solution of unknown concentration, c_u. For the unknown solution

$$A_u = \epsilon c_u l \quad \text{or} \quad \frac{A_u}{c_u} = \epsilon l \tag{3}$$

Since ϵl is the same in Equations (2) and (3),

$$\frac{A_s}{c_s} = \frac{A_u}{c_u} \quad \text{or} \quad c_u = \frac{A_u}{A_s} \times c_s \tag{4}$$

Since A_u and A_s are experimentally determined values and c_s is known, c_u, the unknown concentration, can be calculated from Equation (4). When the absorbing species are colored, this technique is called *colorimetry.*

EQUIVALENT WEIGHT AND NORMALITY / 15.5

The **EQUIVALENT WEIGHT** or an **EQUIVALENT** of a substance is the mass in grams that releases or takes up 1 mole of unit charges (6.02×10^{23} electrons or protons). This definition makes the equivalent weight of a

---

* Called the **MOLAR EXTINCTION COEFFICIENT**.

substance, unlike its atomic or molecular weight, a variable; it depends on the nature of the reaction, so that it is confusing to speak about the equivalent weight of a substance without reference to a specific reaction. The equivalent weight of a substance is determined not by what it is, but by what it does in a reaction. We illustrate with various types of reactions.

Equivalent weights of acids and bases. *One equivalent of an acid is the quantity that transfers (gives up) 1 mole of* H^+ *(1 mole of protons). Correspondingly, 1 eq of a base is the quantity that accepts 1 mole of* H^+. Sodium hydroxide, NaOH, can react with phosphoric acid, H_3PO_4, for example, in any one of three ways:

$$H_3PO_4 + Na^+ + OH^- \longrightarrow Na^+ + H_2PO_4^- + H_2O \qquad (5)$$
$$H_3PO_4 + 2(Na^+ + OH^-) \longrightarrow 2Na^+ + HPO_4^{2-} + 2H_2O \qquad (6)$$
$$H_3PO_4 + 3(Na^+ + OH^-) \longrightarrow 3Na^+ + PO_4^{3-} + 3H_2O \qquad (7)$$

In reaction (5), since 1 mole of H_3PO_4 transfers 1 mole of H^+, there is 1 eq in 1 mole of H_3PO_4. In reaction (6) there are 2 eq, and in reaction (7) there are 3 eq in 1 mole of H_3PO_4 since 1 mole of the acid transfers, respectively, 2 and 3 moles of H^+. The molecular weight of H_3PO_4 is 98.00; the equivalent weights are, therefore,
H_3PO_4, reaction (5):

$$\frac{98.00 \text{ g/mole}}{1 \text{ eq/mole}} = 98.00 \ \frac{\text{g}}{\text{eq}} = 98.00 \ \frac{\text{mg}}{\text{meq}}$$

H_3PO_4, reaction (6):

$$\frac{98.00 \text{ g/mole}}{2 \text{ eq/mole}} = 49.00 \ \frac{\text{g}}{\text{eq}} = 49.00 \ \frac{\text{mg}}{\text{meq}}$$

H_3PO_4, reaction (7):

$$\frac{98.00 \text{ g/mole}}{3 \text{ eq/mole}} = 32.67 \ \frac{\text{g}}{\text{eq}} = 32.67 \ \frac{\text{mg}}{\text{meq}}$$

For NaOH in each of the three reactions, (5), (6), and (7), 1 mole of NaOH accepts one mole of H^+; its equivalent weight is therefore

$$\frac{40.00 \text{ g/mole}}{1 \text{ eq/mole}} = 40.00 \ \frac{\text{g}}{\text{eq}} = 40.00 \ \frac{\text{mg}}{\text{meq}}$$

Equivalent weights of oxidizing and reducing agents. In oxidation–reduction reactions, *1 eq of the oxidizing agent is the quantity that accepts 1 mole of electrons;* correspondingly, *1 eq of a reducing agent is the quantity that loses 1 mole of electrons.* For example, in the reaction between potassium dichromate, $K_2Cr_2O_7$, and sulfur dioxide, SO_2,

$$Cr_2O_7^{2-} + 3SO_2 + 5H^+ \longrightarrow 2Cr^{3+} + 3HSO_4^- + H_2O$$

the partial reactions (page 224) are

reduction: $Cr_2O_7^{2-} + 14H^+ + 6e^- \longrightarrow 2Cr^{3+} + 7H_2O$
oxidation: $SO_2 + 2H_2O \longrightarrow HSO_4^- + 3H^+ + 2e^-$

One mole of $Cr_2O_7^{2-}$ ions accepts 6 moles of electrons, while 1 mole of SO_2 transfers 2 moles of electrons; hence, there are 6 eq in 1 mole of $Cr_2O_7^{2-}$ or in 1 mole of $K_2Cr_2O_7$, 6 eq in 2 moles of Cr^{3+} (3 eq/mole of Cr^{3+}), and 2 eq in 1 mole of SO_2.
Their equivalent weights are then

$$K_2Cr_2O_7 \quad \frac{294 \text{ g/mole}}{6 \text{ eq/mole}} = 49.0 \frac{\text{g}}{\text{eq}}$$

$$Cr_2O_7^{2-} \quad \frac{216 \text{ g/mole}}{6 \text{ eq/mole}} = 36.0 \frac{\text{g}}{\text{eq}}$$

$$SO_2 \quad \frac{64.1 \text{ g/mole}}{2 \text{ eq/mole}} = 32.0 \frac{\text{g}}{\text{eq}}$$

$$Cr^{3+} \quad \frac{52.0 \text{ g/mole}}{3 \text{ eq/mole}} = 17.3 \frac{\text{g}}{\text{eq}}$$

Equivalent weights in ion combinations. In *reactions in which ions combine* to form an insoluble solid (or a soluble but practically nondissociating substance), *the number of equivalents in 1 mole is equal to the total charge of either ion* forming the precipitate (or the weakly dissociated substance). For example, in the reaction

$$Ca^{2+} + SO_4^{2-} \longrightarrow CaSO_4(c)$$

the number of equivalents in 1 mole of $CaSO_4$ is 2, the total charge of either ion forming the precipitate; there are also 2 eq in each mole of Ca^{2+} and SO_4^{2-}. In the reaction

$$2La^{3+} + 3C_2O_4^{2-} \longrightarrow La_2(C_2O_4)_3(c)$$

there are 6 eq $(2 \times 3^+$ or $3 \times 2^-)$ in 1 mole of lanthanum oxalate, $La_2(C_2O_4)_3$; there are also 3 eq in 1 mole of La^{3+} and 2 eq in 1 mole of $C_2O_4^{2-}$.

Example 7 Calculate the equivalent weight of NH_3 in each of the following processes:

(a) $NH_3 + HCl \longrightarrow NH_4^+ + Cl^-$
(b) $NH_3 + 3H_2O \longrightarrow NO_3^- + 9H^+ + 8e^-$

Answer The molecular weight of NH_3 is 17.03 g/mole. (a) Acid–base reaction. One NH_3 molecule accepts 1 proton. Therefore, 1 mole = 1 eq; 17.03 g/eq. (b) This is an oxidation partial reaction. One NH_3 molecule gives up 8 electrons. (The involvement of protons is irrelevant in an oxidation–reduction reaction; we just consider the electrons.) 1 mole = 8 eq:

$$17.03 \frac{g}{\text{mole}} \times \frac{1 \text{ mole}}{8 \text{ eq}} = 2.13 \frac{g}{\text{eq}}$$

What unifies the various definitions of an equivalent is that, in each case, one equivalent supplies or accepts 1 mole of unit charges—in the form of protons (in acid–base reactions), or electrons (in oxidation-reduction reactions), or the ions themselves (in ion combinations). Because electric charge is strictly conserved, the number of moles of charge supplied by one reactant must equal the number of moles of charge accepted by the other reactant. Thus, the system of equivalence is merely a convenient but arbitrary scheme for rebalancing equations such that *1 eq of one substance reacts exactly with 1 eq of another substance to produce 1 eq of each product.* For example, the reaction written as

$$3\text{Ca(OH)}_2(c) + 2\text{H}_3\text{PO}_4 \longrightarrow \text{Ca}_3(\text{PO}_4)_2(c) + 6\text{HOH} \qquad (8)$$
$$\quad 3 \text{ moles} \qquad 2 \text{ moles} \qquad\qquad 1 \text{ mole} \qquad 6 \text{ moles}$$

involves 6 equivalents of each reactant and product; dividing Equation (8) by 6 also yields a balanced equation:

$$\tfrac{1}{2}\text{Ca(OH)}_2(c) + \tfrac{1}{3}\text{H}_3\text{PO}_4 \longrightarrow \tfrac{1}{6}\text{Ca(PO}_4)_2(c) + \text{HOH} \qquad (9)$$
$$\quad 1 \text{ eq} \qquad\qquad 1 \text{ eq} \qquad\qquad 1 \text{ eq} \qquad 1 \text{ eq}$$

Since the coefficients of a balanced equation merely indicate the *relative number* of moles of reactants and products, Equation (9) is as acceptable as Equation (8).

The NORMALITY, N, of a solution is defined as the *number of equivalents per liter of solution.* It is related to the molarity of a solution by the number of equivalents per mole:

$$\text{M} \times \frac{\text{eq}}{\text{mole}} = \text{N}$$

$$\frac{\text{mole}}{\text{liter}} \times \frac{\text{eq}}{\text{mole}} = \frac{\text{eq}}{\text{liter}}$$

Since there are 10^3 milliequivalents (meq) in one equivalent and 10^3 mg in 1 g, the unit mg/meq is equal to g/eq and meq/ml is equal to eq/liter.

Example 8 How many ml of 0.1140 N basic solution are required to react with 28.65 ml of 0.1060 N acid solution?

Answer The number of milliequivalents of acid is

$$28.65 \text{ ml} \times 0.1060 \frac{\text{meq}}{\text{ml}} = 3.037 \text{ meq}$$

Hence, 3.037 meq of base are required. Since the basic solution contains 0.1140 meq/ml, the required volume is given by

$$\frac{3.037 \text{ meq}}{0.1140 \dfrac{\text{meq}}{\text{ml}}} = 26.64 \text{ ml}$$

PRECISION AND ACCURACY / 15.6

The only kind of physical quantity that can be measured with perfect accuracy is a tally of discrete objects, for example, dollars and cents or the number of objects in a museum case. In measuring a quantity capable of continuous variation, for example, mass or length, there is always some uncertainty because the answer cannot be expressed by any finite number of digits. Besides, errors resulting from mistakes made by the experimenter in the construction and use of measuring devices, and other errors over which the experimenter has no control, are inherent in measurements. Therefore, more than one determination of any quantity should be made. The "accepted" value of a quantity is chosen as the most probable value from available data, examined critically for errors.

The PRECISION of a measurement is a statement about the mutual agreement of repeated determinations; it is a measure of the reproducibility of an experiment. The arithmetical average of the values is usually taken as the "best" value. The simplest measure of precision is the AVERAGE DEVIATION, calculated by first determining the average of the series of measurements; then the deviation of each individual measurement from the average is calculated, and finally the deviations, each treated as a positive quantity, are averaged.

Example 9 In a series of determinations, the following values for the electronic charge were obtained: 1.60×10^{-19}, 1.58×10^{-19}, 1.62×10^{-19}, 1.56×10^{-19} C. Calculate the average deviation.

Answer

Individual measurements	Individual deviations from the average
1.60×10^{-19}	0.01×10^{-19}
1.58×10^{-19}	0.01×10^{-19}
1.62×10^{-19}	0.03×10^{-19}
1.56×10^{-19}	0.03×10^{-19}
$4\,)\overline{6.36 \times 10^{-19}}$	$4\,)\overline{0.08 \times 10^{-19}}$
1.59×10^{-19}	0.02×10^{-19}
average 1.59×10^{-19}	average deviation 0.02×10^{-19} C

These results would be reported as $(1.59 \pm 0.02) \times 10^{-19}$ C.

Precise measurements, however, are not necessarily accurate. The ACCURACY expresses the agreement of the measurement with the accepted value of the quantity. Accuracy is expressed in terms of the error, the experimentally determined value minus the accepted value. Significant figures are discussed in Appendix I.14.

Example 10 The accepted value for the electronic charge is 1.60210×10^{-19} C. Calculate the error for the determination of the electronic charge in Example 9.

Answer

$$
\begin{array}{ll}
1.59 \times 10^{-19}\text{ C} & \text{(the determined value)} \\
-1.60 \times 10^{-19}\text{ C} & \text{(the accepted value)} \\
\hline
-0.01 \times 10^{-19}\text{ C} & \text{(the error)}
\end{array}
$$

PROBLEMS

1. Cl⁻ determination. A sample contains 0.0709 g Cl^-; how many ml of 0.10 M $AgNO_3$ should be used for complete precipitation of AgCl?

2. Standard solution. Describe the preparation of 500.0 ml of a 0.02000 M $Cr_2O_7^{2-}$ aqueous solution using $K_2Cr_2O_7$.

3. Titration. 36.72 ml of sulfuric acid reacts with 0.3500 g Na_2CO_3 as shown:

$$CO_3^{2-} + 2H^+ \longrightarrow CO_2 + H_2O$$
$$Na_2CO_3 + H_2SO_4 \longrightarrow Na_2SO_4 + CO_2 + H_2O$$

(a) What is the molarity of the acid? (b) If 20.50 ml of the acid is required to titrate 25.15 ml of NaOH solution, calculate the molarity of the NaOH solution.

$$2NaOH(aq) + H_2SO_4(aq) \rightarrow Na_2SO_4(aq) + 2H_2O$$

4. Titration. A sample contains iron(III) oxide. The oxide is reduced to Fe^{2+}; the Fe^{2+} requires 4.00 ml of 0.0200 M $KMnO_4$ for reaction in an acid solution $(5Fe^{2+} + MnO_4^- + 8H^+ \rightarrow 5Fe^{3+} + Mn^{2+} + 4H_2O)$. Calculate the mass of (a) Fe and (b) of Fe_2O_3 in the sample.

5. Indicators. (a) Select an indicator suitable for determining the pH of solutions with pH around 5. (b) Calculate at 25°C the pH at the equivalence point for the titration of (i) benzoic acid, $HC_7H_5O_2$, with NaOH; (ii) NH_3 with HCl. In each case assume that the concentration of the reaction product at the equivalence point is 0.100 mole/liter. For the benzoate ion $K_b = 1.55 \times 10^{-10}$ and, for the ammonium ion, $K_a = 5.65 \times 10^{-10}$ at 25°C. In each case, select from Table 15.1 (page 301) an indicator suitable for the titration.

6. Volumetric. Find the volume of 0.150 M HI solution required to titrate (a) 30.0 ml of 0.100 M NaOH; (b) 2.16 g of Ag^+ $[Ag^+(aq) + I^-(aq) \rightarrow AgI(c)]$; (c) 190.5 mg Cu^{2+} $[2Cu^{2+}(aq) + 4I^-(aq) \rightarrow 2CuI(c) + I_2(c)]$.

7. Titration. 40.0 ml of NaOH solution reacts with 20.0 ml of 0.150 M sulfuric acid, H_2SO_4, solution forming sodium sulfate, Na_2SO_4. Find the molarity of the NaOH solution.

8. Molarity. A solution contains 59.5 g MnO_4^- in 0.250 liter. The source of MnO_4^- is $KMnO_4$. Calculate the molarity with respect to MnO_4^- and the number of grams of $KMnO_4$ in the solution.

9. Precision and accuracy. Student A determined the molecular weight of an unknown substance in a mass spectrometer (page 37), while student B used a molecular weight analyzer (page 43). The results of replicate experiments are

Student A	Student B
28.046	28.1
28.045	28.4
28.047	28.3
28.047	28.4

For each series, calculate the average ("best") value and the average deviation. If the unknown is ethylene, C_2H_4, calculate the error. Which results are more precise and which are more accurate?

ADDITIONAL PROBLEMS

10. Titration. A sample (0.5760 g) of silver alloy was dissolved and Ag^+ was titrated with 30.20 ml of 0.1000 M Cl^-. [$Ag^+(aq) + Cl^-(aq) \rightarrow AgCl(c)$.] Calculate (a) the mass and (b) the percentage of silver in the alloy.

11. Titration. How many mg of phosphoric acid, H_3PO_4, were in a solution which required 37.60 ml of 0.1200 M NaOH for titration to Na_2HPO_4? [$H_3PO_4(aq) + 2NaOH(aq) \rightarrow Na_2HPO_4(aq) + 2H_2O$.]

12. Titration. Use Fig. 15.2a (page 303) to justify the use of either methyl red or phenolphthalein to determine the end point in the titration of HCl with NaOH. Would the precision of the titration be identical for 0.10 M and for 0.0010 M solutions?

13. Titration. (a) 0.045 g of oxalic acid is titrated with 40.0 ml of $KMnO_4$ solution ($2MnO_4^- + 5H_2C_2O_4 + 6H^+ \rightarrow 2Mn^{2+} + 10CO_2 + 8H_2O$). Calculate the molarity of the permanganate solution. (b) A solution of $Na_2C_2O_4$ is titrated with 20.0 ml of 0.100 M $KMnO_4$ solution. Find the mass of $Na_2C_2O_4$.

14. Volumetric. VINEGAR is an aqueous solution of acetic acid, $HC_2H_3O_2$, with small quantities of nonacidic components. 20.00 ml NaOH is titrated with 32.80 ml 0.0670 M HCl. A 5.00-ml portion of vinegar is diluted with water and titrated with 30.50 ml of the NaOH solution. Calculate (a) the mass of acetic acid, (*i*) per 5 ml, (*ii*) per 100 ml vinegar; (b) the mass percent of acetic acid, if the density of the vinegar is 1.001 g/ml.

15. Analysis. Krypton and fluorine, subjected to an electric discharge at about 90°K and 30 torr, react to form a white solid. The solid hydrolyzes with the formation of HF and Kr; 0.0380 mg of Kr is recovered, and the titration of the fluoride in solution (with thorium nitrate; alizarin sulfonate is used as the indicator) shows the presence of 9.10×10^{-4} mmole F^-. What is the empirical formula of the krypton fluoride?

16. Volumetric. When hydrochloric acid is distilled under a pressure of 760 torr, a distillate containing 20.23% HCl by weight is eventually obtained. If 20.00 g of such a solution is weighed and diluted to 2.000 liters, what is the molarity of the solution?

17. Titration. 12.50 g of sodium thiosulfate, $Na_2S_2O_3$, is dissolved in 500 ml of solution. 30.00 ml of this solution reacts with 23.53 ml of I_2 solution. In turn, 20.11 ml of the I_2 solution is required to titrate a solution containing As_2O_3. Calculate the mass of As_2O_3 in the solution.

$$2Na_2S_2O_3(aq) + I_2(aq) \longrightarrow Na_2S_4O_6(aq) + 2NaI(aq)$$
$$As_2O_3(H_2O)_x(aq) + 5H_2O + 2I_2(aq) \longrightarrow 2H_3AsO_4(aq) + 4HI(aq) + xH_2O$$

18. Precision and accuracy. The International Commission on Atomic Weights has accepted 1.0079 as the atomic weight of hydrogen. Some typical determinations follow: 1.00775, 1.00777, 1.00778, 1.00800. Calculate (a) the average deviation, and (b) the error.

19. Pollution. NH_3 scrubbing is one of the processes being investigated for the removal of SO_2 from stack gases. Analysis of aqueous solutions containing ammonium sulfite is based on the reaction in an acid solution, $SO_3^{2-} + I_2 + H_2O \rightarrow SO_4^{2-} + 2H^+ + 2I^-$. 50 ml of solution is titrated with 2.01 ml 0.10 M I_2 solution. Find the mass in mg of (a) SO_3^{2-} in solution, (b) SO_2 absorbed.

20. Normality. Given a 0.10 M $KMnO_4$ solution. Find its normality with respect to each of the following reactions or partial reactions.

(a) $MnO_4^- + H^+ \rightarrow HMnO_4$
(b) $2MnO_4^- + Ca^{2+} \rightarrow Ca(MnO_4)_2(c)$
(c) $MnO_4^- + 8H^+ + 5e^- \rightarrow Mn^{2+} + 4H_2O$
(d) $MnO_4^- + 1e^- \rightarrow MnO_4^{2-}$
(e) $MnO_4^- + 4H^+ + 3e^- \rightarrow MnO_2(c) + 2H_2O$

21. Normality. 45.0 ml of NaOH solution reacts with 15.0 ml of 0.1000 N sulfuric acid forming Na_2SO_4. Find the normality of the NaOH solution.

22. Colorimetry. What is the effect on the absorbance of doubling the concentration of a solution? What is the effect on the fraction (I/I_0) of the incident radiation transmitted?

23. Normality. 30.0 ml of HCl solution reacts with 45.0 ml of 0.200 N Na_2CO_3 solution ($2HCl + Na_2CO_3 \rightarrow 2NaCl + H_2CO_3$). Find the (a) normality and molarity of the HCl solution; (b) molarity of the Na_2CO_3 solution.

24. Normality. Calculate the mass in mg of Fe^{2+} oxidized by 26.0 ml of 0.150 N $K_2Cr_2O_7$ in acid solution. Find the molarity of the $K_2Cr_2O_7$ solution ($6Fe^{2+} + Cr_2O_7^{2-} + 14H^+ \rightarrow 6Fe^{3+} + 2Cr^{3+} + 7H_2O$).

25. Colorimetry. The complex ion $Co(NH_3)_6^{3+}$ absorbs radiation with wavelength near 4400 Å. It is found that, in a certain apparatus, a 0.100 M solution of $Co(NH_3)_6(NO_3)_3$ absorbs 70.0% and transmits 30.0% of the radiation incident on it. An unknown solution of the same compound, in the same apparatus, absorbs 40.0% of the incident radiation. Calculate the molarity of the unknown solution.

16

GALVANIC CELLS

We saw in Chapter 12 that an electric current can cause chemical reactions. The process can be reversed, and electricity obtained from a chemical reaction. This discovery, by Alessandro Volta in 1796, provided for the first time a source of continuous electric current, and thus made possible the great electrical and electrochemical discoveries of the early nineteenth century. Today, the electrochemical cell, or battery of cells, has been generally superseded by the generator as a source of electricity. However, cells still have a place as portable sources of small amounts of power. They may be making a comeback in the form of fuel cells (page 352). What is more important for chemistry is that the electrical work that can be obtained from a chemical reaction provides a measure of the driving force of the reaction and of its equilibrium constant.

One of the most familiar chemical reactions is the reaction between zinc and a soluble copper salt,

$$Zn(c) + Cu^{2+}(aq) \longrightarrow Cu(c) + Zn^{2+}(aq)$$

This reaction is exothermic ($\Delta H = -52$ kcal). Merely adding a piece of zinc to a $CuSO_4$ solution results in the liberation of heat, but yields no work. The liberated heat can be partially converted to work (as in a steam engine), but this process is notoriously inefficient and does not help us to measure the maximum available work—the quantity in which we are especially interested.

ELECTRICITY FROM A CHEMICAL REACTION / 16.2

To see how a reaction can be harnessed to produce electrical work directly, we first recall that an oxidation–reduction reaction can be

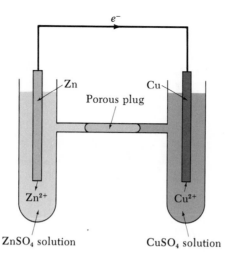

Fig. 16.1. *A galvanic cell with liquid junction.*

resolved into two partial reactions (page 224):

$$Zn(c) \longrightarrow Zn^{2+}(aq) + 2e^- \qquad \text{(oxidation)} \qquad (1)$$
$$Cu^{2+}(aq) + 2e^- \longrightarrow Cu(c) \qquad \text{(reduction)} \qquad (2)$$

If we arrange for these partial reactions to occur in two physically separated places, then the electrons will have to flow from the place where the oxidation occurs (the ANODE) to the place where the reduction occurs (the CATHODE). This flow of electrons constitutes a current, which can be used to do work or light a bulb.

An apparatus in which chemical energy is converted to electrical energy is known as a VOLTAIC CELL or, more often, a GALVANIC CELL, after Luigi Galvani, whose discovery in 1786 of the effects of electricity on frog muscles prepared the way for Volta's work.

The simplest way to separate the partial reactions is to insert the Zn into a solution that does not contain Cu^{2+} and to insert a piece of Cu into a solution containing Cu^{2+} (Fig. 16.1). Specifically, we insert the Zn into a $ZnSO_4$ solution and, in a separate beaker, the Cu into a $CuSO_4$ solution.

A metal consists of ions (Zn^{2+}, Cu^{2+}) and electrons (e^-). The electrons can wander freely through the metal (page 368). We can imagine each of the partial reactions (1) and (2) taking place in either direction. Suppose that the process $Zn(c) \rightarrow Zn^{2+}(aq) + 2e^-$ occurs. The Zn^{2+} ions give the solution a positive charge. The electrons cannot go into the solution—it is not a metal—and they give the zinc a negative charge. These charge buildups stop the process before any detectable amount of Zn has reacted. The same reasoning tells us that the partial reaction $Cu^{2+}(aq) + 2e^- \rightarrow Cu(c)$ also cannot get anywhere. Neither can the reverse reactions. How can we make the cell generate a current? We must do two things:

(a) We connect the Zn and Cu electrodes by a wire (a metallic conductor). The electrons left behind on the anode can then flow through the wire to the cathode and reduce the Cu^{2+} to Cu.

(b) The two solutions must be connected by placing them in contact without allowing them to mix. The positive charge produced in the anode solution by generating Zn^{2+} and the negative charge produced in the cathode solution by removing Cu^{2+} can then cancel each other by the migration of ions from one solution to the other.

Why do the electrons flow from the zinc to the copper, not the other way? We could ask the same question about any chemical reaction: Why does it go this way and not that way? There is a certain "driving force" for any reaction. The driving force for the partial reaction $Zn \rightarrow Zn^{2+} + 2e^-$ is greater than the driving force for the partial reaction $Cu \rightarrow Cu^{2+} + 2e^-$. The first process thus wins and drives the second backwards ($Cu^{2+} + 2e^- \rightarrow Cu$). The driving-force idea will be made more precise as we go through this chapter and the next. We will see that the equilibrium constant is related to the driving force of the reaction.

The meeting place of two solutions is called a LIQUID JUNCTION. There are some advantages in connecting two solutions via a third solution, usually a concentrated solution of KCl or NH_4NO_3. Such a solution is called a SALT BRIDGE. A Zn–Cu cell, modified by inclusion of a salt bridge, is shown in Fig. 16.2.

Figure 16.3 shows an example of a cell with only one solution, needing no salt bridge. Two electrodes are immersed in a solution of $ZnCl_2$. One is a piece of zinc, and the other is a piece of silver with a porous coating of silver chloride. Because of the very low solubility of AgCl, the solution contains only a trace of Ag^+ ion. The partial reactions

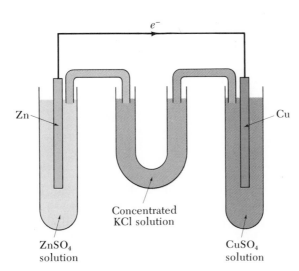

Fig. 16.2. *A galvanic cell with a salt bridge.*

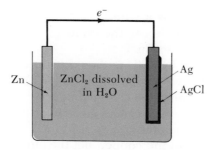

Fig. 16.3. *A galvanic cell without liquid junction.*

are

$$Zn \longrightarrow Zn^{2+} + 2e^-$$
$$2AgCl + 2e^- \longrightarrow 2Ag + 2Cl^-$$

The overall reaction is

$$Zn(c) + 2AgCl(c) \longrightarrow 2Ag(c) + Zn^{2+} + 2Cl^-$$

ELECTRICAL WORK / 16.3

When 1 mole of Zn is converted to Zn^{2+}, 2 faradays (page 228) of charge, or $2 \times 9.65 \times 10^4$ C, pass through the external circuit. The quantity of work that can be obtained as a given amount of charge passes from one point to another is determined by the DIFFERENCE OF POTENTIAL between the two points. Potential is measured in VOLTS (V):

$$\begin{array}{ccc} \text{work} & = & \text{difference of potential} & \times & \text{charge} \\ \text{(in joules)} & & \text{(in volts)} & & \text{(in coulombs)} \end{array}$$

The potential difference between the two electrodes in a galvanic cell is called the ELECTROMOTIVE FORCE, or EMF, of the cell. The emf of the cell in Fig. 16.1 depends on the temperature and the concentrations of the $ZnSO_4$ and $CuSO_4$ solutions. Let us assume that the emf is 1.10 V. This means that when 1 mole of zinc and 1 mole of Cu^{2+} are consumed, and 2 faradays flow through the circuit, the work this electricity will do is

$$1.10 \text{ V} \times 2 \text{ faradays} \times \frac{9.65 \times 10^4 \text{ C}}{\text{faraday}} = 2.12 \times 10^5 \text{ joules}$$

$$= 2.12 \times 10^5 \text{ J} \times \frac{1 \text{ cal}}{4.184 \text{ J}}$$

$$= 5.07 \times 10^4 \text{ cal}$$

A difference of potential is often compared to a difference of hydrostatic pressure. If we have two tanks of liquid at different levels, the liquid can do work in flowing from the higher to the lower tank. The difference of gravitational potential between the two tanks is the work available per gram. In the same way, the difference of electrical potential is the work available per coulomb.

Not all the energy theoretically available from a chemical reaction can actually be realized as work. Some of this energy appears as heat because of the resistance of the solution and of the wires. We are interested, however, not so much in obtaining work from the reaction as in measuring the *maximum* work that can be obtained, for this is what measures the driving force. To find this maximum work, we must measure the potential difference between the electrodes *when no current is flowing through the cell*. The device used for this measurement is called a **POTENTIOMETER**.

We shall not describe the details of the potentiometer. We may regard it as a galvanic cell, the emf of which can be varied continuously and read from a dial. The potentiometer is connected so that it tries to make a current flow through the cell in the direction opposite to the current generated by the cell reaction. The potentiometer is then adjusted until the current is zero. At this point, the known emf of the potentiometer is equal to the emf of the cell, measured with no current flowing. The emf thus determined is a measure of the ideal maximum amount of work that can be obtained from the chemical reaction taking place in the cell.*

CONVENTIONAL NOTATION FOR CELLS / 16.5

One often wishes to describe a galvanic cell without taking the trouble to draw an actual picture of it. The notation used is a rudimentary diagram of the cell. Phase boundaries (between solid and liquid, liquid and gas, or solid and gas) are indicated by |; a liquid junction is represented by ⋮, a salt bridge by ‖. The cells previously discussed may be represented thus (with spaces for filling in the concentrations):

Fig. 16.1: Zn|ZnSO$_4$(___M)⋮CuSO$_4$(___M)|Cu
Fig. 16.2: Zn|ZnSO$_4$(___M)‖CuSO$_4$(___M)|Cu
Fig. 16.3: Zn|ZnCl$_2$(___M)|AgCl|Ag

In the cell description, the *anode* is written on the *left* and the *cathode* on the *right*, a rule that may be remembered by observing the alphabetical order in each of the following pairs: *anode–cathode, oxidation–reduction, left–right*. If we were to write the reaction backwards (Cu + Zn^{2+} → Cu^{2+} + Zn), the cell description would also be reversed: Cu|Cu^{2+}‖Zn^{2+}|Zn. We may not know the *actual* direction of the reaction; the description simply follows the way we choose to write the reaction, or if the description is written first, the equation for the reaction follows the way we write the description.

---

* This zero-current emf is called the **REVERSIBLE EMF**. The reason for the word "reversible" will appear in the next chapter (page 336).

In a galvanic cell, electrons are left behind on the anode by the oxidation (loss of electrons) taking place there. The *anode* pushes these electrons into the external circuit and is therefore regarded as the *negative* electrode—the source of negative charge. The *cathode* sucks electrons in from the external circuit—they are in demand for the reduction process—and is therefore called the *positive* electrode. Dry cells (page 327) are marked "Center terminal is positive"; this means that the center terminal is the cathode.

In an *electrolysis* cell, the cathode still receives the electrons, but this time we are forcibly feeding electrons into the cathode by connecting it to the negative terminal (the electron source) of a generator or a galvanic cell. Therefore, we think of the *cathode* of the electrolysis cell as the *negative* electrode. Electrons are being sucked out of the anode by the positive terminal of the power source, and the *anode* is therefore considered *positive*.

There is a common misconception that "anode" always means "positive electrode" and "cathode" always means "negative electrode." The familiar example of the dry cell shows how wrong this is.

Oxidation takes place at the anode and reduction at the cathode, in a cell of either type. The only difference is in our decision as to what we will call positive and what we will call negative. A table will summarize the usages described in this section:

	Anode	Cathode
Galvanic cell (power source) (e.g., storage cell being discharged)	−, Oxidation	+, Reduction
Electrolysis cell (power consumer) (e.g., storage cell being charged)	+, Oxidation	−, Reduction

THE NERNST EQUATION / 16.7

The emf of a cell measures the driving force of the cell reaction. For the cell of Fig. 16.1, the reaction is

$$Zn(c) + Cu^{2+} \rightleftharpoons Cu(c) + Zn^{2+}$$

Suppose that this reaction has come to equilibrium. According to Le Chatelier's principle (page 200), increasing the concentration of Cu^{2+} shifts the equilibrium to the right, and therefore must have increased the driving force of the left-to-right ("forward") reaction. *The greater is the concentration of a reactant* (left side of equation), *the greater is the driving force of the reaction.* This is still true when the reaction is not in equilibrium—and it has not reached equilibrium as long as the cell can deliver a current. Therefore, increasing the concentration of Cu^{2+} increases the emf of the cell. Conversely, increasing the concentration of Zn^{2+} decreases the driving force of the left-to-right reaction, and therefore decreases the emf of the cell. However, the emf does not depend on how much solid Cu or Zn is present.

The quantitative relation between the emf and the concentrations is given by the NERNST EQUATION (Walther Nernst, 1889), which, for the cell under discussion, is

$$\mathscr{E} = \mathscr{E}^0 - \frac{2.303RT}{2\mathscr{F}} \log \frac{[Zn^{2+}]}{[Cu^{2+}]}$$

$\mathscr{E}$ is the emf of the cell in volts; $\mathscr{E}^0$ is the STANDARD EMF of the cell, a constant at a given temperature for a given cell; R is the gas constant, 8.314 J/deg-mole; T is the absolute temperature (°K); $\mathscr{F}$ is Faraday's constant, 9.6487×10^4 C/faraday; and the logarithm is to base 10. The reader should verify from the equation that changes in the concentrations have the effects described previously.

The general form of the Nernst equation is

$$\mathscr{E} = \mathscr{E}^0 - \frac{2.303RT}{n\mathscr{F}} \log Q$$

where n is the number of moles of electrons (faradays) transferred through the external circuit when the cell reaction takes place as written; Q is the reaction quotient (page 195) for the reaction as written. In the Zn–Cu cell, $n = 2$. The reaction quotient is

$$Q = \frac{[Zn^{2+}]}{[Cu^{2+}]}$$

The solids are, as usual, omitted.

For the Zn–Ag–AgCl cell (Fig. 16.3), the Nernst equation is

$$\mathscr{E} = \mathscr{E}^0 - \frac{2.303RT}{2\mathscr{F}} \log([Zn^{2+}][Cl^-]^2)$$

When gases are involved in the cell reaction, we write the *partial pressure* of each gas (in atmospheres) instead of its concentration. For the reaction

$$Zn + 2H^+ \longrightarrow Zn^{2+} + H_2(g)$$

we write

$$Q = \frac{[Zn^{2+}]p_{H_2}}{[H^+]^2}$$

The relation between partial pressure and concentration is simple (page 195), but the numerical value of Q depends on which is used.

When $Q = 1$, then $\log Q = 0$, and $\mathscr{E} = \mathscr{E}^0$. The standard emf of the cell is the emf when the concentration of each solute species—molecule or ion—appearing in the cell reaction is 1 mole/liter, and each pressure is 1 atm if there are gases. $\mathscr{E}^0$ may be thought of as the emf of the cell under ordinary, average laboratory conditions.

When $T = 298.15°K$ (25°C), then

$$\frac{2.303RT}{\mathscr{F}} = \frac{2.303 \times 8.314 \frac{\text{joules}}{°K\text{ mole}} \times 298.15°K}{9.6487 \times 10^4 \frac{C}{\text{faraday}}}$$

$$= 0.05916 \frac{\text{joule faraday}}{C\text{ mole}} \quad \text{or} \quad \frac{V\text{ faraday}}{\text{mole}}$$

At 25°C, the Nernst equation is thus

$$\mathscr{E} = \mathscr{E}^0 - \frac{0.05916}{n} \log Q$$

Example 1 For the cell

$$Zn(c) \,|\, ZnCl_2(aq,\ 0.100\ M) \,\|\, Cu(NO_3)_2(aq,\ 0.200\ M) \,|\, Cu$$

$\mathscr{E}^0 = 1.1030$ V at 25°C. Calculate $\mathscr{E}$. (See Appendix I.15 for calculations with logarithms.)

Answer

$$\mathscr{E} = \mathscr{E}^0 - \frac{2.303RT}{2\mathscr{F}} \log \frac{[Zn^{2+}]}{[Cu^{2+}]}$$

$$= 1.1030 - \frac{0.05916}{2} \log \left(\frac{0.100}{0.200}\right)$$

$$= 1.1030 + \frac{0.05916}{2} \log 2.00$$

$$= 1.1030 + 0.0089$$

$$= 1.1119\ V$$

We note that $\mathscr{E}$ and $\mathscr{E}^0$ are not greatly different. This is the usual case with ordinary laboratory concentrations in the range 0.01 to 10 M.

Example 2 For the cell $Sn(c) \,|\, SnCl_2(0.10\ M) \,|\, AgCl(c) \,|\, Ag(c)$, the reaction is

$$Sn + 2AgCl \longrightarrow 2Ag + Sn^{2+} + 2Cl^-$$

$\mathscr{E}^0 = 0.359$ V at 25°C. Write the Nernst equation and calculate $\mathscr{E}$.

Answer To find n, we write the partial equations:

$$Sn \longrightarrow Sn^{2+} + 2e^-, \qquad AgCl + e^- \longrightarrow Ag + Cl^-$$

The equation for the overall reaction is obtained by doubling the second partial equation and adding. There are therefore 2 electrons in each partial equation; $n = 2$. The reaction quotient is $Q = [Sn^{2+}][Cl^-]^2$. The Nernst equation is

$$\mathscr{E} = \mathscr{E}^0 - \left(\frac{0.05916}{2}\right) \log[Sn^{2+}][Cl^-]^2$$

In a 0.10 M solution of $SnCl_2$, $[Sn^{2+}] = 0.10$ and $[Cl^-] = 0.20$ mole/liter. Therefore,

$$\mathscr{E} = \mathscr{E}^0 - \left(\frac{0.05916}{2}\right) \log\left[(0.10)(0.20)^2\right]$$

$$= 0.359 - \left(\frac{0.05916}{2}\right) \log(4.0 \times 10^{-3})$$

$$= 0.359 - \frac{0.05916}{2}\left[\log 4.0 + \log(10^{-3})\right]$$

$$= 0.359 - \frac{0.05916}{2}(0.60 - 3)$$

$$= 0.359 + 0.071$$
$$= 0.430 \text{ V}$$

The cell reaction in Example 2 could have been written, just as correctly,

$$\tfrac{1}{2}Sn + AgCl \longrightarrow Ag + \tfrac{1}{2}Sn^{2+} + Cl^-$$

The emf ($\mathscr{E}$ or $\mathscr{E}^0$) is unaffected when the coefficients in the equation are multiplied by a constant. $\mathscr{E}$ is work *per coulomb*, regardless of the number of moles appearing in the equation. The reader can verify (Problem 12, page 332) that the Nernst equation is unaffected by changing the chemical equation in this way. When the equation is reversed, $\mathscr{E}$ and $\mathscr{E}^0$ change sign. If $\mathscr{E}$ comes out to be *negative*, it means that, under the given conditions, the cell reaction is *opposite* to the reaction we assumed.

Example 3 Find $\mathscr{E}$ for the cell

Cu | $CuCl_2$(0.50 M) | Cl_2(1.50 atm) | Pt

given that $\mathscr{E}^0 = +1.018$ V at 25°C.

Answer The anode reaction is the oxidation of Cu, $Cu \rightarrow Cu^{2+} + 2e^-$
The cathode reaction is the reduction of Cl_2, $Cl_2 + 2e^- \rightarrow 2Cl^-$
The overall cell reaction is $Cu + Cl_2 \rightarrow Cu^{2+} + 2Cl^-$ and $n = 2$.

The Nernst equation is

$$\mathscr{E} = \mathscr{E}^0 - \left(\frac{0.05916}{2}\right) \log\left(\frac{[Cu^{2+}][Cl^-]^2}{p_{Cl_2}}\right)$$

Here, $[Cu^{2+}] = 0.50$ mole/liter, $[Cl^-] = 1.00$ mole/liter, and $p_{Cl_-} = 1.50$ atm. Thus

$$\mathscr{E} = \mathscr{E}^0 - \left(\frac{0.05916}{2}\right) \log\left(\frac{0.50 \times 1.00^2}{1.50}\right)$$

$$= 1.018 - \frac{0.05916}{2} \log\left(\frac{1}{3}\right)$$

$$= 1.018 + 0.014 = 1.032 \text{ V}$$

A galvanic cell always requires two electrodes. Oxidation must occur at one electrode (the anode) and reduction at the other (the cathode). There is no way to measure the potential difference between an electrode and a solution without introducing another electrode; what we always measure is a potential difference between two electrodes, which is the emf of an entire cell.

However, it is useful, as a bookkeeping device, to think of a cell as consisting of two half-cells, with each half-cell contributing its share to the emf of the entire cell. A half-cell consists of an electrode and the solution with which it is in contact. *It is possible to assign a potential to each half-cell*, and to calculate from these potentials the emf of the entire cell. Instead of tabulating the standard emf $\mathscr{E}^0$ for every cell, we need tabulate it only for every half-cell.

We can choose any half-cell arbitrarily, call its standard potential zero, and measure the standard potentials of all other half-cells by combining them with this reference half-cell. *The half-cell chosen as the reference is* $Pt \mid H_2(g, 1.0 \text{ atm}) \mid H^+(aq, 1.0 \text{ M})$. The negative ion (X^-) that goes with H^+ can be any ion for which HX is a strong acid: Cl^-, Br^-, I^-, HSO_4^-, ClO_4^-, NO_3^-. See Fig. 16.4.

The standard potential, $\mathscr{E}^0$, of any other half-cell is defined as the standard emf of the cell obtained by combining this half-cell, on the right, with the reference H_2 half-cell on the left. This means we assume that the reference half-cell is the anode[*] $(H_2 \rightarrow 2H^+ + 2e^-)$ and the other half-cell is the cathode. If our assumption is correct—that is, if the reference H_2 half-cell is really the anode—then $\mathscr{E}^0$ is given a $+$ sign. If, on the other hand, our assumption is incorrect—that is, if the reference H_2 half-cell is really the cathode—then $\mathscr{E}^0$ is negative.

Suppose that we want to find the standard potential for the half-cell $Hg^{2+} \mid Hg$. By definition, this potential is $\mathscr{E}^0$ for the cell

$$Pt \mid H_2 \mid H^+(aq) \parallel Hg^{2+} \mid Hg$$

Measurement with a potentiometer gives $\mathscr{E}^0 = 0.851$ V. The cell reaction is

$$H_2(g) + Hg^{2+}(aq) \longrightarrow Hg(l) + 2H^+(aq)$$

The reaction really goes in the direction written. That is why we consider the emf positive. The $H_2 \mid H^+$ half-cell is the anode and the $Hg^{2+} \mid Hg$ half-cell is the cathode. Thus, we say that $\mathscr{E}^0 = +0.851$ V for the half-cell $Hg^{2+} \mid Hg$ or for the partial reaction

$$Hg^{2+} + 2e^- \longrightarrow Hg$$

[*] The words "anode" and "cathode" apply strictly to the electrodes (the metallic conductors). However, it is common practice to refer to the half-cell (electrode + solution) in which oxidation occurs as the anode, and the half-cell in which reduction occurs as the cathode.

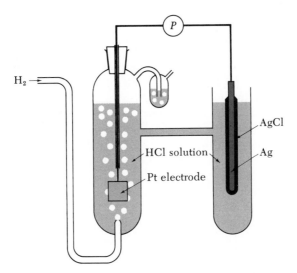

We now find the standard potential for the half-cell $Zn^{2+}(aq)$ | Zn. This is the same as $\mathscr{E}^0$ for the cell

$$Pt \mid H_2(g) \mid H^+(aq) \parallel Zn^{2+}(aq) \mid Zn$$

If we take a potentiometer and measure the emf, we have to reverse the connections. The electrons are trying to come out of the Zn, not the Pt. Therefore, the H_2 | H^+ half-cell is really the *cathode* and the Zn^{2+} | Zn half-cell is the anode. The cell should be written

$$Zn \mid Zn^{2+}(aq) \parallel H^+(aq) \mid H_2(g) \mid Pt$$

for which the cell reaction is

$$2H^+ + Zn \longrightarrow H_2 + Zn^{2+}$$

The half-cell potential for Zn^{2+} | Zn is therefore given a negative sign: $\mathscr{E}^0 = -0.7628$ V. *A negative potential is assigned to any half-cell that acts as an anode when paired with the reference half-cell.*

Table 16.1 gives the standard potentials for a number of common half-cells. We calculate cell emf's from these potentials as follows: For the *reduction,* copy the potential given in Table 16.1, where the partial reactions are written as *reductions.* For the *oxidation,* copy the potential and *change its sign,* for we want the potential corresponding to the *reversed* partial reaction. Then *add* the potentials algebraically. The procedure is summarized by the equation

$$\mathscr{E}^0_{cell} = \mathscr{E}^0_{reduction} + (-\mathscr{E}^0_{oxidation})$$

or

$$\mathscr{E}^0_{cell} = \mathscr{E}^0_{cathode} + (-\mathscr{E}^0_{anode})$$

or

$$\mathscr{E}^0_{\text{cell}} = \mathscr{E}^0_{\text{right}} + (-\mathscr{E}^0_{\text{left}})$$

Note that the $\mathscr{E}^0$'s in the equations are the numbers given in Table 16.1. The *only* sign changes to be made are those specified by the minus signs in the equations.*

The half-cell potentials are usually called **ELECTRODE POTENTIALS** because they specify the potential of the electrode (the metal) relative to the solution.† The proof that the calculation of an emf from electrode potentials is legitimate is left to the reader (Problem 14, page 333).

Example 4 Find $\mathscr{E}^0$ at 25°C for the cell Cu | CuCl$_2$(aq) | Cl$_2$(g) | Pt, in which the reaction is

$$Cu(c) + Cl_2(g) \longrightarrow Cu^{2+} + 2Cl^-$$

Answer First we write the anode and cathode partial reactions. Beside the oxidation, we write $\mathscr{E}^0$ from Table 16.1, and then reverse its sign:

$$Cu \longrightarrow Cu^{2+} + 2e^- \qquad -(+0.3402)$$

We use this awkward form rather than just plain -0.3402 to remind ourselves that the sign has been changed and we must not change it again. Beside the reduction, we write $\mathscr{E}^0$ from Table 16.1 with no change of sign:

$$Cl_2 + 2e^- \longrightarrow 2Cl^- \qquad +1.3583$$

Then we calculate $\mathscr{E}^0$ for the cell by adding the half-cell potentials algebraically:

$$\mathscr{E}^0 = +1.3583 - 0.3402$$
$$= +1.0181 \text{ V}$$

The plus sign indicates that the left electrode is indeed the anode and the reaction goes as written when $\mathscr{E} = \mathscr{E}^0$.

Example 5 Find $\mathscr{E}^0$ at 25°C for a cell in which the reaction is

$$2Ag^+ + Cd(c) \longrightarrow 2Ag(c) + Cd^{2+}$$

Answer We write the partial equations, look up their $\mathscr{E}^0$'s in Table 16.1, record them with a minus in front of the potential for oxidation, and add algebraically:

$2Ag^+ + 2e^- \longrightarrow 2Ag$	(reduction)	$+0.7996$
$Cd \longrightarrow Cd^{2+} + 2e^-$	(oxidation)	$-(-0.4026)$
$2Ag^+ + Cd \longrightarrow 2Ag + Cd^{2+}$		$\mathscr{E}^0 = 1.2022 \text{ V}$

---

* Reversing the sign and adding is, of course, equivalent to subtracting. If you prefer to think of the process as a subtraction, be sure to subtract the $\mathscr{E}^0$'s given in the table, *without change of sign.*

† If we chose a different reference half-cell, all the electrode-solution potential differences would be changed by the same additive constant. The constant makes no difference in the calculation of a cell emf because it would cancel out (Problem 13, page 332).

Table 16.1
Standard Electrode (Reduction) Potentials in Aqueous Solution at 25°C[a]

Element	Partial reaction	$\mathscr{E}^0$, volts
Ag	$Ag^+ + e^- \rightleftharpoons Ag$	+0.7996
Ag	$AgBr(c) + e^- \rightleftharpoons Ag + Br^-$	+0.0713
Ag	$AgCl(c) + e^- \rightleftharpoons Ag + Cl^-$	+0.2223
Ag	$Ag_2CrO_4(c) + 2e^- \rightleftharpoons 2Ag + CrO_4^{2-}$	+0.4463
Ag	$AgI(c) + e^- \rightleftharpoons Ag + I^-$	-0.1519
Ag	$Ag(NH_3)_2^+ + e^- \rightleftharpoons Ag + 2NH_3(aq)$	+0.373
Ag	$Ag(S_2O_3)_2^{3-} + e^- \rightleftharpoons Ag + 2S_2O_3^{2-}$	+0.015
Al	$Al^{3+} + 3e^- \rightleftharpoons Al$	-1.67
Au	$Au^+ + e^- \rightleftharpoons Au$	+1.7
Au	$Au(CN)_2^- + e^- \rightleftharpoons Au + 2CN^-$	-0.60
Au	$Au^{3+} + 3e^- \rightleftharpoons Au$	+1.42
Ba	$Ba^{2+} + 2e^- \rightleftharpoons Ba$	-2.90
Br	$Br_2(l) + 2e^- \rightleftharpoons 2Br^-$	+1.0650
Ca	$Ca^{2+} + 2e^- \rightleftharpoons Ca$	-2.76
Cd	$Cd^{2+} + 2e^- \rightleftharpoons Cd$	-0.4026
Cl	$Cl_2(g) + 2e^- \rightleftharpoons 2Cl^-$	+1.3583
Cr	$Cr^{3+} + 3e^- \rightleftharpoons Cr$	-0.46
Cr	$HCrO_4^- + 7H^+ + 3e^- \rightleftharpoons Cr^{3+} + 4H_2O$	+1.195
Cu	$Cu^{2+} + 2e^- \rightleftharpoons Cu$	+0.3402
F	$F_2(g) + 2e^- \rightleftharpoons 2F^-$	+2.87
Fe	$Fe^{2+} + 2e^- \rightleftharpoons Fe$	-0.409
Fe	$Fe^{3+} + 3e^- \rightleftharpoons Fe$	-0.016
Fe	$Fe^{3+} + e^- \rightleftharpoons Fe^{2+}$	+0.7701
H	$2H^+ + 2e^- \rightleftharpoons H_2(g)$	0.0000
H	$2H_2O + 2e^- \rightleftharpoons H_2(g) + 2OH^-$	-0.8277
Hg	$Hg_2^{2+} + 2e^- \rightleftharpoons 2Hg$	+0.7961
Hg	$Hg_2Cl_2(c) + 2e^- \rightleftharpoons 2Hg + 2Cl^-$	+0.2682
Hg	$Hg^{2+} + 2e^- \rightleftharpoons Hg$	+0.851
I	$I_2(c) + 2e^- \rightleftharpoons 2I^-$	+0.535
K	$K^+ + e^- \rightleftharpoons K$	-2.9241
Li	$Li^+ + e^- \rightleftharpoons Li$	-3.045
Mg	$Mg^{2+} + 2e^- \rightleftharpoons Mg$	-2.375
Mn	$MnO_4^- + 8H^+ + 5e^- \rightleftharpoons Mn^{2+} + 4H_2O$	+1.491
Na	$Na^+ + e^- \rightleftharpoons Na$	-2.7109
Ni	$Ni^{2+} + 2e^- \rightleftharpoons Ni$	-0.23
O	$O_2(g) + 4H^+ + 4e^- \rightleftharpoons 2H_2O$	+1.229
O	$O_2(g) + 2H_2O + 4e^- \rightleftharpoons 4OH^-$	+0.401
Pb	$Pb^{2+} + 2e^- \rightleftharpoons Pb$	-0.1263
Pb	$PbSO_4(c) + 2e^- \rightleftharpoons Pb + SO_4^{2-}$	-0.3563
Pb	$PbO_2(c) + 3H^+ + HSO_4^- + 2e^- \rightleftharpoons PbSO_4(c) + 2H_2O$	+1.685
Pt	$Pt^{2+} + 2e^- \rightleftharpoons Pt$	+1.2
Sn	$Sn^{2+} + 2e^- \rightleftharpoons Sn$	-0.1364
V	$V^{3+} + e^- \rightleftharpoons V^{2+}$	-0.255
Zn	$Zn^{2+} + 2e^- \rightleftharpoons Zn$	-0.7628

[a] Arranged alphabetically by chemical symbols.

The standard electrode potentials in Table 16.1 are sometimes called **REDUCTION POTENTIALS** because they measure the tendency of a partial reaction to occur as a reduction.* The more *positive* the potential, the more the electrode tends to behave as a *cathode;* the more *negative* the potential, the more the electrode tends to be an *anode.* The partial reaction with the more positive potential has the greater driving force to be a reduction; the other partial reaction is pushed backwards as an oxidation. The most positive potentials belong to the partial reactions in which the bext oxidizing agents are reduced; their reduced forms are the poorest reducing agents. For example:

$$F_2 + 2e^- \rightleftharpoons 2F^-,$$
$$\mathscr{E}^0 = +2.87 \text{ V}$$

Reaction has most positive $\mathscr{E}^0$

Reaction has greatest tendency to proceed as a reduction

F_2 is the best oxidizing agent in the table

The reverse reaction (oxidation) has the least tendency to go

F^- is the poorest reducing agent

$$Li^+ + e^- \rightleftharpoons Li,$$
$$\mathscr{E}^0 = -3.05 \text{ V}$$

Reaction has most negative $\mathscr{E}^0$

Reaction has least tendency to proceed as a reduction

Li^+ is the poorest oxidizing agent in the table

The reverse reaction (oxidation) has the greatest tendency to go

Li is the best reducing agent

Consider the electrolysis, at Pt electrodes, of water that contains some sulfuric acid to make it a conductor. The partial reactions are

$$2H_2O \longrightarrow O_2(g) + 4H^+ + 4e^- \quad \text{(oxidation)}$$
$$4H^+ + 4e^- \longrightarrow 2H_2(g) \quad \text{(reduction)}$$
$$\overline{2H_2O \longrightarrow 2H_2(g) + O_2(g)}$$

As soon as current starts to flow, the liberation of H_2 and O_2 converts the apparatus to the galvanic cell (Fig. 16.5)

$$\text{Pt} \mid O_2 \mid H^+ \mid H_2 \mid \text{Pt}$$

for which we calculate the emf:

$$2H_2O \longrightarrow O_2 + 4H^+ + 4e^- \quad -(+1.229 \text{ V})$$
$$4H^+ + 4e^- \longrightarrow 2H_2 \quad \quad 0$$
$$\overline{2H_2O \longrightarrow 2H_2 + O_2} \quad \overline{\mathscr{E} \approx \mathscr{E}^0 = -1.2 \text{ V}}$$

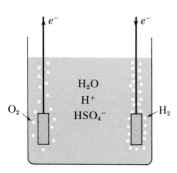

Fig. 16.5. *The cell* Pt|O₂|H⁺|H₂|Pt, *necessarily created when an aqueous solution of* H₂SO₄ *is electrolyzed.*

* Some older books give oxidation potentials, which are simply the negatives of ours.

The negative sign means that the reaction tends to occur in the direction *opposite* to that which we have written. The galvanic cell is trying to make a current flow in the direction opposite to the current with which we are trying to electrolyze the solution. In order to overcome this "back emf," the applied emf must be at least 1.2 V. If the applied emf is increased gradually from zero, practically no current will flow until 1.2 V is attained. Thereafter, the current rises linearly with applied emf (Fig. 16.6). The emf $\mathscr{E}_d$ at which the rise begins is called the DECOMPOSITION POTENTIAL of the solution.

The decomposition potential must be at least equal to the reversible (zero-current) emf of the cell whose reaction opposes the desired electrolysis reaction. However, it is often necessary to increase the applied emf considerably above this minimum value in order to cause a reaction at an appreciable rate. The extra potential required is called the OVERVOLTAGE, and depends on the nature of the electrode surface, the current per unit area, and the composition of the solution. Overvoltages are especially large, up to about 1 V, when gases are involved in the reaction. Another effect of a gas is to cover the electrode with an insulating layer, greatly increasing the resistance.

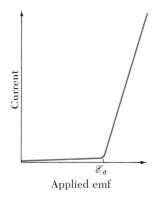

Fig. 16.6. *The dependence of current on emf in an electrolysis cell; $\mathscr{E}_d$, decomposition potential.*

Example 6 Calculate the emf required to bring about the reaction

$$Cu^{2+}(1\ M) + 2Cl^-(1\ M) \longrightarrow Cu(c) + Cl_2(g,\ 1\ atm)$$

with carbon electrodes. Assume that the overvoltage for the evolution of Cl_2 on a carbon surface is (a) 0; (b) 0.25 V.

Answer From Table 16.1, we obtain the following reversible emf for the back reaction:

$$
\begin{array}{ll}
Cu \longrightarrow Cu^{2+} + 2e^- & -(+0.3402) \\
\underline{Cl_2 + 2e^- \longrightarrow 2Cl^-} & \underline{+1.3583} \\
Cu + Cl_2 \longrightarrow Cu^{2+} + 2Cl^- & \mathscr{E}^0 = +1.0181
\end{array}
$$

(a) $\mathscr{E}_d = 1.02$ V
(b) $\mathscr{E}_d = 1.02 + 0.25 = 1.27$ V

Even in the absence of overvoltage, the potential necessary to bring about electrolysis at an appreciable rate may be considerably greater than one would calculate from the $\mathscr{E}^0$'s. Electrolysis causes nonuniformities in the concentration of the electrolyte, always in such a way as to increase the required potential. This effect is called CONCENTRATION POLARIZATION, and can often be diminished simply by stirring the solution.

THE DRY CELL AND THE STORAGE CELL / 16.11

Two of the most familiar galvanic cells used as sources of power are the dry cell and the lead-acid storage cell. Figure 16.7 shows a typical

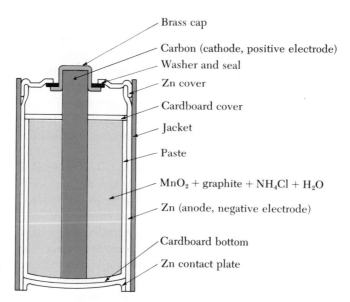

Brass cap
Carbon (cathode, positive electrode)
Washer and seal
Zn cover
Cardboard cover
Jacket
Paste
MnO_2 + graphite + NH_4Cl + H_2O
Zn (anode, negative electrode)
Cardboard bottom
Zn contact plate

Fig. 16.7. *A dry cell as used in flashlights.*

dry cell. The anode is Zn; the cathode is C, which is chemically inert under ordinary conditions. A paste consisting of graphite, MnO_2, NH_4Cl, and H_2O surrounds the cathode. Adjacent to the Zn is a layer of paste, usually made from wheat flour, containing H_2O, NH_4Cl, and $ZnCl_2$, but not MnO_2. The cell is "dry" only in the sense that it contains paste rather than free liquid; it could not function if water were absent. The electrode reactions are complicated and subject to some controversy. One representation is

anode: $$Zn \longrightarrow Zn^{2+} + 2e^-$$
cathode: $$NH_4^+ + MnO_2 + e^- \longrightarrow MnO(OH) + NH_3(aq)$$

The maximum emf is 1.48 V. Because of slow diffusion in the paste, the dry cell is more subject to polarization than are cells with liquid electrolytes. The emf, therefore, drops sharply if large currents are drawn.

Alkaline dry cells use KOH as the electrolyte instead of $NH_4Cl + ZnCl_2$. They give current for a longer time than ordinary dry cells, and the emf drops less when heavy currents are drawn.

The lead–acid storage cell, in its charged state, consists of an electrode of spongy lead and an electrode of finely divided solid PbO_2, each material supported by a grid of Pb–Sb alloy (stronger and more resistant to corrosion than pure Pb). The electrolyte is an aqueous solution of sulfuric acid (Fig. 16.8). The reactions, on discharging, are

$$Pb(c) + HSO_4^- \longrightarrow PbSO_4(c) + H^+ + 2e^-$$
$$\underline{PbO_2(c) + 3H^+ + HSO_4^- + 2e^- \longrightarrow PbSO_4(c) + 2H_2O}$$
$$Pb(c) + PbO_2(c) + 2H^+ + 2HSO_4^- \longrightarrow 2PbSO_4(c) + 2H_2O$$

The cell can be recharged by passing current in the reverse direction,

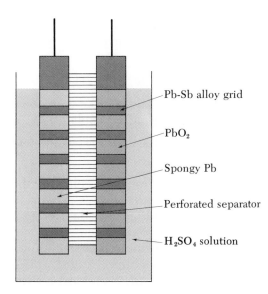

Fig. 16.8. A lead–acid storage cell.

with electrons entering at the Pb electrode and leaving at the PbO_2 electrode. On charging, both electrode reactions are reversed. Sulfuric acid is consumed in the discharging process and regenerated on charging; the density of the solution provides an indication of the degree to which the cell is charged, for the density decreases as the acid content decreases. The solution in a charged cell usually contains 30 to 40% H_2SO_4, corresponding to a density of 1.2 to 1.3 g/ml. The emf of the cell is roughly 2 V.

A **BATTERY** is a number of galvanic cells connected together, in either series (usually) or parallel.

CORROSION / 16.12

Corrosion is any undesired process—usually an oxidation—in which a metal is converted to one of its compounds. We shall consider primarily the corrosion of iron, the most widely used and one of the more easily corroded metals.

A familiar fact about the corrosion, or rusting, of iron is that *both* oxygen and water are necessary. Furthermore, the presence of acids and salts* in the water greatly accelerates corrosion. These observations suggest that electrochemical processes are involved in corrosion. The generally accepted theory is that Fe is oxidized to Fe^{2+} by H_2O or H^+:

$$Fe \longrightarrow Fe^{2+} + 2e^- \tag{3}$$
$$2H^+ + 2e^- \longrightarrow H_2(g) \tag{4}$$

---

* Salts which make the solution basic (for example, Na_2CO_3) are exceptional in that they inhibit corrosion.

If the H^+ concentration is large, bubbles of H_2 appear. In corrosion by natural waters, however, dissolved O_2 is usually reduced instead of H^+:

$$4H^+ + O_2(aq) + 4e^- \longrightarrow 2H_2O \tag{5}$$

The oxygen also oxidizes the Fe^{2+} ions liberated in reaction (3), producing hydrated iron(III) oxide (rust):

$$4Fe^{2+} + O_2 + (4 + 2x)H_2O \longrightarrow 2Fe_2O_3(H_2O)_x + 8H^+ \tag{6}$$

The H^+ consumed in reaction (4) or (5) is regenerated in reaction (6) so that H^+ is a catalyst: It accelerates the reaction without being consumed. CO_2 is nearly always present in natural waters and increases the H^+ concentration by the reaction $CO_2 + H_2O \leftrightharpoons H^+ + HCO_3^-$, thus contributing to corrosion.

The corrosion of an iron or steel pipe is greatly accelerated if it is attached to a copper pipe; reaction (3) occurs on the Fe, reaction (5) mostly on the Cu. Any metal less active (a poorer reducing agent) than Fe will behave in the same way as Cu. Even when only Fe is present, some points on the surface ("anodic sites") will be more chemically reactive than others ("cathodic sites") because of mechanical strain or small variations in composition. Thus, reaction (3) takes place at the anodic sites, which will be corroded, and reaction (5) occurs at the cathodic sites, which will not be corroded. The circuit through the metal must be completed by ionic conduction through the solution, which explains why an electrolyte accelerates corrosion.

If another metal more active (a better reducing agent) than iron — Zn or Mg, for example — is in contact with Fe, the more active metal will be oxidized in preference to the Fe, and the Fe in the vicinity will be protected from corrosion. The Fe is said to be made CATHODIC; the active metal is called a "sacrifical anode," for it is allowed to corrode away in order to protect the Fe. The same result can be achieved by imposing a negative electric potential on the iron and a positive potential on other electrodes placed in the water or the ground nearby. These methods are often used to protect water tanks, boilers, ship hulls, and submerged or buried pipes.

"Galvanized iron" is protected by a coating of Zn, which serves as a sacrificial anode and protects the Fe even if the coating has a hole in it. However, if the Fe is plated with a metal less active than itself, such as Ag or Cu, a small hole will result in very rapid corrosion, for the Fe will behave as a sacrifical anode and will protect the plating — hardly a desirable result. Tin is cathodic (less active) relative to iron in the presence of air, but in the absence of air (as in a can), and especially in the presence of acid (as in many foods), Sn becomes anodic relative to Fe. Thus, Sn is useful for plating Fe, even though the table of electrode potentials makes it appear otherwise. The reasons for this shift in potentials are not fully understood.

Many metals start to corrode by forming a thin, continuous, adherent layer of oxide, which prevents further corrosion. Examples are Al, Co, Cr, and Ni. Such metals are resistant to corrosion under ordinary conditions, even though Al and Cr, for example, are more active than Fe. An especially impervious coating can be produced on many metals, even on Fe, by treating the metal with a strong oxidizing agent, such as concentrated HNO_3 or Pb_3O_4 ("red lead"), or by using it as the anode in an electrolysis with high current density. The metal is said to become PAS-SIVE. Unfortunately, the passivity of Fe is easily destroyed, merely by touching the passive metal with a piece of normal (active) Fe. The resistance of Pt and related elements to electrolytic oxidation is much greater than one would expect from their oxidation potentials; this is also a case of passivity.

PROBLEMS

1. Definitions. Define briefly: (a) difference of potential, (b) electromotive force, (c) salt bridge, (d) anode, (e) positive electrode, (f) reference half-cell, (g) standard electrode potential, (h) decomposition potential, (i) overvoltage, (j) sacrificial anode.

2. Emf and work. The emf of a galvanic cell is 0.80 V. (a) How much work, in joules, is done by this cell when 5.0 C of charge pass from one electrode to the other? (b) When a current of 0.10 A flows, how much power (in watts) is the cell producing? (1 watt = 1 joule/second) (c) How much work, in joules and in calories, does the cell do when 1 faraday passes through the circuit? Assume that the emf of the cell is unaffected by the rate at which current is drawn.

3. Cell construction. For each of the following reactions, draw a diagram of a galvanic cell in which the reaction occurs. Show the composition of the anode, the cathode, and the solution or solutions. Indicate on your diagram the direction of migration of each kind of ion in the cell and the direction of flow of electrons in the external circuit.

(a) $2Cr(c) + 3HgCl_2(aq) \longrightarrow 2Cr^{3+} + 6Cl^- + 3Hg(l)$
(b) $Ni^{2+} + H_2(g) + 2OH^- \longrightarrow Ni(c) + 2H_2O$
(c) $Fe(c) + Hg_2Cl_2(c) \longrightarrow Fe^{2+} + 2Hg(l) + 2Cl^-$

4. Calculation of emf. For each of the following cells, (a) write the equation for the cell process, assuming that the left electrode is the anode; (b) find $\mathscr{E}^0$ at 25°C; (c) find $\mathscr{E}$ at 25°C, using the Nernst equation. Explain the significance of any negative answers in (b) and (c).

(i) $Fe \mid FeBr_2(0.10 \text{ M}) \mid Br_2(l) \mid Pt$
(ii) $Ag \mid AgI(c) \mid HI(0.030 \text{ M}) \mid H_2(0.75 \text{ atm}) \mid Pt$
(iii) $Fe \mid Fe(NO_3)_2(0.10 \text{ M}) \parallel Cu(NO_3)_2(0.050 \text{ M}) \mid Cu$

5. Nernst equation. For cell (ii) of the preceding problem, calculate the pressure of hydrogen that will make the emf of the cell equal to 0 at the given HI concentration.

6. Oxidizing and reducing agents. Select the best oxidizing agent and the best reducing agent in each list: (a) Cl_2, MnO_4^-, Fe^{2+}, Sn. (b) Cd^{2+}, Cd, Al^{3+}, Zn.

7. Decomposition potential. Find the minimum emf necessary to bring about each of the following reactions. Neglect overvoltage.

(a) $Cu(c) + 2H^+(1.0 \text{ M}) \longrightarrow Cu^{2+}(1.0 \text{ M}) + H_2(g, 1.0 \text{ atm})$

(b) $Sn^{2+}(1.0 \text{ M}) + 2Br^-(2.0 \text{ M}) \longrightarrow Sn(c) + Br_2(l)$

8. Corrosion. Which of the following methods of storing used steel wool is best? Which is most conducive to rusting? Explain. (a) Keep it immersed in plain tap water. (b) Keep it immersed in soapy water (a basic solution). (c) Just leave it in the sink.

ADDITIONAL PROBLEMS

9. Cell reaction and standard emf. For each of the following cells (*i–v*), (a) write the partial equations for the anode and cathode processes, assuming that the left electrode is the anode; (b) write the equation for the overall cell reaction; (c) find the standard emf $\mathscr{E}^0$ at 25°C. (d) Does the cell reaction actually occur as written in (b) or in the reverse direction when $\mathscr{E} = \mathscr{E}^0$?

(*i*) $Pt \mid Br_2(l) \mid CuBr_2(aq) \mid Cu$

(*ii*) $Al \mid AlCl_3(aq) \parallel MgSO_4(aq) \mid Mg$

(*iii*) $Fe \mid FeSO_4(aq) \mid PbSO_4(c) \mid Pb$

(*iv*) $Zn \mid ZnCl_2(aq) \parallel FeCl_2(aq), FeCl_3(aq) \mid Pt$

(*v*) $Pt \mid Cl_2(g) \mid KCl(aq) \mid Hg_2Cl_2(c) \mid Hg(l)$

10. Cell reaction and standard emf. An electrolytic cell is constructed as follows: In one beaker, a silver electrode is immersed in an aqueous solution of $Na_2S_2O_3$. In another beaker, a platinum electrode is immersed in an aqueous solution of KOH, and $O_2(g)$ from an outside source is bubbled around the Pt. The two solutions are connected by a salt bridge. (a) Write the equation for the reaction at the anode. (b) Write the equation for the reaction at the cathode. (c) Write the overall equation for the actual reaction. (d) Calculate $\mathscr{E}^0$ for the cell.

11. Pressure and concentration. (a) Calculate $\mathscr{E}^0$ at 25°C for a cell in which the reaction is $2H_2(g) + O_2(g) \rightarrow 2H_2O(l)$. (b) Calculate $\mathscr{E}$ for this cell when the concentrations of $H_2(g)$ and $O_2(g)$ are each 1.00 mole/liter. (First calculate the pressure corresponding to this concentration.) (c) What would $\mathscr{E}^0$ be for this cell if we decided to define $\mathscr{E}^0$ as the emf at unit concentration (1 mole/liter) for gases as well as for solutes?

12. Nernst equation. Suppose that the cell reaction in Example 2, page 320, is written

$$\tfrac{1}{2}Sn + AgCl \longrightarrow Ag + \tfrac{1}{2}Sn^{2+} + Cl^-$$

Write the Nernst equation for the reaction in this form. Show that it is really the same as in Example 2.

13. Electrode potentials. (a) Suppose we were to decide that $\mathscr{E}^0 = 0$ for the half-cell $Pb^{2+} \mid Pb$ instead of the hydrogen half-cell. On the basis of this choice, what would be $\mathscr{E}^0$ for $Cd^{2+} \mid Cd$ and for $Ag^+ \mid Ag$? (b) Calculate $\mathscr{E}^0$ for the cell $Cd \mid Cd^{2+} \parallel Ag^+ \mid Ag$. Compare with the result of Example 5 (page 324).

14. Electrode potentials. Figure 16.9 shows three cells in series. When current flows through this circuit, a reaction occurs at electrode A in cell 1, and the reverse reaction occurs, to the same extent, at electrode A in cell 3. Similarly, the reactions at electrodes B and C in cell 2 are equal and opposite to the reactions at electrodes B and C in cells 1 and 3. (a) The net emf of this arrangement of cells must be zero ($\mathscr{E}_1 + \mathscr{E}_2 + \mathscr{E}_3 = 0$); otherwise, a perpetual-motion machine could be constructed. Explain. (b) Choose A as the reference half-cell. This means that, by definition, $\mathscr{E}_B = \mathscr{E}_1$ and $\mathscr{E}_C = -\mathscr{E}_3$. Show that $\mathscr{E}_2 = \mathscr{E}_C - \mathscr{E}_B$. You have now proved the validity of expressing the emf of a cell as the difference of two electrode potentials, each of which is really the emf of a cell containing the reference half-cell.

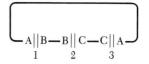

Fig. 16.9. *Three cells with resultant emf equal to zero.*

15. Decomposition potential. Calculate the minimum potential necessary to bring about each of the following reactions. Neglect overvoltage.

(a) $Cd^{2+}(1.0\ M) + H_2O \longrightarrow Cd(c) + 2H^+(1.0\ M) + \frac{1}{2}O_2(1.0\ atm)$
(b) $H_2O \longrightarrow H_2(1.0\ atm) + \frac{1}{2}O_2(1.0\ atm)$

(c) If the potential is gradually increased from 0, which reaction, (a) or (b), will occur first? (d) Would your answer to (c) be changed if the overvoltage for evolution of H_2 is 1.0 V? (e) Would your answer to (c) be changed if the concentration of H^+ were 10^{-7} M instead of 1 M?

16. Dry cell. Why is the dry cell so designed that Zn and MnO_2 do not come into contact? What reaction might occur if they were in contact? How would this reaction affect the usefulness of the cell?

17. Dry cell. The total charge that can be delivered by a 6-inch dry cell, before its emf drops too low, is usually about 35 amp-hours. (One amp-hour is the charge which passes through a circuit when 1 A flows for 1 hour.) Find the mass of Zn consumed when 35 amp-hours of charge are drawn from the cell.

18. Dry cell. In a basic solution, Zn^{2+} is converted to $Zn(OH)_4^{2-}$. Write partial and overall equations for the reactions in an alkaline dry cell.

19. Storage cell. A galvanic cell must be so constructed that not all the reactants can come into contact and react directly. (a) How is this rule violated in the design of the lead–acid storage cell? (b) What reaction, resulting from this violation, helps to account for the fact that a charged storage cell slowly discharges itself on standing idle?

20. Storage cell. The nickel–cadmium storage cell has recently come into use. The reactions are complex, involving mixed oxidation states of Ni, but the discharge process may be represented approximately by

$$Cd(c) + 2Ni(OH)_3(c) \longrightarrow Cd(OH)_2(c) + 2Ni(OH)_2(c)$$

The electrolyte is an aqueous solution of KOH. (a) Write partial equations for the oxidation and reduction processes. (b) Write a description of the cell in the usual notation. (c) Does the emf of this cell change during discharge? Does the composition of the solution change? In what ways are these properties advantageous and disadvantageous, in comparison to the lead–acid cell?

21. Corrosion. Would you recommend plating Fe, to protect it from corrosion, with each of the following metals: Au; Cr; Ba; Pb? Why or why not? (Ignore the cost of the metal in making your decision.)

22. Corrosion. A magnesium bar weighing 5.0 kg is attached to a buried iron pipe by way of an iron wire, to protect the pipe from corrosion. An average current of 0.050 A flows between the bar and the pipe. (a) What reaction occurs at the surface of the bar? Of the pipe? In which direction do electrons flow through the wire? (b) How many years will it take for the Mg bar to be entirely consumed? (One year $= 3.16 \times 10^7$ seconds.) (c) What reaction(s) will occur if the bar is not replaced after the time calculated in (b)?

THE SECOND LAW
OF THERMODYNAMICS

In Chapter 11 (Chemical Equilibrium), we learned that a given chemical system has a preferred state, the state of equilibrium, and we calculated the composition of a mixture at equilibrium, given one magic number, the equilibrium constant. What is there about the reactants and products that determines the value of this constant? In Chapter 5 (First Law of Thermodynamics) and in Chapter 16 (Galvanic Cells), we learned some important ideas about the work and heat involved in chemical reactions. These ideas provide the key to the puzzle of equilibrium constants. We are now in a position to tie them together.

We first review briefly some conventions that were used in Chapters 5 and 16. When a system *absorbs heat* from the surroundings (an *endothermic* process), this heat is considered *positive: $q > 0$*. When heat is *given out* by the system (an *exothermic* process), the heat is considered *negative: $q < 0$*. In the case of work, the convention is that work is *positive ($w > 0$)* when the system *does work* on the surroundings—for example, a gas expands, a cell discharges. When work is *done on* the system—the gas is compressed, the cell is charged—the work is *negative ($w < 0$)*. The sign convention for the emf of a galvanic cell is the same as for work. If $\mathscr{E} > 0$, the cell can do work when the reaction occurs as written: $w > 0$. If $\mathscr{E} < 0$, the reaction is trying to go the other way and work must be done on the cell ($w < 0$) to make the reaction go as written.

The first law of thermodynamics (page 56) is simply the law of conservation of energy: you cannot get something for nothing. There is no perpetual-motion machine that does work with no input of energy.

That still leaves some intriguing possibilities. How about taking in air from the atmosphere, or water from the ocean, on one side of an engine, extracting heat from it, discharging it colder on the other side, and converting the extracted heat to work? This marvelous invention would make coal, oil, and nuclear power obsolete. It would be as good a perpetual-motion machine as the device that creates energy out of nothing. Unfortunately, it does not exist. Mankind has become convinced that it never will exist. This conclusion is the SECOND LAW OF THERMODYNAMICS:

Heat cannot be withdrawn from a reservoir at an initially uniform temperature and converted completely to an equivalent amount of work by a device operating in cycles.

The only way to convert heat to work is to have two different temperatures and convert to work some of the heat that flows from the higher to the lower temperature. It has been proposed that work could be obtained from the ocean by taking advantage of the temperature difference between the surface and the depths. This process is possible, at least in theory. If, however, the ocean were at the same temperature throughout, no work could be obtained by cooling it.

REVERSIBLE PROCESSES / 17.3

In the preceding chapter, we described an ideal way of measuring the emf of a cell. The emf of the cell was balanced by an equal and opposite emf from a potentiometer. We called the result of this measurement the "reversible emf." The word *reversible* has two quite different meanings in chemistry.

The first meaning is the sense with which the reader is probably familiar. A reversible process is one that can be made to occur in either direction, such as the reaction $H_2(g) + I_2(g) \leftrightharpoons 2HI(g)$.

The second meaning is more subtle. A process is said to be THERMODYNAMICALLY REVERSIBLE if it occurs in a delicately balanced way so that an infinitesimal* change in some condition will reverse its direction. A balance provides a familiar example. We adjust the weights until the arm no longer swings either way. The mass of the weights then equals the mass of the object being weighed. A very small change—ideally, an infinitesimal change—in either mass can be detected by movement of the beam. A reversible process cannot be fully realized in practice; it can only be approached as an ideal limit, like a frictionless balance. In such a balance, the infinitely slow motion of the beam, resulting from an infinitesimal mass difference, would be a reversible process.

How can a chemical reaction be carried out reversibly, in the thermodynamic sense? Not by merely mixing the reactants and letting them

---

* "Infinitesimal" means "as close to zero as you want it to be."

react. We must somehow balance an equal opposing force against their tendency to react. By a minute adjustment of the opposing force, the reaction can be made to proceed very slowly in either direction. Our needs are met by a galvanic cell in which the reaction tries to occur, and a potentiometer as the variable opposing force. When we refer to the "reversible emf" of a cell, the word "reversible" is used in the thermodynamic sense: The emf of the potentiometer balances the emf of the cell in the same way that the weights balance the object being weighed. The counterpart of the frictionless balance is a circuit with no resistance and an infinitely sensitive galvanometer (current detector), with which an infinitesimal imbalance of potential could be detected. Naturally, this is only an ideal limit.

In this section, we will confine ourselves to processes that take place with the system and the surroundings always at a single temperature. We can imagine the system immersed in a large tank at constant temperature, such as a tank of ice water (always at 0°C).

Let us assume that we have a system on which work can be done, which can then do work in return until it runs down, which can then have work done on it again, then do work, etc., over and over. It might be a wind-up toy, or a cylinder of gas with a piston, or a storage battery. With the battery in mind, we say that we *charge* the system by doing work w_c on it; this work is *negative* ($w_c < 0$). We *discharge* it by letting it do work w_d for us; this work is *positive* ($w_d > 0$). Every time we discharge the system, we bring it back to the same state as before it was charged. In this cycle of charging and discharging, the net work done by the system is

$$w_n = w_c + w_d$$

Consider two examples:

(a)   If it takes 40 cal to charge the system, then $w_c = -40$ cal; if the system does 50 cal of work on discharge, $w_d = +50$ cal. The net work is $w_n = -40$ cal $+ 50$ cal $= +10$ cal.

(b)   If it takes 42 cal to charge the system ($w_c = -42$ cal) but only 35 cal are obtained on discharge, $w_n = -42$ cal $+ 35$ cal $= -7$ cal. The negative sign means that net work is done *on* the system, not obtained from it.

Suppose we have a system like (a) above, which does positive work when it goes through the charge–discharge cycle. Remember that everything—the system and the reservoir with which it exchanges heat—is at constant temperature. The net effect of the cycle is that the system does work (10 cal in the example) and, by the first law, it must withdraw 10 cal of heat from the reservoir. This is precisely what the second law says cannot be done. *In the complete cycle, the net work cannot be positive.* It can only be zero or negative: $w_n \leqslant 0$.[*] To charge a battery, you must put in at least as much work as you get back by discharging it. System (a) is impossible: it would be a perpetual-motion machine. System (b) is possible: we put in 42 cal and got only 35 back.

The best possible case is attained when the entire process is thermodynamically reversible. Then $w_n = 0$, $w_c + w_d = 0$, and $w_d = -w_c$. The charging and discharging processes involve the same amount of work but with opposite signs: for example, $w_c = -40$ cal, $w_d = +40$ cal, $w_n = -40$ cal $+ 40$ cal $= 0$. No matter how ingenious we are, there is no hope of finding another charging process that will require less work than $-w_c = 40$ cal, or another discharging process that will give more work than $w_d = 40$ cal. We will never actually get the whole 40 cal as work, but it is a goal to push toward.

It seems that the charged system has 40 cal more of something than the discharged system has. This "something" is a measure of the available work. It has the units of energy, but is not equal to the energy. It is another property, the **FREE ENERGY** (G).[†] The word "free" is used in the sense of "available as work," not merely as heat. As with energy and enthalpy, we are concerned only with changes in free energy (ΔG). Calling it a "property"—like energy, volume, temperature—implies that the change in free energy as the system goes from one state to another is determined by the two states, not by the details of the process. *The change in free energy when the system passes from state* d *(discharged) to state* c *(charged) is equal to the work which must be done on the system to take it reversibly from state* d *to state* c (all at constant temperature):

$$\Delta G = G_c - G_d = -w_{c,rev}$$

This work is the same for *any* reversible charging process from d to c; ΔG therefore satisfies the requirement of not depending on the details of the process. If work must be done *on* the system, then $w_{c,rev}$ is *negative* and ΔG is *positive*: for example, $w_{c,rev} = -40$ cal, $\Delta G = -(-40$ cal$) = +40$ cal. This makes sense: We are charging the system and its free energy increases. Conversely, if the system is being discharged, the maximum work that can be obtained is positive, but ΔG should be negative, for the system is doing work, and its ability to do further work is decreasing:

$$\Delta G = G_d - G_c = -w_{d,rev}$$

If $w_{d,rev} = +40$ cal, $\Delta G = -40$ cal.

[*] The sign $\leqslant$ means "less than or equal to"; the sign $\geqslant$ means "greater than or equal to."

[†] G stands for *Gibbs free energy* or *Gibbs function*, after Josiah Willard Gibbs (1839–1903), one of the founders of chemical thermodynamics. The symbol F is also used.

When ΔG is negative, the *decrease* in free energy measures the *maximum work available* from the system when it passes "downhill" from one state to another at constant temperature. When ΔG is positive, the *increase* in free energy measures the *minimum work required* to take the system "uphill" from one state to the other. The work actually obtained will always be less than the maximum; the work actually required will always be more than the minimum.

A chemical reaction can give out (or take in) energy in three forms: heat, pressure–volume work, and electrical work. We now specify that the free-energy change *excludes* the work ($P\,\Delta V$) done by or against a constant external pressure. This means that at constant pressure, the free-energy change is equal to minus the reversible *electrical* work, which is calculated from the reversible electromotive force.

FREE ENERGY AND EMF / 17.5

Let us recall the cell in Fig. 16.1, page 314. For this cell, $\mathscr{E}^0 = 1.10$ V. Then the maximum work available when the cell reaction

$$\mathrm{Zn} + \mathrm{Cu}^{2+} \longrightarrow \mathrm{Zn}^{2+} + \mathrm{Cu}$$

occurs at unit concentrations is

$$1.10 \text{ V} \times 9.65 \times 10^4 \,\frac{\mathrm{C}}{\cancel{\mathrm{faraday}}} \times 2\,\frac{\cancel{\mathrm{faradays}}}{\mathrm{mole\ Zn}} = 2.12 \times 10^5 \text{ joules/mole Zn}$$

This quantity is $-\Delta G$ for the chemical reaction written above. ΔG is the total free energy of 1 mole Zn^{2+} and 1 mole Cu, minus the total free energy of 1 mole Zn and 1 mole Cu^{2+}:

$$\begin{aligned}
\Delta G &= G_{\mathrm{products}} - G_{\mathrm{reactants}} \\
&= G_{\mathrm{Zn}^{2+}} + G_{\mathrm{Cu}} - G_{\mathrm{Zn}} - G_{\mathrm{Cu}^{2+}} \\
&= -2.12 \times 10^5 \text{ joules}
\end{aligned}$$

Here, ΔG is negative, $-212{,}000$ joules, because the cell has done work, and its ability to do further work has declined. The decrease in free energy is not equal to the electrical work actually done when current is drawn from the cell, for the work done is always less than that which might be done, and it is the possibility, not the performance, that is given by the free-energy change.

The reversible emf, $\mathscr{E}$, associated with a reaction is really the free-energy change in different units, and with the sign reversed. Let n be the number of faradays which pass through the circuit when the reaction occurs as written, and let $\mathscr{F}$ be the Faraday constant, 9.65×10^4 C/faraday. Then the free-energy change is

$$\Delta G = -n\mathscr{F}\mathscr{E} \tag{1}$$

$\mathscr{E}$ is in volts (V); ΔG is in joules (J) or, more often, calories (cal) or kilocalories (kcal):

$$\Delta G = -n\ \cancel{\text{faradays}} \times \mathscr{F} \frac{\cancel{\mathrm{C}}}{\cancel{\text{faraday}}} \times \mathscr{E}\ \frac{\mathrm{J}}{\cancel{\mathrm{C}}}$$

$$= -9.65 \times 10^4\ n\mathscr{E}\ \mathrm{J}$$

$$= -9.65 \times 10^4\ n\mathscr{E}\ \mathrm{J} \times \frac{1\ \text{cal}}{4.184\ \mathrm{J}}$$

$$= -2.31 \times 10^4\ n\mathscr{E}\ \text{cal} = -23.1\ n\mathscr{E}\ \text{kcal}$$

When all the coefficients in the chemical equation are multiplied by a constant, $\mathscr{E}$ is unaffected, but ΔG (like ΔH) is multiplied by the constant. Compare the equations

$$2Zn + 2Cu^{2+} \longrightarrow 2Zn^{2+} + 2Cu \qquad n = 2,\ \Delta G = -2\mathscr{F}\mathscr{E} = -2.12 \times 10^5\ \mathrm{J}$$
$$2Zn + 2Cu^{2+} \longrightarrow 2Zn^{2+} + 2Cu \qquad n = 4,\ \Delta G = -4\mathscr{F}\mathscr{E} = -4.24 \times 10^5\ \mathrm{J}$$
$$\tfrac{1}{2}Zn + \tfrac{1}{2}Cu^{2+} \longrightarrow \tfrac{1}{2}Zn^{2+} + \tfrac{1}{2}Cu \qquad n = 1,\ \Delta G = -\mathscr{F}\mathscr{E} = -1.06 \times 10^5\ \mathrm{J}$$

When all concentrations are 1 mole/liter (pressures, 1 atm), $\mathscr{E} = \mathscr{E}^0$. The corresponding free-energy change is known as the STANDARD FREE ENERGY CHANGE, ΔG^0:

$$\Delta G^0 = -n\mathscr{F}\mathscr{E}^0 \tag{2}$$

Example 1 Calculate ΔG^0 in kilocalories at 25°C for the reaction

$$Cu^{2+}(aq) + Ni(c) \longrightarrow Cu(c) + Ni^{2+}(aq)$$

Use data in Table 16.1 (page 325).

Answer We write the two partial reactions, look up $\mathscr{E}^0$ for each, and proceed as on page 324:

$$\begin{array}{rl}
Ni \longrightarrow Ni^{2+} + 2e^- & -(-0.23) \\
Cu^{2+} + 2e^- \longrightarrow Cu & +0.3402 \\
\hline
Cu^{2+} + Ni \longrightarrow Cu + Ni^{2+} & \mathscr{E}^0 = +0.57\ \mathrm{V}
\end{array}$$

When the reaction occurs as written, 2 faradays pass through the circuit: $n = 2$. Then

$$\Delta G^0 = -23.1\ n\mathscr{E}^0\ \text{kcal}$$
$$= -23.1 \times 2 \times 0.57\ \text{kcal}$$
$$= -26\ \text{kcal}$$

Free-energy changes are often recorded as STANDARD FREE ENERGIES OF FORMATION, ΔG_f^0 (analogous to ΔH_f, page 65). ΔG_f^0 is ΔG^0 for the reaction in which *1 mole* of the substance is formed *from its elements* in their stable forms.

Example 2 Calculate ΔG_f^0 in kilocalories per mole for $H_2O(l)$ at 25°C from the data in Table 16.1.

Answer The standard free energy of formation is ΔG^0 for the reaction

$$H_2(g) + \tfrac{1}{2}O_2(g) \longrightarrow H_2O(l)$$

The reaction can be resolved into the partial reactions

$$H_2(g) \longrightarrow 2H^+ + 2e^- \qquad 0$$
$$\underline{2H^+ + \tfrac{1}{2}O_2(g) + 2e^- \longrightarrow H_2O(l) \qquad +1.229}$$
$$H_2(g) + \tfrac{1}{2}O_2(g) \longrightarrow H_2O(l) \qquad \mathscr{E}^0 = +1.229 \text{ V}$$

$$\begin{aligned} \Delta G_f^0 &= -23.1 \; n\mathscr{E}^0 \text{ kcal/mole} \\ &= -23.1 \times 2 \times 1.229 \text{ kcal/mole} \\ &= -56.8 \text{ kcal/mole} \end{aligned}$$

FREE ENERGY AND EQUILIBRIUM / 17.6

What do we mean by saying that a system is in equilibrium? The general idea is familiar: a system is in equilibrium if it will not change its state without external intervention. What is an external intervention? The answer depends on what conditions are held constant as the system finds its equilibrium state. We confine ourselves to one case: the system is closed (no matter enters or leaves) and is at constant temperature and pressure. *A closed system at constant temperature and pressure is in equilibrium when it cannot undergo a change unless electrical work is done on it.* An equilibrium can be shifted by adding or removing matter—but then the system is not closed. It can be shifted by changing the temperature or pressure—but we are now considering the case where temperature and pressure are constant. The changes in free energy as a system is displaced from equilibrium, or as it goes toward equilibrium, are as follows:

Equilibrium → Disequilibrium	Disequilibrium → Equilibrium
More stable state → Less stable state	Less stable state → More stable state
Electrical work must be done on system	System can do electrical work
$w < 0$	$w > 0$
Even for a reversible process, $w_{rev} < 0$	$w_{rev} > 0$
$\Delta G > 0$	$\Delta G < 0$
G increases	G decreases

The state of equilibrium is the state of minimum free energy.

We would expect a process with negative ΔG to happen, for it represents movement toward equilibrium. Such a process is commonly described as "spontaneous," but this word can be misleading, for the process may not actually occur at a measurable rate. The reaction

$$H_2(g) + \tfrac{1}{2}O_2(g) \longrightarrow H_2O(l)$$

disequilibrium state   equilibrium state

has $\Delta G^0 = -57$ kcal at 25°C and is therefore "spontaneous," but the reaction does not occur at room temperature in the absence of a catalyst. The reverse reaction,

$$H_2O(l) \longrightarrow H_2(g) + \tfrac{1}{2}O_2(g)$$
equilibrium state   disequilibrium state

has $\Delta G^0 = +57$ kcal, which means that there is no hope of decomposing water to its elements at 25°C and 1 atm unless work, in the amount of at least 57 kcal for every mole H_2O, is done. This information about free energy does not guarantee that water can be decomposed at all. It can be, by electrolysis, and the electrical work required will be somewhat more than 57 kcal/mole because of the inefficiency to be expected in any process.

The free-energy change in a reaction is thus a measure of the driving force of the reaction. "Drift" might be a better word than "drive," for molecules behave randomly, not purposefully. The only direction in which the reaction can proceed, at constant temperature and pressure, is the direction of lower free energy, unless electrical work is done on the reacting system.

EMF AND EQUILIBRIUM / 17.7

A system in equilibrium can do no work, for its free energy has reached a minimum. If current is drawn from a galvanic cell until the cell reaction has come to equilibrium, the emf must become zero. If the emf were not zero, the cell could do work. *The condition of equilibrium for a reaction is $\mathscr{E}_{cell} = 0$*, or, from Equation (1), $\Delta G = 0$.

The Nernst equation (page 319) is

$$\mathscr{E} = \mathscr{E}^0 - \left(\frac{2.303RT}{n\mathscr{F}} \right) \log Q$$

where Q is the reaction quotient, the same expression that appears in the law of chemical equilibrium. At equilibrium, $\mathscr{E} = 0$ and $Q = K$ (page 195), where K is the equilibrium constant. Therefore,

$$\mathscr{E}^0 - \left(\frac{2.303RT}{n\mathscr{F}} \right) \log K = 0$$

and

$$\log K = \frac{n\mathscr{F}\mathscr{E}^0}{2.303RT} \tag{3}$$

or, at 25°C,

$$\log K = \frac{n\mathscr{E}^0}{0.05916} = 16.90n\mathscr{E}^0$$

Combining Equations (2) and (3) gives another expression for K:

$$\log K = -\frac{\Delta G^0}{2.303RT}$$

For example, in the cell

$$\text{Pt} \mid H_2(g) \mid HCl(aq) \mid AgCl(c) \mid Ag(c)$$

the partial reactions are

$$\frac{1}{2}H_2 \rightleftharpoons H^+ + e^-$$
$$\underline{AgCl + e^- \rightleftharpoons Ag + Cl^-}$$
$$\frac{1}{2}H_2 + AgCl \rightleftharpoons H^+ + Cl^- + Ag$$

For this cell, at 25°C, $\mathscr{E}^0 = 0.2223$ V and $n = 1$; then

$$\log K = \frac{n\mathscr{F}\mathscr{E}^0}{2.303RT} = \frac{0.2223}{0.05916} = 3.758$$

$$K = 10^{3.758} = 10^{0.758} \times 10^3 = 5.73 \times 10^3$$

The equilibrium condition is

$$\frac{[H^+][Cl^-]}{p_{H_2}^{1/2}} = 5.73 \times 10^3$$

The equilibrium constant of any reaction can be determined if we can resolve the reaction into two partial reactions for which the potentials are given in a table. From the resultant emf, $\mathscr{E}^0$, we calculate K as shown above.

Example 3 Find the equilibrium constant at 25°C for the reaction (in aqueous solution)

$$Zn + Cu^2 \rightleftharpoons Zn^{2+} + Cu$$

Answer The partial reactions, with their standard potentials, are

$$
\begin{array}{lll}
Zn \longrightarrow Zn^{2+} + 2e^- & \text{(oxidation)} & -(-0.7628) \\
Cu^{2+} + 2e^- \longrightarrow Cu & \text{(reduction)} & \underline{+0.3402} \\
& & \mathscr{E}^0 = +1.1030 \text{ V}
\end{array}
$$

Thus, $\mathscr{E}^0 = +1.1030$ V and $n = 2$ (2 electrons appear in each partial reaction). Then

$$\log K = \frac{n\mathscr{E}^0}{0.05916} = \frac{2 \times 1.1030}{0.05916} = 37.29$$

$$K = 10^{37.29} = 10^{0.29} \times 10^{37} = 1.9 \times 10^{37} = [Zn^{2+}]/[Cu^{2+}]$$

Example 4 Find the equilibrium constant at 25°C for the reaction

$$Ag_2CrO_4(c) \rightleftharpoons 2Ag^+ + CrO_4^{2-}$$

Answer Inspection of those partial reactions in Table 16.1 involving Ag shows that the following two can be combined to give the desired reaction:

$$Ag^+ + e^- \rightleftharpoons Ag \qquad +0.7996$$
$$Ag_2CrO_4 + 2e^- \rightleftharpoons 2Ag + CrO_4^{2-} \qquad +0.4463$$

We double and reverse the first and add the partial reactions:

$$
\begin{array}{llr}
2Ag \rightleftharpoons 2Ag^+ + 2e^- & \text{(oxidation)} & -(+0.7996) \\
Ag_2CrO_4 + 2e^- \rightleftharpoons 2Ag + CrO_4^{2-} & \text{(reduction)} & +0.4463 \\
\hline
Ag_2CrO_4 \rightleftharpoons 2Ag^+ + CrO_4^{2-} & & \mathscr{E}^0 = -0.3533 \text{ V}
\end{array}
$$

Note that the potential is not affected by doubling the coefficients, or multiplying them by any constant. Here $n = 2$ (2 electrons appear in each partial reaction). Then

$$\log K = \frac{2\mathscr{E}^0}{0.05916} = \frac{2(-0.3533)}{0.05916} = -11.94$$

$$K = 10^{-11.94} = 10^{0.06} \times 10^{-12} = 1.1 \times 10^{-12} = [Ag^+]^2[CrO_4^{2-}]$$

The reader will recognize K as a solubility product (page 287).

Example 5 Find the equilibrium constant at 25°C for the reaction $H_2O \rightleftharpoons H^+ + OH^-$.

Answer The two partial reactions suitable in this case are

$$
\begin{array}{llr}
H_2 \rightleftharpoons 2H^+ + 2e^- & \text{(oxidation)} & 0.0000 \\
2H_2O + 2e^- \rightleftharpoons H_2 + 2OH^- & \text{(reduction)} & -0.8277 \\
\hline
2H_2O \rightleftharpoons 2H^+ + 2OH^- & & \mathscr{E}^0 = -0.8277 \text{ V}
\end{array}
$$

The reaction is double the desired one. We multiply each partial reaction by $\frac{1}{2}$ to make the overall reaction

$$H_2O \rightleftharpoons H^+ + OH^-$$

This does not affect the potentials, but makes $n = 1$ instead of 2, so that

$$\log K = \frac{-0.8277}{0.05916} = -13.99$$

$$K = 1.0 \times 10^{-14} = [H^+][OH^-]$$

We have thus calculated the ion product of water (page 265).

The reactions in Examples 4 and 5 are not oxidation–reduction reactions. Nevertheless, they can be resolved into partial equations that add up to the overall reaction, and the electrode potentials can then be used to calculate $\mathscr{E}^0$, ΔG^0, and K.

PREDICTING THE DIRECTION OF A REACTION / 17.8

Often we wish to know whether a reaction can occur when certain chemicals are present in certain amounts. A rigorous way to answer this question is to find the equilibrium constant, K, for the reaction in question and to calculate the reaction quotient, Q, for the reaction, using the initial concentrations or pressures of the species present. Then if $Q < K$, the reaction may go from left to right; if $Q > K$, from right to left.

Example 6 A solution contains Ag^+ and CrO_4^{2-} ions, with $[Ag^+] = 10^{-1}$ mole/liter and $[CrO_4^{2-}] = 10^{-2}$ mole/liter. Will solid Ag_2CrO_4 precipitate from this solution?

Answer For the reaction

$$Ag_2CrO_4(c) \rightleftharpoons 2Ag^+ + CrO_4^{2-}$$

we have

$$Q = [Ag^+]^2[CrO_4^{2-}] = (10^{-1})^2 \times (10^{-2}) = 10^{-4}$$

Q is greater than $K = 1.1 \times 10^{-12}$ (Example 4), and the reaction will go to the left; in other words, Ag_2CrO_4 will precipitate.

A simpler procedure is often adequate to determine the direction of a reaction. In a large proportion of cases the concentrations are in the range 0.01 to 10 moles/liter—not vastly removed from 1 mole/liter (for pressures, not too far from 1 atm). Thus, when chemicals are mixed together, the chances are that Q is somewhere near 1—neither very large nor very small. However, most equilibrium constants *are* very large or very small, with only an occasional one happening to fall near 1. To decide, therefore, which way a reaction will go, it usually suffices to ascertain whether $K > 1$ or $K < 1$. These cases correspond, respectively, to $\log K > 0$ and $\log K < 0$, and to $\mathscr{E}^0 > 0$ and $\mathscr{E}^0 < 0$ (Equation 3). If $\mathscr{E}^0 > 0$, the reaction goes from left to right; if $\mathscr{E}^0 < 0$, it goes from right to left. The rule becomes unreliable when $\mathscr{E}^0$ is close to 0—and, of course, it applies only when all concentrations are near 1 mole/liter.

Example 7 Will Fe reduce Fe^{3+} to Fe^{2+}?

Answer The reaction in question is $Fe + 2Fe^{3+} \rightarrow 3Fe^{2+}$ which is resolved into partial reactions:

$Fe \longrightarrow Fe^{2+} + 2e^-$	(oxidation)	$-(-0.409)$
$2Fe^{3+} + 2e^- \longrightarrow 2Fe^{2+}$	(reduction)	$+0.7701$
$Fe + 2Fe^{3+} \longrightarrow 3Fe^{2+}$		$\mathscr{E}^0 = +1.179$ V

Since $\mathscr{E}^0 > 0$, the reaction is possible when concentrations are near 1. A practical consequence of this result is that a solution of an iron(II) (Fe^{2+}) salt can be protected from atmospheric oxidation by keeping solid iron in the bottle; any Fe^{3+} formed will be reduced to Fe^{2+}.

Example 8 Will the permanganate ion, MnO_4^-, liberate O_2 from water in the presence of acid (H^+)?

Answer The reaction is

$$4MnO_4^- + 12H^+ \longrightarrow 4\ Mn^{2+} + 5O_2 + 6H_2O$$

which is the resultant of the partial reactions

$(2H_2O \longrightarrow O_2 + 4H^+ + 4e^-) \times 5$	$-(+1.229)$
$(MnO_4^- + 8H^+ + 5e^- \longrightarrow Mn^{2+} + 4H_2O) \times 4$	$+1.491$
$4MnO_4^- + 12H^+ \longrightarrow 4Mn^{2+} + 5O_2 + 6H_2O$	$\mathscr{E}^0 = +0.262$ V

$\mathscr{E}^0$ is positive and the reaction is possible.

Example 9 Will O_2 oxidize gold to $Au(CN)_2^-$ in the presence of cyanide ion (CN^-) and OH^-?

Answer The reactions in question are

$(Au + 2CN^- \rightleftharpoons Au(CN)_2^- + e^-) \times 4$	$-(-0.60)$
$O_2 + 2H_2O + 4e^- \rightleftharpoons 4OH^-$	$+0.401$
$4Au + 8CN^- + O_2 + 2H_2O \rightleftharpoons 4Au(CN)_2^- + 4OH^-$	$\mathscr{E}^0 = +1.00$ V

The reaction goes to the right—a rather surprising result, for Au is notoriously unreactive and does not tarnish in air. The explanation is that the presence of CN^- permits the formation of a complex ion (page 470), $Au(CN)_2^-$, which is much more stable than anything that could be formed from Au, O_2, and H_2O alone.

ENTROPY AND FREE ENERGY / 17.9

The concept of entropy was introduced in Chapter 9 as a measure of the randomness, or disorder, of a system. It is one of the most important, and also one of the most subtle, concepts in physical science. A system tends to arrive at the least orderly configuration—the configuration of greatest entropy—simply because there are many more ways to be disorderly than to be orderly. However, the least orderly configurations have the highest energies; in general, an orderly arrangement (as in a crystal) has low energy, while a disorderly arrangement (as in a gas) has high energy. If we think only about energy and forget entropy, a system at a finite temperature has a greater probability of being in a state of low energy than in a state of high energy. When the pressure is constant, enthalpy plays the role of energy. There are thus two opposing tendencies, toward low enthalpy and high entropy. The system finds a compromise between these two tendencies, and it turns out that free energy is the key to the best compromise.

We recall (page 339) that we calculated $\Delta G^0 = -2.12 \times 10^5$ joules for the reaction

$$Zn + Cu^{2+} \longrightarrow Zn^{2+} + Cu$$

If the concentrations of the ions are 1 mole/liter, $\Delta G = \Delta G^0$. If Zn and Cu^{2+} simply react with each other, not in a galvanic cell, the total heat liberated is 2.17×10^5 joules: $\Delta H = -2.17 \times 10^5$ joules, the negative sign signifying the emission of heat. The maximum work available from the cell is 2.12×10^5 joules. Even in the ideal case of reversible discharge, the maximum available work is not equal to the heat emitted when the reactants merely mix and do no electrical work. The difference,

$$(2.17 \times 10^5) - (2.12 \times 10^5) = 5 \times 10^3 \text{ joules}$$

is the amount of energy that can be obtained only as heat and not as work. When the cell operates with maximum efficiency, 5×10^3 joules is the amount of heat evolved as the cell puts out 2.12×10^5 joules of electrical work. In practice, of course, even more heat will be emitted and less work will be done.

This difference between available heat and available work is related to the entropy change, ΔS. The heat emitted when the maximum work is done is

(heat emitted when no electrical work is done) − (maximum work)
$$= (-\Delta H) \qquad\qquad\qquad\qquad\qquad - (-\Delta G)$$
$$= \Delta G - \Delta H$$
$$= -T \, \Delta S$$

Then the entropy change is given by

$$\Delta S = \frac{\Delta H - \Delta G}{T}$$

For the $Zn + Cu^{2+}$ reaction at 25°C,

$$\Delta S = \frac{-2.17 \times 10^5 - (-2.12 \times 10^5) \text{ joules}}{298°\text{K}}$$
$$= -17 \text{ joules/°K}$$

It is also possible for the maximum work that can be done to be greater than the maximum heat that can be given out; in such a case, heat is absorbed by the cell, and ΔS is positive. An example is the reaction

$$Pb(c) + Hg_2Cl_2(c) \longrightarrow PbCl_2(c) + 2Hg(l)$$

for which $\Delta H = -9.43 \times 10^4$ joules and $\Delta G = -1.034 \times 10^5$ joules at 25°C.

Figure 17.1 illustrates the relations among heat, work, ΔH, ΔG, and ΔS for three ways of carrying out a reaction: simple mixing of the reactants (no electrical work), a galvanic cell from which current is being drawn, and an ideal (reversible) cell. The diagram is for the common case where ΔH, ΔG, and ΔS are all negative.

For a closed system at constant pressure and temperature, the state of equilibrium is the state of minimum G. The equation $\Delta G = \Delta H - T \, \Delta S$, or $G = H - TS$, shows the interplay between the two opposing tendencies toward low enthalpy and high entropy. The higher the temperature, the more sensitive G is to changes in S, and, therefore, the more important is the tendency toward maximum entropy; the lower the tem-

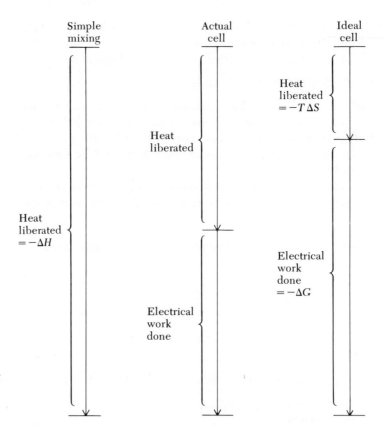

Simple mixing Actual cell Ideal cell

Heat liberated $=-T\Delta S$

Heat liberated

Heat liberated $=-\Delta H$

Electrical work done $=-\Delta G$

Electrical work done

Fig. 17.1. *The division of the enthalpy change between heat and work in three ways of carrying out a chemical reaction.*

perature, the more important is the tendency toward minimum enthalpy.

It is not obvious why $(H - G)/T$ has anything to do with randomness. We offer one comment that may help to make the connection plausible. Heat is random molecular motion; work is ordered motion. $G = H - TS$; for given H and T, the larger S is, the smaller G is. The more randomness the system has, the less of its energy is available in ordered form—work—and the more of its energy is available only in disordered form—heat.

Free-energy changes are most accurately measured and most clearly understood through their relation to emf's. However, the free-energy concept is not dependent on galvanic cells. For processes that do not, or cannot, take place in a cell, free-energy changes can still be calculated from other data.

Example 10 For the process

$$H_2O(c) \longrightarrow H_2O(l)$$

$\Delta H = 1436$ cal/mole and $\Delta S = 5.26$ cal/°K mole. (a) Calculate ΔG for the process at -10°C. Which form, ice or liquid water, is stable at this temperature? (b)

Calculate ΔG for the process at $+10°C$. Which form is stable at this temperature? (c) Calculate the temperature at which $\Delta G = 0$. What is the physical significance of this temperature?

Answer

(a) $\Delta G = \Delta H - T \Delta S$

$\quad\quad = 1436$ cal/mole $- 263°K \times 5.26$ cal/°K mole

$\quad\quad = 53$ cal/mole

ΔG is positive, which means that liquid water has a higher free energy than ice. The stable form is ice.

(b) $\quad \Delta G = 1436$ cal/mole $- 283°K \times 5.26$ cal/°K mole

$\quad\quad = -53$ cal/mole

At this temperature, liquid water has the lower free energy and is more stable.

(c) $\quad \Delta G = \Delta H - T \Delta S = 0$

$$T = \frac{\Delta H}{\Delta S} = \frac{1436 \text{ cal/mole}}{5.26 \text{ cal/°K mole}}$$

$\quad\quad = 273°K \quad$ or $\quad 0°C$

At this temperature, the two forms of water are equally stable and can exist together in equilibrium.

ENTROPY AND HEAT / 17.10

The more efficient a process is, the more work is done by the system and the more heat is absorbed. The most efficient process is a reversible process, in which the heat absorbed is $T \Delta S$. In any irreversible process between the same two states, less heat is absorbed. Therefore, the maximum heat absorbed in an endothermic process is $T \Delta S$:

$$q \leqslant T \Delta S, (q/T) \leqslant \Delta S \tag{4}$$

In words, q/T is less than or equal to ΔS; or, the maximum possible value of q/T is ΔS. In an exothermic process, q is negative; the heat liberated is $-q$, and its *minimum* value is $-T \Delta S$:

$$-q \geqslant -T \Delta S, q \leqslant T \Delta S$$

For example, if $q = -40$ cal and $T \Delta S = -35$ cal, then $40 > 35$, or $-40 < -35$. Thus, the inequality (4) applies to both cases. If a system absorbs heat q_1 at temperature T_1, and then absorbs heat q_2 at temperature T_2, the inequality can be generalized to

$$\frac{q_1}{T_1} + \frac{q_2}{T_2} \leqslant \Delta S$$

The equals sign is for a reversible process:

$$\frac{q_1}{T_1} + \frac{q_2}{T_2} = \Delta S \tag{5}$$

Suppose we have a system that is perfectly insulated from its surroundings. It can exchange no heat with its surroundings. That means $q = 0$ for every process:

$$\Delta S \geqslant q/T, \; \Delta S \geqslant 0$$

The entropy of this system can only increase. An interesting special case of the insulated system is the ISOLATED SYSTEM. Such a system has no interaction with its surroundings. It cannot change its energy no matter what it does; its energy is conserved. Thus, high or low energy is irrelevant to the stability of the states available to an isolated system. The most stable state—the state in which the system is most likely to be found—is simply the state that can be realized in the largest number of ways: the state of *highest entropy*. The system always tends to go from a state of lower entropy to a state of higher entropy. *For any possible process in an isolated system, $\Delta S \geqslant 0$.*

An isolated system is another of those unattainable idealizations. However, there is one genuinely isolated system: the entire universe. *In every possible process, the entropy of the universe increases.* Whenever the entropy of one system decreases, there must be another system whose entropy increases even more. We give two examples:

(a) Water can be frozen by placing it in contact with cooling coils in which liquid ammonia is boiling. The freezing of the water corresponds to a decrease in entropy—the solid is more orderly (has lower entropy) than the liquid—but this effect is more than compensated by the increase in the entropy of the ammonia as it changes from a liquid to a gas.

(b) Among all the products that could possibly be formed from a fertilized egg plus an ample supply of food, only a very minute fraction would be adult organisms. The development of living organisms represents an enormous increase in order—and thus decrease in entropy—but it is possible because the dispersal of radiation from the sun (like the expansion of a gas) involves a far greater increase in entropy.

Rudolf Clausius (1822–1888), who invented and named the concept of entropy, summarized all of thermodynamics in two famous sentences: "The energy of the universe is constant. The entropy of the universe tends toward a maximum."

HEAT ENGINES / 17.11

Most of the energy used by man is obtained by the combustion of fossil fuels and their refined components: coal (C), natural gas (mostly CH_4), liquefied petroleum gas (mostly C_3H_8 and C_4H_{10}), gasoline (C_7H_{16}, C_8H_{18}, etc.), and fuel oil ($C_{14}H_{30}$, $C_{15}H_{32}$, etc.). Hydrogen is an excellent fuel, but it is expensive (it does not occur in large natural deposits like the other fuels) and difficult to transport as a gas (because of its low density) or as a liquid (because of its low boiling point).

Since the development of the steam engine in the eighteenth century, the standard method of extracting work from fuel has been to burn it with air. The heat thus liberated causes the expansion of a *working substance:* water in a steam engine, the products of combustion in an internal-combustion engine. The working substance, in expanding, does work on a piston or a turbine blade. A heat engine is characterized by two temperatures: the temperature of the working substance before it does work (T_h for hot) and its temperature after it does work (T_c for cold). In one cycle or one unit of time, the working substance absorbs an amount of heat q_h at the temperature T_h, does work w, and gives up heat q_c (a negative quantity) as it cools to the temperature T_c. The input to the engine—what we pay for—is the heat q_h put into the working substance. The output heat q_c is wasted as far as work is concerned, being used merely to heat up the atmosphere or the nearest body of water ("thermal pollution"). The EFFICIENCY of the engine is the ratio of useful output (work) to input:

$$\text{efficiency} = \frac{w}{q_h}$$

Sadi Carnot, a French military engineer, set out to improve the efficiency of steam engines by thinking instead of tinkering. In 1824, he showed that there is an upper limit—depending only on the temperatures T_h and T_c—to the efficiency of a heat engine. An ideal engine which would attain this maximum efficiency is now called a CARNOT ENGINE. Nobody has ever built a Carnot engine; it exists only as a concept. It would operate reversibly, in the thermodynamic sense. Equation (5) is therefore applicable, with $\Delta S = 0$ because the working substance returns to the same state after each cycle. By combining this equation with the first law of thermodynamics (Problem 21, page 356), the efficiency can be calculated:

$$\frac{w}{q_h} = \frac{T_h - T_c}{T_h} \tag{6}$$

Thus, if $T_h = 700°K$ and $T_c = 400°K$, the efficiency of a Carnot engine is

$$\frac{w}{q_h} = \frac{700 - 400}{700} = 0.43$$

If T_h is raised to $800°K$, the efficiency is improved to

$$\frac{w}{q_h} = \frac{800 - 400}{800} = 0.50$$

No heat engine operating at the temperatures T_h and T_c can have greater efficiency than a Carnot engine using these same temperatures. It is not just that no one has been able to devise such an engine; rather, the existence of an engine more efficient than the Carnot engine would contradict the second law of thermodynamics.

Carnot's work was of great practical importance in showing how to improve steam engines. To increase the ideal maximum efficiency, increase the difference $T_h - T_c$, which means, in practice, increase T_h. After a while, though, you bump into the law of diminishing returns as more expensive materials are needed to withstand the higher temperatures and as T_h in the denominator gets bigger. Also, there is a theoretical limit to the flame temperature attainable with a given fuel. As if that were not enough, the equilibrium $N_2 + O_2 \rightleftharpoons 2NO$ shifts to the right as the combustion temperature increases, producing more air pollution. People have long hoped for a way to convert the energy of fuels to useful work by some method not subject to the Carnot limitation inherent in any heat engine, and they have long known that the answer was, in principle, a galvanic cell.

FUEL CELLS / 17.12

A galvanic cell in which the reactants are fed continuously to the cell and the products are continuously removed is called a **FUEL CELL**. Any cell process involves the consumption of some materials and the production of others, but in a conventional (nonfuel) cell, the reactants are part of the cell when it is built and they are not replenished, or, in a storage cell, they are periodically replenished by reversing the current. The only satisfactory fuel cells thus far developed require the expensive and cumbersome H_2 as fuel, and they have been used only in special situations where cost is secondary, notably in space vehicles and military equipment. A fuel cell that would burn petroleum products would be of enormous value—it might supersede the internal-combustion engine—and such cells have been built, but they have not reached the stage of practical application.

In a hydrogen fuel cell, the overall reaction is

$$H_2(g,\ 1\ atm) + \tfrac{1}{2}O_2(g,\ 1\ atm) \longrightarrow H_2O(l)$$
$$\Delta H = -68.32\ \text{kcal},\ \Delta G = -56.69\ \text{kcal}$$

The heat produced on merely burning 1 mole H_2, with no work obtained directly, is $-\Delta H = 68.32$ kcal. This is the q_h in the expression for efficiency: the heat that is put into the working substance in a heat engine when the fuel burns. If the reaction can be made to take place in a galvanic cell, the maximum electrical work available (when the cell operates reversibly) per mole H_2 is $-\Delta G = 56.69$ kcal. This is the w in the expression for efficiency: the work done by the device. The maximum efficiency of the cell then is

$$\frac{w}{q_h} = \frac{-\Delta G}{-\Delta H} = \frac{56.69}{68.32} = 0.83$$

352   17 / The Second Law of Thermodynamics

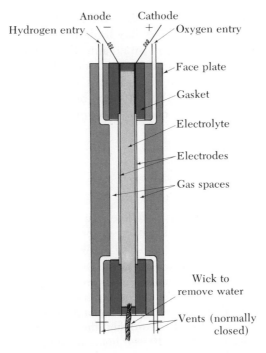

Anode   Cathode

Hydrogen entry   Oxygen entry

Face plate

Gasket

Electrolyte

Electrodes

Gas spaces

Wick to
remove water

Vents (normally
closed)

Fig. 17.2. *A hydrogen–oxygen fuel cell. The electrolyte is a cation-exchange membrane containing mobile* H^+. *The electrodes are of porous platinum.*

Of course, an actual cell will have a lower efficiency because of resistance, polarization, and whatnot, but an engineer is better off working under a ceiling of 83% than under a ceiling of 40 or 50%. The best steam power plants have attained an efficiency near 40%.

Figure 17.2 shows a simplified diagram of a fuel cell similar to those used in spacecraft. If the pressures of the gases are 1 atm and the water has the same vapor pressure as pure water, then the maximum emf of this cell at 25°C can be calculated from Table 16.1:

$$
\begin{array}{lcr}
H_2 \longrightarrow 2H^+ + 2e^- & \text{(oxidation)} & 0 \\
\tfrac{1}{2}O_2 + 2H^+ + 2e^- \longrightarrow H_2O(l) & \text{(reduction)} & +1.229 \\
\hline
H_2 + \tfrac{1}{2}O_2 \longrightarrow H_2O(l) & & \mathscr{E}^0 = 1.229\ \text{V}
\end{array}
$$

In practice, the emf is 0.8 to 1.0 V. The only product should be H_2O. The astronauts can drink this water, and testing its purity provides a diagnosis of the condition of the cell. A number of these cells, typically 30 or 40, are stacked together in series to make a battery yielding 30 V or so. H_2 and O_2 are stored as liquids in insulated containers.

The supply of fossil fuels is rapidly diminishing. The time may not be far off when hydrogen, produced from water by nuclear energy, will be cheaper than oil or natural gas. The hydrogen can be produced by electrolyzing water or by some sequence of reactions requiring only heat (page 431). The energy system of the future may be based on H_2 distributed by pipeline. It is only 20–30% as expensive to pump gas as to transmit electricity over long distances; the ease of using electricity blinds us to the complexity and inefficiency of the system that brings it to us. We may look forward to a time when H_2, instead of natural gas (CH_4), is piped into the home for heating, cooking, refrigeration, and air conditioning*; each building or each neighborhood would have a battery of fuel cells to generate electricity for lights and appliances. Whether H_2 can be used as fuel for automobiles and airplanes, which cannot be connected to a pipeline, is more problematical.

Data in Table 16.1 (page 325) are to be used as needed.

1. Definitions. Define briefly: (a) second law of thermodynamics, (b) reversible process (two senses), (c) free energy, (d) standard free-energy change, (e) free energy of formation, (f) entropy, (g) efficiency, (h) fuel cell.

2. First and second laws. State whether each of the following processes is possible at constant temperature. If it is impossible, which law of thermodynamics does it violate? (a) A gas expands, doing 100 cal of work and absorbing 120 cal of heat. It is then compressed to its initial state; this compression requires 110 cal of work and the gas gives out 120 cal of heat. (b) A battery is charged by doing 10 watt-hours of work on it, while it gives out 1 watt-hour of heat. When the battery is discharged to its initial state, it yields 8 watt-hours of work and gives out 1 watt-hour of heat. (c) A gas is compressed by 15 liter-atm of work and it emits 15 liter-atm of heat. It then expands to its initial state, doing 16 liter-atm of work and absorbing 16 liter-atm of heat.

3. Emf, free energy, and equilibrium constant. For each of the following reactions in aqueous solution (a) find $\mathscr{E}^0$ at 25°C for a cell in which the reaction occurs; (b) calculate ΔG^0 in kcal for the given reaction; (c) calculate the equilibrium constant.

(i) $Cd(c) + Ni^{2+} \rightarrow Cd^{2+} + Ni(c)$
(ii) $H_2(g) + Br_2(l) \rightarrow 2H^+ + 2Br^-$
(iii) $5Sn(c) + 2MnO_4^- + 16H^+ \rightarrow 5Sn^{2+} + 2Mn^{2+} + H_2O$

4. Free energy and equilibrium. For the following process of converting an enantiomer to a racemic mixture (page 496),

$$(+)\text{-}CH_3CHNH_2COOH(aq) \longrightarrow CH_3CHNH_2COOH(aq)$$
$$(+)\text{-alanine} \qquad\qquad\qquad \text{alanine (racemic)}$$

$\Delta G^0 = -360$ cal at 25°C. Calculate K for this reaction.

5. Free energy of formation. Calculate ΔG_f^0 for $AgI(c)$ at 25°C.

6. Form of equation. Compare (a) ΔG, (b) K, and (c) $\mathscr{E}$ for reaction (i) with the corresponding quantity for (ii) and for (iii):

(i) $Zn + 2Ag^+ \leftrightharpoons Zn^{2+} + 2Ag$
(ii) $\frac{1}{2}Zn + Ag^+ \leftrightharpoons \frac{1}{2}Zn^{2+} + Ag$
(iii) $Zn^{2+} + 2Ag \leftrightharpoons Zn + 2Ag^+$

Explain the differing ways in which these three quantities are affected when the equation is written differently.

---

* There is a long-established technology that uses *heat* to operate a *refrigerator*.

7. Direction of reaction. Ascertain, with as little calculation as possible, whether each of the following reactions will go as written or in the reverse direction at 25°C, when all concentrations and pressures are equal to 1:
(a) $2Cr + 3Cu^{2+} \rightarrow 2Cr^{3+} + 3Cu$
(b) $Zn + Mg^{2+} \rightarrow Zn^{2+} + Mg$
(c) $Fe + Cl_2(g) \rightarrow Fe^{2+} + 2Cl^-$
(d) $2Fe^{2+} + Cl_2(g) \rightarrow 2Fe^{3+} + 2Cl^-$
(e) $4Ag + O_2(g) + 2H_2O \rightarrow 4Ag^+ + 4OH^-$
(f) $4Ag + O_2(g) + 2H_2O + 4Cl^- \rightarrow 4AgCl(c) + 4OH^-$

8. Entropy and free energy. For the process S_8 (*rhombic*) $\rightarrow S_8$ (*monoclinic*), $\Delta H = 768$ cal/mole and $\Delta S = 2.08$ cal/°K mole. Calculate ΔG for the process at (a) 25°C, (b) 170°C. What do these results tell about the stability of the two forms of sulfur at each of these temperatures? (c) Calculate the temperature at which $\Delta G = 0$. What is the significance of this temperature?

9. Heat engines. The boiler in a steam engine is at 600°C and the steam exhausted from the turbine is at 150°C. (a) Calculate the maximum possible efficiency of this engine. (b) Suppose that the actual efficiency of the engine is 80% of the theoretical limit. For the reaction $C(graphite) + O_2(g) \rightarrow CO_2(g)$, $\Delta H = -94.05$ kcal. How many kilowatt-hours of work can be produced by this engine when 1 metric ton (1000 kg) of graphite is burned?

10. Fuel cells. For the reaction

$$-195.5 = 23.1 \times 2$$

$$CH_4(g, 1 \text{ atm}) + 2O_2(g, 1 \text{ atm}) \longrightarrow CO_2(g, 1 \text{ atm}) + 2H_2O(l) \quad log.5$$

at 25°C, $\Delta H = -212.80$ kcal and $\Delta G = -195.50$ kcal. (a) Calculate the reversible emf at 25°C of a fuel cell in which this reaction occurs. (b) Calculate the maximum possible efficiency of the cell. (c) Assume that the cell operates with 60% of its maximum efficiency. What is its emf at 25°C?

11. Entropy and free energy. (a) Calculate ΔS for the reaction in Problem 10. (b) Calculate ΔG for the reaction at 300°C. Assume that ΔH and ΔS are independent of temperature.

ADDITIONAL PROBLEMS

12. Reversible and irreversible processes. Classify each of the following processes as reversible (in the thermodynamic sense), nearly reversible, or irreversible. If the process is irreversible, devise a reversible or nearly reversible process by which the system could be brought from the same initial state to the same final state. (a) A bottleful of water evaporates very slowly through a small leak around the stopper. (b) A compressed gas expands in a cylinder fitted with a piston. The expansion is slow because the piston is lubricated by a viscous oil. (c) A gas expands into an evacuated space via a very small hole.

13. Entropy. (a) From information on page 347, calculate the maximum amount of heat (in joules and in calories) absorbed when the reaction

$$Pb(c) + Hg_2Cl_2(c) \longrightarrow PbCl_2(c) + 2Hg(l)$$

occurs as written (1 mole Pb, etc.) in a galvanic cell. (b) Find the entropy change (in cal/deg) in this reaction at 25°C. (c) How much heat (in calories) is absorbed or liberated when the reaction occurs on simple mixing of the reactants?

14. Direction of reaction. Will each of the following reactions proceed as written or in the reverse direction at 25°C when concentrations are about 1 M? (a) $2Ag + Cu^{2+} \rightarrow 2Ag^+ + Cu$. (b) $2Ag + Cu^{2+} + 2Cl^- \rightarrow 2AgCl(c) + Cu$. (c) Would you use your best silver spoon to stir a solution of $Cu(NO_3)_2$? of $CuCl_2$? Explain. (d) What is the molarity of the most concentrated solution of $CuCl_2$ that it would be wise to stir with a silver spoon? Use the Nernst equation.

15. Emf and equilibrium constant. For the reaction $V^{2+} + Md^{3+} \leftrightharpoons V^{3+} + Md^{2+}$, the equilibrium constant is approximately 15 at 25°C. Calculate $\mathscr{E}^0$ for $Md^{3+} + e^- \leftrightharpoons Md^{2+}$ at 25°C. (Mendelevium was the first actinide element found to exist as a stable $2+$ ion in aqueous solution.)

16. Equilibrium constant. Find the equilibrium constant at 25°C for each of the following reactions:
(a) $Ag(c) + H^+ + Cl^- \leftrightharpoons AgCl(c) + \frac{1}{2}H_2(g)$
(b) $Ag(c) + H^+ \leftrightharpoons Ag^+ + \frac{1}{2}H_2(g)$
(c) $AgCl(c) \leftrightharpoons Ag^+ + Cl^-$
Show how the results of (a) and (b) may be used in answering (c). (See page 198.)

17. Equilibrium constant. Find the equilibrium constant at 25°C of each of the reactions in Problem 7.

18. Equilibrium. The reaction $2Fe^{3+} + 2I^- \leftrightharpoons 2Fe^{2+} + I_2(c)$ is carried out in a galvanic cell. (a) Calculate $\mathscr{E}^0$ for this reaction at 25°C. (b) Calculate K for the reaction. (c) KI is added to a solution of Fe^{3+}, initially 0.50 M, until $[I^-] = 1.0$ mole/liter. Calculate $[Fe^{3+}]$ at equilibrium.

19. Emf and efficiency. (a) Calculate the reversible emf at 25°C of a cell in which the reaction is $Pb(c) + Hg_2Cl_2(c) \rightarrow PbCl_2(c) + 2Hg(l)$ from data on page 347. (b) Calculate the maximum efficiency of this cell. (c) Is there anything surprising or impossible about the result of (b)? Explain.

20. Entropy and free energy. For the process $H_2O(l) \rightarrow H_2O(g)$, $\Delta H = 9720$ cal/mole and

$$\Delta S = 26.05 \text{ cal/°K mole} - 2.303R \log P$$

where R is the gas constant (1.987 cal/°K mole) and P is the pressure of $H_2O(g)$ in atm. (a) Write an expression for ΔG in terms of temperature and pressure. (b) Calculate ΔG at 120°C and 1.50 atm. Which state, liquid or gas, is more stable? (c) When $P = 2.00$ atm, at what temperature are the liquid and gas in equilibrium? (d) Calculate the vapor pressure of water at 90°C.

21. Carnot engine. For one cycle in a Carnot engine, $\Delta S = 0$ and $\Delta E = 0$. (a) Write Equation (5) for this process, in terms of q_h, T_h, q_c, and T_c. (b) Write the first law, $\Delta E = q - w$, for this process, in terms of q_h, q_c, and w. (c) Use the result of (b) to eliminate q_c from the equation in (a). Rearrange the resulting equation to obtain Equation (6).

22. Heat engines. The water in a steam engine absorbs energy at the rate of 500 kW and is heated to 400°C in the boiler. The steam leaves the engine at 200°C. (a) What is the maximum possible output of mechanical power from this engine? (b) If the actual efficiency is 75% of the maximum efficiency, what is the rate (in kilocalories/hour) at which waste heat is discharged to the environment? (c) If 10^5 kg of water is available every hour for carrying off this heat, how much will the temperature of the water be raised?

23. Heat engines. Assume that the temperature of the Gulf Stream is 30°C, the flow of water is 2×10^{15} liters/day, and the heat capacity of the water is 1 kcal/deg liter. (a) Calculate the maximum amount of work (in kilocalories) that could be produced each day by a gigantic heat engine that swallows the entire Gulf Stream, uses it to heat the working substance to 15°C, discharges the stream at 15°C, and uses deep water to cool its working substance to 10°C. (b) If world consumption of energy (as work) is 7×10^{16} kcal/year, in how many days would this Gulf Stream engine produce enough work to provide the world's needs for a year?

24. Fuel cells. When a hydrogen–oxygen fuel cell operates reversibly, is heat liberated or absorbed? How much heat is liberated or absorbed per coulomb? Will this amount of heat be increased or decreased in actual (irreversible) operation?

25. Practical cells. The following cell has been proposed for use in electric automobiles:

$$\text{Zn}(c) \,|\, \text{ZnO}(c) \text{ (dispersed in electrolyte)} \,|\, \text{Na}_2\text{SO}_4(aq) \,|\, \text{O}_2(g) \,|\, \text{C} \ (porous)$$

For the cell reaction $\text{Zn}(c) + \tfrac{1}{2}\text{O}_2(g) \rightarrow \text{ZnO}(c)$ at 25°C, $\Delta G^0 = -76.05$ kcal and $\Delta H = -83.17$ kcal. (a) Calculate the standard emf of the cell at 25°C. (b) If the actual emf is 60% of the value calculated in (a), what is the efficiency of the cell? (c) Make the assumption in (b). How many kilowatt-hours of work could be obtained by the oxidation of 1 kg of Zn?

26. Cells and heat engines. Suppose you have a choice between the following devices for powering an automobile:
(*i*) A cell in which the reaction is

$$\text{Li}(l) + \tfrac{1}{2}\text{Cl}_2(g) \longrightarrow \text{LiCl}(l) \qquad \mathscr{E} = 3.3 \text{ V}$$

(*ii*) A gasoline engine using octane (C_8H_{18}) as fuel:

$$\text{C}_8\text{H}_{18}(l) + \tfrac{25}{2}\text{O}_2(g) \longrightarrow 8\text{CO}_2(g) + 9\text{H}_2\text{O}(g), \ \Delta H = -1200 \text{ kcal}$$

The combustion temperature is 2000°K and the exhaust temperature is 500°K. (a) For each device, calculate the maximum work that can be obtained per gram of fuel. In (*i*), fuel includes both Li and Cl_2. (b) Repeat the calculation for (*ii*) but assume that the vehicle must carry O_2 as well as C_8H_{18}.

27. Cells and heat engines. Methanol (CH_3OH, "wood alcohol") has been proposed as a future fuel. It can be produced from coal, wood, garbage, or sewage; it can be burned in an internal-combustion engine (pure or mixed with gasoline) or in a fuel cell. For the reaction

$$\text{CH}_3\text{OH}(l) + \tfrac{3}{2}\text{O}_2(g) \longrightarrow \text{CO}_2(g) + 2\text{H}_2\text{O}(g)$$

$\Delta H = -153$ kcal and $\Delta G^0 = -164$ kcal at 25°C. Calculate the maximum work that can be obtained per gram of methanol, (a) in the engine described in the preceding problem; (b) in a fuel cell at 25°C. (c) If the cell in (b) operates reversibly, does it emit or absorb heat? How much heat per gram of methanol does it emit or absorb? (d) In which direction would your answers to (b) and (c) be changed if the product were $H_2O(l)$ instead of $H_2O(g)$?

28. Fuel production. Page 342 says that H_2O cannot be decomposed to H_2 and O_2 unless at least 57 kcal/mole of electrical work is done. Page 353 says that H_2 can be produced from H_2O by a sequence of reactions requiring only heat. Are these statements in conflict? Explain after reviewing the conditions associated with each.

18

MOLECULAR ORBITAL
THEORY OF BONDING

In this chapter we examine more carefully the formation of the covalent bond. The description of molecules by quantum mechanics can become exceedingly complicated. The best compromise between simplicity and accuracy is probably the molecular orbital (MO) theory, developed about 1932 by Friedrich Hund, Robert Mulliken, Erich Hückel, and John Lennard-Jones.

The force of attraction between bonded atoms is electrostatic. We find this easy to understand for ionic substances, for which we say that the positive ion and negative ion attract each other. The explanation of bonding as the attraction of oppositely charged atoms could be extended to polar molecules like $\overset{\delta+}{H}—\overset{\delta-}{Cl}$, but only with considerable inaccuracy, because the force resulting from the partial charges on the atoms fails to account for the entire bond energy.

The concept fails completely when it confronts a nonpolar molecule like H_2 because the individual atoms bear no net electrical charge. Instead we consider attractions and repulsions among the constituent nuclei and electrons of the entire molecule. In H_2, each of the two nuclei attracts each of the two electrons; the two nuclei repel each other; and the two electrons repel each other. The attractive forces must exceed the repulsive forces. Otherwise the molecule would not exist; the molecule would be less stable than the individual atoms.

BONDING MOLECULAR ORBITALS / 18.2

The MO theory assumes that the covalent bond in a diatomic molecule is derived from the fusion (overlap) of two atomic orbitals, one

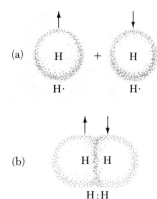

belonging to each of the atoms. This overlap creates a new electron cloud, called a BONDING MOLECULAR ORBITAL, which embraces both atoms. Figure 18.1 depicts the formation of the molecular orbital of an H_2 molecule by the overlap of the $1s$ orbital of the two H· atoms. Note the following: (a) Each electron belongs to the entire molecule rather than to either of the individual atoms; they are said to be DELOCALIZED with respect to the individual atoms. (b) The Pauli exclusion principle applies to molecular orbitals (MO's) as well as to atomic orbitals (AO's). Thus, each MO may have a maximum of two electrons with opposite spins. (c) If electrons were visible, they could be found anywhere within the molecular orbital, but most likely in the region of overlap, that is, between the two nuclei. (d) The strength of the bond results from the fact that *there is a greater electron density between the nuclei than there would be if the two atomic orbitals existed independently.* The more the atomic orbitals overlap, the greater is the charge density between the nuclei, the greater the attractive force in the bond, and the more stable is the bond. This is a crucial point that merits further illustration. Figure 18.2 shows the charge densities of the individual atomic orbitals that would exist if there were no chemical combination of H atoms; it also shows the charge density of the molecular orbital that does exist in H_2. In the molecular orbital the sharing of electrons *increases the negative charge between the nuclei;* the resulting attraction holds the nuclei together.

We are now in a position to visualize the electrostatic nature of the covalent bond. The bond is made up of 2 positive nuclei attracted by a high density of negative charge, as shown in Fig. 18.3a. Such a charge distribution is similar to the alternating Na^+ and Cl^- ions in crystalline NaCl (Fig. 18.3b). There are basic differences, however, between the two. The alternation and attraction of the Na^+ and Cl^- ions extends

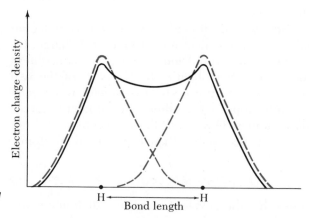

three-dimensionally throughout the entire crystal. In the case of a co-valently bonded molecule, the attraction resides only within the mole-cule. For this reason, the *intra*molecular force of attraction is strong, but the *inter*molecular force of attraction is weak between molecules.

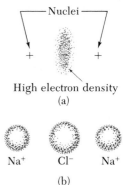

High electron density
(a)

Na⁺   Cl⁻   Na⁺

(b)

Fig. 18.3. *Comparison of the electrostatic nature of (a) a covalent bond, and (b) a segment of an NaCl crystal lattice.*

ANTIBONDING MO'S / 18.3

The mathematical treatment of the MO theory shows that *the number of molecular orbitals formed equals the number of interacting atomic orbitals.* For example, the combination of two atomic orbitals creates two molecular orbitals, one *bonding* and one *antibonding*. The bonding MO is the one discussed in Section 18.2; it has a *lower* energy than the individual AO's from which it is formed. The antibonding or-bital marked with an asterisk, MO*, has a higher energy than the in-dividual AO's. In the MO* there is very little electron density between the nuclei of the atoms, as shown in Fig. 18.4.

This low electron density cannot lessen the repulsion of the nuclei and therefore the MO* has a high energy. But as we shall see, notwith-standing the high energy, MO*'s are capable of housing electrons. The energy of the MO is lower than the energy of the AO's by an amount $-E_{MO}$, whereas the energy of the MO* is greater by approximately the same amount, $+E_{MO}$, as represented in Fig. 18.5. The magnitude of E_{MO} varies, depending on the nature of the interacting atoms.

We can interpret bonding and antibonding molecular orbitals by considering the interacting AO's as each being a standing wave, having a crest and a trough, Fig. 18.6. Conventionally, the crest is assigned a plus (+) sign and the trough is assigned a minus (−) sign. *These signs have nothing to do with electrical charges.* If two waves interact so that the crest (+ sign) of one coincides with the crest (+ sign) of the other,

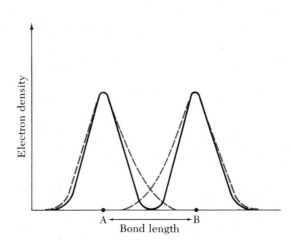

Fig. 18.4. *Plots of electron charge density for an antibonding orbital (*solid black line*) and for separated atomic orbitals (*dashed line*).*

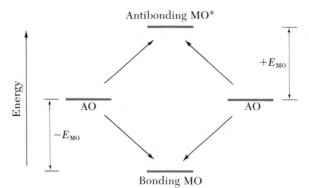

Fig. 18.5. An energy-level diagram showing the combination of two AO's to form two MO's, one bonding MO and one antibonding MO°

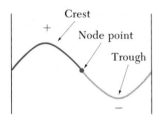

Fig. 18.6. Standing wave showing crest (+) and trough (−).

the new wave is reinforced. The new wave resulting from this type of *in-phase* overlap is analogous to the formation of a bonding molecular orbital. Since the signs of an atomic orbital are assigned in the same way as are those of a wave, we can say that *a bonding molecular orbital results from overlap of atomic orbitals with the same sign.*

If two waves interact so that the crest (+ sign) of one coincides with the trough (− sign) of the other, the new wave is greatly weakened. The new wave resulting from this type of *out-of-phase* interaction is analogous to an antibonding molecular orbital. We can therefore say that *an antibonding molecular orbital results from a combination of atomic orbitals with unlike signs.*

SIGMA (σ) BONDS / 18.4

Schematic representations for bonding MO's are shown in Fig. 18.7 from (a) overlap of two *s* AO's, (b) head-to-head overlap of an *s* and a *p* AO, and (c) head-to-head overlap of two *p* AO's. The three MO's represented in Fig. 18.7 have a high electron density that is distributed symmetrically around an imaginary straight line between the nuclei. This line is called the **BOND AXIS.** Such bonds are referred to as **SIGMA**

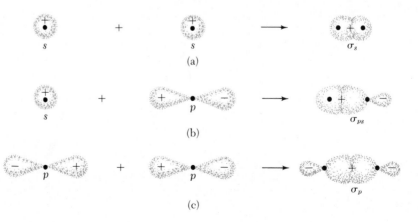

Fig. 18.7. Types of σ-bonding molecular orbitals: (a) σ_s; (b) σ_{ps}; (c) σ_p. The dots represent the nuclei and the + and − represent the signs of the orbitals. Note in each case that like signs overlap.

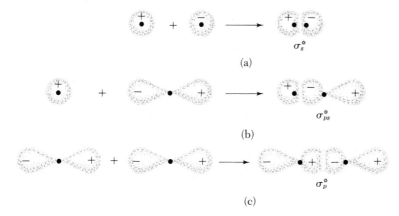

(σ) BONDS. They are more specifically designed σ_s, σ_{sp}, and σ_p depending on the types of AO's from which they are formed. Because of the symmetry of the electron cloud about the bond axis, it is possible to rotate a σ-bonded atom about the axis it makes with its mate without destroying the bond. "Rotation of an atom" means rotation of the electron cloud and any other attached atoms; thus if we rotate C in H_2N—CH_3, we rotate the whole CH_3 group with respect to the H_2N.

If AO's with unlike signs confront each other, as shown in Fig. 18.8a–18.8c, antibonding results. These MO*'s are designated σ^*.

To summarize ($\rightarrow$ stands for "leads to"):

(a) Approach of AO's with like signs $\rightarrow$ overlap $\rightarrow$ high electron density between nuclei $\rightarrow$ attraction $\rightarrow$ bonding MO.

(b) Approach of AO's with unlike signs $\rightarrow$ cancellation $\rightarrow$ almost zero electron density between nuclei $\rightarrow$ repulsion $\rightarrow$ antibonding MO*.

PI (π) BONDS / 18.5

Two p orbitals may also overlap *laterally* (side-to-side) to form a PI (π) BOND (Fig. 18.9). The bonding π MO consists of two electron charge clouds concentrated above and below the axis joining the two nuclei. The MO has a nodal plane (in which the electron density is zero) incorporating the bond axis as shown in Fig. 18.9a. The antibonding MO* is depicted in Fig. 18.9b. *A strong π bond can result*

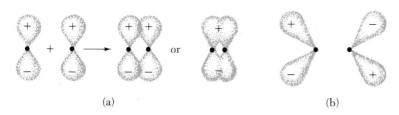

only when the individual p orbitals are parallel. Thus, a π bond can result from overlap of two p_y or two p_z orbitals, but not from a p_y and a p_z. The electron cloud of a π bond is concentrated only in the plane of the component p AO's and therefore the π bond does not exhibit the cylindrical charge symmetry of the σ bond. For this reason π-bonded atoms cannot rotate with respect to each other.

A π bond is rarely formed between atoms unless accompanied by a σ bond. It is the σ bond that supplies most of the electron density along the axis joining the atoms.

MOLECULAR ORBITAL ENERGY LEVELS / 18.6

We now ask the question: Can we set up a series of molecular orbitals from which we can predict the electronic structure of a molecule? Such a procedure would be analogous to what we have learned about setting up a series of atomic orbitals to predict the electronic structure of atoms. We can! The MO theory allows us to do this by making use of relative energy levels of bonding and antibonding molecular orbitals. The energy levels are obtained experimentally. For simplicity, the following application of the MO theory is restricted to the description of bonding in *diatomic* molecules.

A combination of two atomic orbitals gives two molecular orbitals. The one with lower energy is bonding and the one with higher energy is antibonding (Fig. 18.5, page 362). If we combine 2 atoms in the second period of the periodic table, each with five atomic orbitals, $1s$, $2s$, $2p_x$, $2p_y$, and $2p_z$, we get ten molecular orbitals—again the number of molecular orbitals is equal to the number of atomic orbitals. Five of the molecular orbitals are bonding and five are antibonding. Figure 18.10 is an energy-level diagram showing the relative energies of these ten molecular orbitals. Note that an *antibonding* molecular orbital such as the σ_{1s}^* formed from a *lower-energy* atomic orbital has a *lower* energy than a *bonding* molecular orbital such as σ_{2s} formed from a *higher-energy* atomic orbital. The π molecular orbital arising from overlap of the $2p_y$ atomic orbitals has the same energy as the π MO resulting from overlap of the $2p_z$ orbitals. Although the two π molecular orbitals, π_y and π_z, have the same energy, they are not identical. The π_y orbital is oriented along the y axis, and the π_z along the z axis. The same equality of energy exists for the antibonding π_y^* and π_z^* molecular orbitals.

We now ask the important question: How do we use these molecular orbital energy levels to predict the electronic structure of the diatomic molecules? We start with the 2 nuclei and then distribute the total number of electrons into molecular orbitals, applying rules similar to those used for the distribution of electrons into atomic orbitals (page 91). Thus, the molecular orbitals are filled in the order of increasing

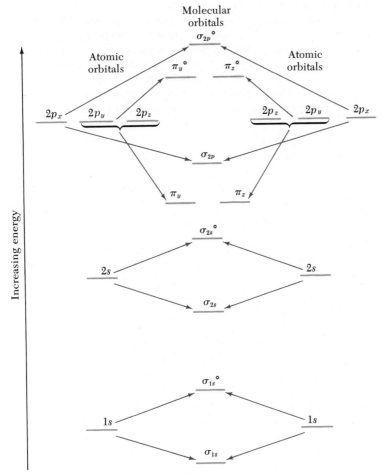

Fig. 18.10. *Energy level diagram showing MO energy levels that arise by overlap of the AO's of the first ten elements.*

energy. Each molecular orbital takes a maximum of 2 electrons, provided the electrons have opposite spins (Pauli's exclusion principle). Hund's rule (page 95) is applicable to molecular orbitals possessing the same energy. *In order for atoms to be bonded there must be an excess of bonding over antibonding electrons.*

Example 1 (a) Use Fig. 18.10 to show the distribution of electrons in (*i*) H_2^+, (*ii*) H_2, (*iii*) H_2^-, (*iv*) He_2. (b) Predict the relative stabilities of the four species.

Answer (a) (*i*) H_2^+ has 1 electron in σ_{1s}. (*ii*) H_2 has 2 electrons in σ_{1s}. (*iii*) H_2^- has 3 electrons, 2 in σ_{1s} and one in σ_{1s}^*. (*iv*) He_2 has 4 electrons, 2 in σ_{1s} and 2 in σ_{1s}^*. These electron distributions are summarized below.

σ_{1s}^*			$\uparrow$	$\uparrow\downarrow$
σ_{1s}	$\uparrow$	$\uparrow\downarrow$	$\uparrow\downarrow$	$\uparrow\downarrow$
Species:	H_2^+	H_2	H_2^-	He_2

(b) The greater the net number of electrons in bonding molecular orbitals, the stronger is the bond. Therefore, the order of bond stability is H_2 (with 2 bonding electrons) $> H_2^+$ and H_2^- (with a net of 1 bonding electron) $> He_2$ (zero net bonding electrons). In fact, He_2 does not exist since there is no advantage for two He atoms to form a bond. Note that the MO theory predicts that both H_2^+ and H_2^- can exist, and indeed they both have been detected. The Lewis shared-pair concept, discussed in Chapter 8, could never accommodate such species; H_2^+ would need a 1-electron bond and H_2^- would need a 3-electron bond.

We now apply this method to several diatomic molecules of second period elements:

Li_2. (atomic number of $Li = 3$) Six electrons (2×3) must be added to the MO's. The four $1s$ "inner-shell" electrons (2 from each atom) fill the σ_{1s} and $\sigma_{1s}{}^*$ molecular orbitals and so contribute nothing to the bond energy. The two $2s$ electrons enter the σ_{2s} bonding molecular orbital, and therefore Li_2 should be stable. The molecule has been detected in the vapor state of lithium. This MO configuration of Li_2 may be represented as

$$Li(1s^2 2s^1) + Li(1s^2 2s^1) \longrightarrow Li_2[(\sigma_{1s})^2(\sigma_{1s}{}^*)^2(\sigma_{2s})^2]$$

N_2. Fourteen electrons are added to the molecular orbitals. The representation is: $(\sigma_{1s})^2(\sigma_{1s}{}^*)^2(\sigma_{2s})^2(\sigma_{2s}{}^*)^2(\pi_y)^2(\pi_z)^2(\sigma_{2p})^2$. Only the electrons in the $n = 2$ energy level appear in the Lewis structural formula. The $(\sigma_{2s})^2$ and $(\sigma_{2s}{}^*)^2$ make no net contribution to the bond energy and therefore are called nonbonding (n) electrons. N_2 has one σ and two π bonds. We relate the MO designation to the structural formula as shown:

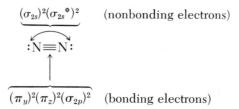

The $(\sigma_{1s})^2(\sigma_{1s}{}^*)^2$ electrons do not participate in bonding and do not appear in the Lewis structural formula.

O_2. The sixteen (8×2) electrons are distributed as shown: $(\sigma_{1s})^2(\sigma_{1s}{}^*)^2(\sigma_{2s})^2(\sigma_{2s}{}^*)^2(\pi_y)^2(\pi_z)^2(\sigma_{2p})^2(\pi_y{}^*)^1(\pi_z{}^*)^1$. Note that in accordance with Hund's rule, the last 2 electrons are distributed, one each, into the equal-energy $\pi_y{}^*$ and $\pi_z{}^*$ MO*'s. These 2 antibonding electrons cancel 2 of the 6 bonding electrons in $(\pi_y)^2(\pi_z)^2$ and $(\sigma_{2p})^2$, leaving a net of 4 bonding electrons in two pairs. Thus the MO theory accurately predicts both the diradical nature and the double-bond character of the O_2 molecule. From this analysis it is evident that neither Lewis formula shown for O_2 (page 126) is adequate by itself. Structure (a) emphasizes the double-bond character, but overlooks the diradical properties. Structure (b) emphasizes the diradical character, but denies the double-bond character.

CO. Figure 18.10 can be used *with caution* to describe the electronic structure of a diatomic molecule composed of 2 different atoms in the second period. The energy levels of the atomic orbitals of the 2 atoms do not coincide, as they do for like atoms, but the order of the energy levels of the molecular orbitals does not change. $:C{\equiv}O:$ has a total of 14 electrons; its electron configuration is isoelectronic with N_2, $(\sigma_{1s})^2(\sigma_{1s}^*)^2(\sigma_{2s})^2(\sigma_{2s}^*)^2(\pi_y)^2(\pi_z)^2(\sigma_{2p})^2$.

METALLIC BONDING / 18.7

We now apply the MO theory to metals. Any theory must explain the typical metallic properties summarized below:

(1) *Metals reflect light and therefore have a lustrous or silvery appearance.* Exceptions are spongy or finely divided forms of metals such as platinum which appears gray to black. Metals in these forms have irregular surfaces which trap the light.

(2) *Metals have high electrical and thermal conductivities.* The electrical conductivity decreases with increase in temperature—a behavior peculiar to metals. Furthermore, the passage of current does not produce any chemical change.

(3) *Metals are malleable (can be flattened) and ductile (can be drawn into a wire).* Especially significant is the fact that the crystalline structure of metals is preserved under moderate deformations.

It has long been recognized that these metallic properties are related in some way to the so-called "free" state of electrons in metals. Metals are composed of atoms that easily ionize. These electrons go into gigantic molecular orbitals which extend over all the atoms in the entire metallic crystal. The units making up the metallic crystal are the positive ions. The following points are helpful in applying MO theory to metallic bonding:

(a) The number of molecular orbitals is exactly equal to the number of AO's which combine to produce them. N atoms with x AO's per atom yield Nx molecular orbitals.

(b) Half are low-energy bonding MO's and half are high-energy antibonding MO*'s.

(c) Within any given shell, as the number of molecular orbitals increases, *the energy difference between the bonding and antibonding molecular orbitals decreases.* For example, the smallest visible piece of lithium contains about 10^{19} atoms; for such a large number, the energy

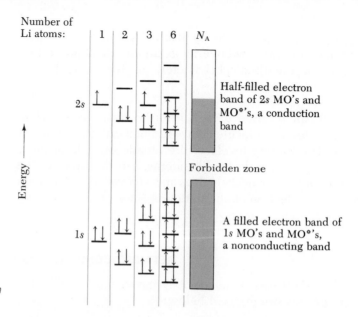

Fig. 18.11. *Electron bands in metallic lithium.*

difference between successive MO and MO* levels is so small that the levels are, for practical purposes, continuous. The group of closely spaced MO's and MO*'s is called an **ELECTRON BAND.** In lithium, one of these bands, originating from the $1s$ atomic orbital, is fully occupied; the other, originating from the $2s$ atomic orbital, is only half-occupied. The energy gap between the two bands is so large that it effectively prevents promotion of electrons from the lower to the higher band. Such energy gaps have come to be called "**FORBIDDEN ZONES.**"

(d) These partially filled electron bands, called **CONDUCTION BANDS,** are responsible for metallic properties. Figure 18.11 summarizes these concepts for a mole (N_A) of Li atoms.

(e) Electrons in the filled shells of an atom, such as $1s$ in Li, will be found in filled electron bands. Only electrons in incompletely filled energy levels, such as $2s$ in Li, are found in conduction bands.

For the Group 2R elements such as Mg ($1s^2 2s^2 2p^6 3s^2$) all electron bands derived from inner shells are filled. The electron band comprising the molecular orbitals from the filled $3s$ AO's is also filled. From these filled bands we might predict that Mg would not be metallic. Yet Mg is a typical metal. The prediction overlooked the fact that the molecular orbitals from the $3s$ AO's and from the $3p$ AO's are very close in energy and blend into a single partially filled electron conduction band. This structure is consistent with the behavior of Mg as a metal.

METALLIC PROPERTIES IN TERMS OF MO BAND THEORY / 18.8

How does an incompletely filled electron band account for metallic properties? In an incompletely filled band an electric current promotes electrons to slightly higher-energy unoccupied molecular orbitals within the band. These promoted electrons, unlike those in filled bands, are free to move in the direction of the applied potential. As a result, the metal conducts electricity. A rise in temperature increases vibration of the positive ions in the crystal lattice; such vibrations interfere with the movement of electrons. As a result, electrical conductivity of metals *decreases* as the temperature *increases*.

Electrons in conduction bands can also absorb thermal energy; the ready transport of energy by these electrons accounts for the high thermal conductivity of metals. The absorption and reemission of photons of visible light by conduction electrons accounts for the high reflectivity of metals. Metals are ductile and malleable because under mechanical stress the positive ions of the crystal can move past each other with very little resistance and without breaking any metallic bonds.

INSULATORS AND SEMICONDUCTORS / 18.9

INSULATORS do not conduct electricity. In insulators the "conduction band" is empty. The forbidden zone below the conduction band represents a large energy barrier, and promotion of electrons to the conduction band normally does not occur.

In SEMICONDUCTORS the conduction band is populated by electrons that are occasionally promoted to it by excitation from lower bands. The forbidden zone is narrower than in the case of insulators, and therefore constitutes less of a barrier to promotion. These substances show

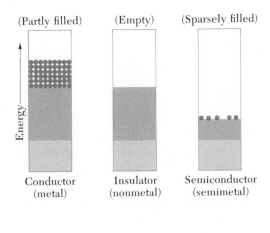

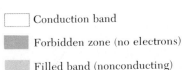

Fig. 18.12. *Band theory of solids. Dots represent electrons in the conduction band.*

meager electrical conductivity. A rise in temperature excites electrons to the conduction band. This enrichment of electrons in the conduction band increases electrical conductivity more than enhanced vibration of the atoms decreases it. The net result is that the electrical conductivity of semiconductors *increases* as the temperature increases. Figure 18.12 illustrates the arrangement of bands and forbidden zones in conductors (metals), insulators (nonmetals), and semiconductors. Some important semiconductors (used in transistors and other "solid-state" devices) are Ge, Si, GaAs, and InSb.

1. Definitions. Define and illustrate each of the following terms: (a) bonding molecular orbital, (b) antibonding molecular orbital, (c) σ bond, (d) π bond, (e) nonbonding electrons, (f) conduction band, (g) forbidden zone, (h) semiconductor.

2. MO designation. (a) Give the MO designation, as was done for O_2 (page 366), for (*i*) Be_2, (*ii*) F_2, (*iii*) He_2^+, (*iv*) HeH^+, (*v*) OF, (*vi*) Ne_2. (b) Which of these species is/are unlikely to exist? Explain.

3. MO theory. (a) Give the molecular orbital designation for C_2, as was done for O_2 (page 366). (b) Should C_2 be paramagnetic?

4. Orbital overlap. Explain why overlap between a p_x and a p_y orbital is nonbonding (neither bonding nor antibonding), whereas a head-to-head overlap of two p_x orbitals can be bonding or antibonding. Draw a schematic representation of each type of overlap indicating the (+) and (−) signs.

5. Metallic bonding. Account for the fact that a diamond or a crystal of NaCl can be shattered by a mechanical blow whereas a piece of gold cannot.

6. Band theory. A small particle of sodium contains 1×10^{-6} mole. (a) How many atoms, (b) how many atomic orbitals, and (c) how many molecular orbitals does this particle contain? (d) Write the symbols showing the distribution of electrons in sublevels for this particle of sodium.

7. MO theory. Which of the following species would you expect to be paramagnetic? (a) He_2^+, (b) NO, (c) NO^+, (d) N_2^{2+}, (e) O_2^-, (f) O_2^{2-}.

8. MO theory. Explain in terms of MO theory (Fig. 18.10) why the first ionization energy of nitric oxide (NO) is less than that of CO.

9. MO theory. Explain why (a) electrons in filled shells have no effect on bonding strengths; (b) there is relatively free rotation about the C—C bond of H_3C—CH_3 but not about the C=C bond of H_2C=CH_2.

10. Sigma and pi bonds. Compare σ and π bonds with respect to (a) mode of formation, (b) rotation about the bond.

11. MO theory. Rationalize the fact that H_2 has a greater ionization energy than an H atom has while the ionization energy of O_2 is less than that of O. (See Fig. 18.10.)

12. Orbital overlap. (Consult page 92 for the schematic representations of the d orbitals.) Sketch a bonding and an antibonding molecular orbital resulting from overlap of a p_y orbital and the d_{xy} atomic orbital.

13. MO theory. In terms of the MO theory, suggest which species of each of the following pairs has the greater electron affinity (page 115) (consult Fig. 18.10): (a) OF or NF, (b) $O_2{}^{2+}$ or $N_2{}^{2+}$, (c) CN or NO.

14. MO theory. Give the distribution of electrons in the molecular orbitals of BN, if in this case π_y, π_z, and σ_{2p} have the same energy.

15. MO theory. When considering the first *three* ionization energies of an O_2 molecule, where would you expect to find the smallest jump in energy?

19

ORBITAL HYBRIDIZATION
AND MOLECULAR SHAPE

We now turn our attention to chemical bonding of molecules with more than two atoms. Later in this chapter we will discuss how the shapes of polyatomic molecules are determined by the kinds of bonds among their atoms. Much of the success of modern chemistry depends on an understanding of the relationship between the physical and chemical properties of substances and the shapes of their molecules. Nowhere are these relationships more important than among the substances of which living organisms are composed.

HYBRIDIZATION OF ATOMIC ORBITALS (AO'S) / 19.1

Any theory that accounts for bonding among more than two atoms must be consistent with the experimentally determined bond energies, lengths, and angles. As an example, let us summarize some of these bond properties for CH_4, methane:

(a) All four C—H bonds are identical; they have the same bond length (1.093 Å) and bond energy (99.4 kcal/mole). Substitution of a Cl atom for any H atom gives only one kind of methyl chloride molecule, CH_3Cl.

(b) All H—C—H bond angles are equal; they are 109.5°.

$$H \underset{C}{\overset{109.5°}{\diagdown\!\!\diagup}} H$$

The 2s and the three 2p orbitals of carbon have different energies and therefore could not form four equivalent bonds with H atoms. The bond from the 2s orbital would be different from the three equivalent bonds from the p orbitals. Carbon, then, must somehow use *four* equivalent atomic orbitals (AO's) for bonding with the H atoms.

sp³ hybrid orbitals. It is assumed that the four equivalent AO's used by carbon are fabricated by blending the 2s and the three 2p orbitals. This blending is called HYBRIDIZATION. Hybridization involves a redistribution of the ground state AO's; it does not result in an increase or decrease in the number of orbitals. We do not attempt to talk about the steps whereby a C atom in its ground state combines with 4 H atoms to form a methane molecule, for no one has found a way to investigate the course of this process experimentally. Instead, we just assume that the 4 unhybridized (ground-state) AO's of the C atom yield 4 equivalent hybridized orbitals. Although the hybridized state of an isolated C is imaginary, it is nevertheless useful in helping us visualize what happens when the molecule is formed. Since these four hybrid AO's are derived from three p AO's and one s AO they are called sp^3 hybrid AO's. The four bonding electrons of carbon are distributed among the sp^3 hybrid AO's according to Hund's rule as shown below:

2p ↑ ↑ __ __ $2sp^3$ ↑ ↑ ↑ ↑

2s ↑↓

1s ↑↓   1s   ↑↓

ground state of carbon sp^3 hybridized state of carbon

Each sp^3 hybrid orbital is shaped like a p orbital, except that its two lobes are of unequal size (Fig. 19.1). In order to simplify the use of orbital pictures for molecules, the small lobes are omitted. The four sp^3 hybrid AO's overlap head-to-head with the 1s AO of each of the four H atoms to form four σ and four σ* molecular orbitals. In general, *bonds formed from hybrid AO's fabricated from s and p atomic orbitals will be sigma bonds.* The 8 electrons needed for bonding (4 from C and 4 from the H atoms) fill the four σ MO's. Each σ MO encompasses only the C atom and the particular H atom—it is said to be a localized two-centered MO. A more sophisticated model involves delocalized multicentered MO's, each of which encompasses the entire molecule. As we shall see later, for some purposes the delocalized description is needed but in general all σ and most π bonds can be assumed to be localized two-centered molecular orbitals.

Since a pair of electrons in one σ bond repels the pair of electrons in each of the other σ bonds, the theory predicts that the four σ bonds formed from the sp^3 hybrid AO's should be separated from each other in space as much as possible. This means that the bond angles must be as large as possible. To best satisfy this need, the axis of each sp^3 orbital should be directed toward the corner of a regular tetrahedron and the bond angle should be 109.5° (Fig. 19.2). The same geometry is pre-

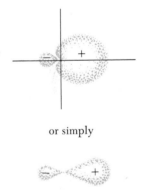

or simply

Fig. 19.1. *Shape of an sp^3 hybrid atomic orbital. This is also the shape of all hybrid orbitals made up from s and p orbitals, the so-called s–p-type hybrid orbitals.*

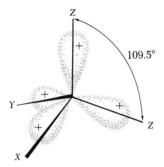

Fig. 19.2 *Orientation of tetrahedral, sp³ hybrid orbitals. (The small lobes are omitted.) Bond to X projects toward the viewer. Bond to Y projects away from viewer. Bonds to each Z are in the plane of the paper.*

dicted by the mathematical process of sp^3 hybridization. The observed bond angle is 109.5°. The success of this prediction strengthens the chemist's belief in the concept of hybridization. Because of this geometry, sp^3 orbitals are called TETRAHEDRAL HYBRID ORBITALS.

sp² hybrid orbitals. The concept of hybridization is also used to account for the three equivalent B—F bonds in boron trifluoride. Each B—F bond length is 1.29 Å, and the bond energy is 154 kcal/mole. All 4 atoms lie in the same plane. The structure has a B atom in the center of an equilateral triangle and a F atom at each of the three corners. The F—B—F bond angle is 120°:

boron trifluoride

The three necessary equivalent AO's are assumed to result from a blending of *one* 2s and *two* of the 2p AO's of boron. The new hybrid AO's are called sp^2:

2p	↑ __ __	2p	
		$2sp^2$	↑ ↑ ↑ __
2s	↑↓		
1s	↑↓	1s	↑↓

ground state of boron sp^2 hybridized state of boron

The observed geometry of the molecule is in accord with the predicted orientation of the hybrid orbitals; the sp^2 AO's are planar and separated by 120° (Fig. 19.3). This is also the arrangement that permits the maximum bond angle (minimum electron repulsion). The sp^2 hybrid AO's always have this triangular, planar orientation and hence are called TRIGONAL HYBRID ORBITALS.

sp hybrid orbitals. A third type of hybridization of s and p atomic orbitals is suggested to explain the bonding in covalent beryllium compounds, such as $BeCl_2$. In the gaseous state it is a linear molecule, Cl—Be—Cl; all 3 atoms lie on a straight line. Each Be—Cl bond has the same bond length and energy, and so they are equivalent. It is necessary, therefore, for the Be atom to provide two hybridized AO's to form the two equivalent σ bonds. It is assumed that the two AO's which hybridize are the 2s and *one* of the 2p orbitals. These two hybrid AO's are designated sp:

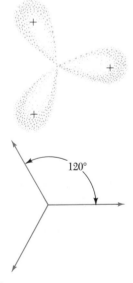

Fig. 19.3. *Planar orientation of trigonal, sp² hybrid orbitals. (The small lobes are omitted.)*

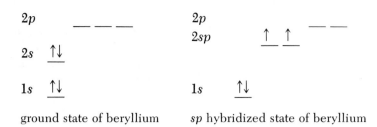

ground state of beryllium   sp hybridized state of beryllium

180°

Fig. 19.4. *Orientation of the two sp hybrid orbitals. (The small lobes are omitted.)*

Since there are only two covalent bonds to the Be atom, the maximum angle separating the axes of the two hybrid AO's should be 180° (Fig. 19.4), which *is* the observed value. Because of this orientation, the *sp* hybrid orbitals are also called DIGONAL HYBRID ORBITALS.

Table 19.1 summarizes the properties of the hybrid AO's formed from *s* and *p* orbitals, the so-called *s–p* type hybrids.

Table 19.1
Properties of s–p-Type Hybrid Atomic Orbitals

Type	Name	s Character	p Character	Bond angle	Geometric arrangement	Type of bond formed
sp	Digonal	$\frac{1}{2}$	$\frac{1}{2}$	180°	Linear	Sigma
sp²	Trigonal	$\frac{1}{3}$	$\frac{2}{3}$	120°	Planar triangular	Sigma
sp³	Tetrahedral	$\frac{1}{4}$	$\frac{3}{4}$	109.5°	Tetrahedral	Sigma

The bond angle can be used as a criterion for the type of AO's used in bonding. For example, in ammonia, NH_3, the three N—H bonds are equivalent, and the H—N—H bond angle is 107°. This is close to the tetrahedral angle of 109.5° and so the N atom in NH_3 is assumed to use sp^3 hybrid orbitals:

2p   ↑ ↑ ↑ $2sp^3$ ↑↓ ↑ ↑ ↑

2s   ↑↓

1s   ↑↓   1s   ↑↓

ground state of nitrogen   sp³ hybridized state of nitrogen

We note that a hybrid AO can also accommodate an unshared pair of electrons. The N atom has available three equivalent $2p$ orbitals with which to form three equivalent N—H bonds. Nevertheless, it is assumed that N uses hybrid AO's. Were N to use *p* orbitals to form the three σ bonds with the H atoms, the bond angle would be approximately 90°, as shown in Fig. 19.5. This is so because each *p* orbital is

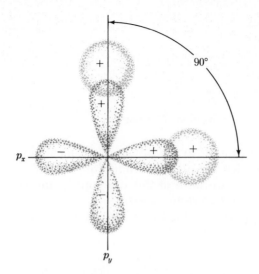

Fig. 19.5. *Representation of two σ bonds, formed from two p orbitals to show 90° bond angle.*

oriented along one of the three axes, x, y, or z, and the angle between any two of these axes is 90°.

The two equivalent O—H bonds in water form an angle of 105°. This value is closer to the sp^3 value of 109.5° than it is to the 90° angle expected if p orbitals were used to form the two σ bonds. The O atom, therefore, is also assumed to use sp^3 hybrid AO's rather than p orbitals:

2p ⇅ ↑ ↑

2s ⇅

1s ⇅

ground state of oxygen

$2sp^3$ ⇅ ⇅ ↑ ↑

1s ⇅

sp^3 hybridized state of oxygen

In most of their molecules, Be, B, C, N, and O use hybrid atomic orbitals to form σ bonds and also to house lone pairs of electrons, when present.

A crude analogy to the concept of hybridization is the mixing of a can of blue paint with cans of white paint. The s orbital is represented by the can of blue paint, and the p orbitals are represented by three cans of white paint. The hybrid orbitals will have lighter shades of blue, depending on the number of white cans mixed with the blue. This mixing is shown in Fig. 19.6. Notice that the fewer the cans of white paint that are mixed, the darker the shade of blue. We can interpret these shades of blue in terms of relative energy levels of the hybrid orbitals. Since the s orbital (can of blue paint) is at a lower energy level than the p orbital (can of white paint), the darker the shade of blue the

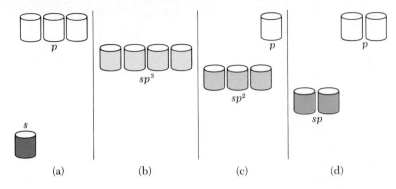

lower is the energy level of the corresponding hybrid orbital. Hence, the sequence of relative energy levels of the *s–p* type of orbital is

$$
\begin{array}{c}
p \\
sp^3 \\
sp^2 \\
sp \\
s
\end{array}
$$

We usually say that for a given atom *the more s character* (or less *p* character) *in an atomic orbital, the lower is the energy of the electrons in the orbital and the closer are the electrons to the nucleus.* Consequently, if the electrons are unshared they become less reactive as the *s* character of their orbital increases. We also find that for a given atom its covalent bond radius decreases as the *s* character of its bond increases.

MULTIPLY BONDED MOLECULES / 19.2

We have seen that the triple bond in N_2 and CO consists of a σ and two π bonds. We will always find that *no more than one bond between 2 atoms can be a σ bond,* and that *all bonds in excess of one must be π bonds.* Thus, single bonds are σ bonds, double bonds are composed of one σ and one π bond, and triple bonds are composed of one σ bond and two π bonds.

Of special interest are the multiply bonded carbon compounds, as typified by ethylene, acetylene, and CO_2. The geometry of these 3 molecules is the clue to the hybridization of the carbon atoms. Ethylene is a planar molecule in which the H—C—H and H—C—C bond angles are approximately 120°:

$$
\begin{array}{ccc}
\text{H} & & \text{H} \\
& \diagdown \,\,120°\,\, \diagup & \\
& \text{C} = \text{C} \,\,120° & \\
& \diagup \quad\quad \diagdown & \\
\text{H} & & \text{H}
\end{array}
$$

Fig. 19.7. *Electronic structures exhibited by carbon: (a) ground state; (b) sp³ hybridization; (c) sp² hybridization; (d) sp hybridization. The distribution of hybridized electrons is shown so that carbon can contribute one electron for each of four covalent bonds.*

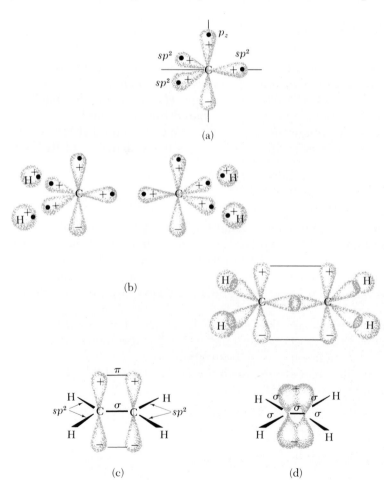

Such geometry implies sp^2 hybridization for each carbon atom. The energy levels of the atomic orbitals of the sp^2 hybridized carbon atom are shown in Fig. 19.7c. The orientation of the sp^2 and p orbitals is shown in Fig. 19.8a.

The bonding of two such units with 4 H atoms is shown in Fig. 19.8b. Two of the three sp^2 hybrid orbitals of each carbon atom overlap with the 1s orbital of each of 4 H atoms. The third sp^2 hybrid orbital overlaps with the sp^2 hybrid orbital of the other carbon atom. This array of σ bonds, called the **SKELETON** of the molecule, is planar (Fig. 19.8c). The p_z orbitals remaining on the adjacent carbon atoms overlap to form a π

Fig. 19.8. *Orbital representation of ethylene: (a) sp² hybrid orbitals and a pz orbital; (b) assembly of two such units plus four H 1s orbitals; (c) skeleton; (d) representation of π bond. (Black dots represent electrons.)*

bond. The π bond is *perpendicular* to the skeleton of the molecule (Fig. 19.8d). The 2 carbon atoms and the 4 hydrogen atoms provide a total of 12 bonding electrons. Ten of these electrons are used for the five σ bonds. The remaining 2 electrons constitute the π bond.

According to MO theory, since twelve AO's are involved in bonding (eight from the 2 C atoms and four from the 4 H atoms), twelve molecular orbitals result; six are bonding and six are antibonding. Five of the bonding and five of the antibonding orbitals are σ. The remaining two are a π and a π^* molecular orbital. The 12 electrons are then used to fill the five σ and the one π bonding molecular orbitals. Each molecular orbital embraces the two bonding atoms (localized MO).

The picture of a triple bond is illustrated with acetylene, C_2H_2. The 4 atoms in acetylene lie on a straight line,

$$H—C{\equiv}C—H$$
acetylene,
a linear molecule

A linear shape indicates that each C uses *sp* hybrid AO's (Table 19.1). The energy levels of an *sp* hybridized C are shown in Fig. 19.7d. Figure 19.9 shows the assembly of 2 H atoms, each with a 1*s* atomic orbital, and 2 C atoms, each with two *sp* hybrid orbitals and a p_y and p_z orbital. Each C atom forms two σ bonds, one by overlapping an *sp* hybrid orbital with the other C atom, and one by overlapping an *sp* orbital with the *s* orbital of a H atom. The p_y orbitals overlap laterally to form a π

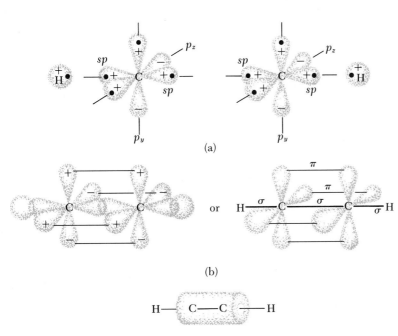

(a)

(b)

(c)

Fig. 19.9. Orbital representation of acetylene; (a) assembly of 2H and 2C atoms; (b) representation showing σ-bonding skeleton and π-bond formation; (c) representation of the triple bond.

bond along the y axis, and the p_z orbitals interact similarly to form a π bond along the z axis (Fig. 19.9b). Therefore, the triple bond comprises one σ and two π bonds. The planes of the two π bonds are at right angles to each other and to the C—C σ bond axis. It is believed, however, that these two π bonds coalesce into a cylindrical shape (Fig. 19.9c).

Carbon dioxide is an example of a molecule with two double bonds to the same carbon atom. It is a linear molecule in which both O=C bonds are equivalent:

$$:\ddot{O}\!=\!\!C\!\!=\!\!\ddot{O}:$$
$$\underset{180°}{}$$

shape of a CO$_2$ molecule

Linearity means the O—C—O bond angle is 180°; we therefore assume that the carbon uses sp hybridized orbitals to form a σ bond to each O atom (Fig. 19.10). The C atom also has two p orbitals, each of which overlaps laterally with a p orbital of an adjacent oxygen atom to give two π bonds. Each π bond holds a pair of electrons, one from each atom, and therefore the octet requirement for C is satisfied. *The separate π bonds are in planes perpendicular to each other.* Since each O atom bonds to only one other atom, there is no bond angle with O at the center to tell us about the kind of AO used by the O atoms. However, we assume that the unshared pairs of electrons are in hybrid orbitals and then both O atoms are sp^2 hybridized.

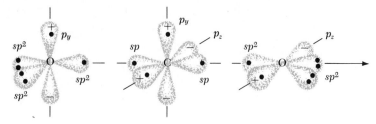

Unshared pairs of electrons

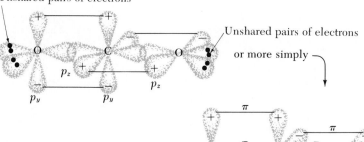

Unshared pairs of electrons

or more simply

Fig. 19.10. *Representation of the construction of a molecule of CO$_2$. (C uses sp hybrid AO's; (O uses sp² hybrid AO's).*

For the 3 molecules just discussed, the geometry was used to indicate the kind of hybridization that should be assumed for the carbon atom. However, it would be preferable if the type of hybridization could be deduced just from the Lewis formula. We do this by assuming that *each unshared and σ-bonded pair of electrons needs a hybrid orbital*. Pi bonds and odd electrons are ignored in counting hybrid orbitals. Then the

number of hybrid AO's = number of (σ bonds + *unshared pairs* of electrons)

Calculated no. of hybrid orbitals	Predicted hybridization
2	sp
3	sp^2
4	sp^3

This assumption applies mainly to elements in the second period of the periodic table, such as Be, B, C, N, and O. For certain compounds, elements in the higher periods of the periodic table may use unhybridized p orbitals for bonding. For example, S must use p orbitals to bond to H in H_2S since the bond angle is 91°.

Table 19.2 summarizes the pertinent information for hybridization of orbitals of carbon.

Table 19.2
Hybridization of Orbitals of Carbon

	sp^3	sp^2	sp
Electron distribution	sp^3 ↑ ↑ ↑ ↑	p ↑ sp^2 ↑ ↑ ↑	p ↑ ↑ sp ↑ ↑
Bonding	σ (single bond)	$\sigma + \pi$ (double bond)	$\sigma + 2\pi$ (triple bond or two double bonds)
Number of hybrid orbitals	4	3	2
Shape Bond angle	Tetrahedral 109°	Trigonal 120°	Digonal 180°
Condition of electrons in hybrid orbitals	s Character ——— increasing ——→ p Character ——— decreasing ——→ Energy ——— decreasing ——→ Stability ——— increasing ——→		

Example 1 Predict the type of hybridized AO's used by the carbon atom in (a) a carbanion, such as $:CH_3^-$; (b) a carbonium ion such as H_3C^+; (c) a free radical such as CH_3; (d) $:C\equiv O:$.

Answer (a) Three σ bonds $+ 1$ unshared pair $= 4$. The C atom uses sp^3 hybridized AO's. (b) Three σ bonds $+ 0$ unshared pairs $= 3$. The C atom uses sp^2 hybridized AO's, and it has one remaining p orbital, with no electrons, perpendicular to the plane of the σ bonds. (c) Since odd electrons do not need a hybrid orbital, the answer is the same as (b), except that the remaining p orbital has 1 electron. (d) One σ bond $+ 1$ unshared pair $= 2$. The C atom uses sp hybridized AO's.

HYBRIDIZATION WITH d ORBITALS / 19.3

Elements of the third and higher periods have d orbitals available for bonding and consequently may exceed the octet. The extra electrons can be in bonds (as in PCl_5 and SF_6) or some can be unshared (as in $:SF_4$ and $:\ddot{C}lF_3$). We shall consider the hybridization of orbitals of central atoms surrounded by five or six pairs of electrons. This discussion is restricted to molecules of *representative elements*. Transition elements are considered in Chapter 22.

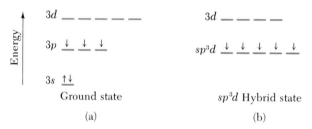

Fig. 19.11. *Electronic arrangement of highest shell of phosphorus: (a) ground state; (b) hybridized sp³d state.*

(a) **Five pairs of electrons on the central atom.** In a molecule of PCl_5 five orbitals are needed to accommodate the five pairs of electrons. It is theorized that P uses the one 3s, three 3p, and *one* 3d orbital to form five sp^3d hybrid orbitals (Fig. 19.11). The PCl_5 molecule is a **TRIGONAL BIPYRAMID** (Fig. 19.12). Three Cl atoms lie at the corners of an equilateral triangle, while a fourth Cl atom lies above and a fifth below the center of the triangle. The five P—Cl bonds are *not* equivalent because the bond angles involving the Cl atoms above or below the triangle are 90° whereas the other bond angles are 120°.

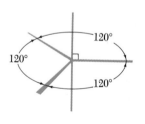

Fig. 19.12. *Orientation of trigonal bipyramidal, sp³d hybrid AO's.*

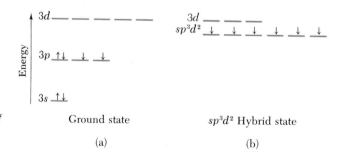

Fig. 19.13. *Electronic arrangement of highest shell of sulfur: (a) ground state; (b) hybridized sp³d² state.*

(b) Six pairs of electrons on the central atom. In a molecule of SF_6, S must provide six AO's to accommodate the six pairs of electrons. It does so by hybridizing its one $3s$, three $3p$, and *two* $3d$ orbitals (Fig. 19.13) to give six equivalent sp^3d^2 hybrid AO's. Each orbital points to the corner of an octahedron (Fig. 19.14), and hence the sp^3d^2 hybrid orbitals are called OCTAHEDRAL. The angle between any two adjacent octahedral hybrid orbitals is 90°.

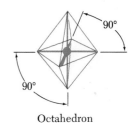

Octahedron

Fig. 19.14. *Representation of octahedral, sp^3d^2 hybrid AO's.*

PROPERTIES OF COVALENT BONDS / 19.4

Chemical bonds are characterized by their length and energy. In this section we are concerned with the structural features that change the normal values of these properties. *Any structural feature that shortens the distance between 2 atoms increases the energy needed to dissociate the atoms.* This happens because the closer the atoms are, the more their orbitals overlap. The generalization is substantiated by the observed distances and energies of the bonds between the 2 carbon atoms in ethane, ethylene, and acetylene:

	H_3C—CH_3	H_2C=CH_2	HC≡CH
Bond length (in Å)	1.54	1.33	1.20
Bond energy (in kcal/mole)	81	141	194

These values also show that the *distance between 2 atoms decreases when the number of bonds between the atoms increases.* Multiple bonding results in a greater electron density between the bonding nuclei. The nuclei are more strongly attracted by this greater electron density and are brought closer together. Multiple-bond covalent radii are shown in Table 19.3 for C, N, O, and S.

Bond polarity may also cause contraction of bond length. For example, the length of the polar Si—F bond, calculated from Table 7.6 (page 112) by adding r_{Si} and r_F, is 1.86 Å. However, the measured bond length in SiF_4 is only 1.54 Å. On the other hand, when a comparison is made of the calculated and observed lengths of a nonpolar bond the correspondence is very good. Thus, in the INTERHALOGEN compound, bromine chloride, Br—Cl, the calculated bond length is

Table 19.3
Multiple Bond Covalent Radii (in Angstrom Units)

Element	Single-bond radius	Double-bond radius	Triple-bond radius
C	0.77	0.665	0.60
N	0.73	0.60	0.55
O	0.74	0.55	
S	1.02	0.94	

2.13 Å; the observed length is 2.14 Å. Coincidental with the contraction of a bond length is an increase in bond energy. The closer the atoms approach each other, the more effectively can their atomic orbitals overlap and the stronger is the bond. Hence, we find that *an increase in polarity of a covalent bond always strengthens the bond.*[†]

RESONANCE AND DELOCALIZED *p* ELECTRONS / 19.5

(a) **Resonance concept.** Let us write a Lewis structural formula for dinitrogen oxide (nitrous oxide), N_2O (page 452). There are two reasonable Lewis formulas for N_2O, both of which fit the octet rule and have the required skeleton. Which formula *if any* is correct? This question might be resolved by comparing the calculated with the observed bond lengths. Formal charges are also shown:

Bond lengths, Å	$:\overset{-}{\overset{..}{N}}\!\!=\!\!\overset{+}{N}\!\!=\!\!\overset{..}{O}:$		$\longleftrightarrow$	$:N\!\!\equiv\!\!\overset{+}{N}\!\!-\!\!\overset{-}{\overset{..}{\underset{..}{O}}}:$	
Calculated	1.20	1.15		1.10	1.47
Observed	1.12	1.19		1.12	1.19

Neither set of predicted bond lengths matches the observed values. Neither formula by itself is correct. This difficulty in assignment of the electronic formula is rationalized by the concept of **RESONANCE**.

According to the resonance concept, the structure of a molecule such as N_2O *cannot be accurately depicted by a single Lewis formula.* Instead, the molecule is depicted by *two or more* formulas, such as those shown above for N_2O, which, taken together, serve as a better description than any single one. Thus, if you ask the question: "What kind of bond exists between the N atoms in N_2O?", each formula, taken by itself, predicts an incorrect bond length. The two formulas taken together, however, imply the answer: "something between a double and a triple bond," which, while not startlingly precise, is somewhat better. The separate formulas are called **CONTRIBUTING (RESONANCE) STRUCTURES**, of which the actual molecule is said to be a **RESONANCE HYBRID**. The word "hybrid" is used in an imaginary, not a biological, sense, and may be exemplified by the description of a rhinoceros offered by a child after a first visit to the zoo as "an animal that looks like a dragon and a unicorn." The dragon and unicorn are mythical, as are the contributing structures; the rhinoceros, like the resonance hybrid, is real. A **DOUBLE-HEADED ARROW**, ↔, is written between the contributing structures to indicate resonance. There are, however, important restrictions on the formulas one may write.

[†] The reader should not form the impression that, since the ionic bond is the extreme case of a polar bond, ionic bonds are stronger than polar covalent bonds. The strongest bond is an optimum blend of covalent and ionic bonding character.

(*i*) The relative positions of all atoms (the skeleton) must be the same in all contributing structures—*only the positions of the electrons may differ.*

(*ii*) There must be the same number of paired and unpaired electrons in all contributing structures, but the number of bonds need not be the same. For example, :Ö::Ö: and :Ö:Ö: are *not* contributing structures of O_2 because the number of paired and unpaired electrons is different. On the other hand, :C̄:::O: and :C: :Ö: are contributing structures of
carbon monoxide since they have the same number of pairs of electrons even though the number of bonds is different.

A resonance hybrid may have more than two contributing structures. If the contributing structures have similar energies, the hybrid has no close resemblance to any of them. If the contributing structures have dissimilar energies, the hybrid "looks" most like the one with the lowest energy.

Orbital hybridization can be incorporated into the concept of resonance. If the bond orbitals in H_2O are made from p orbitals of O, the bond angle should be 90°. If they are made from sp^3 hybrid AO's, the bond angle should be 109.5°. The observed angle, 105°, is between these values. Consequently, the actual AO's used are "resonance hybrids" of p and sp^3 AO's but *closer* to sp^3.

Example 2 Write contributing electronic structures for nitrate ion, NO_3^-, showing formal charges.

Answer The double bond is placed in each of the three equally possible N to O positions:

Observe that the three contributing structures for NO_3^- look equivalent. In each structure the N atom has a single bond to each of 2 O atoms, and a double bond to a third O atom. The three structures taken together have a different connotation than any single one. Any one structure implies, incorrectly, two N—O bond distances of the same length and one shorter N=O bond distance. Actually, all three nitrogen–oxygen bond distances are the same. This experimental fact implies that all 3 oxygen atoms are indistinguishable from each other. No one O atom can therefore be said to be the one that is doubly bonded to the N atom. Instead, *each O atom participates in a double bond*, as is assumed for the resonance hybrid.

(b) **Extended (delocalized) π bonding.** With species like N_2O and NO_3^- we must think of the π MO's as being *delocalized over more than two atoms.*

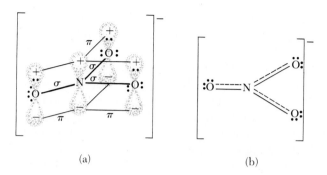

(a)   (b)

We now consider the orbital model for NO_3^-. Since each O—N—O bond angle is 120°, N is assumed to use sp^2 hybrid AO's to form the σ-bonded skeleton. The remaining p orbital of N acts as a pivot for side-to-side overlapping with a p orbital of each O, Fig. 19.15a. These four p AO's engender a delocalized, extended π system made up of four MO's—two bonding and two antibonding. Figure 19.15a denotes the lowest-energy bonding MO which has overlap only between lobes of like signs. For convenience, we depict the delocalized extended π system by drawing dashed lines between the N and each O, Fig. 19.15b. This structure also represents the resonance hybrid.

STRUCTURE OF OXYANIONS; p–d π BONDING / 19.6

Anions of the type $(MO_n)^{x-}$, such as sulfate, SO_4^{2-}, and nitrate, NO_3^-, are called **OXYANIONS**. Their structures are of particular interest because they reveal the role played by π bonding as influenced by the size of the central atom. The oxyanions of the second period elements, B, C, and N,

$$\left[\begin{array}{c} O \\ \overset{\leftarrow 120°}{B} = O \\ O \end{array}\right]^{3-} \quad \left[\begin{array}{c} O \\ \overset{\leftarrow 120°}{C} = O \\ O \end{array}\right]^{2-} \quad \left[\begin{array}{c} O \\ \overset{\leftarrow 120°}{N} = O \\ O \end{array}\right]^{-}$$

| orthoborate anion | carbonate anion | nitrate anion |

have 3 oxygen atoms; each oxygen participates in extended p–p π bonding† with the central atom. The ions are planar and have bond angles of 120°. Each central atom uses sp^2 hybrid orbitals to form the three σ bonds. The remaining p orbital is used to form the π system delocalized over all 4 atoms.

† So named because two p orbitals are involved.

The oxyanions of the third period elements, Si, P, S, and Cl,

$$\left[\begin{array}{c} :\overset{..}{\text{O}}: \\ :\overset{..}{\text{O}}-\underset{\underset{:\overset{..}{\text{O}}:}{|}}{\overset{|}{\text{Si}}}-\overset{..}{\text{O}}: \end{array}\right]^{4-} \left[\begin{array}{c} :\overset{..}{\text{O}}: \\ :\overset{..}{\text{O}}-\underset{\underset{:\overset{..}{\text{O}}:}{|}}{\overset{|}{\text{P}}}-\overset{..}{\text{O}}: \end{array}\right]^{3-} \left[\begin{array}{c} :\overset{..}{\text{O}}: \\ :\overset{..}{\text{O}}-\underset{\underset{:\overset{..}{\text{O}}:}{|}}{\overset{|}{\text{S}}}-\overset{..}{\text{O}}: \end{array}\right]^{2-} \left[\begin{array}{c} :\overset{..}{\text{O}}: \\ :\overset{..}{\text{O}}-\underset{\underset{:\overset{..}{\text{O}}:}{|}}{\overset{|}{\text{Cl}}}-\overset{..}{\text{O}}: \end{array}\right]^{-}$$

| orthosilicate anion | orthophosphate anion | sulfate anion | perchlorate anion |

can have 4 oxygen atoms bonded to the central atom. Each of these oxyanions has a tetrahedral structure, and, hence, the central atom utilizes sp^3 hybrid AO's to form the four σ bonds, one to each atom.

There are two main reasons for this difference in structure between oxyanions of the second and third period elements:

(a) The third period elements are larger and therefore provide more space to surround themselves with four O atoms.

(b) For third period elements, the bond length to O is too long to permit effective overlap of the p orbitals (Fig. 19.16a). Therefore, third period elements cannot form stable p–p π bonds. (Sulfur seems to be an exception; it does form some relatively stable p–p π bonds, as in carbon disulfide, S=C=S.)

The third period elements can acquire more than 8 electrons, by using their available empty $3d$ AO's. As exemplified by S in SO_4^{2-}, a d orbital of S can overlap with a p orbital of O to give a p–d π bond (Fig. 19.16b). Although the third period element could not form a p–p π bond, it can form a p–d π bond, because the d orbital "reaches out" to overlap the p orbital of the O atom. The $3d$ orbital furnished by the S atom is devoid of electrons, whereas the $2p$ orbital of O has 2 electrons. Thus, as a result of p–d π bonding, electron density is delocalized from O to the central atom, in this case to S. Since each O atom can form a p–d bond, the structure of SO_4^{2-} is best represented as possessing a delocalized, extended π system:

$$\left[\begin{array}{c} \text{O} \\ \| \\ \text{O}=\!=\!\text{S}=\!=\!\text{O} \\ \| \\ \text{O} \end{array}\right]^{2-}$$

sulfate anion,
showing delocalized p–d π system

The effectiveness of any kind of π bonding decreases with increasing bond distance. Therefore, as the central atom gets larger, its bond to O gets longer and p–d π bonding becomes less effective.

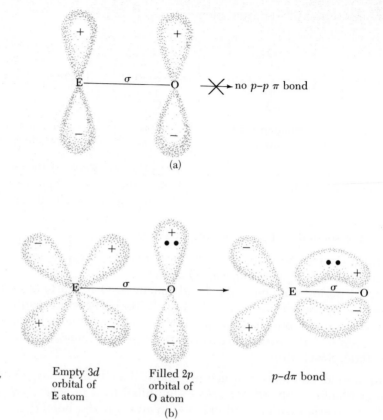

Fig. 19.16. *Pi bonding between a third period element, E, and an O atom: (a) unsuccessful p–p π bond; (b) successful p–d π bond.*

Empty 3d orbital of E atom

Filled 2p orbital of O atom

(b)

no p–p π bond

(a)

σ

p–dπ bond

STRUCTURE AND REACTIVITY OF OXYACIDS AND OXYANIONS / 19.7

The chemist is much concerned with the relationship between the structure and reactivity of substances. A knowledge of such a relationship enables the chemist to make new compounds which have certain desired properties. The Brönsted acid–base reaction is a common and relatively simple reaction; we therefore use it to discuss the relationship of structure and reactivity. We here restrict our discussion to the oxyacids and oxyanions. We will first present the facts about reactivity, and then we will present a broad concept to rationalize the relative reactivities. The acidity of oxyacids follows periodic patterns and is also influenced by the oxidation number of the central atom.

(a) *Acidity of oxyacids increases on proceeding from left to right across a period,* as typified by acids of the elements of the third period.

(b) *Acidity of oxyacids increases on proceeding down a group,* as typified by the group 5R oxyacids

increasing acidity $\longrightarrow$

$Al(OH)_3$	H_4SiO_4	H_3PO_4	H_2SO_4	$HClO_4$
aluminum	silicic	phosphoric↑	sulfuric	perchloric
hydroxide	acid†	acid	acid	acid
(amphoteric)				

H_3AsO_4 increasing
arsenic   acidity
acid

H_3SbO_4

antimonic
acid

(c) *The oxyacids tend to become more acidic the higher the oxidation number of the element*, as illustrated by the chlorine oxyacids:

Acid	Formula	Oxidation number	pK_a
Hypochlorous	$HClO$	+1	7.5
Chlorous	$HClO_2$	+3	2
Chloric	$HClO_3$	+5	−1
Perchloric	$HClO_4$	+7	−11

It is possible to predict relative reactivities of a group of acids or bases provided that only *one* of these factors varies. We can easily compare H_2SO_3 and H_2SO_4 (same central atom but different oxidation numbers) but it is more difficult to compare H_2SeO_4 and H_3PO_4 (different group and different period).

Explanations of the relative acidities of molecules are best given in terms of the stability of the products, more specifically, in terms of the anion, since the H_3O^+ ion is a common product. Thus, *if a conjugate base is more stable (less reactive to H^+) than the conjugate base of a closely related acid, its acid is stronger.* For example, since $HClO_4$ is a stronger acid than $HClO_3$, our theoretical explanation would have to explain the fact that ClO_4^-, the conjugate base of $HClO_4$, is a more stable anion than ClO_3^-, the conjugate base of $HClO_3$.

CHARGE DISPERSAL AND BASICITY / 19.8

We will explain the effects of electronic structure on basicity by the **PRINCIPLE OF CHARGE DISPERSAL.** *Dispersal (delocalization) of the electron density from the donor site of a base stabilizes the base and thus decreases its tendency to accept a proton (weakens its basicity).* As a

† Silicic acid is more likely hydrated silicon dioxide, $SiO_2(H_2O)_n$.

corollary, the conjugate acid of a more stable base tends more to lose a proton to revert to the stable structure, and is thus a stronger acid. These ideas are summarized for the $HClO_4$, ClO_4^- conjugate acid–base pair:

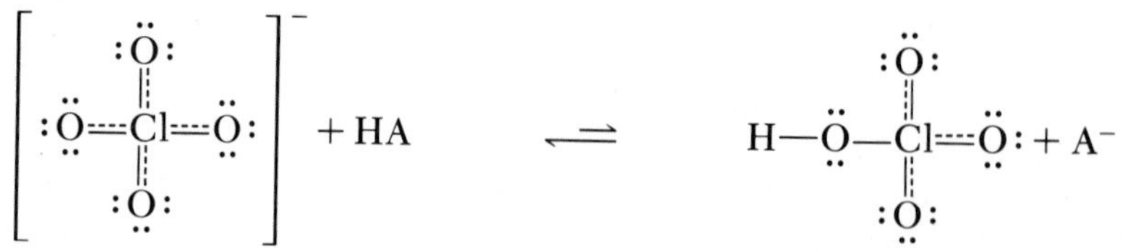

$$\left[\begin{array}{c} :\ddot{O}: \\ \| \\ :\ddot{O}\!=\!\!\!=\!\!Cl\!=\!\!\!=\!\ddot{O}: \\ \| \\ :\ddot{O}: \end{array} \right]^{-} + HA \quad \rightleftharpoons \quad H\!-\!\ddot{O}\!-\!\overset{\displaystyle :\ddot{O}:}{\underset{\displaystyle :\ddot{O}:}{\overset{\|}{Cl}}}\!=\!\!\!=\!\ddot{O}: + A^-$$

Negative charge is widely dispersed,

↓

ion is very stable (unreactive),

↓

tendency to accept H^+ is very low,

↓

basicity is extremely weak.

Tendency to become stable anion by losing H^+ is very high,

↓

molecule is unstable (reactive),

↓

acidity is very strong.

For oxyanions, delocalization (dispersal) of charge results from extended π bonding. For example, in the nitrate ion, a typical oxyanion, the charge is dispersed to all 3 oxygen atoms and *is not just on the O from which the proton was removed*. We could say that each O bears one-third the charge on the anion. This charge delocalization greatly stabilizes the nitrate ion, causing the ion to be a very weak base, and, hence, nitric acid to be a very strong acid. *The more lone O atoms*

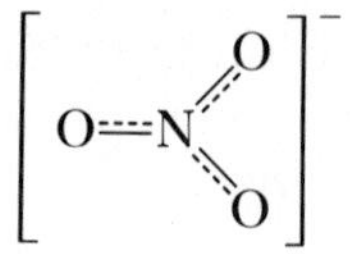

$$\left[O\!=\!\!\!=\!\!N\underset{O}{\overset{O}{<}} \right]^{-} \qquad\qquad \left[O\!=\!\!\!=\!\!N\overset{O}{\underset{\cdot\cdot}{<}} \right]^{-}$$

nitrate ion
The more lone O's bonded
to nitrogen,

↓

the more extended the π bonding,

↓

the more the delocalization of charge,

↓

the weaker the base.

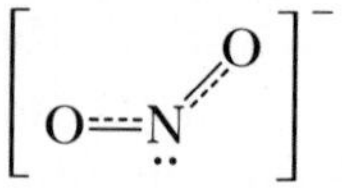

nitrite ion
The fewer lone O's bonded to
nitrogen,

↓

the less extended the π bonding,

↓

the less the delocalization of charge,

↓

the stronger the base.

bonded to a given atom, the more extended is the π system, the more delocalized is the charge, and the more stable is the anion. Thus, NO_3^-, having 3 O atoms participating in extended π bonding, is more stable and therefore less basic than NO_2^-, with only 2 O atoms participating in extended π bonding. (See bottom of page 390.)

The basicity of the chlorine oxyanions also decreases as the number of lone O atoms increases. The oxyanion in which the central atom has the larger oxidation number has the larger number of lone O atoms for participating in extension of the π bond, and therefore is the weaker base:

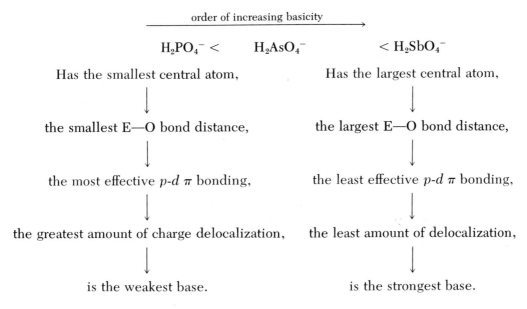

oxidation
number: +7   +5   +3   +1

increasing charge delocalization

decreasing basicity

In this series of anions delocalization of charge is produced by extended p–d π bonding.

In a given periodic group, as the size of the central atom increases, the bond distance to O increases, and the effectiveness of p–d π bonding decreases. This explains the following order of basicity (E represents the central atom):

order of increasing basicity

$H_2PO_4^- <$ $H_2AsO_4^-$ $< H_2SbO_4^-$

Has the smallest central atom,	Has the largest central atom,
↓	↓
the smallest E—O bond distance,	the largest E—O bond distance,
↓	↓
the most effective p-d π bonding,	the least effective p-d π bonding,
↓	↓
the greatest amount of charge delocalization,	the least amount of delocalization,
↓	↓
is the weakest base.	is the strongest base.

Because of the familiar conjugate acid–base pair relationship, we find the following order of acidity:

$$H_3PO_4 > H_3AsO_4 > H_3SbO_4$$

order of decreasing acidity →

Up to now we have only compared oxyanions with identical charge. With anions of different charge, other things being equal, *basicity increases with negative charge* because of the greater attraction for the proton. Phosphate ion, PO_4^{3-}, is a stronger base than hydrogen phosphate, HPO_4^{2-}, which is stronger than dihydrogen phosphate, $H_2PO_4^-$. This order is another way of saying that an anion is a stronger base than its conjugate acid. One can (but with less confidence) generalize about oxyanion bases which have different central atoms. Thus, orthosilicate, SiO_4^{4-}, is a stronger base than phosphate, PO_4^{3-}, which is a stronger base than sulfate, SO_4^{2-}, which is stronger than perchlorate, ClO_4^-. The basicity diminishes as the charge on the anion decreases.

Shapes of molecules are deduced from experimental measurements. They are expressed in terms of bond lengths and bond angles as determined *only* by the position of the *nuclei* of the atoms. Therefore, unshared electrons or electrons in π bonds are not part of the molecular shape, although they do exert an influence. These electrons are shown in Lewis structural formulas but they cannot be "seen" by the instruments used to determine shapes of molecules. The shape of a molecule of the AB_n type is defined by *the σ bonds which project from the central atom*. Since σ bonds result from overlap of atomic orbitals, it follows that the geometry of the molecule is determined essentially by the type of atomic orbitals used by the central atom, A. The σ bond is strongest when the overlap of atomic orbitals is *head-to-head* along the internuclear axis (Fig. 19.17a) rather than off the axis (Fig. 19.17b). Therefore, the σ bond angle is defined by the atomic orbitals of the central atom, A, as in Fig. 19.18.

Fig. 19.17. *Representation of overlap of s–p-type AO's: (a) head-to-head (along axis); (b) off-axis.*

more effective than

Head-to-head overlap

(a)

Off-axis overlap

(b)

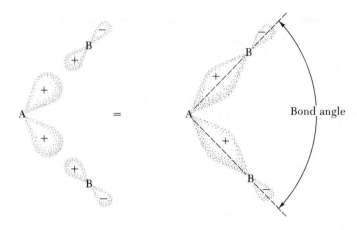

Fig. 19.18. *Representation of a typical B—A—B bond angle formed by overlap of two s-p-type atomic orbitals of A, each with a p orbital of atom B. (Dashed line indicates the bond axis.)*

The tendency for maximum orbital overlap disposes some atoms, as, for example, the N in NH_3 or the O in H_2O, to use an s–p-type hybrid AO rather than an unhybridized p orbital. Sigma bonds formed from s–p-type hybrid orbitals are stronger than those formed from p orbitals.

Another major factor influencing molecular geometry is the tendency to minimize repulsion between pairs of electrons. Orientation of the five types of hybrid atomic orbitals, sp, sp^2, sp^3, sp^3d and sp^3d^2, is such as to maximize the bond angles, thereby minimizing the repulsion. The shapes resulting from these hybrid AO's are summarized in Table 19.4.

EFFECT OF UNSHARED PAIRS OF ELECTRONS / 19.10

In this discussion we assume that the central atom uses hybrid AO's. In the AB_n type of molecule, the replacement of one or more A-B bonds by an unshared pair of electrons alters the shape of the molecule but without changing the hybridized state of the central atom. Remember

Table 19.4
Shapes of Molecules and Ions of Representative Elements with No Unshared Pairs of Electrons on Central Atom

General type	Example	Observed shape	Number of hybrid AO's	Hybrid orbital type
AB_2	$BeCl_2$, $HgCl_2$	Linear	2	sp
AB_3	BF_3, $GaCl_3$	Triangular	3	sp^2
AB_4	CH_4, NH_4^+, BF_4^-	Tetrahedral	4	sp^3
AB_5	$SbCl_5$, PCl_5	Trigonal bipyramidal	5	sp^3d
AB_6	SF_6, AlF_6^{3-}, SiF_6^{2-}	Octahedral	6	sp^3d^2

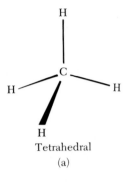

Tetrahedral

(a)

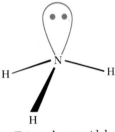

Trigonal pyramidal

(b)

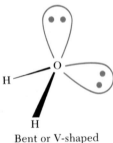

Bent or V-shaped

(c)

Fig. 19.19. *Shapes of (a)
methane, CH_4, AB_4 type; (b)
ammonia, :NH_3, :AB_3 type; (c)
water, $H_2\overset{..}{O}$:,:$\overset{..}{A}B_2$ type.*

that molecular shape is defined by the *positions of nuclei, not of electrons*. Thus, *any* diatomic molecule, like HCl, regardless of the number of unshared pairs, has only one bond axis and is, by definition, linear. With polyatomic molecules we must locate the unshared pairs of electrons on A and consider their effect on the positions of the remaining B atoms. This is best done in terms of electron pair repulsions.

Unshared pairs repel adjacent electron pairs more strongly than do bonding pairs. Consequently, two unshared pairs repel each other more than do two bonding pairs. The repulsion between an unshared pair and a bonding pair is intermediate in magnitude. Electron pair repulsions decrease in the order: unshared pair–unshared pair > unshared pair–bonding pair > bonding pair–bonding pair. Consequently, the presence of unshared pairs causes contraction of angles made by bonding pairs.

We now discuss specific cases of shapes of molecules or ions whose central atom has one or more unshared pairs of electrons.

(a) **Four electron pairs on the central atom.** In this group, there are two types of species, :AB_3, as typified by ammonia, :NH_3, and :$\overset{..}{A}B_2$, as typified by water, :$\overset{..}{O}H_2$. In both molecules, the central atom uses sp^3 tetrahedral hybrid atomic orbitals. All four orbitals are equivalent, and it is therefore immaterial which orbital houses an unshared pair of electrons.

It has been experimentally determined that the molecule :NH_3 is a TRIGONAL (triangular) PYRAMID; the 3 H atoms form the triangular base of the pyramid with the N atom at the apex (Fig. 19.19b.). The $H_2\overset{..}{O}$: molecule is BENT (V-SHAPED), Fig. 19.19c.

(b) **Six electron pairs on the central atom.** In this group are mainly two types, :AB_5, as typified by bromine pentafluoride, :BrF_5, and :$\overset{..}{A}B_4$, as typified by xenon tetrafluoride, $\overset{..}{X}eF_4$. In the AB_6 octahedral molecule (Fig. 19.20a), all six sp^3d^2 hybridized atomic orbitals are equivalent. We should expect that any of the six could accommodate the lone pair of electrons equally well. We find that the shape of :BrF_5 is a SQUARE PYRAMID (Fig. 19.20b); 4 of the fluorine atoms form the square base of the pyramid, and the fifth fluorine atom is at the apex.

With $\overset{..}{X}eF_4$, a new factor appears: the two pairs of electrons can be arranged in *two* different ways about the central atom. The axes of the orbitals with the unshared pairs of electrons can form either an angle of 90° (Fig. 19.20d) or an angle of 180° (Fig. 19.20c). The shape of XeF_4 is square-planar (Fig. 19.20c). The Xe atom is in the center and the F atoms are at the corners of a square; all the atoms are thus in the same plane. Evidently, the two orbitals with the unshared pairs form the maximum angle of 180° with each other, as we would expect from our concept of electron pair repulsion.

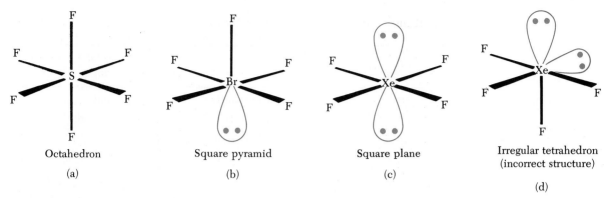

Octahedron	Square pyramid	Square plane	Irregular tetrahedron (incorrect structure)
(a)	(b)	(c)	(d)

Fig. 19.20. *Structures of typical molecules whose central atom has six pairs of electrons: (a) SF_6, AB_6 type; (b) :BrF_5, :AB_5 type; (c) XeF_4, AB_4 type; (d) incorrect structure for XeF_4.*

(c) Five electron pairs on the central atom. This group has mainly three categories of structures: (*i*) :AB_4, as typified by tellurium tetrachloride, :$TeCl_4$; (*ii*) $\ddot{A}B_3$, as typified by chlorine trifluoride, $\ddot{C}lF_3$; (*iii*) :$\ddot{A}B_2$, as typified by xenon difluoride, :$\ddot{X}eF_2$. Of the five types of hybrid atomic orbitals shown in Table 19.4, the sp^3d is unique in that not all five orbitals are equivalent. The three orbitals oriented in a plane toward the corners of a triangle make up the **EQUATORIAL GROUP**, and the two orbitals oriented perpendicular to the plane compose the **AXIAL GROUP** (Fig. 19.21a). The nonequivalence of the bonds is revealed by the bond distances in PCl_5, P—Cl (axial) = 2.19 Å; P—Cl (equatorial) = 2.04 Å. When bonding pairs are replaced by unshared pairs, the observed shapes of the three types of molecule *indicate that the unshared pairs assume equatorial positions*† (Fig. 19.21b–19.21d). Table 19.5 summarizes the typical shapes of molecules and ions whose central atom has at least one unshared pair.

Fig. 19.21. *Shapes of typical molecules with five pairs of electrons on the central atom: (a) PCl_5, AB_5 type; (b) :$TeCl_4$, :AB_4 type; (c) :$\ddot{C}lF_3$, :$\ddot{A}B_3$ type; (d) :$\ddot{X}eF_2$, :$\ddot{A}B_2$ type.*

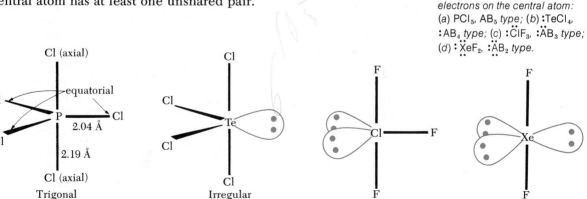

Trigonal bipyramid	Irregular tetrahedron	T-shape	Linear shape
(a)	(b)	(c)	(d)

† The equatorial orbitals have a lower energy since they have more *s* character. They may be sp^2 hybridized and the axial orbitals *pd* hybridized. The unshared pair seeks out the lowest energy orbitals.

Table 19.5
Shapes of Molecules and Ions Having Unshared Pairs on Central Atom

General type	Examples		Shape	Orbital type	Orbital arrangement
:AB_3	:NH_3	H_3O:$^+$	Trigonal pyramid	sp^3	Tetrahedral
:$\ddot{A}B_2$	$H_2\ddot{O}$:	:$\ddot{S}Cl_2$	V shape	sp^3	Tetrahedral
:AB_4	:SF_4	:$TeCl_4$	Irregular tetrahedron	sp^3d	Trigonal bipyramidal
:$\ddot{A}B_3$	:$\ddot{Cl}F_3$		T shape	sp^3d	Trigonal bipyramidal
:$\ddot{A}B_2$	:$\ddot{X}eF_2$	:$\ddot{I}Cl_2^-$	Line	sp^3d	Trigonal bipyramidal
:AB_5	:IF_5	:BrF_5	Square pyramid	sp^3d^2	Octahedral
:$\ddot{A}B_4$	:$\ddot{X}eF_4$	:$\ddot{I}Cl_4^-$	Square plane	sp^3d^2	Octahedral

MOLECULAR SHAPE AND DIPOLE MOMENTS / 19.11

As previously mentioned (page 131), some molecules possess a dipole moment. *The molecular dipole moment is a composite of the individual bond moments and the effect of unshared pairs in hybrid atomic orbitals.* The bond and unshared pair moments can be considered vectors, Appendix I.17, the sum of which gives the molecular dipole moment. The molecular dipole moments, which are experimentally determined, are useful criteria for determining shapes of molecules. For the linear, triangular planar, tetrahedral, trigonal bipyramidal, and octahedral shapes, the sum of the individual bond moment vectors is zero if the central atom is bonded to only one kind of atom or group. This fact is demonstrated for the linear, triangular planar, and tetrahedral structures (Fig. 19.22). For consistency it is assumed that the bonded atom is more electronegative than the central atom. The fact that compounds such as $BeCl_2$, BF_3, CCl_4, gaseous PCl_5, and SF_6 have zero dipole moments supports (but does not prove) the assigned structures. If the bonded atoms are not the same, the vector sum is no longer zero. For example, whereas carbon dioxide, $O{=}C{=}O$, has no dipole moment, carbon oxysulfide, $O{=}C{=}S$, has a finite dipole moment. Bent molecules, such as H_2O, and pyramidal molecules, such as NH_3, have dipole moments, a fact distinguishing them from linear and plane triangular molecules, respectively.

The effect of an unshared pair on the dipole moment depends on the orbital in which it resides. An unshared pair in an *s* orbital, as in H_2S, has no effect on the molecular moment since the *s* orbital is symmetrically disposed around the nucleus. An unshared pair in a *p* orbital has no effect since the two lobes are of equal size and in opposite direction. An unshared pair in an *s–p* type hybrid atomic orbital, as in :NH_3, contributes to the molecular moment, since the two lobes are of unequal

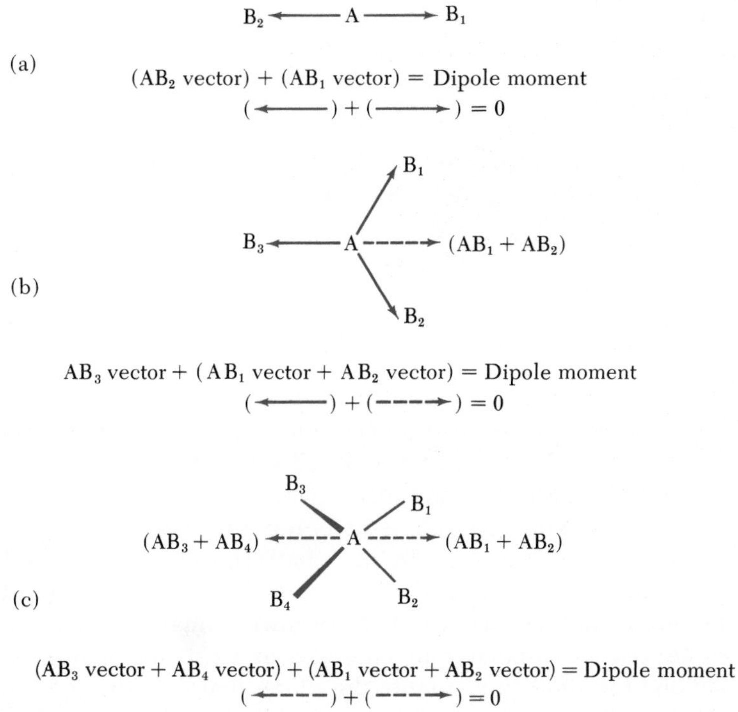

$$B_2 \longleftarrow A \longrightarrow B_1$$

(a)

$$(AB_2 \text{ vector}) + (AB_1 \text{ vector}) = \text{Dipole moment}$$
$$(\longleftarrow) + (\longrightarrow) = 0$$

(b)

$$AB_3 \text{ vector} + (AB_1 \text{ vector} + AB_2 \text{ vector}) = \text{Dipole moment}$$
$$(\longleftarrow) + (\dashrightarrow) = 0$$

(c)

$$(AB_3 \text{ vector} + AB_4 \text{ vector}) + (AB_1 \text{ vector} + AB_2 \text{ vector}) = \text{Dipole moment}$$
$$(\overset{\longleftarrow}{\text{---}}) + (\dashrightarrow) = 0$$

Fig. 19.22. *Vector representation of zero dipole moments for (a) linear, AB_2; (b) triangular planar, AB_3; and (c) tetrahedral, AB_4 molecules. The dashed arrows are resultant vectors.*

size (Fig. 19.19b, page 394). The fact that the dipole moment of NF_3 is very small indicates that the effect of the unshared pair almost cancels the moments of the three N—F bonds.

MOLECULAR SYMMETRY / 19.12

Symmetry is an important aspect of molecular shape. Intuitively, we recognize that a sphere and cube are highly symmetrical bodies. In this section are described the structural features, called SYMMETRY ELEMENTS, which make a molecule symmetrical. The search for symmetry elements sharpens one's ability to visualize the various structures. The use of three-dimensional models is strongly recommended for developing the ability to recognize and visualize symmetry elements. Once the models have been studied, the three-dimensional projectional formulas can be more easily understood. The symmetry elements that we shall consider are planes and points.†

(a) **Symmetry planes.** *A symmetry plane cuts through the molecule so as to divide it into equivalent halves.* This means that a line which starts at any atom, and is drawn perpendicular to the plane and ex-

† There are also proper and improper axes of symmetry which are left for more advanced treatises.

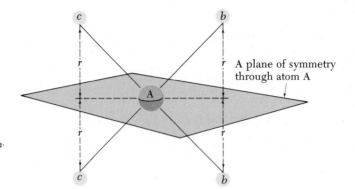

Fig. 19.23. *Examples of a plane of symmetry for a square-planar structure, Ab₂c₂. The lines (r) represent the perpendiculars drawn to the symmetry plane.*

A plane of symmetry through atom A

tended an equal distance on the other side of the plane, terminates at the same kind of atom. Figure 19.23 shows this operation for a square-planar structure of the type Ab_2c_2.

We can consider the plane as being a mirror, so that half the molecule is a mirror image of the other half. In the example cited, the plane passes through an atom, in which case it can be considered as cutting the atom into two halves that are mirror images of each other. Consequently, any planar molecule has at least one plane of symmetry, namely, the molecular plane, the plane that cuts through all its atoms. The structure in Fig. 19.23 has two planes of symmetry, the one shown and the molecular plane. Evidently, a structure can have more than one plane of symmetry. Planes of symmetry may also pass midway between atoms without passing through any atom.

Let us examine some simple types of molecules for symmetry planes. Carbon dioxide, a typical linear molecule, has an infinite number of symmetry planes through the molecular axis, the line extending from one O atom through the C atom to the other O atom. It also has a plane through the carbon atom perpendicular to the molecular axis (Fig. 19.24a). Boron trifluoride, a typical planar triangular molecule, has, in addition to the molecular plane, three planes each passing through the B atom and one of the F atoms (Fig. 19.24b).

(b) Center of symmetry. The *center of symmetry is a point in the center of the molecule to which lines can be drawn from any atom so that, when each line is extended an equal distance past the center, it comes to the same kind of atom.* A simple illustration of a figure with a center of symmetry is a square with points A, A and B, B diagonally opposite each other. If the positions of an A and a B are exchanged, the center of symmetry vanishes:

Two of an infinite number of planes

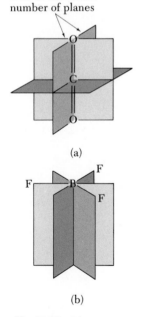

(a)

(b)

Fig. 19.24. *Planes of symmetry in (a) CO₂, and (b) BF₃ (molecular plane not shown).*

B—A
| ↙
A—B

has center of symmetry

B—B
| |
A—A

no center of symmetry

B—A
| |
C—B

no center of symmetry

A molecule can have only one center of symmetry, which may or may not coincide with an atom. *For a molecule to have a center of symmetry, all atoms, with the exception of the one which may coincide with the center, must exist in paired sets.* In carbon dioxide, the C atom is the center of symmetry. Of the typical molecular shapes mentioned in Tables 19.4 and 19.5, only the linear, square-planar, and the octahedral forms, as exemplified by $BeCl_2$, XeF_4, and SF_6, respectively, have centers of symmetry.

Example 3 Determine the number and location of the symmetry elements in (a) $H_2\ddot{O}:$, (b) $:NH_3$, (c) CH_4.

Answer Write the structural formula showing the proper shape, and then look for the elements of symmetry:

(a)

$\overset{\displaystyle \cdot\cdot}{\underset{H \qquad H}{\text{O}}}$

(*i*) Planes of symmetry: There are two present, the molecular plane and the plane through the oxygen atom bisecting the angle.
(*ii*) Center: not present.
(b) Consult Fig. 19.19b (page 394).
(*i*) Planes of symmetry: since there is only 1 nitrogen atom and an odd number of H atoms, any plane of symmetry, if present, must pass through the N atom and either one or all three of the H atoms. Only in this way can the two resulting halves be identical. Since NH_3 is not a planar molecule, no plane can include all 4 atoms. Consequently, we seek planes including the N atom and 1 H atom, and bisecting the angle involving the other 2 H atoms. There are *three* such planes, since any one of the 3 H atoms can be included in a plane.
(*ii*) Center: not present.
(c) To help our discussion, a ball and peg model of CH_4 is shown in Fig. 19.25.
(*i*) Planes of symmetry: Any and all planes must pass through the C atom because only 1 C atom is present. Note that a plane drawn through C, α-H, and β-H bisects the

$\overset{\displaystyle H^{\delta}}{\underset{H^{\gamma}}{\diagdown}}\overset{}{\underset{}{\text{C}}}$

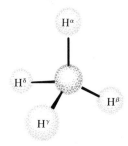

Fig. 19.25. *Ball and peg model for methane. The H atoms are designated with Greek letters.*

angle. The two halves of the molecule are identical, and we have a plane of symmetry. However, there are six combinations for selecting a pair of hydrogens to lie in a symmetry plane with the C atom: α-H, β-H; α-H, γ-H; α-H, δ-H; β-H, γ-H; β-H, δ-H; γ-H, δ-H. Therefore, methane has *six* planes of symmetry.
(*ii*) Center: none present.

PROBLEMS

1. Definitions. Define and illustrate the following terms: (a) hybridization, (b) trigonal bipyramidal shape, (c) octahedral shape, (d) resonance, (e) delocalized π system, (f) p–d π bond, (g) principle of charge dispersal, (h) electron pair repulsion, (i) molecular dipole moment, (j) plane of symmetry, (k) center of symmetry.

2. Miscellaneous. Explain why: (a) N uses sp^3 hybrid AO's in NH_3; (b) some molecules (e.g., O_3) cannot be adequately represented by a single Lewis structural formula; (c) some molecules of the type AB_3 are polar while others are not; (d) NCl_5 has never been observed, while PCl_5 is known; (d) P_2, unlike N_2, is not stable; (e) $H_2C=O$ has a larger dipole than has $Cl_2C=O$.

3. Hybridization. (a) What is the hybridized state of N in each of the following species: (i) $:NH_3$, (ii) NH_4^+, (iii) $H\ddot{N}=\ddot{N}H$, (iv) $HC\equiv N$? (b) Give an orbital description for each species specifying the location of any unshared pairs and the orbitals used for the multiple bonds.

4. Orbital structures. Describe the orbitals (s, p, sp^2, etc.) of the central atom used for bond orbitals and unshared pairs in (a) H_2S (bond angle, 91°); (b) CH_3OCH_3 (C—O—C angle, 110°); (c) S_8 (S—S—S angle, 105°); (d) $BeCl_4^{2-}$; (e) $(CH_3)_2Sn^{2+}$ (C—Sn—C angle, 180°); (f) $(CH_3)_3Sn^+$ (planar with respect to the Sn and 3C atoms); (g) $SnCl_2$ (angle, 95°).

5. Shapes of molecules and ions. Draw the electronic structural formulas and predict the orbital types and the shapes of the following ions and covalent molecules: (a) CCl_4, (b) $HgCl_2$, (c) BF_3, (d) BF_4^-, (e) SeF_2, (f) $SeCl_4$, (g) BrF_3, (h) SbF_6^-, (i) BrF_4^-, (j) ICl_2^-, (k) IO_3^-, (l) XeO_3.

6. Orbital structures. Show the orbital distribution of electrons in the outer energy level of the central atom in (a) SiF_6^{2-}, (b) $AsCl_5$, (c) POF_3 (phosphorus oxyfluoride), (d) ClO_3^-, (e) SeF_4, (f) BrF_3.

7. Basicity and structure. Compare the basicities of the ions in each of the following pairs and explain your choice: (a) SO_4^{2-} and SO_3^{2-}, (b) SeO_3^{2-} and SO_3^{2-}, (c) HCO_3^- and CO_3^{2-}, (d) CO_3^{2-} and NO_3^-.

8. Dipole moments. Indicate which of the following molecules have appreciable dipole moments, and represent the dipole moment by an arrow pointing from positive to negative end of the dipole. In each case, explain your answer. (a) H_2O, (b) $CHCl_3$ (chloroform), (c) PF_5, (d) CCl_2F_2, (e) SF_4, (f) $BClF_2$, (g) HCN, (h) OCS.

9. Symmetry. Indicate the symmetry planes and center (if any) in each of the following molecules or ions: (a) CBr_4, (b) $CHCl_3$, (c) CCl_2F_2, (d) $ClNH_2$, (e) ICl, (f)

10. Resonance. (a) Write two contributing structures consistent with the octet rule for hydrazoic acid, HN_3, for which the sequence of bonded atoms is HNNN. Keep formal charges to a minimum. (b) Locate all formal charges in each structure. (c) Which if any is the correct formula of the molecule? Explain. (d) Predict the two N-to-N bond distances for HN_3 relative to single-, double-, and triple-bond distances.

11. Bond lengths. (a) From the data in Table 7.6 on page 112, calculate the lengths of the following sets of bonds: (*i*) O—F, O—Cl, O—Br, O—I; (*ii*) C—C, C—N, C—O, C—F. (b) What change is observed in the calculated bond length of A—B as (*i*) B increases in atomic number within a family, (*ii*) B increases in atomic number within a period?

12. Multiple bond lengths. Compare the bond lengths and predict the relative bond stabilities for (a) —C≡C—, N≡N; (b) C=O, C=S; (c) —N=O,

—N=C—. Use Table 19.3.

ADDITIONAL PROBLEMS

13. Extended π bonding and resonance. (a) Write the contributing structures of (*i*) ozone, O_3; (*ii*) CO_3^{2-}; (*iii*) formate, HCO_2^-; (*iv*) SO_3; (*v*) diazomethane, H_2CNN. (b) Write structural formulas representing extended π bonding in compounds (*i*)–(*iv*).

14. Hybrid orbitals. (a) What is the hybridized state of each C in each of the following molecules? (*i*) $H_2C=O$; (*ii*) $HC≡N:$; (*iii*) H_3CCH_3; (*iv*) allene, $H_2C=C=CH_2$. (b) Describe the shape of each molecule.

15. Extended π bonding. Account for the fact that the B—F bond distance in BF_3 is shorter than in BF_4^-.

16. Hybridization. (a) Write electronic structural formulas for NO_2, NO_2^+, and NO_2^-. (Assume any electron deficiency is on nitrogen.) (b) Give the hybridization of N in each and give a MO interpretation. (c) Predict the O—N—O bond angle in each case. (d) NO_2 actually has a bond angle of 134°. Explain in terms of resonance theory. (*Hint:* Must the electron deficiency be only on the N atom?)

17. Dipole moment. Account for the fact that the dipole moment of NCl_3 is greater than that of NF_3. (*Hint:* Consider the effect of the unshared pair.)

18. Dipole moment and unshared pairs. After comparing experimental and calculated dipole moments, Charles A. Coulson suggested that in HCl the chlorine atom is sp hybridized. (a) Give the orbital electronic structure for an sp hybridized Cl atom. (b) Which HCl molecule would have a larger dipole moment—the one in which the chlorine uses pure p, or uses sp hybrid orbitals for bonding with the H atom?

19. Dipole moment and structure. The statement was made on page 396 that the zero dipole moment for CCl_4 is consistent with, but does not prove, a tetrahedral structure. Which of the following conceivable structures for CCl_4 would also have a zero dipole moment? (a) Square plane; (b) square pyramid (Cl atoms at the corners of a square base, C at apex); (c) irregular tetrahedron (as in $TeCl_4$).

20. Shape. In their crystalline states, PCl_5 exists as $PCl_4^+PCl_6^-$, and PBr_5 exists as $PBr_4^+Br^-$. (a) Predict the shapes of all these polyatomic ions. (b) Indicate the electronic orbital structure for the P atom in each of its different types of ion. (c) Speculate as to why PCl_5 and PBr_5 do not form the same ionic species.

21. Symmetry. Indicate the kind, number, and location of each symmetry element in the following compounds: (a) PCl_5, (b) SF_6.

22. Symmetry. A cube has at its corners 2 A atoms, 2 B atoms, 2 C atoms, and 2 D atoms. Arrange the 8 atoms so that the cube has (a) a plane of symmetry—locate the plane; (b) a point of symmetry—locate the point.

23. Multiple-bonded molecules. Draw electronic formulas and predict the shapes of the following molecules and ions: (a) nitrosyl bromide, BrNO; (b) nitryl chloride, $ClNO_2$; (c) thionyl chloride, $SOCl_2$; (d) SO_2; (e) phosgene, Cl_2CO.

24. Symmetry. There is evidence that the free radical $H_3C\cdot$ has four planes of symmetry, but the anion $H_3C:^-$ has only three. (a) Illustrate the molecular shape of the radical and of the ion. (b) What is the hybrid state of carbon in the radical? In the ion?

25. π Bonding and shape. (a) In $:N(SiH_3)_3$ the Si—N—Si bond angle is 120°. Describe the hybridization of the orbitals of the nitrogen atom. (b) In the carbon analog, $:N(CH_3)_3$, the C—N—C bond angle is 109°. Describe the hybridization of the orbitals of the nitrogen atom. (c) Account for the difference in the hybridization in the two cases. (d) Suggest an important modification to the determination of the number of hybrid AO's to accommodate the situation in $:N(SiH_3)_3$.

26. Dipole moments. (a) Draw all the structural formulas possible for each of the following species: (*i*) PF_2Cl_3, (*ii*) SF_2Cl_4, (*iii*) $ICl_2Br_2^-$. (b) Explain how knowing the dipole moment could help distinguish among the structures.

27. Multiple bond lengths. Calculate the double bond length in O_2 from the following multiple bond lengths: C=O is 1.21 Å and C=C is 1.33 Å.

28. Extended π bonding and acidity. Account for the fact that (a) nitramide, $H_2\overset{..}{N}NO_2$, and hydrazoic acid, $H—\overset{..}{N}=N=\overset{..}{N}:$, are both stronger acids than ammonia; (b) chloroform, $HCCl_3$, is a stronger acid than fluoroform, HCF_3.

29. Resonance. Contributing structures may not contribute equally to the hybrid. For each pair given below explain why the second contributing structure makes the smaller contribution: (a) $H_2C=CH_2$, $H_2\overset{-}{C}—\overset{+}{C}H_2$; (b) $:C≡\overset{-}{O}:$, $:C=\overset{..}{O}:$; (c) $H—\overset{..}{N}=N=\overset{..}{N}:$, $H—\overset{+}{N}≡\overset{+}{N}—\underset{-2}{\overset{..}{N}}:$; (d) $H—C≡\overset{+}{O}:$, $H—\overset{+}{C}=\overset{..}{O}:$.

30. MO's. (a) How many π MO's are there for NO_3^-? (b) Use signs of the lobes of the interacting p AO's to show the relative energy levels of the lowest and highest energy molecular orbitals.

31. Resonance and polar bonds. Show how HCl, a typical polar molecule, can be considered to be a resonance hybrid of 2 contributing structures.

CHEMICAL KINETICS

Thermodynamics (Chapter 17) can predict the extent of a reaction but the reaction may not actually occur at a measurable rate. An equilibrium constant tells us nothing about the rate of the reaction. The rate of a reaction is the change in concentration of a reactant per unit time. It is also the number of moles of a reactant consumed per unit time. We may write

$$C(graphite) + O_2(g) \longrightarrow CO_2(g) \qquad K = 10^{69}$$

but exposing lumps of graphite to air produces no visible result. If any reaction occurs, its rate is immeasurably slow, even though its equilibrium constant is very large. However, the decomposition of pure hydrogen peroxide,

$$H_2O_2(l) \longrightarrow H_2O(l) + \tfrac{1}{2}O_2(g) \qquad K = 2 \times 10^{20}$$

may occur explosively. Many reactions, such as the rusting of iron in moist air, proceed at intermediate rates.

For experimental convenience, early studies of reaction rates were largely devoted to processes whose reaction times could be measured in minutes or hours. An example is the decomposition of dinitrogen pentoxide dissolved in carbon tetrachloride:

$$2N_2O_5(sol) \longrightarrow 4NO_2(sol) + O_2(g)$$

The data, plotted in Fig. 20.1, show that the concentration of the reactant decreases during the course of the reaction.

Thermodynamic properties describe only the initial and final states of a reaction. From the rate of a reaction, however, we may learn how the reaction goes from the initial to the final state. Some reactions proceed in one step, but, more frequently, the reaction occurs in a sequence of steps (page 414). The single step or the sequence of

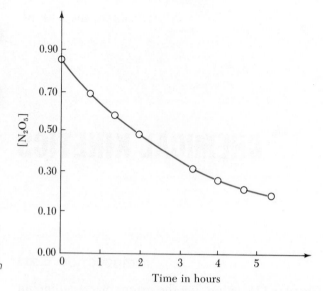

Fig. 20.1. *Decomposition of N_2O_5 in carbon tet.. .'oric'- solution, plotted as concentration versus reaction time.*

steps is called the **MECHANISM OF THE REACTION**. The study of the rates and the mechanisms of chemical reactions is known as **CHEMICAL KINETICS.**

The combustion of graphite is a typical **HETEROGENEOUS REACTION**, a reaction occurring only at the interface (boundary) between two phases. The reaction, $2NO(g) + Br_2(g) \rightarrow 2BrNO(g)$, is a typical **HOMOGENEOUS REACTION**, a reaction occurring in only one phase.

CONDITIONS AFFECTING REACTION RATES / 20.2

The rate of a reaction depends upon the following factors: (1) nature of the reactants, (2) concentration of the reactants, (3) particle size of the solid or liquid in a heterogeneous reaction, (4) temperature, (5) nature of a catalyst if present, and (6) nature of solvent.*

(1) Nature of reactants. Copper or silver strips react very slowly with oxygen even in a flame, while magnesium burns very rapidly. White phosphorus ignites spontaneously in air, whereas red phosphorus does not. Silver ions and chloride ions react very rapidly, $Ag^+ + Cl^- \rightarrow AgCl(c)$, while magnesium ions and oxalate ions react very slowly, $Mg^{2+} + C_2O_4{}^{2-} \rightarrow MgC_2O_4(c)$.

(2) Concentration of reactants. In 1864, Cato Guldberg and Peter Waage recognized that at constant temperature the rate of a homogeneous reaction is usually proportional to some power of the concentration of each reactant. The concentration is expressed in moles of

---

* Influence of solvent will not be considered in this book.

reactant per liter of solution or of gas. This generalization, the LAW OF MASS ACTION, simply states that for the general overall reaction

$$aA + bB \longrightarrow cC$$

the rate of formation of C, or the rate of disappearance of A or B, is proportional to powers of the concentrations of A and B,

$$rate \propto [A]^x[B]^y$$

in which $\propto$ denotes a proportionality. Then we may write

$$rate = k[A]^x[B]^y$$

where k is a constant, called the RATE CONSTANT. The numerical value of the exponents x and y must be *determined experimentally. They cannot be deduced* from the overall reaction or from any theory. Most important, they *need not necessarily be the same as a and b*, the coefficients in the overall reaction. It is possible for one of these exponents to be a fraction, zero, or even negative. For example, if $x = 0$ the rate equation is

$$rate = k[A]^0[B]^y = k \times 1 \times [B]^y = k[B]^y$$

The units of k depend upon the concentration terms and the unit of time. The rate may also be expressed in terms of the symbols

$$-\frac{d[A]}{dt}, \ -\frac{d[B]}{dt}, \ +\frac{d[C]}{dt}$$

in which $-d[A]/dt$ and $-d[B]/dt$ denote, respectively, *the decrease* in the concentration of A and B per unit time; $d[C]/dt$ denotes *the increase* in the concentration of C per unit time. The *plus sign* is always associated with a product because its concentration increases, and the *minus sign* is always associated with a reactant because its concentration decreases.

If experimentally the rate depends on the product of concentration terms, such as

$$rate = \frac{d[A]}{dt} = k[A]^x[B]^y \cdots$$

then the *sum of the powers of the concentration terms in the rate equation* $(x + y + \ldots)$ is called the ORDER OF THE REACTION. For example, the reaction, $2NO + O_2 \rightarrow 2NO_2$, follows the rate equation

$$rate = -\frac{d[NO]}{dt} = k[NO]^2[O_2]$$

The reaction is then described as third order since the sum of the powers is 3. The reaction is said to be *second order* with respect to NO and *first order* with respect to O_2. It is only fortuitous that the orders with respect to the reactants are identical to the coefficients of the reactants. An example of the lack of such a correlation is the reaction in solution between ethyl alcohol and decaborane forming triethyl borate and hydrogen,

$$30C_2H_5OH + B_{10}H_{14} \xrightarrow{\quad} 10B(OC_2H_5)_3 + 22H_2(g)$$

for which the experimentally determined rate equation is

$$-\frac{d[C_2H_5OH]}{dt} = k[C_2H_5OH][B_{10}H_{14}]$$

The exponent for each of the two concentration terms is 1; this reaction is second order, first order with respect to C_2H_5OH and first order with respect to $B_{10}H_{14}$.

The following example will show how rate equations can be derived from experimental data and illustrate some calculations involving the use of these rate equations.

Example 1 For the thermal decomposition of acetaldehyde, CH_3CHO,

$$CH_3CHO(g) \xrightarrow{\quad} CH_4(g) + CO(g)$$

the following data at 800°K are given:

[CH₃CHO]	Rate of decomposition of CH₃CHO
$0.100 \dfrac{\text{mole}}{\text{liter}}$	$9.0 \times 10^{-7} \dfrac{\text{mole}}{\text{liter second}}$
0.200	36.0×10^{-7}
0.300	81.0×10^{-7}
0.400	14.4×10^{-6}

(a) Write the rate equation for the reaction. What is the order of the reaction? (b) Calculate the rate constant for the reaction at 800°K. (c) Calculate the decomposition rate at 800°K at the instant when $[CH_3CHO] = 0.250$ mole/liter.

Answer (a) Note that doubling the concentration of CH_3CHO quadruples the rate while tripling the concentration increases the rate by a factor of 9. In summary,

[CH₃CHO]	Rate
$\times 4 \begin{pmatrix} 0.100 \\ 0.200 \\ 0.300 \\ 0.400 \end{pmatrix} \times 3$	$\times 16 \begin{pmatrix} 9.0 \times 10^{-7} \\ 36.0 \times 10^{-7} \\ 81.0 \times 10^{-7} \\ 144 \;\; \times 10^{-7} \end{pmatrix} \times 9$

The rate of this reaction is proportional to the square of $[CH_3CHO]$, or

$$rate = k[CH_3CHO]^2$$

Since the rate is proportional to the second power of one reactant, the reaction is second order.

(b) Solving for k,

$$k = \frac{rate}{[CH_2CHO]^2} = \frac{9.0 \times 10^{-7} \frac{\cancel{mole}}{\cancel{liter}\ sec}}{(0.100)^2 \frac{mole^{\cancel{2}}}{liter^{\cancel{2}}}} = \frac{36.0 \times 10^{-7}}{(0.200)^2} = \frac{81.0 \times 10^{-7}}{(0.300)^2}$$

$$= \frac{14.4 \times 10^{-6}}{(0.400)^2} = 9.0 \times 10^{-5} \frac{liter}{mole\ second}$$

and k is 9.0×10^{-5} liter/mole-second. Kinetic data with an error less than about 3% are generally considered excellent.

(c) From the previous calculation

$$rate = 9.0 \times 10^{-5} \frac{liter}{mole\ second} [CH_3CHO]^2$$

so that when $[CH_3CHO] = 0.250$ mole/liter

$$rate = 9.0 \times 10^{-5} \frac{\cancel{liter}}{\cancel{mole}\ second} \times (0.250)^2 \frac{mole^{\cancel{2}}}{liter^{\cancel{2}}} = 5.6 \times 10^{-6} \frac{mole}{liter\ second}$$

(3) **Particle size in heterogeneous reactions.** Since heterogeneous reactions occur only at the surface boundary between the reacting phases, the rate of such a reaction is proportional to the surface area. When a given mass is subdivided into smaller particles, the surface area is increased and the rate of reaction increases. Lumps of coal or zinc, for example, are difficult to ignite in air, but, pulverized and dispersed in air, they react explosively.

> **SAFETY NOTE:** Disastrous coal mine explosions do not always result from ignition of a gas. Bituminous ("soft") coal dust suspended in air may be ignited by a match, producing a violent explosion, the rate of pressure rise reaching as much as 40 atm/second. The pressure increase results from the rapid expansion of the gaseous products of the reaction and air, caused by the heat evolved from the combustion reaction. In general, you must regard combustible dusts suspended in air as explosion hazards.

(4) **Effect of temperature.** Temperature has a striking effect on the rate of chemical reactions. Reaction rates negligibly slow at ordinary temperatures may become appreciable and even explosive at elevated temperatures. As a very rough but useful rule, the rate constant is doubled for a rise in temperature of 10 Celsius deg. The effect of temperature on the rate of decomposition of hydrogen iodide is typical; the rate constant increases by a factor of 1.7 for each 10-deg rise in temperature. A rate "constant" is constant only as long as the temperature is constant.

The COLLISION THEORY assumes that molecules must collide to react. The greater the frequency of collisions, the faster the reaction rate, so that the rate is proportional to the number of molecular collisions per second:

rate ∝ number of molecules colliding per liter per second

This assumption accounts for the dependence of the rate on a product (and not a sum) of concentration terms. Let us assume that a molecule of A combines directly with a molecule of B to form AB: A + B → AB. Visualize four molecules each of A and B in a box. In how many ways can the collision A + B occur? One A molecule can collide with any of four B molecules; it has four opportunities for collision. The same is true for each of the four A molecules, making 16 possible collisions in all. Now if there were only two molecules each of A and B in the box, there would be only 2 × 2 or 4 possible collisions. Therefore, assuming the reaction rate is proportional to the number of collisions, the reaction in the box with four molecules each will be 16/4 or 4 times as rapid as in the box with two molecules each. Thus the rate is proportional to the *product* of the concentrations:

rate ∝ [A][B]

The number of colliding molecules calculated from the kinetic theory of gases is enormous, of the order of magnitude of 10^{32} molecules per liter per second at standard conditions. This value does not vary considerably with different gases so that if every collision led to product, all gaseous reactions should proceed at practically the same explosive rate. However, such is not the case. For example, for the same reactant concentration at the same temperature, 300°C, the decomposition rate of gaseous hydrogen iodide is 4.4×10^{-3} mole/liter-hour, while the decomposition rate of gaseous dinitrogen pentoxide is 9.4×10^{5} moles/liter-hour. Clearly, collisions between molecules cannot be the only factor involved in determining the rate of a reaction.

Chemical reactions involve redistribution of atoms, but the redistribution of atoms requires breaking bonds in the reactant molecules, and the formation of bonds in the product molecules. For example, for the reaction $2CN(g) \rightarrow C_2(g) + N_2(g)$ to occur, the bond holding the C atom to the N atom must be ruptured, and bonds must form between C atoms and between N atoms. Since bond breaking is an endothermic process, we postulate that molecules react only if in a collision they possess an energy equal to or greater than a certain critical value. *The critical energy needed for reaction to occur* is called the ENERGY OF ACTIVATION, E_a. *If the colliding molecules have an energy less than E_a, no reaction occurs; if the colliding molecules have an energy equal to or greater than E_a, reaction occurs.* The kinetic (translational) energy is converted

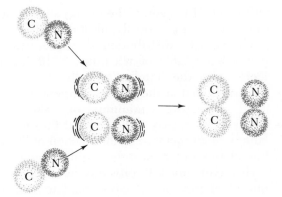

Fig. 20.2. *Chemical reaction pictured in terms of the collision theory.* CN *molecules acquire the energy necessary for the reaction* (E_a) *upon impact; this energy, converted to vibrational energy, weakens the* C—N *bonds.*

to vibrational energy (page 518) upon impact. The atoms in the molecules then vibrate violently so as to weaken the bonds (Fig. 20.2).

At any given temperature not all molecules possess the same kinetic energy (page 20). The speeds of molecules vary from almost zero to practically the speed of light. Figure 20.3 illustrates an energy distribution curve. If we let the area under the curve represent the total number of molecules, then the shaded area represents the number of molecules possessing a velocity of 20×10^4 cm/second or higher. Notice that comparatively few molecules possess the kinetic energy corresponding to 20×10^4 cm/second or higher. Also notice that practically all the molecules possess the kinetic energy corresponding to 2×10^4 cm/second or higher. If the energy of activation for a reaction is 3000 cal/mole, then only a very small fraction of all the colliding molecules

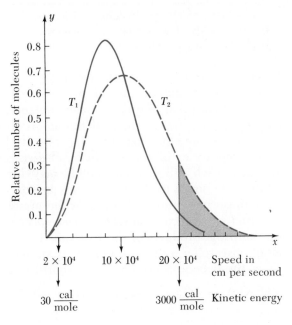

Fig. 20.3. *Kinetic-energy distribution curve (page 21). The y axis gives the relative number of molecules possessing the kinetic energy indicated in calories per mole on the x axis; also included is the corresponding speed in centimeters per second for a gas having a molecular weight of 6 g/mole.* T_2 *is a temperature higher than* T_1; *the kinetic-energy distribution is broader and the average energy higher at the higher temperature.*

will react. However, if the energy of activation requirement is only 30 cal/mole, then practically all of the colliding molecules will react.

The velocity distribution affords an explanation of the effect of temperature on rates of reactions. A 10-deg rise produces a negligible increase in the number of colliding molecules. However, Fig. 20.3 illustrates that at the higher temperature, T_2, the fraction of the molecules possessing a minimum of 3000 cal/mole has increased appreciably. This clearly indicates that the increase in reaction rates is due largely to an increase in the fraction of "activated molecules," the fraction possessing at least the required energy of activation.

However, not all collisions between molecules possessing the required energy of activation lead to reaction. The manner in which they collide is also important. By analogy, the damage ("reaction") resulting from the collision of two automobiles depends not only on their speeds (energy) but also on their relative positions or orientation. Some molecules must be oriented in a very specific manner for reaction to occur (Fig. 20.4). Other molecules may react when colliding in any of a number of random orientations. For example, the combination of 2 H atoms to form an H_2 molecule requires no specific orientation. Since randomness is related to entropy (page 156), this orientation factor is related to the ENTROPY OF ACTIVATION, $S\ddagger$. The more specific the orientation required for the reaction to proceed, the smaller is the entropy of activation, and the slower will be its rate. In summary, the rate is determined by three factors:

Factor	Effect on reaction rate
Greater frequency of collisions	Increase
Higher energy of activation	Decrease
Higher entropy of activation	Increase

E_a is expressed in kcal/mole while $S\ddagger$ is expressed in cal/mole-deg.

More modern theories, developed by Henry Eyring and others, modify the concept that in order to react molecules must "collide." Rather, changes in the arrangement of the atoms in molecules commence as molecules come close together. The chemical change is visualized as a continuous series of changes in bond distances as reactant molecules approach each other. Some bonds may lengthen and finally break, while other bonds may form as atoms approach. Energy changes accompany these continuous changes in the arrangement of atoms. Finally, *the reacting molecules must achieve a specific arrangement* before they can form the products of the reaction. This specific, transient arrangement possessing a definite energy is known as THE TRANSITION STATE. Thus, the decomposition of CN (a short-lived molecule),

$$2CN(g) \longrightarrow C_2(g) + N_2(g) \qquad \Delta H = -20.7 \text{ kcal}$$

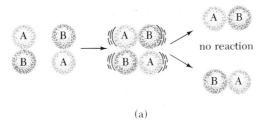

(a)

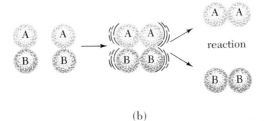

(b)

Fig. 20.4. *A general representation of the possible influence of orientation of molecules on the rate of the reaction 2AB → A₂ + B₂. (a) Orientation that does not lead to reaction; (b) orientation that leads to reaction.*

for which $E_a = 43$ kcal/mole may be visualized as shown in Schemes I and II (distances between atoms are given in angstrom units).

Scheme I. From Reactants to Transition State

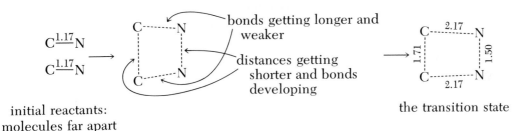

initial reactants: molecules far apart

the transition state

The transition state is not a real substance; it is impossible to isolate a transition state. Nevertheless, it is assumed to possess properties common to real molecules, such as a molecular weight, interatomic distances, a definite enthalpy, and the ability to rotate and vibrate. Once the transition state is formed, it can do only one of two things: either it returns to the initial reactants or it proceeds to form products. It is commonly assumed that once the transition state is formed, it will decompose to the products. In going from the transition state to products, the C—C and N—N bond distances decrease while the C—N bond distance increases further, as shown in Scheme II.

Scheme II. From Transition State to Products

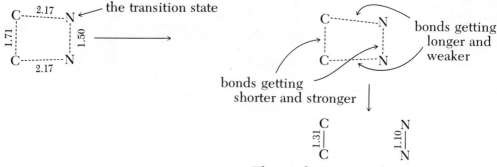

The enthalpy changes accompanying the gradual breaking of the C—N bonds, and the gradual formation of the C—C and N—N bonds, are illustrated in Fig. 20.5. H_r is the enthalpy of the initial reactants. As energy is absorbed, the interatomic distance increases in the C—N molecules, but simultaneously C---C and N---N bonds start to form, evolving energy. However, in forming the transition state more energy is absorbed than evolved; it possesses a larger enthalpy than the initial reactants and corresponds to the state of highest enthalpy, H_{ts}. The difference

$$H_{ts} - H_r = H\ddagger = 43 \text{ kcal/mole}$$

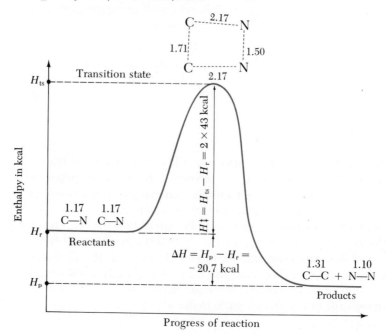

Fig. 20.5. *Reaction-enthalpy diagram. Progress of reaction as the interatomic distances change (not to scale). Distances are in angstrom units.*

is the enthalpy of activation.* As the transition state decomposes to products, the C---N atoms separate even farther, but the energy required to separate them is comparatively small; simultaneously, the C---C and N---N atoms attract each other very strongly and considerable energy is evolved. The product usually possesses a smaller enthalpy, H_p, than the transition state. Note that for this particular reaction, the products have a smaller enthalpy than the reactants,

$$H_p - H_r = -20.7 \text{ kcal}$$

corresponding to ΔH for the reaction. Note also that ΔH *for the reaction depends only on the difference between H_p and H_r and not on H_{ts}.* For an endothermic reaction H_p is larger than H_r.

The double dagger, ‡, is used to indicate differences between the transition state and the reactants:

$H\ddagger$ = enthalpy of activation = $H_{\text{transition state}} - H_{\text{reactants}}$
$S\ddagger$ = entropy of activation = $S_{\text{transition state}} - S_{\text{reactants}}$
$G\ddagger$ = free energy of activation = $G_{\text{transition state}} - G_{\text{reactants}}$

The concept of a transition state permits a scientist to make predictions, a basic aim of science. For example, for the reaction

$(C_2H_5)_3N$ + CH_3I ⟶ $[(C_2H_5)_3NCH_3]^+I^-$
triethylamine   methyl iodide   triethylmethylammonium
                                      iodide

we assume that the transition state is formed by the approach of the N atom to the C atom in CH_3I, the stretching of the C—I bond, and the turning of the H atoms in the methyl iodide molecule, as shown:

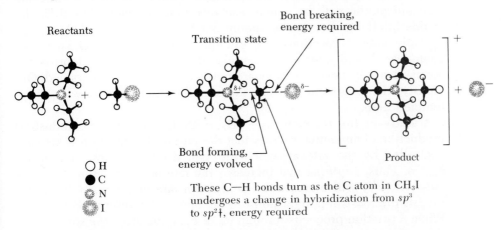

Reactants

Transition state

Bond breaking, energy required

Product

○ H
● C
◉ N
◉ I

Bond forming, energy evolved

These C—H bonds turn as the C atom in CH_3I undergoes a change in hybridization from sp^3 to sp^2†, energy required

---

* It is usually assumed that $H\ddagger$ (the enthalpy of activation) $\approx E_a$ (the energy of activation). It has become customary to refer to enthalpy, rather than energy, of activation and to draw diagrams like Fig. 20.5 in terms of enthalpy. For our purposes in this chapter, *energy* and *enthalpy* may be thought of as synonyms.

† The p orbital lies along the N—C—I axis.

We can now predict that a change in reactants, making it more difficult for the N atom to approach the C atom, will result in an increase in the energy of activation. Thus, for methyl iodide substitute butyl iodide, $(CH_3)_3CI$,

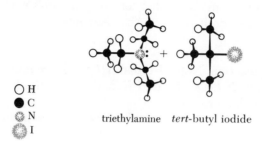

○ H
● C
◉ N
◉ I

triethylamine   *tert*-butyl iodide

It should now be more difficult for the N atom to approach the C atom from which the I atom has to be detached. Consequently, the energy (enthalpy) of activation should increase considerably and the rate of the reaction should be considerably reduced. This conclusion is in agreement with experimental results.

MECHANISM AND RATE EQUATION / 20.4

(a) **The rate-determining step.** Many reactions occur in several steps, each step passing through a transition state. The sum of the equations for all the steps gives the equation for the overall reaction.

The rate equation offers the principal method of determining the mechanism, the path by which reactants go to products. The difficulty of this intellectual endeavor should not be underestimated. Many apparently simple reactions, for example $2H_2 + O_2 \rightarrow 2H_2O$, are very complicated, and after a half-century of labor by many scientists the mechanisms of these reactions are not yet completely understood. Obviously, the treatment of only the simplest mechanisms will be attempted here.

A study of the reaction $N_2 + 3H_2 \rightarrow 2NH_3$, used for the industrial production of ammonia, reveals that the rate of this reaction is actually controlled by the rate at which nitrogen dissociates into atomic nitrogen. Thus, *anything* that increases the rate of the reaction $N_2 \rightarrow 2N$ *automatically* will increase the production rate of ammonia. This also reveals the most important point in the study of reaction mechanisms: When a reaction proceeds through more than one step, *the slowest one determines the rate of the overall reaction and therefore determines the observed rate equation.* The slowest step, the *bottleneck* in the series, is said to be RATE DETERMINING. These ideas may be illustrated by the following analogy:

An assembly line for the production of toasted corn puffs is set up to operate at the following rates:

Step 1. Explosion and toasting of corn kernels (125 pounds/hour)
Step 2. Cooling puffed corn (130 pounds/hour)
Step 3. Packaging (125 pounds/hour)
Step 4. Insertion of prize premium in each package (15 pounds/hour), the bottleneck
Step 5. Sealing package (130 pounds/hour)

Step 4 is the slow step, the measured production rate is only 15 pounds/hour, and improvement in *step 4* will automatically increase the overall production rate.

Although the rates of reactions depend on both $H\ddagger$ and $S\ddagger$, in most (but not all) cases, $H\ddagger$ has a greater influence. For this reason the following discussion assumes that only $H\ddagger$ is decisive in determining relative rates.

Let us consider a reaction R → P, proceeding in two steps:

R ⟶ I	R = reactant
I ⟶ P	I = intermediate
R ⟶ P (overall reaction)	P = product

in which I is an intermediate product or INTERMEDIATE. The intermediate is usually very reactive and short-lived. However, *unlike* transition states, intermediates exist as well-defined molecules, ions, or free radicals. Intermediates *do not* appear in the equation for the overall reaction because they are consumed in ensuing steps.

We can illustrate the important principle, that in a multistep mechanism the slowest step determines the overall rate, by considering two possible reaction-enthalpy diagrams for the mechanism of this reaction.* In both cases (Figs. 20.6a and 20.6b) there is a dip in the enthalpy curve at point I.

In Fig. 20.6a, the enthalpy of the transition state, TS_1, for the first step, R → I, is higher than the enthalpy of the transition state, TS_2, for the second step, I → P. Let us focus our attention on the intermediate. It can undergo one of two possible reactions: return to R or advance to P. But the energy barrier, $H_{-1}\ddagger$, for the return to R is higher than the barrier, $H_2\ddagger$, for the advance to P. The intermediate in this situation races on to products. Very few I molecules go over the higher energy barrier back to R, making the first step practically irreversible. The overall reaction then depends on $H_1\ddagger$ of the first step because it determines the production rate of I from R. The first step is thus the slow step and practically irreversible; the second step is fast.

---

* We disregard those exceptional reactions in which the two steps have transition states with about equal enthalpies.

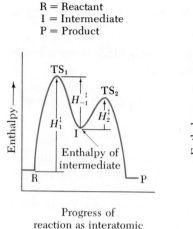

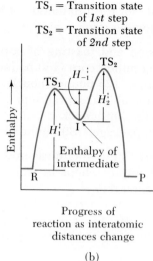

R = Reactant
I = Intermediate
P = Product

TS$_1$ = Transition state of *1st* step
TS$_2$ = Transition state of *2nd* step

Fig. 20.6. *Reaction-enthalpy diagram for two-step reaction.* (a) *First step has higher transition-state enthalpy;* (b) *second step has higher transition-state enthalpy.*

Progress of reaction as interatomic distances change

(a)

Progress of reaction as interatomic distances change

(b)

In Fig. 20.6b, the enthalpy of the transition state, TS_1, for the first step, R → I, is lower than the enthalpy of the transition state, TS_2, for the second step, I → P. Again, focus attention on the intermediate. It can return to R or advance to P. Now, the energy barrier, $H_2\ddagger$, for the advance to P is higher than the barrier, $H_{-1}\ddagger$, for the return to R. The intermediate, in this situation, prefers the reverse pathway to R, making the first step reversible.[*] Very few I molecules advance over the higher barrier to products. This makes the second step, I → P, slow and rate determining. In summary, if TS_1 is higher than TS_2, $H_{-1}\ddagger$ higher than $H_2\ddagger$, then the first step is slow and irreversible, represented as

$$R \xrightarrow{\text{slow}} I$$

$$I \xrightarrow{\text{fast}} P$$

If $TS_2 > TS_1$, $H_2\ddagger > H_{-1}\ddagger$, the first step is fast and reversible and the second step is slow,

$$R \rightleftharpoons I \quad \text{(fast reversible)}$$

$$I \xrightarrow{\text{slow}} P$$

The comparison between $H_{-1}\ddagger$ and $H_2\ddagger$ determines the slow step, another way of saying that the step having the higher transition state enthalpy is the slower step.

[*] It can go either way in the sense of a chemical equilibrium (Chapter 11).

(b) Molecularity of a reaction. The *number* of species (molecules, atoms, free radicals, or ions) that come together to form a transition state is called the MOLECULARITY OF THE REACTION. The molecularity of a reaction must be an integral number. When one molecule produces a transition state the reaction is said to be UNIMOLECULAR. When two or three molecules produce a transition state, the reaction is said to be BIMOLECULAR or TERMOLECULAR, respectively. This concept is therefore theoretical. The probability that four or more species may simultaneously combine to form a transition state is so small that reactions of molecularity higher than three are never postulated. Termolecular reactions are rare. Most reactions are uni- or bimolecular.

We illustrate the concept with the reaction $2NO + O_2 \rightarrow 2NO_2$. We might propose that the mechanism of the reaction is

step 1 $\quad NO + O_2 \longrightarrow NO_3$
step 2 $\quad NO + NO_3 \longrightarrow N_2O_4$
step 3 $\quad\quad N_2O_4 \longrightarrow 2NO_2$

Note that NO_3 and N_2O_4 are intermediates and that the sum of all steps gives the overall reaction. Steps 1 and 2 are bimolecular while step 3 is unimolecular. If the proposed mechanism is a one-step reaction

$$2NO + O_2 \longrightarrow 2NO_2$$

the step is a termolecular reaction. If the proposed mechanism is

step 1 $\quad NO + O_2 \longrightarrow NO_3$
step 2 $\quad NO + NO_3 \longrightarrow 2NO_2$

both forward steps are bimolecular. One could produce other steps that on addition also agree with the overall reaction. But there is another important feature of a proposed mechanism: *a rate equation can be written for any single step in a proposed mechanism from its molecularity.* The number (the coefficient) of each kind of molecule in a proposed step gives us the concentration power (order) for the particular molecule. As we have seen, this is *not* true for the *overall* reaction, which tells us nothing about the orders in the rate equation. If we write

step 1 $\quad NO + O_2 \longrightarrow NO_3$

then we are also saying that the rate of step (1) is given by

$$rate_1 = k_1[NO][O_2]$$

If we write

step 1 $\quad 2NO \longrightarrow N_2O_2$

then we are saying that the rate of step (1) is

$$rate_1 = k_1[NO]^2$$

The clue that is used to propose a mechanism is the subject of Section 20.5.

While a rate equation may be written for each step directly from its molecularity, only the slow step is important. The experimentally observed rate equation tells us much about the slow step and any prior fast step. *If the slow step we write contains only the initial reactants (no intermediates), then the observed rate equation tells us directly the kind and number of molecules of each reactant that are involved in the slow step.* Namely, the number of each kind of molecule is the concentration exponent for the particular molecule in the observed rate equation.

If an intermediate is involved in the rate-determining step, it has to be formed in a prior fast and reversible step (page 416). The observed rate equation then gives a clue to the identity of the reactants involved in the slow step and the preceding fast step. *Any fast step following the slow one may be surmised by using chemical intuition,* subject to the requirement that the sum of all the steps be the same as the overall reaction.

These concepts are illustrated by the following examples. The observed rate equation for the gaseous reaction, $2CN \rightarrow C_2 + N_2$, is

$$rate = -\frac{d[CN]}{dt} = k[CN]^2 \quad \text{(observed)}$$

Since the exponent of the concentration term is two and since only two CN molecules participate in the overall reaction, we may propose that the rate-determining step involves two molecules of CN,

$$CN + CN \longrightarrow \begin{array}{c} C\text{--------}N \\ \vdots \quad\quad \vdots \\ C\text{--------}N \end{array} \longrightarrow C_2 + N_2$$

transition state

The rate equation derived from this proposed bimolecular mechanism

$$rate = -\frac{d[CN]}{dt} = k[CN]^2 \quad \text{(derived)}$$

is in agreement with the observed rate equation. *The rate equation derived from a mechanism must agree with the observed rate equation;* otherwise the proposed mechanism is incorrect. In this case, the overall reaction and the proposed rate-determining step are the same; we conclude that the reaction occurs in a single act. It does not follow that a simple two-molecule overall reaction, for example, $H_2 + Cl_2 \rightarrow 2HCl$, must also occur in a single step; this reaction, in fact, occurs in a sequence of steps (page 425).

We shall now consider a reaction that does not occur in a single step. The observed rate equation shows that the reaction $2NO + O_2 \rightarrow 2NO_2$

is third order:

$$rate = -\frac{d[NO]}{dt} = k[NO]^2[O_2] \quad \text{(observed)} \quad (1)$$

Since the concentration power is two for NO and one for O_2, we might suggest a one-step termolecular mechanism involving two molecules of NO and one O_2 molecule. This termolecular mechanism satisfies the observed overall reaction and the observed rate equation. Since termolecular reactions are rare it is more likely that the reaction proceeds through bimolecular and unimolecular steps. We may then assume that the first step is the fast reversible formation of NO_3 (intermediate). The NO_3 then reacts with a second NO molecule in the slow step:

step 1	$NO + O_2 \rightleftharpoons NO_3$	(fast reversible)
step 2	$NO_3 + NO \longrightarrow 2NO_2$	(slow)
	$\overline{2NO_2 + O_2 \longrightarrow 2NO_2}$	(overall reaction)

Then, the derived rate equation for the slow step is

$$rate = -\frac{d[NO]}{dt} = k_2[NO_3][NO] \quad \text{(derived)} \quad (2)$$

The derived rate Equation (2) cannot be compared directly with the observed rate Equation (1) since the intermediate NO_3 is *not* one of the initial reactants. *Observed rate equations never include the concentration of intermediates.* But it is always possible to substitute the concentrations of reactants for the concentration of the intermediate. Recall that when an intermediate participates in the rate-determining step, it has to be formed in a prior fast and reversible step, in this case, step 1. The equilibrium condition for step (1) is

$$\frac{[NO_3]}{[NO][O_2]} = K$$

so that $[NO_3] = K[NO][O_2]$. We can now substitute the concentration of reactants for $[NO_3]$ in Equation (2), giving

$$rate = -\frac{d[NO]}{dt} = k_2K[NO][O_2][NO] = k[NO]^2[O_2] \quad \text{(derived)}$$

in agreement with the observed rate Equation (1). The constant k in Equation (1) is equal to the product of two constants, k_2 and K. The recent detection of NO_3 strongly supports the bimolecular mechanism.

Example 2 For the oxidation of I^- by H_2O_2 in acidic solutions,

$$H_2O_2 + 2H^+ + 2I^- \longrightarrow I_2 + 2H_2O$$

a proposed mechanism is

$$H_2O_2 + I^- \longrightarrow H_2O + OI^- \quad \text{(slow)}$$

followed by the rapid reactions

$$H^+ + OI^- \longrightarrow HOI$$
$$HOI + H^+ + I^- \longrightarrow I_2 + H_2O$$

For this mechanism to be consistent with kinetic data, what must be the observed rate equation?

Answer The slowest step determines the rate of the overall reaction. Since the slow step involves one molecule of H_2O_2 and one I^- ion, the observed rate equation should be

$$rate = -\frac{d[H_2O_2]}{dt} = k[H_2O_2][I^-]$$

Agreement of a mechanism with the overall reaction and observed rate equation does not necessarily prove that the mechanism is correct. For example, the $H_2 + I_2 \rightarrow 2HI$ reaction was long considered a typical bimolecular reaction involving a four-center transition state

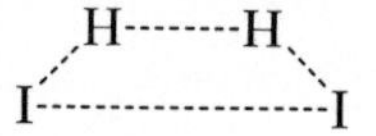

$$
\begin{array}{c}
H\text{-------}H \\
I\text{-------------------}I
\end{array}
$$

It has been recently shown, however, that this reaction occurs through an atomic mechanism:

$$I_2 \rightleftharpoons 2I \quad \text{(fast reversible)} \quad [I]^2/[I_2] = K$$

followed by the termolecular reaction*

$$2I + H_2 \xrightarrow{\text{slow}} 2HI$$

This case serves to emphasize that the proposed mechanism of a chemical reaction is a theoretical pathway consistent with presently known data, and that changes may be necessitated by new experimental or theoretical studies.

CATALYSIS / 20.6

Life as we know it would be impossible without the phenomenon of catalysis. Proteins (page 563) that catalyze biochemical reactions are called ENZYMES. Many industrial processes would also be impossible without catalysts. A CATALYST is a substance that increases the rate of a reaction but is recovered chemically unchanged at the end of the reaction. The catalyst enters into the chemical reaction, but is subsequently regenerated. Typical is the catalytic effect of nitrogen oxide on the rate

---

* The mechanism is more complicated than given here.

of conversion of sulfur dioxide to sulfur trioxide. The reaction

$$2SO_2 + O_2 \xrightarrow{\text{slow}} 2SO_3$$

is slow. However, NO and O_2 react rapidly and NO_2, the product of this reaction, also reacts rapidly with SO_2,

$$2NO + O_2 \xrightarrow{\text{fast}} 2NO_2 \tag{3}$$
$$NO_2 + SO_2 \xrightarrow{\text{fast}} NO + SO_3 \tag{4}$$

Addition of Reaction (3) and twice Reaction (4) yields

$$2SO_2 + O_2 \xrightarrow{\text{fast}} 2SO_3$$

It is evident that we have substituted two fast reactions for a slow one to yield the same overall chemical reaction. This method, the "chamber process," was formerly used to make sulfuric acid, but it has been largely supplanted by the contact process (page 444). However, the method has recently been revived as a potential scheme for removing SO_2 from smokestack gases by converting it to sulfuric acid.

In homogeneous catalysis, the reaction occurs in one phase and the rate depends on the catalyst concentration. The catalyst increases the rate by changing the mechanism of the reaction, thereby decreasing the energy of activation. For example, the homogenous decomposition of acetaldehyde

$$CH_3CHO(g) \longrightarrow CH_4(g) + CO(g)$$

follows the rate equation

$$rate = -\frac{d[CH_3CHO]}{dt} = k_1[CH_3CHO]^2 \quad \text{(observed)}$$

with an energy of activation of 45 kcal/mole, suggesting the mechanism

$$CH_3CHO + CH_3CHO \longrightarrow 2CH_4 + 2CO$$

In the presence of iodine vapor, the observed rate equation is

$$rate = -\frac{d[CH_3CHO]}{dt} = k_2[CH_3CHO][I_2] \quad \text{(observed)}$$

with an energy of activation of 33 kcal/mole, suggesting the mechanism

$$CH_3CHO + I_2 \longrightarrow CH_3I + HI + CO$$
$$CH_3I + HI \longrightarrow CH_4 + I_2$$

At 600°K, k_2 is about 10^7 times larger than k_1.

In heterogeneous catalysis, where the catalyst is usually the surface of a solid, the reaction occurs at the solid surface and the rate depends on the surface area and the number of atoms or "active sites"* per cm².

A general mechanism of surface catalysis involves: (a) diffusion of reactants to the surface of the catalyst, (b) a fast reaction between the molecules of the reactants and the atoms in the surface of the solid catalyst (adsorption), followed by (c) the formation of the transition state (the rate-determining step), which then decomposes rapidly to the catalyst and the products. For example, at about 800°K the homogeneous decomposition of hydrogen iodide,

$$2HI(g) \longrightarrow H_2(g) + I_2(g)$$

follows the rate equation

$$rate = -\frac{d[HI]}{dt} = k_1[HI]^2 \quad \text{(observed)}$$

with an energy of activation of 45 kcal/mole, while in the presence of solid platinum, Pt, the observed rate equation is

$$rate = -\frac{d[HI]}{dt} = k_2[HI][s] \times \text{surface area} = k_3[HI] \times \text{surface area}$$

with an energy of activation of 14 kcal/mole; [s] is the number of active sites per cm². The suggested mechanism is

	H—I | —Pt—	H⋯⋯⋯I \\ / —Pt—	—Pt— + H + I
H—I + —Pt— ⟶	⟶	⟶	
a Pt atom in the surface of the solid	the surface atom adsorbs the HI molecule	the transition state forms	the transition state decomposes

$$2H \xrightarrow[\text{(on surface)}]{\text{fast}} H_2 \qquad 2I \xrightarrow[\text{(on surface)}]{\text{fast}} I_2$$

Negative catalysis. Substances that retard chemical reactions are known as NEGATIVE CATALYSTS, INHIBITORS, or POISONS. They generally act by interfering with the mechanism that leads to a lower energy of activation. Lead atoms, for example, are more strongly adsorbed on platinum than O_2 or hydrocarbon molecules; this interference poisons automobile exhaust catalysts when leaded gasoline is used. Physiological poisons like mercuric chloride and rattlesnake venom react with enzymes, rendering them useless for essential biochemical reactions.

---

* Some places on the surface are more effective in catalysis than others. These active sites are often associated with defects in the crystal lattice.

The decomposition of dinitrogen pentoxide dissolved in carbon tetrachloride,

$$N_2O_5 \longrightarrow 2NO_2 + \tfrac{1}{2}O_2$$

is a first-order reaction:

$$rate = -\frac{d[N_2O_5]}{dt} = k[N_2O_5]$$

This equation leads by integration to

$$\log \frac{[N_2O_5]_0}{[N_2O_5]_t} = \frac{kt}{2.303} \tag{5}$$

in which $[N_2O_5]_0$ is the concentration of N_2O_5 at the start of the experiment $(t = 0)$, and $[N_2O_5]_t$ is the concentration at a time t after the start of the experiment. 2.303 is the conversion factor for changing ln (base e) to log (base 10). Since the ratio appearing in the log is dimensionless, Equation (5) may be rewritten as

$$\log \frac{q_0}{q_t} = \frac{kt}{2.303}$$

in which q_0, the initial quantity of reactant, and q_t, the quantity at time t, are expressible in any convenient quantity unit as moles, grams, number of atoms, number of molecules, etc.

A measure of the rate of a reaction is the HALF-LIFE, *the time required for one-half a given quantity to react;* namely, the time t, denoted $t_{1/2}$, at which $q_t = q_0/2$. Therefore,

$$\log \frac{q_0}{q_0/2} = \log 2 = \frac{kt_{1/2}}{2.303} \tag{6}$$

$$k = \frac{2.303 \log 2}{t_{1/2}} = \frac{0.693}{t_{1/2}}$$

so that the half-life may be evaluated from the rate constant or vice versa. The shorter the half-life, the faster the rate of a reaction. Because of this simple relationship, half-lives are frequently used to express the rates of first-order reactions.

The half-life can be determined from experimental data by plotting, as shown in Fig. 20.7, the quantity present at time t as a function of t; the time required for one-half of any chosen quantity to react is the half-life. For example, the half-life for the decomposition of N_2O_5 in carbon tetrachloride at 30°C is 2.4 hours so that, if we start with 10 g of N_2O_5 at $t = 0$, then, after a period of 2.4 hours, 5.0 g remain; after a second period of 2.4 hours (total of 4.8 hours), 2.5 g remain; after a third period of 2.4 hours (total of 7.2 hours), 1.25 g remain. For each half-life period, the quantity present is reduced by one-half, so that for 3 half-

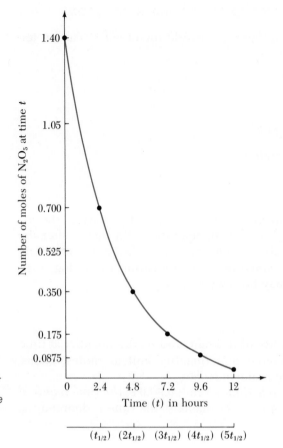

lives $q_t = 10$ g $\times \frac{1}{2} \times \frac{1}{2} \times \frac{1}{2} = 1.25$ g of N_2O_5 remaining. This can be expressed as

$$q_t = 10 \text{ g} \times (0.5)^3 = 1.25 \text{ g}$$

in which 3 is the number of half-lives. Generalizing, we have

$$q_t = q_0(0.5)^{t/t_{1/2}} \tag{7}$$

in which $t/t_{1/2}$ gives the number of half-lives.

Example 3 The half-life for the reaction $N_2O_5 \rightarrow 2NO_2 + \frac{1}{2}O_2$ is 2.4 hours at 30°C. (a) Starting with 10 g, how many grams will remain after a period of 9.6 hours? (b) What period of time is required to reduce 5.0×10^{10} molecules of N_2O_5 to 1.0×10^8 molecules?

Answer (a) 9.6 hours corresponds to four half-lives, 9.6 hours/2.4 hours, whence,

$$q_t = 10 \text{ g} \times 0.5 \times 0.5 \times 0.5 \times 0.5 = 0.63 \text{ g}$$

or

$$q_t = 10 \text{ g} \times (0.5)^{9.6hr/2.4hr} = 10 \text{ g} \times (0.5)^4$$
$$\log q_t = \log 10 + 4 \log 0.5$$

from which

$$q_t = 0.63 \text{ g}$$

(b) Using Equation (7),

$$1.0 \times 10^8 \text{ molecules} = 5.0 \times 10^{10} \text{ molecules} \times (0.5)^{t/2.4hr}$$

and solving,

$$\log \frac{1.0 \times 10^8 \text{ molecules}}{5.0 \times 10^{10} \text{ molecules}} = \frac{t}{2.4 \text{ hours}} \times \log 0.5$$

from which $t = 22$ hours.

CHAIN REACTIONS / 20.8

When a chemical reaction proceeds by a mechanism in which a species consumed in one step is regenerated in a second step, the set of steps is referred to as a CHAIN MECHANISM. Thus, in a CHAIN REACTION each step yields *one* species capable of yielding a product for each *one* species consumed. For example, in the chain reaction

$$:\ddot{C}l\cdot + H_2 \longrightarrow HCl + H\cdot \tag{8}$$
$$H\cdot + Cl_2 \longrightarrow HCl + :\ddot{C}l\cdot \tag{9}$$

each $:\ddot{C}l\cdot$ atom reacts to produce an HCl molecule and an H· atom, which subsequently produces another HCl molecule and regenerates the $:\ddot{C}l\cdot$ atom. The $:\ddot{C}l\cdot$ atom produced in step (9) then starts the cycle all over again.

At ordinary temperatures, hydrogen and chlorine form hydrogen chloride slowly in the dark but explosively in sunlight. The chlorine absorbs radiation (hf) shorter than 4785 Å, producing atomic chlorine,

$$Cl_2 + hf \longrightarrow 2:\ddot{C}l\cdot$$

Without absorption of further light energy, the atoms of chlorine initiate the chain reaction illustrated by reactions (8) and (9). Such chain reactions are fast but not necessarily explosive. They become explosive when the heat evolved in the overall reaction is so large that constant temperature cannot be maintained; as the reaction proceeds, the heat evolved increases the temperature which, in turn, increases the rate of the reaction and the liberation of heat until, finally, the reaction becomes explosive.

In a BRANCHING CHAIN REACTION, *two* or more species capable of producing the reaction are regenerated for every *one* consumed. For example, in

$$O^* + H_2 \longrightarrow H_2O^* \tag{10}$$
$$H_2O^* + O_2 \longrightarrow H_2O + 2O^* \tag{11}$$

for every one "excited" oxygen atom consumed in reaction (10), two are regenerated in reaction (11). The asterisk is usually used to denote excited species. The overall rate then becomes infinite (Problem 25, page 430), and, if the reaction is exothermic, an explosion results even at constant temperature. At about 550°C, for example, the rate of the reaction hydrogen + oxygen → water increases slowly as the concentration of either reactant is increased and then, at the same temperature, the mixture suddenly explodes, signaling the development of the branching chain reaction.

PROBLEMS

1. Definitions. Define and illustrate (a) order of reaction, (b) molecularity, (c) mechanism, (d) enthalpy of activation, (e) entropy of activation, (f) slow step, (g) chain mechanism, (h) reaction-enthalpy diagram, (i) intermediate.

2. Rate equation. The reaction of the reactive fragment OH with H_2S, $4OH(g) + H_2S(g) \rightarrow SO_2(g) + 2H_2O(g) + H_2(g)$, was studied at 27°C:

Reactant concentrations mole/liter		Rate of disappearance of H_2S mole/liter-sec
[OH]	[H_2S]	
1.3×10^{-8}	2.1×10^{-8}	1.4×10^{-6}
3.9×10^{-8}	2.1×10^{-8}	4.2×10^{-6}
3.9×10^{-8}	4.2×10^{-8}	8.4×10^{-6}

(a) Write the rate equation for the reaction. What is the order with respect to OH and to H_2S? What is the order of the reaction? (b) Calculate the rate constant for the reaction at 27°C. (c) Calculate the rate, mole/liter-second, at 27°C at the instant when $[OH] = 1.7 \times 10^{-8}$ and $[H_2S] = 1.0 \times 10^{-8}$. (d) Calculate the rate, mole/second, at 27°C at the instant when $[OH] = 1.7 \times 10^{-8}$ and $[H_2S] = 1.0 \times 10^{-8}$ and the volume of the reacting system is 0.10 liter. (e) At the instant when H_2S is reacting at the rate 5.0×10^{-8} mole/liter-second, what is the rate at which OH is reacting?

3. Temperature. If the rate constant of a reaction is 0.010 liter/mole-second at 30°C, calculate its rate constant at 100°C, assuming the rate doubles for each 10°C rise.

4. Temperature. Explain the increase in rate constants with increasing temperature.

5. Mechanism. Show that the mechanism for the reaction $H_2(g) + 2ICl(g) \rightarrow I_2(g) + 2HCl(g)$ involving two bimolecular reactions,

$$ICl + H_2 \xrightarrow{\text{slow}} HI + HCl$$
$$ICl + HI \xrightarrow{\text{fast}} I_2 + HCl$$

satisfies the observed rate equation

$$-\frac{d[H_2]}{dt} = k[ICl][H_2]$$

6. Mechanism. A plausible mechanism for the decomposition of ozone, $2O_3 \rightarrow 3O_2$, is

$$O_3 \rightleftharpoons O_2 + O \qquad \text{(fast reversible)}$$
$$O + O_3 \longrightarrow 2O_2 \qquad \text{(slow)}$$

(a) What is the molecularity of each step? (b) Derive the rate equation for the overall reaction. (c) Derive another rate equation assuming the first step is slow and the second step is fast.

7. Mechanism. The formation of nitrosyl bromide, $2NO + Br_2 \rightarrow 2BrNO$, follows the rate equation, $rate = k[NO]^2[Br_2]$. Suggest a mechanism involving (a) a termolecular reaction, (b) only bimolecular reactions for this reaction. Derive the rate equation for each mechanism.

8. Catalysis. The catalytic oxidation of vanadium, oxidation state 3, V(III), has been explained in terms of the following mechanism:

$$V(III) + Cu(II) \xrightarrow{\text{slow}} V(IV) + Cu(I)$$
$$Cu(I) + Fe(III) \xrightarrow{\text{fast}} Cu(II) + Fe(II)$$

(a) Write the overall reaction. (b) What species serves as the catalyst? (c) Derive the rate equation for the overall reaction.

9. $t_{1/2}$. The conversion of cyclopropane to propylene is first order with a half-life of 18 minutes at 500°C. (a) What fraction of a given quantity of cyclopropane remains after 54 minutes? (b) Starting with 2.00 g, how many grams remain after (*i*) 72 minutes, (*ii*) 45 minutes? (c) How many minutes will be required to reduce 30.0 mg to 3.75 mg?

10. Transition state. Data are given for two sets of displacement reactions of the type $D + AL \rightarrow DA + L$ (page 256),

		$H\ddagger$, kcal/mole	$S\ddagger$, cal/mole-deg
Set 1:	$OH^- + RF$	9.1	−24
	$I^- + RCl$	14	−24
Set 2:	$Br^- + ROH$	13	−17
	$S^{2-} + ROH$	13	−33

In each set, for which reaction will the rate be faster for the same reactant concentrations and same temperature?

11. Intermediates. Redraw diagrams (a) and (b); on each identify the reactant, the product, the intermediate(s), the transition state(s), and the rate-determining step. In which reaction will the intermediate have a longer life and therefore may be isolated? (Ignore entropy effects.)

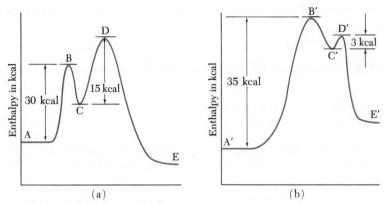

Progress of reaction as interatomic distances change

12. Order. The reaction in methanol, $C_2H_4Br_2 + 3KI \rightarrow C_2H_4 + 2KBr + KI_3$, is first order with respect to $C_2H_4Br_2$ and first order with respect to KI (which exists mainly as KI molecules in this solvent). Write the rate equation. What is the order of the reaction?

13. Rate equation. Data for the reaction in Problem 12 at 60°C are given:

Reactant concentrations mole/liter		Rate of appearance of KI_3 mole/liter-minute
$[C_2H_4Br_2]$	[KI]	
0.500	1.80	0.269
0.500	7.20	1.08
1.50	1.80	0.807

(a) Write the rate equation for the reaction. (b) Calculate the rate constant for the reaction at 60°C. (c) Calculate the rate, mole/liter-minute, at 60°C at the instant when $[C_2H_4Br_2] = 0.75$ and $[KI] = 3.0$. (d) At the instant when $C_2H_4Br_2$ is reacting at the rate of 3.0 mole/liter-minute what is the rate at which KI is reacting and the rate at which KI_3 is forming?

14. Rate equation, catalysis. (a) The equation for an oxidation of hydrogen is $H_2 + CrO_4{}^{2-} + H^+ \rightarrow Cr^{3+} + H_2O$. (*i*) Balance the equation. (*ii*) In the presence of $CuSO_4$ the observed rate equation is

$$rate = -\frac{d[H_2]}{dt} = k\,\frac{[H_2][Cu^{2+}]^2}{[H^+]}$$

What will be the effect on the rate if each of the following, $[H_2]$, $[H^+]$, $[CrO_4{}^{2-}]$, $[Cu^{2+}]$ is individually doubled? (b) The following mechanism has been proposed:

$$Cu^{2+} + H_2 \rightleftharpoons CuH^+ + H^+ \qquad \text{(fast reversible)}$$
$$CuH^+ + Cu^{2+} \longrightarrow 2Cu^+ + H^+ \qquad \text{(slow)}$$
$$2Cu^+ + \tfrac{2}{3}Cr(VI) \longrightarrow 2Cu^{2+} + \tfrac{2}{3}Cr(III) \qquad \text{(fast; details unknown)}$$

(*i*) Write the overall reaction. (*ii*) What substance serves as the catalyst? (*iii*) Derive the rate equation for the overall reaction.

15. Rate equation. Data are given at 660°K for the reaction,

$$2NO + O_2 \longrightarrow 2NO_2$$

Reactant concentrations *mole/liter*		*Rate of disappearance of NO* *mole/liter-second*
[NO]	[O_2]	
0.010	0.010	2.5×10^{-3}
0.010	0.020	5.0×10^{-3}
0.030	0.020	45.0×10^{-3}

(a) Write the rate equation for the reaction. What is the order of the reaction? (b) Calculate the rate constant. (c) Calculate the rate, mole/liter-second, at the instant when $[NO] = 0.015$ and $[O_2] = 0.025$. (d) At the instant when O_2 is reacting at the rate 5.0 mole/liter-second, what is the rate at which NO is reacting and NO_2 is forming?

16. Rate constant. The rate constant for the reaction, $NO_2 + F_2 \rightarrow NO_2F + F$, is 1.4×10^5 ml/mole-second at 50°C. Calculate the number of (a) moles and (b) molecules of NO_2F formed per liter per second when $[NO_2] = [F_2] = 2.8 \times 10^{-6}$ mole/ml. The reaction is first order with respect to NO_2 and to F_2.

17. Kinetics, general. Explain the statements: (a) Only the slow step can be studied by measuring the rate of reaction. (b) A chemical reaction always involves a change in the distance between atoms in molecules. (c) Unreactivity is to be attributed to the forces resisting the deformation and the subsequent breakdown of the initial reactants. (d) Reaction order applies to the experimental rate equation; molecularity applies to a theoretical mechanism. (e) A mechanism is a set of postulates and only that; it is by no means unique.

18. Mechanism. Which reaction, $S^{2-} + CH_3Br \rightarrow CH_3S^- + Br^-$ or $BrO_3^- + 5Br^- + 6H^+ \rightarrow 3Br_2 + 3H_2O$, is more likely to involve a number of steps? Explain.

19. Mechanism. N_2 catalyzes the reaction $2Cl \rightarrow Cl_2$. The reaction $2Cl + N_2 \rightarrow Cl_2 + N_2$ is second order with respect to Cl atoms and first order with respect to N_2. Write the rate equation and suggest a mechanism involving only bimolecular reactions.

20. $H_2 + I_2$ mechanism. Show that the two mechanisms given on page 420 and the one given below for the overall reaction $H_2 + I_2 \rightarrow 2HI$ are consistent with the observed rate equation, $rate = k[H_2][I_2]$.

$$I_2 \rightleftharpoons 2I \qquad \text{(fast reversible)}$$
$$I + H_2 \rightleftharpoons IH_2 \qquad \text{(fast reversible)}$$
$$IH_2 + I \longrightarrow 2HI \qquad \text{(slow)}$$

21. Mechanism. For the following suggested mechanism

$$NO_2 + NO_2 \rightleftharpoons N_2O_4 \qquad \text{(fast reversible)}$$
$$N_2O_4 + F_2 \longrightarrow 2NO_2F \qquad \text{(slow)}$$

what should be the overall reaction and the observed rate equation?

22. Mechanism. At low pressures, the reaction, $2NO_2 + F_2 \rightarrow 2NO_2F$, is first order with respect to NO_2 and to F_2. Suggest a mechanism.

23. Mechanism. For the decomposition reaction, C_4H_8 (cyclobutane)$(g) \rightarrow 2C_2H_4(g)$, the following mechanism is proposed:

$$C_4H_8 + C_4H_8 \rightleftharpoons C_4H_8 + \text{activated } C_4H_8 \quad \text{(fast reversible)}$$
$$\text{activated } C_4H_8 \longrightarrow 2C_2H_4 \quad \text{(slow)}$$

Derive the rate equation. What is the order of the reaction? What is the molecularity of each step?

24. Chain mechanism. The following chain mechanism suggested by Walther Nernst is generally accepted for the reaction between hydrogen and chlorine:

$$Cl_2 \rightleftharpoons 2Cl \quad \text{(fast reversible)}$$
$$H_2 + Cl \longrightarrow HCl + H \quad \text{(slow)} \quad E_a = 6 \text{ kcal}$$
$$H + Cl_2 \longrightarrow HCl + Cl \quad \text{(fast)}$$

Derive the rate equation for the reaction.

25. Chain mechanism. If 1 molecule of HCl is produced in 10^{-10} seconds, while the number of H_2O molecules produced is doubled every 10^{-10} seconds, calculate (a) the number of HCl and of H_2O molecules, and (b) the number of moles of HCl and of H_2O produced in 10^{-7} second.

26. Transition state. For the following exchange reaction, $M(H_2O)_6^{3+} + H_2^{18}O \rightarrow M(H_2O)_5(H_2^{18}O)^{3+} + H_2O$, the following data are given:

Metal ion	$H\ddagger$, kcal/mole	$S\ddagger$, cal/mole-deg
$Al(H_2O)_6^{3+}$	27	28
$Cr(H_2O)_6^{3+}$	27	10

Which ion will exchange faster for the same concentration of reactants at the same temperature?

27. Transition state. Sketch an "enthalpy reaction" diagram for the reaction:

$$K + CH_3I \xrightarrow{\text{slow}} K\text{----}I\text{----}CH_3 \xrightarrow{\text{fast}} KI + CH_3 \quad \Delta H = -20 \text{ kcal}$$
$$\text{(intermediate,}$$
$$\text{lifetime} = 10^{-13} \text{ second)}$$

$H\ddagger$ for first step $= 60$ kcal, $H\ddagger$ for second step $= 5$ kcal.

28. Collision theory. Within the error of calculation, at any given temperature and concentration, the number of colliding molecules is the same for oxygen and for nitrogen. Under comparable conditions will these gases react with any given third gas at the same rate? Explain your answer.

29. Collision theory. It is "estimated that reaction occurred roughly at every collision between sodium, Na, and methyl iodide, CH_3I." What statement can you make about the energy of activation for the gaseous reaction $Na + CH_3I \rightarrow NaI + CH_3$?

30. Energy of activation. The energy of activation for the reaction,

$$O_2 + H \longrightarrow OH + O \qquad \Delta H = +8 \text{ kcal}$$

is 15 kcal/mole, while the energy of activation for the reaction,

$$H + OH \longrightarrow H_2 + O \qquad \Delta H = +7 \text{ kcal}$$

is 17 kcal/mole. At the same concentrations and temperature which reaction is faster? Explain your answer. (Disregard $S\ddagger$.)

31. Catalysis. Show that the mechanism for the acid-catalyzed reaction of substance X is consistent with the observed rate equation, $rate = k[X][CH_3COOH]^{1/2}$:

$$CH_3COOH \rightleftharpoons CH_3COO^- + H^+ \qquad \text{(fast reversible)}$$
$$X + H^+ \rightleftharpoons XH^+ \qquad \text{(fast reversible)}$$
$$XH^+ \longrightarrow \text{products} \qquad \text{(slow)}$$

Would the substitution of a stronger acid for acetic acid affect the rate constant? Explain.

32. k and K. (a) The rate constant for the reaction of methyl radicals with hydrogen iodide, $CH_3 + HI \longrightarrow CH_4 + I$, has been measured at 1100°C with the aid of a mass spectrometer:

$$rate_1 = k_1[CH_3][HI]$$
$$[CH_3] = 10^{-8}, \ [HI] = 2 \times 10^{-8} \text{ mole/liter}$$
$$rate_1 = 2 \times 10^{-7} \text{ mole/liter-second}$$

Calculate k_1. (b) The rate constant for the reverse reaction, $I + CH_4 \longrightarrow CH_3 + HI$, has also been determined:

$$rate_2 = k_2[I][CH_4]$$
$$[I] = 10^{-8}, \ [CH_4] = 3 \times 10^{-6} \text{ mole/liter}$$
$$rate_2 = 3 \times 10^{-4} \text{ mole/liter-second}$$

Calculate k_2. (c) Write the equilibrium condition for $CH_3 + HI \rightleftharpoons I + CH_4$; recall that under these conditions, $rate_1 = rate_2$. Calculate K; the accepted value is about 10^{-1}.

33. Catalysis. The following sequence of reactions has been proposed for large-scale production of hydrogen (page 353):

Reaction	Temperature (°C)
$3I_2(l) + 6LiOH(aq) \longrightarrow 5LiI(aq) + LiIO_3(aq) + 3H_2O(l)$	150
$LiIO_3(c) + KI(aq) \longrightarrow KIO_3(c) + LiI(aq)$	0
$KIO_3(c) \longrightarrow KI(c) + \tfrac{3}{2}O_2(g)$	650
$6LiI(l) + 6H_2O(g) \longrightarrow 6HI(g) + 6LiOH(l)$	500
$6HI(aq) + 3Ni(c) \longrightarrow 3NiI_2(aq) + 3H_2(g)$	150
$3NiI_2(c) \longrightarrow 3Ni(c) + 3I_2(g)$	700

(a) What is the overall reaction? (Ignore differences in physical state.) (b) What are the catalysts? (Make your list as short as possible.)

21

THE REPRESENTATIVE ELEMENTS

The representative elements are those whose d and f subshells are either empty or completely filled. They constitute the R groups in our periodic table.

THE PHYSICAL PROPERTIES OF THE
REPRESENTATIVE ELEMENTS / 21.1

With the exception of the noble gases, which are monatomic, the atoms of all the elements are bonded to each other at ordinary temperatures. Some insights into the nature and periodic variations among the bonding in elements may be gained by examining the heats of fusion and vaporization (ΔH_{fus} and ΔH_{vap}) of the elements. Table 21.1 shows two rectangles per element; the area of the inner (dark) one is proportional to the heat of fusion, and that of the outer (light) one to the heat of vaporization. Of course, the heat of vaporization is *always* the greater of the two, for it represents complete separation into atoms or small molecules, whereas fusion represents only destruction of the crystal lattice.

Note, first, that the ΔH_{vap} values for the elements in the upper right-hand corner of the table are very small compared with those of most of the other elements. This section of the table includes most nonmetals that exist as small molecules, such as Cl_2, O_2, N_2, and P_4. The heats of vaporization are small because only intermolecular forces are disrupted, not the covalent bonds between atoms. ΔH_{vap} for carbon is very high because the element exists as a network structure (graphite or diamond) whose disruption requires the breaking of covalent bonds. These differences among nonmetals arise from the fact that elements of groups 5R, 6R, and 7R require fewer than four covalent bonds for octet completion. These requirements can be met by single bonding (as in

Table 21.1
Heats of Vaporization and Fusion of the Elements[a]

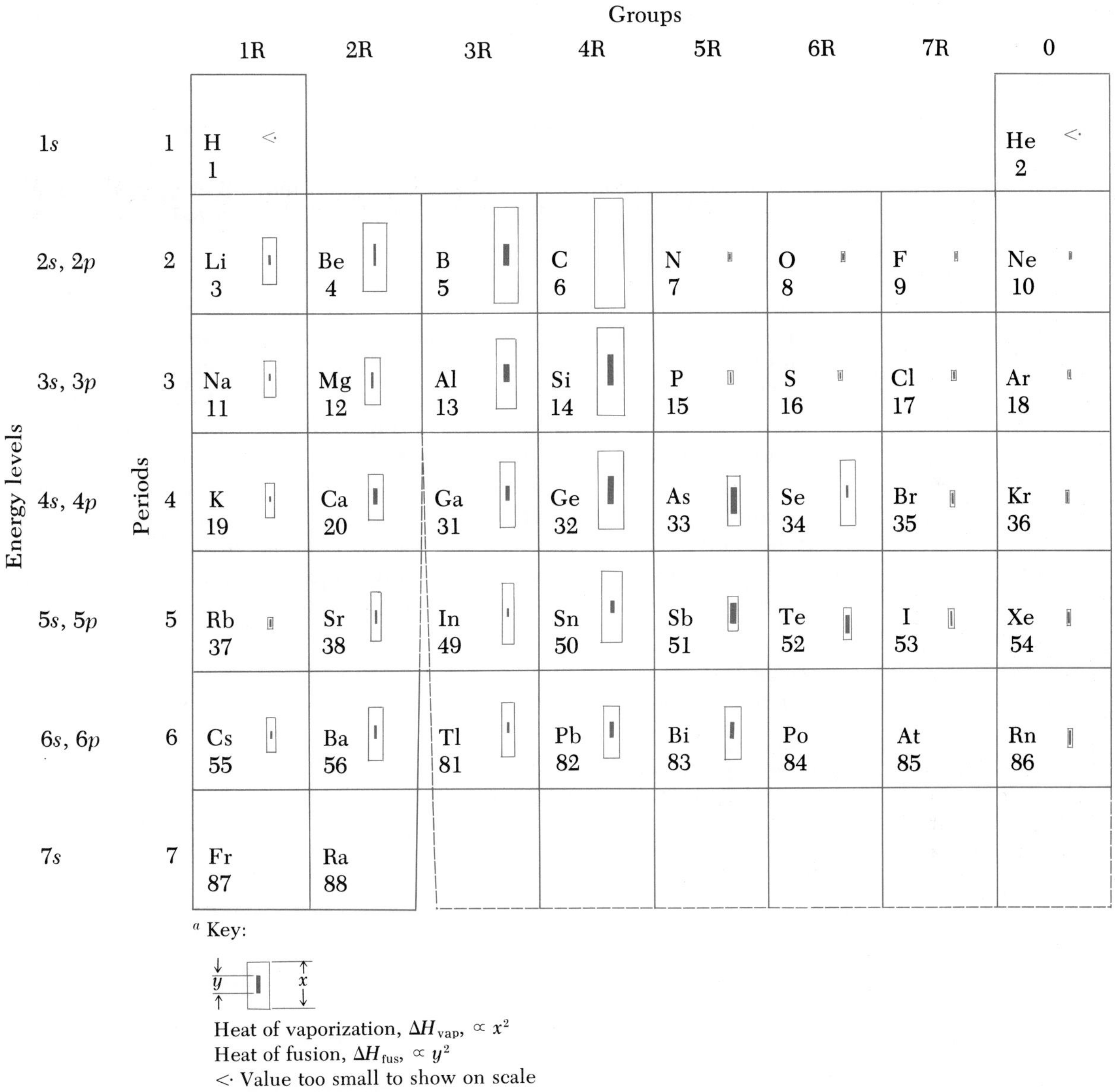

Groups

		1R	2R	3R	4R	5R	6R	7R	0
1s	1	H 1							He 2
2s, 2p	2	Li 3	Be 4	B 5	C 6	N 7	O 8	F 9	Ne 10
3s, 3p	3	Na 11	Mg 12	Al 13	Si 14	P 15	S 16	Cl 17	Ar 18
4s, 4p	4	K 19	Ca 20	Ga 31	Ge 32	As 33	Se 34	Br 35	Kr 36
5s, 5p	5	Rb 37	Sr 38	In 49	Sn 50	Sb 51	Te 52	I 53	Xe 54
6s, 6p	6	Cs 55	Ba 56	Tl 81	Pb 82	Bi 83	Po 84	At 85	Rn 86
7s	7	Fr 87	Ra 88						

Energy levels (left axis) — *Periods* (inner axis)

[a] Key:

Heat of vaporization, ΔH_{vap}, $\propto x^2$
Heat of fusion, ΔH_{fus}, $\propto y^2$
<· Value too small to show on scale

Cl$_2$), by multiple bonding (as in N$_2$), or by cyclic structures (as in S$_8$ and P$_4$). None of these options is available for carbon under ordinary conditions.

It is also noteworthy that the ΔH_{fus} values for metals are especially low; that is, their $\Delta H_{fus}/\Delta H_{vap}$ ratios are lower than those of nonmetals. This circumstance is a reflection of the fact that when metallic solids melt, they preserve much of their metallic character; when they vaporize, metallic character is lost. Hence the energy involved in fusion does not greatly change the type of bonding and is therefore relatively small.

Several elements exist in the form of two or more different substances. This phenomenon is called **ALLOTROPY**. Allotropic forms may differ from each other in chemical bonding, molecular composition, or crystal structure. Only differences in bonding or molecular composition (**PRIMARY ALLOTROPY**) will be considered here.

Primary allotropy is well known in only nine elements: C, O, P, S, As, Se, Sn, Sb, and Te. Table 21.2 shows how they are grouped near the zig-zag line that serves roughly as a borderline between the metallic and the nonmetallic elements (see also Table 7.4, page 109).

The allotropy of carbon, oxygen, phosphorus, and sulfur results from the versatility of their covalent bonding. Carbon occurs as diamond and as graphite (Fig. 21.1). Diamond is extremely hard, in consequence of its stable network covalent structure, which is entirely σ-bonded. Graphite is relatively soft, in part because of the ease with which its π-bonded atomic layers can slip past one another. At ordinary temperatures and pressures, both forms are quite unreactive, and graphite is the form with lower free energy (more stable) by about 0.7 kcal/mole.

Table 21.2
Allotrope Region of Periodic Table[a]

	3R	4R	5R	6R	
		C (diamond; graphite)		**O** (O_2; O_3)	
		Si[b]	**P** (white; red)	**S** (S_2, S_8, "plastic," etc.)	
		Ge[b]	**As**	**Se**	
		Sn	**Sb**	**Te**	

[a] Elements exhibiting primary allotropy shown in bold-face type.

[b] Si and Ge are semimetals (metalloids), and do not exist as allotropes under normal conditions.

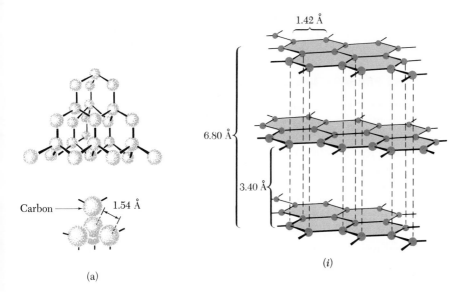

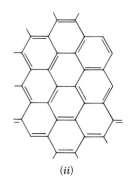

Fig. 21.1. (a) Structure of diamond. (b) Structure of graphite: (i) continuous layers of fused benzenelike rings; (ii) a portion of one layer.

Graphite conducts heat and electricity fairly well, and has an appearance that suggests metallic luster.

The principal primary allotropic forms of oxygen are O_2 (see page 126) and the much less stable ozone, O_3 which spontaneously decomposes to oxygen. The bond angle in ozone is 127°, indicating that

$$\left[\ddot{O}{\cdot}{\cdot}{\cdot}\ddot{O}{\cdot}{\cdot} \longleftrightarrow {\cdot}{\cdot}\ddot{O}{\cdot}{\cdot}{\cdot}{\cdot}\ddot{O} \right] \quad \text{or} \quad {\cdot}{\cdot}\ddot{O}{=}{\cdot}{\cdot}\ddot{O}{\cdot}{\cdot}$$

the central O is sp^2 hybridized. Its reactions in the atmosphere are discussed on page 525.

Phosphorus occurs as tetrahedral P_4 molecules (Fig. 21.2) in phosphorus vapor and the "white" solid form. There are also two solid forms that consist of large molecular networks: the common red (or violet) phosphorus, and the rarer black phosphorus. The black form is a layer lattice that shows some metallic properties (luster, conductivity of heat and electricity); the red form lacks these attributes.

Sulfur exists as (a) cyclic molecules, S_8, in liquid sulfur below about 160°C and in the rhombic and monoclinic crystal forms, (b) long chains, in "plastic sulfur" and the liquid form S_μ, and (c) short chains of various lengths, S_6, S_4, and S_2, in the gaseous state.

Arsenic, selenium, tin, antimony, and tellurium all exist in metallic and nonmetallic forms. For example, Sn occurs as "white tin" which has a metallic luster and is a good conductor of heat and electricity, and "gray tin" in which the atoms are covalently bonded in a diamond-type lattice. Below 18°C, the crystalline white tin slowly changes to the powdery gray tin. This transformation was first observed as blistery outbreaks (called "tin pest") on the surface of tin objects such as organ

Fig. 21.2. Tetrahedral P_4 molecule.

pipes in cold cathedrals. When the conversion is complete, the entire object collapses into a sandy powder.

THE PREPARATION OF THE REPRESENTATIVE ELEMENTS / 21.3

Usually, metals lose and nonmetals gain electrons when they combine with other elements. To recover the elements from their compounds, these changes must be reversed. That is, the metal atoms must regain their lost electrons (reduction), and the nonmetallic atoms must lose their excess (oxidation).

The measure of the thermodynamic ease of reduction of a positive ion is the standard electrode potential (Table 16.1, page 325). Elements with very low $\mathscr{E}^0$'s, therefore, are difficult to prepare by the reduction of their salts. When the $\mathscr{E}^0$ is lower than about -1.5 V, electrolytic reduction is used. An aqueous solvent must be avoided because water is more easily reduced and only H_2 gas would appear at the cathode.

Some typical examples are:

Sodium from fused NaCl

$$2Na^+ + 2e^- \longrightarrow 2Na(l), \qquad \text{at the cathode}$$
$$2Cl^- \longrightarrow Cl_2(g) + 2e^-, \qquad \text{at the anode}$$

Aluminum from Al_2O_3. The energy needed to melt Al_2O_3 and the materials needed to withstand the high temperature would make the cost of electrolytic reduction of fused Al_2O_3 prohibitive. What is needed is a solvent that does not undergo electrolysis itself but permits ionic dissociation of the Al_2O_3. The solvent now used is sodium hexa-fluoroaluminate (CRYOLITE), Na_3AlF_6. Charles M. Hall and Paul Héroult, in 1886, found that Al_2O_3 would dissolve in liquid cryolite (m.p. 1000°C) and that the solution conducted the electric current. The cryolite itself undergoes no appreciable electrolysis because AlF_6^{3-} is resistant to oxidation and Na^+ is resistant to reduction at the voltages used. Molten aluminum accumulates at the cathode, and is tapped off. Carbon is used as the anode and the molten aluminum itself is the cathode. The reactions are:

cathode: $\qquad Al^{3+} + 3e^- \longrightarrow Al(l)$
anode: $\qquad 2O^{2-} \longrightarrow O_2 + 4e^-$

A schematic diagram of the cell appears in Fig. 21.3.

When the reduction potential is less negative than about -1.5 V, any of a number of chemical reducing agents may be used. In many ways, the most attractive one is hydrogen because, as a gas, it can make intimate contact with the surfaces of solids, and because its oxidation product, water, contaminates neither the metal being produced nor the

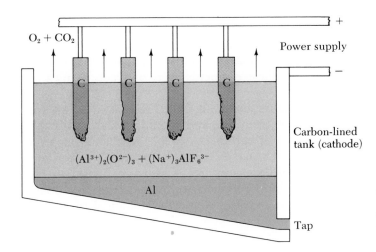

O₂ + CO₂ [annotation, top left]

Power supply

Carbon-lined tank (cathode)

$(Al^{3+})_2(O^{2-})_3 + (Na^+)_3AlF_6^{3-}$

Al

Tap

Fig. 21.3. *Apparatus for electrolytic production of aluminum by the Hall process, using carbon anodes (schematic).*

environment. However, hydrogen is expensive to prepare (by reaction of steam with carbon, $C + H_2O \rightarrow CO + H_2$, or by electrolysis of water) and it is hazardous to handle. The cheapest reducing agent that can be used on a large scale is carbon monoxide, CO. In the presence of a limited supply of air, carbon is oxidized to CO, which, like H_2, is an attractive reducing agent because it is a gas and because its oxidation product, CO_2, is innocuous.

Zinc from zinc sulfide. The sulfide is first "roasted" to the oxide because the oxide is thermodynamically more easily reduced by carbon ($\rightarrow CO_2$) that the sulfide would be ($\rightarrow CS_2$).

$$ZnS + 1\tfrac{1}{2}O_2 \longrightarrow ZnO + SO_2 \quad (\textit{roasting})$$
$$2ZnO + C \longrightarrow 2Zn + CO_2$$

The ease of oxidation of a negative ion to its element stands in just the reverse order of electrode potentials; that is, the *higher* the $\mathscr{E}^0$ (Table 16.1, page 325), the more difficult it is to prepare the element by oxidation of its compounds. For example, a chloride requires a stronger oxidizing agent than a bromide, while F_2 must be prepared electrochemically from a nonaqueous solution. Some nonmetallic elements are obtained directly from natural sources (for example, oxygen and nitrogen from air; sulfur and carbon from underground deposits).

Phosphorus is prepared from phosphates by reduction. The temperature of the process is high enough to convert the phosphate to P_4O_{10} vapor, which is then readily reduced by carbon:

$$
\begin{array}{l}
2Ca_3(PO_4)_2(c) + 6SiO_2(c) \longrightarrow 6CaSiO_3(l) + P_4O_{10}(g) \\
P_4O_{10}(g) + 10C(\textit{graphite}) \longrightarrow 10CO(g) + P_4(g) \\
\hline
2Ca_3(PO_4)_2 + 6SiO_2 + 10C \longrightarrow 6CaSiO_3 + 10CO + P_4
\end{array}
$$

These three elements, together with hydrogen, make up the largest proportion of the mass of living organisms. It is therefore of interest to consider how these elements cycle between living and nonliving forms on the Earth.

Oxygen occurs as O_2, in CO_2, in H_2O, in many organic compounds such as sugars, starches, and proteins, in many dissolved ions such as nitrate (NO_3^-) or carbonate (CO_3^{2-}), and as a major component of the earth's mineral crust. Exchange of oxygen between rock and gaseous or liquid form does occur but is slow compared to other forms of exchange. The most common cycling of oxygen is initiated by the organic process of photosynthesis and its biological reverse, which is respiration or decay:

$$6CO_2(g) + 6H_2O(l) \underset{\substack{\text{respiration} \\ \text{or decay}}}{\overset{\text{photosynthesis}}{\rightleftharpoons}} C_6H_{12}O_6(c) + 6O_2(g), \quad K \approx 0$$

The oxygen cycle is a continual alternation of oxidation and reduction, through both organic and inorganic pathways (Fig. 21.4). The present level in the atmosphere (about 20% by volume) was reached some 450 million years ago, according to our best geological evidence. Some chemists believe that this delicate balance is maintained by the living organisms themselves, and that if life on earth were to cease, the drift toward equilibrium would be favored and oxygen would once again become a trace gas in the atmosphere.

Carbon is present in the atmosphere primarily as carbon dioxide, which constitutes only about 0.03% of the atmosphere by volume. It can be readily seen that the *biological* carbon cycle is intimately related to the oxygen cycle, for whenever oxygen is cycled by life processes, carbon must accompany it (Fig. 21.4). Carbon is incorporated into organic tissue by photosynthesis and released by respiration and decay. However, less than half of the total carbon cycling occurs through biological or other organic pathways. Atmospheric CO_2 dissolves readily in water, and some dissolved molecules escape from the sea to the air. There is little variation in the ratio of dissolved carbon compounds to atmospheric carbon compounds. This geochemical stability is independent of any living process; it is an inherent property of the chemistry of carbon dioxide and water.

Some of the dissolved carbon dioxide reacts with sea water to form carbonates, which settle to the ocean floor as calcium carbonate either in the form of inorganic precipitates (limestone) or as skeletons of various forms of sea organisms. This loss is partially balanced by the action of inland water which slowly dissolves limestone deposits on land and carries the carbonates to sea.

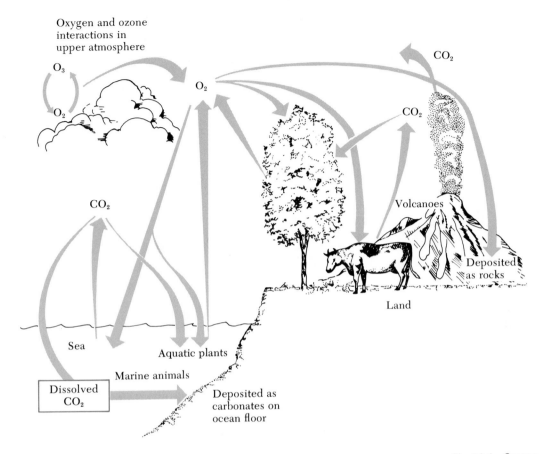

Fig. 21.4. Oxygen–carbon cycle. From A. Turk, J. Turk, J. Wittes, and R. Wittes, Environmental Science. W. B. Saunders Co., Philadelphia, Pennsylvania, 1974.

Currently there is about 15 times as much carbon locked into fossil fuel deposits as there is present in the atmosphere. These deposits were caused by past imbalances in the carbon cycle. The burning of coal and oil since the Industrial Revolution has released much carbon, measurably changing man's environment.

Although N_2 is roughly four times as plentiful in the atmosphere as O_2, it is chemically less accessible to most organisms. Almost every plant and animal can utilize atmospheric O_2, but relatively few organisms can utilize atmospheric N_2 directly. The nitrogen cycle (Fig. 21.5) must therefore provide various bridges between the atmospheric reservoir and the biological community. Lightning, photochemical reactions (page 526), and specialized bacteria and algae (called nitrogen-fixers) transform molecular nitrogen into forms usable by living organisms. Other bacteria and fire provide mechanisms for the return of nitrogen to the atmosphere.

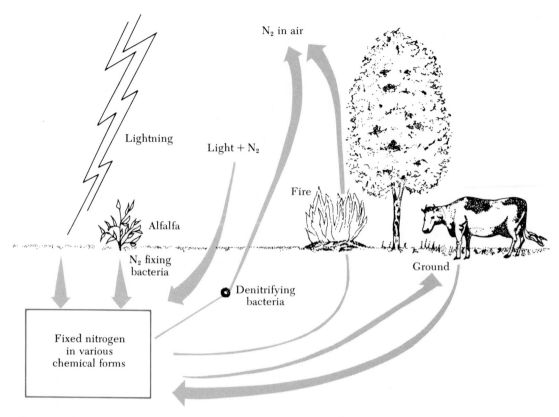

Fig. 21.5. *Nitrogen cycle. From A. Turk, J. Turk, J. Wittes, and R. Wittes, Environmental Science. W. B. Saunders Co., Philadelphia, Pennsylvania, 1974.*

Labels in figure: N₂ in air; Lightning; Light + N₂; Fire; Alfalfa; N₂ fixing bacteria; Denitrifying bacteria; Ground; Fixed nitrogen in various chemical forms

THE HYDRIDES OF THE REPRESENTATIVE ELEMENTS / 21.5

The term **HYDRIDE** is used in two ways. It is the *generic* name for all binary hydrogen compounds, and the *specific* name for the binary hydrogen compounds of the group 1R and 2R elements, for example, lithium hydride and calcium hydride.

Not all the representative elements form stable, well-defined hydrogen compounds. It is observed, for example, that, except for groups 1R and 2R, *as we proceed down the group, the stability of the hydride decreases.* Thus, we find that *no* well-defined hydrogen compounds of the following elements have been isolated in appreciable quantity:

3R	4R	5R	6R
Ga			
In			
Tl	Pb	Bi	Po

We should expect the mode of reaction of hydrides to depend to some extent on the relative electronegativities of hydrogen and the other element. It is evident from Table 8.3 (page 130) that hydrogen is intermediate in electronegativity. Therefore, on the basis of a comparison with hydrogen, we can divide the representative elements into three groups:

(a) Those elements that are less electronegative than hydrogen and, in their hydrides, are therefore positive relative to hydrogen. These are found in the left-hand portion of the periodic table (1R, 2R, bottom of 3R).

(b) Those elements that are more electronegative than hydrogen, and, in their hydrides, are therefore negative. These are found in the right-hand portion of the periodic table (7R, 6R, top of 5R).

(c) Those elements that have approximately the same electronegativity as hydrogen, and have little or no partial charge in their hydrides. These are found in the center region of the periodic table (4R).

The several ways in which hydrides react with water reveal the periodic variations in the bonding and reactivity of these compounds.

(a) **No reaction.** The hydrides of the elements of group 4R and group 5R (except for nitrogen) do not react with water. Included in this class of compounds are phosphine, PH_3, and the hydrocarbons, such as CH_4. Generally, in this group of compounds the central atom has an octet of electrons, and has an electronegativity close to that of hydrogen.

(b) **Basic reaction due to hydride ion.** The hydrides of the elements in groups 1R and 2R (except for Be and possibly Mg) are essentially ionic salts. These elements all have an electronegativity *much less* than that of hydrogen. The hydride ion is very basic, and reacts with water to produce H_2 and OH^-,

$$H{:}^- + H{-}\ddot{O}{-}H \longrightarrow H{-}H + [{:}\ddot{O}{-}H]^-$$

oxidation number	-1	$+1$	$+1$	0	0	$+1$
	$base_1$		$acid_2$	$acid_1$		$base_2$

It is noteworthy that the above reaction is also an oxidation–reduction reaction, as deduced from the change in oxidation numbers shown below each H atom in the equation. In this respect, H^- is a reducing agent. The acceptance of a proton by hydride ion is both an acid–base and a redox reaction.

(c) **Hydrolysis.** Several covalent hydrides react with water to give hydrogen and an oxyacid of the central atom, as typified by the reaction of diborane:

$$\overset{\delta+ \ \delta-}{B_2H_6} + 6H_2O \longrightarrow 6H_2 + 2H_3BO_3, \text{ or } B(OH)_3$$

diborane   boric acid

Generally, during this type of hydrolysis, the more electronegative element combines with the H of H_2O, and the less electronegative element combines with the OH group. In the above example, hydrogen is more electronegative and so H_2 is formed. The hydrides that hydrolyze in this manner contain a central element, less electronegative than hydrogen, that cannot exist as a cation in water because its positive charge would be too high.

(d) **Acidic reaction.** The hydrides of groups 6R and 7R exhibit various degrees of acidity in water. A typical reaction is

$$HCl + H_2O \longrightarrow H_3O^+ + Cl^-$$

Because the group 7R elements are more electronegative than those of group 6R, their hydrides are more acidic. We may also note that acidity *within* the group increases as we go down. Dispersal of electronic charge stabilizes the conjugate base and decreases its tendency to accept a proton. As in all cases, the weaker the base, the stronger its conjugate acid. Thus, going from F^- to I^-, the ionic size increases, charge dispersal increases, basicity decreases, and the acidity of the conjugate acid increases:

$$\xrightarrow{} \text{more acidic}$$

H_2O	HF
H_2S	HCl
H_2Se	HBr
$H_2Te \downarrow$	HI

more acidic

(e) **Basic reaction due to an unshared pair on the central atom.** Ammonia is the only hydride that reacts as a base in water by virtue of an unshared pair of electrons:

$$H_3N: + H_2O \rightleftharpoons NH_4^+ + OH^-$$

This property falls off sharply as we proceed down the group. Thus, the order of basicity of the hydrides of group VA elements is

$$NH_3 > PH_3 > AsH_3 > SbH_3$$

decreasing basicity

$$\xrightarrow{}$$

Ammonia is more basic than water, which in turn is more basic than hydrogen fluoride. Therefore, *basicity due to one or more unshared pairs of electrons on the central atom decreases* on proceeding across a period from left to right.

(f) Lewis acids. The known hydrides of group 3R possess electron-deficient atoms, and so behave as Lewis acids. Of particular interest are their reactions with ionic hydrides:

$$\mathrm{AlH_3} \quad + \quad \mathrm{Li^+H^-} \longrightarrow \mathrm{Li^+} \left[\begin{array}{c} \mathrm{H} \\ \mathrm{H\!:\!\ddot{A}l\!:\!H} \\ \mathrm{H} \end{array} \right]^-$$

aluminum hydride (Lewis acid)	lithium hydride (Lewis base)	lithium aluminum hydride (complex)

$$\tfrac{1}{2}\mathrm{B_2H_6} + \mathrm{Na^+H^-} \longrightarrow \mathrm{Na^+} \left[\begin{array}{c} \mathrm{H} \\ \mathrm{H\!:\!\ddot{B}\!:\!H} \\ \mathrm{H} \end{array} \right]^-$$

diborane   sodium borohydride

Sodium borohydride, a typical salt, is widely used as a reducing agent.

THE OXIDES OF THE REPRESENTATIVE ELEMENTS / 21.6

(a) Basic and acidic anhydrides. All elements except some of the noble gases can form oxides; almost all elements form more than one. With metallic elements oxygen tends to form ionic oxides, while with nonmetallic elements it forms covalent oxides. Ionic oxides react with water to give basic hydroxides; these oxides are known as **BASIC ANHYDRIDES**, as typified by calcium oxide, CaO,

$$\mathrm{Ca^{2+}\!:\!\ddot{O}\!:^{2-}} + \quad \mathrm{O} \longrightarrow \underbrace{\mathrm{Ca^{2+}} + 2\mathrm{OH^-}}$$

a basic anhydride   basic hydroxide

Covalent oxides react with water to give oxyacids; these oxides are known as **ACID ANHYDRIDES**, as typified by sulfur trioxide, SO_3,

$$\mathrm{SO_3 + H_2O \longrightarrow H_2SO_4}$$
an acid anhydride

This reaction may be considered a Lewis acid–base attack by the electron-rich oxygen of water on the electron-deficient sulfur atom:

$$\underset{\text{(Lewis base)}}{\overset{\delta+H}{\underset{H}{\diagdown}}\!\!\ddot{O}\!:^{\delta-}} + \underset{\text{(Lewis acid)}}{\overset{:\ddot{O}:}{\underset{:\ddot{O}:}{S^{\delta+}\!\!=\!\!O\!:^{\delta-}}}} \longrightarrow \mathrm{H^+} + \left[\begin{array}{c} :\ddot{O}: \\ | \\ \mathrm{H\!-\!\ddot{O}\!-\!S\!-\!\ddot{O}:} \\ | \\ :\ddot{O}: \end{array} \right]^-$$

Some metallic oxides are AMPHOTERIC; they react with both acid and base:

$$ZnO + 2H^+ \longrightarrow Zn^{2+} + H_2O$$
$$ZnO + 2OH^- + H_2O \longrightarrow Zn(OH)_4^{2-}$$

(b) Periodicity of acidic properties. The following periodic trends are observed.

(1) *When the oxidation state is equal to the group number, the general trend within a period, from left to right, is from basic through amphoteric to acidic oxides.* The trend also holds for the hydroxy compounds as typified by the elements of the third period:

NaOH	Mg(OH)$_2$	Al(OH)$_3$	SiO$_2$(H$_2$O)$_x$	(HO)$_3$PO	(HO)$_2$SO$_2$	HOClO$_3$
strong base	weak base	amphoteric	very weak acid	moderate acid	strong acid	very strong acid

(2) *The general trend within a group, downward, is toward increased basicity and decreased acidity of oxides and of hydroxy compounds* (page 388).

(3) *For any element, the oxide and the hydroxy compound tend to become more acidic the higher the oxidation state of the element* (page (388).

(c) Acid anhydrides and water; the preparation of acids. The reactions of acidic anhydrides with water are used to prepare acids.

Phosphoric acid is made by reacting its acid anhydride, P_4O_{10},† with water,

$$P_4O_{10} + 6H_2O \longrightarrow 4H_3PO_4$$

Sulfuric acid, one of the most important industrial chemicals, is prepared by a similar method. The acid anhydride of H_2SO_4 is sulfur trioxide, SO_3. Direct oxidation of sulfur produces sulfur dioxide, SO_2, which can be subsequently oxidized in the presence of a vanadium pentoxide (V_2O_5) catalyst to SO_3 (see page 420 for discussion of catalysis). The hydration reaction of SO_3 is very exothermic. The heat liberated vaporizes the water, and the steam cloud carries away much of the SO_3 in the form of a sulfuric acid mist. For more effective results, the SO_3 is dissolved in concentrated H_2SO_4 to form disulfuric acid,§ $H_2S_2O_7$, while water is added continuously to maintain a constant con-

---

† In recognition of its common name, phosphorus pentoxide, this formula may be written $(P_2O_5)_2$.

§ Obsolete name: pyrosulfuric acid; see Appendix II.

centration of H_2SO_4:

$$S_8 + 8O_2(g) \longrightarrow 8SO_2(g)$$

$$2SO_2(g) + O_2(g) \rightleftharpoons 2SO_3(g) \qquad \text{[contact process (surface catalysis)]}$$

$$SO_3(g) + H_2SO_4 \rightleftharpoons H_2S_2O_7, \text{ or } HO-\overset{\overset{\displaystyle O}{|}}{\underset{\underset{\displaystyle O}{|}}{S}}-O-\overset{\overset{\displaystyle O}{|}}{\underset{\underset{\displaystyle O}{|}}{S}}-OH$$

$$H_2S_2O_7 + H_2O \longrightarrow 2H_2SO_4$$

(d) Oxidation of nonmetallic oxides; preparation of nitric acid. Nitrogen dioxide does not yield an oxyacid in which nitrogen has the same oxidation number. Instead, one of its molecules oxidizes another (**SELF-REDOX**) in water to give a mixture of nitric and nitrous acids:

$$\begin{array}{cccc} 2NO_2 + 2H_2O \longrightarrow & H_3O^+ + NO_3^- & + HNO_2 \\ +4 & +5 & +3 \end{array}$$

This reaction is a key step in the industrial preparation of nitric acid as shown in the following sequence of steps:

$$\begin{array}{ll} N_2 + 3H_2 \longrightarrow 2NH_3 & \text{(Haber process)} \\ 4NH_3 + 5O_2 \longrightarrow 4NO + 6H_2O & \text{(Ostwald process)} \\ 2NO + O_2 \longrightarrow 2NO_2 & \\ 2NO_2 + 2H_2O \longrightarrow H_3O^+ + NO_3^- + HNO_2 & \end{array}$$

PEROXIDES / 21.7

In substances containing the **PEROXIDE BOND**, $-\overset{..}{\underset{..}{O}}-\overset{..}{\underset{..}{O}}-$, or the peroxide ion, O_2^{2-}, the oxidation number of oxygen is -1. **HYDROGEN PEROXIDE**, H_2O_2, is a dense liquid that is miscible with water in all proportions. The hydrogen peroxide solution that is sometimes used as an antiseptic is about 3% H_2O_2 in water; a more concentrated form, sometimes used as an oxidant, is 30% H_2O_2. Hydrogen peroxide decomposes exothermically:

$$H_2O_2(l) \longrightarrow H_2O(l) + \tfrac{1}{2}O_2(g) \qquad \Delta H = -23.5 \, \frac{\text{kcal}}{\text{mole}}; \qquad K = 2 \times 10^{20}$$

Many substances, such as metals and metallic oxides, effectively catalyze this decomposition; therefore, small quantities of impurities may encourage pure H_2O_2 or concentrated solutions to decompose explosively.

> **SAFETY NOTE:** Solutions of H_2O_2 should be kept clean, cold, and vented to avoid buildup of internal pressure (leave the stopper loose).

H_2O_2 typically functions as a strong oxidizing agent, although in some circumstances it behaves as a mild reducing agent:

$$\underset{\text{oxidizing agent}}{H_2O_2(aq)} + SO_3^{2-}(aq) \longrightarrow H_2O(l) + SO_4^{2-}(aq)$$

$$\underset{\text{reducing agent}}{5H_2O_2(aq)} + 2MnO_4^-(aq) + 6H^+(aq) \longrightarrow 5O_2(g) + 2Mn^{2+}(aq) + 8H_2O(l)$$

Hydrogen peroxide is weakly acidic in water solutions, $K_a \approx 10^{-12}$.

$$H_2O_2(aq) \rightleftharpoons H^+(aq) + \underset{\text{hydroperoxide ion}}{HO_2^-(aq)}$$

Metallic peroxides of the alkali metals, $(M^+)_2O_2^{2-}$, and of most of the alkaline earth metals, $M^{2+}O_2^{2-}$, are well known. Examples are Na_2O_2, Rb_2O_2, and BaO_2. These peroxides readily react with water or acids to form hydrogen peroxide:

$$Na_2O_2(c) + 2H_2O(l) \longrightarrow 2NaOH(aq) + H_2O_2(aq)$$

Since some peroxides, notably Na_2O_2 and BaO_2, are easily formed simply by heating the metal in air, the hydrolysis of such peroxides affords an easy route for the laboratory preparation of H_2O_2. A particularly convenient procedure is the reaction of BaO_2 with sulfuric acid:

$$BaO_2(c) + H_2SO_4(aq) \longrightarrow BaSO_4(c) + H_2O_2(aq)$$

The insoluble $BaSO_4$ can be filtered off and the remaining liquid phase is an aqueous solution of hydrogen peroxide.

PROPERTIES OF THE HALIDES / 21.8

A review of the periodicity of the chemical properties of the halides involves complexities of dealing with the four common halogens, fluorine, chlorine, bromine, and iodine, and in some cases, more than one oxidation state of the other element, as in PCl_3 and PCl_5.

Note from Table 8.3 (page 130) that the electronegativities of the elements increase toward the right of the periodic table. Each halogen is therefore the most electronegative element in its period. The ionic character of the halides EX, therefore, increases as the electronegativity of the element E *decreases*, and as that of the halogen X *increases*. Thus, for example, BCl_3 is a typical covalent substance, $BeCl_2$ is a polar substance with considerable covalent character, and LiCl is ionic. Also, $AlCl_3$ is largely covalent while AlF_3 is ionic. The halides of the nonmetals are typically gases such as BCl_3, or low-boiling covalent liquids such as NCl_3 and CCl_4. Furthermore, as the oxidation state of the element increases, the corresponding halide behaves more like that of a nonmetal. We shall illustrate this fact below, when comparing the behavior of $SbCl_3$ and $SbCl_5$.

The halides of most nonmetallic elements hydrolyze to give the oxy-acid of the element and the corresponding hydrohalogen acid. The oxidation numbers of the element and of the halogen are retained:

$$BCl_3 + 6H_2O \longrightarrow H_3BO_3 + 3H^+ + 3Cl^-$$
$$PBr_3 + 6H_2O \longrightarrow H_2PHO_3 + 3H^+ + 3Br^-$$
$$PCl_5 + 9H_2O \longrightarrow H_3PO_4 + 5H^+ + 5Cl^-$$

The reaction with PBr_3 is used to prepare hydrobromic acid. In the above examples, the halogen is more electronegative than the central atom. If the halogen is less electronegative, the hydrolysis proceeds as shown for nitrogen triiodide, where N is more electronegative than I:

$$NI_3 + 3H_2O \longrightarrow NH_3 + 3HOI$$
$$NH_3 + HOI \rightleftharpoons NH_4^+ + OI^-$$

As the metallic properties of the central atom increase, as is the trend on proceeding down a group, hydrolysis is incomplete and oxyhalides are formed, as shown for antimony(III) chloride:

$$SbCl_3 + 3H_2O \rightleftharpoons SbOCl + 2H_3O^+ + 2Cl^-$$
$$\text{antimony}$$
$$\text{oxychloride}$$

Antimony(V) chloride reacts with water as a typical nonmetallic halide:

$$SbCl_5 + 4H_2O \longrightarrow H_3SbO_4(aq) + 5HCl(aq)$$

Carbon tetrachloride, CCl_4, is one of the few nonmetallic halides that resist hydrolysis. It is immiscible with water and relatively inert. An even more inert halide is sulfur hexafluoride, SF_6.

In contrast to CCl_4, silicon tetrachloride, $SiCl_4$, as well as $GeCl_4$ and $SnCl_4$, reacts vigorously with water,

$$SiCl_4(l) + 4H_2O + 4H_2O \longrightarrow Si(OH)_4 + 4HCl$$
$$\text{"silicic acid"}$$

This sharp difference between the behavior of CCl_4 and the other chlorides does *not* arise from thermodynamic differences, because they are *all* thermodynamically unstable by comparison with their hydrolysis products. For example:

$$CCl_4(l) + 2H_2O(l) \rightleftharpoons CO_2(g) + 4HCl(g), \qquad K = 10^{66}$$

Instead, the difference arises because CCl_4 is kinetically stable, a circumstance resulting from the fact that carbon has no d orbitals available for binding to accommodate reaction intermediates.

Lead(IV) chloride, $PbCl_4$, is very unstable and at room temperature decomposes spontaneously to give lead(II) chloride:

$$PbCl_4 \longrightarrow PbCl_2 + Cl_2$$

Lead(IV) bromide and iodide are not known.

As we proceed down groups 3R, 4R, and 5R, we find that the higher oxidation states of the metal become less stable, and the lower oxidation states become more stable. This trend is illustrated in the following sequence for group 4R:

$$\underset{\text{unknown}}{\cancel{CCl_2}} \qquad \underset{\text{stable}}{CCl_4}$$

$$\underset{\substack{\text{known, but very}\\\text{unstable}}}{2SiCl_2} \longrightarrow \underset{\text{stable}}{SiCl_4 + Si}$$

$$\underset{\text{unstable}}{2GeCl_2} \longrightarrow \underset{\text{stable}}{GeCl_4 + Ge}$$

$$\underset{\text{stable}}{Cl_2 + SnCl_2} \rightleftharpoons \underset{\text{stable}}{SnCl_4}$$

$$\underset{\text{stable}}{Cl_2 + PbCl_2} \longleftarrow \underset{\text{unstable}}{PbCl_4}$$

Among the group 3R elements, the +3 oxidation state is the stable one for boron, aluminum, gallium, and indium. However, thallium, the heaviest element in the group, exists mainly in the Tl^+ state.

The tendency for the lower oxidation state to become more stable than the higher oxidation state as the atomic number increases within a group in the periodic table is a general phenomenon. For example, we find that it is more difficult to reduce phosphoric acid to phosphorous acid than to reduce arsenic acid to arsenious acid, as indicated by the following standard electrode potentials:

$$\underset{\text{phosphoric acid}}{H_3PO_4} + 2H^+ + 2e^- \longrightarrow \underset{\text{phosphorous acid}}{H_2PHO_3} + H_2O \qquad \mathscr{E}^0 = -0.28 \text{ V}$$

$$\underset{\text{arsenic acid}}{H_3AsO_4} + 2H^+ + 2e^- \longrightarrow \underset{\text{arsenious acid}}{HAsO_2} + 2H_2O \qquad \mathscr{E}^0 = +0.56 \text{ V}$$

ELECTRON-DEFICIENT MOLECULES / 21.9

The binary compounds of boron are interesting because they illustrate alternative types of bonding. In the trifluoride, BF_3, the boron atom has only 6 electrons in its valence shell. However, the incomplete octet of the boron atom can be compensated in part by extended p–p π bonding between B and F,

The compound BH_3, which is analogous to BF_3, is highly reactive; until recently it was not even known to exist. H, unlike F, lacks the extra electrons needed for H—B π bonding. The known compounds of boron and hydrogen, called BORANES, therefore exhibit ELECTRON DEFICIENCY. The simplest of these compounds is DIBORANE, B_2H_6. A conventional structural formula for B_2H_6 would require 7 bonds, and hence 14 electrons. But the total number of available electrons is only 12 (3 for each B and 1 for each H atom). The observed structure is:

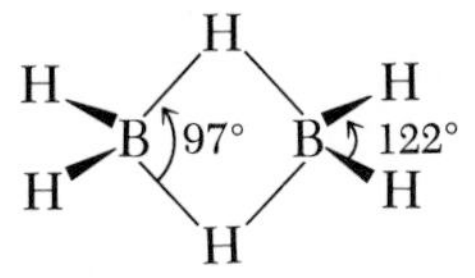

The four terminal B—H bonds

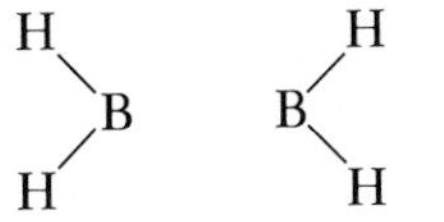

are ordinary, each formed by sharing a pair of electrons between 2 atoms. These account for 8 electrons, leaving only 4 for the bonding of the 2 H atoms bridging the B atoms

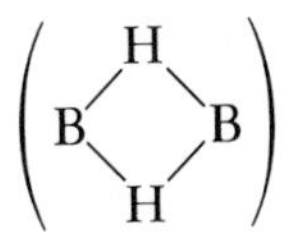

or 2 electrons for each

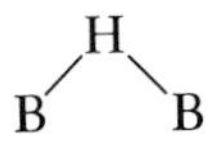

linkage. The bond is formed by overlap of the $1s$ orbital of the H atom with some sort of s–p-type hybrid atomic orbital of each B atom. Such bonding of three nuclei with 2 electrons is called a THREE-CENTER BOND.

THE ZINC SUBGROUP (GROUP 2T) / 21.10

Our periodic table (page 107) includes zinc, cadmium, and mercury among the transition elements, but this classification is awkward, because their d orbitals are completely filled and therefore are not involved in chemical bonding.

$_{30}$Zn, Ar core $3d^{10}4s^2$
$_{48}$Cd, Kr core $4d^{10}5s^2$
$_{80}$Hg, Xe core $4f^{14}5d^{10}6s^2$

Thus, zinc might be likened to calcium ($_{20}$Ca, Ar core $4s^2$) and so be considered a representative element. However, the repulsions exerted by the $3d$ electrons of Zn on its $4s$ electrons do not fully nullify the greater attraction of the $_{30}$Zn nucleus (compared to the $_{20}$Ca nucleus) for the $4s$ electrons. (Refer back to page 111 for a more complete discussion of this effect.) As a result, the metallic radius of Zn is smaller than that of Ca, its ionization energy is greater, its electronegativity is higher, it is a poorer conductor of electricity, and, in general, it is chemically less metallic. The differences between barium (Ba, Xe core $6s^2$) and mercury are even more profound because Hg has both f and d filled subshells intervening before the $6s^2$ electrons. Consequently, mercury exhibits some remarkably covalent behavior. It is the only metal that is liquid at ordinary temperatures (hence its other name, quicksilver). It readily forms covalent bonds with carbon and so can enter biological systems, often with severely toxic effects. It was assumed for many years that mercurial wastes from industry would settle harmlessly and permanently into the bottom muds of rivers and lakes. The conversion of such wastes to covalent compounds like dimethyl mercury, CH_3—Hg—CH_3, by microorganisms, and the resulting incorporation of mercury into plant and animal food chains, showed how wrong this assumption was.

SOME COMMON OXIDES AND THEIR ROLE
AS ENVIRONMENTAL POLLUTANTS / 21.11

(a) **Oxides of carbon.** The most prevalent oxides of carbon are CO and CO_2, although others are known. All have multiple bonds. The formulas of three of them are

$$:C{\equiv}O:$$ carbon monoxide

$$:\ddot{O}{=}C{=}\ddot{O}:$$ carbon dioxide

$$:\ddot{O}{=}C{=}C{=}C{=}\ddot{O}:$$ carbon suboxide (tricarbon dioxide)

Of these, CO_2 represents the most highly oxidized condition of carbon and it is, therefore, the ultimate oxidation product and also the major product obtained when carbonaceous matter is burned in an abundant supply of air or oxygen. When the supply of oxygen is insufficient to provide complete conversion to CO_2, the less highly oxidized product CO is formed, together with traces of C_3O_2.

CARBON DIOXIDE is a colorless, odorless, noncombustible gas with a slight acid taste. The gas is not toxic in low concentrations, but *it is toxic in high concentrations.* Human beings lose consciousness when exposed for only a few minutes to air containing 10% CO_2.

The rate of production of CO_2 by the combustion of fossil fuels has increased sharply during the last century. CO_2 absorbs photons in the infrared region (page 519). But this is the region in which most of the Earth's heat is radiated into space. There is therefore some concern that an increase in the concentration of atmospheric CO_2 may bring about a warming of the Earth's climate, but we do not yet know enough to make reliable quantitative predictions.

Carbon dioxide is soluble in water (90 ml CO_2 gas at 1 atm in 100 ml water at 20°C), but only about 1% of it reacts to form the hydrate, carbonic acid.

$$CO_2 + H_2O \underset{\sim 1\%}{\rightleftharpoons} \begin{matrix} H \\ \diagdown \\ O \\ \diagup \\ \\ O \\ \diagup \\ H \end{matrix} C=O(aq) \quad \text{or} \quad H_2CO_3(aq)$$

carbonic acid (unstable)

CARBON MONOXIDE is an odorless, colorless, tasteless, flammable, highly toxic gas. Mixtures of CO in air in the concentration range between 12.5 and 74% CO by volume are explosive.

> **SAFETY NOTE:** Carbon monoxide is insidious because it gives little sensory warning of its presence; a concentration of 0.1% by volume produces unconsciousness in 1 hour and death in 4 hours. Therefore, the combustion products of any carbonaceous fuel should be regarded as a possible source of CO and should not be allowed to accumulate without control in occupied spaces.

More than 90% of the carbon monoxide in the Earth's atmosphere comes from natural sources, specifically from the oxidation of methane which is emitted by decaying matter. The world-wide atmospheric concentration of CO is kept relatively low (0.1–0.5 ml/10^6 ml of air) and stable by oxidation aided by microorganisms in ordinary soil. However, the global averages do not apply to cities, where 95 to 98% of the CO comes from combustion of fuel, and where the prevailing levels can be 50 to 100 times the world-wide concentration. Cigarette smoke, too, has a high CO concentration resulting from the incomplete combustion of the tightly packed tobacco. The adverse effects of CO stem from its ability to combine with hemoglobin and thereby impair the ability of hemoglobin to transport oxygen in the blood.

Carbon monoxide is difficult to detect or to dispose of because it is comparatively unreactive. (Note that CO is isoelectronic with the very stable molecule N_2.) Cigarette filters do not remove CO. One method of eliminating it from automobile exhaust is to oxidize it to CO_2, possibly with the aid of a catalyst.

(b) Oxides of nitrogen. Nitrogen oxides, like those of carbon, have multiple bonds. Oxides corresponding to every oxidation state of nitrogen from +1 to +6 are known.

Oxide		Oxidation number of N
Dinitrogen oxide, nitrous oxide	N_2O	+1
Nitrogen oxide, nitric oxide	NO	+2
Dinitrogen trioxide	N_2O_3	+3
Nitrogen dioxide	NO_2	+4
Dinitrogen tetroxide	N_2O_4	+4
Dinitrogen pentoxide	N_2O_5	+5
Nitrogen trioxide (unstable)	NO_3	+6

DINITROGEN OXIDE, N_2O (page 384), called "laughing gas," is prepared by heating ammonium nitrate to about 200°C. This is a self-redox reaction.

$$NH_4NO_3 \; (ionic \; crystal) \longrightarrow 2H_2O + N_2O(g)$$
$$ -3 \;\; +5 +1$$

N_2O is a colorless gas (b.p. −88°C) with a slightly sweet odor and taste, and has anesthetic properties. It is stable at ordinary temperatures, but will support combustion:

$$H_2(g) + N_2O(g) \longrightarrow H_2O(g) + N_2(g)$$

NITROGEN OXIDE, $:\ddot{N}{=}\dot{O}: \leftrightarrow :\dot{N}{=}\ddot{O}:$, is prepared industrially by the high temperature oxidation of N_2 or NH_3. It may be conveniently prepared in the laboratory by reduction of sodium nitrite with ferrous sulfate:

$$NO_2^- + Fe^{2+} + 2H^+ \longrightarrow NO(g) + Fe^{3+} + H_2O$$
$$+3 \phantom{NO_2^- + Fe^{2+} + 2H^+ \longrightarrow NO(} +2$$

NO is a colorless paramagnetic gas (note the odd number of electrons). It occurs as a by-product of all combustion processes that occur in air, and rapidly undergoes further oxidation to NO_2,

$$NO(g) + \tfrac{1}{2}O_2(g) \longrightarrow NO_2(g)$$

If NO loses 1 electron, it forms the relatively stable nitrosyl ion, NO^+, which is isoelectronic with the stable gases N_2 and CO. Known compounds of this ion are nitrosyl perchlorate, $NO^+ClO_4^-$, and nitrosyl hydrogen sulfate, $NO^+HSO_4^-$.

NITROGEN DIOXIDE, $:\ddot{O}{-}\dot{N}{=}\ddot{O}:$, a reddish-brown gas, is made industrially by oxidation of NO, and may be prepared in the laboratory by gently heating lead nitrate:

$$NO + \tfrac{1}{2}O_2 \longrightarrow NO_2(g)$$
$$2Pb(NO_3)_2(c) \rightleftharpoons 2PbO(c) + 4NO_2(g) + O_2(g)$$

It is paramagnetic, and its free-radical character predisposes it to combine with itself to produce the lighter-colored dinitrogen tetroxide. The two gases exist in equilibrium at ordinary temperatures (page 127):

$$2NO_2(g) \rightleftharpoons N_2O_4(g)$$

NO_2 is a deadly poison; its great danger is often unappreciated. Initial exposure may cause inflammation of the lungs with only slight pain, but death may occur a few days later from the resulting edema. Continuous exposure to concentrations as low as 500 ppm (0.05% by volume) may be fatal in 48 hours. NO_2 is of great significance in air pollution because it is the principal oxide of nitrogen formed in air during combustion or in secondary reactions from other oxides, and because it is involved in a series of atmospheric reactions leading to smog formation. Cigarette smoke (filtered or unfiltered) contains about 300 parts per million of NO_2; pipe tobacco smoke and cigar smoke contain higher concentrations. NO_2 is not removed by filters.

If NO_2 loses an electron, it forms the relatively stable NITRONIUM ION, NO_2^+, which is isoelectronic with CO_2. Salts of this ion include nitronium perchlorate, $NO_2^+ClO_4^-$, and nitronium hydrogen sulfate, $NO_2^+HSO_4^-$.

DINITROGEN TRIOXIDE, $:\ddot{O}{=}\ddot{N}{-}\ddot{O}{-}\ddot{N}{=}\ddot{O}:$, is formed by cooling an equimolar mixture of NO and NO_2 to about $-20°C$:

$$:\ddot{O}{=}\ddot{N}\cdot\,(g) + \cdot\ddot{O}{-}\ddot{N}{=}\ddot{O}:(g) \longrightarrow N_2O_3 \text{ (blue liquid)}$$

Note that this reaction is the combination of two substances with unpaired electrons to form a nonradical product. N_2O_3 is the acid anhydride of nitrous acid:

$$N_2O_3(g) + H_2O(l) \longrightarrow 2HNO_2(aq) \longrightarrow H_2O + NO + NO_2$$
$$\text{nitrous acid}$$
$$\text{(unstable)}$$

The conjugate base of HNO_2 is the nitrite ion:

$$HNO_2 + OH^- \longrightarrow H_2O + NO_2^-$$
$$\text{nitrite ion}$$

DINITROGEN PENTOXIDE is represented in the vapor state by the structural formula

In the solid form, however, it is an ionic salt of nitronium ion, $NO_2^+NO_3^-$. The solid melts at about 30°C and decomposes readily:

$$N_2O_5(c) \longrightarrow 2NO_2(g) + O_2(g)$$

N_2O_5 is best prepared by reaction of NO_2 with ozone:

$$2NO_2(g) + O_3(g) \longrightarrow N_2O_5(c) + O_2(g)$$

N_2O_5 is the acid anhydride of nitric acid:

$$N_2O_5(c) + H_2O(l) \longrightarrow 2HNO_3(aq)$$
$$\text{nitric acid}$$

Nitric acid is a strong acid, a strong oxidizing agent, and an effective nitrating agent (page 502). The oxidizing action of HNO_3 is evidenced by its ability to dissolve metallic sulfides that are insoluble in HCl by oxidizing the S^{2-} to S,

$$3PbS(c) + 8HNO_3(aq) \longrightarrow 3Pb(NO_3)_2(aq) + 3S(c) + 2NO(g) + 4H_2O$$

(c) **Oxides of sulfur.** The known oxides are those with oxidation numbers for sulfur from 1 to 8 (except 5), but the oxides of most general interest are SO_2 and SO_3.

SULFUR DIOXIDE,

(delocalized p–p π bonding)

is prepared industrially by burning sulfur or sulfides:

$$S + O_2 \longrightarrow SO_2$$
$$CuS + O_2 \longrightarrow Cu + SO_2$$

SO_2 is a toxic, colorless, nonflammable gas that most people can detect in air by taste in concentrations around 1 ppm by volume. In concentrations of 3 ppm or more it has a pungent, irritating odor. It is probably the most significant single air pollutant in urban atmospheres in winter, when large quantities of it are produced by burning sulfur-containing fuels.

SO_2 is the anhydride of sulfurous acid:

$$SO_2(g) + H_2O(l) \longrightarrow H_2SO_3(aq)$$
$$\text{sulfurous acid (unstable)}$$

The corresponding anions are hydrogen sulfite, HSO_3^-, and sulfite, SO_3^{2-}. Sulfur dioxide can serve either as a reducing agent or as an oxidizing agent, although the former is its more usual role:

As reducing agent:

$$SO_2(g) + 2Fe^{3+}(aq) + 2H_2O \longrightarrow SO_4^{2-}(aq) + 2Fe^{2+}(aq) + 4H^+(aq)$$

As oxidizing agent:

$$SO_2(g) + 2H_2S(g) \longrightarrow 2H_2O(l) + 3S(c)$$

The latter equation represents a process that can be used to remove sulfur-containing substances from natural gas by converting them to useful sulfur.

SULFUR TRIOXIDE, SO_3, is prepared by oxidation of SO_2 (page 445). At room temperature it is a solid that may exist in any of several different crystal modifications. It is the anhydride of sulfuric acid (see also page 445).

$$SO_3(g) + H_2O(l) \longrightarrow H_2SO_4(l)$$

The corresponding anions are hydrogen sulfate, HSO_4^-, which is itself a fairly strong acid, and sulfate, SO_4^{2-}. Sulfates and hydrogen sulfates in aqueous solution are not particularly effective oxidizing agents, but hot concentrated sulfuric acid is moderately effective, as evidenced by its ability to oxidize bromides to bromine and iodides to iodine:

$$2NaBr(c) + 2H_2SO_4(l) \longrightarrow Br_2(g) + SO_2(g) + Na_2SO_4 + 2H_2O(l)$$
$$8KI(c) + 5H_2SO_4(l) \longrightarrow 4I_2(g) + H_2S(g) + 4K_2SO_4 + 4H_2O(l)$$

Because sulfuric acid will not oxidize fluorides or chlorides, it is convenient to prepare HCl and HF by the action of concentrated H_2SO_4 on their respective salts. HBr and HI, on the other hand, cannot be prepared in this way, because their salts become oxidized, as shown in the above equations.

(d) Oxides of phosphorus. The common oxides are phosphorus(III) oxide, P_4O_6, and phosphorus(V) oxide, P_4O_{10}.

Phosphorus(III) oxide is produced when white phosphorus is burned in a limited supply of air:

$$2P_4(c) + 6O_2(g) \longrightarrow 2P_4O_6(c)$$

It is the anhydride of phosphorous acid, a diprotic acid:

$$P_4O_6 + 6H_2O \longrightarrow 4H_2PHO_3,$$
$$\text{phosphorous acid}$$

When phosphorus burns in an excess of oxygen, the principal product is phosphorus(V) oxide, P_4O_{10}. This is the anhydride of phosphoric acid, H_3PO_4, an acid of intermediate strength (page 246). P_4O_{10} is a very effective dehydrating agent.

Both P_4O_6 and P_4O_{10} have molecular structures in which phosphorus uses sp^3 hybrid orbitals (Fig. 21.6).

The oxyanions corresponding to P_4O_6 include $HPHO_3^-$ and PHO_3^{2-}. The (V) oxyanions include phosphate, PO_4^{3-}, the protonated phosphate ions HPO_4^{2-} and $H_2PO_4^-$, as well as other ions with various patterns of

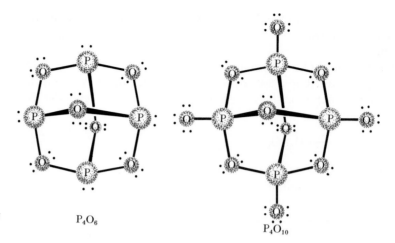

Fig. 21.6. *Structures of phosphorus oxides.*

P_4O_6

P_4O_{10}

bonding to oxygen and various degrees of polymerization, such as:

diphosphate

cyclic triphosphate

Various phosphates have been incorporated in detergent compositions to contribute to water softening and cleaning effectiveness. However, phosphates are also essential plant nutrients, and are frequently in short supply in natural waters. This deficiency serves as a limit to the growth of algae. The limiting barrier to aquatic plant blooms may thus be removed by supplying phosphates or other needed nutrients, with results that far exceed unsophisticated expectations. In many areas of the world, especially in the great rivers and lakes of the tropical and subtropical regions, aquatic weeds have multiplied explosively as a result of phosphates from waste disposal, and have interfered with fishing, navigation, irrigation, and the production of hydroelectric power.

THE NOBLE GASES (GROUP 0) / 21.12

The apparently inert chemical behavior of the noble gases (s^2p^6 configuration) gave the original and sustaining support to the theoretical significance of the stable electronic octet. The octet concept, however, conditioned subsequent generations of chemists to believe that these elements could not react, and, therefore, that investigations of their chemistry would be futile. Even the periodic "group 0" (implying zero valence) designation reinforced this doctrine. In 1962, however, Neil Bartlett noted that oxygen reacts with platinum(VI) hexafluoride, a deep-red vapor at ordinary temperatures, to give an ionic solid, dioxygenyl hexafluoroplatinate(V), $(O_2)^+(PtF_6)^-$. Since the first ionization energy of O_2, 281.3 kcal/mole, is practically the same as that of xenon, it seemed reasonable to attempt the reaction with xenon. The PtF_6 vapor, mixed with an equimolar amount of xenon, immediately formed a yellow solid. In this reaction, xenon transfers an electron to PtF_6,

$$:\overset{..}{\underset{..}{Xe}}: \quad + \quad PtF_6 \longrightarrow [:\overset{..}{Xe}\cdot]^+[PtF_6]^-$$

colorless   red in   yellow solid

gas   vapor state

Also, xenon tetrafluoride, XeF_4, a yellow crystalline solid, was prepared by direct interaction of xenon and fluorine:

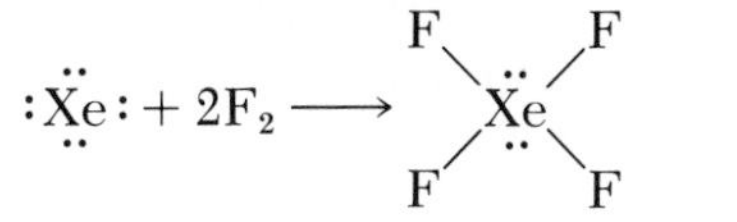

$$:\overset{..}{\underset{..}{Xe}}: + 2F_2 \longrightarrow \quad$$

yellow crystalline
solid

In XeF_4, xenon has 12 outer electrons, and uses sp^3d^2 hybrid orbitals to form a square-planar molecule (page 395). Numerous other investigations soon followed. Among the new noble gas compounds prepared were XeF_2, XeF_6, $XeOF_4$, XeO_3, Na_4XeO_6, and KrF_2.

One recent practical use of such reactions has been the removal of radioactive noble gases (such as ^{222}Rn and ^{135}Xe) from the gaseous effluents of nuclear power plants by converting them to nonvolatile ions and compounds. However, attempts to synthesize compounds of the three lightest noble gases, He, Ne, and Ar, have thus far been unsuccessful.

PROBLEMS

1. Definitions. Define and illustrate the following terms: (a) allotropy, (b) basic anhydride, (c) acid anhydride, (d) oxyanion, (e) self-redox.

2. Allotropy. Allotropy is not observed among the halogens. Offer an explanation for this fact.

3. Chemical properties of elements. Write a balanced chemical equation for each of the following reactions: (a) Reaction of Rb with cold water. (b) Combination of Ba with F_2. (c) Burning of Si_2H_6 in oxygen. (d) Combination of nitrogen with calcium. (e) Oxidation of SO_2 to SO_3. (f) Burning of PH_3 in excess oxygen. (g) Combination of Al with Br_2.

4. Basic and acidic anhydrides. Complete and balance the following equations. If no reaction occurs, write "no reaction." Indicate the physical states of the products.

(a) K_2O (*ionic crystal*) $+ H_2O(l) \longrightarrow$
(b) Al_2O_3 (*ionic crystal*) $+ HBr(aq) \longrightarrow$
(c) Al_2O_3 (*ionic crystal*) $+ Na^+OH^-(aq) + H_2O \longrightarrow$
      (*Hint:* Al_2O_3 is amphoteric.)
(d) $Cl_2O_7(l) + H_2O(l) \longrightarrow$

5. Carbon oxides. Write the balanced equation for the complete oxidation of carbon suboxide (tricarbon dioxide).

6. Nitrogen oxide. The decomposition of $(NH_4)^+(NO_3)^-$ is a self-redox reaction. Write the partial ionic equations, identify the oxidation and the reduction, and show that the equation on page 452 is the sum of the partial equations.

7. Nitric acid. (a) Under some conditions, zinc reduces dilute nitric acid to nitrogen. Write the partial ionic equations and the overall ionic equation for this reaction. (b) Support or refute the statement, "Concentrated nitric acid is always a more powerful oxidizing agent than dilute nitric acid."

8. Phosphorus oxides. The molecular formulas of the (III) and (V) oxides of phosphorus differ numerically from those of the corresponding nitrogen oxides. Explain these differences in terms of the relative tendencies of these two elements to participate in multiple bonding.

9. Hydrogen peroxide. Write the partial ionic equations and the balanced overall equation for the oxidation of Fe^{2+} to Fe^{3+} in acidic hydrogen peroxide solution.

10. Halides. The decomposition of $PbCl_4$ and the self-redox of $SiCl_2$ are both redox reactions. For each reaction, identify all changes in oxidation numbers and state what is being oxidized and what reduced.

11. Allotropy. If Si and Ge had not yet been discovered, would you predict, from their positions in the periodic table, that they would exhibit allotropy? Justify your answer. (Si and Ge in fact normally occur only in a diamondlike structural form, and are both semiconductors. However, denser forms of both elements have recently been produced at very high pressures.)

12. Acid anhydrides. Does the oxidation number of the central atom change when an anhydride combines with water to form an acid? Justify your answer for the anhydrides SO_2, SO_3, N_2O_3, N_2O_5, and P_4O_{10}.

13. Preparation of acids. Write equations for the preparation of (a) HI from PI_3; (b) perchloric acid, $HClO_4$, from its acid anhydride.

14. Properties of oxides. Classify each of the following oxides as basic, acidic, or amphoteric: (a) Ga_2O_3, (b) TeO_3, (c) BaO, (d) Cs_2O, (e) As_2O_5.

15. Oxides. Imagine that you wish to extinguish a magnesium fire in air. (a) Write the equation or equations for the combustion. (b) Give your recommendations for or against each of the following agents for extinguishing the fire. Support your answer, if necessary, with chemical equations: (*i*) water, (*ii*) sand, (*iii*) N_2, (*iv*) Al_2O_3 powder, (*v*) Ar.

16. Oxides of nitrogen. (a) Write an equation, using Lewis (electronic) formulas, for the dimerization of NO_2. Does every atom in your formula of N_2O_4 obey the octet rule? (b) Write an equation, using Lewis formulas, for the combination of NO with NO_2 to form N_2O_3 (page 453). Use arrows to indicate the electron shifts which you assign to obtain the correct formula for N_2O_3. (c) Write the Lewis formula for an N_2O_3 structure that would be formed if unpaired electrons on N in NO and NO_2 were to couple to produce an N—N bond. Explain why this structure would be less stable than that given in your answer to part (b). (Bond energies, kilocalories per mole, N—N, 39; N—O, 53; N=O, 145.)

17. Self-redox. Write the partial ionic equations for the reaction of NO_2 in water to yield nitric and nitrous acids. Specify which partial reaction is oxidation and which is reduction.

18. Self-redox. The reaction of NO_2 in water (see Problem 17) may also yield a mixture of nitric acid and nitric oxide under some conditions. Write the partial ionic equations and the balanced overall equation for the reaction that gives these products.

19. Hydrogen peroxide. The decomposition of H_2O_2 is a self-redox reaction. Write the partial ionic equations and state which is the oxidation and which is the reduction reaction.

20. Halides. On page 448, the properties of the halides are discussed in relation to their positions in the periodic table and the oxidation states of the elements with which they are combined. Illustrate these trends with the aid of a chart based on the periodic table.

21. Sulfur oxides. Write formulas for all the sulfur oxides for which the oxidation states of S were given on page 454.

22. Carbon oxide. Pentacarbon dioxide is a linear molecule with double bonds. Write its structural formula.

23. Aluminum. Aluminum does not react with pure water, but it does react in basic solution with the evolution of H_2 gas. (a) Show from the table of reduction potentials (page 325) that the reaction between Al and H_2O is thermodynamically "spontaneous." (b) Explain why no reaction is observed in pure water. (c) Explain why the reaction occurs in basic solution.

24. Halides. You are asked to select one of three chlorine compounds, Li^+Cl^-, CCl_4, or ClO_2, to be used as an agent for destroying fungus growths by oxidation. Which compound would make the best choice? Justify your selection.

25. Group 2T. Would you expect organo-cadmium compounds containing C—Cd bonds to exist? Defend your answer.

26. Oxides. Chlorine oxides in which Cl has an even oxidation number undergo self-redox. In basic solution, the products are oxyanions such as ClO_2^-, ClO_3^-, and ClO_4^-. Write the balanced ionic equations for the self-redox of (a) ClO_2, (b) Cl_2O_6.

27. Hydrides. Complete and balance the following equations. If no reaction occurs, write "no reaction." Indicate aggregation states of the products.

(a) BaH_2(ionic crystal) $+ H_2O(l) \longrightarrow$
(b) $C_2H_6(g) + H_2O(l) \longrightarrow$
(c) $AlH_4^- + H_3O^+ \longrightarrow$
(d) $HBr(g) + H_2O(l) \longrightarrow$
(e) $(H_3C)_3N{:}\ (g) + H_2O(l) \longrightarrow$

28. Hydrides. Ionic hydrides, such as CaH_2, give a basic reaction in water; covalent hydrides like HCl are acidic in water. Does it follow that NH_3, which is basic in water, is more like an ionic hydride than is CH_4, which gives no basic reaction? Explain.

29. Hydrides. LiH melts at 689°C. Would you expect liquid LiH to conduct an electric current? Write the equations for the reaction at each electrode. Write the equations for the electrode reactions in the electrolysis of the solution formed by mixing water with LiH. (Assume that the electrodes are Pt in all cases.)

30. The DIAGONAL RULE. This rule states that metallic elements of the second period resemble the metals "diagonally to the right" (that is, metals of the next higher group in the third period) more than they do the other metals of their own group. (a) To what three pairs of metallic elements does this rule apply? (b) The more concentrated the positive charge on a real or hypothetical cation, the greater is its ability to distort and attract the negative charge from an anion and thus engender a polar covalent bond. This concentration of charge is proportional to the ionic charge divided by the cube of the ionic radius (Table 7.6, page 112). Show that this relationship leads to the diagonal rule.

22

TRANSITION ELEMENTS; COMPLEX IONS AND MOLECULES

In the broadest sense, the **TRANSITION ELEMENTS** are those which have partly filled d or f subshells in any of their common oxidation states. Used in this way, the definition includes the lanthanides and actinides, which we have designated as inner transition elements. The subgroup 2T, comprising zinc, cadmium, and mercury, is thus excluded, as discussed on page 449. Refer now to the inside front cover and note that the two lower sections of the periodic table comprise 58 elements, 55 of which (excluding the 3 in the zinc group) are transitional. Thus, more than half of all the elements (55 out of 103) are transitional. If you could see a display of all the elements divided into two groups, "transition" and "other," the "others" would include hard and soft solids, colorless and colored gases, and two liquids; furthermore, they would exhibit a very wide range of chemical reactivities. The transition elements, on the other hand, would be a much more homogeneous group; they are *all* what we think of as "typical" metals: more or less hard, high melting, and good conductors of heat and electricity. Their ability to form alloys with each other further attests to their similarity. They almost all exhibit various oxidation states, always including some compounds that are paramagnetic and colored.

Most of the transition elements do not react with strong acids, such as HCl and H_2SO_4. Some do have negative standard reduction potentials for the reaction $M^{n+} + ne^- \rightarrow M$, and liberate hydrogen from hydrochloric acid. These include Mn, Cr, and the iron triad. Silver, gold, the palladium triad, and the platinum triad, the so-called noble metals, are especially inert to acids, both to the nonoxidizing species, such as hydrochloric and hydrofluoric acids, and to the oxidizing acids, such as nitric acid.

Table 22.1

Common Oxidation Numbers of Transition Elements[a]

	3T	4T	5T	6T	7T	——8T——			1T
4	Sc	Ti	V	Cr	Mn	Fe	Co	Ni	Cu
	3	4, 3	**5**, 4, 3, 2	**6**, 3, 2	**7**, 4, 2	**3, 2**	**3, 2**	**2**	**2**, 1
5	Y	Zr	Nb	Mo	Tc	Ru	Rh	Pd	Ag
	3	4	**5**, 4	**6**, 4	7, 4	8, 6, 4, 3	**3**	**2**	1
6	La	Hf	Ta	W	Re	Os	Ir	Pt	Au
	3	4	**5**, 4	**6**	7, 4	8, 6, 4, 3	4, 3	4, 2	3, 1

[a] The most important values are in boldface type.

The maximum oxidation number exhibited by a transition element is generally the group number. Thus, chromium and manganese have maximum oxidation numbers of +6 and +7, respectively. The group 1T elements exhibit the oxidation number +1, but Cu(II), Ag(II), and Au(III) states are known. Table 22.1 lists the *commonly* observed oxidation states. The most important ones are in boldface type.

In groups 4T through 7T the most stable oxidation state usually corresponds to the group number. This is especially true for the heavier elements in each of these groups. Within each of these groups, as the atomic number increases, the higher oxidation states become more prevalent and the lower oxidation states less prevalent. The most common oxidation states, other than the one corresponding to the group number, are the +2 and +3 states. In these, and the rarely observed +1 state, the element usually forms ionic bonds. For example, manganese(II) sulfate, silver(I) nitrate, and copper(II) sulfate are salts. In the higher oxidation states, covalent bonding is observed. For example, titanium(IV) chloride is a liquid; vanadium(V) fluoride is a volatile white solid, m.p. 19.5°C; and osmium (VIII) oxide, OsO_4, is a very toxic liquid. Stable +1 oxidation states occur only among the group 1T elements.

In water, elements in unstable oxidation states often undergo self-redox to give two more stable oxidation states, one higher and one lower than the original. The behavior of Mn(III) is typical (the oxidation numbers are shown):

$$2Mn^{3+} + 2H_2O \longrightarrow Mn^{2+} + MnO_2(c) + 4H^+$$

oxidation number: +3   +2   +4

In many compounds, the existence of complexes serves to maintain various oxidation states that would not otherwise be stable.

One of the most noteworthy chemical properties of the transition elements is their ability to form COMPLEXES.* A complex is a molecule or ion composed of several parts, each of which has some independent existence in solution. We would not count SO_4^{2-} as a complex ion, for example, because S^{6+} and O^{2-}, or other such pairs, do not exist in aqueous solution.

A complex consists of a central atom that is a Lewis acid, surrounded by several Lewis bases, as follows:

| Lewis acid (electron acceptor) a positive metal ion, usually transitional Ag^+, Fe^{2+}, Co^{3+}, Pt^{4+}, etc. | + | Lewis bases (electron donors) also called LIGANDS; anions or molecules with lone pairs of electrons: $:CN:^-$, $:\ddot{O}H^-$, $:\ddot{F}:^-$, $H_2\ddot{O}:$, $:NH_3$, etc. | $\longrightarrow$ | Complex ion or molecule $Ag(NH_3)_2^+$, $Fe(CN)_6^{4-}$, $Cu(H_2O)_4^{2+}$, $Pt(NH_3)_2Cl_4$ |

Note that the charge on the complex is simply the sum of the charges on the central atom and on the ligands:

$$[\overset{+1}{Ag}(\overset{0}{NH_3})_2]^+ \qquad 1 + 2(0) = +1$$

$$[\overset{+2}{Fe}(\overset{-1}{CN})_6]^{4-} \qquad 2 + 6(-1) = -4$$

$$[\overset{+2}{Cu}(\overset{0}{H_2O})_4]^{2+} \qquad 2 + 4(0) = +2$$

$$[\overset{+4}{Pt}(\overset{0}{NH_3})_2\overset{-1}{Cl_4}]^0 \qquad 4 + 2(0) + 4(-1) = 0$$

When no other ligand is present, most transition-metal cations in aqueous solutions are complexed with water molecules, $[M(H_2O)_x]^{n+}$, where x is usually 3 to 6.

COORDINATION NUMBER / 22.3

The number of atoms attached to the central atom, called the COOR-DINATION NUMBER, depends on the charge on the cation, the charge on the ligand, the relative sizes of cation and ligand, and the repulsion among ligands. The interplay of these various factors cannot be evaluated quantitatively, and, hence, reliable predictions cannot be made. Fortunately, from experience, we can make a useful but not universal generalization—*a cation usually has a coordination number that is twice its charge.* Thus, some typical values are 2 for Ag^+, 4 for Cu^{2+}, 6

---

* See Appendix II for a discussion of the nomenclature of complexes.

for Co^{3+}, 4 for Ni^{2+}, and 6 for Fe^{3+}. Some important exceptions are 6 for Fe(II), Co(II), Sn(II), Sn(IV), and Pt(IV). The same cation can have more than one coordination number, as in FeF_6^{3-} and $FeCl_4^-$. However, the coordination number is rarely more than 3 times the charge of the cation.

Alfred Werner in 1893 proposed a rather simple structural theory to explain the unusual properties of complexes. We use for illustration the several compounds with the general formula $PtCl_4(NH_3)_n$, where n can be 2, 3, 4, 5, or 6. One mole of $PtCl_4(NH_3)_6$ reacts with 4 moles of silver nitrate to give 4 moles of insoluble AgCl:

$$PtCl_4(NH_3)_6 + 4Ag^+ \longrightarrow 4AgCl(c) + Pt(NH_3)_6^{4+}$$

What, then, is the valence of Pt? Is it 4 as if in $PtCl_4$? If so, what do we do with the six NH_3's? More puzzling yet, one mole of $PtCl_4(NH_3)_5$ reacts with only *three* moles of $AgNO_3$:

$$PtCl_4(NH_3)_5 + 3Ag^+ \longrightarrow 3AgCl(c) + [Pt(NH_3)_5Cl]^{3+}$$

Thus, one of the four chlorines remains with the Pt. Most startling is $PtCl_4(NH_3)_2$, which does not react with $AgNO_3$ at all. Werner rationalized such behavior by suggesting that the Pt had two kinds of valences. The first was the ordinary "ionic" valence, such as +1 for Na in NaCl. The "ionic" valence in all the Pt compounds cited above is thus +4 (which we now recognize as the oxidation number). The second was an additional or "auxiliary" valence, which for our platinum compounds is 6 (we now recognize this as the coordination number.) Thus, the various complexes may be formulated as follows:

$$[Pt(NH_3)_6]^{4+}(Cl^-)_4 + 4Ag^+ \longrightarrow 4AgCl$$
$$[Pt(NH_3)_5Cl]^{3+}(Cl^-)_3 + 3Ag^+ \longrightarrow 3AgCl$$
$$[Pt(NH_3)_4Cl_2]^{2+}(Cl^-)_2 + 2Ag^+ \longrightarrow 2AgCl$$
$$[Pt(NH_3)_3Cl_3]^+Cl^- + Ag^+ \longrightarrow AgCl$$
$$[Pt(NH_3)_2Cl_4]^0 + Ag^+ \longrightarrow \text{no reaction}$$

The species containing the Pt are the complexes; the first four are ions, the last is a neutral molecule. In all cases the oxidation number (Werner's "ionic valence") of Pt is +4, and the coordination number (Werner's "auxiliary valence") is 6.

Werner's ideas of the nature of the bonding within the complexes were rudimentary in comparison with our present concepts, but his geometry was good. He suggested that a definite spatial array is associated with each coordination number. We now know that this sup-

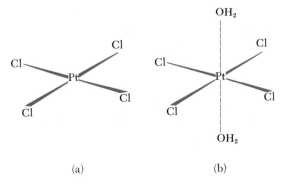

Fig. 22.1. (a) A typical square-planar complex, PtCl$_4^{-2}$. (b) The square-planar complex, represented as an incomplete octahedron. The "missing" parts are two molecules of water, one above and one below the square PtCl$_4$ plane.

(a)　　　　　　　　　　(b)

position is in accord with the wave-mechanical model for chemical bonding. The coordination numbers most frequently encountered are 6, 4, and 2. The six-coordinate complexes are all octahedral (page 395). The four-coordinate complexes are usually tetrahedral, but some are **SQUARE-PLANAR**. The square-planar complex has the cation at the center and the ligands each at one of the corners of the square (Fig. 22.1a). The square is the shape for Pt(II) and Au(III), as in PtCl$_4^{2-}$ and AuCl$_4^-$, and occurs frequently among Ni(II) and Cu(II) complexes. Recent findings seem to indicate that square-planar complexes are really cases of distorted octahedra, in which 2 loosely held molecules of water occupy the two vacant positions (Fig. 22.1b).

A coordination number of 2 is observed especially for Cu, Ag, and Au in their +1 oxidation state; for example, CuCl$_2^-$, Ag(NH$_3$)$_2^+$, and AuCl$_2^-$. The shape of these complexes is linear. Coordination numbers of 5 are rare and of 3 are very rare. The few well-defined complexes with coordination number 5 are trigonal bipyramids (page 395), for instance, iron pentacarbonyl, Fe(CO)$_5$.

BONDING IN TRANSITION-METAL COMPLEXES / 22.6

We have already seen that bonding between atoms can be viewed as two extremes: (1) electrostatic interaction between oppositely charged ions (or between an ion and a dipolar molecule); and (2) an overlap of atomic orbitals to give a covalent bond. Most often, there is a blending of the two extremes, so that many bonds have both ionic and covalent character. The same blending of bonding types prevails for the transition-metal complexes.

Let us first make the extreme "ionic" assumption, namely, that the bonding between the metal ion and the ligands is a purely electrostatic interaction between point charges. This viewpoint was first explored by physicists in the early 1930's, in connection with the spectra of ionic crystals, and the ideas about bonding that are derived from it are called **CRYSTAL-FIELD THEORY**.

Consider, for example, a Fe(CN)$_6^{4-}$ complex ion (Fig. 22.2). As we

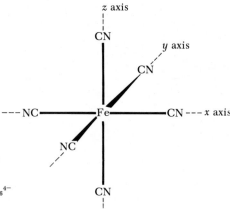

Fig. 22.2. *Shape of the Fe(CN)$_6^{4-}$ complex ion.*

have already learned, the six CN$^-$ ligands are spread apart from each other as widely as is geometrically possible, which is to say they are at the corners of an octahedron, where *all* the C—Fe—C angles are 90°. The six CN$^-$ ligands may therefore be considered to be along the *x*, *y*, and *z* axes. Now recall the shapes and spatial orientations of the five *d* orbitals that surround the nucleus of the Fe atom (see page 92). Two of the orbitals lie along the same axes as the ligands; these axial orbitals are the d_{z^2} orbital, which is oriented along the *z* axis, and the $d_{x^2-y^2}$ orbital which is oriented along the *x* and *y* axes (Fig. 22.3). The other three orbitals, d_{xy}, d_{yz}, and d_{xz}, lie *between* the axes. The two axial (d_{z^2} and $d_{x^2-y^2}$) orbitals will therefore be *closer* to the ligands than the other three orbitals are, and we now have *two sets* of *d* orbitals that are no longer equivalent to each other in energy. Furthermore, the

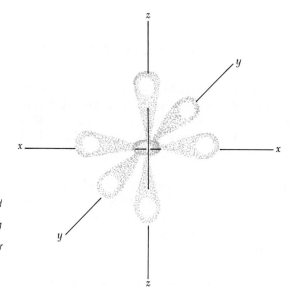

Fig. 22.3. *The d_{z^2} and $d_{x^2-y^2}$ orbitals lie along the x, y, and z axes. The other three d orbitals, which lie between the axes, are not shown.*

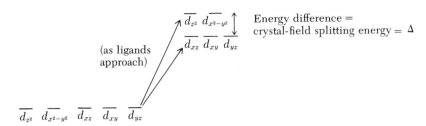

Fig. 22.4. Crystal-field splitting energy.

electrons in the two axial orbitals, being closer to the negative ligands, would be more strongly repelled by them, and would therefore be at a higher energy level than the electrons in the other three orbitals. The d orbitals are then said to be split into two groups at different energy levels, and the energy difference between them is called the CRYSTAL-FIELD SPLITTING ENERGY, symbolized by Δ, as shown in Fig. 22.4.

The energy difference Δ is often in the range corresponding to the frequencies of visible light. This means that visible light is absorbed as the d electrons are promoted to higher energy levels. The energy of the absorbed radiation is also inversely proportional to the wavelength and, in fact, Δ is calculated from this relationship:

$$\Delta = \Delta E = hf = \frac{hc}{\lambda}$$

The longer the wavelength (λ) of the absorbed light, the lower is the energy of the electronic excitation. Colorless complexes usually absorb in the ultraviolet region. Colored substances absorb in the visible region, 400 to 750 nm (page 516). Hence, it takes less energy to excite a colored substance than a colorless substance that absorbs in the ultraviolet. Color appears when the energy level of the excited state is close to the ground-state energy level.

If all visible light is absorbed, the substance appears black. If all but one color is absorbed, the substance will have that color. More frequently, however, only one color is absorbed, in which case the substance has the complementary* color.

Studies of the absorption spectra of complex ions show that the value of Δ depends in large part on the nature of the ligand. The splitting power for some of the more common ligands decreases in the order: $CN^- > NO_2^- > NH_3 > H_2O > OH^- > F^- > Cl^- > Br^- > I^-$. Thus, at the head of the list CN^-, with the greatest splitting power, is said to be a very *strong* ligand. At the other end, I^- is a very *weak* ligand.

Distribution of electrons in the d orbitals; weak and strong crystal-field complexes. The distribution of electrons in an atom is influenced by two factors: (1) the tendency for electrons to resist pairing (Hund's

---

* Complementary colors are two colors that, when combined, produce white or nearly white light. Examples are red + green, orange + blue, and yellow + violet.

Table 22.2
Electronic Distribution in Strong- and Weak-Field Complex Ions

No. of d electrons	Typical ion	Weak field	Strong field — Lower energy	Strong field — Higher energy
d^1	Ti^{3+}	↑ _ _ _ _	↑ _ _	_ _
d^2	V^{3+}	↑ ↑ _ _ _	↑ ↑ _	_ _
d^3	Cr^{3+}, V^{2+}	↑ ↑ ↑ _ _	↑ ↑ ↑	_ _
d^4	Cr^{2+}	↑ ↑ ↑ ↑ _	↑↓ ↑ ↑	_ _
d^5	Fe^{3+}, Mn^{2+}	↑ ↑ ↑ ↑ ↑	↑↓ ↑↓ ↑	_ _
d^6	Fe^{2+}, Co^{3+}	↑↓ ↑ ↑ ↑ ↑	↑↓ ↑↓ ↑↓	_ _
d^7	Co^{2+}	↑↓ ↑↓ ↑ ↑ ↑	↑↓ ↑↓ ↑↓	↑ _
d^8	Ni^{2+}	↑↓ ↑↓ ↑↓ ↑ ↑	↑↓ ↑↓ ↑↓	↑ ↑
d^9	Cu^{2+}	↑↓ ↑↓ ↑↓ ↑↓ ↑	↑↓ ↑↓ ↑↓	↑↓ ↑

rule), and (2) the tendency for electrons to occupy orbitals of lower energy before entering orbitals of higher energy.

If Δ is small, the resistance to being paired predominates and the electron enters the slightly higher energy orbital. Such an arrangement is called a **WEAK FIELD COMPLEX**. If Δ is large, pairing occurs in the lower-energy orbital to give a **STRONG FIELD COMPLEX**. Table 22.2 shows the two types of electronic arrangements for ions with different numbers of d electrons. When there are 4, 5, 6, or 7 d electrons, the two states have different numbers of unpaired electrons, so that the electronic structures can be deduced from the paramagnetic behavior of the complexes. Other properties, such as color and chemical reactivity, are also influenced by the electron distribution pattern.

Recall (page 115) that the orbital energies for ions are not the same as those for atoms. Generally, the ionic orbital energies are in the order $3d < 4s$, and $4d < 5s$.

Example 1 $Fe(CN)_6{}^{3-}$ and $FeF_6{}^{3-}$ have paramagnetic behavior due to 1 and 5 unpaired electrons respectively. Explain.

Answer First determine the number of d electrons in the cation of the complex. In this case the cation is Fe^{3+}, which is a d^5 ion. $Fe(CN)_6{}^{3-}$ with one odd electron is a strong field complex whereas $FeF_6{}^{3-}$ is a weak field complex.

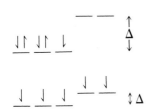

Large energy difference (strong field complex). Electrons enter lower level first, hence there is 1 unpaired electron.

Small energy difference (weak field complex). Electrons resist pairing, hence there are 5 unpaired electrons.

Transition-metal complexes may also be interpreted from the viewpoint of molecular orbital theory, which assumes *that the cation and ligand coordinate by overlap of atomic orbitals to form σ bonds.* Both the molecular orbital and crystal-field theories predict an arrangement of the same number of electrons in orbitals with energies in the same order, even though they consider the nature of the bonding in the complex from different points of view.

STEREOISOMERISM / 22.7

The reaction of ammonia with the tetrachloroplatinate(II) ion, $[PtCl_4]^{2-}$, produces a neutral molecule, $Pt(NH_3)_2Cl_2$. A compound with the same molecular formula is also obtained from the action of Cl^- on the tetraammineplatinum(II) cation, $Pt(NH_3)_4^{2+}$. However, the two compounds differ somewhat in color and solubility:

$$PtCl_4^{2-} + 2NH_3 \longrightarrow Pt(NH_3)_2Cl_2 + 2Cl^-$$
orange–yellow

$$Pt(NH_3)_4^{2+} + 2Cl^- \longrightarrow Pt(NH_3)_2Cl_2 + 2NH_3$$
canary yellow

Alfred Werner suggested that the two compounds differed in the spatial arrangement of the ligands about the Pt. Molecules differing in the spatial arrangements of atoms, rather than in their sequential order, are called **STEREOISOMERS**. Werner reasoned that if these four-coordinate complexes were square-planar, a complex with two different pairs of ligands (MA_2B_2) could have two different spatial arrangements. The one having like groups adjacent to each other is called the *cis*-isomer. The one with the like groups diagonally opposite is called the *trans*-isomer. These are shown in Fig. 22.5. That the canary-yellow compound is the *trans*-isomer is confirmed by the observation that it has a zero dipole moment. The other isomer has a measurable dipole moment, and, hence, is *cis*. These *cis-trans* isomers are called **GEOMETRIC ISOMERS**. Such isomers differ in many physical and chemical properties. A unique difference has been reported in that *cis*-$Pt(NH_3)_2Cl_2$ inhibits tumor growth, whereas the *trans*-isomer does not.

$$\begin{array}{c} H_3N \qquad NH_3 \\ \diagdown \quad / \\ Pt \\ / \quad \diagdown \\ Cl \qquad Cl \end{array}$$
cis isomer
$\mu \neq 0$

$$\begin{array}{c} Cl \qquad NH_3 \\ \diagdown \quad / \\ Pt \\ / \quad \diagdown \\ H_3N \qquad Cl \end{array}$$
trans isomer
$\mu = 0$

Fig. 22.5. *Geometric isomerism in a square-planar complex, MA_2B_2.*

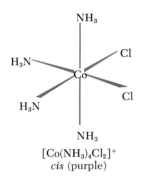

$$[Co(NH_3)_4Cl_2]^+$$
cis (purple)

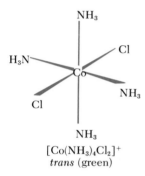

$$[Co(NH_3)_4Cl_2]^+$$
trans (green)

Fig. 22.6. *Geometric isomerism in an octahedral complex, MA₄B₂.*

Geometrical isomerism with respect to the central atom is *impossible* in tetrahedral complexes. The observation of geometrical isomers is, therefore, a way of deciding that a certain four-coordinate complex is square-planar rather than tetrahedral.

Geometrical isomerism is also observed among octahedral complexes. Werner prepared two isomers of $[Co(NH_3)_4Cl_2]^+$, one *cis* and one *trans*, shown in Fig. 22.6. The difference in color is probably due to the dependence of the crystal-field splitting energy, Δ, on the variation in spatial orientation of the two kinds of ligands. An MA_3B_3 type of complex, such as $Co(NH_3)_3(NO_2)_3$, also exists as *cis-trans* isomers.

THE STABILITY OF COMPLEX IONS / 22.8

We have learned that a Cl^- ligand bonded to a Pt^{4+} ion cannot be removed by converting it to the white insoluble AgCl with the aid of Ag^+ ions. Evidence of this kind indicates that complex ions may be very stable. Everyone knows that gold does not dissolve in aerated water, but it will dissolve if cyanide ion, CN^-, is added to form a complex.

The stability of any substance may be related to the fraction that would survive when it reached equilibrium with its decomposition products; this attribute is called the **THERMODYNAMIC STABILITY**. Thus, water is thermodynamically stable with respect to H_2 and O_2:

$$H_2O(l) \rightleftharpoons H_2(g) + \tfrac{1}{2}O_2(g) \qquad K \approx 10^{-48} \text{ at } 25°C$$

but hydrogen peroxide is thermodynamically unstable with respect to H_2O and O_2:

$$H_2O_2(l) \rightleftharpoons H_2O(l) + \tfrac{1}{2}O_2(g) \qquad K \approx 10^{20} \text{ at } 25°C$$

A substance may also be considered stable because it reacts very slowly; this attribute is called **KINETIC STABILITY**. We shall consider each type in turn, with specific reference to complexes.

(a) The **THERMODYNAMIC STABILITY** of a complex may be expressed by the equilibrium constant for its decomposition. Thus, the dissociation of the complex ion $Ag(NH_3)_2^+$ may be represented as

$$Ag(NH_3)_2^+ \rightleftharpoons Ag^+ + 2NH_3$$

The equilibrium constant for this reaction is called the **DISSOCIATION CONSTANT** of $Ag(NH_3)_2^+$,

$$\frac{[Ag^+][NH_3]^2}{[Ag(NH_3)_2^+]} = K$$

Like solubility products (page 344), these constants are calculated from standard electrode potentials (Table 22.3). Table 22.4 gives the dissociation constants of several complex ions.

Table 22.3
Standard Reduction Potentials for Complex Ions in Water at 25°

Half-reaction	$\mathcal{E}^0$, volts
$Ag(CN)_2^- + e^- \rightleftharpoons Ag + 2CN^-$	-0.395
$Ag(S_2O_3)_2^{3-} + e^- \rightleftharpoons Ag + 2S_2O_3^{2-}$	$+0.015$
$Ag(NH_3)_2^+ + e^- \rightleftharpoons Ag + 2NH_3(aq)$	$+0.373$
$Ag^+ + e^- \rightleftharpoons Ag$	$+0.7996$
$AuBr_4^- + 3e^- \rightleftharpoons Au + 4Br^-$	$+0.87$
$AuCl_4^- + 3e^- \rightleftharpoons Au + 4Cl^-$	$+1.00$
$Au^{3+} + 3e^- \rightleftharpoons Au$	$+1.42$
$Hg(CN)_4^{2-} + 2e^- \rightleftharpoons Hg + 4CN^-$	-0.37
$HgI_4^{2-} + 2e^- \rightleftharpoons Hg + 4I^-$	-0.04
$HgBr_4^{2-} + 2e^- \rightleftharpoons Hg + 4Br^-$	$+0.21$
$HgCl_4^{2-} + 2e^- \rightleftharpoons Hg + 4Cl^-$	$+0.38$
$Hg^{2+} + 2e^- \rightleftharpoons Hg$	$+0.851$

Example 2 Find the dissociation constant of $Ag(NH_3)_2^+$ at 25°C.

Answer We must find $\mathcal{E}^0$ for the reaction

$$Ag(NH_3)_2^+ \rightleftharpoons Ag^+ + 2NH_3$$

by combining $\mathcal{E}^0$'s for half-reactions which add to the desired reaction:

$Ag \rightleftharpoons Ag^+ + e^-$ (oxidation)	$-(+0.7996)$
$Ag(NH_3)_2^+ + e^- \rightleftharpoons Ag + 2NH_3$ (reduction)	$+0.373$
$Ag(NH_3)_2^+ \rightleftharpoons Ag^+ + 2NH_3$	$\mathcal{E}^0 = -0.427$ V

Table 22.4
Dissociation Constants of Complex Ions in Water at 25°C

Complex ion	K
$Ag(NH_3)_2^+$	6.0×10^{-8}
$Cd(CN)_4^{2-}$	1.25×10^{-17}
CdI_4^{2-}	3×10^{-7}
$Cd(NH_3)_4^{2+}$	2.57×10^{-7}
$Co(NH_3)_6^{2+}$	1×10^{-5}
$Co(NH_3)_6^{3+}$	10^{-34}
$Cu(CN)_2^-$	8.7×10^{-17}
$Cu(NH_3)_4^{2+}$	2.14×10^{-13}
$Ni(NH_3)_4^{2+}$	1.12×10^{-8}
$Zn(NH_3)_4^{2+}$	3.47×10^{-10}
$Zn(OH)_4^{2-}$	4.5×10^{-16}

Now

$$\mathscr{E}^0 = \frac{2.303RT}{n\mathscr{F}} \log K$$

$$= \frac{0.05916}{n} \log K \text{ (at 25°C)}$$

Here, $n = 1$ faraday/mole. (One faraday passes through the circuit when the reaction occurs as written.) Then

$$\log K = \frac{n\mathscr{E}^0}{0.05916}$$

$$= \frac{1(-0.427)}{0.05916} = -7.22$$

$$K = 10^{0.78} \times 10^{-8} = 6.0 \times 10^{-8}$$

These constants can be used in typical calculations of equilibrium concentrations:

Example 3 A solution is prepared by dissolving 0.20 mole $AgNO_3$ (ionic salt) and 0.50 mole NH_3 in H_2O to make 1.00 liter of solution. Find the concentration of Ag^+ in the solution.

Answer The process may be represented as

	Ag^+	$+$	$2NH_3$	$\rightleftharpoons$	$Ag(NH_3)_2^+$
Initial molar concentrations	0.20		0.50		0
Change	$-y$		$-2y$		$+y$
Concentrations at equilibrium	$0.20 - y$		$0.50 - 2y$		y

But practically *all* the Ag^+ reacts. Therefore, y is almost 0.20, and, at equilibrium,

$$[NH_3] = 0.50 - 2y \approx 0.50 - 2(0.20) \approx 0.10$$
$$[Ag(NH_3)_2^+] = y \approx 0.20$$
$$[Ag^+] = 0.20 - y = \text{a small number}$$

Then, for the reaction

$$Ag(NH_3)_2^+(aq) \rightleftharpoons Ag^+(aq) + 2NH_3(aq)$$

we have the equilibrium condition

$$\frac{[Ag^+][NH_3]^2}{[Ag(NH_3)_2^+]} = 6.0 \times 10^{-8}$$

$$\frac{[Ag^+] \times 0.10^2}{0.20} = 6.0 \times 10^{-8}$$

$$[Ag^+] = 1.2 \times 10^{-6} \text{ mole/liter}$$

The small value of $[Ag^+]$ justifies our assumption.

When a complex ion is formed, the concentration of the uncomplexed (actually hydrated) metal ion is greatly diminished. The result is that salts of this metal become more soluble, as equilibria of the type $AgCl(c) \rightleftharpoons Ag^+ + Cl^-$ are shifted to the right, transferring more of the metal ions to the complex:

$$Ag^+ + 2NH_3 \rightleftharpoons Ag(NH_3)_2{}^+$$

Example 4 Calculate the solubility of AgCl in 0.20 M NH_3. The reaction is

$$AgCl(c) + 2NH_3(aq) \rightleftharpoons Ag(NH_3)_2{}^+(aq) + Cl^-(aq)$$

Answer
First obtain K:

$AgCl(c) \rightleftharpoons Ag^+ + Cl^-$	$K_1 = 1.74 \times 10^{-10*}$
$Ag^+ + 2NH_3 \rightleftharpoons Ag(NH_3)_2{}^+$	$K_2 = \dfrac{1}{6.0 \times 10^{-8}}$
$AgCl(c) + 2NH_3(aq) \rightleftharpoons Ag(NH_3)_2{}^+(aq) + Cl^-(aq)$	$K = K_1K_2 = 2.9 \times 10^{-3}$

Then, the solubility of AgCl(c) is equal to $[Cl^-]$, since AgCl is the only source of Cl^-. Let the solubility of AgCl $= [Cl^-] = y$

$$AgCl(c) + 2NH_3(aq) \rightleftharpoons Ag(NH_3)_2{}^+(aq) + Cl^-(aq)$$

Initial concentration	0.20	0	0
Change	$-2y$	$+y$	$+y$
Equilibrium concentration	$0.20 - 2y$	y	y

Equilibrium condition:

$$\frac{[Ag(NH_3)_2{}^+][Cl^-]}{[NH_3]^2} = \frac{y \times y}{(0.20 - 2y)^2} = 2.9 \times 10^{-3}$$

$$\frac{y^2}{(0.20 - 2y)^2} = 2.9 \times 10^{-3}$$

$$\frac{y}{0.20 - 2y} = \sqrt{29 \times 10^{-4}} = 5.4 \times 10^{-2}$$

$$y = \text{solubility of AgCl} = 9.7 \times 10^{-3} \text{ mole/liter}$$
(compared to 1.3×10^{-5} mole/liter in pure water)

(b) **KINETIC STABILITY** refers to the *rate* at which any substance reacts. A complex ion that reacts rapidly is called **LABILE**; one that reacts very slowly is **INERT**. Kinetic stability and thermodynamic stability do not necessarily correlate with each other. For example, $Co(NH_3)_6{}^{3+}$ is thermodynamically unstable toward acid: the equilibrium constant for the reaction

$$Co(NH_3)_6{}^{3+} + 6H_3O^+ \rightleftharpoons Co(H_2O)_6{}^{3+} + 6NH_4{}^+$$

is about 10^{25}, yet it persists for weeks in an acid medium.

* See Table 14.2, page 289.

The octahedral complexes of the first transition series are fairly labile except for those of Cr^{3+} and Co^{3+}. The inert complexes of these two cations were among those used by Werner in his historic researches. The octahedral complexes of these cations are inert when they are strong field. Cr(III) is a d^3 ion and Co(III) is a d^6 ion. Therefore it appears that when the three lower-energy d orbitals of a strong-field complex are half-filled or filled, a special stability is imparted to the complex. The electron arrangements are shown in Table 22.2.

1. Definitions. Define and illustrate the following terms: (a) complex ion, (b) ligand, (c) strong-field complex, (d) weak-field complex, (e) geometrical isomerism, (f) kinetic stability, (g) thermodynamic stability.

2. Transition elements. How does the definition, "Transition elements are those whose neutral atoms have partly filled d or f subshells" differ from that given at the beginning of this chapter? Which definition includes more elements? Name the elements covered only by the broader definition.

3. d electrons. Give the number of d electrons in (a) Ru(0), (b) Mn(VII), (c) Zn(IV), (d) Ir(IV), (e) Ag(II).

4. Oxidation number. What is the oxidation number of manganese in each of the following molecules or ions? (a) MnO_4^-, (b) Na_2MnO_4, (c) $MnPt(IV)Cl_6(H_2O)_6$, (d) MnO_2, (e) Mn_3O_4.

5. Werner theory. (a) Add two more equations to the series on page 463, in which the complex is (i) $[Pt(NH_3)Cl_5]^x$ and (ii) $[PtCl_6]^x$. If a positive ion is needed to achieve neutrality, use K^+. (b) Construct a parallel series of equations which include all possible Pt(II) complexes with NH_3 and Cl^- ligands. State which, if any, of these complexes can exhibit geometrical isomerism and draw the structural formulas of the isomers.

6. Crystal-field theory. Draw the electronic configurations for the strong and weak-field octahedral complexes of (a) Mn(III); (b) Ni(III).

7. Geometrical isomerism. Draw the structural formulas of the geometrical isomers of the square-planar complexes (a) $[PtNH_3Cl_2Br]^-$; (b) $[PtNH_3ClBrF]^-$.

8. Crystal-field splitting. $Mn(H_2O)_6^{3+}$ absorbs in the visible region at a shorter wavelength than does $Cr(H_2O)_6^{2+}$. Explain the significance of this finding in terms of the crystal-field splitting energy and account for the difference in absorption in terms of the oxidation number of the metal.

9. Crystal-field theory. Give the distribution of the d electrons of the central atom according to crystal-field theory for (a) $Co(NH_3)_6^{3+}$, diamagnetic; (b) CoF_6^{3-}, paramagnetic; (c) $Co(NH_3)_6^{2+}$.

10. Stereoisomerism. Draw the formulas for the two geometrical isomers of $Co(NH_3)_3(NO_2)_3$.

11. Complex-ion calculations. Find the concentration of Cd^{2+} in a solution in which the other equilibrium concentrations are $[I^-] = 0.10$ mole/liter, and $[CdI_4^{2-}] = 0.20$ mole/liter.

12. Stability. Explain the following in terms of thermodynamic and kinetic stability. (a) HgI_4^{2-} has a dissociation constant of 5.3×10^{-31}. When radioactive iodide ($^{131}I^-$) is added to a solution of this complex ion, $Hg^{131}I_4^{2-}$ can be isolated within a short period of time. (b) $Fe(CN)_6^{4-}$ has $K = 10^{-35}$. After 100 hours in an aqueous solution of $^{14}CN^-$, less than 2% of the cyanide in the complex has exchanged.

13. Reactions. Write balanced ionic equations to represent each of the following observations. (a) SO_2 gas turns the orange solution of acidified $Cr_2O_7^{2-}$ to the green Cr^{3+}. (b) H_2S gas decolorizes a purple solution of MnO_4^-, leaving the brown-black precipitate MnO_2. (c) MnO_2 is dissolved by acidified H_2O_2, leaving a pale pink solution of Mn^{2+}. (d) The addition of KNO_2 to a pink solution of Co^{2+} acidified with acetic acid, $HC_2H_3O_2$, decolorizes the solution, leaving a yellow precipitate of $K_3Co(NO_2)_6$. (e) Black CoS is dissolved by a solution of NaCN. (f) $Co(OH)_2$ is precipitated on the addition of NaOH to a solution of Co^{2+}, but not on the addition of concentrated NH_3. (g) Zirconium dissolves in a mixture of HF and HNO_3, but not in either acid individually.

14. Crystal-field theory. Given the following standard reduction potentials:

(*i*) $Co(H_2O)_6^{3+} + e^- \longrightarrow Co(H_2O)_6^{2+} \quad \mathscr{E}^0 = +1.84$ V
(*ii*) $Co(NH_3)_6^{3+} + e^- \longrightarrow Co(NH_3)_6^{2+} \quad \mathscr{E}^0 = +0.1$ V

(a) Explain the significance of these $\mathscr{E}^0$ values in terms of relative stabilities of the Co(II) and Co(III) states in these complexes. (b) Offer an explanation in terms of the crystal field theory for the differences in the $\mathscr{E}^0$ values.

15. Ligand strength. (a) Draw the orbital structures for $:C\equiv N:^-$ and NO_2^-. (b) Do any other ligands in the series shown on page 467 have π bonds? (c) Draw a conclusion about the crystal-field effect of ligands that possess π bonds. (d) About where would CO fit into the series?

16. Complex ions. A compound X contains 30.3% Na, 18.4% Fe, 23.7% C, and 27.6% N. When it is electrolyzed with a mercury cathode, metallic Na dissolves in the mercury. HCN (a deadly gas) is slowly released when X is heated with a strong acid. However, medical records indicate that several would-be suicides have swallowed a few grams of X without apparent ill effects. (a) Write the structural formula for X. (b) Offer explanations for the action of strong acid and the failures at suicide. (c) Name the compound (see Appendix II).

17. Stability of complexes. Starting with a clear solution of silver nitrate, it is possible to produce and dissolve consecutively three precipitates in the following reactions, all of which are carried out in sequence in the same beaker:

(*i*) Add a solution of NaCl → white precipitate.
(*ii*) Add conc. NH_3 → white precipitate dissolves to yield a clear solution.
(*iii*) Add a solution of KBr → light yellow precipitate.
(*iv*) Add a solution of sodium thiosulfate, $Na_2S_2O_3$ → yellow precipitate dissolves to yield a clear solution.
(*v*) Bubble in H_2S (*g*) → black precipitate.
(*vi*) Add concentrated NaCN solution → black precipitate dissolves to yield a clear solution.
(a) Write a balanced chemical equation for each reaction. (b) List the three complex ions in the order of increasing stabilities. Explain your order.

18. Oxidizing oxyanions. (a) Ammonium salts of oxidizing anions decompose on heating. The "volcano" experiment involves heating a mound of the orange solid $(NH_4)_2Cr_2O_7$. Among the products are a large volume of the green solid Cr_2O_3 and N_2 gas. Write an equation for the reaction. (b) Under some conditions, the decomposition yields NH_3 gas. Would you conclude from this observation that the final oxidation state of some of the Cr is lower or higher than it is in Cr_2O_3? Explain.

19. Polyanions. A test for PO_4^{3-} involves the formation of a yellow precipitate of ammonium molybdophosphate, $(NH_4)_3PMo_{12}O_{40}(HNO_3)_2H_2O$, on the addition of nitric acid and ammonium molybdate, $(NH_4)_6Mo_7O_{24}(H_2O)_4$. No redox occurs. Determine the oxidation number of Mo in the polyanions $PMo_{12}O_{40}^{3-}$ and $Mo_7O_{24}^{6-}$.

20. Complex ions. (a) Calculate the dissociation constant of $Ag(S_2O_3)_2^{3-}$. (b) Calculate the solubility of AgBr in 0.25 M $Na_2S_2O_3$. (See Tables 14.2 and 22.3.) The reaction is

$$AgBr(c) + 2S_2O_3^{2-} \longrightarrow Ag(S_2O_3)_2^{3-} + Br^-$$

21. Nomenclature. (See Appendix II.) Give a systematic name for (a) $[Co(NH_3)_5Cl]Cl_2$, (b) $Na_3[Co(NO_2)_6]$, (c) $K_2[PtCl_4Br_2]$, (d) $Ni(CO)_4$, (e) $K_2[Fe(CN)_5NO]$.

23

ORGANIC CHEMISTRY

WHAT IS ORGANIC CHEMISTRY? / 23.1

The adjective "organic" implies a product that originates from living organisms. Until the middle of the nineteenth century, most chemists thought that the atoms in organic molecules were bound together not merely by electrical forces but also by some mysterious vital force. As life must come from previous life, so it was believed that an organic chemical must come from another organic chemical, and ultimately from a living organism, the only source of the vital force. However as chemists began to synthesize organic chemicals from inorganic compounds, the "vital force" theory gradually fell into disrepute.

Most of these organic compounds contained carbon. For this reason, even after the demise of the vital force theory, it was natural to retain the term "organic chemistry" to refer to the chemistry of carbon compounds. This definition is not precise, however; organic chemists do not usually concern themselves with carbonate rocks, such as limestone, $CaCO_3$. Some definitions require that a compound have at least one C—H or C—halogen bond to be classified as organic, but even this would incorrectly exclude carbon disulfide, CS_2, and urea, $(H_2N)_2C{=}O$. Thus, there is no exact definition of "organic chemistry."

Carbon is the key element in the classification because we find so many carbon-containing compounds. There are so many because carbon atoms form stable bonds to each other, forming continuous or branched chain compounds, as well as cyclic compounds:

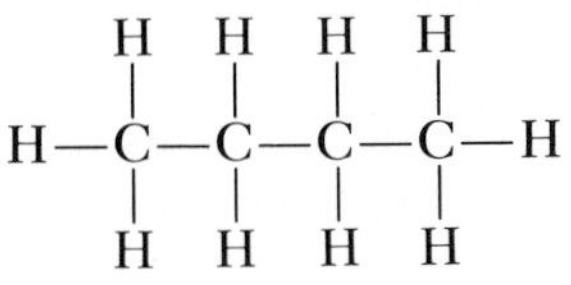

$CH_3CH_2CH_2CH_3$
n-butane
(continuous chain)

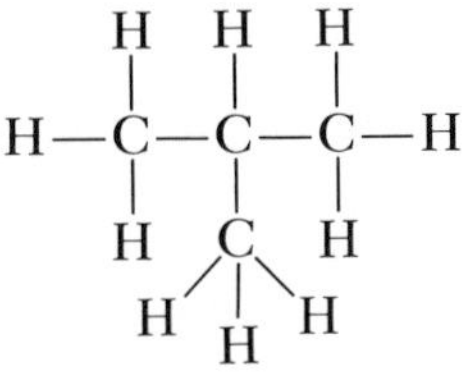

$(CH_3)_2CHCH_3$
isobutane
(branched chain)

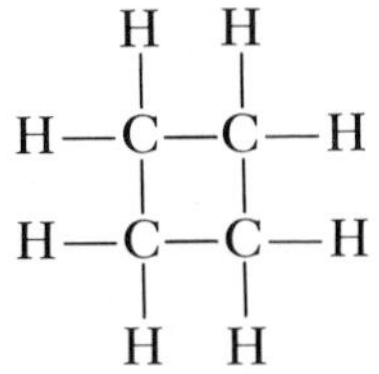

C_4H_8 or $(CH_2)_4$
cyclobutane
(cyclic)

Compounds are known with more than a hundred carbon atoms bonded in sequence. The number of organic compounds is further increased by the ability of C to exist in sp, sp^2, and sp^3 hybrid states and to bond to practically every element in the periodic table except the noble gases. The most common partners are H, O, N, F, Cl, Br, I, S, P, and Si.

Carbon ($1s^22s^22p^2$) does not usually participate in ionic bonding. The C^{4+} ion, which would be isoelectronic with helium ($1s^2$), does not exist in chemical systems because it would require a very high ionization energy (3412 kcal/mole) to remove the four electrons. Thus, even CF_4 is covalent, not ionic. The C^{4-} ion, isoelectronic with neon ($1s^22s^22p^6$), is thought to exist only in beryllium carbide, Be_2C, and aluminum carbide, Al_4C_3, in which its attraction by the nearby Be^{2+} or Al^{3+} ions suffices to overcome the mutual repulsions of its four extra electrons.

There are compounds in which a carbon atom participates in both ionic and covalent bonding. When methyl chloride, CH_3Cl, reacts with potassium, potassium methide, CH_3K, is formed,

$$\underset{\substack{\text{methyl} \\ \text{chloride}}}{H-\overset{\overset{\displaystyle H}{|}}{\underset{\underset{\displaystyle H}{|}}{C}}-Cl} + 2K\cdot \longrightarrow \underset{\substack{\text{potassium} \\ \text{methide}}}{H-\overset{\overset{\displaystyle H}{|}}{\underset{\underset{\displaystyle H}{|}}{C}}:^-K^+} + K^+Cl^-$$

Potassium methide is a crystalline, ionic solid in which there are three covalent C—H bonds, and one ionic bond between the $[:CH_3]^-$ anion and the K^+ cation. Such carbon anions, called CARBANIONS, are almost invariably bonded only to the highly electropositive elements of group 1R or 2R. With the less electropositive metallic elements, carbon forms a covalent bond. For example, dimethylmercury, $H_3C-Hg-CH_3$, is a typical covalent compound (b.p. 96°C). Carbanions are formed as very reactive intermediates in some organic reactions.

The cationic analog of CH_3^- is CH_3^+. Such a species, called a CARBONIUM ION, has only a sextet of electrons about the carbon atom and, as expected, reacts readily as a Lewis acid to complete the octet. Carbonium ions are also formed as very short-lived intermediates during the course of certain organic reactions. Some carbonium ions have been identified in solution under especially favorable conditions. A typical preparation of a carbonium ion is the following electrophilic displacement (page 258) on F by the Lewis acid SbF_5 as shown:

$$CH_3-\underset{\underset{H}{|}}{\overset{\overset{\displaystyle F \;\;\;\;}{\cdot\cdot}}{C}}-CH_3 + SbF_5 \longrightarrow CH_3\overset{+}{C}HCH_3 + SbF_6^- \tag{1}$$

isopropyl fluoride	a Lewis acid	isopropyl cation a carbonium ion

The most typical bonding of carbon is covalent, using sp^3 (tetrahedral), sp^2 (trigonal), or sp (linear) hybridized states. The particular hybridization sets a limit on the number of σ bonds (4 for sp^3, 3 for sp^2, and 2 for sp). The deficit between the number of σ bonds and the total covalency of 4 for carbon is made up by π bonds:

Hybridization	σ bonds	π bonds	Bond multiplicity
sp^3	4	0	4 single bonds, $-\overset{\overset{\displaystyle |}{}}{\underset{\underset{\displaystyle |}{}}{C}}-$
sp^2	3	1	2 single + 1 double bond, $-\underset{\underset{\displaystyle |}{}}{C}=$
sp	2	2	1 single + 1 triple bond, $-C\equiv$, or 2 double bonds, $=C=$.

ALKANE HYDROCARBONS*; ISOMERISM AND HOMOLOGY / 23.3

HYDROCARBONS are compounds that contain only carbon and hydrogen. A subdivision of hydrocarbons is the ALKANES. The carbon atoms in alkanes have no multiple bonds and form only chains, not cycles.

The major sources of alkanes are petroleum and natural gas. Petroleum is separated by distillation into various fractions differing in boiling point range. Each fraction is a mixture containing alkanes with several different numbers of C atoms. The important alkane fractions are (a) gas (1–4 C atoms), (b) gasoline (5–10 C atoms), (c) kerosene (11 and 12 C atoms), (d) heating oil (13–17 C atoms), (e) lubricating oil (18–25 C atoms), and (f) petroleum jelly (26–38 C atoms). Fractions with 5 carbons or more may also contain cyclic hydrocarbons.

A manufacturer of pure hydrocarbons offers for sale 18 different alkanes called OCTANES, whose molecular formula is C_8H_{18}. Seventeen of them are clear, colorless liquids at room temperature. The eighteenth is a white crystalline solid. We determine their boiling points, and find that they range from 99° to 126°C; some of the boiling points are so close to others that they cannot easily be distinguished from each other. Now we try a somewhat more heroic experiment. We take 18 identical new automobiles with empty gasoline tanks and add one of

---

* See Appendix II for a discussion of nomenclature of the alkanes and some of their simple derivatives.

the octanes to each. Of course, this is a bit difficult with the solid octane, but we can liquefy it by dissolving it in a little commercial gasoline. We find that some of the octanes serve very well as gasolines; others cause some engine "knock"; some knock so badly that we cannot drive the automobile.

Clearly, what we have here are 18 different substances (ISOMERS) that are distinct in their properties despite their common molecular formula. As the number of C atoms in the alkane increases, the number of isomers increases sharply. This comes about because branching can become more extensive. We will illustrate this with one of the octanes ("isooctane"), which serves as the standard of 100-octane gasoline. Imagine that we construct a model of this molecule from atomic models by starting with methane, CH_4, and adding C and H atoms with the provision that with each addition *we maintain the proper covalencies* for C and H. We remove one H, shown as $\boxed{H}$, and add $\textcircled{C}H_3$ as shown:

$$H_3C-\boxed{H} \xrightarrow{-\boxed{H}} H_3C$$

$$H_3C \xrightarrow{+\textcircled{C}H_3} H_3C-\textcircled{C}H_3 \quad\Bigg\} \quad (1)$$
$$\text{ethane}$$

The net addition is thus CH_2.
A similar second step gives

$$\overset{\boxed{H}}{\underset{}{H_3C-CH_2}} \xrightarrow[+\textcircled{C}H_3]{-\boxed{H}} \overset{\textcircled{C}H_3}{\underset{}{H_3C-CH_2}} \quad (CH_3CH_2CH_3)$$
$$\text{propane}$$

Continuing, we have

removed in (4)

$$\overset{CH_3}{\underset{\boxed{H}}{H_3C-CH}} \xrightarrow{(3)} \overset{CH_3}{\underset{\textcircled{C}H_3}{H_3C-C\boxed{H}}} \xrightarrow{(4)} \overset{CH_3}{\underset{CH_3\ H}{H_3C-C-\textcircled{C}\boxed{H}}} \overset{H}{} \xrightarrow{(5)}$$

$\llcorner$removed in (5)

removed in (6)

$$\overset{H_3C\ \ H\ \ H}{\underset{H_3C\ \ H\ \ H}{H_3C-C-C-\textcircled{C}-\boxed{H}}} \xrightarrow{(6)} \overset{H_3C\ \ H\ \ H}{\underset{H_3C\ \ H\ \boxed{H}}{H_3C-C-C-C-\textcircled{C}H_3}} \xrightarrow{(7)} \overset{CH_3}{\underset{CH_3\ \ \ \textcircled{C}H_3}{H_3C-C-CH_2-CH-CH_3}}$$

$\llcorner$removed in (7)   "isooctane"

Note the following:

(a) We started with CH_4 and added CH_2 seven times, thus, $CH_4 + 7(CH_2) \rightarrow C_8H_{18}$.

(b) We had no choice in steps (1) and (2) because all the hydrogens that we might have removed were equivalent; we cannot distinguish one from another. In steps (3) through (7), however, there were other possibilities than the ones we chose to construct the model we wanted. For example, instead of step (3), we could have replaced an H from an end carbon as shown in step (3'):

$$
\underset{\underset{H}{|}}{\overset{\overset{H}{|}}{CH_3CH_2C}}\boxed{H} \xrightarrow{(3')} \underset{\underset{H}{|}}{\overset{\overset{H}{|}}{CH_3CH_2C}}\textcircled{C}H_3
$$

These other ways could have led to any of the other seventeen isomers, two of which are

$$
CH_3CH_2CH_2CH_2CH_2CH_2CH_2CH_3 \qquad \underset{\underset{H_3C}{|} \quad \underset{CH_3}{|}}{\overset{\overset{H_3C}{|} \quad \overset{CH_3}{|}}{CH_3-C-C-CH_3}}
$$

n-octane   tetramethylbutane

These compounds, which differ among each other in the *sequence of linkages of atoms*, are called **STRUCTURAL ISOMERS**.

All alkanes have one common feature: their C and H content can be expressed by the general formula C_nH_{2n+2}, where n is a whole number. A series of compounds, describable by a general formula, is called a **HOMOLOGOUS SERIES**. Among carbon compounds the molecular formulas of successive members of such a series differ from each other by CH_2.

The alkanes with continuous (unbranched) carbon chains are called *n*-alkanes (*n* stands for "normal"). Table 23.1 gives the names and properties of the first ten of these. Note that the first three, methane, ethane, and propane, are not designated n because they have no branched isomers from which they need be distinguished.

Removing an H atom from an alkane leaves an *alkyl* group. To name the group, replace the suffix *ane* by the suffix *yl*. Some common alkyl groups are: H_3C- (methyl), $H_3C\underset{\underset{H}{|}}{\overset{\overset{H}{|}}{C}}-$ (ethyl), $H_3CCH_2\underset{\underset{H}{|}}{\overset{\overset{H}{|}}{C}}-$ (*n*-propyl),

and $H_3C-CH-CH_3$ (isopropyl). Refer to Appendix II.
|

Table 23.1
Names and Physical Properties of Normal Alkanes, C_nH_{2n+2}

Name	Molecular formula	Condensed formula	Boiling point, °C	Melting point, °C
Methane	CH_4	CH_4	−161.5	−182.5
Ethane	C_2H_6	CH_3CH_3	− 88.6	−183.3
Propane	C_3H_8	$CH_3CH_2CH_3$	− 42.1	−189.9
n-Butane	C_4H_{10}	$CH_3CH_2CH_2CH_3$	− 0.5	−138.4
n-Pentane	C_5H_{12}	$CH_3(CH_2)_3CH_3$	36.1	−129.7
n-Hexane	C_6H_{14}	$CH_3(CH_2)_4CH_3$	68.7	− 95.3
n-Heptane	C_7H_{16}	$CH_3(CH_2)_5CH_3$	98.4	− 90.6
n-Octane	C_8H_{18}	$CH_3(CH_2)_6CH_3$	125.7	− 56.8
n-Nonane	C_9H_{20}	$CH_3(CH_2)_7CH_3$	150.8	− 53.5
n-Decane	$C_{10}H_{22}$	$CH_3(CH_2)_8CH_3$	174.1	− 29.7

Example 1 Write the structural formulas for the isomers of C_5H_{12}.

Answer C_5H_{12} fits the formula C_nH_{2n+2} and is therefore the formula for one or more alkanes. We systematically select the different ways of arranging 5 C atoms. We simplify the formulas by omitting H atoms. First, we have 5 consecutive C atoms:

(I) $-\overset{|}{\underset{|}{C}}-\overset{|}{\underset{|}{C}}-\overset{|}{\underset{|}{C}}-\overset{|}{\underset{|}{C}}-\overset{|}{\underset{|}{C}}-$. Next, we take 4 C atoms, and add a fifth C atom as a

branch. Adding it to an end C (C_1 or C_4) gives us compound (I) again, with 5 consecutive C atoms. Therefore, we add the fifth C to either C_2 or C_3, which are

equivalent, giving us (II) $-\overset{|1}{\underset{|}{C}}-\overset{|2}{\underset{|}{C}}-\overset{|3}{\underset{|}{C}}-\overset{|4}{\underset{|}{C}}-$. We then take 3 consecutive carbons,
$\qquad -\underset{|}{C}-$

$\overset{1}{C}-\overset{2}{C}-\overset{3}{C}-$, and add 2 more C atoms. Again, we cannot add a C to the end carbons, C_1 or C_3, because this would give a 4-carbon sequence. The remaining C

atoms must be added to C_2 to give (III) $-\overset{|}{\underset{|}{C}}-$ with the structure. We could have added the

two C atoms as a 2-carbon sequence to give $-\overset{1}{C}\overset{2}{-C}-C-$ ... $-\overset{3}{C}-$ ... $-\overset{4}{C}-$. But now we actually

have a four-carbon chain with a C on C_2. This compound is identical to (II),

We see that bending bonds does not result in different isomers:

$$CH_3—CH_2—CH_2—CH_2—CH_3$$

$$\begin{array}{ll} CH_2—CH_2—CH_2—CH_3 & CH_2—CH_2 \\ | & | \quad | \\ CH_3 & CH_3 \quad CH_2—CH_3 \end{array}$$

All represent the same compound, n-pentane, because all the carbon sequences are continuous, not branched.

THE SHAPES OF ALKANES; CONFORMATION / 23.4

The bond angle corresponding to sp^3 hybridization is 109.5°. CH_4 is thus a tetrahedral molecule. A chain of σ-bonded carbon atoms, which maintains the tetrahedral C—C—C angles, is therefore shaped as a series of 109.5° zig-zags. But these σ-bonded atoms can rotate more or less freely about the bond. This means that a chain of singly bound C atoms in an alkane can be arranged in any zig-zag shape in space as long as the tetrahedral bond angles are preserved. Two of these arrangements, called **CONFORMATIONS**, are shown in Fig. 23.1 for models of the carbon skeleton of a continuous 8-carbon chain. Note that the linear structural formula (Fig. 23.1a) represents only the projection of one particular conformation; such a formula does not mean that the chain is actually a straight line in space. Conformations are rarely isomers because they are usually rapidly interconverted, without breaking chemical bonds, and therefore cannot be isolated and stored as distinct substances. Thus, the conformations shown in Fig. 23.1a and 23.1b are carbon skeletons of the same isomer of C_8H_{18}.

Let us look more carefully at the structure of CH_3CH_3. Its entropy, measured thermochemically, is lower than the value calculated from spectral data. The calculations *assumed free rotation* about the C—C σ bond to give an infinite number of conformations having the same energy and, therefore, an equal chance of existence. There is an in-

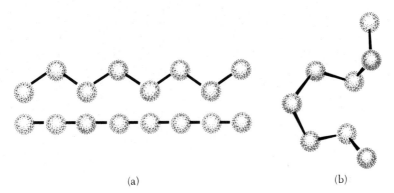

(a)   (b)

Fig. 23.1. *Model of carbon skeleton of n-octane. (a) Conformation whose projection is linear. (b) Conformation that does not have a linear projection.*

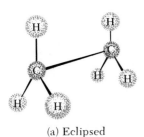

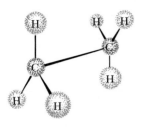

(a) Eclipsed

(b) Staggered

Fig. 23.2. *Conformations of ethane:* (a) *eclipsed,* (b) *staggered.*

finite number of conformations because there is an infinite number of angles through which one of the CH₃ groups could be rotated. We will consider only the two extreme cases.

If we sight along the C—C bond in Fig. 23.2a, the H's on the front C eclipse those on the back C. In Fig. 23.2b, the H's are said to be **STAGGERED**. A rotation of 60° would convert the **ECLIPSED** into the staggered conformation.

To explain the discrepancy between the calculated and observed values, it was reasoned that the two conformations actually do *not* have the same energy. It was estimated that the staggered conformation is more stable than the eclipsed conformation by 3 kcal/mole.* Therefore, at a given instant, a quantity of ethane will have more staggered-like than eclipsed-like molecules. Although rotation about the C—C σ bond occurs, it is not completely random or "free" because of the energy barrier of 3 kcal/mole required to pass from a staggered through an eclipsed to another staggered conformation, as shown in the energy diagram, Fig. 23.3. Thus the actual lack of complete randomness causes a loss in entropy. Hence, the observed value for the entropy was less than the calculated value.

Fig. 23.3. *Energy changes during rotation of 120° about carbon–carbon single bond of ethane.*

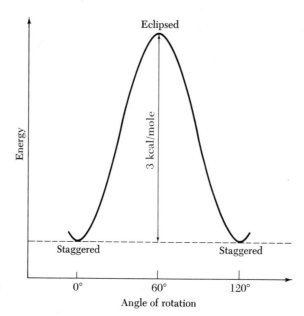

* The repulsive force between the pair of electrons in one bond and those in nearby bonds is greater in the eclipsed conformation because of the greater proximity of the electrons.

Hydrocarbons like ethene, $H_2C=CH_2$, that possess one carbon-to-carbon double bond ($C=C$), belong to the **ALKENE** homologous series. The suffix *-ene* indicates the presence of the $C=C$ bond. Since the alkenes have two fewer H atoms than the corresponding alkanes, their general formula is C_nH_{2n}, where n is a whole number larger than 1.

The presence of a double bond usually increases the opportunities for isomerism. Thus, although there are only two butane isomers, there are four isomeric butenes. Isobutene is branched, and we might expect two *n*-butenes as shown:

$$(H_3C)_2C=CH_2 \qquad \overset{1}{H_2}\overset{2}{C}=\overset{3}{C}H\overset{}{C}H_2\overset{4}{C}H_3 \qquad \overset{1}{H_3}\overset{2}{C}CH=\overset{3}{C}H\overset{4}{C}H_3$$

isobutene   1-butene   2-butene

(For naming alkenes, see Appendix II.)

Why then are there four butene isomers? The reason is that in 2-butene, the two CH_3 groups may be on the same side (**CIS**) of the double bond, or on opposite sides (**TRANS**). Therefore there are two 2-butenes, the *cis-* and *trans-*isomers, rather than the single 2-butene we originally proposed.

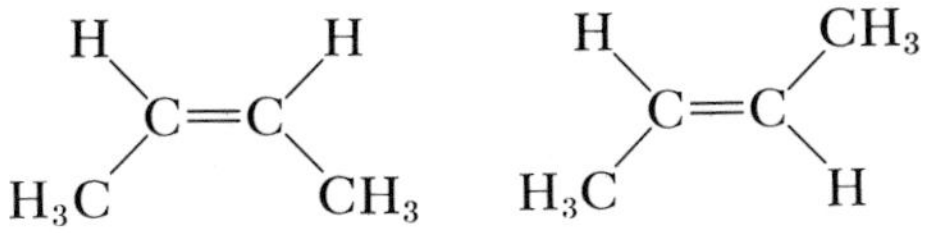

cis-2-butene   *trans*-2-butene

This is a type of **GEOMETRICAL ISOMERISM** (page 469).

To be considered isomers, molecules must maintain their identity when separated from each other. Rotation about a double bond has an activation energy (rotation barrier) of more than 30 kcal/mole, and is therefore more restricted than rotation about a single bond. The reason for this difference is that rotation involving a π bond would destroy the lateral overlap of the p orbitals, as illustrated in Fig. 23.4.

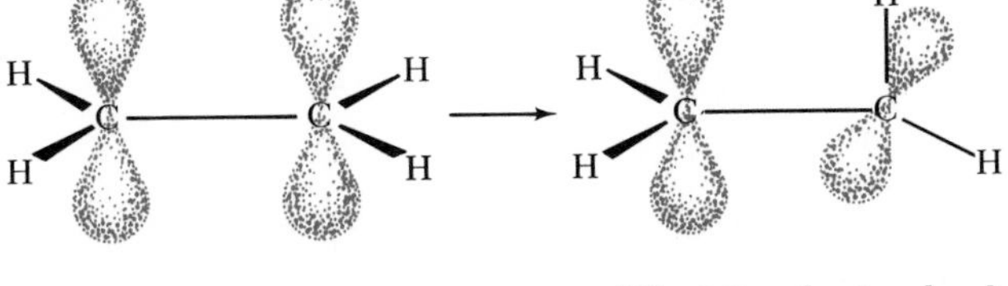

90° rotation about a π bond
(no p-orbital overlap)

Fig. 23.4. *Rotation about a π bond.*

In the case of 1-butene,

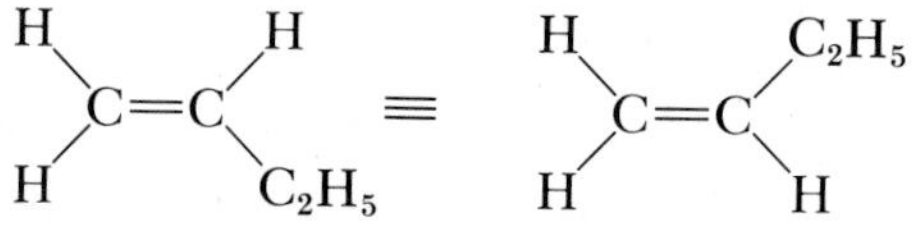

$$\underset{H}{\overset{H}{>}}C=C\underset{C_2H_5}{\overset{H}{<}} \quad \equiv \quad \underset{H}{\overset{H}{>}}C=C\underset{H}{\overset{C_2H_5}{<}}$$

1-butene (no geometrical isomerism)

the two structures can be superimposed by flipping one on the other without breaking chemical bonds. The lack of free rotation about the double bond of 1-butene, therefore, has no structural consequences, and no geometrical isomerism exists for this compound. *In order for geometrical isomerism to exist, each doubly bonded carbon atom must have two different atoms or groups attached to it.* The two sets of atoms or groups may, however, be identical, as in the isomers of $ClHC=CHCl$.

Acetylene (ethyne), $HC\equiv CH$, is the first member of the **ALKYNE** homologous series, whose general formula is C_nH_{2n-2}. The ending *-yne* indicates the presence of the $-C\equiv C-$ bond.

Molecules can have more than one double bond. Thus, hydrocarbons having two double bonds are called **DIENES**. One of the more important dienes is 1,3-butadiene,

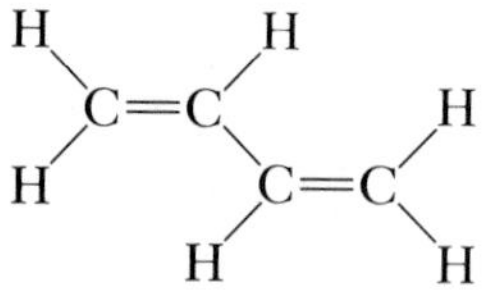

$$\underset{H}{\overset{H}{>}}C=C\underset{H}{\overset{H}{<}}\,\underset{H}{\overset{}{}}C=C\underset{H}{\overset{H}{<}}$$

1,3-butadiene

which serves as starting material for preparing a variety of polymers.

CYCLOALKANES / 23.6

Carbon atoms may also form cyclic compounds. Since the minimum number of carbon atoms needed to form a cycle is 3, the lowest member of this series is cyclopropane,

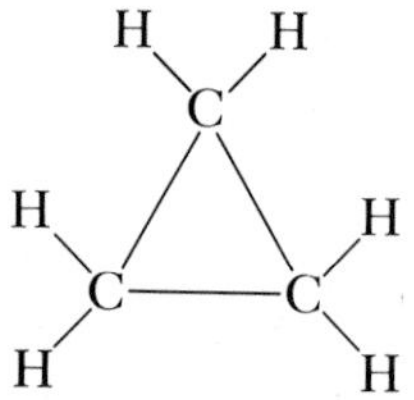

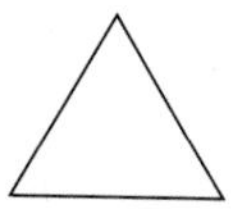

cyclopropane   abbreviated formula

To name this group of hydrocarbons, *cyclo-* is prefixed to the name of the alkane with the corresponding number of carbons. The cycloalkanes have the general formula C_nH_{2n}, and therefore are isomeric with the alkenes. For example, cyclobutane (C_4H_8) and 1-butene are isomers.

BENZENE AND AROMATIC COMPOUNDS / 23.7

Cyclic compounds can also possess double bonds. Of particular interest is the cyclic triene called **BENZENE**. Benzene is a volatile liquid (b.p. 80°C), first isolated in 1825 by Michael Faraday by heating soft coal in the absence of oxygen. Its molecular formula is C_6H_6, and the structure suggested by August Kekulé in 1866 is

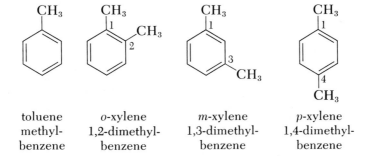

Kekulé formula for benzene   abbreviated formula

The ring system is an integral unit of a very large number of hydrocarbons which are referred to as **AROMATIC*** compounds. Coal tar, a distillate from soft coal, is a source of many aromatic compounds, for example,

common name:	toluene	o-xylene	m-xylene	p-xylene
systematic name:	methyl-benzene	1,2-dimethyl-benzene	1,3-dimethyl-benzene	1,4-dimethyl-benzene

Note the use of the prefixes, *o-* for *ortho-*, *m-* for *meta-*, and *p-* for *para-*, to indicate the 1,2-, 1,3-, and 1,4-disubstituted benzene, respectively. There are many aromatic hydrocarbons with fused benzene rings—rings with 2 C atoms in common—as in naphthalene, $C_{10}H_8$, a white solid (m.p. 80°C) used as a moth repellent:

---

* This class of compounds was originally called aromatic because many of their derivatives have pleasant odors. Many of them are also toxic, some are carcinogenic, and therefore should not be inhaled.

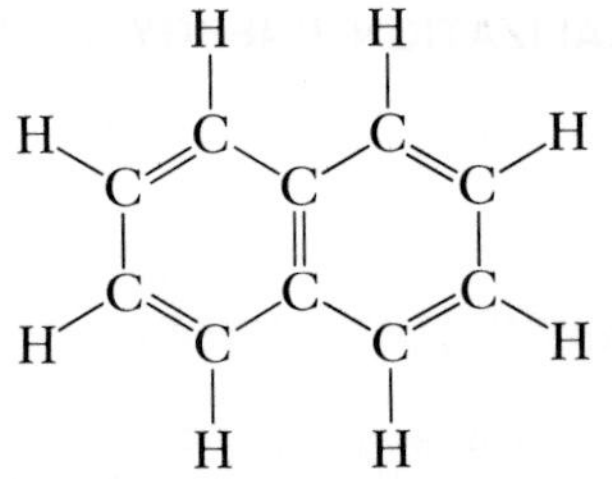

or

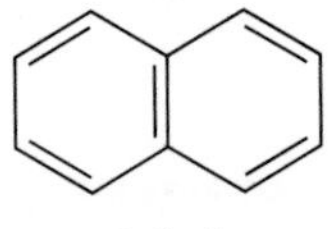

naphthalene

Graphite is composed of layers of *fused benzene rings* (page 435).

Aromatic hydrocarbons are used extensively in the manufacture of plastics, drugs, dyes, insecticides, explosives, and a host of other common materials. A few examples of compounds derived from aromatic hydrocarbons are

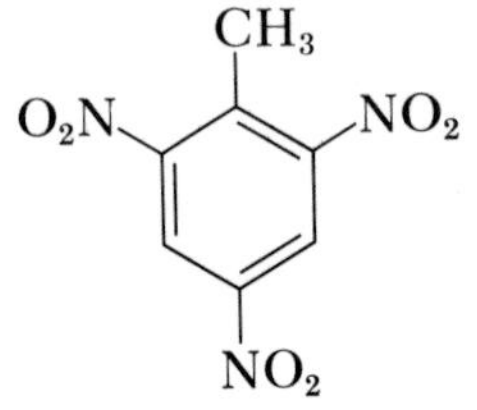

TNT	aspirin	DDT
(an explosive)	(a drug)	(an insecticide)

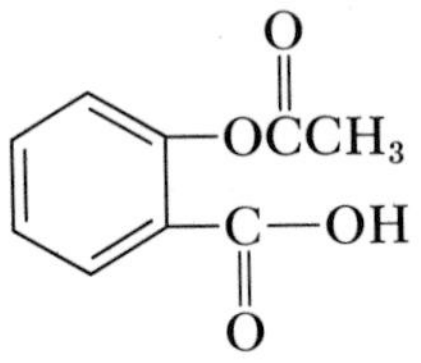

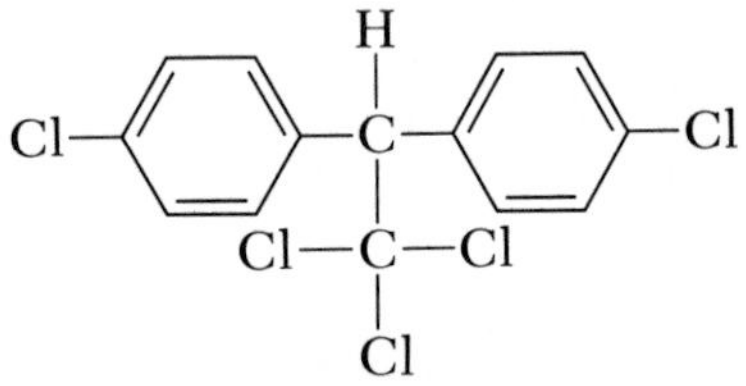

Table 23.2 summarizes the hydrocarbon series referred to in the preceding sections. The meaning of the last column will be explained in Section 23.13.

Table 23.2
Hydrocarbon Homologous Series

Series name	General formula	Typical example	Formula	Chemical behavior
Alkane	C_nH_{2n+2}	Ethane	H_3C-CH_3	Saturated
Alkene	C_nH_{2n}	Ethene	$H_2C=CH_2$	Unsaturated
Alkyne	C_nH_{2n-2}	Ethyne	$H-C\equiv C-H$	Unsaturated
Alkadiene	C_nH_{2n-2}	1,3-Butadiene	$H_2C=CH-CH=CH_2$	Unsaturated
Cycloalkane	C_nH_{2n}	Cyclohexane	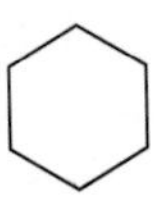	Saturated
Benzenoid	C_nH_{2n-6}	Benzene		Aromatic

(a) **Structure of benzene.** Benzene is much less reactive than typical alkenes. This property has led to the idea that there is a stability inherent in this ring of 6 carbon atoms with three alternating double bonds. The Kekulé structural formula for benzene shows three fixed double bonds. A second Kekulé formula could have been drawn with the double bonds alternating the other way; these are contributing structures (page 384), *neither of which is correct:*

contributing structures (Kekulé)

These formulas are incorrect because they imply that the bonds in benzene should alternate in length between 1.34 Å for a C=C bond and 1.54 Å for a C—C bond. Experimentally, all the carbon-to-carbon bond distances are 1.40 Å (Fig. 23.5a).

We will now use an orbital picture to obtain a single structural formula consistent with the geometrical and chemical properties of benzene. Benzene is a planar molecule. Each C—C—C and H—C—C bond angle is 120° (Fig. 23.5a). This shape indicates that each C atom is using sp^2 hybridized atomic orbitals to form three σ bonds for the carbon–hydrogen skeleton (Fig. 23.5b). Each C atom utilizes 3 of its 4 outer electrons to fabricate its three σ bonds. The fourth electron must reside in the remaining p orbital. Each carbon atom has such a p orbital, and the six p orbitals are parallel and adjacent to each other (Fig. 23.5c). A given p orbital can overlap at the same time with each parallel p orbital adjacent to it (Fig. 23.5d), so that *the 6 p electrons are delocalized in an extended, cyclic π system encompassing all 6 carbons.* The electron distribution has the shape of a pair of lumpy doughnuts, one above and the other below the σ-bonded carbon skeleton. Since the electron density is somewhat greater at the carbon atoms and somewhat less between the carbons, the doughnuts are depicted with lumps at the carbon atoms (Fig. 23.5e).

This model for benzene predicts that the bonding is uniform between any pair of carbon atoms, and hence each carbon-to-carbon bond should have the same length. The model also predicts the carbon-to-carbon bond length to be less than a C—C bond length and greater than a C=C bond length, in agreement with experiment. The delocalized extended π system is represented by a circle in the center of the hexagon (Fig. 23.5f). *The extended π system is not a single molecular orbital.* The six p atomic orbitals interact by lateral overlap to give six π molecular orbitals, three bonding and three antibonding. The 6 p elec-

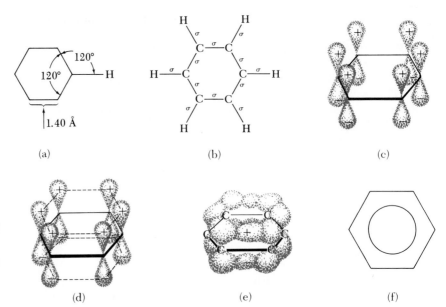

Fig. 23.5. Molecular orbital representation for benzene: (a) structure showing geometry; (b) σ-bonded skeleton; (c) p orbitals; (d) cyclic overlap of p orbitals; (e) extended π bond; (f) accepted shorthand notation.

(a)

(b)

(c)

(d)

(e)

(f)

trons of benzene fill the three bonding molecular orbitals, each of which is delocalized over all 6 carbon atoms. The charge cloud (shown in Fig. 23.5e) actually represents the most stable of the three bonding molecular orbitals.

(b) Delocalization energy. Electrons may be found anywhere in an extended π bond. Since the delocalized electrons can move in a larger volume, *the repulsive forces between them are weakened.* As a result, the real molecule with delocalized p electrons has less energy than we would assign to the fictitious contributing structures, and therefore is more stable:

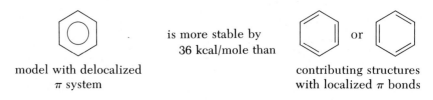

model with delocalized π system

is more stable by 36 kcal/mole than

contributing structures with localized π bonds

The difference in energy between the localized (Kekulé) and delocalized models is called **DELOCALIZATION ENERGY.** *The more extensive the delocalization of p electrons, the greater is the delocalization energy and the more stable is the molecule.* The term *aromatic* is now used to describe those substances with large delocalization energies. All aromatic substances have p orbitals overlapping laterally in a cyclic fashion. They are all *less* reactive chemically than isomers without this property. For example, benzene is less reactive than its nonaromatic isomer

$$H-C\equiv C-\underset{\underset{\displaystyle H}{|}}{\overset{\overset{\displaystyle H}{|}}{C}}-\underset{\underset{\displaystyle H}{|}}{\overset{\overset{\displaystyle H}{|}}{C}}-C\equiv C-H.$$

An H atom of a hydrocarbon can be replaced by another atom or group of atoms not of the hydrocarbon variety. The resulting compounds are called **SUBSTITUTED** hydrocarbons, and the atom or group is known as a **FUNCTIONAL GROUP**. The more common functional groups are listed in Table 23.3. The symbol R is used to represent any carbon-containing group. The chemistry of these substituted hydrocarbons is characteristic mainly of the functional group.

The reactions of the various functional groups constitute a large part of the subject matter of organic chemistry, and any systematic treatment would be too voluminous for this chapter. However, we shall cite some properties and reactions of several classes of compounds of general interest.

With **ALCOHOLS**, ROH, we could consider the R to have replaced an H of water, and, indeed, alcohols show some similarities to water in their chemical behavior. For example, ethanol, like water, has a weakly acidic H atom, and therefore reacts with metallic sodium with the evolution of hydrogen:

$$C_2H_5OH(l) + Na(c) \longrightarrow C_2H_5O^-Na^+(alcoholic) + \tfrac{1}{2}H_2(g) \quad \text{(slower)}$$

$$HOH(l) + Na(c) \longrightarrow HO^-Na^+(aq) + \tfrac{1}{2}H_2(g) \quad \text{(faster)}$$

Alcohols can form H bonds among themselves and with water. Therefore, alcohols are more soluble in water than are the corresponding alkanes, since alkanes have practically no ability to H-bond.

Ethyl alcohol, or ethanol, is sometimes called grain alcohol because much of it is made by fermentation of sugar or starch in the presence of an enzyme. Consequently, it is found in wines, beers, and "hard" liquors. The body can tolerate moderate amounts of ethanol, but excessive consumption causes addiction, leading to breakdown of liver tissue (cirrhosis) and brain tissue (delirium tremens), conditions that can be fatal.

Other common alcohols are methyl alcohol, CH_3OH (methanol or wood alcohol), ethylene glycol, $HOCH_2CH_2OH$, used as an antifreeze, and glycerol (glycerine), $HOCH_2CHOHCH_2OH$, an important trihydroxy compound.

Table 23.3
Common Functional Groups in Organic Compounds

Group	General formula	General name	Example Formula	Example Name
—Cl	R—Cl	Chloride	CH_3Cl	Methyl chloride (chloromethane)
—Br	R—Br	Bromide	CH_3Br	Methyl bromide (bromomethane)
—OH	R—OH	Alcohol	CH_3CH_2OH	Ethyl alcohol (ethanol)
—O—	R—O—R	Ether	$CH_3CH_2OCH_2CH_3$	Diethyl ether
—NH$_2$	RNH$_2$	Amine	$CH_3CH_2CH_2NH_2$	n-Propylamine
$-N\big\langle{}^{O}_{O}$	RNO$_2$	Nitro compound	$C_6H_5NO_2$	Nitrobenzene
$-C\!=\!O$ $\;\;\mid$ $\;\;H$	$R\!-\!C\!=\!O$ $\;\;\;\;\;\mid$ $\;\;\;\;\;H$	Aldehyde	$CH_3CH_2C\!=\!O$ $\;\;\;\;\;\;\;\;\;\;\mid$ $\;\;\;\;\;\;\;\;\;\;H$	Propionaldehyde
$-C\!=\!O$	$\;\;\;\;\;R$ $\;\;\;\;\;\mid$ $R\!-\!C\!=\!O$	Ketone	$\;\;\;\;\;\;\;O$ $\;\;\;\;\;\;\;\|$ $CH_3\!-\!C\!-\!C_2H_5$	Methyl ethyl ketone
$-C\!=\!O$ $\;\;\mid$ $\;\;OH$	$R\!-\!C\!=\!O$ $\;\;\;\;\;\mid$ $\;\;\;\;\;OH$	Carboxylic acid	$\;\;\;\;\;\;\;O$ $\;\;\;\;\;\;\;\|$ $CH_3\!-\!C\!-\!OH$	Acetic acid
$\;\;O$ $\;\;\|$ $-C\!-\!OR'$	$\;\;\;\;\;O$ $\;\;\;\;\;\|$ $R\!-\!C\!-\!OR'$	Ester	$\;\;\;\;\;\;\;O$ $\;\;\;\;\;\;\;\|$ $CH_3\!-\!C\!-\!OC_2H_5$	Ethyl acetate
—O—O—	R—O—O—R'	Peroxide	$C_2H_5\!-\!O\!-\!O\!-\!C_2H_5$	Diethyl peroxide

ETHERS, ROR, would result from replacing both H's of H_2O by an R group. They may be formed by intermolecular dehydration of alcohols. Diethyl ether, called "ether," an anesthetic, is made this way from ethanol:

$$C_2H_5O\overline{H\;\;+\;\;HO}C_2H_5 \xrightarrow[\text{heat}]{Al_2O_3} C_2H_5OC_2H_5 + H_2O$$

$$\text{ethanol} \hspace{5cm} \text{diethyl ether}$$

Since ethers do not have OH groups, they do not react with Na, nor do they undergo H-bonding. Because they have an $-\overset{\cdot\cdot}{\underset{\cdot\cdot}{O}}-$ atom with unshared electrons they behave as Brönsted bases toward strong acids, and as Lewis bases (nucleophiles) toward Lewis acids:

$$C_2H_5\overset{..}{\underset{..}{O}}C_2H_5 + H:\overset{..}{\underset{..}{I}}: \longrightarrow [C_2H_5\overset{\overset{\displaystyle H}{\displaystyle |}}{O}C_2H_5]^+ + \quad :\overset{..}{\underset{..}{I}}:^-$$

$$\text{base}_1 \qquad \text{acid}_2 \qquad\qquad \text{acid}_1 \qquad\qquad \text{base}_2$$

$$\text{diethyloxonium ion}$$

$$C_2H_5\overset{..}{\underset{..}{O}}C_2H_5 + BF_3 \longrightarrow C_2H_5\overset{\displaystyle +}{\overset{..}{\underset{\underset{\underset{F\;\;|\;\;F}{\displaystyle F}}{\displaystyle B^-}}{O}}}C_2H_5$$

$$\text{Lewis base}\quad\text{Lewis acid}$$

ALDEHYDES and **KETONES** both possess the **CARBONYL**, $\rangle C{=}O$, group. An aldehyde has at least one H atom bonded to the carbonyl carbon; a ketone has two R groups attached to this carbon:

$$\underset{\text{an aldehyde}}{R{-}\overset{\overset{\displaystyle H}{\displaystyle |}}{C}{=}O} \qquad \underset{\text{formaldehyde}}{H{-}\overset{\overset{\displaystyle H}{\displaystyle |}}{C}{=}O} \qquad \underset{\text{a ketone}}{R{-}\overset{\overset{\displaystyle R}{\displaystyle |}}{C}{=}O} \qquad \underset{\text{acetone}}{H_3C{-}\overset{\overset{\displaystyle CH_3}{\displaystyle |}}{C}{=}O}$$

Formalin, familiar to biology students as a preservative, is a 40% water solution of formaldehyde.

These compounds can be prepared by oxidation of the appropriate alcohols. If the OH group of the alcohol is attached to a C bearing 2 H atoms, the alcohol can be oxidized by reagents such as potassium permanganate, $KMnO_4$, and potassium dichromate, $K_2Cr_2O_7$, to an aldehyde,

$$CH_3{-}\overset{\overset{\displaystyle H}{\displaystyle |}}{\underset{\underset{\displaystyle H}{\displaystyle |}}{C}}{-}O \;\; +[O]^* \longrightarrow \underset{\substack{\text{acetaldehyde}\\(\text{a typical aldehyde})}}{CH_3{-}\overset{\overset{\displaystyle H}{\displaystyle |}}{C}{=}O} + H_2O$$

If only 1 H atom is attached to the C atom with the OH group, the alcohol is oxidized to a ketone,

$$CH_3{-}\overset{\overset{\displaystyle CH_3}{\displaystyle |}}{\underset{\underset{\displaystyle H}{\displaystyle |}}{C}}{-}O{-}H +[O] \longrightarrow \underset{\substack{\text{acetone}\\(\text{a typical ketone})}}{CH_3{-}\overset{\overset{\displaystyle CH_3}{\displaystyle |}}{C}{=}O} + H_2O$$

$$\underset{\substack{\text{2-propanol}\\(\text{isopropyl alcohol})}}{}$$

* To simplify writing the above equations, [O] is used to indicate the oxidizing agent.

An alcohol with no H atoms on the carbon bearing the OH group resists oxidation under the usual conditions,

$$CH_3-\underset{\underset{CH_3}{|}}{\overset{\overset{CH_3}{|}}{C}}-OH + [O] \longrightarrow \text{no reaction}$$

2-methyl-2-propanol

Oxidation is a useful method for the synthesis of ketones since ketones resist further oxidation. It has only limited utility for the preparation of aldehydes since the aldehyde obtained is readily oxidized further to a carboxylic acid,

$$H-\underset{\underset{H}{|}}{\overset{\overset{H}{|}}{C}}-\overset{\overset{H}{|}}{C}=O + [O] \longrightarrow H-\underset{\underset{H}{|}}{\overset{\overset{H}{|}}{C}}-\overset{\overset{OH}{|}}{C}=O$$

acetaldehyde   acetic acid

The oxidation of appropriate alcohols to aldehydes or ketones can be reversed. Thus, aldehydes and ketones add hydrogen in the presence of platinum to give the corresponding alcohol,

$$CH_3\underset{\underset{H}{|}}{\overset{}{C}}=O + H_2 \xrightarrow{Pt} CH_3-\underset{\underset{H}{|}}{\overset{\overset{H}{|}}{C}}-OH$$

CARBOXYLIC or FATTY ACIDS, $R-\overset{\overset{O}{\|}}{C}-OH$, are weak acids ($K_a \approx 10^{-5}$) which react with hydroxides to form salts:

$$CH_3COOH(l) + Na^+OH^-(aq) \longrightarrow CH_3COO^-Na^+(aq) + H_2O(l)$$

acetic acid   sodium acetate

The salt of a long-chain fatty acid is called a SOAP (page 177). Some important naturally occurring fatty acids are

$CH_3(CH_2)_{14}COOH$ palmitic acid
$CH_3(CH_2)_{16}COOH$ stearic acid
$CH_3(CH_2)_7CH{=}CHCH_2COOH$ oleic acid
$CH_3(CH_2)_4CH{=}CHCH_2CH{=}CHCH_2COOH$ linoleic acid

Carboxylic acids react with alcohols and eliminate water to form ESTERS. The reaction is catalyzed by strong acids such as HCl.

$$CH_3CO\overset{\text{}}{|}OH + H\overset{\text{}}{|}OC_2H_5 \underset{\underset{\text{Hydrolysis}}{\overset{H^+}{\longleftarrow}}}{\overset{\overset{H^+}{\longrightarrow}}{\text{Esterification}}} CH_3C\overset{\overset{O}{\diagup\!\!\diagup}}{-}OC_2H_5 + H_2O$$

ethyl acetate
(an ester)

This reaction is *not* analogous to the reaction between a strong acid and a strong base, like $ClH + HO^-Na^+ \rightarrow Na^+Cl^- + H_2O$. Instead, the oxygen eliminated in esterification comes from the carboxylic acid, not from the hydroxy group of the alcohol. This conclusion is derived from experiments in which an alcohol "labeled" with ^{18}O isotope, $R^{18}OH$, is esterified with an unlabeled acid, $R'COOH$. The eliminated water does *not* contain ^{18}O, but the ester does, showing that the oxygen in H_2O must have come from the carboxylic acid:

$$R'CO\overline{|OH + H|}{}^{18}OR \longrightarrow R'C\!\!\overset{O}{\underset{}{\diagup}}\!\!-^{18}OR + H_2O$$

An ester is hydrolyzed in the presence of hydroxide ion. The process is called SAPONIFICATION because the carboxylate salt is a soap when R' is a long chain of carbon atoms:

$$\underset{\text{ester}}{R'COOR} + Na^+OH^- \longrightarrow \underset{\text{soap}}{R'COO^-Na^+} + \underset{\text{alcohol}}{ROH}$$

Natural FATS are solid esters and fatty OILS are liquid esters of the trihydroxy alcohol glycerol. Natural fats and oils contain two or three different carboxylic acid components, each with an even number of carbon atoms ranging from 4 to 26. The saponification of such esters, therefore, yields a mixture of soaps:

$$
\begin{array}{l}
R-C\!\!\overset{O}{\underset{}{\diagup}}\!\!-O-CH_2 \\[4pt]
R'-C\!\!\overset{O}{\underset{}{\diagup}}\!\!-O-CH + 3Na^+OH^- \longrightarrow \underset{\text{soap}}{RCOO^-Na^+} + \underset{\text{soap}}{R'COO^-Na^+} \\[4pt]
R''-C\!\!\overset{O}{\underset{}{\diagup}}\!\!-O-CH_2
\end{array}
$$

$$+ \underset{\text{soap}}{R''COO^-Na^+} + \underset{\text{glycerol}}{HOC\!-\!C\!-\!COH}$$

The esters of saturated fatty acids (such as stearic acid) (R = alkyl) are called SATURATED FATS; esters of fatty acids having at least one C=C bond are usually liquids at ambient temperatures and are called UNSATURATED OILS. When the carboxylic acids have two or more double bonds, their esters are called POLYUNSATURATED OILS.

AMINES are derived from NH_3 by replacing at least one H by a hydrocarbon group. Some examples are

$$\langle\bigcirc\rangle-NH_2 \qquad CH_3CH_2\overset{\overset{\displaystyle H}{|}}{N}CH_3 \qquad CH_3\overset{\overset{\displaystyle CH_3}{|}}{N}CH_3$$

aniline   ethylmethylamine   trimethylamine

Amines are bases that react with acids to form alkylammonium salts, in a manner analogous to the behavior of ammonia:

$$CH_3NH_2 + HBr \longrightarrow CH_3NH_3^+Br^-$$
methylamine   methylammonium bromide

$$NH_3 + HBr \longrightarrow NH_4^+Br^-$$
ammonia   ammonium bromide

CHIRALITY; OPTICAL ISOMERISM / 23.10

Lactic acid, $CH_3CH(OH)COOH$, isolated from sour milk and muscle tissue, has the remarkable ability to rotate a plane of polarized light* to the right (clockwise from observer's viewpoint). It is called (+)-lactic acid. The (+) indicates right-rotating. Another kind of lactic acid, (−)-lactic acid, synthesized in the laboratory, rotates the plane of polarized light to the left (counterclockwise). Because of this optical activity, these forms of lactic acid are called OPTICAL ISOMERS. Their other physical properties, such as boiling point, melting point, and solubilities in different solvents, are identical. An equal mixture of these isomers, called a RACEMIC FORM, no longer rotates a plane of polarized light and is therefore optically inactive.

To understand the basis of optical isomerism, construct a ball-and-peg model of lactic acid, Fig. 23.6a. By exchanging a pair of groups, we

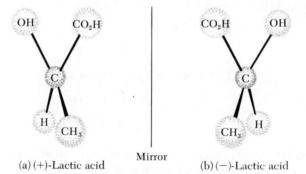

Fig. 23.6. *Nonidentical mirror images in the tetrahedral system.*

(a) (+)-Lactic acid   Mirror   (b) (−)-Lactic acid

* Light waves are assumed to vibrate in an infinite number of planes, all of which are at right angles to the direction in which the light travels. When ordinary light is passed through a Nicol prism, composed of calcite (a crystalline form of $CaCO_3$), or through Polaroid, it emerges vibrating in only one plane. Such light is PLANE-POLARIZED. The rotation of the plane polarized light is observed in an instrument called a POLARIMETER.

get a second structure, Fig. 23.6b, which is not identical with the first one, because they do not match atom for atom in space. They are indeed isomers, interchangeable only by breaking and reforming bonds, and not by mere rotation about bonds. They are called ENANTIOMERS. A mirror placed between the enantiomers shows them to be mirror images of each other. Isomers which are not identical with their mirror images are called CHIRAL.* Chiral molecules do not possess a plane or center of symmetry (page 397). A molecule with a plane and/or a center of symmetry is ACHIRAL; it has an identical mirror image and shows no optical activity. For example, if the CH_3 group of lactic acid is replaced by an H, we get glycolic acid, $CH_2(OH)COOH$. A ball-and-peg model of this compound, Fig. 23.7, is now identical with its mirror image; it has a plane of symmetry bisecting the H—C—H bond angle and it is achiral.

The relationships between chirality and symmetry elements are

Symmetry designation	Relationship to mirror image	Symmetry elements
Achiral (symmetric)	Identical	Plane and/or point of symmetry present
Chiral (dissymmetric)	Not identical	Plane and point of symmetry absent

Lactic acid is only one of many molecules having carbon as the central tetrahedral atom. To be chiral, the molecule must possess a tetrahedral C atom attached to *four different atoms or groups*. In the case of lactic acid, the four groups or atoms are COOH, OH, CH_3, and H. It is always possible to arrange the four different groups or atoms about a CHIRAL CENTER to give two structures—the optical isomers. The two different spatial arrangements are called CONFIGURATIONS. Since optical isomers, like geometrical isomers, differ because of a different spatial arrangement of atoms, they are also stereoisomers.

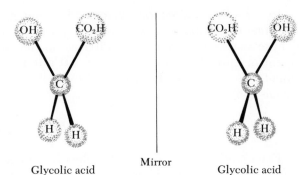

Glycolic acid   Mirror   Glycolic acid

Fig. 23.7. *Identical mirror images.*

* Formerly called DISSYMMETRIC.

Achiral reagents, catalysts, or solvents react in an identical fashion with enantiomers. For example, HI, an achiral compound, reacts identically in every respect with (+)-lactic acid and with (−)-lactic acid:

$$2 \ CH_3CH(OH)COOH + 2 \ HI \longrightarrow 2 \ CH_3CH_2COOH + 2 \ H_2O + I_2$$
(+) or (−)-lactic acid

However, with chiral substances, enantiomers react at different rates. For example, enzymes are chiral biochemical catalysts. They may catalyze the reaction of one enantiomer, leaving the other enantiomer unchanged. Considerations of molecular chirality are thus of great importance in many chemical reactions, especially those involving living organisms.

Example 2 Which of the following compounds can exist as stereoisomers? Indicate the type of stereoisomers, geometric or optical. (a)

$$\begin{array}{cc} H_3C & Cl \\ \diagdown & \diagup \\ C=C & \\ \diagup & \diagdown \\ H & Cl \end{array}$$ (b)

$$H_3C-\underset{\underset{Cl}{\mid}}{\overset{\overset{H}{\mid}}{C}}-CH_2CH_3;$$ (c) $CH_3CH_2CH{=}CHCH_3$; (d) HCDClBr (D is deuterium).

Answer (a) Cannot. Has no chiral center; multiple-bonded atoms cannot be chiral centers. With 2 Cl's on the same double-bonded C, the compound cannot have a geometric isomer. (b) Can have an optical isomer. The C bonded to the Cl is a chiral center; the four different groups are H_3C, H, Cl, and CH_2CH_3. Note that we consider the entire group attached, e.g., H_3C and CH_2CH_3, and not just the adjacent atom which happens to be a C in both of these groups. (c) Has geometric isomers. Each double-bonded carbon is attached to a pair of different groups or atoms; H, CH_3 are attached to one C and H, CH_2CH_3 are attached to the other. (d) Has optical isomers. The C is a chiral center with 4 different atoms; H and D are different.

REACTIONS OF ORGANIC COMPOUNDS

In the following sections we consider some general types of reactions of organic compounds which are typical of all covalent substances. These reactions may be classified as (a) decomposition, (b) displacement or substitution, (c) addition, (d) elimination, and (e) rearrangement. We have already described examples of displacement reactions (page 256).

DECOMPOSITION / 23.11

Fragmentation of a covalent molecule into two or more smaller particles without the involvement of other reactants is called a **DECOMPOSITION** reaction. Heat or radiation causes the decomposition by rupturing the weaker bonds. Bond cleavage can occur in either of two ways. (a) *One of the atoms retains both electrons:*

$$A:\!\{B \longrightarrow A:+ B \quad \text{or} \quad A\}\!:B \longrightarrow A + :B$$

as illustrated by:

$$H_3N:B(CH_3)_3 \xrightleftharpoons{\text{heat}} H_3N:+ B(CH_3)_3$$

This type of reaction is the reverse of a Lewis acid–base reaction. (b) *Each atom takes 1 electron:*

$$A:B \longrightarrow A\cdot + \cdot B$$

as illustrated by the decomposition of dinitrogen tetroxide,

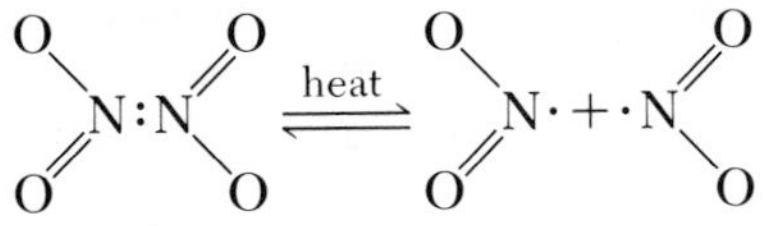

dinitrogen   nitrogen
tetroxide   dioxide

This reaction leads to the formation of free radicals. Most free radicals are reactive species that rapidly undergo further reaction. Free radicals are frequently needed to initiate reactions. Compounds with weak covalent bonds, such as the bond between less active metals and carbon, or the peroxide, —O—O—, bond, are frequently used as free radical sources, as shown:

$$H_3C:Hg:CH_3 \xrightarrow{\text{heat}} \cdot Hg\cdot + 2H_3C\cdot$$
dimethylmercury

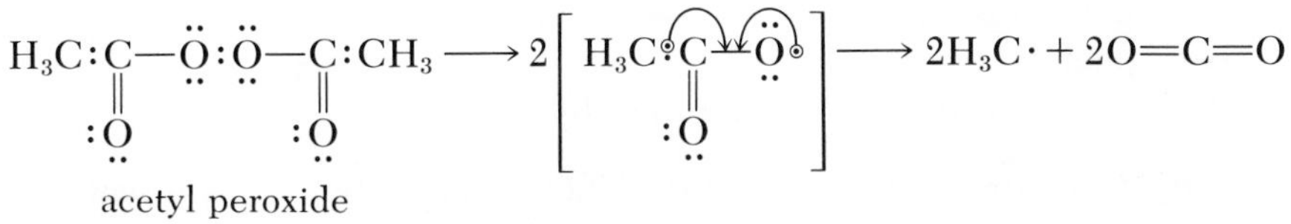

acetyl peroxide

To understand the course of these reactions (and those discussed later in the chapter), it is necessary to keep track of the assignment of electrons. To help in this "electron bookkeeping," the chemist frequently uses curved arrows to indicate the disposition of electrons, as shown above.

DISPLACEMENT OR SUBSTITUTION ON CARBON / 23.12

(a) **Nucleophilic displacement on a σ-bonded carbon atom.** These reactions are important in the synthesis of organic compounds, a few of which are illustrated:

$$H:\overset{..}{\underset{..}{O}}:^- + (CH_3)_2CH\overset{\frown}{(\overset{..}{\underset{..}{Br}}:)} \longrightarrow (CH_3)_2CH\overset{..}{\underset{..}{O}}H + :\overset{..}{\underset{..}{Br}}:^-$$

an alcohol

$$:N\equiv C:^- + (CH_3)_2CH\overset{\frown}{(\overset{..}{\underset{..}{Cl}}:)} \longrightarrow (CH_3)_2CHC\equiv N: + :\overset{..}{\underset{..}{Cl}}:^-$$

a cyanide

$$H_3N: + (CH_3)_2CH\overset{\frown}{(\overset{..}{\underset{..}{Br}}:)} \longrightarrow (CH_3)_2CHNH_3^+ + :\overset{..}{\underset{..}{Br}}:^-$$

an ammonium salt

These reactions go nearly to completion; in each case the reaction proceeds so that the stronger *Brönsted* base displaces the weaker *Brönsted* base. For this reason, in most bimolecular reactions on carbon that go to completion, the leaving groups are the extremely weak bases, such as I⁻, Br⁻, and Cl⁻. These anions are referred to as "good leaving groups."

These reactions can proceed by two different mechanisms. *The rate equation is one of the best ways of elucidating the mechanism.* One group of nucleophilic displacements as typified by

$$H:\overset{..}{\underset{..}{O}}:^- + CH_3CH_2\overset{\frown}{(\overset{..}{\underset{..}{Br}}:)} \longrightarrow CH_3CH_2:\overset{..}{\underset{..}{O}}H + :\overset{..}{\underset{..}{Br}}:^-$$

has a second-order rate equation

$rate = k[OH^-][CH_3CH_2Br]$

This one-step reaction is termed the S_N2 reaction: S for substitution (displacement), N for nucleophilic, and 2 for bimolecular. It is bimolecular because the slow step—the only step—involves a collision of two species.

A second kind of reaction, typified by

$$(CH_3)_3C:\overset{..}{\underset{..}{Br}}: + :\overset{..}{\underset{..}{O}}:H^- \longrightarrow (CH_3)_3C:\overset{..}{\underset{..}{O}}:H + :\overset{..}{\underset{..}{Br}}:^- \qquad \text{(in water)}$$

has a first-order rate equation,

$rate = k[(CH_3)_3CBr]$

and is designated the S_N1 reaction: S for substitution, N for nucleophilic, and 1 for unimolecular. The S_N1 mechanism is typical of a second broad type of displacement reaction; *the leaving group is ejected first, and then combination with the displacing group occurs.* The first step, which is the slow one, is the ionization of the alkyl halide to form a reactive carbonium ion:

step (1) $(CH_3)_3C\!:\!\overset{..}{\underset{..}{Br}}\!:$ $\xrightarrow[\text{(aq)}]{\text{slow}}$ $(CH_3)_3C^+(aq) + :\overset{..}{\underset{..}{Br}}:^-(aq)$

carbonium ion

In step (2) the carbonium ion reacts with the solvent, in this case H_2O:

step (2) $(CH_3)_3C^+ + H_2\overset{..}{\underset{..}{O}}:$ $\xrightarrow{\text{fast}}$ $(CH_3)_3C\overset{..}{O}H_2{}^+$

conjugate acid of the alcohol

The OH^-, acting as a Brönsted base, now plays a role:

step (3) $\underset{\text{acid}_1}{(CH_3)_3COH_2{}^+} + \underset{\text{base}_2}{OH^-}$ $\xrightarrow{\text{fast}}$ $\underset{\text{base}_1}{(CH_3)_3COH} + \underset{\text{acid}_2}{H_2O}$

The addition of Ag^+ increases the rate of those S_N1 reactions in which halide ions are leaving groups. This observation is consistent with the proposed mechanism; the Ag^+ ion facilitates the ionization step by forming AgBr,

$(CH_3)_3C\!:\!\overset{..}{\underset{..}{Br}}: + Ag^+ \longrightarrow (CH_3)_3C^+ + Ag\!:\!\overset{..}{\underset{..}{Br}}: (c)$

A third general type of mechanism first involves bond formation with the nucleophile, followed by ejection of the leaving group:

step (1) $X: + A:L \longrightarrow X:A:L$

intermediate

bond forming (association)

step (2) $X:A:L \longrightarrow X:A + :L$

bond breaking (dissociation)

This type is not observed for nucleophilic displacement on a σ-bonded carbon because the first step would lead to a C with five σ bonds. However, displacement on silicon

$(CH_3)_3SiCl + OH^- \xrightarrow{H_2O} (CH_3)_3SiOH + Cl^-$

and displacement on boron

$(CH_3)_2BCl + OH^- \xrightarrow{H_2O} (CH_3)_2BOH + Cl^-$

do occur by this mechanism. Silicon can acquire five σ bonds by using an empty d orbital. Boron has only a sextet of electrons and is "eager" to acquire another pair by bonding with a nucleophile. In nucleophilic displacement of ligands of complex ions all three types of mechanisms are observed.

(b) **Other substitution reactions.** When the displacing agent is a free radical,

$$R\cdot + A:L \longrightarrow R:A + \cdot L$$

the group or atom displaced must also be a free radical. Such displacements usually occur as *intermediate* steps in reactions. This type of reaction is illustrated by the light-induced chlorination of methane,

$$CH_4 + Cl_2 \xrightarrow{\text{light}} CH_3Cl + HCl$$

The mechanism of this reaction involves the following steps:

step 1
$$:\overset{..}{\underset{..}{Cl}}:\overset{..}{\underset{..}{Cl}}: \xrightarrow{\text{light}} 2:\overset{..}{\underset{..}{Cl}}\cdot$$

step 2
$$:\overset{..}{\underset{..}{Cl}}\cdot + H\!\!-\!\!CH_3 \longrightarrow :\overset{..}{\underset{..}{Cl}}:H + \cdot CH_3$$

step 3
$$H_3C\cdot + :\overset{..}{\underset{..}{Cl}}\!\!-\!\!\overset{..}{\underset{..}{Cl}}: \longrightarrow H_3C:\overset{..}{\underset{..}{Cl}}: + \cdot\overset{..}{\underset{..}{Cl}}:$$

The first step furnishes the free radicals, which in this case are the chlorine atoms. The next two steps are free radical displacements. The Cl· generated in step 3 can react as shown in step 2 and, therefore, steps 2 and 3 constitute a chain reaction (page 425). Chain reactions are characteristic of free radical processes. The sum of the chain reaction steps gives the overall chemical equation.

Of considerable importance are displacements by electrophiles (Lewis acids) on aromatic compounds such as benzene. Two of the more common examples are bromination and nitration. The formula for the specific catalyst needed is shown above the arrow:

Bromination:
$$C_6H_6 + Br_2 \xrightarrow{\text{FeBr}_3} C_6H_5Br + HBr$$
bromobenzene

Nitration:
$$C_6H_6 + HNO_3 \xrightarrow{\text{H}_2\text{SO}_4} C_6H_5NO_2 + H_2O$$
nitrobenzene

The mechanism of nitration is typical of such electrophilic displacements. The electrophile (Lewis acid) is the nitronium ion, NO_2^+, formed as shown:

$$HNO_3 + 2\ H_2SO_4 \longrightarrow H_3O^+ + 2HSO_4^- + NO_2^+$$
nitronium ion

Benzene has an electron-rich extended cyclic π system and therefore behaves as a Lewis base. The NO_2^+, a Lewis acid, bonds to one of the sp^2 carbons of benzene:

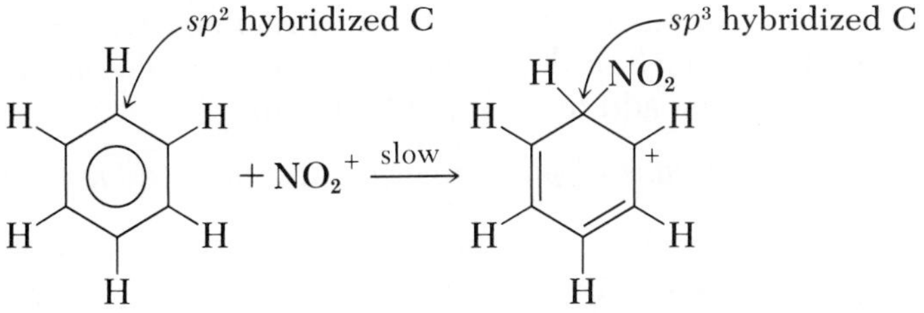

sp² hybridized C

sp³ hybridized C

Lewis base   Lewis acid   unstable carbonium ion intermediate

The carbon to which the Lewis acid bonds becomes sp^3 hybridized, since it now has four σ bonds. As a result, the extended π system is disrupted and most of the delocalization energy (page 490) is lost. Therefore, this step is endothermic and rate-controlling. The unstable intermediate carbonium ion reestablishes the stable cyclic π system by transferring a proton to a base to give the product, as shown:

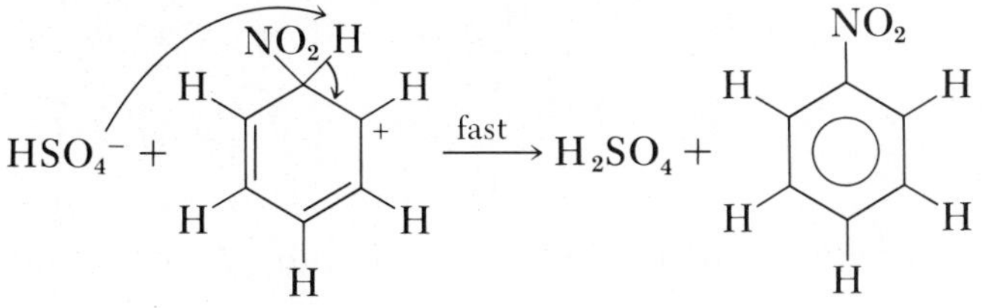

base₁   acid₂   acid₁   base₂

ADDITION REACTIONS / 23.13

Compounds with multiple bonds are capable of reacting with other molecules so as to acquire only σ bonds. A few such ADDITION reactions are illustrated for alkenes, alkynes, and carbonyl compounds. Ethylene, a typical alkene, reacts with chlorine readily at room temperature:

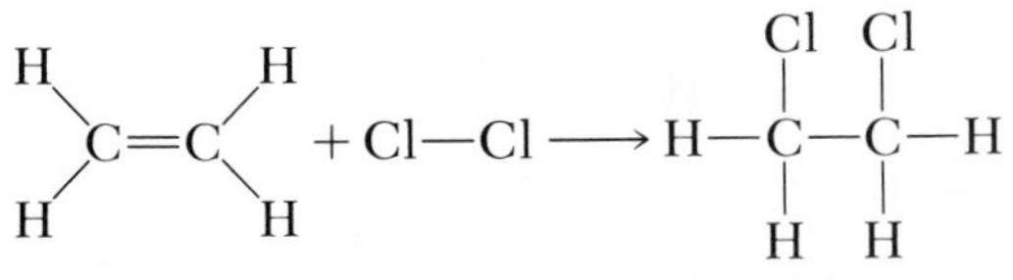

1,2-dichloroethane

Alkynes readily add 2 moles of halogen, as exemplified by the addition of bromine to propyne,

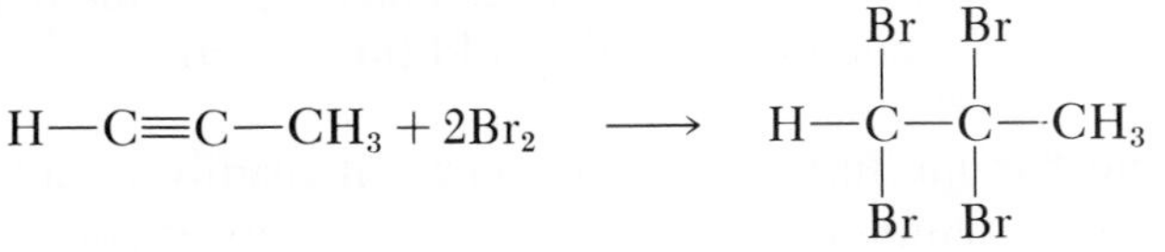

propyne (colorless)   (red-brown)   1,1,2,2-tetrabromopropane (colorless)

The decolorization of the red-brown solution of bromine, usually in a solvent such as carbon tetrachloride, is used as a test for alkenes and alkynes. Other reagents that add to multiple bonds include

(a) *Hydrogen in the presence of platinum, palladium, or nickel,*

$$HC\equiv CCH_2CH_3 + H_2 \xrightarrow{\text{Pt}} \underset{\text{H}}{\overset{\text{H}}{\diagup}}C=C\underset{\text{CH}_2\text{CH}_3}{\overset{\text{H}}{\diagdown}}$$

1-butyne   1-butene

$$\underset{\text{H}}{\overset{\text{H}}{\diagup}}C=C\underset{\text{CH}_2\text{CH}_3}{\overset{\text{H}}{\diagdown}} + H_2 \xrightarrow{\text{Pt}} H-\overset{\overset{\text{H}}{|}}{\underset{\underset{\text{H}}{|}}{C}}-\overset{\overset{\text{H}}{|}}{\underset{\underset{\text{H}}{|}}{C}}-CH_2CH_3$$

1-butene   *n*-butane

$$CH_3CH_2\underset{\underset{O}{\|}}{C}CH_2CH_3 + H_2 \longrightarrow CH_3CH_2\underset{\underset{OH}{|}}{\overset{\overset{H}{|}}{C}}CH_2CH_3$$

(b) *Dry hydrogen halides,* such as hydrogen chloride gas, HCl, add to alkenes and alkynes:

$$\underset{\text{H}}{\overset{\text{H}}{\diagup}}C=C\underset{\text{H}}{\overset{\text{H}}{\diagdown}} + HCl \longrightarrow H-\overset{\overset{\text{H}}{|}}{\underset{\underset{\text{H}}{|}}{C}}-\overset{\overset{\text{H}}{|}}{\underset{\underset{\text{H}}{|}}{C}}-Cl$$

chloroethane

(c) *Water* adds to alkenes and alkynes in the presence of dilute sulfuric acid. With alkenes the product is an alcohol,

$$\underset{\text{H}}{\overset{\text{H}}{\diagup}}C=C\underset{\text{H}}{\overset{\text{H}}{\diagdown}} + HOH \xrightarrow[\text{H}_2\text{SO}_4]{\text{dilute}} H-\overset{\overset{\text{H}}{|}}{\underset{\underset{\text{H}}{|}}{C}}-\overset{\overset{\text{H}}{|}}{\underset{\underset{\text{H}}{|}}{C}}-OH$$

ethanol

Molecules undergoing addition reactions are said to be UNSATURATED. Saturated molecules, such as alkanes and cycloalkanes, do not have multiple bonds and so do not undergo the addition reaction (Table 23.2, column 5).

Although benzene has multiple bonds, it does not readily undergo addition reactions. For example, under ordinary conditions, it does not react with H_2, although it will form cyclohexane (C_6H_{12}) at high temperature and high pressure of H_2. As we have seen, the typical reaction of benzene is substitution. The resistance to addition reactions reflects the tendency to preserve the stable delocalized π system. This resistance to addition is another aspect of aromaticity.

The reverse of addition is **ELIMINATION**. Elimination reactions are one way of introducing multiple bonds into a molecule, as shown:

$$\underset{\substack{| \quad | \\ H \quad OH}}{H-\overset{\displaystyle H}{\underset{\displaystyle H}{C}}-\overset{\displaystyle H}{\underset{\displaystyle OH}{C}}-H} \xrightarrow[\text{H}_2\text{SO}_4]{\text{conc.}} \underset{H}{\overset{H}{>}}C=C\overset{H}{\underset{H}{<}} + H_2O$$

$$H-\overset{\displaystyle H}{\underset{\displaystyle H}{C}}-\overset{\displaystyle H}{\underset{}{C}}-\underset{\substack{Br}}{\overset{\displaystyle H}{C}}-H + K^+OH^- \xrightarrow{\text{ethanol}} H-\overset{\displaystyle H}{\underset{\displaystyle H}{C}}-C=\overset{\displaystyle H}{C}-H + K^+Br^- + H_2O$$

1-bromopropane   propylene

The oxidation of alcohols to carbonyl compounds is a kind of elimination reaction.

The conversion of one structural isomer into another is known as a **REARRANGEMENT REACTION**. Many such reactions involve organic compounds; some require a catalyst, others proceed under the influence of heat or radiation. A simple example of a rearrangement is the conversion of 1,4-pentadiene to 1,3-pentadiene. The structure of the latter compound permits extended π bonding, and it is therefore more stable than the former (page 490). This enhanced stability is the thermodynamic drive that favors the rearrangement:

$$\underset{H}{\overset{H}{>}}C=C\overset{H}{\underset{H}{-}}\overset{H}{\underset{H}{C}}-C\overset{H}{\underset{H}{=}}C\overset{H}{\underset{H}{<}} \xrightarrow[\text{or base}]{\text{acid}} \underset{H}{\overset{H}{>}}C=C-C=C-\overset{H}{\underset{H}{C}}-H$$

1,4-pentadiene   1,3-pentadiene
(no extended π bonding)   (extended π bonding)

PROBLEMS

1. Definitions. Define and illustrate the following terms: (a) isomerism, (b) homologous series, (c) conformation, (d) configuration, (e) geometrical isomerism, (f) optical isomerism, (g) aromaticity, (h) functional group, (i) unsaturation, (j) delocalization energy, (k) chirality, (l) rearrangement, (m) elimination reaction, (n) addition reaction, (o) structural isomerism.

2. General. Discuss the correctness or incorrectness of the following statements: (a) The compound whose formula is C_3H_6 must be an alkene. (b) In order to have different conformations a molecule must have at least 4 σ-bonded atoms in sequence, —A—B—C—D—. (c) Molecules with carbon-to-carbon multiple bonds readily undergo addition reactions. (d) All bonds to carbon are covalent. (e) The compound whose molecular formula is C_4H_8O must have a C=C bond or be a cyclic compound.

3. Conformation. How many different substances do the following structural formulas represent?

$$CH_3—CH—CH_2—CH—CH_2—CH_3 \qquad CH_2—CH—CH_2—CH—CH_3$$
$$\qquad\quad | \qquad\qquad\quad | \qquad\qquad\qquad\quad | \quad\ | \qquad\qquad\quad |$$
$$\qquad\quad CH_3 \qquad\qquad CH_3 \qquad\qquad\ CH_3\ \ CH_3 \qquad\qquad CH_3$$

$$\qquad\quad CH_3 \qquad\qquad\qquad CH_3 \qquad\qquad CH_3 \qquad\quad CH_3$$
$$\qquad\quad | \qquad\qquad\qquad\quad | \qquad\qquad\quad | \qquad\qquad\ |$$
$$CH_3—CH—CH—CH_2—CH_2 \qquad CH_2—CH—CH_2—CH$$
$$\qquad\qquad\quad | \qquad\qquad\qquad\qquad | \qquad\qquad\qquad\ |$$
$$\qquad\qquad\ CH_3 \qquad\qquad\qquad\quad CH_3 \qquad\qquad\quad CH_3$$

4. Homologous series. Given the compounds (a) C_4H_{10}, (b) C_4H_8, and (c) C_4H_6, (*i*) Name and write a general formula for all the homologous series (when more than one is possible) to which each of these hydrocarbons belongs (see Table 23.2, page 488). (*ii*) Write a structural formula consistent with each series.

5. Isomerism. Write structural formulas and give the name of the homologous series or type of functional group(s) present for (a) 5 isomers of C_6H_{14}; (b) 6 isomers of C_4H_8; (c) 4 noncyclic isomers of C_3H_6O (The C=C—O—H grouping is rarely stable and should not be used in your answer.); (d) three isomers of C_3H_4; (e) 4 isomers of C_3H_9N; (f) 4 aromatic isomers of C_7H_7Cl; (g) 5 noncyclic, non-alkene isomers of $C_3H_6O_2$.

6. Conformations. Which of the following compounds can exist in different conformations? (a) CH_3Cl; (b) $HOOH$; (c) $H_2C=CH_2$; (d) $CH_3CH_2CH_3$; (e) H_3COH; (f) $HOCl$.

7. Isomer formation. With the aid of equations show that at least two isomers are possible from each of the following reactions and indicate the type of isomerism: (a) $CH_3CH=CH_2 + HBr$; (b) $CH_3CHBrCH_2CH_3 + KOH$ (3 isomers); (c) one mole of each $CH_3C\equiv CCH_3 + H_2(Pt)$; (d) $CH_3CH_2CH_3 + Cl_2$ in

$$\overset{\displaystyle O}{\overset{\|}{}}$$

light; (e) $CH_3\overset{\displaystyle O}{\overset{\|}{C}}CH_2CH_3 + H_2$; (f) monochlorination (substitution) of naphthalene (page 488).

8. Functional group isomerism. (a) Draw the structural formulas for the four alcohols with the molecular formula $C_4H_{10}O$. (b) Which alcohol has optical isomers? (c) Write the structural formulas for the three ethers with the molecular formula $C_4H_{10}O$. (d) Give a simple chemical test to distinguish alcohols from ethers.

9. Oxidation of alcohols and aldehydes. There are four isomeric alcohols with the molecular formula C_4H_9OH. (a) Which isomer does not undergo oxidation? (b) Which isomer is oxidized to a ketone? Draw the structure of the ketone. (c) Two isomers are oxidized to aldehydes. Give the equations for these reactions. (d) Give the structures of the carboxylic acids resulting from the oxidation of the aldehydes formed in part (c).

10. Ether formation. (a) Give the equation for the conversion of methanol to the corresponding ether. Name the ether formed. (b) How many ethers are possible when a mixture of CH_3OH and CH_3CH_2OH reacts with concentrated sulfuric acid? (c) Would you choose the method in part (b) to synthesize methyl ethyl ether, $CH_3OCH_2CH_3$? Explain.

11. Stereoisomerism. Which of the following compounds can exhibit geometric and/or optical isomerism? State which in each case. (a) $H_2C{=}C(Cl)CH_3$; (b) $ClFC{=}CHCl$; (c) $CH_3CH_2CH{=}CHCH(CH_3)_2$; (d) $CH_3CHClCOOH$; (e) $HC{\equiv}CCH{=}CHCl$; (f) $ClCH{=}CHCHClCH_3$; (g) cyclohexene; (h) $H_3CN{=}NCH_3$; (i) $CH_3SiHClBr$; (j) $[NH(CH_3)(C_2H_5)(C_3H_7)]^+Cl^-$

12. Decomposition. (a) At 60° to 100°C the following reaction occurs:

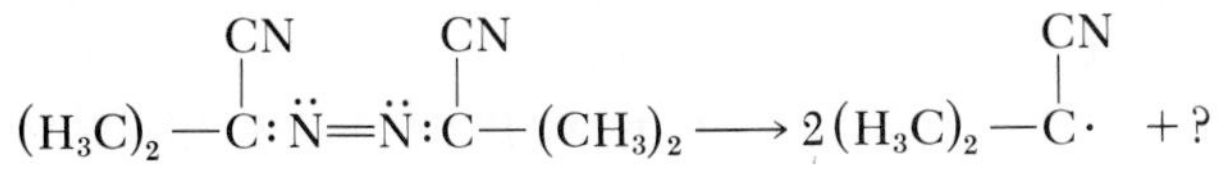

$$(H_3C)_2{-}\underset{\underset{CN}{|}}{C}{:}\ddot{N}{=}\ddot{N}{:}\underset{\underset{CN}{|}}{C}{-}(CH_3)_2 \longrightarrow 2\,(H_3C)_2{-}\underset{\underset{CN}{|}}{C}\cdot\ +?$$

What is the formula of the missing molecule? (b) When acetyl peroxide, $H_3C\underset{\underset{O}{\|}}{C}{-}O{-}O{-}\underset{\underset{O}{\|}}{C}CH_3$, is heated, one of the products isolated is ethane. Suggest a mechanism for the formation of ethane. What gaseous substance is also produced?

13. Nucleophilic displacement on carbon. Write an equation for the reaction of ethyl bromide with each of the following nucleophiles: (a) $H\ddot{S}{:}^-$; (b) $CH_3\ddot{O}{:}^-$; (c) $:H^-$; (d) $:CH_3^-$; (e) $CH_3\ddot{N}H_2$.

ADDITIONAL PROBLEMS

14. Optical isomerism. Write a structural formula for the simplest alkane exhibiting optical isomerism.

15. Redox equations. Use the ion-electron method to write a balanced ionic equation for the oxidation of CH_3CHO to CH_3COOH by acidified MnO_4^- which is reduced to Mn^{2+}.

16. Isomerism. Distinguish between the following types of isomerism: (a) stereo and structural; (b) geometrical and optical.

17. Carboxylic acids. Write balanced equations for (a) the formation of potassium oleate; (b) the formation of calcium palmitate; (c) the hydrolysis of methyl acetate; (d) the saponification of glyceryl trilinoleate.

18. Chemical tests. Give a simple test to distinguish between (a) the processes of addition and substitution of Cl_2 with a hydrocarbon; (b)

H_2C=CHC_3H_7 and cyclopentane; (c) CH_3CH_2OH and CH_3OCH_3; (d) CH_3CH_2CHO and CH_3CCH_3.
$$\overset{\parallel}{O}$$

19. Acetylene. Acetylene is produced when calcium carbide, CaC_2, reacts with water. The other product is $Ca(OH)_2$. Write an equation for the reaction. What category of reaction is this? How is CaC_2 related chemically to acetylene?

20. Displacement on carbon. Account for the fact that $(CH_3)_3CBr$ reacts with very dilute OH^- in CH_3OH to give $(CH_3)_3COCH_3$ and not $(CH_3)_3COH$.

21. Synthesis. Show steps in the synthesis of CH_2BrCH_2Br from CH_3CH_2OH.

22. Nucleophilic displacement. Ethanol, C_2H_5OH, a typical alcohol, reacts with hydrogen bromide to give ethyl bromide, C_2H_5Br. Ethanol does not react with NaBr. (a) Why cannot Br^- replace OH^-? (b) Suggest a mechanism for the reaction with HBr. (*Hint:* Alcohols are weak bases.)

23. Reactions of alkanes. (a) Write the equation for the chlorination of ethane, C_2H_6, to give ethyl chloride, C_2H_5Cl. (b) Ethyl chloride can be chlorinated to give two isomers, A and B, of formula $C_2H_4Cl_2$. A can be chlorinated to give two isomers of $C_2H_3Cl_3$, while B gives only one isomer of $C_2H_3Cl_3$ on chlorination. Write the equations for the chlorination of ethyl chloride. (c) Give the structural formulas of A and B.

24. Substitution in alkanes. (a) What two monobromo substitution products arise from the reaction of $(CH_3)_3CH$ with Br_2 in sunlight? (b) The major product is $(CH_3)_3CBr$. What conclusion can you draw about the reactivity of the two different sets of H's in $(CH_3)_3CH$ toward bromination? (c) Give an equation for the preparation of the compound in part (b) by an addition reaction.

25. Esterification. Use structural formulas to complete the equations:
(a) $CH_3CH_2COOH + CH_3OH \rightarrow$
(b) $CH_3COOH + CH_3CH_2CH_2OH \rightarrow$
(c) $HCOOH + CH_3CH_2OH \rightarrow$

26. Elimination. The three mechanisms proposed for the reaction

$$\overset{H}{\underset{\displaystyle |}{\overset{\displaystyle |}{-C}}}\overset{Br}{\underset{\displaystyle |}{\overset{\displaystyle |}{-C-}}} + OH^- \longrightarrow\; >C=C< + H_2O + Br^-$$

are (*i*) a one-step reaction; (*ii*) a two-step reaction with a carbanion intermediate; (*iii*) a two-step reaction with a carbonium ion intermediate. (Bond breaking is slower than bond formation.) (a) Show steps for each mechanism. (b) Give the rate equation expected for each mechanism. (c) Which mechanism can be verified by a study of isotope effects? (A C—H bond is broken at a faster rate than is a C—D bond.)

27. Addition. (a) Write an equation for the addition of Br_2 to H_2C=CH_2. (b) Write equations for a simplified two-step mechanism, initiated by the formation of a carbonium ion with a C—Br bond, and completed by bonding with Br^-, formed in the first step.

28. Basicity and *s* character. (a) In what kind of hybrid orbital do we find the unshared pair of electrons of nitrogen in each of the following compounds, given in order of decreasing basicity?

$$CH_3CH_2\ddot{N}H_2 > CH_3\underset{\underset{\displaystyle H}{|}}{C}{=}\ddot{N}H > CH_3C{\equiv}N\colon$$

(i)   (ii)   (iii)
ethylamine   ethylimine   acetonitrile

(b) In terms of your answer to part (a), suggest a relationship between basicity and the *s* character of the orbital holding an unshared pair.

29. Acidity, basicity, and delocalization. When an atom with an unshared pair of electrons is attached to a multiple-bonded atom ($-\overset{..}{A}-X{=}Y$ or $-\overset{..}{A}-C_6H_5$), it will hybridize so as to have the unshared pair in a *p* orbital that can overlap with the adjacent π system. Use this knowledge to explain the following: (a) aromatic amines such as aniline, $C_6H_5NH_2$, are less basic than aliphatic amines such as *n*-hexylamine, *n*-$C_6H_{13}NH_2$; (b) aromatic hydroxyl compounds such as phenol, C_6H_5OH, are more acidic than alcohols such as *n*-hexyl alcohol, $C_6H_{13}OH$; (c) amides such as acetamide, $CH_3\underset{\underset{\displaystyle O}{\|}}{C}-\ddot{N}H_2$, are less basic than amines such as ethylamine, $C_2H_5NH_2$; (d) carboxylic acids, RCOOH, are more acidic than alcohols, ROH.

30. Nomenclature. Give the IUPAC name for

(a) $CH_3-\underset{\underset{\displaystyle CH_3}{|}}{CH}-CH_2-CH_3$

(b) $CH_3-\underset{\underset{\displaystyle CH_3}{|}}{CH}-CH_2-\underset{\underset{\displaystyle CH_3}{|}}{CH}-CH_3$

(c) $CH_3-CH_2-\overset{\overset{\displaystyle CH_3}{|}}{\underset{\underset{\displaystyle CH_3}{|}}{C}}-CH_3$

(d) $CH_3-CH_2-\overset{\overset{\displaystyle CH_3}{|}}{\underset{\underset{\displaystyle CH_2}{|}}{C}}-CH_3$
          CH_3

(e) $CH_3-CH_2-\underset{\underset{\displaystyle CH_2}{|}}{\overset{\overset{\displaystyle CH_3}{|}}{C}}-CH_2-CH_3$
          CH_3

(See Appendix II.)

31. Optical activity. Account for the stereochemical results in each step of the following sequence:

$$CH_3CHO \xrightarrow[\text{H}_2\text{O}]{\text{D}_2(\text{Pt})} CH_3-\underset{\underset{\displaystyle D}{|}}{\overset{\overset{\displaystyle H}{|}}{C}}-OH \xrightarrow[\text{alcohol oxidase}]{\text{the enzyme}} CH_3CHO + CH_3-\underset{\underset{\displaystyle D}{|}}{\overset{\overset{\displaystyle H}{|}}{C}}-OH$$

chiral but
optically inactive

unreacted optically active

32. Configuration and mechanism. Given the following stereochemical observation:

$$\text{HO}^- + \text{H}\cdots\overset{\displaystyle \text{CH}_2\text{CH}_3}{\underset{\displaystyle \text{H}_3\text{C}}{\text{C}}}\text{—Br} \xrightarrow{\text{S}_\text{N}2} \text{HO—}\overset{\displaystyle \text{CH}_2\text{CH}_3}{\underset{\displaystyle \text{CH}_3}{\text{C}}}\cdots\text{H} \qquad \text{no}\left(\text{H}\cdots\overset{\displaystyle \text{CH}_2\text{CH}_3}{\underset{\displaystyle \text{CH}_3}{\text{C}}}\text{—OH} \right)$$

optically active   optically active

(a) In an $S_\text{N}2$ reaction does the nucleophile approach the carbon atom from the same side of the C—Br bond or the opposite side of this bond? (b) Offer a reason for the kind of approach in terms of electrostatic theory.

24

RADIATION AND MATTER

INTRODUCTION / 24.1

Photons impinging on matter may be absorbed, transmitted, reflected, refracted, or diffracted. Among all these phenomena, the one that interests us the most is absorption, because the wavelength pattern of the absorbed photon is most characteristic of the structure of the absorbing molecule. This pattern is called the **ABSORPTION SPECTRUM**.

MOLECULAR ENERGIES / 24.2

The absorption of a photon by a molecule is quantized. This means that a molecule absorbs only radiation of certain specific wavelengths; all other wavelengths are transmitted without any change in the molecule. Photons can be absorbed from almost every region of the electromagnetic spectrum. Photons from a given region of the spectrum will excite a particular molecule in a specific way. To understand the relationship between the wavelength of absorbed radiation and the resulting excitation, we must first consider the varieties of molecular energy.

A molecule can be considered to possess several distinct kinds of energy, each of which can be changed by the absorption of a photon:

E_{nuclear}, associated with the composition of its nuclei.
$E_{\text{electronic}}$, associated with the distribution of its electrons
$E_{\text{translational}}$, associated with its kinetic energy.
$E_{\text{rotational}}$, associated with its rotational motion.
$E_{\text{vibrational}}$, associated with the modes of vibration of its atoms.
$E_{\text{nuclear spin}}$, associated with the spin of its nuclei.
$E_{\text{electron spin}}$, associated with the spin of its electrons.

We will consider only the electronic, vibrational, and nuclear spin energies. The differences between the excited- and ground-state en-

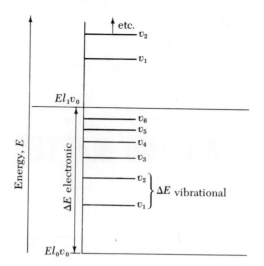

Fig. 24.1. Schematic (not to scale) comparison of energy levels of electronic states El_0 and El_1 and vibrational states v_0, v_1, v_2, etc. $\Delta E_{\text{nuclear spin}}$ is very much smaller than $\Delta E_{\text{vibrational}}$.

ergies, designated ΔE, are vastly different for these kinds of energies. Figure 24.1 compares electronic and vibrational energy levels.

The energy differences between the first excited state and ground state decrease in the order $\Delta E_{\text{electronic}} > \Delta E_{\text{vibrational}} \gg \Delta E_{\text{nuclear spin}}$ (not shown in Fig. 24.1). Any ΔE is related to the frequency, f (or ν), and wavelength, λ, of the absorbed photon by

$$\Delta E = hf = hc/\lambda$$

$\Delta E_{\text{electronic}}$, having the largest magnitude, requires photons of the highest energy and, therefore, the shortest wavelength. At the other extreme, $\Delta E_{\text{nuclear spin}}$ requires photons of the smallest frequency and, therefore, the longest wavelength. In Table 24.1, the regions of the

Table 24.1
Effect of Radiation on Molecules[a]

f, sec^{-1}	6×10^{16}	3×10^{16}	8×10^{14} 4×10^{14}		1×10^{12}	3×10^8
λ, nm	5	15	4×10^2 8×10^2		3×10^4	10^9
	X rays	Far uv[b]	Near uv[b]	Visible	Ir	Microwave	Radiowave
E, kcal per mole of photons	5800	2300	82	36	1	10^{-4}
	Breaking of bonds	Excitation of σ electrons	Excitation of π and unshared electrons	Vibrational excitation	Rotational and electron spin excitation	Nuclear spin excitation

[a] The boundaries are approximate.
[b] "Near" means near to the visible region; "far" means far from the visible region; uv means ultraviolet.

electromagnetic spectrum and the corresponding changes in molecular energies are listed.

The nature of the radiation can be expressed in terms of frequency (in sec⁻¹), wavelength (in nanometers, nm), or wave number, defined as the reciprocal of the wavelength, $1/\lambda$ (in cm⁻¹). The wave number is the number of waves per centimeter and, therefore, the smaller the wave number the longer the wavelength.

GENERAL FEATURES OF SPECTRA / 24.3

A spectrophotometer is a device that directs photons of a known single frequency into the material being studied. The instrument then senses and records whether these photons are absorbed or transmitted. The instrument automatically varies the frequency of the photons. As a result, the record of absorbed (or transmitted) radiation also varies continuously. If this record is transcribed on a moving paper, a graph is generated automatically, with absorption or transmission plotted against wavelength. Figure 24.2 is a schematic representation of a spectrophotometer.

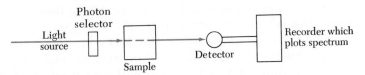

Fig. 24.2. *Schematic representation of a spectrophotometer.*

A typical absorption "band" appears in Fig. 24.3.† This spectrum shows complete transmission (or absence of absorption) of photons from 450 to 300 nm. Some absorption begins at wavelengths shorter than 300 nm and reaches maximum absorption at 250 nm, the minimum

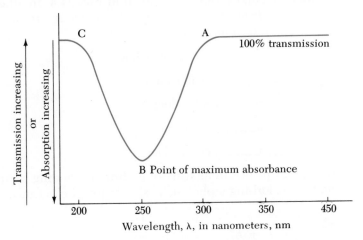

Fig. 24.3. *A typical absorption band; a plot of absorption and transmission versus wavelength. The point of maximum absorption, 240 nm, is called a "peak," even though this point is a "valley." Some spectra are recorded so that the point of maximum absorption is a peak rather than a valley. In such cases increasing absorption would be ascending rather than descending.*

† The explanation of why bands are broad rather than sharp is complex and will not be given here.

point in the curve. The wavelength 250 nm is recorded as λ_{max} for the substance. At 200 nm transmission is again complete.

(a) **Types of electronic states.** Electronic excitations are classified on the basis of the state of the electron being excited. Only an outer electron is excited by photons in the ultraviolet region. Hence, we omit reference to those electrons present in the filled inner energy levels and consider only sigma (σ), pi (π), and unshared pairs of electrons (n). These three categories of electron pairs are illustrated in formaldehyde:

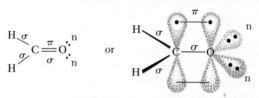

(b) **Types of electronic excitations.** The nature of electronic excitations is best explained by MO theory (page 359). This theory assumes the existence of bonding and antibonding molecular orbitals. Corresponding to the σ and π bonding MO's, there are the sigma antibonding, σ^*, and pi antibonding, π^*, MO's. Since unshared electrons are not involved in bonding, the MO's housing such electrons are called **NONBONDING MOLECULAR ORBITALS**. All molecules have at least one σ bonding MO and, hence, will also have at least one σ^* antibonding MO. If the molecule does not have a multiple bond, it will not possess either a π bonding or π^* antibonding molecular orbital.

The antibonding orbitals have higher energies than the nonbonding molecular orbitals, which in turn have higher energies than the bonding molecular orbitals, as shown in Fig. 24.4. In all excitations *the electron goes from a lower occupied molecular orbital into a higher unoccupied molecular orbital.*

_____ σ^* _____	sigma antibonding
_____ π^* _____	pi antibonding
_____ n _____	nonbonding
_____ π _____	pi bonding
_____ σ _____	sigma bonding

(Energy increasing →)

Fig. 24.4. *Representation of relative energy levels of the five types of molecular orbitals, σ, π, n, π^*, and σ^*.*

Sigma and pi electrons usually go from a bonding to an antibonding orbital on absorbing a photon, as shown for pi electrons:

$$\pi^*_{MO} \underline{\qquad} $$
$$\pi_{MO} \underline{\;\uparrow\downarrow\;} \quad + \quad hf \quad \longrightarrow \quad \pi^*_{MO} \underline{\;\uparrow\;} \quad \pi_{MO} \underline{\;\uparrow\;}$$

ground state + photon $\longrightarrow$ excited state

An unshared electron is excited from a nonbonding molecular orbital to an antibonding molecular orbital. Since the sigma antibonding molecular orbital, σ^*, is at a higher energy than the pi antibonding molecular orbital, π^*, more energy is required to excite an electron from a given molecular orbital into a σ^* than into a π^* antibonding molecular orbital.

Since the relative energies of molecular orbitals is $n > \pi > \sigma$, the least energy is required to excite an electron from an n molecular orbital and the greatest energy is required to excite an electron from a σ molecular orbital to a given antibonding orbital. These electronic excitations are summarized schematically in the energy-level diagram in Fig. 24.5. Only the $n \rightarrow \pi^*$, $\pi \rightarrow \pi^*$, and $n \rightarrow \sigma^*$ electronic excitations correspond to an energy transition small enough to occur in the near ultraviolet or visible regions—the regions detectable by ordinary uv spectrophotometers. Species such as methane, CH_4, without π bonds or without at least 1 unshared electron, absorb in the far ultraviolet region and, hence, their uv spectra cannot be detected in an ordinary spectrophotometer. Table 24.2 lists some common groups and indicates their λ_{max} and the type of electronic excitation.

By writing* for the excited electron in the π^* molecular orbital and removing an electron dot from the formula we indicate the $\pi \rightarrow \pi^*$ and the $n \rightarrow \pi^*$ transition in acetone, $(CH_3)_2C{=}O$, as follows:

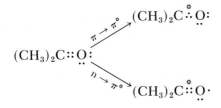

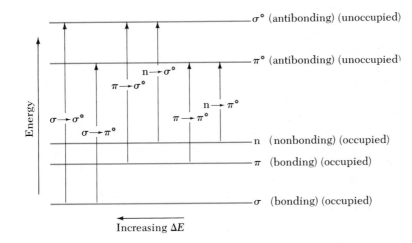

Fig. 24.5. *Schematic representation of electronic excitations showing relative magnitudes of ΔE.*

Table 24.2
Some Common Groups Absorbing in Near-Uv Range

Atom or group	Typical compound	Type of excitation	λ_{max} (nm)
$-\ddot{\text{O}}-$ [a]	CH_3OH	$n \longrightarrow \sigma^*$	183
$-\ddot{\text{S}}-$	$C_6H_{13}SH$	$n \longrightarrow \sigma^*$	224
$-\ddot{\text{C}}\text{l}\colon$ [a]	CH_3Cl	$n \longrightarrow \sigma^*$	173
$-\text{Br}\colon$	CH_3Br	$n \longrightarrow \sigma^*$	204
$-\ddot{\text{I}}\colon$	CH_3I	$n \longrightarrow \sigma^*$	258
$-\dot{\text{N}}-$	$(CH_3)_3N$	$n \longrightarrow \sigma^*$	227
$>C=C<$ [a]	$H_2C=CH_2$	$\pi \longrightarrow \pi^*$	171
$-C\equiv C-$ [a]	$HC\equiv CH$	$\pi \longrightarrow \pi^*$	173
$>C=\ddot{\text{O}}\colon$	$(CH_3)_2CO$	$\pi \longrightarrow \pi^*$	189
$-\text{N} \diagup^{\text{O}} \diagdown_{\text{O}}$	CH_3NO_2	$n \longrightarrow \pi^*$ $\pi \longrightarrow \pi^*$ $n \longrightarrow \pi^*$	279 201 274

[a] These are listed because they absorb at just slightly higher frequencies than the near-uv region.

(c) **Color.** When electronic excitation occurs in the visible region, the absorbing substance is colored. The color of the substance is the complement of the color absorbed. Some complementary pairs are yellow-violet, orange-blue, and red-green. A photon absorbed in the visible region has less energy than a photon absorbed in the ultraviolet region; therefore,

$$\Delta E \underset{\text{visible}}{\text{electronic}} < \Delta E \underset{\text{ultraviolet}}{\text{electronic}}$$

$\Delta E_{\text{electronic}}$ depends on the energy gap between the bonding or nonbonding MO from which the excited electron leaves and the antibonding MO* to which it goes. As this gap becomes narrower, $\Delta E_{\text{electronic}}$ becomes smaller. For a $\pi \rightarrow \pi^*$ transition, this narrowing occurs as the length of the alternating double-single bond sequence is increased (Fig. 24.6a, page 518). This extension increases the delocalization of the π electrons. In summary, the *more delocalized the electrons, the smaller the energy and hence the longer the wavelength of the light needed for excitation.*

We illustrate this phenomenon with vitamins A_1 and A_2, both of which are yellow:

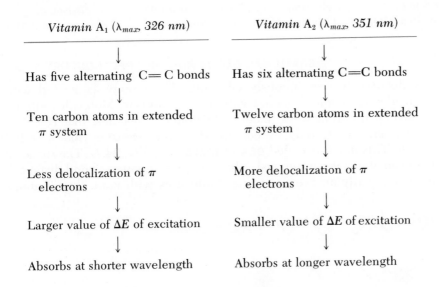

The extended π system arises from the overlapping lateral p orbitals as shown in Fig. 24.6b. The relationship between the number of alternating C=C groups and the absorbed wavelength, λ_{max}, is given in the comparison below:

Vitamin A_1 (λ_{max}, 326 nm)	*Vitamin A_2 (λ_{max}, 351 nm)*
↓	↓
Has five alternating C=C bonds	Has six alternating C=C bonds
↓	↓
Ten carbon atoms in extended π system	Twelve carbon atoms in extended π system
↓	↓
Less delocalization of π electrons	More delocalization of π electrons
↓	↓
Larger value of ΔE of excitation	Smaller value of ΔE of excitation
↓	↓
Absorbs at shorter wavelength	Absorbs at longer wavelength

All organic dyes possess alternating multiple bonds. These multiple bonds may also be, for example, C=O, C≡N, N=O, C≡C, and N=N.

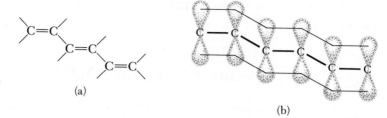

Fig. 24.6. (a) Some alternating double bonds; (b) overlapping p orbitals.

(a)

(b)

(d) Photochemistry. The study of chemical changes involving electronically excited molecules is called **PHOTOCHEMISTRY**. One such change is the rearrangement between *trans-* and *cis-*isomers, crucial in the biochemical process of vision.

In our discussion of geometrical isomerism (page 485) it was noted that rotation about a double bond could not occur readily because it would necessitate destroying the π bond formed by the overlap of p orbitals. However, on excitation, one electron undergoes a $\pi \rightarrow \pi^*$ transition. The π^* electron cancels the bonding effect of the π electron. The π bond is now broken and rotation can occur to give some of the other isomer:

*trans-*isomer
(2 electrons in π MO)

1 electron (°) in π^* molecular orbital
1 electron (·) in π molecular orbital

*cis-*isomer
(2 electrons in π MO)

VIBRATIONAL EXCITATION; IR SPECTROSCOPY / 24.5

Molecules are continuously vibrating. The energy associated with molecular vibrations is called **VIBRATIONAL ENERGY**. Molecules vibrate in several different ways (modes). Diatomic molecules, like H—H or H—Cl can vibrate only by compression and extension, like a coiled spring. This mode is called **BOND STRETCHING** (Fig. 24.7). The distance between atoms is continuously changing; the reported "bond distance" is thus actually an average value. Molecules with more than 2 atoms

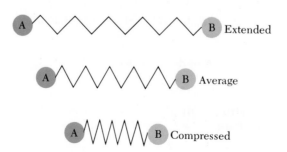

A ⟋⟍⟋⟍⟋⟍⟋⟍ B Extended

A ⟋⟍⟋⟍⟋⟍ B Average

A ⟋⟍⟋⟍⟋⟍⟋ B Compressed

Fig. 24.7. Bond stretching in a diatomic molecule, AB.

Table 24.3
Some Characteristic Infrared Absorption Frequencies (in Wave Numbers)

Bond	Frequency, cm^{-1}	Bond	Frequency, cm^{-1}
—C—H	2850–2960	—C≡C—	2100–2260
—C—D	~2200	—C≡N	2000–2300
=C—H	3010–3100	—C—O—	1000–1300
≡C—H	3300	>C=O	1700–1750
—C—C—	600–1500	—O—H	3590–3650 (sharp)
>C=C<	1620–1680	—O—H (H-bonded)	3200–3400 (broad)

have bond angles which are also continuously changing. Therefore bond angle, like bond distance, is also an average, not a constant, value. The several modes of molecular vibration that involve changes of bond angles are referred to as **BENDINGS**.

A given bond, such as —O—H, has roughly the same stretching frequency in different molecules. Similarly, the same combination of 3 or more atoms, such as —NO_2, —NH_2, or —CH_3, has roughly the same bending frequency in different molecules. *When exposed to infrared radiation, an absorbing species gives an absorption band at a wavelength characteristic of each of its stretching and bending modes.* Table 24.3 shows the absorption bands associated with various vibrational modes of some common bonds. Since many of the absorption bands are attributable to the presence of certain specific groups, infrared spectra are useful in elucidating the structure of unknown compounds. The overall infrared spectrum of a pure substance is a uniquely characteristic property and, like a fingerprint, can be used as an identification. The presence of absorptions that cannot be attributed to the substance under examination thus indicates the presence of impurities. Figure 24.8 shows the spectrum of acrylonitrile, $H_2C=CHC≡N$, with some structural assignments.

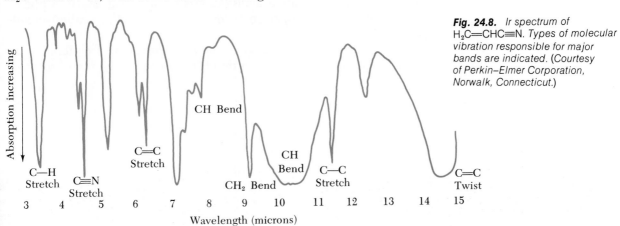

Fig. 24.8. *Ir spectrum of $H_2C=CHC≡N$. Types of molecular vibration responsible for major bands are indicated. (Courtesy of Perkin–Elmer Corporation, Norwalk, Connecticut.)*

The energy levels of protons and neutrons in atomic nuclei are quantized like the energy levels of electrons in atoms. As with electrons, one way to classify these energy states is by spin orientation. There are two spin orientations corresponding to quantum numbers of $+\frac{1}{2}$ ($\uparrow$) and $-\frac{1}{2}$ ($\downarrow$). However, spin pairing ($\uparrow \downarrow$) occurs so that nuclei with an *even number of protons and an even number of neutrons*, for example $^{12}_{6}C$, $^{16}_{8}O$, and $^{32}_{16}S$, do not have a net nuclear spin. On the other hand, $^{1}_{1}H$, $^{2}_{1}D$, $^{31}_{15}P$, $^{15}_{7}N$, $^{19}_{9}F$, and $^{15}_{8}O$ are examples of nuclei having spin. Of particular importance is the nucleus of $^{1}_{1}H$ that has two nuclear spin states, $+\frac{1}{2}$ and $-\frac{1}{2}$. Under ordinary conditions, these spin states have practically the same energy (see Fig. 24.9a). However, in a strong external magnetic field these spin states differ (Fig. 24.9b) in energy by an amount, ΔE, whose magnitude varies with the magnetic field strength.

Photons of the proper frequency can excite the H nucleus (proton) of a compound in a magnetic field from the more stable to the less stable state. This phenomenon, often called "changing the spin," occurs in the *radiowave region* of the electromagnetic spectrum. The technique for causing such excitations, developed by Felix Bloch, Edward Purcell, and Herbert Gutowsky in the mid-1940's, is the basis for NUCLEAR MAGNETIC RESONANCE (nmr) spectroscopy. It is a powerful method for elucidating molecular structures. Practically all organic compounds contain H atoms. Therefore, proton studies are of special value to the organic chemist. Fortunately, ^{12}C and ^{16}O, the other prevalent elements in organic compounds, have no nuclear spin and so do not complicate the nmr spectrum of ^{1}H.

The compound to be studied is placed in the homogeneous field of a strong magnet and radiowaves are passed through the sample. The frequency, f, of the absorbed photons is proportional to the magnetic field strength H, or $f \propto H$. Because it is easier, most instruments operate by keeping f constant and varying the field strength. A typical nmr absorption band, called a SIGNAL, is shown in Fig. 24.10. A higher field strength is called UPFIELD; a lower field strength is called DOWNFIELD.

If all H's in a molecule gave a signal at the same frequency, nmr spectroscopy would have very little utility. Fortunately, *the absorption*

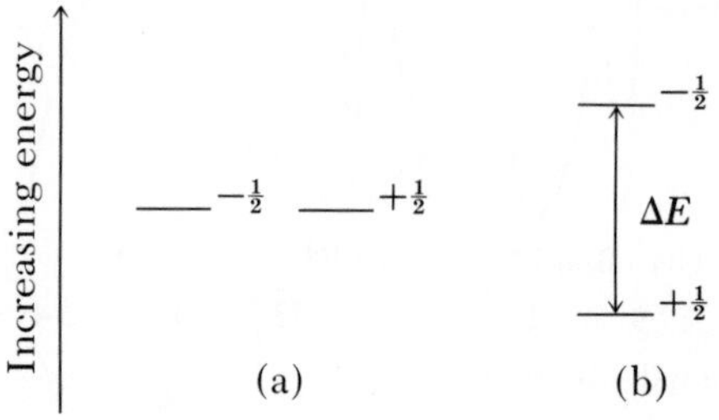

Fig. 24.9. *Relative energy levels of nuclear spin. (a) Under ordinary conditions; (b) in a strong external magnetic field.*

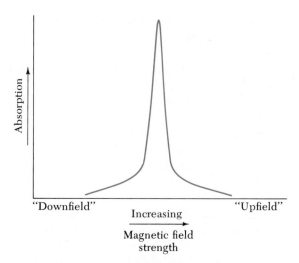

Fig. 24.10. *A typical nmr absorption signal.*

by a proton depends on the bonding environment. To enter the excited state at a given frequency, *all* protons must "feel" the same magnetic field strength. The magnetic field sensed by a proton is *not necessarily what is applied by the magnet in the instrument.* This difference exists because the electrons in the bond holding the H and the electrons in π bonds in the vicinity of the H *induce* their own small magnetic fields. This induced field may either *oppose or reinforce the applied field.* The field sensed by the proton is then the *resultant* of the applied and induced fields. In Fig. 24.11 a vector is used to indicate the field strength needed to make the frequency, f, just sufficient to change the spin of a proton. In Fig. 24.11b, an induced field reinforces the applied field so that to attain the field strength needed to change the spin, a smaller field must be applied. Such a deshielded proton absorbs more "downfield" (Fig. 24.10). In Fig. 24.11c, the induced field opposes the applied field so that to attain the needed field strength a stronger external field strength must be applied. Such a shielded proton

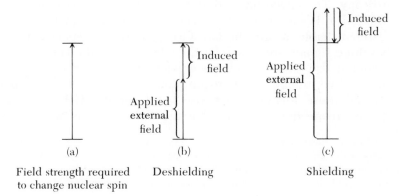

Fig. 24.11. *Effect of field induced by electron clouds on absorption of a photon by a proton. (a) No induced field; (b) induced field reinforces applied field, deshielding; (c) induced field opposes applied field, shielding.*

absorbs more "upfield." The variation in the applied magnetic field strength as influenced by the environment of the proton is called a CHEMICAL SHIFT.

A few important generalizations can be made about molecular structure and the relative shift:

(a) All equivalent H atoms in a given molecule have the same chemical shift.

(b) Strongly electronegative atoms such as F, O, N, Cl, and Br withdraw electron density from nearby H atoms causing such H's to have signals more downfield. For example, the H's in dimethyl ether, CH_3OCH_3, show a signal more downfield than the signal for the H's in ethane, CH_3CH_3.

(c) Electropositive atoms such as Si have the opposite effect to that of an electronegative atom. Thus, the protons in tetramethyl silane, $(CH_3)_4Si$ (b.p. 25°C), are among the most shielded protons known. They produce a signal far upfield. The signal from $(CH_3)_4Si$ is used as the zero reference point for measuring chemical shifts of different protons.

(d) Because of the π electrons, H atoms in the following groups,

$$-\overset{|}{\underset{|}{C}}=O, \quad H-\overset{|}{C}=\overset{|}{C}-, \quad H-\hexagon, \quad \text{and} \quad H-C\equiv C-, \quad \text{have a more}$$

downfield chemical shift.

The nmr spectrum gives several important kinds of information:

(a) The *number of signals* (peaks) tells us how many kinds of protons (protons with different chemical environments) are present in a molecule.†

(b) The *position* (*chemical shift*) of the signal informs us about the bonding environment of each proton.

(c) The relative *area under each signal* tells us the relative number of each kind of proton in the molecule.

The low-resolution spectra of the isomers, dimethyl ether, Fig. 24.12a, and ethanol, Fig. 24.12b, illustrate the application of nmr spectroscopy. The spectrum of dimethyl ether shows only one signal because all 6 H's are equivalent. The 3 H's in a given CH_3 group are indistinguishable, as are the two CH_3 groups. The spectrum of ethanol has three signals—one each for the CH_3, CH_2, and OH protons. The instrument also gives the relative areas under the peaks. Each area gives the relative number of hydrogen atoms associated with the peak. For ethanol these relative areas are $3:2:1$ for the CH_3, CH_2, and OH groups, respectively. Notice that the CH_2 signal is farther downfield

† Neighboring protons, mainly those on adjacent atoms (H—X—Y—H), interact magnetically with each other so as to increase the number of possible protonic spin states. Therefore, with an instrument of sufficiently high resolving power it is possible to obtain a number of closely spaced signals in place of one absorption signal.

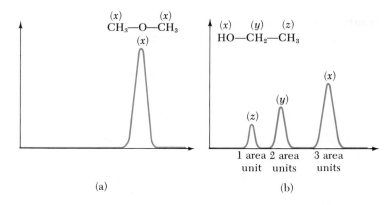

Fig. 24.12. *Low-resolution nmr spectrum of (a) dimethyl ether* CH_3OCH_3; *(b) ethanol,* CH_3CH_2OH.

than the CH_3 signal because of the electron-withdrawing effect of the adjacent highly electronegative O atom.

MASS SPECTRAL ANALYSIS / 24.7

On absorbing a high-energy photon, a given molecule (called the "parent") may eject an electron and become a positive ion. Ionization of vapor molecules can also be brought about by an electron beam. An energetic electron from the beam knocks an electron off the molecule:

$$R:Q \quad + \quad e^- \quad \rightarrow \quad R{\cdot}Q^+ \ + 2e^- \ \text{(ionization)}$$

| parent molecule | electron from beam | parent ion (a radical) |

Most parent molecules possess only paired electrons so that the parent ion arising from this ionization step is a radical.

This is the first step in the application of mass spectrometry to the determination of the structure of compounds. The theory of mass spectrometry has been discussed (page 36) and the reader may wish to review it. Only a few micrograms of a compound need be vaporized and ionized. Most parent ions are formed in highly excited vibrational states and therefore undergo fragmentation:

$$R{\cdot}Q^+ \quad \text{(parent ion)}$$

$$R^+ + {\cdot}Q \quad \text{or} \quad R{\cdot} + Q^+$$

R^+ and Q^+ are fragment ions which can undergo further fragmentation to give a mixture of positive ions of varying masses.

The parent and all fragment cations are separated and detected according to their mass/charge (m/e) value. For most ions the charge is $+1$, hence m/e is simply the mass of the ion. Neutral particles are not detected. The mass spectrum is recorded as a series of sharp peaks, the position of each peak representing the mass of an ion. The height of each peak is a measure of the relative abundance of the corresponding

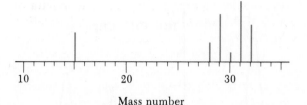

Fig. 24.13. *The mass spectrum of methanol, CH₃OH.*

Mass number

ion. The mass spectrum for methanol, CH_3OH, is shown in Fig. 24.13.

The chemist assigns a structure to each ion, as we now illustrate for the four most prominent peaks from methanol, those of masses 32, 31, 29, and 15.† The mass number of the parent ion is 32; the other three values, 31, 29, and 15, result from fragment ions.

$$H_3C\ddot{O}H + e^- \longrightarrow [H_3C\dot{O}H]^+ \text{ (mass 32; } [4 \times 1] + 12 + 16) + 2e^-$$

parent　　　　　　　　　parent ion

$$[H_2\overset{\textstyle\frown}{\text{(H)}}\dot{C}-\ddot{O}H]^+ \longrightarrow [H_2C=\ddot{O}H]^+ \text{ (mass 31)} + H\cdot$$

parent ion　　　　　　fragment ion(1)

$$[HC=O\overset{\frown}{\text{(S H)}}]^+ \longrightarrow [HC\equiv\ddot{O}]^+ \text{ (mass 29)} + H_2$$

fragment ion(1)　　fragment ion(2)

$$[H_3C:\ddot{O}H]^+ \longrightarrow H_3C^+ \text{ (mass 15)} + :\ddot{O}H$$

parent ion　　　fragment ion(3)

Note that two of the ions appear to come directly from fragmentation of the parent ion, while one of them can only come from further dissociation of a fragment ion.

The reader should ask why we assumed that the peak for mass 31 arose from a loss of an H· from the carbon atom to give $H_2C=\ddot{O}H^+$ rather than from the oxygen atom to give $H_3C-\ddot{O}:^+$. There are two reasons. First, the O—H bond (111 kcal) is stronger than the C—H bond (99 kcal), and hence is more difficult to break. Second, the fragment $H_2C=\ddot{O}H^+$ has an octet of electrons on the C and O whereas $H_3C-\ddot{O}:^+$ has only a sextet of electrons on the O atom. In general, when more than one structure can be assigned to an ion, *the more stable one is chosen.*

What is the utility of mass spectrometry to the chemist? The energy of the electron beam can be so controlled that the spectrum of a particular compound is highly reproducible. Therefore, the mass spectrum can

† Some small peaks will be found at mass of one unit more than each of these four mass numbers because of the presence of some ¹³C in all carbon-containing compounds.

be used to identify a compound by comparing it with the spectra of known samples until a match is found. Also, since the fragmentation patterns of a given class of compounds, such as alcohols, ethers, ketones, and carboxylic acids, are reasonably consistent, one can deduce the presence of certain functional groups in a compound. Therefore, mass spectroscopy can be used to establish the structure of a new compound. A simple but typical structural assignment is shown in Example 1.

Example 1 A gaseous organic compound has prominent peaks with the following masses: 26, 25, 24, and 13. What is the structure of the compound?

Answer The assumption is made that the parent ion is the one with the largest mass.† In this case the parent ion has mass 26. Therefore, the compound has a molecular weight of 26. The peaks at 25 and 24 must represent fragments formed by a stepwise loss of H atoms. These 2 H atoms account for two mass units, leaving a mass of 24 for one of the fragments. Since the compound is organic, it contains at least 1 C atom (mass 12). The remaining mass of 12 cannot be attributed to any atom or group of atoms other than another C atom. The ion with mass 24 must be C_2^+. The molecular formula of the parent is then C_2H_2 (26 g/mole) and the compound could only be acetylene, HC≡CH. The ion of mass 13 must be CH^+. The equations for the reactions are shown:

$$H—C≡C—H + e^- \longrightarrow [H—C\overset{.}{=}C—H]^+ \text{ (mass 26)} + 2e^-$$
$$[H—C\overset{.}{=}C:H]^+ \longrightarrow [H—C≡C]^+ \text{ (mass 25)} + H^{\cdot}$$
$$[H:C≡C]^+ \longrightarrow [\cdot C≡C]^+ \text{ (mass 24)} + H^{\cdot}$$

The CH^+ ion could come from either $C_2H_2^+$ or from C_2H^+.

ATMOSPHERIC RADIATION CHEMISTRY / 24.8

Chemical reactions induced by high-energy photons are especially important in atmospheric chemistry. Besides sustaining plant and animal life because of the presence of O_2 and CO_2, the atmosphere plays the essential role of preventing the most energetic rays of the sun from reaching and injuring living organisms. Most of the sun's radiation is absorbed by molecules in the atmosphere which are thereby ionized or fragmented. The following reactions are illustrative:

$$O_2 + hf \longrightarrow O_2^+ + e^-$$
$$N_2 + hf \longrightarrow N_2^+ + e^-$$
$$:\overset{..}{O}—\overset{..}{O}: + hf \longrightarrow 2 :\overset{..}{O}\cdot$$

O_2 is ionized by far ultraviolet light, while the ionization of N_2 requires solar X rays. That the ionization energy of N_2 is greater than

---

† This assumption is not always true. Some parent ions may be too unstable to be found in appreciable abundance in the spectrum so that the peak at the highest mass number may not represent the parent ion.

that of O_2 is understandable when we recall (page 366) that 2 un-paired electrons in the ground state of O_2 are in higher-energy an-tibonding MO's ($2\pi_y{}^*$, $2\pi_z{}^*$), whereas all the outer electrons of N_2 are paired in lower-energy bonding MO's. It therefore takes photons of higher energy to ionize N_2.

Another prominent ion in the atmosphere is NO^+. It is believed to be produced in either of two ways, neither of which requires the ioniza-tion of nitrogen oxide, NO. Each path involves a two-step sequence:

path 1 $\quad N_2 + hf \longrightarrow N_2{}^+ + e^-$
$\quad\quad\quad\quad O + N_2{}^+ \longrightarrow NO^+ + N \quad\quad$ (a displacement)

(Atomic oxygen is present because of fragmentation of O_2.)

path 2 $\quad O + hf \longrightarrow O^+ + e^-$
$\quad\quad\quad\quad O^+ + N_2 \longrightarrow NO^+ + N \quad\quad$ (a displacement)

Oxygen atoms can also absorb photons to give excited atoms:

$O + hf \longrightarrow O^*$

The emission of radiation from these excited O atoms is the major visible component of "airglow," the faint, diffuse radiation that is barely visible to the human eye on moonless nights at points far re-moved from city lights. These reactions which occur in the outer part of the atmosphere do not remove the lower-energy ultraviolet photons (210 to 290 nm). Even these photons could cause severe tissue damage leading to skin cancer. As these photons approach the earth, they hit a region of the atmosphere (stratosphere) where there is a relatively high concentration of ozone. Ozone absorbs these photons, and reverts to ox-ygen by a two-step reaction:

$O_3 + hf$ (near uv) $\longrightarrow O_2 + O +$ heat
$\quad\quad\quad\quad O_3 + O \longrightarrow 2O_2{}^*$

Supersonic jet airplanes fly in the stratosphere. It is argued that the exhaust gases (oxides of nitrogen) of these planes would decompose O_3, removing its beneficial action of filtering out deleterious low-energy ultraviolet photons.

PROBLEMS

1. Definitions. Define and illustrate the following terms: (a) absorption spectrum, (b) molecular energies, (c) vibrational mode, (d) nuclear magnetic resonance, (e) chemical shift, (f) parent ion, (g) fragmentation.

2. Molecular excitations. In which region(s) of the electromagnetic spec-trum are the photons causing the following excitations usually found? (a) Elec-tronic (σ electrons); (b) electronic (π electrons); (c) vibrational; (d) nuclear spin.

3. Color. Given a red, a yellow, a blue, a green, a violet, and an orange dye, list them in the order of increasing wavelength (decreasing energy) of the absorbed photon.

4. Infrared absorption. What absorption bands would you look for in order to distinguish between the isomers in each of the following pairs by infrared spectroscopy?

(a) CH_3CH_2OH and CH_3OCH_3
(b) $H_2C=CHCH_2OH$ and $(CH_3)_2C=O$
(c) $HC≡CCH_2CH_2CH_3$ and $H_2C=CHCH_2CH=CH_2$
(d) $HC≡CCH_2CH_2=CH_3$ and $H_2C=CHCH_2CH=CH_2$

5. Nuclear magnetic resonance. Give the number of signals observed in the low-resolution nmr hydrogen spectrum of each of the following compounds. Indicate the relative areas if more than one signal is present. (a) CH_3CH_3; (b) $CH_3CH_2CH_3$; (c) $CH_3CH_2CH_2CH_3$; (d) $(CH_3)_3CH$; (e) $(CH_3)_2CHCH_2CH_3$; (f)

$$(CH_3)_2C=O; \text{ (g) } CH_3CH_2\overset{\overset{\displaystyle H}{|}}{C}=O;$$ (h) $H_2C=CH_2$; (i) $CH_3CH=CH_2$; (j) $CH_3CH=CHCH_3$; (k) $C_6H_5NO_2$.

6. Chemical shift. Sketch a low-resolution proton nmr curve for each of the following compounds showing the *relative* chemical shifts and indicating the relative areas: (a) CH_3CH_2Cl; (b) $(CH_3)_3COCH_3$; (c) $DOCH_2CH_2Br$. (Signals for D are not observed in proton spectra.)

7. Electronic excitation. List the possible types of electronic excitations in the order of decreasing ΔE for each of the following compounds: (a) CH_3CH_3; (b) CH_3CH_2Br; (c) $H_2C=CH_2$; (d) $H_2C=CHOCH_3$; (e) $H_2C=NH$.

8. Atmospheric chemistry. Supply the formula for the missing species in each of the following balanced equations:

(a) $N_2 + O_2^+ \longrightarrow NO + ?$
(b) $N^+ + O_2 \longrightarrow O^+ + ?$
(c) $? + O_3 \longrightarrow SO_2 + O_2$

9. Mass spectrometry. Suggest the structure of each of the following compounds from the peaks in its mass spectrum, as given below. Indicate the formula for each fragment ion. Assume that the highest mass number corresponds to the parent ion. Compounds (b)–(f) contain at least one C atom in each. (a) 18, 17, 16; (b) 16, 15, 14, 13, 12, 1; (c) 44, 28, 16, 12; (d) 27, 26, 14, 13, 12, 1; (e) 30, 29, 28, 16, 14, 13, 12, 1; (f) 34, 33, 32, 31, 19, 15, 14, 1.

10. Electron bookkeeping. Using structural formulas, rewrite the following balanced equations, showing the fate of all outer electrons, and state which products are reactive intermediates.

(a) 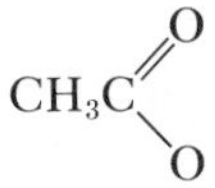 $CH_3C\overset{\displaystyle O}{\underset{\displaystyle O}{\diagdown\!\!\!\diagup}} \longrightarrow CO_2 + CH_3$

(b) $(CH_3)_4Pb \longrightarrow Pb + 4CH_3$
(c) $[(CH_3)_2C=O]^+ \longrightarrow CH_3CO^+ + CH_3$
(d) $CH_3I + e^- \longrightarrow CH_3 + I^-$
(e) $H_2O_2 + Fe^{2+} \longrightarrow Fe^{3+} + OH + OH^-$
(f) $HCCl_3 + OH^- \longrightarrow CCl_2 + H_2O + Cl^-$

11. Molecular excitations. For each of the following pairs, which would require a higher energy photon? (a) $\pi \rightarrow \pi^*$ transition in $H_2C{=}O$ or excitation of the $C{=}O$ stretch in $H_2C{=}O$; (b) $\pi \rightarrow \pi^*$ transition in $H_2C{=}CHCH_2CH{=}CH_2$ or in $H_2C{=}CHCH{=}CH_2$; (c) excitation of the C-to-O stretch in $H_2C{=}O$ or in $H_3C{-}OH$; (d) $\pi \rightarrow \pi^*$ or $n \rightarrow \pi^*$ transition for the $-N{=}\ddot{O}$: (nitroso) group.

12. Uv. (a) From the values given in Table 24.2 give the trend which prevails for λ_{max} versus position in a group in the periodic table. Suggest a reason. (CH_3F has a λ_{max} less than that of CH_3Cl.) (b) Give the trend which prevails for λ_{max} versus group number in the periodic table (compare F, O, and N). Suggest a reason.

13. Ir. From the values given in Table 24.3 answer the following. (a) Compare the stretching frequency of a bond involving an H atom such as C—H and O—H to one with no H atom such as C—C and C—O. (b) What is the effect of bond multiplicity on stretching frequency, for example, C—C versus C=C versus C≡C? (c) What is the effect on the stretching frequency of replacing an atom by a heavier isotope, for example, O—H versus O—D? (d) What is the effect on C—H stretching frequency of changing the hybrid state of the C?

14. Atmospheric chemistry. In terms of MO theory account for the fact that the reaction

$$N_2^+ + O_2 \longrightarrow N_2 + O_2^+$$

occurs readily, whereas the reaction

$$NO^+ + N_2 \longrightarrow NO + N_2^+$$

does not occur readily.

15. Atmospheric chemistry. Predict which reaction has the more negative ΔH and briefly explain your choice:

(a) $He^+ + O \longrightarrow He + O^+$ or $He^+ + N \longrightarrow He + N^+$
(b) $N_2^+ + O_2 \longrightarrow NO^+ + NO$ or $N_2 + O_2^+ \longrightarrow NO^+ + NO$
(c) $N^+ + O_2 \longrightarrow NO + O^+$ or $N + O_2^+ \longrightarrow NO + O^+$

16. Ultraviolet absorption. Select the molecule in each pair absorbing at the longer wavelength in the ultraviolet region and explain your choice.

(a) $(CH_3)_3N$ or $(CH_3)_3P$
(b) $CH_3CH_2OCH_2CH_3$ or $CH_2{=}CHOCH_2CH_3$
(c) $CH_2{=}CHCH_2CH{=}CH_2$ or $CH_2{=}CHCH{=}CHCH_3$

17. Nmr. Give the number of signals observed in the low-resolution nmr hydrogen spectrum of each of the following compounds and indicate the relative area for each signal. (a) $p\text{-}C_6H_4(NO_2)_2$; (b) $m\text{-}C_6H_4(NO_2)_2$; (c) $o\text{-}C_6H_4(NO_2)_2$; (d) cyclohexene; (e) cyclopentene.

18. Ir. Use Table 24.3 to account for the following observations: (a) A concentrated solution of ethanol, CH_3CH_2OH, in CCl_4 has a broad band at 3350 cm^{-1}, but on increased dilution this band disappears and a sharp band at 3640 cm^{-1} appears. (b) When the infrared spectrum of ethylene glycol,

H_2COHCH_2OH, is taken in CCl_4 a broad O—H stretching band appears at about 3350 cm^{-1} regardless of the concentration of the solution.

19. Fragmentation. Write the electronic structural formula of the missing species in the following equations:

(a)　　$CH_3CH_2CH_2 \xrightarrow{hf} ? + \cdot CH_3;$　　(b)　　$H_2CN_2 \xrightarrow{hf} ? + N_2$

20. Fragmentation. Irradiation of acetone, $(CH_3)_2C{=}O$, yields ethane, CH_3CH_3, and carbon monoxide, CO. Give a three-step sequence, the first two steps of which involve fragmentation, for the formation of these products.

21. Stability of carbonium ions. Carbonium ions (page 478) are important intermediates in many organic reactions. Mass spectrometry can be used to elucidate the relative stabilities of tertiary, R_3C^+, secondary, R_2CH^+, and primary, RCH_2^+, carbonium ions. What conclusion about relative stabilities can be drawn from the fact that the mass spectrum of 2,2,3-trimethylbutane, $(CH_3)_3CCH(CH_3)_2$, shows three fragment ions of masses 57 (97), 43 (8), and 15 (1)? The relative abundances are in parentheses.

22. Electronic excitation. 1,3-Butadiene, $H_2C{=}CHCH{=}CH_2$, absorbs in the ultraviolet at a longer wavelength than does 1-butene, $H_2C{=}CHCH_2CH_3$. This phenomenon can be explained in terms of MO theory. Butadiene has four π molecular orbitals, two bonding and two antibonding, each with a different energy. (a) Show the electron distribution of the four π electrons of butadiene in these energy levels. (b) Compare this distribution of electrons in molecular orbitals to that of 1-butene. (c) Explain the difference in ΔE of absorption of the two compounds.

23. Excitation and geometry. Although the initially formed electronically excited state of formaldehyde, $H_2C{=}O$, has the same shape as the ground state, it quickly relaxes to a pyramidal structure. (a) Draw a reasonable electronic structure for "pyramidal" $H_2C{=}O$ and compare it with the planar excited form. (b) Should the planar ground state or the excited pyramidal state have (*i*) the longer C—O bond length, (*ii*) the larger dipole moment?

24. Mass spectrometry. Identify each of the following fragment ions and explain why it is a prominent peak:

(a)　　From ethanol, $CH_3CH_2\ddot{\underset{\cdot\cdot}{O}}H$, the ion of mass 31.
(b)　　From butylamine, $CH_3CH_2CH_2CH_2\ddot{N}H_2$, the ion of mass 30.
(c)　　From acetic acid, $CH_3CO\ddot{\underset{\cdot\cdot}{O}}H$, the ion of mass 43.

25

NUCLEAR CHEMISTRY

In this chapter we consider the structure of the nucleus. We shall examine in some detail the nature of the bond (force) holding nucleons together in nuclei, and the properties of the nuclei such as radioactivity, artificial transmutations, fission, and fusion. Where possible, this discussion is in terms of principles previously applied to the properties of atoms and molecules.

RADIOACTIVITY / 25.1

Henri Becquerel in 1896 accidentally discovered that uranium compounds emit a radiation similar in nature to X rays. *Elements such as uranium, which spontaneously emit energy without the absorption of energy, are said to be naturally* RADIOACTIVE.

Experiments using electric or magnetic fields showed that the emission is composed of three distinct types. In Fig. 25.1, an electric field, applied to a narrow parallel beam of the emission from a naturally radioactive source, splits the beam into three beams, labeled ALPHA (α), BETA (β), and GAMMA (γ). The α beam, deflected to the negative plate, must be composed of positively charged particles; the β beam, more sharply deflected to the positive plate, must be composed of lighter negatively charged particles. The γ beam, undeflected, must be electrically neutral. The α particle, also designated as $^4_2\text{He}^{2+}$, carries two unit positive charges and possesses the mass of a helium nucleus. The electric charge and mass of a β particle are those of an electron, also designated as $_{-1}^{0}\beta$. The properties of γ rays are similar to those of X rays, except that their wavelengths are shorter. Gamma rays are photons several million times higher in frequency than visible light (Fig. 6.7, *see color plate following page 78*).

It was also discovered that radioactive isotopes (radioisotopes) frequently produce other elements. Rutherford found that radium, for

530

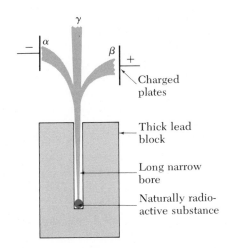

Fig. 25.1. *The behavior of α, β, and γ emissions in an electric field; apparatus is enclosed in a vacuum.*

γ

α

β

+

−

Charged plates

Thick lead block

Long narrow bore

Naturally radioactive substance

example, spontaneously emits α particles and radon, a radioactive gas. To account for these observations, Rutherford and Frederick Soddy in 1902 proposed that radioactivity is the result of a spontaneous change of the atoms of one element into the atoms of another element.* These **NUCLEAR REACTIONS** (transmutations) involve a change in the atomic number or the mass number (or both) of the radioisotope. The transmutation of radium may be represented by the equation

$$\underset{\text{α particle}}{^{226}_{88}\text{Ra} \longrightarrow \ {}^{4}_{2}\text{He}} + {}^{222}_{86}\text{Rn} \tag{1}$$

in which the subscript is the atomic number, and the superscript is the mass number—the number of nucleons (page 79) in the nucleus.

In chemical reactions, atoms are rearranged; they are not created or destroyed. In nuclear reactions, nucleons are rearranged; they are not created or destroyed. Hence, in nuclear reactions, mass numbers are conserved: *The sum of the mass numbers of reacting nuclei and particles must equal the sum of the mass numbers of the nuclei and particles produced.* The conservation of charge requires *the sum of the atomic numbers of reacting nuclei and particles to equal the sum of the atomic numbers of the products*, illustrated in Equation (1). In the reaction of an α particle with $^{27}_{13}\text{Al}$, in which a neutron is produced,

$$\underset{\text{neutron}}{^{27}_{13}\text{Al} + {}^{4}_{2}\text{He} \longrightarrow \ {}^{1}_{0}n} + \ ?$$

the product nucleus must be $^{30}_{15}\text{P}$. In the reaction

---

* The theory had to contend with a century of work that established the idea of unalterable atoms that could *not* be transmuted into other atoms. "Soddy turned to his colleague and blurted: 'Rutherford, this is transmutation!' Rutherford rejoined: 'For Mike's sake, Soddy, don't call it *transmutation*. They'll have our heads off as alchemists.' Rutherford and Soddy were careful to use the term 'transformation' rather than 'transmutation.'" (*Scientific American*, August 1966, page 91.)

$$^{235}_{92}\text{U} \longrightarrow \ _{-1}^{\ 0}\beta + ?$$

the product must be $^{235}_{93}\text{Np}$.

Although neutrons, *outside the nucleus*, disintegrate into protons and electrons, they are *not* composed of protons and electrons. Experiments by Robert Hofstadter on electron scattering by protons and by neutrons, analogous to the experiments by Rutherford on the scattering of α particles by nuclei, suggest that the proton consists of a neutral core surrounded by two positively charged clouds, and that a neutron possesses a similar structure except that the inner cloud is negative. In one, the two clouds add up to one unit positive charge; in the other, the two clouds cancel each other electrically. This view is consistent with the dyon* model of particles. Dyons are *hypothetical* subparticles (suggested by Paul Dirac and not yet definitely observed in nature) with assigned properties. The model predicts that the proton is composed of one dyon with a charge of $\frac{1}{3}e^-$ and two dyons, each with a charge of $\frac{2}{3}e^+$, yielding a net charge of $\frac{3}{3}e^+$. The neutron should consist of two dyons, each with a charge $\frac{1}{3}e^-$, and one dyon with a charge of $\frac{2}{3}e^+$, yielding a net charge of zero.

The energy of nuclear reactions, ΔE, calculated from the difference between the masses of products and reactants in accordance with the Einstein law (page 69), is in agreement with the measured values. For the transmutation of radium, Equation (1), the masses (g/mole) of the atoms are

$$^{226}_{88}\text{Ra} = 226.0254 \qquad \begin{array}{l} ^{4}_{2}\text{He} = 4.0026 \\ \underline{^{222}_{86}\text{Rn} = 222.0176} \\ \phantom{^{222}_{86}\text{Rn} = }226.0202 \end{array}$$

so that Δm, the difference in mass between the products and the initial reactants, is

$$\Delta m = 226.0202 - 226.0254 = -0.0052 \ \frac{\text{g}}{\text{mole}}$$

and

$$\Delta E = \Delta mc^2 = -5.2 \times 10^{-3} \ \frac{\text{g}}{\text{mole}} \left(3.0 \times 10^{10} \ \frac{\text{cm}}{\text{second}} \right)^2$$

$$= -4.7 \times 10^{18} \ \frac{\text{ergs}}{\text{mole}}$$

---

* So called because the predicted particle carries two (*dy*) charges, an electric charge and a magnetic charge (*one pole*). The dyon has also been dubbed the *quark*.

or in kcal/mole

$$\Delta E = \frac{-4.7 \times 10^{18} \frac{\text{ergs}}{\text{mole}}}{4.18 \times 10^{10} \frac{\text{ergs}}{\text{kcal}}} = -1.1 \times 10^8 \frac{\text{kcal}}{\text{mole}}$$

The measured value is $\Delta E = -1.1 \times 10^8$ kcal/mole.

Upon passage through matter, the energy of the emissions from radio-active nuclei is consumed by the ionization or excitation of atoms, molecules, or ions and by the fragmentation of chemical bonds (page 523). Alpha particles usually cannot penetrate paper more than a few sheets in thickness. The walls of the ordinary glass beaker generally can stop β particles. X and γ rays are highly penetrating and are stopped only by thick layers of lead or concrete. Distance from the source is also important in minimizing exposure since the intensity varies inversely with the square of the distance.

NUCLEAR STRUCTURE / 25.3

In certain respects, the structure of the nucleus is analogous to the electronic structure of atoms. Quantum mechanics shows that the energy states (levels) of a nucleon in the nucleus are quantized and characterized by four quantum numbers. The Pauli principle also applies to nucleons: No two protons and no two neutrons can possess the same four quantum numbers. One of these corresponds to m_s for electrons. Protons and neutrons may thus have two spin orientations (page 96), corresponding to quantum numbers of $+\frac{1}{2}$ and $-\frac{1}{2}$. Therefore, protons and neutrons, like electrons, possess a characteristic (but much smaller) permanent magnetic moment. However, spin pairing, $\uparrow\downarrow$, occurs and, hence, nuclei with *an even number of protons* and *an even number of neutrons*, for example, ^4_2He, $^{12}_6\text{C}$, $^{16}_8\text{O}$, $^{32}_{14}\text{Si}$, do not have a permanent magnetic moment. In ^2_1D, however, the spins of the proton and neutron are parallel and deuterium nuclei possess a magnetic moment; these nuclei, like electrons, may be made to orient with respect to an applied external magnetic field (page 520).

Nucleons may be assigned to nuclear orbitals, analogously to the assignment of electrons to atomic orbitals (page 91). Just as closed (filled) sublevels, e.g., He $1s^2$, Ne $1s^22s^22p^6$ possess relative atomic (chemical) stability, closed sets of nuclear energy levels ("shells") also correspond to great nuclear stability. Nuclei containing 2, 8, 20, 28, 50, 82, or 126* nucleons of each kind are particularly stable. For example, the ^4_2He nucleus contains 2 protons and 2 neutrons corresponding to a

---

* Commonly referred to as "magic numbers." Recent theoretical studies predict that the bounds of a reasonably stable region of nuclei may range from about $^{298}_{114}\text{X}$ to about $^{310}_{126}\text{X}$.

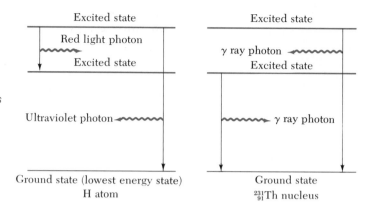

Fig. 25.2. (Not to scale.) The excited states of the H atom may be produced by electron beam bombardment (page 87); the excited states of Th are produced when Np emits an α particle. A photon is produced when an electron or a nucleon returns to a lower energy state (shown in the diagram symbolically with an arrow). The energy of the photon is determined by the difference in the energies of the two states.

closed shell. This accounts for the observation that many naturally radioactive nuclei spontaneously emit the extremely stable α particles.* The less stable nuclei ^1_1H (proton), ^2_1D (deuteron), ^3_1T (triton), ^3_2He, or ^6_3Li, however, are never seen as products of natural radioactive decompositions. Of the isotopes of oxygen, ^{14}O, ^{15}O, ^{16}O, ^{17}O, ^{18}O, ^{19}O, and ^{20}O, the most stable is ^{16}O; its nucleus contains 8 protons and 8 neutrons, corresponding to a closed shell. $^{40}_{20}Ca$ forms the next closed shell. Although this model ("nuclear shell model") does not always yield correct predictions, it is nevertheless very useful in correlating much information about nuclei.

A nucleus remaining after the emission of an α or β particle often emits a γ ray (photon). In terms of the energy-level model of nuclei, the nucleus is in an excited state and emits a photon in a manner analogous to the emission of a photon by an excited hydrogen atom (Fig. 25.2).

THE STABILITY OF NUCLEI / 25.4

The energy required to separate a molecule in the gaseous state into its component atoms equals the energy evolved when these atoms recombine at the same temperature. Similarly, the **BINDING** (bond) **ENERGY OF A NUCLEUS** is the energy required to separate the nucleus into its component nucleons or the energy evolved when the nucleons recombine to form the nucleus. However, the heats of nuclear reactions are about a million times larger than the heats of chemical reactions. The nature of the force holding nucleons together must be fundamentally different from the electrostatic force involved in atomic (chemical) bonding. As in chemical bonding, gravitational and magnetic forces are too weak to be significant. Further, the force cannot be electrical. For example, the deuteron, composed of one type of charged particle (the

---

* This does not mean that α particles exist in the nucleus; rather, the probability that the four nucleons will "stick and escape" is comparatively high.

proton) and a neutral particle, cannot be held together by electric forces. An even more striking illustration is the great stability of the nucleus of $^3_2$He in which the electrostatic interaction of the two protons is repulsive. In 1935, Hideki Yukawa suggested that *a particular particle oscillating between nucleons* with practically the speed of light *is the force* that holds the nucleons together. He further predicted that this particle should have a mass about 275 ± 25 times the electron mass and may be electrically neutral, positive, or negative. These particles, discovered later, are called "pi-mesons"* or "pions," designated as π^0, π^+, π^-. As a result of the continuous pion transfer, protons are changed to neutrons and neutrons are changed to protons.

At any given instant, the nucleus has a fixed number of nucleons with a definite positive charge equal to the atomic number. At any given instant, two nucleons are sharing the same exchange particle. The sharing of pions by nucleons is analogous to the sharing of electrons by bonded atoms in molecules; see Fig. 25.3. The two attractive forces are alike in that they drop off sharply beyond a certain critical distance, about 10^{-8} cm between bonded atoms and about 10^{-13} cm between adjacent nucleons. As with chemical bonding, the closer the nucleons, the stronger the nuclear bond. However, a small separation of nucleons beyond 10^{-13} cm leads to fragmentation of the nucleus.

This theory explains $_{-1}\beta$ decay, which is a paradoxical phenomenon, since electrons do not exist in the nucleus. Beta particles result from the decay of an *ejected* π^-. For every ejected π^-, a nuclear neutron is converted to a nuclear proton. The ejected π^- undergoes a series of transformations, the net result of which is the formation of a $_{-1}\beta$ particle and 3 neutrinos.

$$\text{nuclear neutron} \longrightarrow \text{nuclear proton} + \text{ejected } \pi^-$$
$$\text{ejected } \pi^- \longrightarrow {}_{-1}^0\beta + 3{}_0^0\nu \text{ (neutrino)}$$

The existence of the neutrino, massless† and chargeless, was predicted by Wolfgang Pauli and Enrico Fermi in 1931; the prediction has been verified experimentally. The pion results from a transition from an excited to a lower nuclear state similar to a transition that creates a photon.

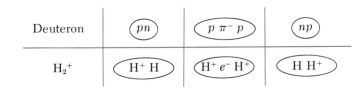

Fig. 25.3. *The deuterium nucleus and the H_2^+ ion illustrate the analogy between electron sharing in molecules and pion sharing in nuclei.*

---

* From the Greek *mesos*, "intermediate"; the mass of the meson is between that of an electron and a nucleon.

† The mass is too small to be distinguished experimentally from zero.

Radioactivity. Natural nonradioactive (stable) nuclei, plotted in Fig. 25.4, fall within a narrow region of stability. In the light elements the ratio N/Z is close to 1. However, in the heavier elements the ratio of neutrons to protons increases to about 1.5. Apparently, relatively more neutrons are needed in the heavier nuclei to balance the electrostatic repulsion between the protons. All nuclei with atomic numbers greater than 83 are radioactive; of these, only $^{235}_{92}U$, $^{238}_{92}U$, and $^{232}_{90}Th$ occur on the Earth in relatively large amounts. The only stable nuclei that contain fewer neutrons than protons are ^1_1H and ^3_2He. It is of interest to note that about 86% of the nuclei in the Earth's crust (excluding the waters and the atmosphere) have even mass numbers.

Nuclei outside of the stability band spontaneously transform (decay) to nuclei closer to or within the stability band. Nuclei to the right of the stability band have a relative excess of protons and transform by $_{+1}^{0}\beta$,

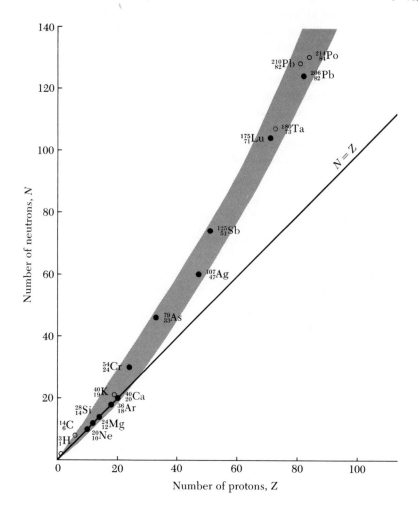

Fig. 25.4. *The stability region, approximately represented by the shaded area. The solid line locates nuclei with equal numbers of protons and neutrons. Several stable nuclei are given to serve as reference points. A few naturally occurring radioactive nuclei are indicated by open circles.*

$^1_1$H, or $^4_2$He emission or by electron capture:

$$^{78}_{35}\text{Br} \longrightarrow {}^{78}_{34}\text{Se} + {}^{0}_{+1}\beta \text{ (positron emission)}$$
$$^{43}_{21}\text{Sc} \longrightarrow {}^{42}_{20}\text{Ca} + {}^{1}_{1}\text{H} \text{ (proton emission)}$$
$$^{11}_{6}\text{C} \longrightarrow {}^{7}_{4}\text{Be} + {}^{4}_{2}\text{He} \text{ (alpha emission)}$$
$$^{7}_{4}\text{Be} + e^- \text{ (orbital electron)} \longrightarrow {}^{7}_{3}\text{Li} \qquad \text{(electron capture)}$$

Stable nuclei do not capture orbital electrons; for example, the highly endothermic process $^1_1$H + orbital electron $\rightarrow {}^1_0 n$ does not occur.

Nuclei to the left of the stability band have a relative excess of neutrons and decay by $_{-1}^{0}\beta$ or $^1_0 n$ emission:

$$^{87}_{36}\text{Kr} \longrightarrow {}^{87}_{37}\text{Rb} + {}_{-1}^{0}\beta, \qquad {}^{87}_{36}\text{Kr} \longrightarrow {}^{86}_{36}\text{Kr} + {}^1_0 n$$

$^{235}_{92}$U, $^{238}_{92}$U, and $^{232}_{90}$Th, the parents, respectively, of three natural radioactive series, have an excess of protons and decay mostly by α-particle emission. The successive transformations of the ^{238}U series are shown in Fig. 25.5.

Artificial transmutations. Everyone is more or less familiar with the dreams, shattered by Dalton, of the alchemists who attempted to transform a cheap metal into gold. With the postulation of the nuclear theory of the atom, artificial transmutation came to appear feasible. The first induced transmutation was demonstrated in 1919 by Rutherford, who exposed nitrogen to α particles from radium and detected the production of protons:

$$^{14}_{7}\text{N} + {}^{4}_{2}\text{He} \longrightarrow {}^{18}_{9}\text{F} \longrightarrow {}^{17}_{8}\text{O} + {}^{1}_{1}\text{H}$$

With the removal of the particle source, the reaction stops.

While treating light elements such as boron or aluminum with α particles, Irène and Fréderic Joliot-Curie detected in 1934 the emission of positrons and neutrons. More important, they observed that the emission of positrons does not stop, although the emission of neutrons does stop, upon removal of the α particle source. The reactions are

$$^{10}_{5}\text{B} + {}^{4}_{2}\text{He} \longrightarrow {}^{14}_{7}\text{N} \longrightarrow {}^{13}_{7}\text{N}^* + {}^{1}_{0} n$$

followed by

$$^{13}_{7}\text{N}^* \longrightarrow {}^{13}_{6}\text{C} + {}_{+1}^{0}\beta$$

in which $^{13}_{7}$N* is the first radioisotope artificially produced. An asterisk is sometimes used to indicate a radioisotope.

Before it can be captured by a nucleus, a positively charged particle must possess sufficient kinetic energy to overcome the repulsive force that develops as the positive particle approaches the positive nucleus. Particle accelerators, such as cyclotrons, were invented to increase the kinetic energy of charged particles to levels required for capture by nuclei with high atomic numbers. With the accelerators available in 1934, it was not possible to induce radioactivity in the elements beyond

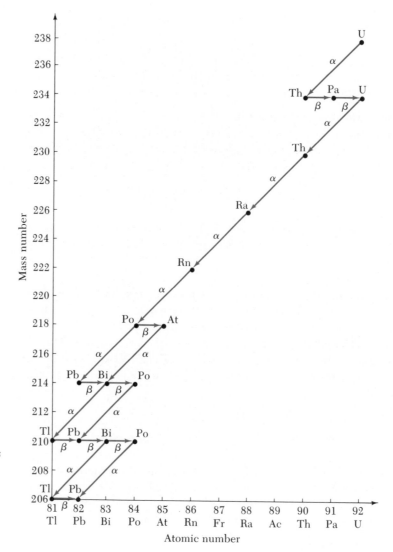

Fig. 25.5. *Uranium-238 radioactive series. $^{238}_{92}U$ is the parent; $^{206}_{82}Pb$, nonradioactive, is the end product. The radioisotopes emit either α or $_{-1}\beta$ particles; occasionally, a fraction of the nuclei of an element, for example, ^{218}Po, emits α particles, while the remainder emit $_{-1}\beta$ particles.*

potassium. Since the neutron is electrically neutral, Enrico Fermi reasoned that its entry into a nucleus would not be opposed by repulsive forces. It should be possible to transform all known elements by exposure to neutron sources.† He was correct: Practically all known elements have been transformed, and a number of transuranium elements have been synthesized:

$$^{238}_{92}U + ^1_0n \longrightarrow ^{239}_{92}U^{\ast} \longrightarrow ^{239}_{93}Np^{\ast} + ^0_{-1}\beta$$
$$^{239}_{93}Np^{\ast} \longrightarrow ^{239}_{94}Pu^{\ast} + ^0_{-1}\beta$$

† The half-life of free neutrons is 12 minutes. This time is ample for using them in nuclear reactions.

Nuclear reactors, in which controlled fission reactions (below) are carried out, are excellent sources of neutrons. These neutrons are used to bombard atomic nuclei to synthesize useful isotopes of which ^{14}C is typical:

$$^{14}_{7}N + ^{1}_{0}n \longrightarrow ^{14}_{6}C^{*} + ^{1}_{1}H$$

A solution of a nitrate salt is run through the reactor. Exposure to neutrons converts the nitrogen in the NO_3^- to $^{14}_{6}C$, creating ^{14}C-labeled carbonate ion, $^{14}CO_3^{2-}$, which is recovered and sold as solid $Ba^{14}CO_3$. Other isotope syntheses are the preparation of radioactive $^{60}_{27}Co$, used in cancer therapy, and tritium:

$$^{59}_{27}Co + ^{1}_{0}n \longrightarrow ^{60}_{27}Co^{*}, \qquad ^{6}_{3}Li + ^{1}_{0}n \longrightarrow ^{3}_{1}H^{*} + ^{4}_{2}He$$

Fission. It is instructive to plot, as in Fig. 25.6, the average energy released per nucleon in the formation of a nucleus from nucleons. The greater the energy released per nucleon, the greater the stability of the nucleus.† It is evident that the very light and the very heavy nuclei are least stable compared to nuclei in the vicinity of $^{55}_{26}Fe$. We might expect that spontaneous changes should include the fusion of very light nuclei into heavier nuclei, and the splitting (fission) of a very heavy nucleus into two nuclei. These reactions should be rich energy sources, compared with chemical changes.

Although spontaneous fission has been detected in $^{232}_{90}Th$ and heavier elements, the rate is too slow for any practical purpose because of the

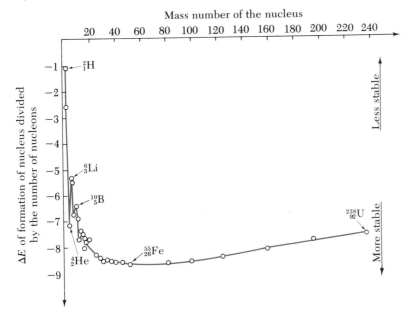

Fig. 25.6. *The average energy (in MeV) evolved per nucleon in the formation of nuclei from the separate nucleons.*

† The energy differences are so large that ΔG and ΔE are practically the same. The nucleus of lowest energy is the most stable.

high energy of activation (roughly 1.4×10^8 kcal/mole) required to produce the deformation of the nucleus (a slight separation of nucleons) necessary for fission. The energy of activation is supplied by a neutron which forms an unstable intermediate nucleus, for example,

$$^{235}_{92}U + ^1_0n \longrightarrow {}^{236}_{92}U \longrightarrow {}^{137}_{53}I + {}^{97}_{39}Y + 2^1_0n \tag{2}$$

The situation here, illustrated in Fig. 25.7, is exactly comparable to a two-step chemical reaction (page 415).

When Fermi claimed in 1934 the synthesis of transuranium elements by neutron bombardment of uranium, many investigators doubted this interpretation of the chemical analytical results. In particular, Ida Noddack suggested that the uranium nuclei were being split into smaller nuclei. Practically every laboratory engaged in the study of nuclear reactions undertook an analytical investigation of the possible products.

Among the products of the neutron bombardment of uranium, a radioactive element was found that behaved chemically like the elements of Group 2R of the periodic system (strontium, barium, and radium). Among these, only radium had been known to be radioactive, and therefore the new radioactive element was assumed to be radium. However, because its radioactivity differed from that of natural radium, it was thought to be some other isotope of radium. In 1939, Otto Hahn and Fritz Strassman attempted to isolate this "radium" by the same chemical methods used in the original discovery of radium by the

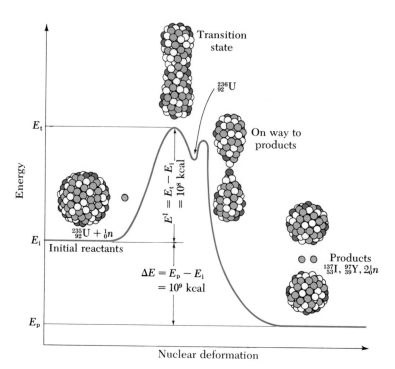

Fig. 25.7. *The change in energy with deformation of a nucleus. For a chemical reaction, the horizontal axis would be changed to molecular deformation.*

Curies—namely, the introduction of a larger quantity of barium as a "carrier" to help separate the small quantity of radium from all other impurities, followed by a final careful separation of the radium from the barium by fractional crystallization. Barium is particularly convenient to help carry a trace of radium out of a mixture because the properties of the two elements and their compounds are so much alike; for example, a precipitate of $BaCO_3$ from a solution will also carry with it any trace of radium as $RaCO_3$. Of course, the same similarity that helps the crude separation makes the final separation of the desired component from the carrier difficult. Hahn and Strassman, however, found that the final separation of the new "radium" from the barium was not just difficult, but impossible. They finally showed that the new material could not be separated from barium for the simple, but unexpected, reason that *it was an isotope of barium!* It was then concluded that the uranium nucleus had divided into two fragments and it was immediately realized that a tremendous amount of energy would accompany this division, a conclusion verified by the large readings produced by this energy in an ionization chamber. Similar readings obtained by previous investigators had been dismissed as "electronic noises." Nuclear fission was thus discovered by chemists.

From the principle of conservation of charge, the other nucleus formed with $_{56}$Ba must be $_{36}$Kr; a typical fission reaction* is

$$^{235}_{92}U + ^{1}_{0}n \longrightarrow ^{236}_{92}U \longrightarrow ^{142}_{56}Ba + ^{91}_{36}Kr + 3^{1}_{0}n$$
$$\Delta E = -4.6 \times 10^9 \text{ kcal/mole}$$

Since more than one neutron is produced for each one consumed and since a ^{236}U nucleus undergoes fission in about 10^{-9} second, it becomes possible to develop a branching chain reaction (page 425) leading to a nuclear explosion (miscalled an "atomic explosion"), the release of unprecedented quantities of energy in extremely small periods of time. However, through the use of "chain breakers," materials such as cadmium that act as neutron scavengers, the number of neutrons available for the fission of $^{235}_{92}$U may be decreased by about $\frac{2}{3}$. The rate of the fission reaction is thereby decreased, an explosion is prevented, and the heat of the reaction may be utilized as a practical energy source.

The first controlled and sustained fission reaction was operated successfully in 1942 at the University of Chicago with comparatively impure ^{235}U and "chain breakers." The first branching chain fission reaction was successfully carried out in a test explosion with pure ^{239}Pu in 1945 near Alamogordo, New Mexico.

Fission fragments, the waste products from nuclear reactors, emit $_{-1}^{0}\beta$ particles, neutrons, and γ radiation and will remain lethal for at least 500 years.

---

* See Equation (2), page 540; fission may occur in many ways depending upon the energy of the intermediate formed.

Fusion. Tremendous amounts of energy can be released by the fusion of small nuclei into larger ones. The largest energy release occurs with the lightest elements. Two of the more important reactions are

$$^2_1H + ^3_1H \longrightarrow ^4_2He + ^1_0n \qquad \Delta E = -4.2 \times 10^8 \text{ kcal/mole}$$
$$^2_1H + ^2_1H \longrightarrow ^3_2He + ^1_0n \qquad \Delta E = -6.9 \times 10^7 \text{ kcal/mole}$$

Energy from the fusion of one pair of nuclei induces the fusion of other pairs. Such a sustained reaction results in an explosion, specifically described as THERMONUCLEAR.

A tank of hydrogen does not explode to form deuterium because the rate of the reaction is infinitesimally small. The rate is proportional to the number of collisions between the nuclei, but the number of collisions is essentially zero because of repulsive electrostatic forces. To increase the number of collisions, the temperature of the reactants must be increased to 10^6 to 10^7 °C. At these temperatures, light elements are completely stripped of orbital electrons forming an ionized PLASMA, a gaseous system containing equivalent numbers of positive ions and electrons. For the hydrogen isotopes, the probability of reaction per collision appears to be highest for

$$^3_1H + ^3_1H \longrightarrow ^4_2He + 2^1_0n$$

and least for

$$^1_1H + ^1_1H \longrightarrow ^2_1H + ^0_{+1}\beta$$

The hydrogen bomb makes use of nuclear fusion. The high temperature needed to activate the reacting atoms is supplied by a fission bomb. As yet, ways of controlling nuclear fusion for the benefit of man have not been devised. One of the major stumbling blocks is the inability to handle the plasma. Hopefully, research to harness nuclear fusion will be successful. The burgeoning consumption of energy will require, in the future, that nuclear fission and fusion supplement the burning of fossil fuels. These exothermic nuclear processes are often referred to as "combustion" or "burning."

Stellar energy. The mean temperature of the sun, 4×10^6 °K, suffices for fusion between ^1H, ^2H, and ^3He,

$$^1_1H + ^1_1H \longrightarrow ^2_1H + ^0_{+1}\beta \text{ (slow step)}$$
$$^2_1H + ^1_1H \longrightarrow ^3_2He$$
$$^3_2He + ^3_2He \longrightarrow ^4_2He + 2^1_1H$$

Once the sun has synthesized ^3He and ^4He, further nuclear reactions involving these nuclei may lead to the production of ^{12}C, which then starts a carbon–nitrogen–oxygen cycle:

$$^{12}_{6}C + ^{1}_{1}H \longrightarrow ^{13}_{7}N \longrightarrow ^{13}_{6}C + ^{0}_{+1}\beta$$
$$^{13}_{6}C + ^{1}_{1}H \longrightarrow ^{14}_{7}N$$
$$^{14}_{7}N + ^{1}_{1}H \longrightarrow ^{15}_{8}O \longrightarrow ^{15}_{7}N + ^{0}_{+1}\beta$$
$$^{15}_{7}N + ^{1}_{1}H \longrightarrow ^{12}_{6}C + ^{4}_{2}He$$

The last reaction replenishes ^{12}C and closes the cycle; ^{12}C serves as a catalyst. The overall reaction is the fusion of 4 protons:

$$4^{1}_{1}H \longrightarrow ^{4}_{2}He + 2^{0}_{+1}\beta \qquad \Delta E = -6 \times 10^{8} \text{ kcal/mole}$$

Other reactions, still unknown, also contribute to the energy output of the sun.

RATE OF RADIOACTIVE DECAY / 25.6

It is impossible to predict when, if ever, an atom or molecule will react. However, the chemist is almost never concerned with the behavior of a single particle. Even in a small mass of radioactive material, there are many trillions of atoms. For example, in 1 mg of ^{238}U, there are 2.5×10^{18} atoms. With such large numbers of atoms or molecules, it is possible to make precise predictions about reaction rates. Experiment shows that radioactive decay rates are directly proportional to the number of atoms present. For example, if 10^{20} atoms have a decay rate of 10^{3} atoms per minute, then 4×10^{20} atoms will have a decay rate of 4×10^{3} atoms/minute; radioactive decay is a typical first-order reaction:

$rate = k \times N$ where N = number of atoms

The time it takes for one-half the atoms to disintegrate, the so-called half-life (page 423), is independent of the number of atoms. The half-life is a characteristic of each radioisotope. The shorter the half-life, the faster is the rate of decay. The half-lives of some important radioisotopes are given in Table 25.1.

Table 25.1
Half-Lives of Some Radioisotopes

Isotope	Decay reaction	$t_{1/2}$
$^{3}_{1}H$	$^{3}_{1}H \longrightarrow ^{3}_{2}He + _{-1}\beta$	12.5 years
$^{14}_{6}C$	$^{14}_{6}C \longrightarrow ^{14}_{7}N + _{-1}\beta$	5.73×10^{3} years
$^{32}_{15}P$	$^{32}_{15}P \longrightarrow ^{32}_{16}S + _{-1}\beta$	14.3 days
$^{59}_{26}Fe$	$^{59}_{26}Fe \longrightarrow ^{59}_{27}Co + _{-1}\beta$	45.1 days
$^{60}_{27}Co$	$^{60}_{27}Co \longrightarrow ^{60}_{28}Ni + _{-1}\beta$	5.2 years
$^{90}_{38}Sr$	$^{90}_{38}Sr \longrightarrow ^{90}_{39}Y + _{-1}\beta$	27.7 years
$^{131}_{53}I$	$^{131}_{53}I \longrightarrow ^{131}_{54}Xe + _{-1}\beta$	8.1 days

The half-life of $^{15}_{8}O$ is 2.0 minutes. Therefore, given 8 mg of this isotope to start with, 4 mg would remain after 2.0 minutes, 2 mg would remain after 4.0 minutes, and 1 mg would remain after 6.0 minutes. The relation between q_t, the quantity at time t, and q_0, the initial quantity, is (page 424)

$$q_t = q_0(0.5)^{t/t_{1/2}} \tag{3}$$

in which $t_{1/2}$ is the half-life and $t/t_{1/2}$ gives the number of half-lives. In Equation (3), rates of decay may be substituted for q_t and q_0.†

Example 1 Radon "seeds," radon collected from radium bromide and sealed in capillary tubing, are used for cancer therapy. The half-life for radon, $^{222}_{86}Rn \rightarrow$ $^{218}_{84}Po + ^{4}_{2}He$, is 3.825 days. (a) Starting with 0.0200 mg, how many milligrams will remain after 11.475 days? (b) A radon seed initially emitted 7.0×10^4 α particles per second; it now emits 2.1×10^4 α particles per second. What is the age of the seed?

Answer (a) Substituting the given data into Equation (3),

$$q_t = 0.0200 \text{ mg } (0.5)^{11.475 \text{ days}/3.825 \text{ days}} = 0.0200 \text{ mg } (0.5)^{3.000}$$
$$= 0.0200 \text{ mg} \times 0.125 = 0.00250 \text{ mg}$$

The quantity left is 2.50×10^{-3} mg.
(b) Using the rates in place of q_t and q_0, Equation (3) gives

$$2.1 \times 10^4 \frac{\text{particles}}{\text{second}} = 7.0 \times 10^4 \frac{\text{particles}}{\text{second}} (0.5)^{t/3.8 \text{ days}}$$

and solving:

$$\log(2.1 \times 10^4) = \log(7.0 \times 10^4) + \frac{t}{3.8 \text{ days}} \times \log(5 \times 10^{-1})$$

$$4.322 = 4.845 + \frac{t}{3.8 \text{ days}} (-0.301)$$

$$0.523 = \frac{0.301t}{3.8 \text{ days}}$$

from which $t = 6.6$ days. The age of the seed is 6.6 days.

Radiochemical dating. Radiochemical dating has made important contributions to paleoanthropology and geology. The full significance of the discoveries of hominid fossils and tools at Olduvai Gorge (Tanzania), for example, could not be realized until the ages (1.7×10^6 years) of the beds at the Gorge were accurately determined. Other

† Units used in connection with radioactivity are summarized in Appendix I.13.

radiochemical measurements show that man lived in South America many centuries before 11,000 B.C. In radiochemical dating, ages are *not* measured directly; rather, the number of disintegrations per unit time per gram of material (the ACTIVITY) is measured, from which ages are calculated on the basis of one or more assumptions.

Dating with $^{14}_{6}C^*$, as conceived by Willard Libby, is based on the following reasoning. Natural carbon contains a small fraction of $^{14}_{6}C^*$ whose activity, as found in living plants and animals and in the air, is approximately constant at 14 disintegrations per minute per gram (d/minute-g) of carbon. The concentration of ^{14}C remains constant because it is continuously replenished by the bombardment of ^{14}N in the upper atmosphere by cosmic-ray neutrons: $^{14}_{7}N + ^{1}_{0}n \rightarrow ^{15}_{7}N \rightarrow ^{14}_{6}C^* + ^{1}_{1}H$. It is assumed that the $^{14}_{6}C^*$ activity in the air was the same in ancient times.† This $^{14}_{6}C^*$ is then incorporated into living tissues. Then, so long as a plant or an animal is alive, the activity of $^{14}_{6}C^*$ per gram of carbon remains constant, but upon death the $^{14}_{6}C^*$ continues to decay without replenishment and the activity decreases. From the known half-life, 5.73×10^3 years, for $^{14}_{6}C^* \rightarrow _{-1}^{0}\beta + ^{14}_{7}N$, the age of archeological samples may then be determined.

Example 2 The activity of the hair of an Egyptian woman is 7.50 d/minute-g carbon. What is the age of the hair?

Answer From the statement of the problem,

$$q_0 = 14 \ \frac{d}{minute\text{-}g} \quad \text{and} \quad q_t = 7.50 \ \frac{d}{minute\text{-}g}$$

Using Equation (3),

$$7.50 \ \frac{\cancel{d}}{\cancel{minute\text{-}g}} = 14 \ \frac{\cancel{d}}{\cancel{minute\text{-}g}} \ (0.5)^{t/5730 \ yr}$$

from which $t = 5.2 \times 10^3$ years; the age of the hair is, therefore, 5.2×10^3 years.

Neutron activation analysis. New methods of quantitative analysis based on the formation of radioactive nuclei by neutron capture have been developed. The sample is exposed to neutrons, each constituent element forms a specific radioisotope with a characteristic half-life, and from the activity of the radioisotope, the mass of the element from which it was formed may be calculated. The sensitivity of the method, about one part or less per billion, makes it very useful for the nondestructive analysis of trace elements in complex biological systems, in archeological samples, in meteorites, and in lunar rocks.

---

† Recently, comparison of ^{14}C dates with tree-ring dates has shown that this assumption is incorrect. The ^{14}C dating method can still be used, but corrections must be applied to the calculated ages.

1. Nuclear chemistry. Pick the true statements: (a) The internal structures of nuclei affect the chemical properties of atoms and molecules. (b) The principles of spectroscopy are applicable to the smallest nucleus and to the largest molecule. (c) In the atom, electrons are distributed in atomic orbitals; in the nucleus, nucleons are distributed in nuclear orbitals. (d) Electrostatic force is the same between all electron pairs; nuclear force is the same between all nucleon pairs. (e) Nuclear force is similar to electrostatic force. (f) Unlike the electronic charge, a nuclear charge may be altered. (g) The total number of nucleons in a nucleus can be changed in a transmutation. (h) Hydrogen can be "burned" to helium. (i) Nucleons are structureless (have no internal structure). (j) It is definitely proven that electrons are structureless. (k) The total charge of a nucleus always equals the sum of Z protons and N neutrons. (l) The sum of protons and neutrons is changed during fission. (m) Confined electrons and nucleons can only have quantized energies. (n) The pion is the "glue" that keeps nucleons together. (o) The proton is a structureless point charge.

2. α, β, γ. How is the number of protons in a nucleus changed when (a) a γ ray, (b) a β particle, (c) an α particle, is emitted from the nucleus?

3. Nuclear reactions. Balance the following nuclear reactions (indicate symbol where possible, mass number, and atomic number for ?):

$$^1H + {}^1H \longrightarrow {}_{+1}^{\ 0}\beta + ?$$
$$^{236}_{92}U \longrightarrow {}^{95}_{39}Y + ? + 3{}_0^1n$$
$$^{252}_{98}Cf + {}^{10}_5B \longrightarrow ? + 5{}_0^1n$$
$$^{37}_{17}Cl + {}_0^0\nu(\text{neutrino}) \longrightarrow ? + {}_{-1}^{\ 0}\beta \ (\text{reaction by which neutrinos are detected})$$
$$^{127}_{52}TeO_4{}^{2-} \longrightarrow ?O_4{}^- + {}_{-1}^{\ 0}\beta$$
$$^{235}_{92}U \longrightarrow 7\alpha + 4{}_{-1}^{\ 0}\beta + ?$$
$$^{252}_{98}Cf + {}^{11}_5B \longrightarrow {}^{257}_{103}Lr + 6?$$
$$^{238}_{92}U + 37{}_0^1n \longrightarrow ? \longrightarrow 14{}_{-1}^{\ 0}\beta + ?$$

4. ΔE. Calculate ΔE for the proposed basis of an absolutely clean source of nuclear energy: $^{11}_5B + {}_1^1H \to {}^{12}_6C \to 3{}_2^4He$. Atomic masses: $^{11}B = 11.00931$, $^4He = 4.00260$, $^1H = 1.00783$.

5. β decay. Positrons do not exist as particles in nuclei. Explain $_{+1}\beta$ emission (by analogy to $_{-1}\beta$ emission).

6. Abundance. ISOBARS are atoms that have the same mass number but different atomic numbers. For each pair pick the more abundant isobar: (a) $^{14}_6C$, $^{14}_7N$; (b) $^{32}_{15}P$, $^{32}_{16}S$; (c) $^{16}_8O$, $^{16}_9F$.

7. Nuclear spin. Pick the nuclei that should have permanent magnetic moments: ^7_4Be, ^8_4Be, $^{15}_7N$, $^{20}_9F$, $^{20}_{10}Ne$, ^2_1H.

8. $t_{1/2}$. 2.000 picogram (pg) of ^{33}P decays by $_{-1}^{\ 0}\beta$ emission to 0.250 pg in 75.9 days. Find the half-life of ^{33}P.

9. Dating. A sample of bristlecone pine wood of age 7000 ± 100 years, known by counting growth rings, possesses an activity of 6.6 d/minute-g of carbon. Calculate the age of the wood sample from radiochemical evidence. Does this finding have any bearing on the assumption that the rate of production of $^{14}_6C^*$ in ancient times was the same as it is today?

10. Nuclear chemistry. Pick the true statements: (a) Nuclei have a fixed (rest) mass, charge, spin, and magnetic moment. (b) In principle, it is impossible to construct a periodic table for nuclei. (c) In reactions between nuclei as between molecules, the slow step is generally the fusion of the nuclei followed by a rapid dissociation process. (d) Nuclear "combustion" requires less energy (is easier) to start than molecular combustion. (e) Fractional charge units such as one-third electronic charge have been definitely discovered in nature. (f) In a molecule, the nucleus of one atom "knows" nothing of the nuclear forces exerted by the nucleus of the atom next to it. (g) It should be possible for a proton to disintegrate into 2 positrons and an electron, thus conserving electronic charge. (h) Two electrons, or 2 protons, or 2 neutrons in the same atom cannot have identical quantum numbers.

11. Radioactivity. *Scientific American*, April 1903: "Professor Curie has announced to the French Academy of Sciences that radium possesses the extraordinary property of continuously emitting heat without combustion, without chemical change of any kind, and without change in its molecular structure. Radium, he states, maintains its own temperature at a point 1.5°C above the surrounding atmosphere. Despite this constant activity, the salt apparently remains just as potent as it was at the beginning." Would you, at the present time, accept this as a correct statement? Justify your position.

12. ΔE. Calculate Δm, the difference in mass between the final and initial nuclei in grams per mole for the emission of a γ ray,

$$^{19}_{8}O \text{ (excited state)} \longrightarrow {}^{19}_{8}O \text{ (ground state)} + \gamma \ (1.06 \times 10^8 \text{ kcal/mole})$$

13. Nuclear structure. Draw energy-level diagram(s) to illustrate the statement: Electromagnetic radiation emitted by nuclei, atoms, or molecules in reverting to lower energy states is packaged in quanta.

14. Nuclear structure. Rutherford's experiments (page 78) show that nuclei repel positively charged particles, yet it is accepted almost as fact that nuclei contain protons. In the absence of experimental information, would you predict that such nuclei should fly apart?

15. Nuclear reaction. Pick the nuclear reaction most likely to occur:

(a) $^{35}_{15}P \rightarrow {}^{0}_{+1}\beta + {}^{35}_{14}Si$
(b) $^{38}_{18}Ar \rightarrow {}^{0}_{+1}\beta + {}^{38}_{17}Cl$
(c) $^{41}_{21}Sc \rightarrow {}^{0}_{+1}\beta + {}^{41}_{20}Ca$

16. ΔE. (a) Show that the energy of a $_{-1}^{0}\beta$-decay nuclear reaction, for example,

$$^{12}_{5}B \longrightarrow {}^{12}_{6}C + {}^{0}_{-1}\beta$$

expressed in grams per mole is equal to the difference between the atomic masses of $^{12}_{6}C$ and $^{12}_{5}B$. (b) Show that the energy of a $_{+1}^{0}\beta$-decay nuclear reaction, for example,

$$^{10}_{6}C \longrightarrow {}^{10}_{5}B + {}^{0}_{+1}\beta$$

expressed in grams per mole is equal to the difference between the atomic masses of $^{10}_{5}B$ and $^{10}_{6}C$ plus the mass of 2 electrons.

17. Nuclear energy. Calculate (a) ΔE for the reaction ${}^6_3\text{Li} + {}^2_1\text{H} \rightarrow 2{}^4_2\text{He}$; (b) the heat evolved in kilocalories when 60.00 g ${}^6\text{Li}$ reacts with ${}^2\text{H}$ as in (a). Atomic masses: ${}^6\text{Li} = 6.0151$, ${}^2\text{H} = 2.0141$, ${}^4\text{He} = 4.0026$. (c) A suggestion for harnessing fusion reactions involves a two-step reaction: ${}^2_1\text{H} + {}^3_1\text{H} \rightarrow {}^4_2\text{He} + {}^1_0 n$ and ${}^6_3\text{Li} + {}^1_0 n \rightarrow {}^4_2\text{He} + {}^3_1\text{H}$. (*i*) What is the overall reaction? (*ii*) This reaction is referred to as "catalytic" burning of Li. What substance acts as the catalyst?

18. Protection. Alpha particles can be stopped by paper or by the outer skin layers. Does it follow that an α emitter, internally located, presents no hazard to an animal?

19. Fission. Let us assume that the neutrons produced by ${}^{235}\text{U}$ fission travel an average distance of 5 cm before being captured by ${}^{235}\text{U}$. The density of ${}^{235}\text{U}$ is 19 g/ml. Will a sphere of pure ${}^{235}\text{U}$ with a mass of (a) 10 g, (b) 5 kg, explode on exposure to a stray neutron? Explain your answers.

20. Fission. Is the fission reaction

$${}^{232}_{90}\text{Th} \longrightarrow {}^{175}_{71}\text{Lu} + {}^{41}_{19}\text{K} + 16{}^1_0 n$$

endothermic or exothermic? Atomic masses: ${}^{41}\text{K} = 40.9618$, ${}^1_0 n = 1.00867$, ${}^{175}\text{Lu} = 174.9408$, ${}^{232}\text{Th} = 232.0387$.

21. Rate. The half-life of ${}^{35}\text{S}$ is 87.2 days and its rate constant is 7.95×10^{-3}/day. If a sample containing ${}^{35}\text{S}$ has an activity of 10^5 d/minute, what is the number of atoms, the number of moles, and the number of picograms (Appendix I.3) of ${}^{35}\text{S}$ in the sample?

22. Dating. Will the production of ${}^{14}\text{C}^*$ in the atmosphere by other than natural means affect radiocarbon dating (a) of existing relics; (b) of matter now living that may become future relics? If so, will the error in the age be positive or negative?

23. Dating. The overall equation for the decomposition of ${}^{238}\text{U}$ $(t_{1/2} = 4.50 \times 10^9$ years$)$ is

$${}^{238}_{92}\text{U} \longrightarrow {}^{206}_{82}\text{Pb} + 8\alpha + 6_{-1}^{\ 0}\beta$$

In a rock from the mid-Atlantic rise, the atomic ratio ${}^{206}\text{Pb}/{}^{238}\text{U}$ (corrected for ${}^{206}\text{Pb}$ not generated by ${}^{238}\text{U}$ in the rock) is 1.128. (a) For every 1 mole of ${}^{238}\text{U}$ now present in the rock, how many moles were present when the rock was formed? (b) Calculate the age of the rock. (c) See Fig. 25.5 (page 538). Does the leakage of radon from uranium minerals introduce a positive or a negative error in the age of the mineral?

24. Dating. (a) The activity of ${}^{14}_6\text{C}$ in the linen wrappings of the Book of Isaiah of the Dead Sea Scrolls, found in 1947, is about 11 dpm. Find the approximate age of the Book.

25. Dating. Present instrumentation detects a minimum of 4.8×10^3 atoms of tritium per mole of ${}^1\text{H}$. Ground water contains 4.8×10^6 tritium atoms per mole of ${}^1\text{H}$; $t_{1/2}$ for ${}^3\text{H} = 12.3$ years. Calculate the maximum age of samples that can be dated by the tritium method. (The supply of ${}^3\text{H}$ in the atmosphere, like that of ${}^{14}\text{C}$, is maintained by cosmic-ray bombardment: ${}^{14}_7\text{N} + {}^1_0 n \rightarrow {}^{12}_6\text{C} + {}^3_1\text{H}$.)

26. Pollution. Neutron activation is used for the quantitative determination of many elements present in minute amounts in environmental samples. After exposure to neutrons a standard arsenic sample ($^{75}_{33}\text{As} + ^1_0n \rightarrow ^{76}_{33}\text{As} \rightarrow ^{76}_{32}\text{Ge} + ^0_{+1}\beta$) has an activity of 5.0×10^7 counts/mg of arsenic. A 1.0-g sample exposed to the same neutron source for the same period and counted in an identical manner acquires an activity of 20 counts. Calculate the grams of As per gram of sample and the grams of As per 10^6 grams of sample.

27. Solvated electrons. Electrons produced by ionizing radiation are delocalized in a molecular orbital over a number of solvent molecules (e.g., water, ammonia). (a) Label the conjugate acid–base pairs in the reaction $e^-(aq) + \text{H}_3\text{O}^+ \leftrightharpoons \text{H}_2\text{O} + \text{H}$. (b) The rate equation for the reaction $e^-(aq) + \text{CH}_3\text{I}(aq) \rightarrow \text{CH}_3(aq) + \text{I}^-(aq)$ is

$$\text{rate} = 1.7 \times 10^{10} \; \frac{\text{liter}}{\text{mole sec}} \; [\text{CH}_3\text{I}][e^-(aq)]$$

What is the CH_3I concentration in an aqueous solution when the measured rate is 6.6×10^{-6} mole/liter-second and $[e^-(aq)] = 2.0 \times 10^{-8}$ mole/liter?

26

POLYMERS AND BIOCHEMICALS

In Chapter 1 "molecules" were described as electrically neutral individual particles of matter whose atoms are linked by chemical bonds. The idea that matter is composed of molecules became quite useful in the study of gases during the eighteenth and nineteenth centuries. Gases that lend themselves to experimental study rarely exceed about 300 in molecular weight, and their molecules are rarely larger than about 10 Å in diameter. Substances whose molecules do not exceed these limits can generally be characterized by their vapor pressures or by their melting, boiling, or sublimation temperatures. The molecular weights of these substances can readily be determined from their vapor densities or from the colligative properties of their solutions. Moreover, such measurements can be made with laboratory equipment that has been available at least since the beginning of modern chemistry. As a result of this happy combination of concept and method, the first 150 years of modern chemistry were largely directed to the study of materials of low molecular weight. During their investigations, however, chemists sometimes encountered materials that did not respond to studies of these kinds, materials that decomposed instead of melting or vaporizing, materials that were gummy, gluey, waxy, or resinous, materials that were insoluble in most solvents, and that, when they did dissolve, yielded solutions whose colligative properties were too small to measure. Chemists found these materials so difficult to work with that they usually threw them into the trash can. It gradually came to be recognized, however, that some naturally occurring substances of unquestioned importance to man, like cellulose, rubber, and protein, resembled the sticky residues that plagued experimenters. During the latter part of the nineteenth century, chemists began to realize that what these classes of substances had in common was the fact that their molecules were very large. It was evident that entirely new methods of investigation would have to be developed to study such materials.

Substances of high molecular weight can often be characterized by their decomposition products. Thus, if natural rubber is heated, the hydrocarbon isoprene, C_5H_8, distills off. If starch is chewed, the sweet taste of glucose is detected. If egg albumin (a protein) is boiled in dilute sulfuric acid, the amino acids leucine, alanine, serine, glutamic acid, methionine, and some 13 others are produced (Table 26.3, page 564). It is reasonable to assume that these fragments detected in the decomposition of large molecules are indeed their units of structure. Thus, starch is called a POLYSACCHARIDE (many sugar units) and, in general, materials whose molecules are made up of repetitions of individual units are called POLYMERS (Greek, "many parts"). The unit substance of which a polymer is made is called a MONOMER. Molecules composed of at least two different kinds of monomeric units are called COPOLYMERS. When the number of units in a polymer is small and specifically known, it is often specified by use of the appropriate Greek prefix (DImer, TRImer, and so on). When polymerization is so extensive that molecular weights reach the range of 10^4 to 10^6 g/mole, the substance is called a HIGH POLYMER, and its constituent particles are called MACROMOLECULES.

THE DECOMPOSITION OF NATURAL HIGH POLYMERS; REPEATING UNITS / 26.2

The structure of isoprene (2-methyl-1,3-butadiene) produced by the destructive distillation of natural rubber is

isoprene

Natural rubber itself is a long chainlike structure of the type

(For a more accurate rendition of bond angles, see Table 26.1.) The formula of this macromolecule can be written by repetition of any one

Table 26.1
Some Natural Polymers and Their Decomposition Products[a]

Polymer and source	Formula	Decomposition product(s)

Cellulose
(primary
structural material
of plants).

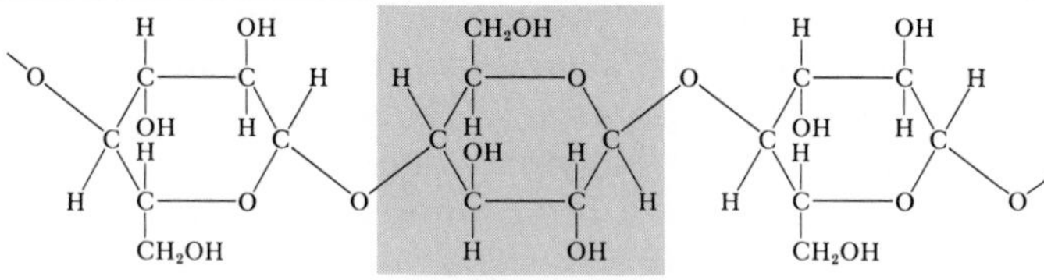

H_2O (+acid) →

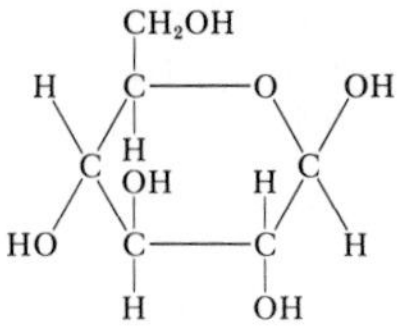

Glucose

Chitin
(protective
shell of insects,
crustaceans, etc)

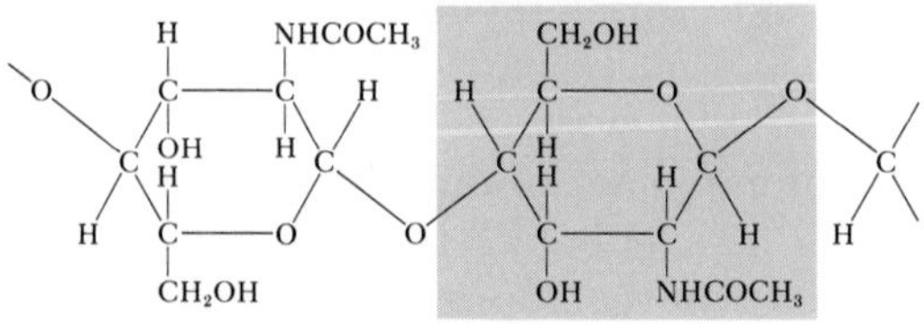

H_2O (+acid) →

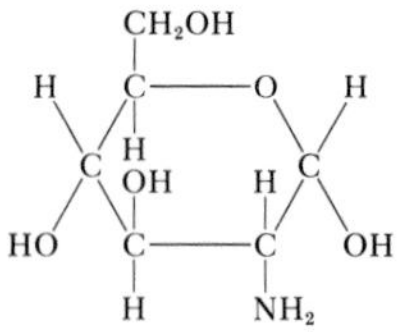

Aminoglucose

Natural rubber

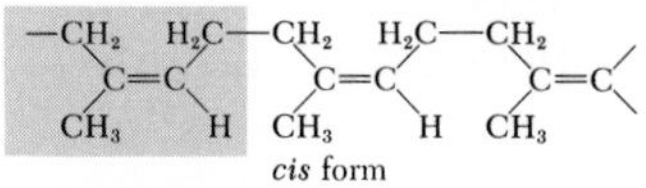

$-CH_2$ H_2C-CH_2 H_2C-CH_2
C=C   C=C   C=C
CH_3 H CH_3 H CH_3
cis form

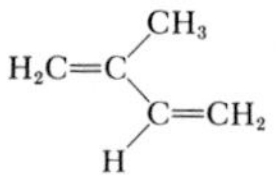

heat →

Gutta-percha
(related to natural
rubber)

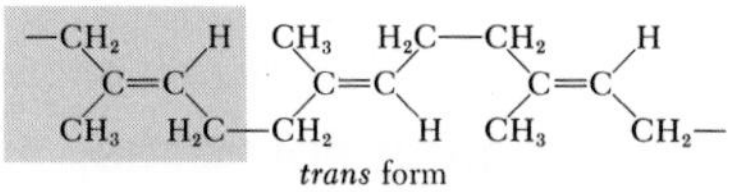

$-CH_2$ H CH_3 H_2C-CH_2 H
C=C   C=C   C=C
CH_3 H_2C-CH_2 H CH_3 CH_2-
trans form

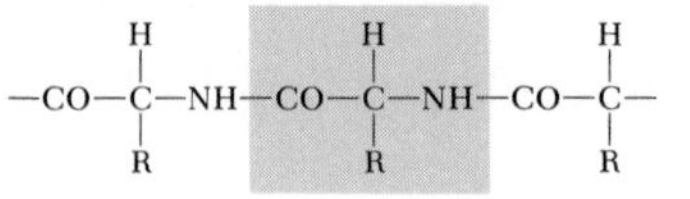

heat →

$H_2C=C$
   CH_3
   $C=CH_2$
   H
Isoprene

Silk
(a protein)

$-CO-\overset{H}{\underset{R}{C}}-NH-CO-\overset{H}{\underset{R}{C}}-NH-CO-\overset{H}{\underset{R}{C}}-$

H_2O (+acid) →

Amino acid

R—H, CH_3, CH_2OH, or any of about 12 other groupings.

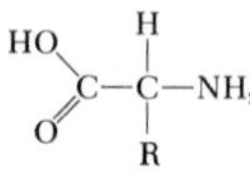

[a] The shaded portion is the repeating unit.

of the portions *A*, *B*, *C*, or *D*. However, only unit *A* has the same atomic skeleton as isoprene, the observed decomposition product. Furthermore, the treatment of isoprene with appropriate reagents yields a rubbery solid. The repeating unit of natural rubber is, therefore, unit *A*. In general, the repeating unit is identified by study of the decomposition or synthesis of the polymer.

Table 26.1 shows the formulas of some natural organic polymers and of their decomposition products. The identification of the decomposition product, however, is far from a full description of the polymer, for the following reasons:

The mode of linkage of the monomer will influence the properties of the macromolecule. Note (Table 26.1) that in natural rubber the isoprene units are linked in the *cis* form; in gutta-percha the linkage is *trans*.

The sequence and extent of monomeric linkages also are important determinants of polymer structure. Cellulose and starch are both glucose polymers, but the rigid cellulose consists of linear macromolecules of about 3500 glucose units each, whereas the more pliable starch is a lower polymer (about 500 glucose units per linear chain of cornstarch) with different structural linkage of the monomeric groups and more extensive chain branching.

The shape of the macromolecules and their orientation in space with respect to each other are also critical factors affecting polymer structure. Raw and cooked egg white (albumin) differ in the shape of their constituent protein molecules.

Examination of Table 26.1 shows two kinds of relationships between monomer and polymer. In rubber and gutta-percha, the repeating unit of the macromolecule is identical with the decomposition product in atomic composition and skeletal structure. If the reaction could be reversed, a simple ADDITION of isoprene units to each other would form the polymer.

In cellulose, chitin, and silk, however, decomposition is accompanied by the uptake of water. A reversal of the process, if it could be effected, would require the CONDENSATION* of monomers with elimination of water.

Polymers of the first type are called ADDITION POLYMERS; those of the second type are called CONDENSATION POLYMERS.

THE PRODUCTION OF SYNTHETIC HIGH POLYMERS BY CONDENSATION / 26.3

The controlled synthesis of high polymers by condensation methods reached a high degree of industrial success as early as the 1930's, largely as a result of the pioneer work of Wallace H. Carothers. A condensation polymer must be made from monomers that contain more than one functional group so as to enable intermolecular reactions to proceed continuously.

To illustrate this point, consider the reaction of an amine with a carboxylic acid:

$$CH_3-CH_2-NH_2 + \quad \overset{\displaystyle O}{\underset{HO}{\diagup}}\!\!\diagdown C-CH_2-CH_3 \longrightarrow CH_3-CH_2-\underset{\underset{H}{|}}{N}-\underset{\underset{O}{\|}}{C}-CH_2-CH_3 + H_2O$$

| an amine | a carboxylic acid | | an amide |

This condensation does not lead to a polymer because the product has no NH_2 or COOH group with which to continue the reaction. However, the reaction of a diamine with a diacid leaves two reactive groups, and

---

* Condensation is the union of two or more molecules with elimination of a smaller unit like H_2O or NH_3.

the condensation can continue indefinitely to form a polymer:

$$H_2N-CH_2-CH_2-NH_2 + \quad \underset{HO}{\overset{O}{\underset{}{C}}}-CH_2-\overset{O}{C}{\overset{}{\underset{}{OH}}} \quad \xrightarrow{-H_2O}$$

a diamine   a dicarboxylic acid

$$H_2N-CH_2-CH_2-\underset{H}{\overset{}{N}}-\underset{O}{\overset{}{C}}-CH_2-C\overset{O}{\underset{OH}{}} \quad \xrightarrow[-H_2O]{H_2N-CH_2-CH_2-NH_2}$$

$$H_2N-CH_2-CH_2-\underset{H}{N}-\underset{O}{C}-CH_2-\underset{O}{C}-\underset{H}{N}-CH_2-CH_2-NH_2$$

condensations continue
and polymer is formed
$$\xrightarrow{-nH_2O} \left(\underset{H}{N}-CH_2-CH_2-\underset{H}{N}-\underset{O}{C}-CH_2-\underset{O}{C} \right)_n$$

repeating unit of polymer

Carothers succeeded in producing condensation polymers by heating the salts of diamines with dicarboxylic acids at 200° to 250°C, with elimination of water. These substances are called NYLONS. The popular hosiery polymer is Nylon-66, the copolymer of a diamine and a dicarboxylic acid each having 6 carbon atoms per molecule. (The numeral 66 indicates that each monomer molecule has 6 carbon atoms.)

$$H_2N-(CH_2)_6-NH_2 + HOOC-(CH_2)_4-COOH \xrightarrow{-H_2O}$$

$$\cdots -CO-NH-(CH_2)_6-NH-CO-(CH_2)_4-CO-NH-\cdots$$

the shaded portion is the repeating
unit of Nylon-66

A Nylon can also be made by the polymerization of a single amino acid, such as 6-aminohexanoic acid, which is polymerized to the material called Nylon-6:

$$H_2N-\underset{H}{\overset{H}{C}}-\underset{H}{\overset{H}{C}}-\underset{H}{\overset{H}{C}}-\underset{H}{\overset{H}{C}}-\underset{H}{\overset{H}{C}}-COOH \xrightarrow{polymerization} \cdots NH-CO-(CH_2)_5-NH-CO-\cdots$$

6-aminohexanoic acid   polymer (Nylon-6)
(shaded portion in the repeating unit)

The single numeral 6 indicates that there is but one monomer and that it has 6 carbon atoms per molecule.

In the case of addition polymerization it is necessary for the addition product of two monomers to be capable of undergoing further addition, and for this process to continue to macromolecule formation. Otherwise, of course, the addition stops after a single step. Such termination occurs, for example, after the chlorination of ethylene; the product is saturated and therefore cannot undergo further addition or chain lengthening (page 503). It is found that addition polymerizations proceed by a path involving the initial formation of some reactive species, such as free radicals or ions, and by the addition of the reactive species to another molecule with the *regeneration of the reactive feature*. In this way the addition reactions may proceed continuously. Pioneer studies in this field were carried out by Hermann Staudinger in the 1920's. An example of free radical addition polymerization (the curved arrows represent electron shifts that occur during reaction) is shown below. The hydroxyl radical reacts at the carbon-to-carbon double bond to produce a new free radical. The unshared electron in the dimer is associated with the carbon that bears the CN group:

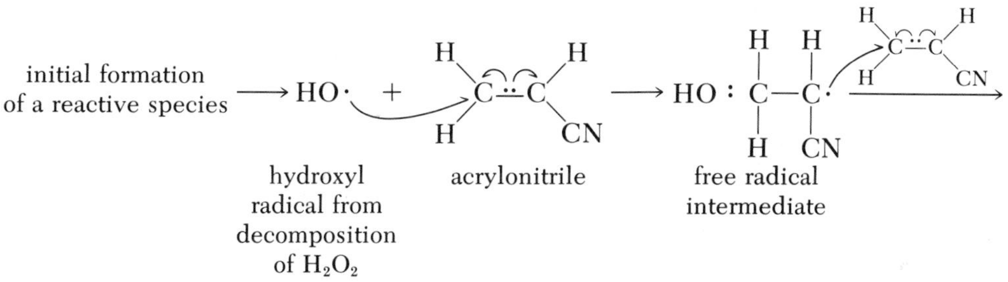

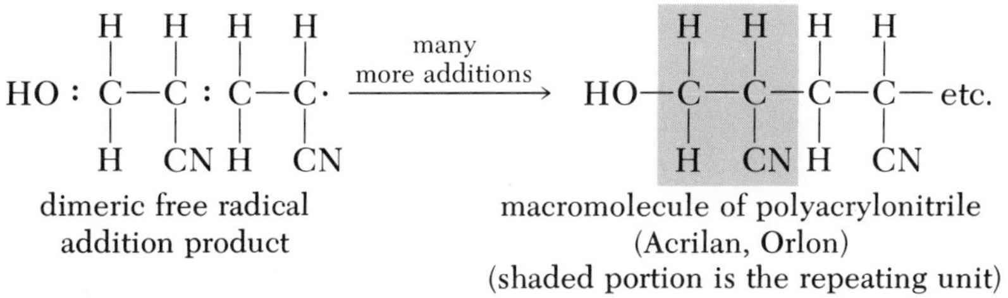

initial formation of a reactive species

hydroxyl radical from decomposition of H_2O_2

acrylonitrile

free radical intermediate

dimeric free radical addition product

macromolecule of polyacrylonitrile
(Acrilan, Orlon)
(shaded portion is the repeating unit)

Recall that a dash is equivalent to a pair of dots in a structural formula. Therefore, bonds may be represented as:

single   C—C   or   C:C
double   C=C   or   C⋯C   or   C::C

In these equations, curved arrows represent bonds being made or broken.

In a polymer made up of long, chainlike molecules, the side groups may be oriented in either an orderly or a random pattern with respect to the chain. There are two types of orderly arrangements: (a) all the side groups may lie on the same side of the chain; (b) the side groups may alternate with respect to the chain. These patterns are illustrated for polypropylene in Fig. 26.1. The structural difference between natural rubber (*cis*) and gutta-percha (*trans*), shown in Table 26.1, is a consequence of restricted rotation around double bonds.

We have seen that molecules must provide two reaction sites to produce a condensation polymer. If more than two sites per molecule are available, then the chain may branch and a three-dimensional polymeric network structure may be formed. Chains connected to each other by occasional bridges are called **CROSS-LINKED POLYMERS**.

An example of a branched modification of a linear polymer is amylopectin, a type of starch. The difference due to branching is shown in Fig. 26.2.

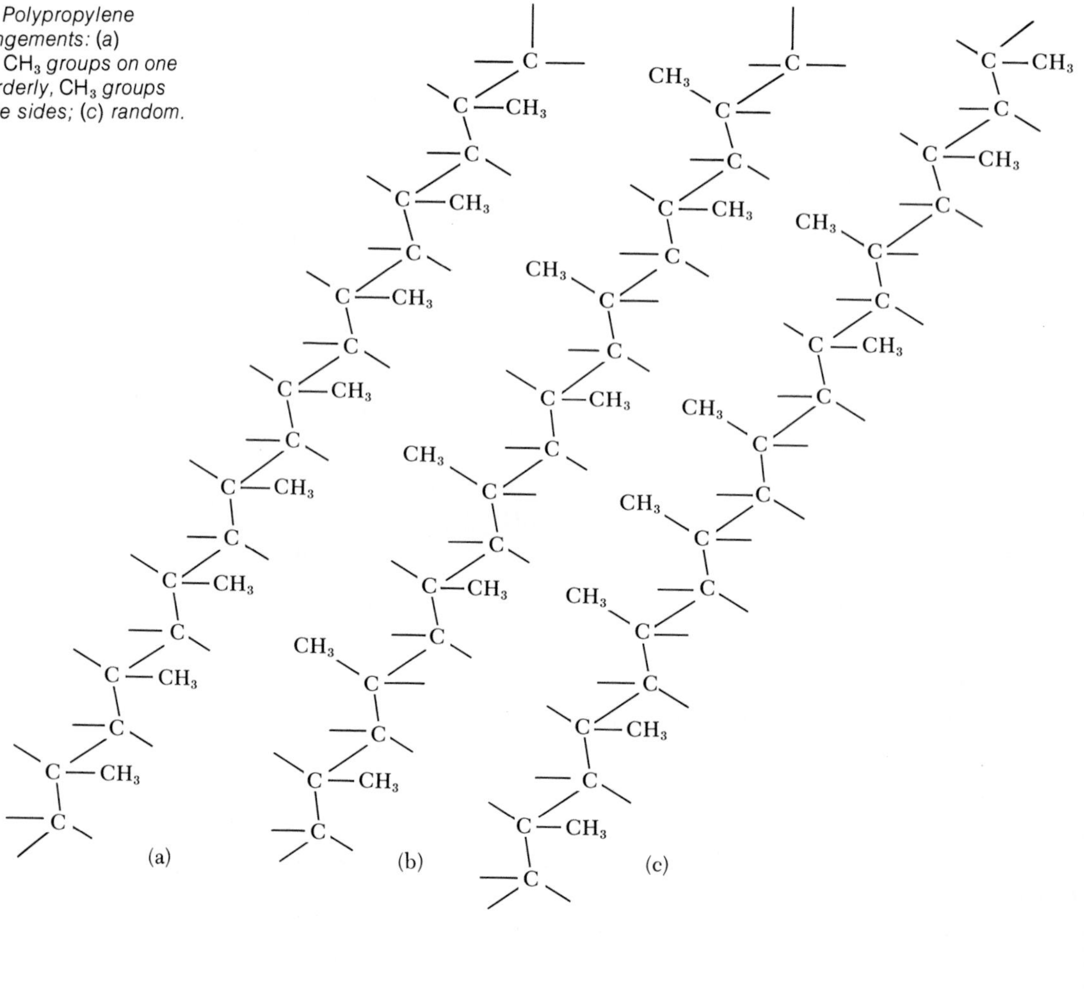

Fig. 26.1. *Polypropylene chain arrangements: (a) orderly, all CH$_3$ groups on one side; (b) orderly, CH$_3$ groups on alternate sides; (c) random.*

(a) (b) (c)

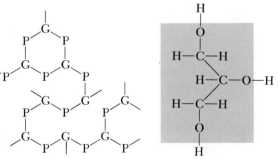

(a)

(b)

Fig. 26.2. (a) Straight-chain amylose starch (structural skeleton only); (b) branched-chain amylopectin starch (structural skeleton only).

A typical cross-linking bridge in protein is the diamino acid, cystine,

$$\underset{\displaystyle \text{COOH}}{\overset{\displaystyle \text{NH}_2}{\text{CH}}}-\text{CH}_2-\text{S}-\text{S}-\text{CH}_2-\underset{\displaystyle \text{COOH}}{\overset{\displaystyle \text{NH}_2}{\text{CH}}} \qquad \text{cystine}$$

which has four sites for reaction (shown in boldface) and can serve as a ladder rung between protein chains.

A good example of difference in molecular requirements for linear and network polymers is the formation of polyesters from dihydroxy or trihydroxy alcohols. As was shown on page 554 for diamines, reaction of a dihydroxy alcohol with a diacid gives a linear copolymer. If a trihydroxy alcohol is used instead, a network copolymer is formed. An example is the class of **ALKYD** resins (Fig. 26.3), produced from glycerol and phthalic acid, and used as ingredients of paints and varnishes.

Fig. 26.3. Network copolymer of glycerol and phthalic acid (alkyd resin). Water is eliminated in the condensation.

Copolymer

P represents shaded portion of phthalic acid structure

G represents shaded portion of glycerol structure

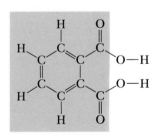

Glycerol

Phthalic acid

Polymeric substances display wide varieties of physical and chemical properties; this is to be expected from the many possible kinds of composition and arrangement. The broad range of fibrous, adhesive, plastic, filmy, foamy, rubbery materials attests to this versatility. What generalizations can be made regarding the relationships between properties and structure? Although the problems are complicated, considerable advances have been made in recent years.

The chemical reactivity of a polymer is, in large measure, the reactivity of its molecular components. Natural rubber, for example, undergoes deterioration when ozone attacks the double bonds of the polymer chain; a saturated hydrocarbon chain, like polyethylene, is resistant to such attack. Celluloses offer their hydroxyl groups to a variety of reagents and thus make it possible to modify properties. Reaction with nitric acid produces NITROCELLULOSE, from which propellant (guncotton) and plastic (Celluloid) products are formed. Reaction with acetic acid produces CELLULOSE ACETATE, which can be fabricated into films, sheets, and other useful forms:

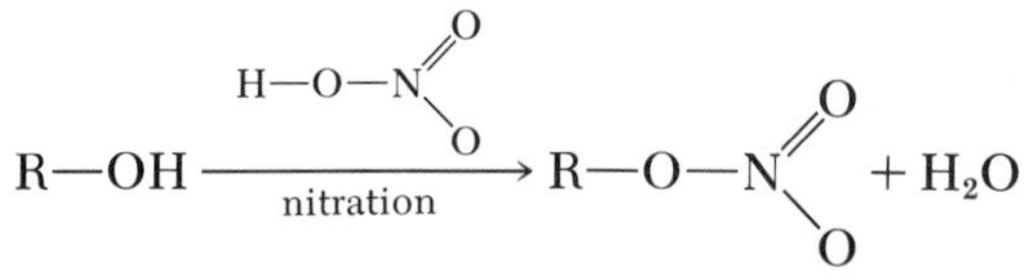

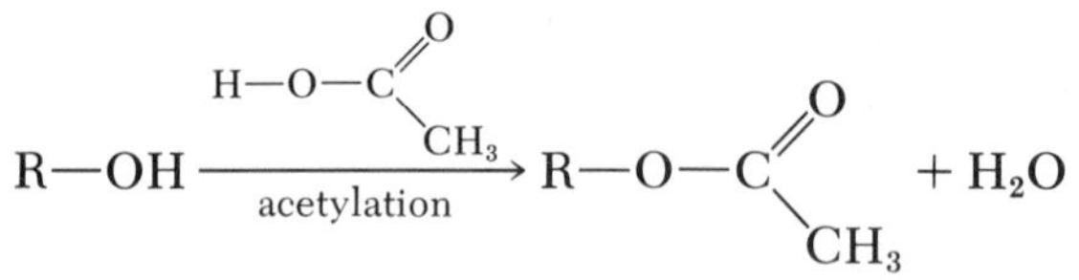

More drastic chemical differences yield wider differences in chemical properties. Especially notable is the family of silicone polymers, in which the macromolecular chains contain —Si—O— linkages,

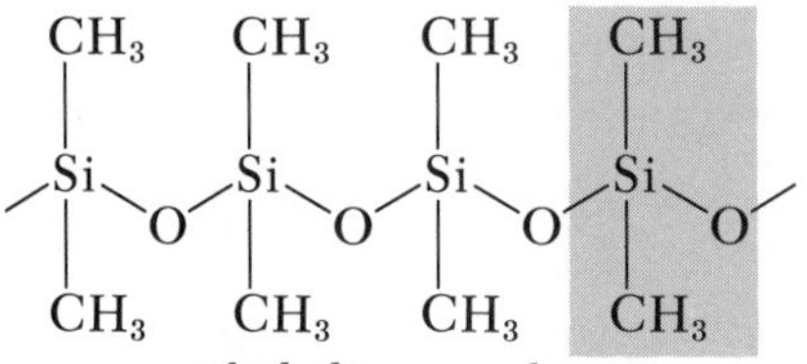

methylsilicone polymer
(the shaded portion is the repeating unit)

The great thermal stability of the O—Si bond, owing in part to the p–d π bonding* of O to Si, makes it possible to use silicone products at high temperatures. The hydrocarbon side chains contribute oily or

---

* Reminder: this refers to π bonding between p and d orbitals.

lubricating properties. There have been, of course, attempts to extend the range of "inorganic" polymers by using other varieties of linkage for the chain backbones. Potential candidates include boron–carbon, boron–oxygen, arsenic–oxygen, and beryllium–oxygen. The problems are difficult, in part because our understanding of inorganic linkages, which may involve d-orbital π bonding and other less well-recognized interactions, lags behind our knowledge of the simpler σ bonds and σ–π multiple bonds typical of common organic molecules. Another difficulty is the annoying (to the polymer chemist) tendency of inorganic systems to cyclize in units of relatively low molecular weight. The tendency to form multiple bonds decreases with increasing atomic number; as a result it becomes more difficult to use the addition-polymerization techniques that work so well with organic monomers like ethylene. Despite such difficulties, however, many inorganic polymers have been prepared (Fig. 26.4).

The properties of a polymer are also determined by the form and arrangement of its macromolecules. The critical factors are the molecular weight (which is a function of the degree of polymerization), the extent of branching, cross-linking, or network structuring, the steric disposition of the monomeric units, and the degree and kind of crystallinity of the macromolecules. Certainly there is room enough for variation even without alteration of chemical functional groups. Some of these relationships are shown in Table 26.2.

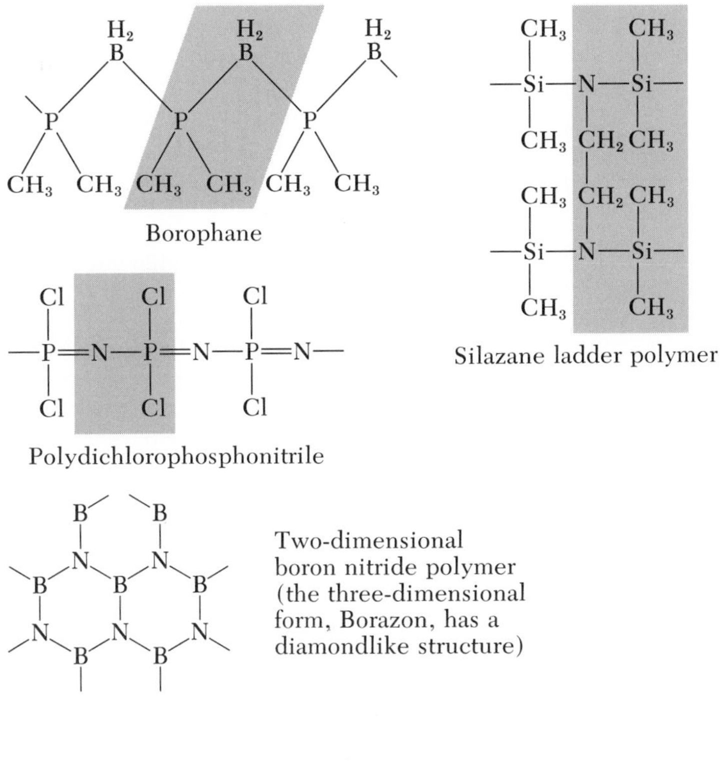

Borophane

Polydichlorophosphonitrile

Silazane ladder polymer

Two-dimensional boron nitride polymer (the three-dimensional form, Borazon, has a diamondlike structure)

Fig. 26.4. *Some inorganic polymers. The shaded portions are the repeating units.*

Table 26.2
Properties and Molecular Makeup of Polymers

Physical nature of polymer	Molecular requirements

Hard
(difficult to scratch)
	High molecular weight
	High crystallinity[a]
	Cross-linking or network structure
	   Example: phenol–formaldehyde copolymer
	     (Bakelite)

Strong
(cannot easily be pulled
apart)
	High molecular weight
	Cross-linking or network structure
	Partly crystalline—crystallites embedded in
	  amorphous matrix that acts like a cement
	   Example: polymethylmethacrylate (Lucite,
	     Plexiglas) reinforced with glass fiber

Fibrous	High molecular weight
	Linear macromolecules
	Long parallel arrangements of crystalline and
	  amorphous regions
	   Example: polyacrylonitrile (Acrilan, Orlon)

Leathery	High molecular weight
	Linear macromolecules with slight degree of
	  cross linking; fragments of the chains are
	  free to move under stress
	Low crystallinity
	   Example: vinyl chloride–vinyl acetate co-
	     polymer (vinyl floor covering)

Rubbery	Linear macromolecules with little cross link-
	  ing; entire chains are free to move under
	  stress, but rotation is restricted by C=C
	  double bonds
	High molecular weight
	Low crystallinity (but crystallinity increases
	  with elongation)
	   Example: polybutadiene (Buna rubber)

Soft, waxy	Low molecular weight ($< 10,000$) Low crystallinity *Example:* polyvinyl acetate chewing gum

[a] "Crystallinity" is used here in the sense of "orderly arrangement." Polymer structures range from amorphous (random, or noncrystalline) to highly ordered ("crystalline").

The strength and thermal stability of three-dimensional network polymers are not attributed to stronger chemical bonds, nor even to a greater number of bonds. Instead, these properties result from the fact that the network polymer is better able than a linear polymer to preserve its structural integrity even after some bonds are broken. This advantage derives from the differences in geometrical structure between the two types of polymers, as shown below:

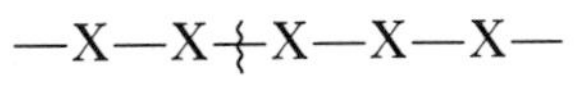

$$-X-X \;{+}\; X-X-X-$$

linear polymer: breaking a bond severs the chain.

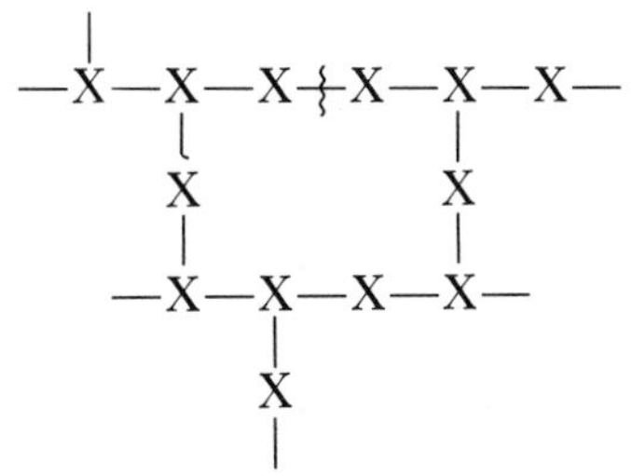

$$
\begin{array}{c}
\mid\\
-X-X-X\;{+}\;X-X-X-\\
\mid\qquad\qquad\mid\\
X\qquad\qquad X\\
\mid\qquad\qquad\mid\\
-X-X-X-X-\\
\mid\\
X\\
\mid
\end{array}
$$

network polymer: breaking a bond does not sever the chain.

Shifts in properties also accompany the changes in molecular arrangements produced by a rise in temperature. A polymer may progress, on heating, from a rigid glassy state, through a partly flexible leathery condition, to a rubbery condition, and finally to a flowing viscous liquid. Mechanical deformation also changes the properties of a polymer. Perhaps the most striking instance is the stretching of rubber. In the unstretched form, rubber molecules undergo random motion; on stretching this motion is restricted, the entropy is reduced, and the molecules assume an orderly, linear arrangement. Stretching is thus analogous to freezing, an exothermic process. The release of energy is familiar to anyone who has stretched a wide rubber band, touched it immediately to his lips, and felt its sudden warmth.

ION-EXCHANGE RESINS / 26.7

An **ION-EXCHANGE RESIN** is a three-dimensional polymer containing electrically charged sites of one particular charge sign, and mobile ions of the other charge sign. In a **CATION EXCHANGER** the immobile molecular network is anionic; the mobile ions are cations. In an **ANION EXCHANGER** the charge signs are reversed. Figure 26.5 is a schematic representation of a cation exchanger. Ion exchange can be used, as the name implies, to substitute one ion for another in a solution. For example, if water containing Ca^{2+} ions flows through a column of cationic exchanger containing Na^+ ions, displacement occurs, and Na^+ ions appear in the eluent. This particular process is called **WATER SOFTENING**, because it prevents any subsequent precipitation of an insoluble calcium soap (page 495). The action is an equilibrium that may be represented as

$$2[(Na^+)_x \, resin^{x-}] + xCa^{2+} \rightleftharpoons [(Ca^{2+})_x(resin^{x-})_2] + 2xNa^+$$

A cation-exchange resin treated with acidified water becomes populated with $H^+(aq)$ ions; an anion-exchange resin in basic solution takes up $OH^-(aq)$ ions. When water containing salt impurities is treated

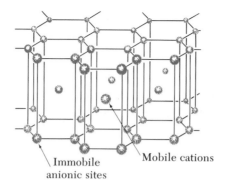

Fig. 26.5. *Cation exchanger (schematic). Some of the atoms are charged.*

Immobile anionic sites

Mobile cations

in series with two resins that have been thus loaded with H^+ and OH^-, the cation and anion impurities are removed and are replaced by H^+ and OH^-, respectively. This combination of substitutions amounts to the replacement of salts by water, and thus constitutes a **DEMINERALIZING** action, comparable to distillation. The removal of $CaCl_2$ from water may be represented as

$$2[(H^+)_x \, resin^{x-}] + 2[(OH^-)_x \, resin^{x+}] + xCa^{2+} + 2xCl^- \longrightarrow$$
$$[(Ca^{2+})_x (resin^{x-})_2] + 2[(Cl^-)_x \, resin^{x+}] + 2xH_2O$$

BIOCHEMICAL POLYMERS / 26.8

Biochemistry is the chemistry of organisms. An **ORGANISM** is anything that is alive (a "living organism"), or if not, was once alive (a "dead organism"). All organisms are made up in large part of polymers; as far as we know, polymeric material is essential to life.

Some organisms may be considered to be on the very borderline of life. A **VIRUS** (Fig. 26.6), for example, consists of particles several

Fig. 26.6. *Crystalline tobacco necrosis virus, magnification 160,000 times [J. Ultrastructure Research, Vol. 2, page 8 (1958); courtesy of Dr. R. W. G. Wyckoff].*

hundred angstrom units in length or diameter; these particles can reproduce themselves in a suitable environment but they do not ingest food or carry on any other metabolic processes. The reproductive mechanisms of these and all organisms, however, involve processes that can occur only with macromolecular materials.

In this section we consider the two types of biochemical polymers that are most universally characteristic of living things: proteins and nucleic acids. Proteins serve structural and catalytic functions in biochemical systems; nucleic acids serve functions in protein synthesis and in reproduction.

(a) PROTEINS (Greek, *proteos*, "primary") are condensation polymers of amino acids, and have molecular weights in the range of 6000 to 50,000.

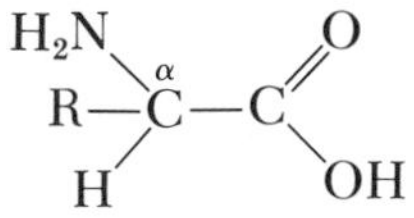

$$\underset{H}{\overset{H_2N}{\underset{|}{R-\overset{\alpha}{C}}}}-\overset{O}{\overset{\|}{C}}\begin{subarray}{l}\\OH\end{subarray}$$

General formula for an α-AMINO ACID. R *may be hydrogen, an alkyl group, or some other organic group. The* α-*carbon atom refers to the one adjacent to the* COOH *group. When all four groups attached to the* α *carbon are different, the molecule is chiral. See page 496 and Table 26.3.*

The condensation of two or more molecules of an amino acid involves the formation of PEPTIDE LINKAGES in which 1 molecule of water is eliminated between one NH_2 and one COOH group. (See also Table 26.1, page 552, showing the degradation of silk.)

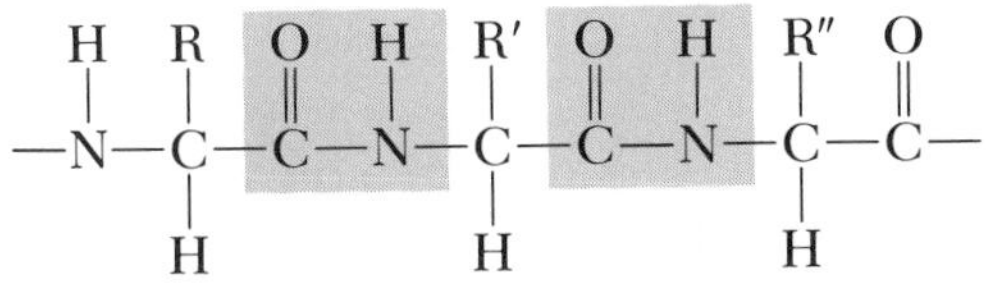

$$-\overset{H}{\underset{H}{\overset{|}{N}}}-\overset{R}{\underset{|}{\overset{|}{C}}}-\overset{O}{\overset{\|}{C}}-\overset{H}{\underset{|}{\overset{|}{N}}}-\overset{R'}{\underset{H}{\overset{|}{C}}}-\overset{O}{\overset{\|}{C}}-\overset{H}{\underset{|}{\overset{|}{N}}}-\overset{R''}{\underset{H}{\overset{|}{C}}}-\overset{O}{\overset{\|}{C}}-$$

shaded portions are peptide linkages

Practically all of the chiral amino acid portions of protein in living organisms on earth, so far as we know, have the same "left-handed" configurations; the two possibilities are shown below:

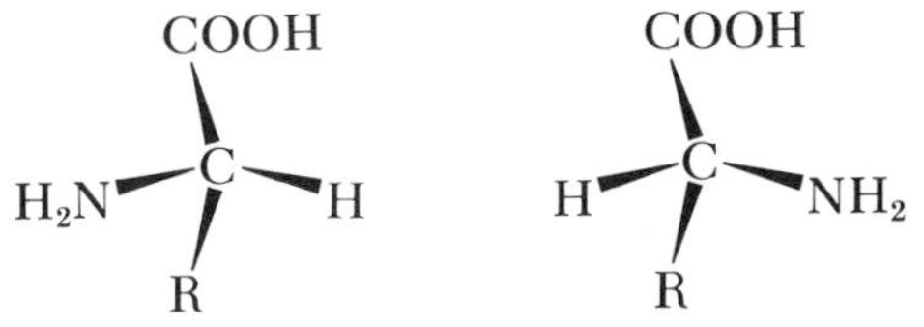

$$\underset{R}{\overset{COOH}{\underset{|}{\overset{|}{C}}}}$$ H_2N — C — H

$$\underset{R}{\overset{COOH}{\underset{|}{\overset{|}{C}}}}$$ H — C — NH_2

(left-handed)   (right-handed)

Table 26.3
Some Common Amino Acids from Proteins

Name	Abbreviation	Formula	Name	Abbreviation	Formula
Glycine	Gly	H_2NCH_2COOH	Cystine	CysS—SCys	$\underset{\underset{NH_2}{\mid}}{HOOCCHCH_2}SSCH_2\underset{\underset{NH_2}{\mid}}{CHCOOH}$
Alanine	Ala	$CH_3-\underset{\underset{NH_2}{\mid}}{CHCOOH}$	Serine	Ser	$HOCH_2\underset{\underset{NH_2}{\mid}}{CHCOOH}$
Valine	Val	$(CH_3)_2CH\underset{\underset{NH_2}{\mid}}{CHCOOH}$	Threonine	Thr	$CH_3\underset{\underset{OH}{\mid}}{CH}-\underset{\underset{NH_2}{\mid}}{CHCOOH}$
Leucine	Leu	$(CH_3)_2CHCH_2\underset{\underset{NH_2}{\mid}}{CHCOOH}$	Histidine	His	(ring structure) $HC\,\overset{N-CH}{\underset{\underset{H}{N}}{}}C-CH_2\underset{\underset{NH_2}{\mid}}{CHCOOH}$
Isoleucine	Ileu	$CH_3CH_2\underset{\underset{CH_3}{\mid}}{CH}-\underset{\underset{NH_2}{\mid}}{CHCOOH}$	Aspartic acid	Asp	$HOOCCH_2\underset{\underset{NH_2}{\mid}}{CHCOOH}$
Proline	Pro	$\overset{H_2C-CH_2}{\underset{\underset{H}{N}}{H_2C\quad CHCOOH}}$	Glutamic acid	Glu	$HOOCCH_2CH_2\underset{\underset{NH_2}{\mid}}{CHCOOH}$
Phenylalanine	Phe	(C₆H₅)—$CH_2\underset{\underset{NH_2}{\mid}}{CHCOOH}$	Lysine	Lys	$H_2NCH_2CH_2CH_2CH_2\underset{\underset{NH_2}{\mid}}{CHCOOH}$
Tyrosine	Tyr	HO—(C₆H₄)—$CH_2\underset{\underset{NH_2}{\mid}}{CHCOOH}$	Arginine	Arg	$H_2N\underset{\underset{NH}{\parallel}}{C}NHCH_2CH_2CH_2\underset{\underset{NH_2}{\mid}}{CHCOOH}$
Cysteine	CysH	$HSCH_2\underset{\underset{NH_2}{\mid}}{CHCOOH}$			

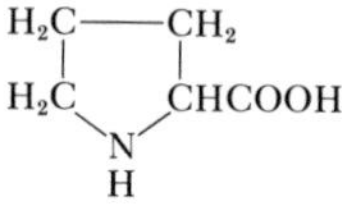

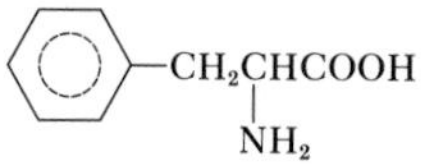

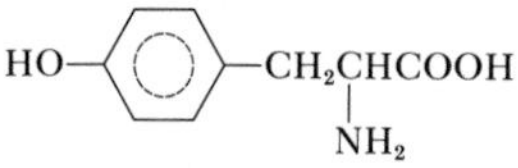

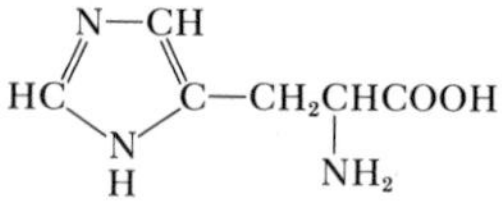

In fact, we cannot use the right-handed acids in our bodies; if they were the only type available to us, we could not survive. We do not know, however, why our amino acids are left-handed, nor what amino acid configurations (if any) may exist in extraterrestrial organisms.

A protein may be characterized according to its *primary* and *secondary* structures. The "primary structure" of a protein is that which can be depicted by the usual structural formula of the organic chemist. Such a formula shows the sequence of atomic linkages in the molecule, without regard to conformation (page 483) or to relatively weak

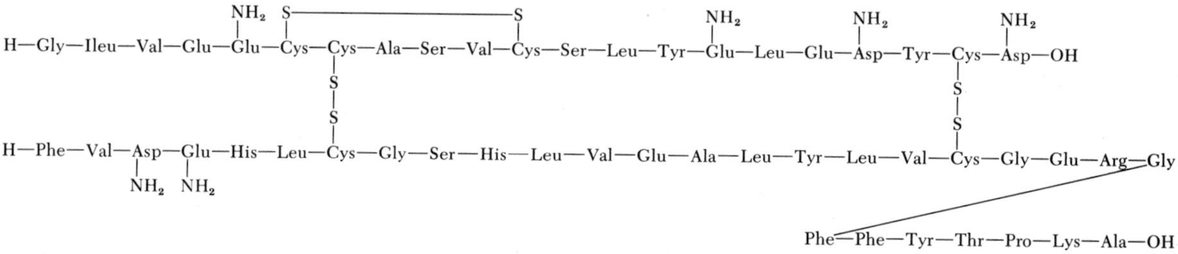

Fig. 26.7. *Amino acid sequence in beef insulin, determined in 1955 by Frederick Sanger.*

linkages such as hydrogen bonding. The first formulation of this kind was reported in 1955 by Frederick Sanger for the protein INSULIN. Figure 26.7 depicts the primary structure of insulin.

The secondary structure of a protein is determined by the spatial arrangement of the polypeptide chain. Evidence obtained mainly from X-ray diffraction patterns (Linus Pauling, 1951, and others) has shown that the chain is typically wound into a helix. The helical form is maintained by hydrogen bonds located at spaced intervals, as shown in Fig. 26.8. The entire structure is called the ALPHA-HELIX.* Other secondary structures of proteins include pleated sheets and random coils.

Proteins that are catalysts in biochemical reactions are called ENZYMES. They are very highly specific, each enzyme being capable of catalyzing only a particular reaction of a particular substance. This high specificity implies that steric (spatial) effects, in addition to interactions of functional groups, are critical in enzyme action, because there are many more possible shapes of molecules than there are different types of chemical bonds. Figure 26.9 shows a model of an enzyme-catalyzed

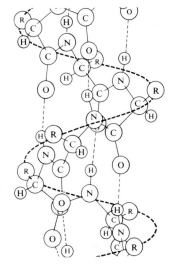

Fig. 26.8. Part of the alpha-helix of a protein molecule, showing hydrogen bonding,

$$\text{C}=\text{O}\cdots\text{H}-\text{N}$$

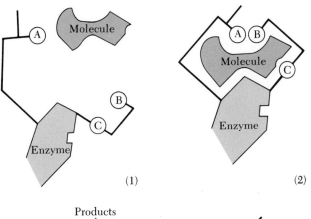

(1)

(2)

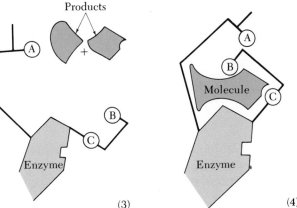

(3)

(4)

Fig. 26.9. Schematic representation of the mode of enzyme action [D. E. Koshland, Jr., Science 142, 1533 (1963)]. (1) Molecule and enzyme. (2) Molecule binds to enzyme via group C forming a complex; the binding induces proper alignment of catalytic groups A and B. (3) Reaction ensues yielding products and the original enzyme. (4) Illustrates that molecules which are either too large or too small may be bound but fail to react because of improper alignment of the catalytic groups.

* The "alpha" refers to the alpha-carbon atom, to which are attached the groups that are linked by H bonds.

decomposition of a molecule. Under the influence of the approaching molecule, the enzyme bends so as to grip the molecule in a perfect fit. A complex is formed in which chemical bonds within the molecule are weakened so that decomposition occurs more rapidly than it would without the enzyme.

(b) NUCLEIC ACIDS are substances that determine protein synthesis and cell reproduction, and are thus the genetic determinants of all living organisms. Although their name refers to the nuclei of cells, they also occur in bacterial cells (which have no nuclei) and in viruses (which have no cells).

DEOXYRIBONUCLEIC ACID (DNA) is a polymer containing up to 15,000 monomeric units. Each unit contains a phosphoric acid portion, a carbohydratelike portion (deoxyribose), and one of four possible organic bases. The four bases, all nitrogen compounds, are ADENINE, CYTOSINE, GUANINE, and THYMINE. Formulas appear in Fig. 26.10.

The condensation of the monomeric units involves elimination of water to form the bond.

shaded area represents the chain formed by elimination of H_2O molecules

DNA is a double-stranded helix which is held in this configuration mainly by hydrogen bonds between pairs of nitrogen bases. This model (Fig. 26.11) was proposed in 1953 by James D. Watson and Francis H. C. Crick, and confirmed in 1973 by an X-ray technique capable of "seeing" individual atoms.

Another nucleic acid is RIBONUCLEIC ACID (RNA), which is similar to

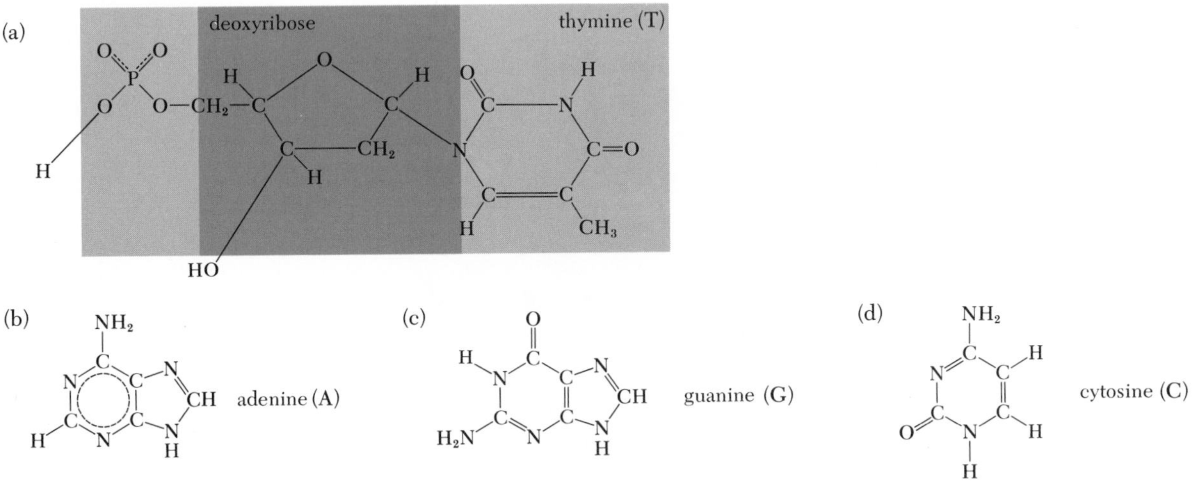

Fig. 26.10. (a) Thymine monomer. The entire colored portion is the repeating unit and the H and OH outside the colored portion are used in the linkage. (b), (c), and (d) are other nitrogen bases.

DNA except that the carbohydrate portion is ribose instead of deoxyribose. The difference between the two is an atom of oxygen.

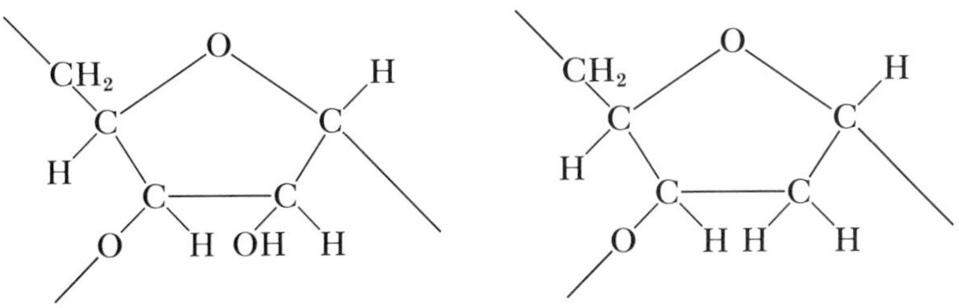

ribose portion
of RNA

deoxyribose portion
of DNA

How do nucleic acids determine cell reproduction? Each time a cell (or virus) reproduces itself, it transmits the ability to continue the reproduction. The reliability of this transmittal accounts for the continuity of species. It is therefore reasonable to assume that some sort of durable pattern is preserved or reconstructed with each duplication. A "pattern" is a physical entity whose size, shape, and makeup carry information, such as a punched card, a dress pattern, or a perforated metal plate that establishes the positions of bolt holes. In the cell nucleus, the reproduction pattern is the DNA molecule. The sequence of bases embodies the pattern, or code, for the synthesis of proteins. For brevity, let us designate the bases by their initials, A, G, C, and T, as shown in Fig. 26.10. Now if we imagine that a single DNA strand were isolated, and stretched out along a straight line, we would have a structure such as

$$-\text{sugar}-\text{PO}_4-\text{sugar}-\text{PO}_4-\text{sugar}-\text{PO}_4-\text{sugar}-\text{PO}_4-$$

A	G	C	T

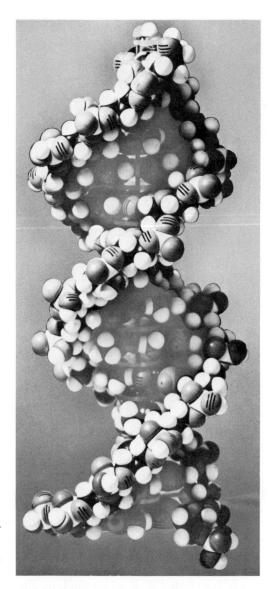

Fig. 26.11. *DNA helix (model). Courtesy of Professor M. H. F. Wilkins, Medical Research Council, Biophysics Unit, Kings College, London.*

Reading from left to right, we see that this particular stretch of DNA gives the base sequence, —AGCT—. But, as we have seen, the DNA molecules consists of two strands; these are arranged spatially so that the bases from the two separate chains are brought into very close proximity. It turns out, however, that the structures of the individual bases are such that not all combinations of pairs are allowed; in fact the only pairs that are allowed are A–T and G–C (other combinations simply won't fit in a hydrogen-bonding pattern). We can therefore extend our schematic diagram one step further:

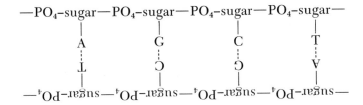

Thus, the two strands of DNA are not identical but they complement each other in a *fixed relationship of base pairing.* A DNA molecule reproduces itself when its two strands separate, and each strand then serves as the template for the formation of a new strand with the complementary bases. This pairing gives rise to two DNA molecules, each identical with the parent. DNA is used by the cell ultimately to direct the synthesis of proteins. It is the specific *sequence* of bases in the DNA that contains the information which determines the sequence of amino acids in the protein. For example, the sequence of bases —ATCCGTAACG— is different from the sequence —ATGGG-TATCC—; the proteins that are made from the information contained in such stretches of DNA will differ in their amino acid sequence simply because the DNA base sequences are different.

Protein synthesis does not occur by direct interaction of amino acids with DNA, but rather by a complex transfer mechanism. A portion of the DNA determines the sequence of bases in a smaller single-stranded RNA molecule, called MESSENGER RNA. Again, this sequence does not duplicate the DNA bases, but complements them in a fixed pairing relationship; the crucial point is that the *information* is carried by this transfer. The messenger RNA then carries the information to structures (called RIBOSOMES) outside the cell nucleus where protein synthesis takes place. The actual synthesis of proteins is carried out with the aid of yet another RNA called TRANSFER RNA. This synthesis is shown schematically in Fig. 26.12.

To visualize the process, start at the lower right-hand side of Fig. 26.12a, where it says "amino acid$_1$." This amino acid interacts with a molecule of ATP (page 64) and becomes activated so that it can complex with a molecule of transfer RNA that *has the proper sequence of bases to accept this particular amino acid.* The messenger RNA (complexed to the ribosome) then undergoes base pairing with the complementary transfer RNA.

Now look at Fig. 26.12b, which represents a view of the process after several amino acids have interacted. The amino acids become linked to each other by peptide bonds to form the protein chain. The transfer RNA's are released and are reused. (They can be said to be "recycled.") What is most significant about the entire transformation is that the "base-language" of DNA and RNA has been translated to the "amino acid-language" of proteins.

(a)

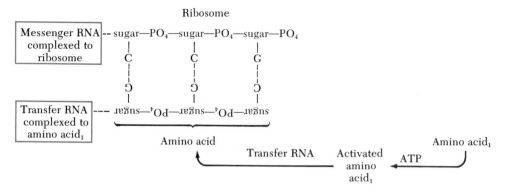

(b)

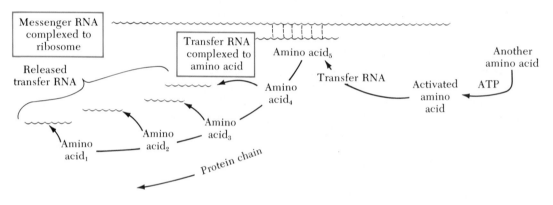

Fig. 26.12. *Protein synthesis.*

Biochemists necessarily assume as their working hypothesis that the laws of physical systems apply equally well to living organisms. Their success has been taken by some to prove that living beings can be interpreted, at least in principle, by the laws of physics and chemistry. Some theoretical physicists disagree with this proposition, arguing that the description of a living system necessarily transcends the physical and chemical laws that describe its atomic constituents. The subject has been a matter of serious debate among philosophers of science.

PROBLEMS

1. Definitions. Define and illustrate the following terms: (a) monomer, (b) polymer, (c) copolymer, (d) tetramer, (e) addition polymer, (f) condensation polymer, (g) ion-exchange resin, (h) protein, (i) enzyme.

2. Monomeric units. Neoprene, a synthetic rubber introduced commercially in 1931, is able to withstand ozone and oils more effectively than natural rubber. It is made from "chloroprene," CH_2=CCl—CH=CH_2 and is known to have the following structure:

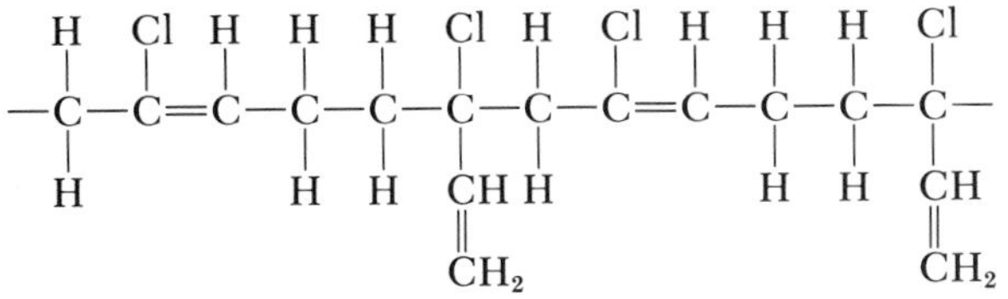

$$
\begin{array}{c}
 \overset{H}{\underset{H}{\vert}} \overset{Cl}{\vert} \overset{H}{\vert} \overset{H}{\vert} \overset{H}{\vert} \overset{Cl}{\vert} \overset{H}{\vert} \overset{Cl}{\vert} \overset{H}{\vert} \overset{H}{\vert} \overset{H}{\vert} \overset{Cl}{\vert}
\end{array}
$$

—C—C=C—C—C—C—C—C=C—C—C—C—

H   H H CH H   H H CH

CH_2CH_2

(a) Write the structures of all the geometrically possible repeating units. (b) Write the structure of the actual monomeric unit of the rubber.

3. Polymerization. Early attempts at polymerization of isoprene failed to produce a product with the properties of natural rubber. Write the structural formula of an isoprene polymer that might be produced by random polymerization.

4. Network structure. Draw a section of the glycerol–phthalic acid network copolymer of Fig. 26.3, showing all the atoms in each monomeric unit in place of the symbols P and G.

5. Properties of polymers. When polyethylene is subjected to intense ionizing radiation from a nuclear pile, it becomes much stronger, more difficult to melt, and less soluble. Account for these effects. (Refer to Table 26.2, page 560.)

6. Condensation polymers. Write a structural formula for a condensation polymer formed from each of the following monomeric units. Indicate which portion of each formula is the repeating unit.

(a) Copolymer of   (terephthalic acid)

with HO—CH_2—CH_2—OH (ethylene glycol).

(b) Copolymer of

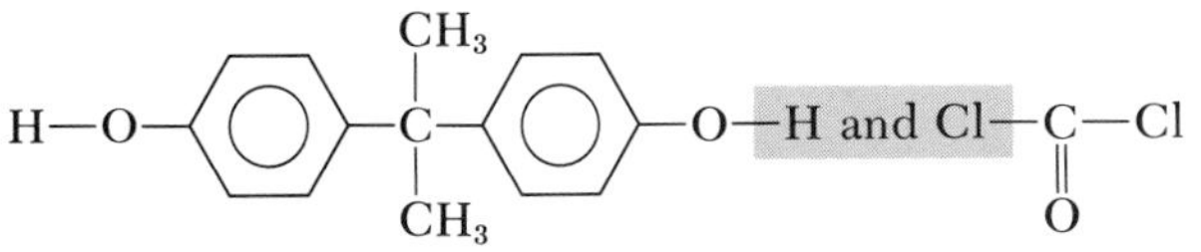

H—O— ⟨◯⟩ —C(CH$_3$)(CH$_3$)— ⟨◯⟩ —O—H and Cl—C(=O)—Cl

a phenol derivative   phosgene

with elimination of HCl as shown, to form **POLYCARBONATE PLASTIC**.

7. Addition polymerization. (a) Write the structural formula for the addition product of the vinyl chloride radical to a molecule of vinyl chloride:

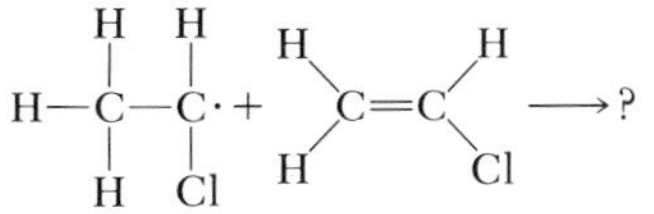

H—C(H)(H)—C(H)(Cl)· + C(H)(H)=C(H)(Cl) ⟶ ?

vinyl chloride
radical

(b) Write the structural formula for a portion of a molecule of polyvinyl chloride. Indicate the repeating unit.

8. Peptides. A DIPEPTIDE is formed from 2 amino acid molecules. Write structural formulas for the four possible dipeptides formed from a mixture of alanine and glycine.

9. Polybutadiene. The structural formula of butadiene is

$$
\begin{array}{c}
\text{H} \\ \diagdown \\ \quad\quad \text{C}=\text{C} \\ \diagup \quad\quad\quad \diagdown \\ \text{H} \quad\quad\quad\quad \text{C}=\text{C} \\ \quad\quad \diagup \quad\quad\quad \diagdown \\ \quad \text{H} \quad\quad\quad\quad \text{H}
\end{array}
$$

(a) Draw the structural formula of *cis*-polybutadiene and of *trans*-polybutadiene, assuming in each case that the carbon skeleton is unbranched. Show which portion of each formula is the repeating unit. (b) Under the influence of ultraviolet radiation, an added "sensitizer" such as phenyl disulfide

$$ \text{C}_6\text{H}_5 - \text{S} - \text{S} - \text{C}_6\text{H}_5 $$

is converted into radicals which can add to the double bonds in the polymer to form transitory radical structures that are freely rotating. When the sensitizer radicals subsequently become detached, the double bonds reform. Under these conditions, irradiation of *cis*-polybutadiene converts it to a mixture of *cis* and *trans* forms. Show these transformations with the aid of structural formulas. Explain why the transitory radical derivative of polybutadiene is freely rotating.

10. Addition polymers. Write a structural formula for an addition polymer formed from each of the following monomeric units. Indicate which portion of each formula is the repeating unit.

(a)
$$
\begin{array}{c}
\text{F} \\ \diagdown \\ \quad \text{C}=\text{S} \\ \diagup \\ \text{F}
\end{array}
$$

(b)
$$
\begin{array}{c}
\text{H} \quad\quad \text{H} \quad\quad\quad\quad \text{O} \\ \diagdown \quad\quad | \quad\quad\quad \diagup\!\!\!\diagup \\ \text{C}=\text{C}-\text{C} \\ \diagup \quad\quad\quad\quad\quad\quad \diagdown \\ \text{H} \quad\quad\quad\quad\quad \text{OC}_2\text{H}_5
\end{array}
$$

(c) Copolymer of
$$
\begin{array}{c}
\text{H} \quad\quad \text{H} \\ \diagdown \quad\quad \diagup \\ \text{C}=\text{C} \\ \diagup \quad\quad \diagdown \\ \text{H} \quad\quad \text{H}
\end{array}
\quad \text{and} \quad
\begin{array}{c}
\quad\quad \text{O} \\ \quad\quad \| \\ \text{CF}_3-\text{C}-\text{CF}_3.
\end{array}
$$

11. Cellulose derivatives. In nitration and acetylation of cellulose the free hydroxyl groups are converted to $-\text{ONO}_2$ and $-\text{OCOCH}_3$ groups, respectively. Draw the structures of the monomeric units of cellulose nitrate and cellulose acetate.

12. Proteins. (a) Nylon-66 (page 554) is characterized by amide linkages. Argue for or against the proposition that Nylon-66 is a synthetic protein. (b) Nylon-6 (page 554) is an amino acid polymer. Argue for or against the proposition that Nylon-6 is a synthetic protein.

13. Insulin. Using the information in Fig. 26.7 (page 564), write the portion of the structural formula for insulin that includes the first four amino acids starting with glycine. Do not use the amino acid abbreviations; write your formula completely, using a symbol for each atom.

14. Network polymers. Is the greater stability of network polymers compared with linear polymers analogous to the higher boiling point of water as compared with hydrogen fluoride (page 141)? Defend your answer.

15. DNA. Experiments have been carried out in which bacteria are nourished and reproduce in a nutrient medium in which all the nitrogen is the ^{15}N isotope. It is found that 50% of the nitrogen in the DNA of the first new generation of bacteria is ^{15}N. (a) How does this evidence support the double-helix structure of DNA? (b) If the first new generation of bacteria were transferred to a nutrient medium containing only ^{14}N, what isotopic makeups would you expect to find in the DNA of the second new generation?

16. Enzyme. Starch (amylose), $(C_6H_{10}O_5)_x$, is a polymer of the sugar glucose, $C_6H_{12}O_6$, linked by —O— bonds; each starch molecule may contain 200 to 1000 glucose units. A mixture of amino acids added to a solution of amylose produces no detectable reaction. One molecule of the enzyme amylase, containing the same amino acids, however, breaks (hydrolyzes) 4000 —O— bonds per second. Estimate the mass in picograms of glucose produced by an amylase molecule in 1 day and the heat in calories liberated to a cell by the oxidation of the glucose:

$$C_6H_{12}O_6(aq) + 6O_2(g) \longrightarrow 6CO_2(g) + 6H_2O(l) \qquad \Delta H = -686 \text{ kcal}$$

17. Genetic continuity. Imagine that you purchase a dress pattern. Is it reasonable to say that this item contains *information?* Now if you use this pattern to make a dress, could you say that the information has been transmitted from the pattern to the dress material? Is it reasonable to call the transfer of information *communication?* Which process involves a greater increase of entropy, the cutting of a fabric to match a dress pattern, or the cutting of a fabric by a child playing randomly with scissors? Would you agree that the increase of information in a system decreases the entropy? Is it therefore reasonable to consider that a molecule of DNA communicates with other molecules, thereby transferring information, and thus preserves the character of genetic continuity as a low-entropy process?

APPENDIX I

SYSTEMS OF MEASUREMENT AND REVIEW OF PHYSICAL CONCEPTS

MATTER / I.1

The tendency to maintain a constant velocity is called INERTIA. Thus, unless acted on by an unbalanced force, a body at rest will remain at rest; a body in motion will remain in motion with uniform velocity.

MATTER is anything that exhibits inertia; the quantity of matter is its MASS. Mass is difficult to define without falling into the circle of defining mass in terms of force and force in terms of mass. However, there is no practical difficulty in deciding when two masses are equal, for their weights are then equal; an instrument that verifies this condition is a BALANCE.

MOTION / I.2

MOTION is the change of position or location in space. Motions of objects may be classified as follows:

TRANSLATIONAL MOTION occurs when the center of mass of an object changes its location. *Example:* an arrow in flight.

ROTATIONAL MOTION occurs when some points of a moving object remain stationary. *Examples:* a spinning top (the points along its axis of rotation are stationary); a rotating molecule (the center of mass remains stationary).

DISTORTION is change of shape, or the motion of the points of an object relative to one another. *Example:* a sagging metal bar.

VIBRATION is periodic distortion and recovery of original shape. *Examples:* a struck tuning fork; a vibrating molecule.

THE METRIC SYSTEM / I.3

The metric system, established by international treaty at the Metric Convention in Paris in 1875, has since been extended and improved.

The currently established official metric system is called the International System of Units (Système International d'Unités, abbreviated SI).

The metric system has seven fundamental units, from which all others can be derived:

Quantity	Unit	Abbreviation
length	meter	m
mass	kilogram	kg
time	second	s, or sec
electric current	ampere	A, or amp
temperature	Kelvin	K, or °K
luminous intensity	candela	cd
amount of substance	mole	mol

Larger or smaller units in the metric system are expressed by the following prefixes:

Multiple or fraction	Prefix	Symbol
10^{12}	tera	T
10^{9}	giga	G
10^{6}	mega	M
10^{3}	kilo	k
10^{-1}	deci	d
10^{-2}	centi	c
10^{-3}	milli	m
10^{-6}	micro	μ
10^{-9}	nano	n
10^{-12}	pico	p
10^{-15}	femto	f
10^{-18}	atto	a

We will now consider several fundamental and derived physical quantities which the chemist frequently uses.

LENGTH / I.4

One meter, m, is 1,650,763.73 wavelengths in vacuum of the orange-red radiation of the krypton-86 atom at the triple point of nitrogen ($-210°C$).

Length conversions:

from	to	multiply by
angstrom, Å	centimeter	10^{-8} cm/Å (exactly)
angstrom, Å	nanometer	10^{-1} nm/Å (exactly)
inch, in.	centimeter	2.54 cm/in. (exactly)

The INERTIAL MASS, m, of a body is determined by the force, f, required to produce a given acceleration, a:

$$f = ma \quad \text{or} \quad m = f/a$$

The GRAVITATIONAL MASS of a body is that property which determines its weight. Experimentally, the inertial mass and gravitational mass of a given body are exactly identical.

One kilogram, kg, is the mass of Prototype Kilogram Number 1 (a platinum–iridium alloy) kept at the International Bureau of Weights and Measures, Sèvres, France.

Mass conversions:

from	to	multiply by
gram, g	kilogram	10^{-3} kg/g (exactly)
pound (avoirdupois), lb	kilogram	4.536×10^{-1} kg/lb
kilogram, kg	pound (avoirdupois)	2.205 lb/kg

TEMPERATURE / I.6

If two bodies, A and B, are in contact, and there is a spontaneous transfer of heat from A to B, then A is said to be *hotter* or at a higher temperature than B. When, despite an available path, there is no net heat transfer between two bodies, they are said to be at the same temperature. TEMPERATURE is thus the property of a body that predisposes it to lose heat: the higher the temperature, the greater the tendency for heat to flow away from the body.

Many properties of substances change with changes in temperature. Examples are density, color, ability to conduct electricity, and ability to stimulate nerve impulses. A set of values in which temperature is related to some measured property is called a TEMPERATURE SCALE. A device used to obtain such measurements is a THERMOMETER. Additional discussion of temperature scales appears in Chapter 2.

The standard unit of temperature is the Celsius (formerly Centigrade) degree, °C, or the Kelvin degree, officially designated the kelvin, K, although °K is more often used.

Kelvin and Celsius scales. The Kelvin scale is fixed by reference to the triple point (page 160) of water, to which is assigned the temperature 273.16K (exactly). Each Kelvin degree is then defined as exactly 1/273.16 of the triple point of water. The Celsius scale is fixed by its relationship to the Kelvin scale.

International practical temperature scale (Celsius). This scale defines six temperatures, all in degrees Celsius, for example:

Temperature	°C
boiling point of oxygen	−182.97
triple point of water	+0.01
boiling point of water	+100.0

ABSOLUTE ZERO (page 11), 0°K, is therefore −273.15°C.

Fahrenheit scale. The relationship to the Celsius scale is:

temperature, °C = $\frac{5}{9}$(temperature, °F − 32)

VOLUME / I.7

Volume is based on length; hence, the metric unit is 1 cubic meter, m^3. The metric system also recognizes the LITER, l, which is 10^{-3} m^3. The liter was formerly defined as the volume of 1 kg of pure water at the temperature of its maximum density (3.98°C) at 1 standard atmosphere. Since the liter is an approved metric unit, so are its multiples and fractions, of which the most common is the milliliter, ml.

Volume conversions:

from	to	multiply by
liter, l	cubic meter	10^{-3} m³/l (exactly)
milliliter, ml	liter	10^{-3} l/ml (exactly)
milliliter, ml	cubic centimeter	1 cm³/ml (exactly)
quart (U.S. liquid), qt	liter	0.9464 l/qt
cubic foot (U.S.), ft³	liter	28.316 l/ft³

DENSITY / I.8

Density is mass per unit volume. The metric unit is therefore kilograms per cubic meter, kg/m^3, or the more usual expressions, grams per cubic centimeter, g/cm^3, or grams per milliliter, g/ml.

An earlier expression of density is specific gravity. The unit is obsolete and should no longer be used. The following explanation is offered as a guide:

$$\text{specific gravity} = \frac{\text{density of a given substance}}{\text{density of a standard substance}}$$

The temperatures of both substances must be specified; they are frequently, but not always, the same. For solids and liquids the standard substance is usually water; for gases it is usually air, or sometimes hydrogen. Thus, the specific gravity of carbon tetrachloride, $1.594^{20°/4°}$, means that the density of this liquid at 20°C is 1.594 times as great as the density of water at 4°C. Since the density of water at 4°C is 1.0000 g/ml, the specific gravity$^{x°/4°}$ is numerically equal to the density in grams per milliliter at $x°$.

FORCE AND WEIGHT / I.9

FORCE is that which changes the velocity (that is, the state of rest or motion) of a body; it is defined as:

force = mass × acceleration

The metric unit is the NEWTON, N, whose dimensions are kg m/sec². A newton is thus the force needed to change the velocity of a mass of 1 kg by 1 m/second in a time of 1 second. However, the older unit, the DYNE (defined below) is still very much in use.

The WEIGHT of a body is the force exerted on it by the Earth's gravity, and is therefore equal to the mass of the body times the acceleration due to gravity:

weight = mass × acceleration due to gravity

Since the Earth's gravity is not the same everywhere, the weight of any given body is not a constant. However, at any given spot on Earth, gravity is constant and therefore weight is proportional to mass. When a balance tells us that a given sample (the "unknown") has the same weight as another sample (the "weights," as given by a scale reading or by a total of counterweights), it also tells us that the two sets of masses are equal. The balance is therefore a valid instrument for measuring the mass of an object when no other forces but gravity are exerted (no magnetic forces, for example).

Force conversion:

from	to	multiply by
dyne	newton, N	10^{-5} newtons/dyne (exactly)
lb	newton, N	4.4482 N/lb

Pressure is force per unit area, and therefore has the dimensions (kg m/sec²)/m², or kg/sec²-m = newtons/m². The metric unit of pressure is the BAR, defined as 10⁵ newtons/m², or 10⁶ dynes/cm².

Chemists also measure pressure in terms of the heights of liquid columns, especially water and mercury. This usage is not completely satisfactory because the pressure exerted by a given column of a given liquid is not a constant, but depends on the temperature (which influences the density of the liquid) and the location (which influences gravity). Such units are, therefore, not part of the metric system, and attempts are being made to discourage their use. However, habit is hard to change, and the units described below are used so frequently in books, journals, industrial literature, and the like, that most chemists are more familiar with them than they are with bars and millibars.

The pressure of a liquid or a gas depends only on the depth (or height), and is exerted equally in all directions. At sea level, the pressure exerted by the earth's atmosphere varies around 14.7 lb/in.². This value is equivalent to the pressure exerted by a column of mercury about 76 cm high, or of water about 34 ft high.

One STANDARD ATMOSPHERE (atm) = the pressure exerted by exactly 76 cm (= exactly 760 mm) of mercury at 0°C (density 13.5951 g/cm³) and at standard gravity, 980.665 cm/sec².

One TORR* = the pressure exerted by exactly 1 mm of mercury at 0°C and standard gravity.

Pressure conversions:

from	to	multiply by
atmosphere	bar	1.10325 bar/atm
bar	dyne/cm²	10⁶ dyne/cm²-bar (exactly)
pound(force)/in.²	newton/m²	6894.76 newtons in.²/lb-m²
atmosphere	torr	760 torr/atm (exactly)
mm of mercury at 0°C and standard gravity	torr	1 torr/mm mercury (exactly)
atmosphere	lb/in.²	14.6959 lb/in.²-atm

A BAROMETER is a device that measures atmospheric pressure. The original and simplest form was invented by Torricelli in 1643. It is made by inverting a tube longer than 76 cm filled with mercury into a dish of mercury (Fig. I.1). The atmosphere will support only that height of mercury which exerts an equivalent pressure; any excess mercury will fall into the reservoir and leave a space in the upper part of the

* Named after Evangelista Torricelli.

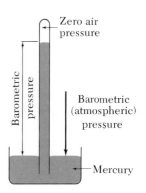

Fig. I.1. Mercury barometer (*schematic*).

tube with zero air pressure. Note that the height in millimeters is numerically equal to the pressure in torr units only if the mercury is at 0°C and standard gravity. Otherwise, appropriate corrections must be made.

ENERGY is the capacity to do work on a body or to transfer heat to a body. WORK is the activity required to overcome inertia. Any moving object can do work by virtue of its motion because, by collision, it can overcome the inertia of another object; this capacity is called KINETIC ENERGY. Energy of motion may be either translational, rotational, or vibrational. The energy that a moving body loses in a *perfectly elastic* collision is transmitted as kinetic energy to another object. The magnitude of the kinetic energy, for a body of mass m and velocity u, is

$$\text{kinetic energy} = \tfrac{1}{2}mu^2$$

The energy of a body may stem from some attribute other than motion. A compressed spring, for example, may do work when it expands, a stick of dynamite when it explodes. In every case in which we attribute a quantity of energy to an object, we imagine that an event could occur in which this energy may be transmitted in some way by the object to another body.

The metric unit of energy is the product of the units of force and distance, or (kg m/sec²) × m, which is kg m²/sec²; this unit is called the JOULE, J. The joule is thus the work done when a force of 1 newton acts over a distance of 1 meter.

Work may also be expressed in terms of pressure and volume (Fig. I.2):

$$\text{work} = \text{pressure} \times \text{volume change}$$

Work may also be done by moving an electric charge in an electric field. A unit of electrical work that is recognized in the metric system is the ELECTRON VOLT, eV, which is the kinetic energy acquired by 1 electron (1.60×10^{-19} coulomb) when it is accelerated in an electric field produced by a potential difference of 1 volt:

(a) $1.60 \times 10^{-19} \, \dfrac{\text{coulomb}}{\text{electron}} \, \dfrac{\text{joule}}{\text{volt coulomb}} \times 10^7 \, \dfrac{\text{ergs}}{\text{joule}} = 1.60 \times 10^{-12} \, \dfrac{\text{erg}}{\text{electron volt}}$

(b) $1.60 \times 10^{-12} \, \dfrac{\text{erg}}{\text{electron volt} \left(4.18 \times 10^{10} \, \dfrac{\text{ergs}}{\text{kcal}} \right)} = 3.83 \times 10^{-23} \, \dfrac{\text{kcal}}{\text{electron volt}}$

There is, therefore, 1.60×10^{-12} erg or 3.83×10^{-23} kcal in 1 electron volt. Then, the kinetic energy acquired by *1 mole of electrons* (1 faraday) when they are accelerated by a potential difference of 1 volt is

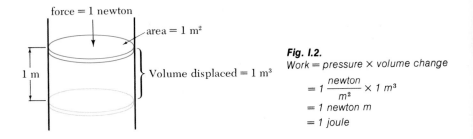

force = 1 newton

area = 1 m²

1 m

Volume displaced = 1 m³

Fig. I.2.
Work = pressure × volume change

$$= 1 \frac{newton}{m^2} \times 1\ m^3$$
$$= 1\ newton\ m$$
$$= 1\ joule$$

(a) $1.60 \times 10^{-12} \dfrac{\text{erg}}{\text{electron volt}} \times 6.02 \times 10^{23} \dfrac{\text{electrons}}{\text{faraday}} = 9.63 \times 10^{11} \dfrac{\text{ergs}}{\text{faraday volt}}$

(b) $3.83 \times 10^{-23} \dfrac{\text{kcal}}{\text{electron volt}} \times 6.02 \times 10^{23} \dfrac{\text{electrons}}{\text{faraday}} = 23.1 \dfrac{\text{kcal}}{\text{faraday volt}}$

The most widely used conversion factor obtained from these relationships is

$$1 \frac{\text{electron volt}}{\text{atom}} = 23.1 \frac{\text{kcal}}{\text{mole}}$$

Energy conversions:

from	to	*multiply by*
erg	joule	10^{-7} joule/erg (exactly)
calorie (thermochemical)*	joule	4.184 joules/cal (exactly)
kilocalorie, kcal, also called kilogram-calorie, or Calorie (capital C, used in expressing food energies for nutrition)	calorie	10^3 cal/kcal (exactly)
kilocalorie	joule	4.184×10^3 joules/kcal (exactly)
liter atmosphere	joule	101.325 joules/1-atm
liter atmosphere	calorie	24.2173 cal/l-atm
liter atmosphere	liter torr	760 torr/atm (exactly)
electron volt, eV	joule	1.6021×10^{-19} joule/eV
electron volt	calorie	3.8291×10^{-20} cal/eV
electron volt	erg	1.6021×10^{-12} erg/eV
million electron volts, MeV	electron volt	10^6 eV/MeV (exactly)
British thermal unit, btu	calorie	252 cal/btu
electron volt per molecule	calorie per mole	$2.3061 \times 10^4 \dfrac{\text{cal molecule}}{\text{eV mole}}$
electron volt per molecule	kcal per mole	$23.061 \dfrac{\text{kcal molecule}}{\text{eV mole}}$
coulomb volt, CV	joule	1 CV/joule (exactly)
kilowatt-hr	kcal	860.4 kcal/kwatt-hr
kilowatt-hr	joule	3.6×10^6 joule/kwatt-hr (exactly)

*The "small calorie" or "gram-calorie" is the quantity of heat required to warm 1 g of water from 3.5° to 4.5°C. The "normal calorie" involves the temperature change from 14.5° to 15.5°C, and the "mean calorie" is 1/100 the heat needed to warm 1 g of water from 0° to 100°C. All of these units are nearly the same. In this book, we use the thermochemical calorie.

One may do work on an object and yet fail to convert such work into equivalent energy of motion, electricity, magnetism, radiation, or chemical or physical change. For example, one may bend an iron bar back and forth several times, or stir a liquid or a gas in a confined space, or force an electric current through a copper wire and, after all such expenditures of energy, observe that the object retains substantially its original form and position in space. We assume that this energy has been conserved, and exists as an undifferentiated internal energy called HEAT, which is retained by the object in question. It is believed that heat is the energy of random motion of all the elementary particles of a body. The traditional unit of heat is the CALORIE, cal, or KILOCALORIE, kcal, but neither of these is part of the metric system.

POWER is the amount of energy delivered per unit time. A common unit is the watt. Watt-hours and kilowatt-hours are units of energy (see page 581).

Power conversions:

from	to	multiply by
watt	kwatt	10^{-3} kwatt/watt (exactly)
watt	amp-volt	1 amp-volt/watt (exactly)
kwatt	erg/sec	10^{10} erg/sec-kwatt (exactly)

Units Derived from the International System (SI Derived Units)

Quantity	Name of unit	Symbol	Definition in SI Units
length	angstrom	Å	10^{-10} m or 0.1 nm
volume	liter	l	dm^3 or 10^{-3} m^3
pressure	standard atmosphere	atm	1.01325 bar or 101325 N/m^2
	torr	torr	1.33322×10^{-3} bar or 133.322 N/m^2
energy	calorie	cal	4.184 J

ELECTRIC CURRENT AND DERIVED ELECTRICAL QUANTITIES / I.12

When electric current flows along parallel conductors, a magnetic force is produced between them. The metric unit of electric current, the AMPERE, A, is the constant current which, if maintained in each of two parallel conductors of infinite length and 1 meter apart in a vacuum, would produce between them a force of 2×10^{-7} newton/meter of conductor length (exactly).

The magnitude of the current which can pass through a given section of matter depends not only on the potential difference between the two points of reference but also on the resistance imposed by the matter. The unit of resistance is the ohm, Ω, and its relationship to potential differences and current is called OHM'S LAW:

$$\text{electric current (amperes)} = \frac{\text{potential difference (volts)}}{\text{resistance (ohms)}}$$

Therefore, an ohm is equivalent to a volt per ampere, or V/amp. The reciprocal of the resistance is called the CONDUCTANCE (unit: reciprocal ohm, or mho, Ω^{-1}).

To account for observed electrostatic interactions, it is postulated that two kinds of electric charge exist; they are called POSITIVE and NEGATIVE. Unlike charges attract, and like charges repel each other. The magnitude of electrostatic force is given by Coulomb's law:

$$F = \frac{1}{D} \frac{q_1 q_2}{d^2}$$

where F is the force of attraction or repulsion, q_1 and q_2 are the magnitudes of two electric charges separated by distance d, and D is the dielectric constant (page 220).

RADIOACTIVITY / I.13

The CURIE, Ci, the standard used in measuring the activity of all radioactive substances, has been fixed at 3.7×10^{10} disintegrations per second (dps). All human beings are continuously exposed to radiation (exclusive of "fallout") originating within as well as outside the body. This natural source averages about 2 to 3 dps.

The ROENTGEN, R, the standard used in measuring the intensity of X or γ radiations, is defined as the quantity of X or γ radiation that produces 1.61×10^{12} ion pairs in 1 g of air, equivalent to the absorption of 84 ergs/g of air. The RAD (radiation absorbed dose) is the dosage of any nuclear emanation equivalent to the absorption of 100 ergs/g of any material.

The REM (roentgen equivalent man) is the standard unit of the biological effect on man of exposure to radiation. It measures the biological effectiveness of different kinds of radiation by accounting for both the radiation dosage absorbed by the body and its biological damage: $rem = rad \times rbe$ (relative biological effectiveness). $rbe = 1$ for X rays, γ rays, and electrons; 10 for neutrons, protons, and α particles; 20 for high-speed nuclei. Typical natural background radiation levels in the United States are about 0.125 rem/year, including about 0.025 rem received from radioactive carbon and potassium in our bodies. The Atomic Energy Commission (1972) dosage limits are 0.500 for any one individual per year and 0.170 rem/year for the general population. A chest X ray may deliver a dosage of 0.500 rem.

Information obtained from measurements may be communicated to human beings (through sensory routes) or to machines. Any measurement culminates in a sensation; the sensing element may be the human eye or other organ, or an inanimate instrument. In any event, the sensor must be stimulated by a quantity of energy (such as light reflected from an object being observed) if it is to receive the information. As a result, the process of measurement disturbs to some degree the matter being measured. For measurements of ordinary bodies, the magnitude of this disturbance is too small to be of consequence, but for measurements of very small bodies like individual electrons the inherent uncertainty introduced is relatively great.

Significance of measurement figures. Some uncertainties in measurement are inevitably introduced through factors such as human error, malfunction of measuring devices, and experimental bias (error in one direction, as by a ruler whose "inches" are too short). Information is useful only to the extent that one can be confident of its validity. To insure such utility, each figure or digit in the numerical expression of a measurement should be significant. A SIGNIFICANT FIGURE may be defined as a number that we believe to be correct within some specified or implied limit of error. Thus, if the height of a man, expressed in significant figures, is written as 5.78 feet, it is assumed that only the *last figure* may be in error. Clearly, any uncertainty in the first or second figure would remove all significance from the last figure. (If one is uncertain of the number of feet, it is idle to speak of inches.) If we have reason to believe that the last figure will be in doubt by a *specified* amount, we may so indicate by expressions such as 5.78 ± 0.01 feet.

Value	*Number of significant figures*
0.0301 pound	3
4.290 hours	4
1.030 mg	4
0.001030 g	4
5000 miles	Ambiguous—the zeros may have measurement significance or may be merely spacers to indicate the magnitude of the first digit
5.00×10^3 miles	3—the zeros clearly have measurement significance; they are not needed to indicate magnitude
6 pencils	If this is used in the sense of a tally ("Tom has 6 pencils"), it is an exact number, and the concept of significant figures has no relevance; if it is used as a measurement ("The average student uses 6 pencils per semester"), it has 1 significant figure
1 foot = 12 inches	These are exact numbers which express a definition, not measurements

To count the significant figures in a number, read the number from left to right and count all the digits starting with the first digit that is not zero. The position of the decimal point should be ignored because it is determined by the particular units employed, and not by the precision of the measurement. Thus, the measurements 12.2 cm and 122 mm are identical and therefore all the corresponding figures are equally significant. Various illustrative examples are given in the table on the preceding page.

We must guard against introduction of uncertainty by arithmetical procedures. The following rules will be helpful.

Rule 1. In addition or subtraction, any figure in the answer is significant only if *each number* in the problem contributes a significant figure at that decimal level. Therefore, the value which terminates at the highest decimal level (that is, the level of greatest magnitude) will determine how far the significant figures should be carried in the answer:

$$
\begin{array}{r}
308.7812 \\
0.00034 \\
\underline{10.31} \\
319.09
\end{array}
$$

Rule 2. When a number is "rounded off" (nonsignificant figures discarded), the last significant figure is unchanged if the next figure is less than 5, and is increased by 1 if the next figure is 5 or more:

$$4.6349 \longrightarrow 4.635 \quad \text{(four significant figures)}$$
$$4.6349 \longrightarrow 4.63 \quad \text{(three significant figures)}$$
$$2.8150 \longrightarrow 2.82 \quad \text{(three significant figures)}$$

Rule 3. In multiplication and division, the number of significant figures in the answer is the same as that in the quantity with the fewest significant figures:

$$\frac{3.0 \times 4297}{0.0721} = 1.8 \times 10^5$$

Rule 4. In a multistep computation, it will be convenient first to determine the number of significant figures in the answer by rules 1–3 above, and to round off each number that contains excess significant figures to one more significant figure than necessary. Then round off the answer to the correct number of significant figures. This procedure will preserve significance with minimum labor.

Example 1 Evaluate the expression

$$V = 4.3 \times \frac{311.8}{273.1} \times \frac{760}{784 - 2}$$

Answer There are two significant figures in the number 4.3 and therefore the answer will have two significant figures. Therefore, round off according to rules 1 and 2, to one extra significant figure. Note that the presence of only one significant figure in 2 does not mean that there is only one significant figure in the answer because $784 - 2 = 782$, which has three significant figures. Thus,

$$V = 4.3 \times \frac{312}{273} \times \frac{760}{782}$$

Solve and round off to two significant figures,

$$V = 4.8$$

Measurement information, to be complete, must specify the units employed, which are called DIMENSIONAL UNITS. Like other algebraic terms, these may have positive or negative exponents. Thus, the unit of area, square feet, is expressed as ft^2. A unit of speed may be written *miles/hour* or *miles hr^{-1}*. A unit of frequency, the number of times some event occurs *per second*, is *sec^{-1}*.

EXPONENTS AND LOGARITHMS / I.15

In the expression x^n, x is called the BASE, and n the EXPONENT. The expression x^{-n} is the reciprocal of x^n,

$$x^{-n} = \frac{1}{x^n}$$

Thus,

$$4^{-1} = \tfrac{1}{4}$$

$$2^{-3/2} = \frac{1}{2^{3/2}} = \frac{1}{\sqrt{2^3}} = \frac{1}{\sqrt{8}} = \frac{1}{2\sqrt{2}} = \frac{\sqrt{2}}{4}$$

Any number (except 0) to the 0th power is 1:

$$(\tfrac{1}{2})^0 = 1^0 = 8^0 = 1000^0 = (10^{23})^0 = e^0 = x^0 = 1$$

When exponential expressions having the same base are multiplied, the exponents are added; when such expressions are divided, the exponents are subtracted:

$$10^2 \times 10^3 = 10^5, \qquad 10^6 \times 10^{-4} = 10^2$$

$$\frac{10^5}{10^2} = 10^3, \qquad \frac{10^5}{10^{11}} = 10^{-6}$$

When a power is raised to a power, the exponents are multiplied:

$$(10^8)^3 = 10^{24} \quad \text{and} \quad (10^{-1})^2 = 10^{-2}$$

Addition and subtraction of numbers in exponential notation:
Step 1. Convert all numbers to the same exponent of 10.
Step 2. Add or subtract as required.
Step 3. Adjust to the proper number of significant figures.
Step 4. Convert to an exponent of 10 such that the decimal point in the coefficient follows the first numeral.

Example 2 Add the following numbers in exponential notation: 3.48×10^5, 1.23×10^6, -0.78×10^4.

Answer

Numbers to be added	Convert to same exponent of 10	
3.48×10^5	3.48×10^5 ⎫	
1.23×10^6	12.3×10^5 ⎬ step 1	
-0.78×10^4	-0.078×10^5 ⎭	
	15.702×10^5	step 2
	15.7×10^5	step 3
	1.57×10^6	step 4

A logarithm is an exponent:

$$N = a^x \qquad (a \text{ is the BASE}; x \text{ is the EXPONENT})$$
$$\log_a N = x \qquad (a \text{ is the BASE}; x \text{ is the LOGARITHM})$$

The base of logarithms used as an aid to ordinary computations is 10. Then

$$10{,}000 = 10^4$$

and

$$\log_{10} 10{,}000 = \log_{10} 10^4 = 4$$

When the base is not specified, 10 is generally understood (except in purely mathematical writings, including tables of integrals, which use natural logarithms exclusively):

$$\log 0.001 = \log 10^{-3} = -3$$

Because logarithms are exponents, logarithms of products are added and logarithms of quotients are subtracted:

$$\log(a \times b) = \log a + \log b$$
$$\log\left(\frac{a}{b}\right) = \log a - \log b$$

Logarithms that cannot be expressed as integral exponents are found in tables or on slide rules. Logarithmic tables present logarithms of numbers between 1 and 10; a decimal point is assumed to follow the first digit of the number and to precede the first digit of the logarithm. Thus, we find that the logarithm of 191 is 281; this means that

$$\log 1.91 = 0.281$$

A number that is not between 1 and 10 may be written in proper exponential form, and the logarithm obtained as follows:

$$
\begin{aligned}
\log 7040 &= \log(7.040 \times 10^3) \\
&= \log 7.040 + \log 10^3 \\
&= 0.8476 + 3 = 3.8476
\end{aligned}
$$

$$
\begin{aligned}
\log 0.000625 &= \log(6.25 \times 10^{-4}) \\
&= \log 6.25 + \log 10^{-4} \\
&= 0.7959 + (-4) = -3.204
\end{aligned}
$$

The procedure can be reversed. The number, N, whose logarithm has a given value, x, is the ANTILOGARITHM of that value:

$$
\left.
\begin{aligned}
10^x &= N \\
\log N &= x \\
\text{antilog } x &= N
\end{aligned}
\right\} \quad (x \text{ is the LOG}; N \text{ is the ANTILOG})
$$

Rules for significant figures in logarithms and antilogarithms. The number of *decimal places* (digits after the decimal point, including zero) in the logarithm (x) should be equal to the number of significant figures in the number (N).

(*i*) $\log 1.21 = 0.083$

Note that 1.21 has three significant figures, and its logarithm has 3 decimal places. Note also that the logarithms given in the previous examples conform to this rule.

The number of significant figures in the antilogarithm is equal to the number of decimal places in the logarithm. Some examples follow:

(*ii*) $\quad \log 100 = \log 10^2 = 2$
$\quad\quad$ antilog $2 = 10^2 = 100$

We assume above that 2 and 100 are exact numbers.

(*iii*) antilog $6.3909 =$ antilog$(6 + 0.3909)$
$\quad\quad\quad\quad\quad\quad\quad = $ antilog $6 \times$ antilog $0.3909 = 10^6 \times 10^{0.3909}$
$\quad\quad\quad\quad\quad\quad\quad = 10^6 \times 2.460$ (from log table, page 605)
$\quad\quad\quad\quad\quad\quad\quad = 2.460 \times 10^6$

Note that only four significant figures are permitted.

(*iv*) antilog$(-0.0079) =$ antilog$(0.9921 - 1)$
$\quad\quad\quad\quad\quad\quad\quad\quad = $ antilog $0.9921 \times$ antilog(-1)
$\quad\quad\quad\quad\quad\quad\quad\quad = 9.820 \times 10^{-1}$
$\quad\quad\quad\quad\quad\quad\quad\quad = 0.9820$

Note that four significant figures are permitted because there are four figures after the decimal point in −0.0079.

(v) antilog(−9.42) = antilog 0.58 × antilog(−10)
$$= 3.8 \times 10^{-10}$$

Note that only two significant figures are permitted.

It is also possible to work with the exponential forms in antilog calculations, as shown below:

(vi) antilog(−11.05) = $10^{-11.05}$ = $10^{(0.95-12)}$ = $10^{0.95} \times 10^{-12}$
$$= (\text{antilog } 0.95) \times 10^{-12}$$
$$= 8.9 \times 10^{-12}$$

Logarithms to the base e (where $e = 2.71828 \cdots$) are **NATURAL LOGARITHMS** and are given the symbol ln. Thus:

$$\ln e^x = x$$
$$\ln e^{0.2} = 0.2$$

The relation between the logarithms of the two bases, e and 10, is

$$\ln x = 2.303 \log x$$

To solve for the value of e^x, for example, $e^{-3.0}$, the following procedure is recommended:

(a) Take the natural logarithm of $e^{-3.0}$, $\ln e^{-3.0} = -3.0$;
(b) Convert to the base 10 by dividing by 2.303:

$$\frac{-3.0}{2.3} = -1.3$$

(c) Take the antilogarithm:

$$-1.3 = 0.7 - 2$$
$$\text{antilog}(-1.3) = \text{antilog } 0.7 \times \text{antilog}(-2)$$
$$= 5 \times 10^{-2}$$

Therefore,

$$e^{-3.0} = 5 \times 10^{-2}$$

The notation exp(x) is often used to represent e^x.

Values of exponential functions and natural logarithms are given in tables in standard reference books. However, these tables are usually less adequate than the tables of logarithms to the base 10.

APPROXIMATE SOLUTIONS TO QUADRATIC EQUATIONS / I.16

A quadratic equation is one in which the highest exponent to which a variable is raised is 2. Any quadratic equation may be written as

$$ax^2 + bx + c = 0$$

The equation has two solutions, given by

$$x = \frac{-b \pm \sqrt{b^2 - 4ac}}{2a}$$

If the term in x or the term in x^2 (not both) is very small compared with the constant, c, it may be dropped:

If $bx \ll c$, then $\qquad$ If $ax^2 \ll c$, then
$ax^2 + c \approx 0$ $\qquad\qquad\quad$ $bx + c \approx 0$

$$x^2 \approx -\frac{c}{a} \qquad\qquad\qquad x \approx -\frac{c}{b}$$

$$x \approx \sqrt{-\frac{c}{a}}$$

The case that frequently arises in problems of ionic equilibrium (Chapter 14) is that in which the x term is dropped. For example,

$$\frac{x^2}{0.10 - x} = 2.6 \times 10^{-4}$$

Solve by using the quadratic formula:	Drop the x term. Then:
$x = 5.0 \times 10^{-3}$	$\dfrac{x^2}{0.10} = 2.6 \times 10^{-4}$ $x = 5.1 \times 10^{-3}$

We see that the two answers nearly agree. But how can we know this without solving the equation both ways (which doesn't save any work)? The idea is this: Make the simplifying assumption and solve the equation. Then note whether or not the assumption is true (in this case, whether 5.1×10^{-3} is much less than 0.10; it is). If it is, then the equation $x^2/(0.10 - x) = 2.6 \times 10^{-4}$ and the equation $x^2/0.10 = 2.6 \times 10^{-4}$ are essentially the same, because $0.10 - x \approx 0.10$. Therefore, they have essentially the same solution. The answer can also be checked by the usual method of substituting it for x in the original equation to see whether an identity is obtained.

Let us now try this method in a case where it does *not* work:

$$\frac{x^2}{0.10 - x} = 2.6 \times 10^{-2}$$

We drop the x term. Then,

$$\frac{x^2}{0.10} = 2.6 \times 10^{-2}$$

$$x = 5.1 \times 10^{-2}$$

We see that 5.1×10^{-2} ($= 0.051$) is *not* small compared to 0.10. Therefore, the x term may *not* be dropped and the quadratic formula must be used. The correct solution thus obtained is

$$x = 4.0 \times 10^{-2}$$

A **VECTOR** is a quantity that has both magnitude and direction; examples are velocity, displacement, and force. By contrast, a **SCALAR** quantity, such as mass or volume, has magnitude but no direction.

Consider a vector represented by a line from point 0 to point A, designated $\overrightarrow{0A}$, and another from the same origin to point B, designated $\overrightarrow{0B}$:

The lengths and directions of the two vectors could represent, for example, the magnitudes and directions of two different forces exerted on a body at point 0, such as a force of gravity, $\overrightarrow{0A}$, and a force exerted by a magnetic field, $\overrightarrow{0B}$.

To add the two vectors, we construct a parallelogram, as follows:

Then the vector, $\overrightarrow{0C}$, drawn from 0 to the opposite vertex of the parallelogram, is the **VECTOR SUM** (or **RESULTANT**) of $\overrightarrow{0A} + \overrightarrow{0B}$. This sum has physical meaning because the actual object will move just as if it were impelled by a force exerted in the direction $\overrightarrow{0C}$ and whose magnitude is represented by the length of the $\overrightarrow{0C}$ line vector.

APPENDIX II

NOMENCLATURE

NOMENCLATURE OF INORGANIC COMPOUNDS / II.1

BINARY COMPOUNDS, VALENCES

The name of a **BINARY COMPOUND**, a compound composed of two elements, is usually derived from the names of the elements (see back cover); usually, the metallic or electropositive element is written first and the second element is given the suffix *-ide*. There are four systems of nomenclature:

(a)   When there is more than one atom of an element in the chemical formula, a Greek prefix, *di-*, *tri-*, *tetra-*, *penta-*, *hexa-*, etc., is used to indicate the number of atoms. Examples are tin dichloride, $SnCl_2$; trisodium phosphide, Na_3P; and trioxygen, O_3 (see below). However, when there is only one compound of the two elements, the prefix is often omitted, as in sodium phosphide, Na_3P, or barium chloride, $BaCl_2$.

(b)   When an element can exhibit more than one oxidation number or valence, Roman numerals are used to denote the oxidation number in the particular compound.

(c)   When an element can exhibit more than one valence, the higher state may be denoted by the suffix *-ic* and the lower by *-ous*. However, this method is now obsolete.

(d)   Traditional names are used for common compounds such as water (dihydrogen oxide) and ammonia (trihydrogen nitride). Sometimes these names designate the mineral source of the compound.

	(a)	(b)	(c)	(d)
BN	boron nitride			borazon
Na_3P	trisodium phosphide			
	sodium phosphide			
$SnCl_2$	tin dichloride	tin(II) chloride	stannous chloride	
$SnCl_4$	tin tetrachloride	tin(IV) chloride	stannic chloride	
HCl	hydrogen chloride			
Fe_3O_4	triiron tetroxide	iron(II,III) oxide	ferroso-ferric oxide	magnetite
O_3	trioxygen			ozone

Table II.1
Names and Common Valences of Some Elements and Groups

Name	Valence	Illustrative compound Ionic	Illustrative compound Nonionic
Metals and cations			
Ammonium	1	NH_4Cl	—
Copper(I) (cuprous)	1	—	Cu_2O
Copper(II) (cupric)	2	$CuCl_2$	—
Hydrogen	1	—	H_2O
Mercury(I) (mercurous)	1	—	Hg_2S
Mercury(II) (mercuric)	2	HgF_2	$HgCl_2$
Potassium	1	K_2SO_4	—
Silver	1	$AgNO_3$	AgI
Sodium	1	$NaCl$	—
Barium	2	$BaCO_3$	—
Cadmium	2	CdF_2	—
Calcium	2	CaO	—
Cobalt	2	$Co(NO_3)_2$	$CoCl_2$
Iron(II) (ferrous)	2	$FeSO_4$	$FeBr_2$
Iron(III) (ferric)	3	$Fe_2(SO_4)_3$	$FeCl_3$
Lead	2	PbF_2	$Pb(C_2H_3O_2)_2$
Magnesium	2	$MgSO_4$	—
Nickel	2	NiO	$NiCl_2$
Strontium	2	$SrCl_2$	—
Tin(II) (stannous)	2	$SnSO_4$	$SnCl_2$
Tin(IV) (stannic)	4	SnO_2	$SnCl_4$
Zinc	2	ZnF_2	$ZnBr_2$
Aluminum	3	Al_2O_3	AlI_3
Antimony	3	—	SbI_3
Bismuth	3	BiF_3	$BiCl_3$
Chromium	3	$Cr_2(SO_4)_3$	$CrCl_3$
Nonmetals and anions			
Acetate	1	$NaC_2H_3O_2$	$HC_2H_3O_2$
Arsenite	1	$KAsO_2$	—
Bicarbonate	1	$NaHCO_3$	—
Bromide	1	$NaBr$	HBr
Bromate	1	$NaBrO_3$	$HBrO_3$
Chloride	1	$NaCl$	HCl
Chlorate	1	$KClO_3$	$HClO_3$
Cyanide	1	KCN	HCN
Dihydrogen phosphate	1	NaH_2PO_4	—
Fluoride	1	KF	$(HF)_x$
Hydride	1	LiH	CH_4
Hydroxide	1	$NaOH$	HOH
Iodide	1	KI	HI
Nitrate	1	$NaNO_3$	HNO_3
Nitrite	1	$NaNO_2$	HNO_2
Permanganate	1	$KMnO_4$	$HMnO_4$
Carbonate	2	$CaCO_3$	—
Chromate	2	K_2CrO_4	—
Dichromate	2	$K_2Cr_2O_7$	—
Hydrogen phosphate	2	Na_2HPO_4	—
Oxygen (oxide)	2	Na_2O	HgO
Oxygen (peroxide)	2	Na_2O_2	H_2O_2
Sulfate	2	Na_2SO_4	H_2SO_4
Sulfide	2	K_2S	CdS
Sulfite	2	Na_2SO_3	—
Arsenate	3	K_3AsO_4	H_3AsO_4
Phosphate	3	Na_3PO_4	H_3PO_4

Further information needed for such nomenclature is given in Table II.1, which includes common valences of various elements and groups.

BINARY ACIDS are compounds of hydrogen and one other element. Type formulas are HX or H_2X. They are named by bracketing the root, derived from the name of the element X, by the prefix *hydro-* or *hydr-* and suffix *-ic* followed by the word *acid*. The common binary acids are restricted mainly to the elements of the halogen and oxygen families (page 442), since the water solutions of the hydrogen compounds of other elements are not sufficiently acidic to warrant the use of the name acid. The corresponding negative ion is named by adding the suffix *-ide* to the root as shown in the tabulation below. Hydrocyanic acid, HCN, is included because it is named as if it were a binary acid. In hydrazoic acid, the root *azo* means nitrogen.

HF	*Hydrofluoric acid*	F^-	Fluoride ion
HCl	*Hydrochloric acid*	Cl^-	Chloride ion
HBr	*Hydrobromic acid*	Br^-	Bromide ion
HI	*Hydriodic acid*	I^-	Iodide ion
H_2S	*Hydrosulfuric acid*	S^{2-}	Sulfide ion
H_2Se	*Hydroselenic acid*	Se^{2-}	Selenide ion
H_2Te	*Hydrotelluric acid*	Te^{2-}	Telluride ion
HCN	*Hydrocyanic acid*	CN^-	Cyanide ion
HN_3	*Hydrazoic acid*	N_3^-	Azide ion

TERNARY COMPOUNDS

Ternary compounds, which are composed of three elements, are usually characterized by the presence of a group; the name is therefore usually derived from the name of the element and the group. Groups containing oxygen are named according to the number of oxygen atoms by a system that is related to the names of the corresponding acids. We will therefore first describe the nomenclature of oxyacids.

OXYACIDS have the general formula H_mXO_n. They possess at least one OH group attached to a central atom. However, only the OH protons are acidic (that is, can be donated to bases). Therefore, only these H's precede the central atom in the formula; the others follow it:

$HC_2H_3O_2$	acetic acid	(monoprotic)
H_2PHO_3	phosphorous acid	(diprotic)

Other examples appear below.

The central element may exist in different oxidation states, and there may be an acid corresponding to each state. The older system of nomenclature uses various prefixes and suffixes to denote these conditions. These designations, together with the newer IUPAC (International Union of Pure and Applied Chemistry) names, are given below.

The oxidation states are designated by the prefixes *hypo-* and *per-*, and by the suffixes -*ous* and -*ic*, according to the following schedule. Note that when there is only one oxidation state, the suffix -*ic* is used. When there is more than one oxidation state, the listings, reading down, are in the order of increasing oxidation states (more oxygen atoms).

Number of oxidation states	Prefix/suffix	Example	Oxidation number of central atom	IUPAC name
One	_____ic	Boric acid, H_3BO_3	+3	Same
Two	_____ous	Sulfurous acid, H_2SO_3	+4	Same
	_____ic	Sulfuric acid, H_2SO_4	+6	Same
Three	hypo_____ous	Hypophosphorous acid, HPH_2O_2	+1	Phosphinic acid
	_____ous	Phosphorous acid, H_2PHO_3	+3	Phosphonic acid
	_____ic	Phosphoric acid, H_3PO_4	+5	Same
Four	hypo_____ous	Hypochlorous acid, $HClO$	+1	Same
	_____ous	Chlorous acid, $HClO_2$	+3	Same
	_____ic	Chloric acid, $HClO_3$	+5	Same
	per_____ic	Perchloric acid, $HClO_4$	+7	Same

Oxyacids in the same oxidation state may differ in water content, and the prefixes *ortho-*, *pyro-*, and *meta-* are used in order of decreasing hydration. When no prefix is used, *ortho-* is sometimes understood, as in phosphoric acid, H_3PO_4.

Prefix	Meaning	Examples	Names Common	IUPAC
Ortho	Contains most water possible	H_3PO_4	Orthophosphoric acid	Phosphoric acid
Pyro	Intermediate water content, an acid that appears to be formed by the removal of 1 H_2O molecule from 2 acid molecules	$H_2S_2O_7$ $H_4P_2O_7$	Pyrosulfuric acid Pyrophosphoric acid	Disulfuric acid Diphosphoric acid
Meta	Contains least water possible (but must retain 1 hydrogen)	HPO_3 HBO_2	Metaphosphoric acid Metaboric acid	Same Same

These relationships are further illustrated by the following equations for the phosphoric acids:

$$2H_3PO_4 \longrightarrow H_2O + H_4P_2O_7$$
ortho   pyro

$$H_3PO_4 \longrightarrow H_2O + HPO_3$$
ortho   meta

The anions and salts of the oxyacids are named by retaining the prefixes but changing the suffixes as follows:

Oxyacid	Anion or salt
_____ic	_____ate
H_2SO_4, sulfuric acid	$SO_4{}^{2-}$, sulfate ion
	Na_2SO_4, disodium sulfate, or sodium sulfate
	$FeSO_4$, iron(II) sulfate
	$Fe_2(SO_4)_3$, iron(III) sulfate
H_3PO_4, phosphoric acid	$PO_4{}^{3-}$, phosphate ion
	Na_3PO_4, trisodium phosphate
_____ous	_____ite
H_2SO_3, sulfurous acid	$SO_3{}^{2-}$, sulfite ion
	Na_2SO_3, disodium sulfite, or sodium sulfite
$HClO_2$, chlorous acid	$NaClO_2$, sodium chlorite
per_____ic	per_____ate
$HClO_4$, perchloric acid	$NaClO_4$, sodium perchlorate
hypo_____ous	hypo_____ite
$HClO$, hypochlorous acid	$NaClO$, sodium hypochlorite

More than one anion may be derived from an acid H_mXO_n when m is greater than 1 because it is possible for the molecule to lose more than 1 hydrogen atom. These anions are differentiated by indicating the number of hydrogen atoms. For example, $H_2PO_4{}^-$ is named dihydrogen phosphate and $HPO_4{}^{2-}$ is named monohydrogen phosphate, although in practice the prefix mono- is frequently omitted. When no more than 1 H atom can be present in the anion, the prefix bi- is often used; $HSO_4{}^-$ and $HCO_3{}^-$ are commonly called bisulfate and bicarbonate, respectively. Thus,

Na_3PO_4	Trisodium phosphate
Na_2HPO_4	Disodium hydrogen phosphate
NaH_2PO_4	Monosodium dihydrogen phosphate
Na_2PHO_3	Disodium phosphite or disodium phosphonate
$NaHPHO_3$	Sodium hydrogen phosphite or sodium hydrogen phosphonate
$NaHSO_4$	Sodium bisulfate or sodium hydrogen sulfate
$NaHCO_3$	Sodium bicarbonate or sodium hydrogen carbonate

COMPLEX COMPOUNDS

The rules presented in this section apply only to the naming of the simple coordination compounds—those with no multidentate* ligands and with only 1 central atom.

* A multidentate ligand, exemplified by ethylenediamine, $H_2NCH_2CH_2NH_2$, has more than one bonding site.

(1) Naming the ligand. Anionic ligands are usually named by adding the letter *o* to the stem name, for example, Cl^- (*chloro*), CN^- (*cyano*), SO_4^{2-} (*sulfato*), NH_2^- (*amido*), $—NO_2^-$ (*nitro*), $—ONO^-$ (*nitrito*), and $S_2O_3^{2-}$ (*thiosulfato*). Molecular ligands are given special names, for example, H_2O (*aquo*), NH_3 (*ammine*), CO (*carbonyl*), NO (*nitrosyl*), and PH_3 (*phosphine*).

(2) Naming compounds with a complex ion. Four rules are observed:

(a) If the compound is ionic, the cation is mentioned first, whether it is the complex ion or not.

(b) In naming the complex, the order is negative ligand, neutral ligand, and then central atom, followed by a Roman numeral in parentheses to indicate the oxidation number.

(c) If the complex is an anion, the suffix *-ate* is appended to the name of the central atom. The oxidation number then follows this suffix.

(d) The number of each kind of ligand is specified, using the Greek prefixes *di-*, *tri-*, *tetra-*, *penta-*, and *hexa-*.

The examples below illustrate these rules:

$[Co(NH_3)_6]Cl_3$	Hexaamminecobalt(III) chloride
	or hexaamminecobalt trichloride
$K_3[Fe(CN)_6]$	Potassium hexacyanoferrate(III)
	or tripotassium hexacyanoferrate
$[Cr(H_2O)_4Cl_2]Cl$	Dichlorotetraaquochromium(III) chloride
$Na_3[Ag(S_2O_3)_2]$	Sodium dithiosulfatoargentate(I)

NOMENCLATURE OF ORGANIC COMPOUNDS / II.2

Many organic compounds are given **COMMON** or **TRIVIAL** names. However, as the number of carbon atoms in a molecule increases, the possibilities for isomerism increase. For example, there are 9 isomers for heptane, and 18 for octane. It would be obviously foolhardy to attempt to give each isomer an unambiguous trivial name. Chemists have developed a method of nomenclature that assigns a systematic, unique name to each possible compound. The system presently used is called the IUPAC system.

(1) Naming saturated compounds. The procedure will be applied to the naming of

$$
\begin{array}{c}
\text{CH}_3 \\
|\\
\text{CH}_2 \\
|\\
\underset{\underset{\textcircled{\scriptsize 7}}{}}{\text{CH}_3}\!-\!\underset{\underset{\textcircled{\scriptsize 6}}{}}{\text{CH}_2}\!-\!\underset{\underset{\textcircled{\scriptsize 5}}{}}{\text{CH}_2}\!-\!\underset{\underset{\textcircled{\scriptsize 4}}{}}{\text{CH}}\!-\!\underset{\underset{\textcircled{\scriptsize 3}}{}}{\text{CH}}\!-\!\text{CH}_3 \\
|\\
\textcircled{\scriptsize 2}\,\text{CH}_2 \\
|\\
\textcircled{\scriptsize 1}\,\text{CH}_3
\end{array}
$$

(a) **Selecting and naming the longest chain.** Name the compound as a DERIVATIVE of the alkane (saturated, noncyclic hydrocarbon) represented by the longest chain or continuous sequence of carbon atoms. In the above compound, the longest chain contains 7 carbons as numbered. The compound is therefore named as a heptane, and is considered to be a *derivative* of this parent compound. The name must end in the suffix -*ane* to indicate that the compound is a member of the alk*ane* series. Except for the first four members of the alkane series (page 479), the prefix of the name is derived from the Greek word for the number of carbon atoms in the chain. Thus, for example, *hept-* indicates 7 carbons.

(b) **Numbering the longest carbon chain.** Number the carbon atoms in the longest chain, as illustrated, so that the branching groups are on the lower- rather than the higher-numbered carbon atoms.

(c) **Locating the branching groups.** The location of the branching hydrocarbon groups is indicated by the number of the carbon atom of the parent chain to which they are attached. Thus, in our illustrative compound, we find a CH_3 group on C-3 and a CH_3CH_2 group on C-4.

(d) **Naming the branched groups.** Each branched group is named according to the number of carbons in its longest chain. To name the group, replace the suffix -*ane* by -*yl*. Thus, CH_3 is called meth*yl* since it stems from meth*ane*, and CH_3CH_2 is named eth*yl* from eth*ane*.

Some other prefixes are

Fluoro	F—		
Chloro	Cl—		
Bromo	Br—		
Iodo	I—		
Hydroxy	HO—	Phenyl	
Mercapto	HS—		
Amino	H_2N—		
Nitro	O_2N—		

(e) **Arranging the names of the branched groups.** The names of these groups are arranged in alphabetical order and each is prefixed with the number of the carbon atom to which each group is attached. Thus, the name of the illustrative compound is 4-ethyl-3-methylheptane.

The rules of punctuation are: (*i*) words are run together; (*ii*) a dash is used to separate a number from a word; and (*iii*) a comma is used to separate numbers.

The following examples further illustrate the method:

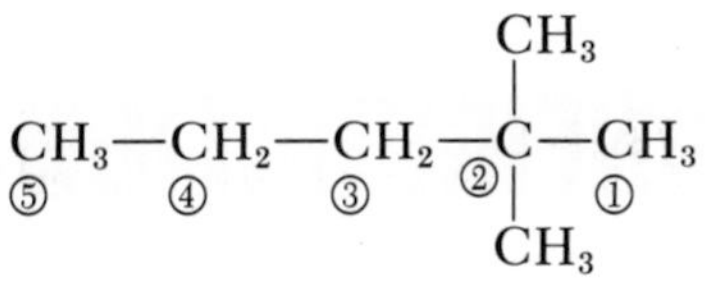

$$CH_3-CH_2-CH_2-\underset{\underset{CH_3}{|}}{\overset{\overset{CH_3}{|}}{C}}-CH_3$$
⑤　④　③　②｜　①

2,2-dimethylpentane

$$CH_3-\underset{\underset{①}{CH_3}}{\overset{}{CH}}-\underset{\underset{③}{CH_3}}{\overset{\overset{CH_2}{|}}{\underset{②}{C}}}\underset{}{\overset{\overset{H}{}}{\underset{⑤ CH_2}{\overset{}{\underset{|}{C}}}}}-CH_3$$
①　②｜　③｜　④｜
　　CH_3　CH_3 ⑤ CH_2
　　　　　　　⑥ CH_3

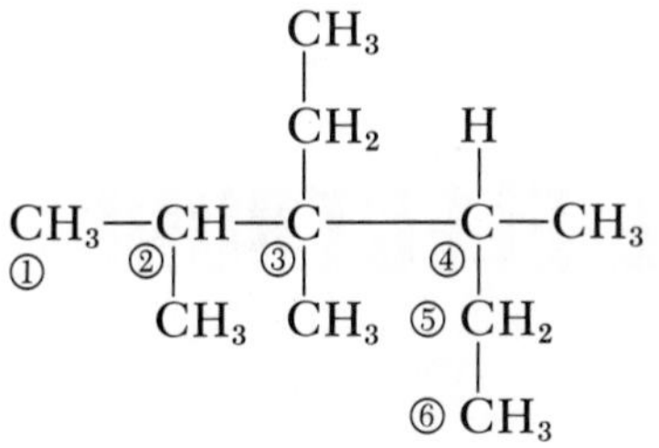

3-ethyl-2,3,4-trimethylhexane

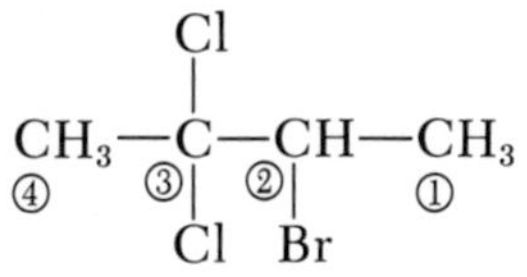

$$CH_3-\underset{\underset{Cl}{|}}{\overset{\overset{Cl}{|}}{C}}-\underset{\underset{Br}{|}}{CH}-CH_3$$
　　　④　③｜　②｜　①
　　　　　Cl　Br

2-bromo-3,3-dichlorobutane

Note the repetition of numbers when a group appears twice on the same carbon, and the use of *di-* and *tri-* to indicate the number of times the group appears as a branch.

(2) **Naming unsaturated compounds.** The C=C double bond is named by the suffix -*ene* and the C≡C triple bond by the suffix -*yne*. The name is prefixed with the number of the first multiply bonded carbon atom in the chain. Examples are:

$$CH_2-CH=CH-CH_2-CH_3$$
2-pentene

$$HC≡C-CH_2-CH_3$$
1-butyne

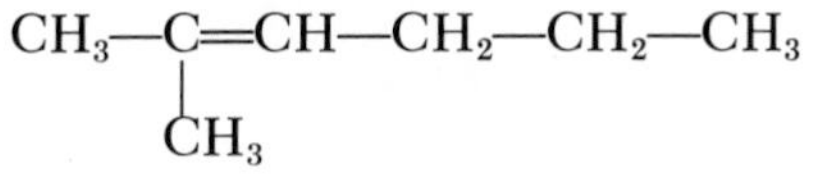

$$CH_3-\underset{\underset{CH_3}{|}}{C}=CH-CH_2-CH_2-CH_3$$
2-methyl-2-hexene

APPENDIX III

THE ELECTRONIC CONFIGURATIONS OF THE ELEMENTS

Abbreviations: $K = 1s^2$ *Note:* Underlined subshells are partly filled
$L = 2s^2 2p^6$
$M = 3s^2 3p^6 3d^{10}$
$N = 4s^2 4p^6 4d^{10} 4f^{14}$

Representative elements, atomic numbers 1–20, 30–38, 48–56, 80–88: electron enters an s or p subshell.
Transition elements, atomic numbers 21–29, 39–47, 72–79: electron enters a d subshell.
Inner transition elements, atomic numbers 57–71, 89–103: electron enters an f subshell.

$_1$H	$\underline{1s^1}$	$_{15}$P	$KL\ 3s^2\underline{3p^3}$	$_{29}$Cu	$KLM\ \underline{4s^1}$
$_2$He	$1s^2$	$_{16}$S	$KL\ 3s^2\underline{3p^4}$	$_{30}$Zn	$KLM\ 4s^2$
$_3$Li	$K\ \underline{2s^1}$	$_{17}$Cl	$KL\ 3s^2\underline{3p^5}$	$_{31}$Ga	$KLM\ 4s^2\underline{4p^1}$
$_4$Be	$K\ 2s^2$	$_{18}$Ar	$KL\ 3s^2 3p^6$	$_{32}$Ge	$KLM\ 4s^2\underline{4p^2}$
$_5$B	$K\ 2s^2\underline{2p^1}$	$_{19}$K	$KL\ 3s^2 3p^6\underline{4s^1}$	$_{33}$As	$KLM\ 4s^2\underline{4p^3}$
$_6$C	$K\ 2s^2\underline{2p^2}$	$_{20}$Ca	$KL\ 3s^2 3p^6 4s^2$	$_{34}$Se	$KLM\ 4s^2\underline{4p^4}$
$_7$N	$K\ 2s^2\underline{2p^3}$	$_{21}$Sc	$KL\ 3s^2 3p^6\underline{3d^1}4s^2$	$_{35}$Br	$KLM\ 4s^2\underline{4p^5}$
$_8$O	$K\ 2s^2\underline{2p^4}$	$_{22}$Ti	$KL\ 3s^2 3p^6\underline{3d^2}4s^2$	$_{36}$Kr	$KLM\ 4s^2 4p^6$
$_9$F	$K\ 2s^2\underline{2p^5}$	$_{23}$V	$KL\ 3s^2 3p^6\underline{3d^3}4s^2$	$_{37}$Rb	$KLM\ 4s^2 4p^6\underline{5s^1}$
$_{10}$Ne	KL	$_{24}$Cr	$KL\ 3s^2 3p^6\underline{3d^5}4s^1$	$_{38}$Sr	$KLM\ 4s^2 4p^6 5s^2$
$_{11}$Na	$KL\ \underline{3s^1}$	$_{25}$Mn	$KL\ 3s^2 3p^6\underline{3d^5}4s^2$	$_{39}$Y	$KLM\ 4s^2 4p^6\underline{4d^1}5s^2$
$_{12}$Mg	$KL\ 3s^2$	$_{26}$Fe	$KL\ 3s^2 3p^6\underline{3d^6}4s^2$	$_{40}$Zr	$KLM\ 4s^2 4p^6\underline{4d^2}5s^2$
$_{13}$Al	$KL\ 3s^2\underline{3p^1}$	$_{27}$Co	$KL\ 3s^2 3p^6\underline{3d^7}4s^2$	$_{41}$Nb	$KLM\ 4s^2 4p^6\underline{4d^4}5s^1$
$_{14}$Si	$KL\ 3s^2\underline{3p^2}$	$_{28}$Ni	$KL\ 3s^2 3p^6\underline{3d^8}4s^2$	$_{42}$Mo	$KLM\ 4s^2 4p^6\underline{4d^5}5s^1$

$_{43}$Tc KLM $4s^24p^64d^65s^1$

$_{44}$Ru KLM $4s^24p^64d^75s^1$

$_{45}$Rh KLM $4s^24p^64d^85s^1$

$_{46}$Pd KLM $4s^24p^64d^{10}$

$_{47}$Ag KLM $4s^24p^64d^{10}5s^1$

$_{48}$Cd KLM $4s^24p^64d^{10}5s^2$

$_{49}$In KLM $4s^24p^64d^{10}5s^25p^1$

$_{50}$Sn KLM $4s^24p^64d^{10}5s^25p^2$

$_{51}$Sb KLM $4s^24p^64d^{10}5s^25p^3$

$_{52}$Te KLM $4s^24p^64d^{10}5s^25p^4$

$_{53}$I KLM $4s^24p^64d^{10}5s^25p^5$

$_{54}$Xe KLM $4s^24p^64d^{10}5s^25p^6$

$_{55}$Cs KLM $4s^24p^64d^{10}5s^25p^66s^1$

$_{56}$Ba KLM $4s^24p^64d^{10}5s^25p^66s^2$

$_{57}$La KLM $4s^24p^64d^{10}5s^25p^65d^16s^2$

$_{58}$Ce KLM $4s^24p^64d^{10}4f^25s^25p^66s^2$ (?)

$_{59}$Pr KLM $4s^24p^64d^{10}4f^35s^25p^66s^2$ (?)

$_{60}$Nd KLM $4s^24p^64d^{10}4f^45s^25p^66s^2$

$_{61}$Pm KLM $4s^24p^64d^{10}4f^55s^25p^66s^2$ (?)

$_{62}$Sm KLM $4s^24p^64d^{10}4f^65s^25p^66s^2$

$_{63}$Eu KLM $4s^24p^64d^{10}4f^75s^25p^66s^2$

$_{64}$Gd KLM $4s^24p^64d^{10}4f^75s^25d^15p^66s^2$

$_{65}$Tb KLM $4s^24p^64d^{10}4f^95s^25p^66s^2$ (?)

$_{66}$Dy KLM $4s^24p^64d^{10}4f^{10}5s^25p^66s^2$ (?)

$_{67}$Ho KLM $4s^24p^64d^{10}4f^{11}5s^25p^66s^2$ (?)

$_{68}$Er KLM $4s^24p^64d^{10}4f^{12}5s^25p^66s^2$

$_{69}$Tm KLM $4s^24p^64d^{10}4f^{13}5s^25p^66s^2$

$_{70}$Yb $KLMN$ $5s^25p^66s^2$

$_{71}$Lu $KLMN$ $5s^25p^65d^16s^2$

$_{72}$Hf $KLMN$ $5s^25p^65d^26s^2$

$_{73}$Ta $KLMN$ $5s^25p^65d^36s^2$

$_{74}$W $KLMN$ $5s^25p^65d^46s^2$

$_{75}$Re $KLMN$ $5s^25p^65d^56s^2$

$_{76}$Os $KLMN$ $5s^25p^65d^66s^2$

$_{77}$Ir $KLMN$ $5s^25p^65d^76s^2$

$_{78}$Pt $KLMN$ $5s^25p^65d^96s^1$

$_{79}$Au $KLMN$ $5s^25p^65d^{10}6s^1$

$_{80}$Hg $KLMN$ $5s^25p^65d^{10}6s^2$

$_{81}$Tl $KLMN$ $5s^25p^65d^{10}6s^26p^1$

$_{82}$Pb $KLMN$ $5s^25p^65d^{10}6s^26p^2$

$_{83}$Bi $KLMN$ $5s^25p^65d^{10}6s^26p^3$

$_{84}$Po $KLMN$ $5s^25p^65d^{10}6s^26p^4$

$_{85}$At $KLMN$ $5s^25p^65d^{10}6s^26p^5$

$_{86}$Rn $KLMN$ $5s^25p^65d^{10}6s^26p^6$

$_{87}$Fr $KLMN$ $5s^25p^65d^{10}6s^26p^67s^1$

$_{88}$Ra $KLMN$ $5s^25p^65d^{10}6s^26p^67s^2$

$_{89}$Ac $KLMN$ $5s^25p^65d^{10}6s^26p^66d^17s^2$

$_{90}$Th $KLMN$ $5s^25p^65d^{10}6s^26p^66d^27s^2$

$_{91}$Pa $KLMN$ $5s^25p^65d^{10}5f^26s^26p^66d^17s^2$

$_{92}$U $KLMN$ $5s^25p^65d^{10}5f^36s^26p^66d^17s^2$

$_{93}$Np $KLMN$ $5s^25p^65d^{10}5f^46s^26p^66d^17s^2$

$_{94}$Pu $KLMN$ $5s^25p^65d^{10}5f^66s^26p^67s^2$

$_{95}$Am $KLMN$ $5s^25p^65d^{10}5f^76s^26p^67s^2$

$_{96}$Cm $KLMN$ $5s^25p^65d^{10}5f^76s^26p^66d^17s^2$

$_{97}$Bk $KLMN$ $5s^25p^65d^{10}5f^96s^26p^67s^2$ (?)

$_{98}$Cf $KLMN$ $5s^25p^65d^{10}5f^{10}6s^26p^67s^2$

$_{99}$Es $KLMN$ $5s^25p^65d^{10}5f^{11}6s^26p^67s^2$

$_{100}$Fm $KLMN$ $5s^25p^65d^{10}5f^{12}6s^26p^67s^2$

$_{101}$Md $KLMN$ $5s^25p^65d^{10}5f^{13}6s^26p^67s^2$

$_{102}$No $KLMN$ $5s^25p^65d^{10}5f^{14}6s^26p^67s^2$

$_{103}$Lr $KLMN$ $5s^25p^65d^{10}5f^{14}6s^26p^66d^17s^2$

APPENDIX IV

FUNDAMENTAL CONSTANTS

c	Speed of light *in vacuo*	2.997925×10^{10} cm/second
e	Charge on the electron	1.60219×10^{-19} C
$\mathscr{F}$	Faraday's constant	9.64846×10^{4} C/mole
h	Planck's constant	6.6262×10^{-27} erg second/particle
k	Boltzmann's constant	1.38066×10^{-16} erg/°K molecule
N_A	Avogadro's number	6.02205×10^{23} particles/mole
R	Ideal gas constant	8.2057×10^{-2} liter atm/°K mole
		8.31441 joules/°K mole
		1.98717 cal/°K mole
T_{ice}	Ice point	$273.1500°K = 0°C$
V_0	Molar gas volume	22.4138 liters/mole (standard conditions)

From *Dimensions/NBS*, Jan. 1974, pp. 4–5.

VAPOR PRESSURE
OF WATER

Temperature, °C	Pressure, torr	Temperature, °C	Pressure, torr
0	4.6	24	22.4
5	6.5	25	23.8
10	9.2	26	25.2
11	9.8	27	26.7
12	10.5	28	28.3
13	11.2	29	30.0
14	12.0	30	31.8
15	12.8	35	42.2
16	13.6	40	55.3
17	14.5	60	149.4
18	15.5	80	355.1
19	16.5	100	760.0
20	17.5	110	1075
21	18.7		
22	19.8		
23	21.1		

APPENDIX VI

ABBREVIATIONS

ampere, amp or A
angström, Å
atmosphere, atm
calorie, cal
coulomb, coul or C
cubic centimeter, cm^3
cubic foot, ft^3
degree Celsius (Centigrade), °C
degree Fahrenheit, °F
degree Kelvin, °K or K
disintegrations per minute, dpm
electron volt, eV
energy, E
enthalpy, H
entropy, S
foot, ft
free energy, G

gram, g
Hertz, Hz $(=sec^{-1})$
hour, hr
inch, in.
joule, J
kilocalorie, kcal
kilogram, kg
liter, l
minute, min
mole, mol
molal (mol/kg solvent), m
molar (mol/1), M
ohm, Ω
pound, lb
second, s or sec
volt, V
watt, W

LOGARITHMS

Logarithms to the Base 10

N	0	1	2	3	4	5	6	7	8	9
10	0000	0043	0086	0128	0170	0212	0253	0294	0334	0374
11	0414	0453	0492	0531	0569	0607	0645	0682	0719	0755
12	0792	0828	0864	0899	0934	0969	1004	1038	1072	1106
13	1139	1173	1206	1239	1271	1303	1335	1367	1399	1430
14	1461	1492	1523	1553	1584	1614	1644	1673	1703	1732
15	1761	1790	1818	1847	1875	1903	1931	1959	1987	2014
16	2041	2068	2095	2122	2148	2175	2201	2227	2253	2279
17	2304	2330	2355	2380	2405	2430	2455	2480	2504	2529
18	2553	2577	2601	2625	2648	2672	2695	2718	2742	2765
19	2788	2810	2833	2856	2878	2900	2923	2945	2967	2989
20	3010	3032	3054	3075	3096	3118	3139	3160	3181	3201
21	3222	3243	3263	3284	3304	3324	3345	3365	3385	3404
22	3424	3444	3464	3483	3502	3522	3541	3560	3579	3598
23	3617	3636	3655	3674	3692	3711	3729	3747	3766	3784
24	3802	3820	3838	3856	3874	3892	3909	3927	3945	3962
25	3979	3997	4014	4031	4048	4065	4082	4099	4116	4133
26	4150	4166	4183	4200	4216	4232	4249	4265	4281	4298
27	4314	4330	4346	4362	4378	4393	4409	4425	4440	4456
28	4472	4487	4502	4518	4533	4548	4564	4579	4594	4609
29	4624	4639	4654	4669	4683	4698	4713	4728	4742	4757
30	4771	4786	4800	4814	4829	4843	4857	4871	4886	4900
31	4914	4928	4942	4955	4969	4983	4997	5011	5024	5038
32	5051	5065	5079	5092	5105	5119	5132	5145	5159	5172
33	5185	5198	5211	5224	5237	5250	5263	5276	5289	5302
34	5315	5328	5340	5353	5366	5378	5391	5403	5416	5428
N	0	1	2	3	4	5	6	7	8	9

N	0	1	2	3	4	5	6	7	8	9
35	5441	5453	5465	5478	5490	5502	5514	5527	5539	5551
36	5563	5575	5587	5599	5611	5623	5635	5647	5658	5670
37	5682	5694	5705	5717	5729	5740	5752	5763	5775	5786
38	5798	5809	5821	5832	5843	5855	5866	5877	5888	5899
39	5911	5922	5933	5944	5955	5966	5977	5988	5999	6010
40	6021	6031	6042	6053	6064	6075	6085	6096	6107	6117
41	6128	6138	6149	6160	6170	6180	6191	6201	6212	6222
42	6232	6243	6253	6263	6274	6284	6294	6304	6314	6325
43	6335	6345	6355	6365	6375	6385	6395	6405	6415	6425
44	6435	6444	6454	6464	6474	6484	6493	6503	6513	6522
45	6532	6542	6551	6561	6571	6580	6590	6599	6609	6618
46	6628	6637	6646	6656	6665	6675	6684	6693	6702	6712
47	6721	6730	6739	6749	6758	6767	6776	6785	6794	6803
48	6812	6821	6830	6839	6848	6857	6866	6875	6884	6893
49	6902	6911	6920	6928	6937	6946	6955	6964	6972	6981
50	6990	6998	7007	7016	7024	7033	7042	7050	7059	7067
51	7076	7084	7093	7101	7110	7118	7126	7135	7143	7152
52	7160	7168	7177	7185	7193	7202	7210	7218	7226	7235
53	7243	7251	7259	7267	7275	7284	7292	7300	7308	7316
54	7324	7332	7340	7348	7356	7364	7372	7380	7388	7396
55	7404	7412	7419	7427	7435	7443	7451	7459	7466	7474
56	7482	7490	7497	7505	7513	7520	7528	7536	7543	7551
57	7559	7566	7574	7582	7589	7597	7604	7612	7619	7627
58	7634	7642	7649	7657	7664	7672	7679	7686	7694	7701
59	7709	7716	7723	7731	7738	7745	7752	7760	7767	7774
60	7782	7789	7796	7803	7810	7818	7825	7832	7839	7846
61	7853	7860	7868	7875	7882	7889	7896	7903	7910	7917
62	7924	7931	7938	7945	7952	7959	7966	7973	7980	7987
63	7993	8000	8007	8014	8021	8028	8035	8041	8048	8055
64	8062	8069	8075	8082	8089	8096	8102	8109	8116	8122
65	8129	8136	8142	8149	8156	8162	8169	8176	8182	8189
66	8195	8202	8209	8215	8222	8228	8235	8241	8248	8254
67	8261	8267	8274	8280	8287	8293	8299	8306	8312	8319
68	8325	8331	8338	8344	8351	8357	8363	8370	8376	8382
69	8388	8395	8401	8407	8414	8420	8426	8432	8439	8445
70	8451	8457	8463	8470	8476	8482	8488	8494	8500	8506
71	8513	8519	8525	8531	8537	8543	8549	8555	8561	8567
72	8573	8579	8585	8591	8597	8603	8609	8615	8621	8627
73	8633	8639	8645	8651	8657	8663	8669	8675	8681	8686
74	8692	8698	8704	8710	8716	8722	8727	8733	8739	8745
N	0	1	2	3	4	5	6	7	8	9

N	0	1	2	3	4	5	6	7	8	9
75	8751	8756	8762	8768	8774	8779	8785	8791	8797	8802
76	8808	8814	8820	8825	8831	8837	8842	8848	8854	8859
77	8865	8871	8876	8882	8887	8893	8899	8904	8910	8915
78	8921	8927	8932	8938	8943	8949	8954	8960	8965	8971
79	8976	8982	8987	8993	8998	9004	9009	9015	9020	9025
80	9031	9036	9042	9047	9053	9058	9063	9069	9074	9079
81	9085	9090	9096	9101	9106	9112	9117	9122	9128	9133
82	9138	9143	9149	9154	9159	9165	9170	9175	9180	9186
83	9191	9196	9201	9206	9212	9217	9222	9227	9232	9238
84	9243	9248	9253	9258	9263	9269	9274	9279	9284	9289
85	9294	9299	9304	9309	9315	9320	9325	9330	9335	9340
86	9345	9350	9355	9360	9365	9370	9375	9380	9385	9390
87	9395	9400	9405	9410	9415	9420	9425	9430	9435	9440
88	9445	9450	9455	9460	9465	9469	9474	9479	9484	9489
89	9494	9499	9504	9509	9513	9518	9523	9528	9533	9538
90	9542	9547	9552	9557	9562	9566	9571	9576	9581	9586
91	9590	9595	9600	9605	9609	9614	9619	9624	9628	9633
92	9638	9643	9647	9652	9657	9661	9666	9671	9675	9680
93	9685	9689	9694	9699	9703	9708	9713	9717	9722	9727
94	9731	9736	9741	9745	9750	9754	9759	9763	9768	9773
95	9777	9782	9786	9791	9795	9800	9805	9809	9814	9818
96	9823	9827	9832	9836	9841	9845	9850	9854	9859	9863
97	9868	9872	9877	9881	9886	9890	9894	9899	9903	9908
98	9912	9917	9921	9926	9930	9934	9939	9943	9948	9952
99	9956	9961	9965	9969	9974	9978	9983	9987	9991	9996
N	0	1	2	3	4	5	6	7	8	9

$$E = E^\circ - \frac{2.303RT}{nF} \log Q$$

$$R = 9.6487 \times 10^{-4} \, C/\text{Far.}$$

ANSWERS

CHAPTER 2

2.
(a) 373°K
(b) 546°K
(c) 0°K
(d) 195.0°K
(e) −272.7°C
(f) 1.1×10^5 °C
(g) 0°C

3.
(a) 250 ml
(b) 1.00 liter
(c) 4.2×10^5 liters

4. (b) 18.4 liters

5. (a) 0.020 ml/°K = V/T

6.
(a) 3.71 liters
(b) p_{N_2}, 450 torr (initial and final); p_{O_2}, 310 torr (initial), 0 torr (final)

7.
(a) 5.0×10^{10} molecules
(b) 2.0×10^{-5} ml

8. 24.6 liters

9. 59.6 ml

14.
(a) 1.41 liters
(b) 160 cm

16.
(a) 193.7 ml
(b) 6.3 ml

18. 0.227 g/liter

19. 0.919 g/liter

20. 227 liters

21.
(a) 1×10^{-4} liters
(b) 6×10^{10} liters

23.
(a) 62.4 liters torr/°K mole
(b) 1.36×10^{-25} liter atm/°K molecule
(c) 1.99 cal/°K mole

CHAPTER 3

2. 25.5%, 0.342

3. 40.8%, 20.4 g

4.
(a) 4.00
(b) 3.00 liters

6. 1 to 8

7. $PtCl_2N_2H_6$

8.
(a) 54.9 g, 3.01×10^{23} molecules
(b) 454 moles, 2.73×10^{26} molecules

9.
(a) 231 g/mole
(b) 2.61×10^{17} molecules

10. (a) $C_8H_{10}N_4O_2$

11. 12.01

12. 16.3 g

13.
(a) 0.554 g
(b) 2.00 g
(c) 88.87%

14.
(a) 41.6%
(b) 50.0%

15.
(a) 8.3 g
(b) HS

17. 15, 10 liters

18. E_2, F, E_2F_3, EF_2

19.
(a) 2.48 moles, 1.49×10^{24} atoms
(b) 1.78, 1.07×10^{24}
(c) 0.263, 1.58×10^{23}

20.
(a) (i) 1.56
   (ii) 50.0
   (iii) 4.15
   (iv) 8.30 moles O
(b) (i) 0.705
   (ii) 25.0
   (iii) 4.15
   (iv) 8.30 moles Cl_2

22. 1.0078

23. 1.5×10^3 atoms/liter

24. 1:3

25.
(b) 195, 35.5, $PtCl_4$, PtCl, $PtCl_3$, $PtCl_2$, Cl_2

27. 10.81

28. $C_6H_{12}O_3NaSN$

29.
(a) XeF_6
(b) CH_2, C_2H_4

30. 299 g/mole

31. $C_{27}H_{46}O$

32. 238.062995, 238.074228

33. $HBrO_4$

CHAPTER 4

5.
(a) $\dfrac{\text{foot}}{12 \text{ inches}}$

(b) $\dfrac{\text{liter}}{10^3 \text{ ml}}$

(c) $\dfrac{2 \times 6 \times 10^{23} \text{ molecules } B_5H_9}{5 \text{ moles } B_2H_6}$

(d) (i) $\dfrac{426 \text{ g I(ClO}_4)_3}{3 \text{ moles AgClO}_4}$

   (ii) $\dfrac{426 \text{ g I(ClO}_4)_3}{2 \times 22.4 \times 10^3 \text{ ml (sc) } I_2}$

   (iii) $\dfrac{426 \text{ g I(ClO}_4)_3}{3 \times 235 \text{ g AgI}}$

   (iv) $\dfrac{426 \text{ g I(ClO}_4)_3}{2 \text{ moles } I_2}$

(e) $\dfrac{79.904 \text{ g Br}^-}{187.78 \text{ g AgBr}}$

6. 7.7 g

7.
(a) 35.0 liters
(b) 19.2 liters

8. 1.07×10^{25} molecules

9.
(b) 2.5 g
(c) 5.5 g

10.
(a) 1.7×10^{-9} g
(b) 4.1×10^{-8} ml

11. 98%

14.
(a) 2.80 liters
(b) 8.21 liters

15.
(a) 215 g
(b) 165 g

16.
(a) 58.6 g
(b) 335 liters
(c) 2.69 g

17.
(a) 1.04 g
(b) 53.9 g

18.
(a) 16.8 liters
(b) 22.0 g

19. 28.5 mg

21.
(a) 96.0%
(b) 8.0 g, 24.0 g, 1:3

22. HO_2, HO, HO_2 (H_2O_4), HO_3 (H_2O_6), HO_4 (H_2O_8)

23. 90.2%

CHAPTER 5

2. 397 cal

3.
(a) 40 cal
(b) (d) 539 cal
(c) 499 cal

4.
(a) (b) 1050.8 cal
(c) 84.4 cal
(d) 966.4 cal

5.
(a) (b) −41 kcal
(c) −1.0 kcal
(d) (e) −42 kcal

6. 5.00 kcal

8.
(a) 68.3 kcal
(b) (*i*) 2.9 kcal
(*ii*) 1.3 kcal

9. 3.0 kcal

10.
(a) −48.5 kcal
(b) (*i*) 0.388 cal
(*ii*) 97 deg

11.
(a) −221.6 kcal
(b) 2 bonds, 111 kcal

12.
(a) −47 kcal
(b) 355 kcal

13. −136 kcal

16. 2×10^{-11} g

18. 209 J

19.
(a) 0
(b) (c) +50 cal

21. 2.7 Cal/day

22.
(a) −1346 kcal
(b) −216 kcal

23. −543.8 kcal

25. 106 kcal

26. 141 kcal

27. 4×10^{17} tons, 2×10^{-14}%

28.
(a) 1.626 kcal
(b) 1.311 kg

29. 0.8225 kcal/deg

30.
(a) 5.0×10^{16} kcal
(b) 1.4×10^{21} kcal

CHAPTER 6

7. -21.79×10^{-16} erg/atom

8. 12.5 kcal/mole

9. 3; 7.57×10^{-13}, 4.086×10^{-12}, 1.634×10^{-11} erg/photon

10. 1

11.
(a) 2, 1; 4, 0; 3, 2; 6, 0
(b) 10s, 18p, 12d

12.
(a) Ag, Na
(b) Co^{4+}

13. a, c, d

15. 1.6×10^{-19}C

16.
(a) 1.60×10^{-19}C
(b) 10, 1, 8, 1, 15

17. 5.48×10^{-4} g/mole, 9.10×10^{-28} g/electron

19. 7.829×10^{-12} erg

20. 1.3×10^{-19} erg/photon

21. 20.43×10^{-12} erg, 9.722×10^{-6} cm

24. $_{23}V^{3+}$ KL $3s^2 3p^6 3d^2$

25. KL $3s^2 3p^6 3d^5 4s^1$

26.
(a) 5
(b) 1
(c) 0
(d) 4

CHAPTER 8

14. −144 kcal/mole

22. −183.1 kcal/mole

CHAPTER 9

3. 2 atoms/unit cell

14.
(a) 4 atoms/unit cell
(b) 19.3 g/cm³
(c) 1.44 Å

16.
(a) 2 atoms/unit cell
(b) 8
(c) 10.8 cm³/mole of atoms
(d) 3 Å³
(e) 1.43 Å

CHAPTER 10

2.
(a) 0.840 mole/liter
(b) 0.704 mole/liter
(c) 0.555 mole/liter

3.
(a) 0.050 mole
(b) 16.7 ml

4.
(a) 0.17
(b) 8.9 moles/liter

5.
(a) 10.0 mmole
(b) 365 mg
(c) 125 ml

6.
(a) 35%
(b) 4.9 moles/liter

8. 50 torr

9.
(a) −0.612°C
(b) 7.88 atm

10. 74 g/mole

11.
(a) 7.7×10^3 g/mole
(b) −0.0012°C

12.
(a) 4.250 g
(b) 19.61 g
(c) 10.21 g

13.
(a) 18.4%
(b) 19.3%

14. 0.1001 M

15.
(a) 15%
(b) 7.7 moles/liter
(c) 26 ml

16. 131 g

17. 39.6 g

21. 1.203 g/ml

22.
(a) 63.3 ml
(b) 50.1 ml
(c) 109.5 ml
(d) 3.9 ml less

23.
(a) 226 torr
(b) 0.690

24. 115, 62 g/mole

25. (a) 172 g, 334 g

26.
(a) 300 g/mole
(b) $C_{24}H_{12}$

27.
(a) 6.45×10^4 g/mole
(b) 2.9×10^{-5} °C

CHAPTER 11

4. 0.68

5.
(a) 0.0178 mole/liter
(b) 0.0289 mole/liter
(c) 0.53

6. 3.5

7. 0.050 mole/liter

8.
(a) 0.0664 mole
(b) 30.6%

9.
(a) 0.062
(b) 23%

10.
(a) 0.022 mole/liter, 0.53 atm
(b) 7.0×10^{-3} mole/liter

11. (a) 399, 21%

12. 8.5×10^{-5}

13. 0.0115, 1.5×10^{-3}, 8.5×10^{-3} mole/liter, 4.2

14.
(a) 0.61, 0.78 mole/liter, 42 atm
(b) 2.00
(d) 1.41, 1.19 mole/liter, 78 atm, 1.86

15.
(a) 6.8×10^{-3}, 1.75×10^{-2} mole/liter
(b) 4.5×10^{-2}

16. 0.011

17.
(b) 400
(c) 0.50%

18. 5.5×10^{-4} mole

19.
(a) 8.28×10^{-4} mole/liter
(b) 1.46×10^{-4}, 6.82×10^{-4} mole/liter, 82%

20.
(a) 0.0114 mole/liter
(b) 1.61×10^{19}

21.
(a) 9.4×10^{-42}
(b) 0.014

22. 7.99×10^{-6} mole/liter

23. (a) 59%

24.
(a) 0.019
(b) 27.4%

25. 4×10^{-5} mole/liter

CHAPTER 12

2. 2

4.
(a) 2.90×10^{-3} mole
(b) 0.0927 g
(c) 0.0649 liter
(d) 1118 C
(e) 0.310 A

5. $+3$

6. 96.6 Ω^{-1} cm^2 faraday^{-1}

7. 2.3%

8. $-0.0372°C$; experimental, $-0.03556°C$

9.
(a) 2.2

10. 30 atm

13. 0.303 g Zn, 3.01 g $Hg(NO_3)_2$

14.
(a) 9.57 hours
(b) 0.180 g

15.
(a) 3
(b) -3

16.
(a) 2.06 kg
(b) 1.00×10^3 liters

17. (c) 2.41×10^7 C

18. (a) 0.38 kg

19.
(a) 1.19 g
(b) 49%
(c) 4.61×10^{-3} mole

20. 0.99996

22. 5.5×10^{-8} Ω^{-1} cm^{-1}

23. 1.3×10^{-5} mole/liter

24.
(a) 0.998, 0.981
(b) 0.012, 0.15

CHAPTER 14

2.
(a) 3.30
(b) 6.89

3. 7.472

4. 1.5×10^{-5}, 4.84

5.
(a) 0.046
(b) 2.03

6. 11.77

7. 11.59

8. 2.97

9. 2.20×10^{-10}

10. 7.0×10^{-4} mole/liter

11.
(a) 9.07
(b) 8.98
(c) 1.0

12. 7.1×10^{-8}

13.
(a) 1.5×10^{-5}
(b) 1.1×10^{-9} mole/liter

15.
(a) 4×10^{-8}
(b) 4×10^{-20} mole/liter

16.
(a) 1.00×10^{-8}, 6.00
(b) 3.0×10^{-5}, 3.3×10^{-10}
(c) 1.0×10^{-7}, 1.1×10^{-8}
(d) 2.0×10^{-11}, 10.70

17. 7.435

18. 3.05

19. 4.80

20.
(a) 0.27
(b) 2.26

21. 6.71 ml

22.
(a) 2.12×10^{-3}
(b) 1.91×10^{-3} mole/liter

23.
(d) $(\sqrt{K_w}/K_a + 1)^{-1}$
(e) 0.9943, 4.9×10^{-3}

24. 8.68

25.
(a) 10^{-8}
(b) 10^{-7}

26.
(a) 2.5×10^{-10}
(b) 5.45

27.
(a) 1.30
(b) 6.24×10^{-8} mole/liter

28.
(a) (b) 0.024
(c) 6.22×10^{-8}
(d) 1.2×10^{-18} mole/liter

29. (b) 6.89

30.
(a) 5.68×10^{-11}
(c) 7.10

32.
(a) 5.57×10^{-3}
(b) 3.57×10^{-7} mole/liter

33. 1.7×10^{-3}

34. 3.66

35. 0.59 mole/liter

36.
(a) Formic
(b) 1.27

37. 7.67

38. 512 ml, 488 ml

39.
(a) 1.03×10^{-4}
(b) 5.3×10^{-8} mole/liter

40.
(a) 6.7×10^{-8}
(b) 2.6×10^{-4} g/100 ml

41.
(a) 7.5×10^{-3}, 3.1×10^{-3} mole/liter

42. -2.0

43. (a) 3.8

CHAPTER 15

1. 20.0 ml

2. 2.942 g in 500 ml

3.
(a) 0.08992 M
(b) 0.1466 M

4.
(a) 22.3 mg
(b) 32.0 mg

5.
(b) (*i*) 8.60
   (*ii*) 5.12

6.
(a) 20.0 ml
(b) 133 ml
(c) 40.0 ml

7. 0.150 M

8. 2.00 M, 79.0 g

9.
(a) 28.046, 0.001, −0.008
(b) 28.3, 0.1, +0.2

10. 325.8 mg, 56.56%

11. 221.1 mg

13.
(a) 5.00×10^{-3} M
(b) 670 mg

14.
(a) 201 mg, 4.02 g
(b) 4.02%

15. KrF_2

16. 0.05549 M

17. 0.2005 g

18.
(a) 0.00009
(b) −0.0001

19.
(a) 16 mg
(b) 13 mg

20.
(a) (b) (d) 0.10 N
(c) 0.50 N
(e) 0.30 N

21. 0.0333 N

23.
(a) 0.300 N, M
(b) 0.100 M

24. 218 mg, 0.0250 M

25. 0.0424 M

CHAPTER 16

2.
(a) 4.0 J
(b) 0.08 W
(c) 7.7×10^4 J, 1.8×10^4 cal

4.
(i) (b) 1.474 V (c) 1.545 V
(ii) (b) 0.1519 V
(c) −0.0246 V
(iii) (b) 0.749 V (c) 0.740 V

5. 0.11 atm

7.
(a) 0.340 V
(b) 1.18 V

9.
(c) (i) −0.7248 V
(ii) −0.71 V
(iii) +0.053 V
(iv) +1.5329 V
(v) −1.0901 V

10. (d) 0.386 V

11.
(a) 1.229 V
(b) 1.291 V
(c) 1.291 V

13.
(a) −0.2763 V, +0.9259 V
(b) 1.2022 V

15.
(a) 1.63 V
(b) 1.23 V

17. 43 g

22. (b) 25 years

CHAPTER 17

3.
(i) (a) 0.17 V (b) −7.8 kcal
(c) 5.6×10^5
(ii) (a) 1.0650 V
(b) −49.12 kcal
(c) 1.0×10^{36}
(iii) (a) 1.627 V (b) −375 kcal (c) 10^{275}

4. 1.84

5. −15.8 kcal

8. 148 cal/mole
(b) −154 cal/mole
(c) 96°C

9.
(a) 0.515
(b) 3.8×10^3 kWhr

10.
(a) 1.060 V
(b) 0.919
(c) 0.64 V

11.
(a) −58.0 cal/deg
(b) −180 kcal

13.
(a) 9.1×10^3 joules, 2.2×10^3 cal
(b) 7.3 cal/°K
(c) 2.25×10^4 cal liberated

14. (d) 3×10^{-2} M

15. −0.185 V

16.
(a) 1.75×10^{-4}
(b) 3.0×10^{-14}
(c) 1.7×10^{-10}

17.
(a) 10^{81}
(b) 3.1×10^{-55}
(c) 5.6×10^{59}
(d) 7.7×10^{19}
(e) 1×10^{-27}
(f) 1×10^{12}

18.
(a) 0.235 V
(b) 8.8×10^{7}
(c) 5.3×10^{-5} mole/liter

19.
(a) 0.536 V
(b) 110%

20.
(b) -205 cal/mole
(c) $394°K$
(d) 0.70 atm

22.
(a) 149 kW
(b) 3.3×10^{5} kcal/hr
(c) 3.3°C

23.
(a) 5.2×10^{14} kcal/day
(b) 135 days

24. 0.0603 cal/C liberated

25.
(a) 1.65 V
(b) 55%
(c) 0.81 kWhr

26.
(a) (*i*) 1.8 kcal/g (*ii*) 7.9 kcal/g
(b) 1.8 kcal/g

27.
(a) 3.58 kcal/g
(b) 5.12 kcal/g
(c) 0.34 kcal/g absorbed

CHAPTER 19

27. 1.09 Å

CHAPTER 20

2.
(a) 1st, 1st, 2nd
(b) 5.1×10^{9} liters/mole-second
(c) 8.7×10^{-7} mole/liter-second
(d) 8.7×10^{-8} mole/second
(e) 2.0×10^{-7} mole/liter-second

3. 1.3 liters/mole-second

6.
(a) unimolecular (forward), bimolecular

9.
(a) 1/8
(b) (*i*) 0.125 g, (*ii*) 0.354 g
(c) 54 min

12. 2nd

13.
(b) 0.299 liter/mole-minute
(c) 0.673 mole/liter-minute
(d) 9.0, 3.0 moles/liter-minute

15.
(b) 2.5×10^{3} liter²/mole² second
(c) 1.4×10^{-2} mole/liter-second
(d) 10 moles/liter-second

16. 1.1×10^{-3} mole/liter-second, 6.6×10^{20} molecules/liter-second

23. 1st; bimolecular, unimolecular

25. (a) 10^{3}, 10^{301} molecules; 1.7×10^{-21}, 10^{277} moles

CHAPTER 22

11. 6×10^{-4} mole/liter

20.
(a) 5×10^{-14}
(b) 0.11 mole/liter

CHAPTER 25

1. (b) (c) (d) (f) (h) (k) (m) (n)

4. -2.01×10^{8} kcal

6. N, S, O

7. ^{7}Be, ^{15}N, ^{20}F, ^{2}H

8. 25.3 days

9. 62×10^{2} years

10. (a) (f) (h)

12. -4.92×10^{-3} g/mole

15. (c)

17.
(a) -5.17×10^8 kcal
(b) 5.16×10^9 kcal

20. endothermic

21. 1.8×10^{10} atoms, 3.0×10^{-14} mole, 1.1 pg

23.
(a) 2.128 moles
(b) 4.90×10^9 years

24. 2000 years

25. 123 years

26. 4.0×10^{-10} g/g, 4.0×10^{-4} g/10^6 g

27. (b) 1.9×10^{-8} mole/liter

CHAPTER 26

16. 0.10 pg; 3.9×10^{-10} cal

INDEX

Hard and soft acids and bases, 259–261

Hard water, *see* Water

Heat, 57–73, 582
  entropy and, 349–350
  of formation, 65–66
  of fusion, 154 *fig*, 155 *tbl*, 433 *tbl*
  of solution, 220
  specific, 61
  of sublimation, 154 *fig*
  of vaporization, 154 *fig*, 155 *tbl*, 433 *tbl*

Heat capacity, 61

Heat engines, 350–352

Heavy rare-earth elements, 108

Heisenberg, W., 85 *ftn*

Helium atom, spectrum of, *facing* 79 *fig*

Helium molecule, 365 *prb*

n-Heptane, 482 *tbl*

Hero's engine, 11 *ftn*

Héroult, P., 436

Herz, G., 100 *ftn*

Hess's law, 64–66

Heterogeneity, 4

n-Hexane, 482 *tbl*

Higgins, W., 30

High polymers, 551

History of chemistry, 1–2

Hofstadter, R., 532

Homogeneity, 4

Homologous series, 481, 485, 486, 488 *tbl*

Hückel, E., 359

Hund, F., 95 *ftn*, 359

Hund rule, 95

Hybrid orbital number, 381, 381 *tbl*

Hybrid orbitals, 373–377, 381 *tbl*

Hybridization, 373, 376–377
  of atomic orbitals, 372–377
  of *d* orbitals, 382–383
  of *s* and *p* orbitals, 373–377

Hydration, 173
  energy of, 220
  of ions, 219

Hydrazine, 250

Hydrazoic acid, 145 *prb*

Hydride ion, 252

Hydrides, 440–443
  basic reactions of, 441–442
  hydrolysis of, 441–442
  as Lewis acids, 443

Hydrocarbons, 479–491
  substituted, 491

Hydrogen
  as fuel, 353–354
  production of, 431 *prb*

Hydrogen atom
  energy levels in, 86–88, 87 *tbl*, 88 *fig*
  spectrum of, 84, 88 *fig*, *facing* 79 *fig*

Hydrogen bonding, 139–141
  by ions, 219
  solubility and, 176

Hydrogen difluoride ion, 140–141

Hydrogen fluoride, 139–141, 248

Hydrogen half-cell, 323 *fig*

Hydrogen halides, 504

Hydrogen iodide, formation of, 193–194, 420, 429 *prb*

Hydrogen ion, 222

Hydrogen peroxide, 445–446

Hydrogen species, theory of, 365 *prb*

Hydrogen sulfide, ionization of, 250, 283–284, 291

Hydrogenation, 504

Hydrolysis, 250
  of anions, 250–252
  of cations, 252–254
  of nonmetal halides, 258

Hydronium ion, 222

Hypochlorous acid, 257

Ideal gas, 17–19

Ideal gas behavior, deviations from, 22–24

Ideal gas law, 13–15

Indicators, 300–304, 301 *tbl*

Inertia, 574

Infinitesimal, 336 *ftn*

Infrared spectroscopy, 518–519, 519 *tbl*

Inhibitor, 422

Inner transition elements, 108

Insulator, 369, 369 *fig*

Insulin, 564 *fig*

Intensive properties, 2

Interfacial tension, 177

Interhalogen compound, 383

Intermediate, 415

Ion combinations, equivalent weights in, 307

Ion pairs, 235

Ion-electron method, 224–227

Ion-exchange resins, 239 *prb*, 561–562, 562 *fig*

Ionic bond, 120–121

Ionic conduction, 218

Ionic equations, net, 222–223

Ionic equilibrium calculations, 265–296

Ionic radius, 111–113, 112 *tbl*

Ionic solids
  formation of, 121–122
  properties of, 128–129, 129 *tbl*

Ionic solutions, 216–240

Ionization
  degree of, 236, 271
  of water, 265–266
  of weak acids and bases, 269–278

Ionization constant, 270 *tbl*

Ionization energy, 113–114, 113 *tbl*

Ions, 4, 218–223
  positive, 76–77 *fig*

Iron
  complexes of, 465–466
  corrosion of, 329–330
  galvanized, 330

Iron(III) chloride, 253

Iron(III) halides, as catalysts, 502

Iron(III) hydroxide, 254

Isobar, 546 *prb*

Isoelectronic ions, 120

Isolated system, 350

Isomerism, 480–481
  geometrical, 485
  optical, 496–498
  structural, 481

Isoprene, 551, 552 *tbl*

Isotopes, 38, 79, 495

Joliot-Curie, I. and F., 537

Joule, J. P., 56

Joule (unit), 580

Kekulé, A., 487

Kelvin, Lord, 12 *ftn*

Kelvin scale, 12, 577

Kerosene, 479

Ketones, 493

Kinetic energy, 580

Kinetic molecular hypothesis, 17–19

Kinetic stability, 470, 473

Kinetics, 403–431

Kossel, W., 119

Lactic acid, 496

Lambert-Beer law, 305

Langmuir, I., 119

Lanthanoids, 108

Lattice defects, 153

Lattice energy, 122, 220

Lavoisier, A., 1

Law, *see* name of specific law

Lead, red, 331

Le Chatelier principle, 199–202
  for heterogeneous reactions, 203–204
  for solution reactions, 205

Length, 575

Lennard-Jones, J., 359

Leucippus, 28

Level, 90

Leveling effect, 245–246, 252

Lewis, G. N., 81 *ftn*, 119, 254

Lewis acids and bases, 254–256, 493, 499, 502–503

Lewis formulas, 123–124, 135–137

CONVERSION FACTORS[°]

Length 10^6 μm/meter
2.54 cm/in.
1.61×10^5 cm/mile
10^{-8} cm/Å

Energy 1.60×10^{-12} erg/eV
10^7 ergs/joule
4.184 joules/cal
4.184×10^7 ergs/cal
24.22 cal/liter-atm
860.4 kcal/kwatt-hour

Mass   28.35 g/oz (avdp)
453.59 g/lb (avdp)
10^6 g/metric ton

Time 3.16×10^7 sec/year

Area   6.45 cm²/in.²

Volume   16.4 cm³/in.³
946.4 cm³/qt

Pressure   760 torr/atm
1 torr/mm mercury

[°] These and other units are more fully discussed in Appendix I.